Environmental Engineering
Analysis and Practice

Environmental Engineering

Analysis and Practice

BURGESS H. JENNINGS

Professor of Mechanical Engineering
Northwestern University

Past President of the
American Society of Heating, Refrigerating,
and Air-Conditioning Engineers

INTERNATIONAL TEXTBOOK COMPANY

Scranton, Pennsylvania

COPYRIGHT ©, 1970, 1958, 1956, 1949, 1944, 1939
BY INTERNATIONAL TEXTBOOK COMPANY

All rights reserved. No part of the material protected by this copyright notice may be reproduced or utilized in any form or by any means, electronic or mechanical, including photocopying, recording, or by any informational storage and retrieval system, without written permission from the copyright owner. Printed in the United States of America by the Haddon Craftsmen, Inc., Scranton, Pennsylvania. Library of Congress Catalog Card Number: 73-105074.

Standard Book Number 7002 2259 6

Preface

More than a quarter century has elapsed since the late Samuel R. Lewis and the present author brought out *Air Conditioning*. Subsequent revisions, carrying through four editions and a name change to *Air Conditioning and Refrigeration*, have altered the text to the point that it bears little or no similarity to the first impression. In the ten years since the last edition, so many changes have occurred in engineering practice and in teaching patterns, with greater emphasis on fundamentals, that it seemed desirable not merely to provide a new edition, but to consider this volume almost as a new book, even though it retained portions of material from *Air Conditioning and Refrigeration*. The new title exemplifies the thinking involved.

Previous volumes were rich in tabular material to provide data and make it possible to solve a variety of real problems. The author is still of the opinion that no one is a master of a subject area unless he is able to express his mastery quantitatively. This aspect of the book has not changed but with the desirability of providing more theory and emphasizing fundamental aspects of the subject area, less space has remained for tabular and descriptive material in the new text. With the greater facility that students now have in the use of mathematics and computer technology, no effort has been made to avoid rigorous treatment of subject matter where this appeared to be desirable.

In early days, man's environment required merely providing enough heat to give reasonable comfort in winter. Then, with the advent of refrigeration, comfort was available in summer as well. Now, man's environment must also consider special control chambers to take him to outer space and even under the sea. This means the confrontation of extremely low temperatures, or high temperatures, along with both low and high pressures.

The importance of refrigeration and its extension into cryogenics has become so well recognized that an understanding of low-temperature engineering is almost a present-day necessity. Two chapters on cryogenics with extensive data have been prepared for this volume emphasizing fundamental aspects of liquefaction, gaseous separation, and cryogenic usage.

The author's claim to originality in this text must rest largely on the method of presentation and the selection of material since the technical literature has been most freely used, particularly the research publications of the American Society of Heating, Refrigerating, and Air-Conditioning Engineers and its predecessor societies, ASRE and ASHVE. Individuals, too numerous to mention, have also most generously provided information, help, and advice during preparation of this manuscript. Specific acknowledgement must, however, be given to Mr. Victor J. Johnson of the National Bureau of Standards for his inestimably valuable aid in providing cryogenic data from the extensive files of the Bureau of Standards.

This text has purposely been made to give a broad coverage of its field and contains more material than would be required for a class meeting three periods per week for a semester. However, by judicious selection of material, a one-semester course can easily be planned. This merely requires the elimination of certain chapters or topics which appear to have least importance for the objectives of the given teacher of a class.

Problems are provided at the end of each chapter and, in most cases, answers are given to the problems so that the student himself can check the accuracy of his solution. However, a few problems are purposely given without answers because many teachers have indicated a preference for a few problems of this type. The increased application of metric-system units, outside of purely scientific areas, has prompted the author to employ the CGS (centimeter, gram, second) system along with conventional engineering units at a number of places throughout the text. The large amount of tabular data provided should make the book valuable as a reference book both in and outside of class. It is also felt that the book can readily be used for self-instruction by engineers wishing to develop or review their background in this field.

B. H. JENNINGS

Evanston, Illinois
December, 1969

Contents

	Symbols and Abbreviations	ix
1.	Basic Concepts and Instrumentation	1
2.	Thermodynamics, Steam, and Gas Properties	32
3.	Air and Humidity Calculations	85
4.	Heat Transfer and Transmission Coefficients	134
5.	The Heating Load	188
6.	Cooling Systems and Cooling Load	213
7.	Steam Heating	254
8.	Hot Water Heating	289
9.	Heat-Transfer Elements and Combustion	326
10.	Physiological Reactions to the Environment	374
11.	Warm Air Heating	396
12.	Fluid Flow, Duct Design, and Air-Distribution Systems	409
13.	Fans and Air Distribution	461
14.	Principles of Refrigeration	484
15.	Refrigerants and Refrigeration Systems	512
16.	Refrigerant Equipment and Arrangement	553
17.	Cryogenics and Gas Liquefaction	594
18.	Gas Separation and Cryogenic Systems	642
19.	Controls and Control Systems	671
20.	Panel (Radiant) Heating and the Heat Pump	690
21.	The Cleaning of Air	715
22.	Industrial Air Conditioning and Food Preservation	731
	Appendix	749
	Index	755

Symbols and Abbreviations

SYMBOLS

α linear coefficient of expansion
A area [sq ft, sometimes sq in.]
A_i inside area [sq ft or sq in.]
A_o outside area [sq ft or sq in.]
a thermal conductance of an air space [Btu per hr sq ft deg F]
β_v volume coefficient of expansion
b bypass factor of coil
C specific heat, not specified (but usually at constant pressure) [Btu per lb deg F]
C thermal conductance [Btu per hr sq ft deg F]
C_p specific heat at constant pressure [Btu per lb deg F]
C_v specific heat at constant volume [Btu per lb deg F]
Δ Greek letter *delta* (often used to signify "change in")
ΔP pressure difference
Δt temperature difference [deg F]
D diameter of pipe or conduit [ft, rarely in.]
d density [lb per cu ft]
d_a density of air [lb per cu ft]
d_s density of saturated water vapor at dry-bulb air temperature [lb per cu ft]
d actual water-vapor density in air [lb per cu ft]
dt differential temperature difference [F]
dx differential length of path through a material, in direction of heat flow
η_v volumetric efficiency
E evaporative heat loss

e, e_s, e_r absorptivity or emissivity factor
F, F_G shading or radiation factor for solar energy
f film coefficient in heat transfer [Btu per hr sq ft deg F]
f friction factor in fluid flow
f_i film or surface coefficient for inside wall [Btu per hr sq ft deg F]
f_o film or surface coefficient for outside wall [Btu per hr sq ft deg F]
f_s specific humidity correction factor
g gravitational constant (32.2)
H heat demand [Btuh]
H height [ft]
h height of fluid in column [in. or ft]
h_{fgw} heat of vaporization of one pound of steam at wet-bulb temperature
h_L head loss [in. of water]
h_v velocity head [in. of water]
$\mathbf{h}$ enthalpy [Btu per lb]
$\mathbf{h}_a$ enthalpy of one pound of dry air [Btu]
$\mathbf{h}_f$ enthalpy of liquid [Btu per lb]
$\mathbf{h}_{fg}$ enthalpy of vaporization [Btu per lb]
$\mathbf{h}_g$ enthalpy of dry saturated vapor
$\mathbf{h}_x$ enthalpy of wet vapor at any quality x
k coefficient of diffusion
k ratio of specific heats (C_p/C_v)

Symbols and Abbreviations

k	thermal conductivity [Btu in. per hr sq ft deg F]	T, T_f	temperature in degrees Fahrenheit absolute
L	length of lagging measured along axis of pipe [ft]	T_a	air temperature in degrees Fahrenheit absolute ($t_a + 460$)
μ	absolute viscosity [usually in centipoises]	T_c	temperature in degrees centigrade absolute
M	metabolic heat	t_c	temperature in degrees centigrade [C]
m	hydraulic radius	t_d	dry-bulb temperature [F]
M	molecular weight	t_{dp}	dew-point temperature [F]
ν	kinematic viscosity	t_f	temperature in degrees Fahrenheit [F]
N	revolutions per minute	t_{fg}	average flue-gas temperature [F]
n	ratio of roof area (A_r) to ceiling area (A_c)	t_i	inside temperature [F]
ϕ	relative humidity [per cent]	t_o	outside temperature [F]
P, p	pressure [psf or psi]	t_w	wet-bulb temperature [F]
P_a	pressure of "dry" air (partial)	U	over-all coefficient of heat transfer [Btu per hr sq ft deg F]
P_B, p_B	barometric pressure [in. Hg or lb per sq in.]	u	internal energy [Btu per lb]
Pv	flow work [ft-lb]	V, v	velocity [fpm or fps]
p_s	actual pressure of saturated water at dry-bulb temperature [in. Hg or psi]	V_e	equivalent wind velocity [mph]
p	actual pressure of water vapor in air [psi, in. Hg, or psf]	V_m	velocity [fpm]
Q, q	heat added or rejected or transferred [Btu, Btuh, or Btu per min]	v	specific volume [cu ft per lb]
		v	volume [cu ft]
ρ	density [lb per cu ft]	v_f	specific volume of liquid [cu ft per lb]
R	gas constant ($1544/M$ for a given gas, when P is given in lb per sq ft)	v_g	specific volume of saturated vapor [cu ft per lb]
R_1	gas constant ($10.7/M$ for a given gas, when P is given in lb per sq in.)	W	mass (weight) of a body [lb]
		W, W_s	specific humidity (pounds or grains of water vapor associated with one pound of dry air)
R	Reynolds number		
R	thermal resistance to heat flow [hr sq ft deg F per Btu]	W	work [ft-lb or Btu]
r_1 and r_2	radii to innermost and outermost section of lagging [ft or in.]	X	weight fraction of air that is to be bypassed through a conditioner [expressed as a decimal]
S	stored energy	x	quality of vapor
S	surface or area [sq ft or sq in.]	x	thickness of wall [in. or ft]
s_f	entropy of liquid	Y	weight fraction of air which passes through a conditioner [expressed as a decimal]
s_g	entropy of dry saturated vapor		
θ	time [sec, min, or hr]		

ABBREVIATIONS

NOTE: American Standards Association abbreviations are used when applicable; and, following ASA practice, the same abbreviation is used for both singular and plural expressions.

abs	absolute	ASHRAE	American Society of Heating, Refrigerating and Air-Conditioning Engineers; formerly:
a-c	alternating-current		
AGA	American Gas Association		
air hp	air horsepower	ASHVE	American Society of Heating and Ventilating Engineers
API	American Petroleum Institute		

Symbols and Abbreviations

ASRE	American Society of Refrigerating Engineers	KE	kinetic energy
avg	average	kw	kilowatt
bhp	brake horsepower	kwhr	kilowatthour
Btu	British thermal unit	lb	pound
Btuh	British thermal units per hour	log mtd	logarithmic mean temperature difference
C	degrees centigrade	l-p	low-pressure
cfh	cubic feet per hour	Ltd	lesser temperature difference
cfm	cubic feet per minute	Mbh	1000 British thermal units per hour
cfs	cubic feet per second	min	minimum
cm	centimeter	min	minute
COP	coefficient of performance	mph	miles per hour
cu ft	cubic feet	MRT	mean radiant temperature
cu in.	cubic inch	mtd	mean temperature difference
db	dry bulb	μ	micron
d-c	direct-current	mv	millivolt
deg or °	degree	OD	outside diameter
diam	diameter	oz	ounce
EDR	equivalent direct radiation	PD	piston displacement
EHDR	enthalpy–humidity-difference ratio	PE	potential energy
eq	equation	psi	pounds per square inch
ET	effective temperature	psia	pounds per square inch absolute
ETL	equivalent total length	rel hum	relative humidity (ϕ)
F	degrees Fahrenheit	rpm	revolutions per minute
fpm	feet per minute	sec	second
fps	feet per second	SH	sensible heat
ft	foot	shp	shaft horsepower
ft-lb	foot-pound	SHF	sensible heat factor (same as SHR)
gal	gallon	SHR	sensible heat ratio (same as SHF)
gpm	gallons per minute	sp gr	specific gravity
Gtd	greater temperature difference	SPL	static pressure loss
HCOP	heating coefficient of performance	SPR	static pressure regain
HHV	higher heating value [Btu per lb]	sq in.	square inch
h-p	high-pressure	sq ft	square foot
hp	horsepower	std	standard
hp-hr	horsepower-hour	temp	temperature
hr	hour	UMRT	unheated mean radiant temperature
H.R.T.	horizontal-return tubular (boiler)	v	volt
IBR	Institute of Boiler and Radiator Manufacturers	vol	volume
ID	inside diameter	w	watt
in.	inch	whr	watthour
in. Hg	inches of mercury	wt	weight

1

Basic Concepts and Instrumentation

1–1. HEATING AND AIR CONDITIONING

The practice of heating and ventilating has made it possible for man to exist under forbidding climatic conditions. (The term *heating*, as used here, means the maintenance of a space at a temperature above that of its surroundings, while *ventilation* means the supplying of atmospheric air, and the removal of inside air, in sufficient amounts to provide satisfactory living conditions.) Ever since early man in his cave huddled close to a fire, and worried about the removal of smoke, he has used the combustion of fuel to aid him in adjusting to a harsh environment. Heating and ventilating methods have changed greatly since those early days of prehistoric man, but the fundamental problem remains; in temperate climates, heating and ventilating in winter are necessary to life as we know it.

Early civilizations had their origin in tropical areas, where heating requirements were slight or even nonexistent; but residents of tropical areas were (and still are) faced with the opposite problem—how to devise satisfactory cooling methods in order to keep body temperature at sufficiently low levels. Under both warm and cold climatic conditions, a balance must be maintained between the individual and his environment. The objective in heating or cooling for comfort is to provide an atmosphere having such characteristics that the occupants of a space can effectively lose enough heat to permit proper functioning of the metabolic processes in their bodies and yet not lose this heat at so rapid a rate that the body lowers in temperature. The regulatory mechanisms of the human body endeavor to keep body temperature at or about 98.6 F, which is the normal temperature for human beings. In hot summer weather, dissipation of body heat may be difficult, whereas in winter the heat dissipation is not difficult and in fact must be controlled so as not to be excessive. Outdoors, the customary heavier clothing of winter reduces the rate of heat loss. However, it is also desirable and usually necessary to keep the indoor temperature in winter within a suitable comfort range so that body functions are properly carried out without taxing the heat regulatory system.

Heating. Early heating involved the use of fireplaces or room stoves. These were not efficient in the use of fuel, since much of the heat from combustion went directly up the chimney. Also, the space being heated was drafty, because of the excessive amount of cool air brought in and wasted up the chimney, and the occupants were therefore frequently overheated on one side and too cool on the other. The individual stoves in each room were an improvement over open fires, but in addition to the inconvenience of operating such separate heaters, the stoves had certain of the disadvantages associated with fireplaces.

Toward the middle of the nineteenth century, *central heating systems* began to come into use. In such systems, a furnace, which is placed in a convenient location, frequently in the basement of a building, burns fuel and the resultant heat generated is carried by a suitable medium to other parts of the building. The transfer mediums usually employed are air, steam, or hot water. *Gravity warm-air systems*, which depend on differences in density of the air for circulation of the heated air inside the building, have been very successful in small homes. In larger buildings, *forced-air systems*, also known as *mechanically-circulated-air systems* are frequently employed. With forced-air systems, it is possible to clean and humidify the air delivered, and to control accurately the amount of air circulated. *Steam systems* are of numerous types, from the simple one-pipe systems to the more complex vapor and vacuum systems. *Hot-water heating* can make use of natural-circulation effects or can function in forced-circulation hot-water systems, which employ pumps to force the hot water through small pipes to radiators or convectors located in the rooms to be heated. Industrial heating systems most usually employ steam or hot water.

Air Conditioning. The term air conditioning, although it has been used in many different concepts, implies the creation and maintenance of an atmosphere having such conditions of temperature, humidity, air circulation, and purity as to produce desired effects upon the occupants of that space, or upon the materials that are handled or stored there. The simultaneous control of these four factors within required limits, when directed toward human comfort and health or when industrially directed toward conditions permitting the best product yield during manufacture and storage, can rightly be called air conditioning. Air conditioning is independent of time or season and can function effectively under all extremes of weather.

Complete air conditioning of spaces for comfort, and for industrial control of product, developed largely during the last fifty years. The term air conditioning was probably first employed to signify the process of humidifying the air in textile mills to control the static-electricity effects and to reduce breaking of the fibers. In winter, when the air was heated and dry, static charges produced on the moving threads caused trouble in the looms, and the dry threads also became brittle and frequently broke. Humidifying (that is, adding moisture to the air) reduced or eliminated these difficulties and gave a real impetus to the development of air conditioning in industry.

It was eventually thought that if air could be moistened in winter, so as to permit delicate manipulation of materials in process of manufacture, it might be possible to dehumidify the air in summer so as to control the absorption of moisture and the consequent variation of size and weight of moisture-absorbing materials in process of manufacture. With vegetal and animal materials, weight change due to loss or gain of moisture can be very great. Some types of cotton, for example, may change in weight as much as 25 per cent under conditions of extreme humidity variations in the atmosphere. Also, in certain printing and lithographing processes where the paper is subjected to several impressions for different colors, often applied at intervals of many hours or days, the size of the paper changes as its moisture content varies, resulting in imperfect registry and blurring. Satisfactory color printing in many cases can be done only when the relative humidity of the air remains constant.

Moisture can easily be introduced into the air for humidifying, by evaporating moisture into the air, but the problem of removing surplus moisture (dehumidification) is more difficult. Dehumidification can be accomplished by the use of desiccant materials which can be periodically reactivated. More frequently, however, dehumidification is brought about by using refrigeration to chill the air to a sufficiently low temperature that the excess water vapor can be removed by condensation. The resulting water can then be withdrawn from the system. Water-removal by condensation from air is illustrated by the familiar formation of moisture on the cold surfaces of a glass of ice water.

The creation of better comfort conditions for human beings thus evolved from industrial air conditioning. It was obvious that systems similar to cooling systems for industrial purposes could also be designed primarily for the comfort of human beings. For comfort, the lowering of temperature is a most important function of the air-conditioning system whether or not dehumidification is also necessary. A refrigeration system is thus an adjunct of any air-conditioning system whenever temperatures below those of the surrounding atmosphere are required.

1–2. REFRIGERATION AND CRYOGENICS

Control of the environment, whenever cooling is required, involves the use of refrigeration in one form or another. Comfort air conditioning does not require extremely low temperatures for the cooling process and the equipment seldom has to operate at temperatures lower than 40 F. However, in food processing and storage, temperatures down to −40 F may be required. Industrial processes may employ temperatures as low as −150 F. Temperatures in these ranges can be reached by conventional refrigeration equipment.

However, in the last decade or so, man's environment has been extended to outer space where temperatures below −400 F may be encountered. With the need of duplicating such temperatures in the laboratory and working with appropriate media at low temperatures, the area of cryogenics has become extremely important. The term *cryogenics* is usually applied to the temperature regime below −240 F while the term *refrigeration* is applied to cold production above −240 F. One very important cryogenic area involves the liquefaction of gases. For example, by liquefying air, it can readily be separated into its major components, oxygen, nitrogen, argon, and even neon. Liquid oxygen and liquid hydrogen represent basic reactants for rocket engines and gaseous oxygen is required for the numerous basic oxygen furnaces of the steel industry. In addition to use of cryogenic temperatures for space and industrial use, extensive research is being carried out making use of low temperatures to advance the scientific frontiers of knowledge.

Refrigeration and cryogenics are thus intimately associated with man and his environment and need to be thoroughly understood. It is significant that every aspect of heating, air conditioning, refrigeration, and cryogenics involves the application of fundamental principles drawn from the concepts of thermodynamics, fluid flow, and heat transfer coordinated to adapt mediums to serve the environmental requirements for occupants or materials in a space, whether for the comfort of individuals, for the preservation and creation of products, or for the production of power. The development of interrelated principles toward these desired ends will constitute the subject matter of much of this text.

1-3. RELATION OF PROPERTIES OF MATERIALS TO INSTRUMENTATION

The reproducible response of materials to changes in temperature and pressure can often furnish a basis on which to design measuring devices.

Thermal expansion with temperature increase is a useful property of matter which is used in instrumentation. Most materials increase in size (length and volume) as the temperature increases. Over small temperature ranges (not greatly in excess of 250 deg on the Fahrenheit scale) the change in length is essentially a linear function of temperature and can be expressed by the relation

$$L_t = L_o[1 + \alpha(t - t_o)] \qquad (1-1)$$

$$\Delta L = L_t - L_o = L_o \alpha(t - t_o) \qquad (1-2)$$

Basic Concepts and Instrumentation

where L_o = original length of object at temperature t_o;
L_t = length of object at temperature t;
t = temperature, in degrees Fahrenheit;
t_o = datum or reference temperature, in degrees Fahrenheit;
α = linear coefficient of expansion, in in./in. °F, ft/ft °F, 1/°F;
ΔL = increase in length, in consistent units of inches or feet.

Values of linear expansion coefficients are tabulated in Table 1–1 for a variety of substances. The volume coefficient of expansion is closely equal to 3α over moderate temperature changes. Volume coefficients of expansion for some liquids are listed in Table 1–2. For gases, which undergo relatively-large volume change under temperature variation, computation methods are discussed in section 2–8.

Example 1–1. A steam pipe of ordinary steel is 80 ft long at 60 F when installed. Find the increase in the length of this pipe when it is carrying steam at 215 F.

Solution: From Table 1–1, α is read as 0.0000064. Then, by equation 1–2,
$\Delta L = (80)(0.0000064)(215 - 60) = 0.079$ ft
$= 0.95$ in. *Ans.*

The familiar mercury-in-glass thermometer consists of a small glass bulb, and a stem in which there is a passage of capillary size. Liquid mercury in the bulb expands when the bulb warms, and the expansion is in evidence as the mercury thread moves upward in the capillary. Proper calibration of thermometers from standardized temperature points is accomplished by accurately etching and marking the stem between these points to indicate temperature values on a selected scale.

Another example of expansion effect in instrumentation is the use of the bimetal strip, which consists of two dissimilar metals fused lengthwise to each other. Under temperature increase, the side of the strip made from the metal with the greater coefficient of expansion elongates more than the other side of the strip, which causes the strip to bend out of a straight-line position. The resultant side motion can be used to close electric contacts for relay operation, or the motion can be used in other ways to produce other desired effects.

In the thermostat of Fig. 1–1 can be seen the wide, centrally-located, U-shaped bimetal strip which, in response to temperature changes, moves in or out so as to make or break an electric circuit. Adjustment of the thermostat is brought about by moving the bimetal strip about its fulcrum anchor near the bottom. This is done by adjusting the temperature dial *1* and its cam *3*. When a high temperature is to be maintained, the low point of the cam moves the strip far to the left so that only a small temperature drop is required to make the U-shaped strip contract sufficiently to close the contact points of the electric circuit at *7*. When the space temperature rises a desired amount, the resultant effect is to spring the

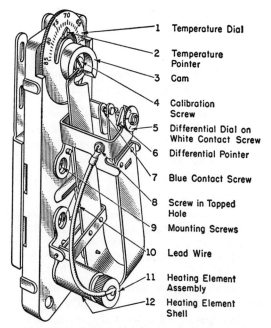

Fig. 1-1. Internal mechanism of bimetal-type open-contact thermostat, with cover removed. (Courtesy Minneapolis-Honeywell Regulator Co.)

bimetal strip ends further apart and thereby to open the circuit. In many types of heating devices it is possible that the burner (or steam valve) may operate for too long a period before the thermostat reacts to stop the burner (or close a valve). To reduce this possibility, units like the one shown in Fig. 1-1 have an internal heater element. When the thermostat circuit is actuated, electric energy from the circuit warms the heater and the inside of the thermostat case so that the control circuit opens sooner and acts to prevent overheating of the space. If the duration of the heating cycle is too long or too short, adjustment can be made by turning the dial at 5.

As is true with many such thermostats, particularly those for home use, low voltage (20 to 24 v) is employed in its operating circuit. The reduced voltage is obtained from a small transformer, and the control action of the thermostat acts through a relay to stop or start the line-voltage motors of the heating system. For units of this type, ingenious holding circuits with double contacts are employed to keep the contacts from arcing and burning as they close and open. Other designs use a magnetic catch which rapidly closes the circuit and which releases only when a predetermined temperature differential has been established.

Another example of a bimetal control device is the limit control, illus-

Basic Concepts and Instrumentation

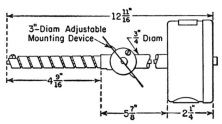

Fig. 1-2. Warm-air limit-control for insertion in furnace-duct or plenum chamber. (Courtesy Mercoid Corporation.)

trated in Fig. 1-2. The particular control shown is used to prevent overheating of a warm-air furnace operated by an automatic burner. It is designed for insertion in the hot-air outlet duct leading from the furnace. As the temperature in the duct rises, the coiled bimetal element reacts by twisting about its central axis until, when the predetermined limit temperature is reached, the coil has twisted sufficiently to trip open the circuit. This action stops the burner. The limit setting of the control can easily be adjusted by pressing the knurled knob on the front of the control either to the left (for a lower temperature) or to the right (for a higher temperature). The adjustable range for this control is from 70 to 310 F, and with its mercury trip-switch it can be used for either line-voltage (110 v) or reduced-voltage applications. Devices similar to this but with the bimetal coil mounted in a leakproof separable well or shield are also applicable for limit control of water temperatures in systems or in boilers alone.

Invar and brass have been used in some bimetallic designs. Invar, which is an alloy of iron containing 36 per cent nickel, has an extremely low coefficient of thermal expansion, and when used with brass in a bimetallic strip the high coefficient of expansion of the brass in contrast to that of the invar makes possible a relatively large travel for small temperature change.

Among other characteristics of matter which are useful in measurement and instrumentation are the thermoelectric effect (thermocouple), which will be described later; the elasticity of metals, which is used in certain pressure-measuring instruments; and the variation in electrical resistance of metals to temperature, which is utilized for thermometric devices.

1-4. TEMPERATURE SCALES AND TEMPERATURE DEVICES

Temperature is a measure of the relative hotness of a body, and may be expressed on any suitable arbitrary scale. The most-used datum points for thermometric scales are the melting point of ice, 32° on the Fahrenheit scale (0° on the centigrade scale), and the boiling point of water at atmospheric pressure, 212° on the Fahrenheit scale (100° on the centigrade scale).

The temperature ranges thus expressed represent 180 Fahrenheit degrees and 100 centigrade degrees, and having defined a degree or unit on either of these arbitrary scales they may be extended above and below the steam and ice points. To change from one scale to the other the following obvious relations can be used:

$$t_f = \frac{9}{5} t_c + 32 = 1.8 t_c + 32 \qquad (1\text{-}3)$$

$$t_c = \frac{5}{9}(t_f - 32) = \frac{t_f - 32}{1.8} \qquad (1\text{-}4)$$

where t_f and t_c represent temperatures expressed in degrees Fahrenheit and centigrade respectively.

Consideration of the laws which govern the behavior of gases and the thermodynamic work scale render the concept of an absolute zero of energy and temperature a reasonable conclusion. It can be shown by calculation that the absolute zero of temperature occurs at 459.69 deg below zero on the Fahrenheit scale and 273.16 deg below zero on the centigrade scale. Absolute temperature can be found on the Fahrenheit and centigrade scales by the relationships

$$T_f = t_f + 459.69 \text{ °R (degrees Rankine)} \qquad (1\text{-}5)$$
or
$$T_f = t_f + 460 \text{ (approx) °R} \qquad (1\text{-}6)$$
and
$$T_c = t_c + 273.16 \text{ °K}$$

where T_f represents degrees Fahrenheit absolute (degrees Rankine), and T_c degrees centigrade absolute (degrees Kelvin). In most calculations the value 460 is employed instead of the more exact 459.69.

The main use that will be made of the absolute-temperature scale in this work will be in relationships dealing with gases such as air and low-pressure superheated steam.

Mention has already been made of the mercury-in-glass thermometer, which has wide utility. Mercury thermometers can be used to temperatures which approach 1000 F, particularly when the stem is filled with nitrogen or another inert gas. However, the temperature range on the low side is limited by the freezing point of mercury, which is −39.6 F. For ranges below this point it is customary to use thermometers filled with colored alcohol or pentane.

Usually mercury thermometers are calibrated for complete immersion in the medium whose temperature is being measured, but they are often used under conditions of partial immersion. Partial immersion means that only the bulb and a portion of the stem are in the medium being investigated and that the rest of the stem is in a colder or warmer atmosphere. Partial immersion leads to inaccurate indications of the thermometer and it is therefore customary to make a *stem correction*. Stem-correction data can be determined by observing the number of degrees of mercury thread exposed outside of the medium being measured and by finding the temperature of the stem itself. The temperature of the stem can be determined approximately by tying an auxiliary thermometer to the exposed stem and insulating the bulb and stem at the point of attachment. The temperature t of the medium being measured to a close approximation is then, in degrees Fahrenheit,

$$t = t_1 + 0.000088 E(t_1 - t_s) \tag{1-7}$$

where t_1 = temperature indicated by the thermometer, in degrees Fahrenheit;

t_s = temperature of stem as indicated by the auxiliary thermometer, in degrees Fahrenheit;

E = number of Fahrenheit degrees of emergent mercury thread outside the thermometer well or outside the fluid medium under measurement;

0.000088 = difference between the coefficients of expansion of mercury and representative glass.

Thermometric elements are also made by filling a temperature-actuated bulb with a suitable fluid such as mercury, methane gas, aniline, or the like, and connecting this bulb through a capillary tube to a pressure-responsive gage. Under proper calibration the pressure readings of the gage can be related directly to the increase in the volume of the fluid, and thus to the temperature of the bulb and of the medium being measured. Such instruments have an essentially uniform scale for temperature indication. Vapor-pressure instruments are built in a similar manner, but depend on the vapor pressure of the fluid in the bulb being transmitted through the capillary tube to an indicating pressure gage calibrated to read in temperature. The relation of temperature to the vapor pressure of a fluid at saturation is not a linear function, so these instruments have nonuniform divisions on their temperature scales. Remote-reading and recording-type gages are frequently one or the other of these two types. The bulb of such an instrument is immersed in the hot medium whose temperature is being controlled, and under variations in temperature of the medium the contents of the bulb expand or contract. The resulting pressure changes then actuate a control element or switch.

Thermocouples. The simplest thermocouple circuit consists of two wires of dissimilar metals, with junctions made at both ends. If each junction is maintained at a different temperature, it is found that a measurable electric current will flow in the completed circuit, or if one of the wires is broken a potential difference (small voltage) of measurable amount will exist at the point of breakage in the circuit. Figure 1-3a is a diagram of such a circuit. A thermoelectric pyrometer makes use of this effect, and serves as a temperature-indicating device.

In Fig. 1-3b, a simple thermocouple pyrometer circuit is shown. Here one junction, in this case the hot one, is indicated at H, and the other junction, the cold one, is indicated at C. The thermocouple element is the

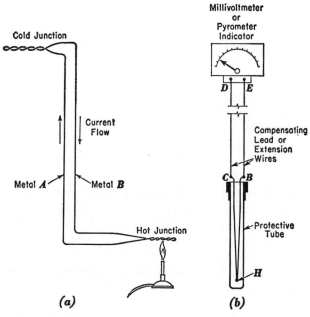

FIG. 1-3. (a) Basic thermocouple circuit. (b) Thermocouple circuit with temperature-indicating device.

portion CHB, and from C and B lead wires complete a circuit to an indicating instrument, which in most cases is essentially a high-resistant voltmeter. One of the temperature-responsive junctions is frequently enclosed in a protective tube. When C and H are at different temperatures a current flows through this circuit, and the reading of the instrument can then be used to indicate the temperature difference between the hot and cold junction. In this simple circuit the wire CH is of one metal, and BH of a dissimilar metal. The cold junction is located at C, although by use of compensating lead wires it can be transferred to the instrument location at D or E if this is desired. In a simple instrument of this type, it is very desirable that the temperature of the cold junction remain essentially constant, as the reading of the instrument is always a measure of the temperature differential above that of the cold junction. The compensating lead wires are made of one of the metals used in the couple, or of a metal with similar thermoelectric characteristics.

Basic Concepts and Instrumentation

Various metal combinations are used in thermocouples. In the temperature range to 600 F, and in refrigeration work, copper and constantan are frequently used. Constantan is an alloy containing approximately 60 per cent copper and 40 per cent nickel, and it develops approximately 0.024 mv per Fahrenheit degree of temperature difference existing between the two junctions. For a range to 1600 F, iron and constantan can be employed, and for a range to 2000 F, chromel and alumel. Chromel is an alloy of 10 per cent chromium and 90 per cent nickel; alumel is an alloy of 2 per cent aluminum and 98 per cent nickel. For higher temperature ranges, and also in certain scientific work, "noble" metals are often used, of which platinum and platinum-rhodium represent a frequently employed combination.

Couple junctions can be joined by welding, brazing, soldering, or, for very temporary work, sometimes even by twisting, provided the surfaces are clean and corrosion is not a factor.

The electromotive effects for various thermocouple combinations are plotted in Fig. 1-4.

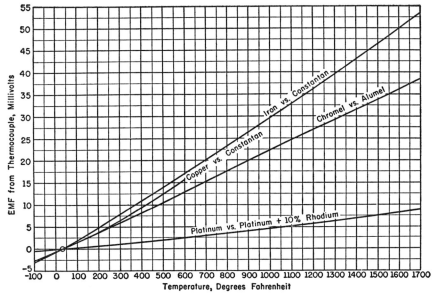

FIG. 1-4. Electromotive forces developed by dissimilar metals used in thermocouple circuits, referred to a 32 F datum.

Thermocouple circuits, in addition to serving as temperature indicators, have also been employed in control devices. One such device is illustrated in Fig. 1-5. Here the hot junction and the cold junction are shown, and the complete electric circuit from the hot and cold junction includes an operating electric coil. When the junctions are at different temperatures

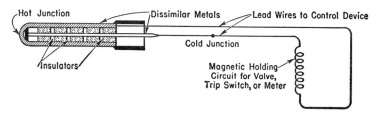

Fig. 1-5. Thermocouple circuit used as control device.

a current flows in the circuit, and under sufficient temperature difference the current is strong enough to operate switches, control valves, or other devices in the circuit. One use which is made of such a circuit is in connection with gas burners, where the hot junction is placed adjacent to the pilot light. As long as the pilot is burning and the junction is hot, the electric circuit holds open the safety trip in the main gas-supply circuit, so that when the thermostat calls for gas flow and active combustion, the gas valve can open and combustion can proceed following ignition by the pilot flame. On the other hand, if the pilot goes out, the gas-valve safety trip releases, and this prevents the gas valve from opening and loading the furnace with unignited gas. The same thermocouple principle is applicable to many other uses.

In scientific work, and in other types of work requiring precision measurement, it is often desirable to eliminate the electrical resistance of the circuits, and for this a potentiometer-type of instrument is often employed. In Fig. 1-6, the circuit of such an instrument is shown. An electromotive force is generated from the junctions C and H. By means of the battery, a current is caused to flow through the circuit $RGABDE$, and by adjusting the position of the movable slide B the potential drop along the slide wire from D to B can be changed. When the potential between B and D is exactly the same as the thermocouple voltage, the galvanometer will not indicate current flow if the key at K is closed; on the other hand, if the voltages are unbalanced the galvanometer will indicate current flow, and the slide position on the wire DBA is then adjusted until current flow ceases. Because the voltage of a dry- or wet-cell battery is not constant, most instruments are provided with a standard (cadmium) cell. Standard cells are precision-made to produce a constant voltage if used so that a minimal current of short duration is taken from them. To check the battery circuit, the key J is momentarily closed to indicate whether the proper current flows through DE and also through the slide wire DBA. If the current is not as desired, the galvanometer deflects and the resistance R in the battery circuit is then adjusted until the desired current flows through the slide wire. Thus readings of the slide wire are then automatically re-evaluated in terms of a standard current flow, and from the

Basic Concepts and Instrumentation

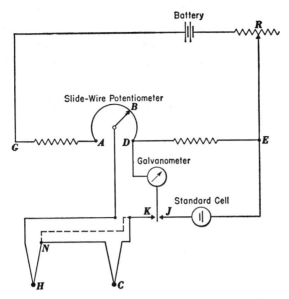

FIG. 1-6. Potentiometer circuit for thermocouple.

basic calibration of the instrument the slide-wire contact-setting B can be used to indicate the electromotive force, or temperature difference of the external thermocouple circuit.

In some cases, the junction C is immersed in an ice and water bath and enclosed in an insulated container so that a fixed datum of 32.0 F is maintained by the melting ice. In this way a constant reference point is obtained, in terms of which the other couple can be evaluated. More frequently, however, in industrial instrumentation, a datum point which is essentially room temperature is used as the reference point, and the partially dotted circuit HNK is substituted for the circuit $HNCK$. Under these conditions, the second junction of the thermocouple circuit is either at N or K. If the same metal runs from H to K, the second junction will be at the instrument point K, whereas if the lead wires are of a different, noncompensating metal, the second junction point would be at N.

1-5. PRESSURE

Unit pressure is defined as force per unit area. The total pressure acting on an area represents the force acting over the whole area and is equal to the product of the unit pressure and the area. Pressure is nearly always measured relative to some datum of pressure, usually that of the atmosphere, and the measurement is really an indication of how much the pressure is greater or less than that of the atmosphere. Pressures measured in this way are known as *gage pressures*. Thus the gage on a steam boiler indicating 150 psi signifies that the pressure in the boiler is 150 psi higher than atmospheric (barometric) pressure. If the barometric pressure is 14.7 psi, the pressure exerted by the steam in the boiler is really 150 + 14.7 = 164.7 psi. This total, or real, pressure is known as the *absolute pressure*.

When a pressure measured is less than atmospheric it is usually called a *vacuum*, and the absolute pressure is found by subtracting the vacuum pressure from the barometric (atmospheric) pressure.

Figure 1-7 is a diagram illustrating gage and absolute pressures at standard barometric pressure.

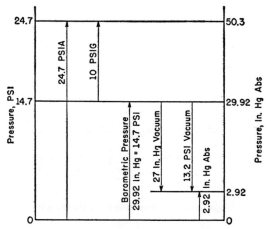

Fig. 1-7. Diagrammatic representation of barometric, gage, and absolute pressures.

A simple method of measuring pressure is to balance a column of liquid against the pressure and then measure the height of the column. It is well-known from the principles of hydrostatics that the unit pressure exerted by a column of liquid is a function of its height and its density and is independent of the cross section of the column. For example, a cubic foot of 68 F water weighs 62.33 lb. If this water is in a cubical container the pressure on the bottom of the container is 62.33 psf, or 0.433 psi. The same unit pressure applies to every vertical section in the cube, and this would be true whether the cube were cut in half to make two containers or reduced any number of times. Water itself can be used as a measuring material, but mercury, with its greater density, is a more common measuring medium. Mercury has a density of 0.4912 lb per cu in. at 32 F, and 0.4893 lb per cu in. at 70 F. It is thus obvious that

1 in. Hg = 0.491 psi at 32 F   1 psi = 2.036 in. Hg at 32 F
1 in. Hg = 0.489 psi at 70 F   1 psi = 2.044 in. Hg at 70 F

Basic Concepts and Instrumentation

To change fluid pressure units, this obvious relation is applicable:

$$h_1 d_1 = h_2 d_2 \qquad (1\text{-}8)$$

where h_1 and h_2 are the respective heights of the two fluids in question, measured in consistent units (feet, inches, etc.), and where d_1 and d_2 are the respective densities or relative densities of the two fluids in question, measured in consistent units (lb per cu ft for each, or specific gravity for each, etc.).

Example 1-2. An oil of specific gravity 0.8 shows a height of 18 in. when the pressure in an air chamber is measured. To what pressure is this equivalent in inches of water? (Specific gravity of water is unity.)

Solution: Let subscript 1 stand for water, and 2 for the oil, in equation 1-8, which reads

$$h_1 d_1 = h_2 d_2$$

Then
$$(h_1)(1) = (18)(0.8)$$

and
$$h_1 = 14.4 \text{ in.} \qquad Ans.$$

Example 1-3. In some air-flow measurements it is desired to find what height of an imaginary column of air of uniform density, 0.07 lb per cu ft, is equivalent to a pressure of 14 in. of water. (Water density is 62.3 lb per cu ft at 70 F.)

Solution: Use $h_1 d_1$ for the air and $h_2 d_2$ for the water:

$$h_1 d_1 = h_2 d_2$$

Then
$$h_1(0.07) = 14/12\,(62.3)$$

and
$$h_1 = 1038 \text{ ft} \qquad Ans.$$

Example 1-4. Express one standard atmosphere, 14.7 psi (more exactly, 14.696), as (a) feet of water and (b) inches of mercury.

Solution: When, in equation 1-8, h_1 and d_1 are expressed in units of feet and pounds per cubic foot respectively, the resultant units are

$$\text{ft} \times \frac{\text{lb}}{\text{ft}^3} = \frac{\text{lb}}{\text{ft}^2} = \text{lb per sq ft}$$

a) Set the pressure of one atmosphere, in pounds per square foot, to $h_1 d_1$ in pounds per square foot:

$$(144)(14.696) = h_1 d_1 = h_1(62.3)$$

Therefore
$$h_1 = 33.96 \text{ (say 34) ft of 70 F water} \qquad Ans.$$

b) The density of mercury at 70 F is 0.4893 lb per cu in. and at 32 F is 0.4912 lb per cu in. Use hd in units of

$$\text{in.} \times \frac{\text{lb}}{\text{in.}^3} = \frac{\text{lb}}{\text{in.}^2}$$

Then
$$14.696 = hd = h \times 0.4893$$

and
$$h = 30.03 \text{ in. of 70 F mercury} \qquad Ans.$$
$$= 29.92 \text{ in. of 32 F mercury} \qquad Ans.$$

Manometers, shown at the top of Fig. 1-8, consist of a tube bent in the shape of the letter U and about half-filled with mercury or other suitable indicating liquid—water, carbon tetrachloride, or colored kerosene. When the ends of the tube are connected to different pressure regions, the liquid surfaces will stand at different levels in the two legs, and the difference in height of liquid will represent the difference in pressure of the two regions,

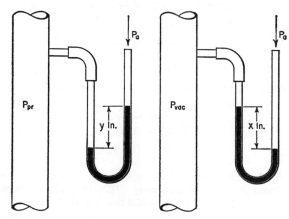

U-Tube Manometer for Pressure and Vacuum Measurement

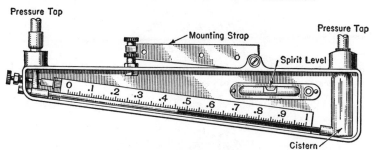

Inclined Draft Gage for Measuring Small Pressure Differences

FIG. 1-8. Manometric pressure-measuring devices.

expressed in inches of mercury or inches of the particular indicating fluid. Frequently the atmospheric (barometric) pressure is one of the pressure regions employed.

At the bottom of Fig. 1-8 is shown a draft-gage type of manometer. Draft gages are used to measure small pressure differences, such as those that exist in duct systems or in combustion chambers and stacks of furnaces. One of the tubes is inclined at an angle so that for a given vertical displacement of measuring fluid a much larger travel occurs on the inclined leg, and the reading scale can be read to smaller unit divisions. In these instruments, use is frequently made of low-density colored petroleum oil.

Basic Concepts and Instrumentation 17

A barometer measures atmospheric pressure. Mercury barometers are made by filling with mercury a tube sealed at one end and, after removing extraneous gases and sealing the other end, inserting the mercury tube upside down in a mercury cistern. Atmospheric pressure supports the column and prevents the mercury from falling out of the tube. The height of the supported mercury column is a measure of the atmospheric pressure. The space above the mercury in the sealed top of the tube is almost a perfect vacuum except for the slight vapor pressure of mercury existing at ordinary temperatures.

Barometric pressure in a given locality varies from day to day, but only to a slight extent. Standard atmosphere, by definition, is a column of mercury 29.921 in. high, measured at 32 F. This is equal to 14.6959 psi, or 2116.2 psf. Barometric pressure decreases with altitude, and a range of barometric pressures has been adopted as standard by the National Advisory Committee for Aeronautics. The following table shows several of these pressures.

ELEVATION (FT)	STANDARD ATMOSPHERE	
	In. Hg	PSI
0	29.921	14.6959
1000	28.85	14.17
2000	27.82	13.67
3000	26.82	13.17
4000	25.84	12.69
5000	24.90	12.23
10,000	20.58	10.11

The Bourdon-tube gage represents a very common type of pressure-measuring instrument used in engineering practice. The Bourdon tube, one form of which is illustrated in Fig. 1-9, is usually made of brass or steel and has a flattened cross section resembling an ellipse or, in some cases, a rectangle with rounded corners. The tube is bent in the form of a circular arc covering from 100 deg to some 300 deg. The open end of the tube is attached to the case of the instrument and connected to the source of pressure. The other end is sealed and free to move. When pressure is applied to a bent tube of this type, it tends to increase in volume, and in so doing its radius of curvature also increases so that the circular arc tends to straighten. The movement of the free end is attached through a suitable linkage to operate an indicating pointer. Bourdon tubes can measure partial vacuums as well as positive (gage) pressures. In the case of a partial vacuum, the tube tends to contract in depth and its radius of curvature decreases.

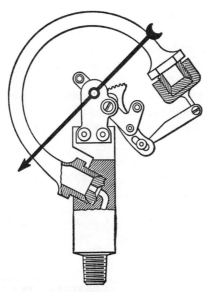

Fig. 1-9. Bourdon-tube pressure-gage element.

Bourdon-tube types of control instruments are manufactured which are responsive to the changes in liquid or vapor volume in a closed system. The motion of the free end of the tube actuates a switch or other control mechanism upon changes in pressure (temperature) of the medium in the control bulb. Temperature-responsive bulbs, when filled with liquid or gas, cause the tube to move in a linear path under changes in temperature.

1-6. THERMOELECTRIC EFFECTS[1]

Earlier in this chapter the thermocouple was described as a temperature-measuring instrument. At that point, no discussion was given to other thermoelectric phenomena; however, this area is so extensive, involving instrumentation, the creation of refrigeration, and the production of power, that it merits discussion in depth.

Historically a German, Thomas Seebeck, in 1821 discovered that an electromotive force was produced when a junction between two dissimilar metals was heated (thermocouple effect), while the other wire (metal) junctions were unheated. Following this in 1834, Jean Peltier in France discovered that when an electric current was passed through a junction connection of two dissimilar metals, a cooling effect was produced. If the current was then reversed, a heating effect was produced at that junction and with two junctions in a complete circuit, heat was produced at

[1]Some readers may wish to defer studying section 1-6 until energy relationships and systems of units in Chapters 2 and 4 have been considered.

TABLE 1-1
Linear Coefficients of Thermal Expansion per Degree Fahrenheit

Material	Coefficient α [1/(deg F)]	Temperature Range Applicable (deg F)
Aluminum	0.0000123	32–212
	0.0000143	68–570
Brass	0.0000106	32–212
Brick	0.0000053	68
Fireclay	0.0000061	600
Bronze	0.0000102	32–212
Carbon, graphite	0.0000044	104
Concrete	0.0000056	68
Copper	0.0000093	32–212
	0.0000098	77–570
Glass		
Jena	0.0000045	32–212
Plate	0.0000049	32–212
Pyrex	0.0000020	70–880
Quartz	0.0000003	60–1800
Gold	0.0000082	32–212
Granite	0.0000046	68
Ice	0.0000283	0–30
Inconel (0.8 Ni, 0.14 Cr)	0.0000089	100–1400
Invar	0.0000005	68
Iron, cast	0.0000059	104
Iron, wrought	0.0000063	0–212
Lead	0.0000139	32–212
Marble	0.0000065	60–212
Masonry	0.000002	68
	0.000004	68
Monel metal (.67 Ni, .33 Cu)	0.0000078	70–212
	0.0000089	70–1100
Paraffin	0.0000723	60–100
Silver	0.0000105	68
Steel	0.0000064	32–212
	0.0000068	32–392
	0.0000072	32–572
Stainless	0.0000054	68–392
Tin	0.0000149	64–212
Wood		
Oak	0.0000027	36–92
Across fiber	0.0000302	36–92
Pine	0.0000030	36–92
Across fiber	0.0000189	36–92
Zinc	0.0000165	32–212

TABLE 1-2

VOLUME COEFFICIENTS OF THERMAL EXPANSION AND SPECIFIC VOLUMES OF LIQUIDS

MATERIAL	VOLUME COEFFICIENT β_v [1/(DEG F)]	APPLICABLE TEMPERATURE RANGE (DEG F)	SPECIFIC VOLUME AND DENSITY AT TEMPERATURE INDICATED			
			Cu Ft per Lb	Cu Cm per Gram	Lb per Cu Ft	Deg F
Alcohol, ethyl............	0.000562	80–115	0.02029	1.267	49.27	68
50% water (weight)...	0.000413	32–102	0.01753	1.094	57.05	68
Alcohol, methyl.........	0.000630	32–142	0.02153	1.2566	49.62	60
50% water (weight)...	....	....	0.01746	1.0887	57.27	60
Benzene................	0.000650	52–176	0.01782	1.1123	56.12	32
Calcium chloride, 40% solution in water....	0.000235	68	0.01148	0.7164	87.13	68
Mercury...............	0.000101	32–212	0.00118	0.0735	848.7	32
			0.00118	0.0738	845.6	68
			0.00119	0.0741	842.9	100
			0.00120	0.0749	833.5	212
Petroleum (sp gr 0.899)..	0.00044	75–248	0.01784	1.1138	56.05	60
Sea water..............	....	....	0.01565	0.9756	63.90	60
Sodium chloride 20% solution in water....	0.00020	32–85	0.01396	0.8712	71.65	68
Water.................	....	....	0.01602	1.0001	62.422	32
	....	....	0.01602	1.0000	62.426	39.6
	....	....	0.01604	1.0009	62.35	60
	....	....	0.01605	1.0018	62.31	68
	0.00022	60–160	0.01613	1.0070	61.99	100

one junction while cooling was produced at the other. In England, William Thompson (Lord Kelvin) in 1855 developed a theory applicable to the Seebeck and Peltier discoveries and indicated that a third characteristic, namely an unusual type of heating or cooling of the wires of a thermocouple circuit occurred when a temperature gradient existed along the conductor links of a circuit and an electric current passed through the system. The theory that Lord Kelvin developed has recently been rigorously proved by the methods of modern statistical thermodynamics. It is interesting that Kelvin was able to develop his material without the benefit of this modern theory.

Let us refer to Fig. 1–10, where a circuit is shown consisting of two dissimilar conductors, a and b, connected as shown. At the end points, C and D, provision can be made for connecting a battery or a potentiometer (E).

First, if the potentiometer is connected at D and C and the junctions A and B are maintained at different temperatures, a voltage (potential) is produced and can be measured by the potentiometer. Calling this poten-

Basic Concepts and Instrumentation

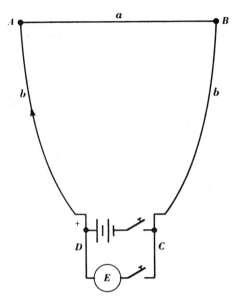

FIG. 1-10. Thermoelectric circuit arranged to show Seebeck or Peltier effects.

tial, E, the Seebeck coefficient or thermoelectric power is defined as

$$\alpha_{ab} = \frac{E}{T_A - T_B} \tag{1-9}$$

In differential notation and generalized

$$\alpha = \frac{dE}{dT} \tag{1-10}$$

Let us now consider what will happen if the junctions are brought to the same temperature and if the potentiometer is disconnected and the battery switch closed at C. Under this circumstance, a current I will flow around the circuit. This will result in heating at one junction, let us say B, while cooling occurs at the same rate at A. If the amount of this cooling is designated as q, we can then define the Peltier coefficient as

$$\pi = \frac{q}{I} \tag{1-11}$$

It should be observed that this heating is not the same type of heating that occurs in an ordinary electric circuit, namely the I^2R loss, where R represents the resistance of the circuit and when R is measured in ohms and I is in amperes, the resulting heat loss is in watts.

To define the Thompson coefficient, we have to envisage that in addition to the current flow, I, there is a temperature gradient dT/dx which produces heating or cooling in amount, q_τ, per unit length. In terms of these parameters, the Thompson coefficient, τ, can be stated.

$$\tau = \frac{q_\tau}{I \dfrac{dT}{dx}} \qquad (1\text{-}12)$$

$$q_\tau = \tau\, I\, \frac{dT}{dx} \qquad (1\text{-}13)$$

As has been mentioned, Lord Kelvin first developed relationships between the thermoelectric variables. The proof, which is very long, has been confirmed by the Russian scientist A. F. Ioffe (ref. 1) and can also be found in some thermodynamics texts (ref. 2). Before presenting the equations, it should be noted that both the Seebeck coefficient, α, and the Peltier coefficient, π, must relate to the junctions or branches of the circuit. The matter is simplified if we define the coefficients in difference form, namely $\alpha = \alpha_a - \alpha_b$ and $\pi = \pi_a - \pi_b$. There is theoretical and experimental justification for doing so and on this basis the thermoelectric equations appear

$$\pi = (\alpha_a - \alpha_b)\, T = \alpha T \qquad (1\text{-}14)$$

$$\tau_a - \tau_b = T\, \frac{d\alpha}{dT} = \frac{d\,(\alpha_a - \alpha_b)}{dT} \qquad (1\text{-}15)$$

Compared to the Seebeck and Peltier effects, the Thompson gradient effect is small, so small that for almost all thermoelectric purposes it can be disregarded, and with this the case, equations 1–11 and 1–14 become the significant parameters. Combine the latter equations and there results

$$q = \alpha\, I\, T \qquad (1\text{-}16)$$

Except for making thermocouples for temperature measurement, little use was made of thermoelectric phenomena until the last quarter of a century when it was found that certain semiconductors possessed much greater Seebeck coefficients than did metals. In particular, bismuth telluride was found suitable as an element for the thermoelectric circuit. Since that time, other semiconductors have been investigated and research is continuing to find better semiconductor formulations. The manner in which the semiconductors operate to produce high Seebeck coefficients can be explained using theory from modern solid state physics [3]. The theory will not be presented here beyond stating that there are two types of thermoelectric materials usually called n and p elements. The n type has a Seebeck coefficient which is negative and possesses an excess of electrons, the p type has a positive Seebeck coefficient and has a deficiency of electrons. The elements are arranged with a large number of them

Basic Concepts and Instrumentation

constituting a module and with the space between them filled by insulating material. A single thermoelectric unit is shown in Fig. 1–11. Note that the n and p materials are connected by a copper strap. Four connections are shown but there are only two junctions, the upper one which is shown

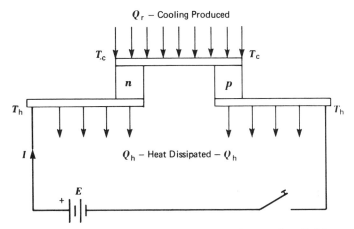

FIG. 1–11. Thermoelectric element arranged to produce Peltier cooling at the cold surface, T_c.

as the cold one, absorbing heat, and the lower one dissipating heat. The insertion of the copper as an extraneous material does not inhibit the performance of the junctions and makes a usable geometry for the system. A number of couples of the type shown can be placed in series to form a row and many rows can be placed beside each other to form a plate, capable of absorbing heat on the cold-plate side to produce refrigeration while its other side dissipates heat. Manufacturers make their modules in a form and shape to suit a user and can provide heat-transfer fins or such liquid-cooling passages as may be desired by a purchaser on proper sides of a module. Values of α for dissimilar metal combinations do not exceed 0.00005 volts per Centigrade degree while semiconductor n and p pairs have α values ranging from 0.0002 to slightly better than 0.00025 volt per C degree.

Heat Balance of a Couple. The proper direction of current flow, n to p, for the top couple to be a cooling (heat absorbing) device is shown in Fig. 1–11. However, independent of the direction of current flow, the semiconductors have electrical resistance and I^2R heat is generated in them. In addition to this, heat conduction takes place through the elements from the hot side of the module to the cold side. Thus the designer would wish the element to be long in the direction of heat flow to minimize heat conduction from the hot to the cold side but any increased length in turn increases the electrical resistance which leads to increased I^2R loss, so a compromised length and cross section for the element must be chosen. In addition, the cost of semiconductor material is high so that the net result is that elements of short length, usually 0.5 in. or less are chosen.

The I^2R loss (heat) generated within an element flows in both directions and we can assume for analysis that half of it dissipates to the cold side and half to the hot side. Thus, for the cold side, there is received by conduction, heat amounting to

$$C(T_h - T_c) + \frac{1}{2}I^2R = C\Delta T + \frac{1}{2}I^2R \tag{1-17}$$

while for the hot side, the heat leaving by conduction is

$$C\Delta T - \frac{1}{2}I^2R$$

After equilibrium is reached for the cold side, the Peltier cooling must offset, Q_r, the heat absorbed (useful cooling) plus the heat conduction in the legs of the couple and

$$\alpha_{pn} I T_c = Q_r + C\Delta T + \frac{1}{2}I^2R \tag{1-18}$$

$$Q_r = \alpha_{pn} I T_c - C\Delta T - \frac{1}{2}I^2R \tag{1-19}$$

The power, w, required from the battery (source) is represented by the product of current, I, and the voltage, E, used to force the current through the resistance offered by the semiconductors n and p (along with their contact resistances at the copper plates) and against the Seebeck voltage produced by the couple itself; thus,

$$w = IE = I(IR + \alpha_{pn}\Delta T)$$

$$w = I^2R + \alpha_{pn} I\Delta T \tag{1-20}$$

$$w = I^2R + (\alpha_p - \alpha_n) I (T_h - T_c) \tag{1-21}$$

The performance of a refrigeration system is expressed as the ratio of useful cooling to the energy required to produce that cooling and is usually expressed as *coefficient of performance*, COP. Using equations 1–19 and 1–20

$$\text{COP} = \frac{Q_r}{w} = \frac{\alpha_{pn} I T_c - C\Delta T - \frac{1}{2}I^2R}{\alpha_{pn} I\Delta T + I^2R} \tag{1-22}$$

$$\text{COP} = \frac{(\alpha_p - \alpha_n) I T_c - C(T_h - T_c) - \frac{1}{2}I^2R}{(\alpha_p - \alpha_n) I (T_h - T_c) + I^2R} \tag{1-23}$$

Basic Concepts and Instrumentation

In the preceding equations note that

$\alpha = \alpha_{pn} = \alpha_p - \alpha_n$ = difference between the values of the Seebeck coefficients for the p and n materials, in volts per degree Centigrade (or Kelvin)

I = current in amperes

T_h = temperature of the hot junction (hot-plate side), in degrees Kelvin

T_c = temperature of the cold junction (cold-plate side), in degrees Kelvin

$\Delta T = T_h - T_c$

R = electrical resistances of the elements in series including contact resistances at the copper strap connectors, in ohms. The resistance of the copper straps can be considered negligible, $R = R_p + R_n$

C = thermal conductance of both elements, $C_p + C_n$, in watts per Centigrade (Kelvin) degree, $C = C_p + C_n$

Thermoelectric Performance. The electrical resistances of the legs of a couple can be written

$$R_n = \rho_n \frac{L_n}{A_n} \quad \text{and} \quad R_p = \rho_p \frac{L_p}{A_p} \tag{1-24}$$

and the thermal conductances of the legs are similarly

$$C_p = k_p \frac{A_p}{L_p} \quad \text{and} \quad C_n = k_n \frac{A_n}{L_n} \tag{1-25}$$

and to be consistent with the preceding equations of this article, we should express equations 1-24 and 1-25 in the following units

ρ_p, ρ_n = electrical resistivities, in ohm centimeters

k_p, k_n = thermal conductivities in (watts) × (centimeters) per (square centimeter) × (degree Centigrade)

A_p, A_n = square centimeters

L_p, L_n = length of elements in centimeters

Note that the reciprocal of electrical resistance is electrical conductance, and likewise the reciprocal of thermal conductance is thermal resistance. It is obvious from an observation of equation 1-22 that the best performance of a couple unit will arise when $R = R_p + R_n$ and $C = C_p + C_n$

are minimal and when α_{pn} is a maximum. It is possible to maximize the first and second goals by arrangement of system geometry to meet the following relationship

$$\frac{A_p}{L_p} \times \frac{L_n}{A_n} = \left(\frac{k_n \, \rho_p}{k_p \, \rho_n}\right)^{\frac{1}{2}} \quad (1\text{--}26)$$

It will be assumed that the design parameters which follow are based on this geometry layout.

In order to find the maximum value of coefficient of performance with respect to current flow I, the procedure to follow is to differentiate equation 1-22 with respect to I, equate the result to zero and solve for I. This laborious procedure will show that for maximum performance the maximum current flow is

$$I_{\text{maxCOP}} = \frac{(\alpha_{pn})(T_h - T_c)}{R(\sqrt{1 + Z T_m} - 1)} \quad (1\text{--}27)$$

where

$$Z = \frac{(\alpha_{pn})^2}{C\,R} = \frac{(\alpha_p - \alpha_n)^2}{C\,R} \quad (1\text{--}28)$$

and

$$T_m = \frac{T_h + T_c}{2}$$

$$\text{COP}_{\max} = \frac{T_c}{T_h - T_c} \cdot \frac{\sqrt{1 + Z T_m} - T_h/T_c}{\sqrt{1 + Z T_m} + 1} \quad (1\text{--}29)$$

If the conditions of equation 1-26 apply for Z

$$Z = \frac{(\alpha_{pn})^2}{(\sqrt{k_p \, \rho_p} + \sqrt{k_n \, \rho_n})^2} = \left(\frac{\alpha_p - \alpha_n}{\sqrt{k_p \, \rho_p} + \sqrt{k_n \, \rho_n}}\right)^2 \quad (1\text{--}30)$$

The term Z, which has units of $1/°K$, is called the *figure of merit* of a thermoelectric assembly. It can be seen that it is independent of the temperature range and depends only on the characteristics of the couple materials; Seebeck coefficients, thermal conductivities, and electrical resistivity.

Reference to equation 1-18 will show that the maximum temperature difference ΔT is reached when the load Q_r approaches zero and

$$\Delta T = T_h - T_c = \frac{\alpha I_c T - \frac{1}{2} I^2 R}{C} \quad (1\text{--}31)$$

Differentiate equation 1-31 with respect to current I and equate to zero to find the maximum, this shows that

Basic Concepts and Instrumentation

$$I_{max} = \frac{\alpha T_c}{R} \tag{1-32}$$

$$\Delta T_{max} = \tfrac{1}{2} \frac{\alpha^2 T_c^2}{RC} \tag{1-33}$$

and by use of equations 1-24, 1-25, 1-26, and 1-30, it can be shown that

$$Z = \frac{\alpha^2}{(\sqrt{k_p \rho_p} + \sqrt{k_n \rho_n})^2} = \frac{\alpha^2}{RC} \tag{1-34}$$

and

$$\Delta T_{max} = \tfrac{1}{2} Z T_c^2 \tag{1-35}$$

To find the maximum cooling (heat pumped) by a couple differentiate equation 1-19 with respect to I, equate to zero to find a maximum value and again

$$I_{max} = \frac{\alpha T_c}{R} \tag{1-36}$$

Substitution into equation 1-19 shows

$$Q_{r(max)} = \tfrac{1}{2} \frac{\alpha^2 T_c^2}{R} - C\Delta T \tag{1-37}$$

substitute equation 1-33 in the above and

$$Q_{r(max)} = C\Delta T_{max} - C\Delta T = C(\Delta T_{max} - \Delta T) \tag{1-38}$$

The preceding equations presented give enough information on which to lay out and design thermoelectric modules.

The use of thermoelectric refrigeration for special purposes is growing but for medium and large capacities other types of refrigeration will continue to be used. Semiconductor materials are expensive and Z values would need to be appreciably higher to make thermoelectric refrigeration fully competitive.

Example 1-5. A thermoelectric cooling device uses bismuth telluride (Bi_2Te_3) elements for which the elements have Seebeck coefficients of $+240$ and -230 microvolts per °C, the resistivity is 0.0007 (ohm) (cm) and the thermal conductivity is 0.018 (watt) (cm) per (sq cm) (°C), and is essentially the same for both p and n materials, with all properties measured at 20 C. Assume that the elements are 1.3 cm long and 1.0 cm in diameter. Compute the basic characteristics of this couple to use in a design to produce a temperature of -15 C with 30 C ambient with 200 watts of cooling needed.

Solution:

$$R_p = \rho_p \frac{L_p}{A_p} = 0.0007 \frac{1.3}{\frac{\pi}{4}(1.0)^2}$$

$$= (0.0007)(1.655) = 0.001158 \text{ ohm}$$

$$R = R_p + R_n = 0.001158 + 0.001158 = 0.00232 \text{ ohm}$$

$$C = k_p \frac{A_p}{L_p} + k_n \frac{A_n}{L_n} = 2k \frac{A}{L} = 2 \left[(0.018) \frac{1}{1.655} \right]$$

$$= 0.0217 \text{ watt per }°C$$

By equation 1-28

$$Z = \frac{[240 \times 10^{-6} - (-230 \times 10^{-6})]^2}{(0.0217)(0.00232)} = 0.0044 \text{ per deg K}$$

The maximum temperature difference possible by equation 1-35 is

$$\Delta T_{\max} = (\tfrac{1}{2})(0.0044)(273.16 - 15)^2 = 146.6 \text{ deg}$$

Current for maximum cooling by equation 1-36 is

$$I = \frac{(470 \times 10^{-6})(258.16)}{0.00232} = 52.3 \text{ amp}$$

For the actual temperature lift find the maximum heat delivery by equation 1-38

$$Q_{r(\max)} = 0.0217 (146.6 - 45) = 2.2 \text{ watts}$$

To remove 200 watts would require

$$\frac{200}{2.2} = 91 \text{ couples}$$

The total resistance for 91 couples is

$$91 \times R = 91 \times 0.00232 = 0.211 \text{ ohm}$$

The voltage required, $E = IR$, becomes

$$V = (52.3)(0.211) = 11.04 \text{ volts}$$

For this small voltage the couples could justifiably be arranged in series.

This design does not represent the optimum performance which could have been found by using equation 1-27 as a starting point to find the current, I, to give the optimum COP. The resulting lower current would have given reduced capacity per couple and a designer is faced with the dilemma of choosing higher efficiency and reduced capacity or higher capacity and lower performance. If continuous service is required the first choice may be more important while for intermittent service the latter may be a better approach.

This example illustrates the methods to be followed in design layout and many other items can desirably be computed one of which is the actual COP. This is left as a problem for student solution. It should also be noted that the contact resistances of the couples should also be included in the value of R. Great care must be taken to keep the contact resistance at a minimum. It is possible to have this as low as some 5 per cent but it may rise to 20 per cent or more with careless soldering or with any inadequate attachment procedures. Figure 1-12 reproduced from

Basic Concepts and Instrumentation

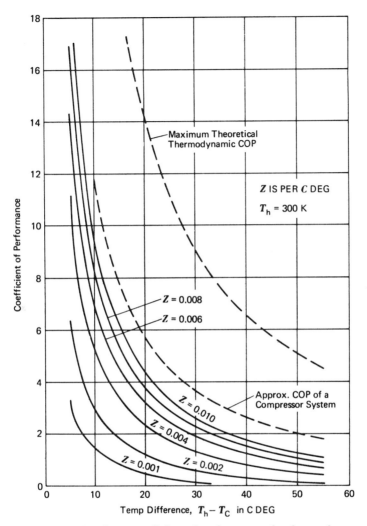

FIG. 1-12. Maximum coefficient of performance of a thermoelectric couple as a function of temperature difference and figure of Merit, Z. (From Reference 4, by permission.)

Reference 4 is a plot of the performance that can be envisaged for different Z values.

PROBLEMS

1-1. Compare the relative increase in length of a 10-ft length of copper pipe imbedded in concrete at 70 F, when the pipe, while carrying water, gradually warms to 140 F. (Obviously the increase in length is not sufficient to break or disturb the relative bond between the tube and the concrete.) *Ans.* $0.078 - 0.047 = 0.031$ in.

1-2. Write equation 1–1 in the form $L_t = L_o[1 + \alpha(\Delta t)]$ and then cube both sides of the equation. Define the volume coefficient of expansion (β_v) in an equation of the form $L^3_t = V_t = V_o(1 + \beta_v \Delta t)$ and show that for moderate temperature changes and representative coefficients of linear expansion, $\beta_v = 3\alpha$ for all practical purposes.

1-3. Change 20 C to degrees Fahrenheit and to degrees Fahrenheit absolute.
Ans. 68 F, 528 F abs

1-4. Change 14 F and 200 F to degrees centigrade. *Ans.* −10 C, 93.3 C

1-5. A mercury-in-glass thermometer is used to measure the temperature of the gases flowing through an insulated metal duct. The thermometer indicates a temperature of 640 F and has an emergent mercury thread of 100 deg showing. The ambient temperature at the stem of the thermometer is 110 F. Find the probable true temperature that would be indicated by this thermometer if completely immersed.
Ans. 644.7 F

1-6. A mercury-in-glass thermometer inserted in a thermometer well is being used to measure the temperature in a steam line. The thermometer reading is 400 F and the emergent thread of mercury outside the well is 70 deg. If the ambient temperature is 90 F at the thermometer stem, what is the probable true temperature in the steam line?
Ans. 401.9 F

1-7. An underground steam pipe, made of steel, runs 2000 ft underground. In summer, with no steam flowing, the pipe can be considered to be at a ground temperature of 60 F. In winter the pipe carries saturated steam at 115 psia at a temperature of 338.1 F. Compute the increase in pipe length in winter, which must be absorbed by expansion devices, over the summer length. *Ans.* 3.8 ft

1-8. Compare the millivoltage generated by thermocouples employed to measure the temperature of hot gases at 500 F, when the cold junction of the thermocouple is at 100 F and when the couples are (a) copper-constantan, (b) iron-constantan, and (c) chromel-alumel. *Ans.* (a) 11.1; (b) 12; (c) 9.2

1-9. Thermopiles consist of several thermocouple circuits connected in series. The cold junctions are brought out to a common region of low or standard temperature, while the hot junctions, in a separate compartment, are exposed to an unknown temperature, to a source of radiant energy, or the like. A particular thermopile has 50 iron-constantan thermocouple circuits in series. This thermopile shows 250.0 mv on its potentiometer when exposed to a source of radiation with the cold junctions held at 32 F. Find (a) the EMF per couple and (b) the temperature in the hot compartment of the thermopile.
Ans. (a) 5 mv; (b) 200 F

1-10. (a) For the thermocouple control device in Fig. 1–5, using an iron-constantan couple, find the EMF developed when the hot junction is 600 F and the cold junction is 100 F. (b) If the electrical resistance of the circuit is 0.1 ohm, what is the probable current flowing? *Ans.* (a) 15 mv; (b) 0.15 amp

1-11. Transform a pressure of 12 psig into inches of mercury at (a) 32 F and (b) 70 F. *Ans.* (a) 24.4; (b) 24.5

1-12. What is the pressure, in inches of water, of an air column 900 ft high? Assume constant air density at 0.078 lb per cu ft and water density at 62.3 lb per cu ft.
Ans. 13.5 in.

1-13. Convert a mercury column 20 in. high, measured at 70 F, into pounds per square inch, and feet of water, at 68 F. *Ans.* 9.78 psi, 22.6 ft

Basic Concepts and Instrumentation

1-14. Assume that in problem 1-13 the barometric pressure is 29.92 in. Hg at 32 F. Find answers expressed as absolute pressures. *Ans.* 24.47 psia, 56.5 ft

1-15. A less commonly used method of changing centigrade to Fahrenheit temperatures employs the following sequence of operations: To the temperature in degrees centigrade, 40 is added and the result is multiplied by 1.8. From this product, 40 is subtracted and the resulting number represents the temperature in degrees Fahrenheit. By simple proof, show that this method is the equivalent of equation 1-3.

1-16. Making use of calculated data available in example 1-5, (a) compute the power in watts required for the production of 200 watts of cooling and (b) find the actual coefficient of performance. *Ans.* (a) 577; (b) 0.349

1-17. (a) Refer to example 1-5 and compute the current flow for optimum coefficient of performance under the design conditions indicated. (b) For this current flow find the cooling produced with each couple. (c) Find the number of couples required for 200 watts of cooling. (d) Find the COP.
Ans. (a) 18.5 amp; (b) 0.87; (c) 230; (d) 0.735

1-18. A doped bismuth antimony telluride p element has a Seebeck value of 250 microvolts per °C and a doped bismuth selenium telluride n element has a coefficient of -280 microvolts per °C at $\rho = 0.0006$ (ohm) (cm) at 20 C. The corresponding k values are 0.012 and 0.013 (watt) (cm) per (sq cm) (°C) respectively. Making use of these data compute the significant items as found in example 1-5. Round elements are 1×1.3 cm long and -20 C is required in a 20 C space.
Ans. $Z = 0.0093$ per °K, $\Delta T = 298$, 67 amp, 3.9 watts, 52 couples

1-19. Rework example 1-5 assuming a contact resistance of 10 per cent is added for the total resistance with no other change made.

1-20. Prove by algebraic computation that the two sides of equation 1-34 in the text are really equivalent.

REFERENCES

1. A. F. Ioffe, "Semiconductor Thermoelements and Thermoelectric Cooling." Translated from the Russian (London: Infosearch Ltd., 1957).
2. M. W. Zemansky, *Heat and Thermodynamics*, 5th ed. (New York: McGraw-Hill Book Company, Inc., 1957), pp. 409-420.
3. H. J. Goldsmid, *Thermoelectric Refrigeration* (New York: Plenum Press, 1964).
4. ASHRAE, *Handbook of Fundamentals,* Chapter 2 (New York, 1967).
5. L. A. Stabler, "A Primer of Thermoelectric Refrigeration," *ASHRAE Journal* (August, 1959), p. 60.

2

Thermodynamics, Steam, and Gas Properties

2–1. THE FIRST LAW OF THERMODYNAMICS

Thermodynamics (that branch of science which deals with energy and its transformations), heat transfer, and fluid flow are the basic disciplines underlying air conditioning and refrigeration. The first law of thermodynamics is fundamentally equivalent to the law of conservation of energy, which states that energy can neither be created nor destroyed but can change in form. When energy does change from one form to another, it always transforms in definite fixed ratios. The first law in its simplest differential form can be written

$$dQ = du + d\mathbf{W} \qquad (2\text{–}1)$$

which states that heat Q added to a system or process transforms to internal energy, u, and to work, $\mathbf{W}$. Since this is an equation, any of these terms can be positive, zero or negative and it is customary to give Q a positive sign for heat added to a system and $\mathbf{W}$ is positive when work is being delivered from or produced by the system. The internal energy, u, is a thermodynamic property really defined by the equation and measures the thermal energy stored in the medium of the system. Heat, Q, is not a thermodynamic property but is energy in transition flowing under the impetus of a temperature gradient. Likewise work, $\mathbf{W}$, is energy in transition and we can express work as the product of force F acting through distance, x, in the direction of the force, or

$$d\mathbf{W} = F dx \qquad (2\text{–}2)$$

$$\mathbf{W}_{1-2} = \int_{1}^{2} F dx \qquad (2\text{–}3)$$

and if F is constant

$$\mathbf{W}_{1-2} = F(x_2 - x_1) \qquad (2\text{–}4)$$

Work for example is done when a mass of one pound exerting a one-pound gravitational force is being raised vertically ten feet and for this case since the force is constant, we could say that ten foot-pounds of work

Thermodynamics, Steam, and Gas Properties

have been done. Note that work in this case created potential energy of ten foot pounds stored in gravitational form and this stored energy could produce work if one had the mass to drive a suitable mechanism as it returned to its original datum, or if under free fall it transferred its potential energy to equivalent kinetic energy, KE.

Consider the case of a gas in a cylinder having work on it as the gas is compressed into smaller volume. Here, if P represents unit pressure say in pounds force per sq ft (psf) and if A is the face area of the piston in square feet,

$$F = PA$$

and

$$dW = Fdx = PAdx = PdV \tag{2-5}$$

The term dV appears since area, A, sweeping through distance dx necessarily produces a volume change dV. Equation 2-1 is thus often written in the form

$$dQ = du + PdV \tag{2-6}$$

Note that, for the case just described, since the gas volume is decreased, dV is minus in sense thus indicating that work is done on and not by the system in equation 2-6.

Consider in connection with equation 2-5 the case of a cylinder filled with a gaseous medium in which the piston is locked, so that the volume cannot change, and heat being added. In this case the heat addition to the working medium must all reappear as increase in internal energy. When matter, either gaseous, liquid or solid, does not change in phase, heat addition to it would be evidenced by a rise in its temperature as energy is stored in increased activity of its constituent molecules. For example, a bar of iron heated to redness differs from the cold bar only in that its molecular activity is greater and the greater internal energy that the bar possesses is evidenced by its greater temperature.

When a solid is heated it gradually rises in temperature until a point is reached when the molecular activity becomes so great that the substance can no longer exist as a solid and gradually begins to change over to liquid form. During this process a large amount of energy is required to bring about the change in state from solid to liquid, and the temperature remains constant during the process. This energy is known as the *heat of fusion*. If more heat is then added to the resulting liquid, the temperature will continue to rise until a temperature is reached at which the liquid begins to change into the vapor or gaseous state. Here again the temperature remains constant during vaporization, and much energy, called the *heat of vaporization*, must be added in order to effect this change of state.

Consider the familiar ice, water, steam process as an example. If ice is heated, when its temperature reaches 32 F it starts to change to liquid and the temperature will stay constant at 32 F until all the ice in contact with the water has melted. To accomplish this change, the heat of fusion of ice amounting to 144 Btu per pound (more exactly, 143.35) must be supplied. (The Btu, discussed later, is a unit of energy almost exactly equal to 778 ft-lb.) If the water were now further heated, at atmospheric pressure, it would not boil or change into vapor until a temperature of 212 F was reached. To effect this change from liquid to vapor, 970.3 Btu per pound of water would have to be supplied. This large amount of energy is required to break up the molecular bond conditions in the liquid state as compared to the vapor state, and to supply the energy required as the increased volume of the steam over that of the liquid makes room for itself in expanding against the surrounding atmospheric pressure. The temperature remains constant during this vaporization. If the resultant vapor is further heated out of contact with the liquid, there will be a continuous temperature rise above the saturation or boiling temperature and such vapor is called superheated. If the whole process is reversed and heat is removed, exactly the same changes will occur, but in reversed order. During condensation, which will take place at a constant temperature of 212 F, 970.3 Btu per pound of vapor condensed must be removed to permit this change; and during freezing, which will take place at 32 F, 143.3 Btu per pound of water must be removed.

This form of energy which is stored in the molecules or atoms of a substance is known as *internal energy*. It is a form of thermal energy and may be loosely and inaccurately called "heat." It is preferable always to call the energy, which is associated with the substance itself, internal energy, and to reserve the term "heat" for the energy transfer occurring whenever differences in temperature exist. *Heat* will thus be restricted to describe the energy flow, or energy in transition, which results from the driving action of a temperature gradient or difference.

In the last century much experimental work was conducted to determine the relationship of mechanical work to heat and internal energy. The historic work of James Prescott Joule in 1843 to 1850, and later that of Henry A. Rowland, showed that an invariant relation held between these forms of energy. Both Joule and Rowland demonstrated by doing work on water, such as by intensively paddling or churning it, that exactly the same results could be obtained as if the water were heated by fire. From such experiments it was found that almost exactly 778 ft-lb are equivalent to one British thermal unit (Btu). The Btu was defined as the amount of heat required to raise the temperature of one pound of water through one degree Fahrenheit (from 62 F to 63 F). Although the specific heat of water is approximately unity, it is not exactly unity over a range of

Thermodynamics, Steam, and Gas Properties   35

temperature, and so the Btu is more accurately defined as 1/180 the amount of heat required to raise one pound of water from 32 F to 212 F. The Btu is 778.26 ft-lb, based on the International Steam Table Calorie. Based on other evaluations of fundamental constants, it often appears as 778.16.

2–2. THE OPEN SYSTEM AND THE STEADY-FLOW ENERGY EQUATION

Our considerations, up to this point, have tacitly considered the first law in relation to a closed system, that is one in which a medium participated in energy interchanges but was of itself invariant in mass. Of much greater interest for thermodynamic application is a so-called open system, which is one into which mass can flow at a given time rate or from which, except for periods of variable mass storage in the system, mass at an equivalent time rate of flow leaves the system. The medium flowing into the system has associated with it various forms of energy, as is similarly the case for the medium leaving the system. Energy also can be delivered, or abstracted from, the flowing medium as it passes through the system, with this interchange most commonly occurring as heat or as work (electricity) through shafts or wires leading to the external environment. For purposes of this discussion, we can define a system as any convenient region that we wish to isolate for purposes of analysis and then consider the mass and energy interchanges which take place at its boundaries.

Fortunately, many real processes involve essentially steady flow, that is the time rate of mass flow into and the time rate of mass flow from a system are equal, and we can expect the law of continuity of mass as being applicable for the flow patterns of the system. Many illustrations of steady-flow systems exist. For example, in a water pump, as the fluid flows through the pump at an essentially constant rate the water absorbs energy from the work transmitted into the system by the pump shaft. A steam boiler, after it has reached equilibrium operation, is essentially a steady-flow device receiving a constant supply of feed water per hour and sending out an equivalent weight of steam per hour. Heat from the burning fuel flows into the water, changing the condition of the water to that of the delivered steam at a greater energy level. Many devices in engineering employ processes which are steady-flow in character, or which approach this condition so closely as to permit treatment by steady-flow methods. Other examples of such equipment are steam turbines, nozzles, centrifugal compressors, and even reciprocating machines, such as steam engines and piston-type compressors.

When the laws of the conservation of energy and continuity of mass flow are applied to a steady-flow system or process, a useful equation can be developed which is known as the steady-flow energy equation. To develop the equation, imagine a device of any kind (boiler, pump, compressor, etc.) to which a fluid is supplied. In the device (Fig. 2–1), work can be added or removed, as through a power shaft. Heat can be added, as from a fire or from steam coils, or removed, as by a refrigeration evaporator. Finally the fluid departs, with energy different from that which it possessed when it entered the device.

Considering the energy associated with each pound of fluid, it is found that at entry each pound has:

1. Potential energy, in amount $(1)(Z_1)$ ft-lb above a convenient datum plane.
2. Kinetic energy, in amount $\dfrac{(1)V_1^2}{2g}$ ft-lb, dependent on its velocity at point (1).
3. Internal energy, in amount $(1)(u_1)$ Btu, or $778u_1$ ft-lb.
4. Flow work, in amount P_1v_1 ft-lb.

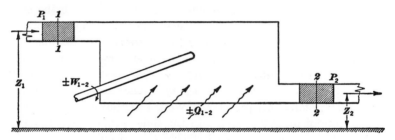

FIG. 2–1. Steady-flow energy device.

Flow work, which may also be called the *work of intrusion*, or the *work of extrusion*, is the energy required to force each pound of the fluid, having a specific volume of v_1 cu ft per lb, into the system at the constant pressure P_1 psf existing during inlet steady-flow conditions (or to force each pound from the system under conditions of v_2 and P_2). Proof:

Work = force × distance
Work = pressure × area × distance
Area × distance = volume

Therefore   Work = Pv

This proof is true under the steady pressure P existing when the volume v of one pound of the fluid is being forced into or from the device by the succeeding portions of fluid.

This flow work has little significance except for steady-flow processes. Its use will be illustrated in some later examples.

The fluid, having entered the device, bringing with it the forms of energy heretofore enumerated, can now do *mechnical work* in amount

$-W_{(1-2)}$ or mechanical work can be added to the fluid in amount $+W_{(1-2)}$. The subscripts *1–2* under W simply signify that work occurs between the fluid inlet *1* and outlet *2* points. In the case of a steam engine, work energy leaves the fluid through the driven shaft in amount $-W_{(1-2)}$. In the case of a compressor, work is supplied to the fluid system through the driven shaft in amount $+W_{(1-2)}$. In the engine example, some of the work $-W_{(1-2)}$ generated in the cylinder may be dissipated as friction in the bearings; not all of it will appear as delivered energy from the shaft. Such a condition also exists in an opposite sense for a compressor.

Heat in amount $+Q_{(1-2)}$ Btu per lb, or $+778 Q_{(1-2)}$ ft-lb per pound of fluid flowing, can be added in certain devices, as from a fire in a boiler, or perhaps may leave, in amount $-778 Q_{(1-2)}$ ft-lb per lb, by radiation or conduction to cooling coils, colder surroundings, or the like.

From the law of conservation of energy, the energy in the leaving fluid must equal the energy in the entering fluid plus any energy supplied, or minus any energy diminution, in the device itself. Thus, in mathematical form, for one pound of fluid the energy equation appears as follows:

$$Z_1 + \frac{V_1^2}{2g} + 778 u_1 + P_1 v_1 \pm W_{(1-2)} \pm 778 Q_{(1-2)}$$
$$= Z_2 + \frac{V_2^2}{2g} + 778 u_2 + P_2 v_2 \quad (2\text{-}7)$$

The term $u + (Pv/778)$ is of particular interest and has been named *enthalpy*. This composite term, consisting of the sum of internal energy and flow work, is the term which is tabulated in most tables of vapor properties. The conventional symbol for enthalpy is **h**.

$$\mathbf{h} = u + \frac{Pv}{778} \text{ Btu per lb} \quad (2\text{-}8)$$

where u = internal energy, in Btu per pound;
P = absolute pressure, in pounds per square foot;
v = specific volume, in cubic feet.

If equation 2–7 is divided through by 778, and if the enthalpy h is substituted for $u + (Pv/778)$, there results

$$\frac{Z_1}{778} + \frac{V_1^2}{2g(778)} + \mathbf{h}_1 \pm \frac{W_{(1-2)}}{778} \pm Q_{(1-2)} = \frac{Z_2}{778} + \frac{V_2^2}{2g(778)} + \mathbf{h}_2 \quad (2\text{-}9)$$

Equation 2–9 is more convenient than equation 2–7 for steady-flow processes involving steam or a refrigerant vapor.

The general steady-flow energy equation 2–7 or 2–9 is applied to a given process by eliminating negligibly small or irrelevant terms to obtain the simplest possible form of equation.

Apply equation 2–9 first to a steam boiler. Considering the $Z/778$ terms, it can be shown that where steam is concerned, small changes in potential energy are negligible, since the enthalpy term values have magnitudes about 1000 Btu per lb (778,000 ft-lb), and thus differences in elevation up to 100 ft (100 ft-lb per lb) are relatively negligible. Considering the kinetic-energy terms, it will be realized that they are small and that entering and leaving values will practically cancel out in well-designed feed and discharge pipes. The work term also does not appear, as no shafts supply or remove energy from a boiler. There remains the term $+Q_{(1-2)}$, for heat added, and the equation appears as follows:

$$\mathbf{h}_1 + Q_{(1-2)} = \mathbf{h}_2$$

Thus, for a boiler,

$$Q_{(1-2)} = \mathbf{h}_2 - \mathbf{h}_1 \text{ Btu per lb} \qquad (2\text{--}10)$$

is the heat added and is equal to the difference in enthalpies of the leaving steam ($\mathbf{h}_2$) and the feed water entering ($\mathbf{h}_1$).

It should be noted that this equation does not apply to a nonsteaming boiler, such as would be encountered when a cold boiler is being warmed up for service. It happens that for this case the heat added equals the difference in the u's based on each composite pound of water-steam at the end of the heating period and each composite pound of water-steam at the beginning of the heating period.

Let us apply the energy equation 2–7 to a fan. The Z terms are relatively negligible for slight changes in elevation of air. Work is supplied in amount $+\mathbf{W}_{(1-2)}$. The heat flow $Q_{(1-2)}$ is usually negligible unless the surrounding air is much hotter or colder than the circulated air. Evaluating, we have

$$\mathbf{W}_{(1-2)} = \frac{V_2^2 - V_1^2}{2g} + P_2 v_2 - P_1 v_1 + 778(u_2 - u_1) \text{ ft-lb}$$

The work applied to a fan appears primarily as the change in kinetic energy of the leaving and entering air at the points of measurement and as the increase in the flow-work term. The frictional losses, and a portion of the compression work, appear as an increase in the u, or internal energy, term. The slight increase in temperature of the air because of a change in u is difficult to measure or even detect, and in fan testing the other two terms, which might be called the useful output of the fan, are the ones considered. Thus, per pound of air,

$$\mathbf{W}_{(1-2)} = \frac{V_2^2 - V_1^2}{2g} + (P_2 - P_1) v_1 \text{ ft-lb} \qquad (2\text{--}11)$$

As the specific volume of the air, v_1, does not change appreciably in going through a fan, v_1 may be called equal to v_2 and the equation appears as above. The $(P_2 - P_1)$ term must be expressed in consistent units, usually

Thermodynamics, Steam, and Gas Properties

pounds per square foot of pressure change, and represents what is called the static-pressure change in the fan. Actually, in fan work it is customary to measure this change in pressure units of inches of water. The equation given, with the kinetic-energy term present, is called the *total fan work*. If the kinetic-energy term is removed, because the practical application of the fan prevents any possible utilization of the kinetic energy in the delivery, the equation represents the *static fan work*.

These two illustrations show how the steady-flow energy equation is used. For any practical problem, the method of use is to eliminate irrelevant or negligible terms and then to analyze what remains. Further use of this equation will be made in other parts of the book.

2-3. ENERGY AND POWER

Before further thermodynamic relationships can be developed consideration should be given to describing logical and workable systems of units. In the world today there are two major systems of units in use, the English (British) system used in engineering practice largely throughout the English-speaking world, and the metric (meter, gram second) system almost universally used for scientific activity throughout the world and, in the non-English speaking areas, it is also used in daily life. The English system uses the pound, foot, second for its fundamental units of mass, length, and time. It is a consistent and workable system but it is extremely confusing to those accustomed only to the metric system because its nondecimal character makes it less flexible and it still carries with it many archaic vestiges of former ages. For example, there are three kinds of pounds representative of mass measure, namely, the common avoirdupois pound, the troy pound used in coinage, and its equivalent the apothecary pound used in medicine and in the drug industry. In the avoirdupois pound there are 16 ounces, in the troy and apothecary pound there are 12 ounces. Neither the avoirdupois pound nor the avoirdupois ounce equals its troy (apothecary) equivalent. In fact, the only common unit in all three pounds is the *grain* and 7000 grains equal 1 lb avoirdupois while 5760 grains equal 1 lb troy (or apothecary).

Fortunately engineering practice is concerned only with the avoirdupois pound and this is rigorously and legally defined in terms of an internationally recognized standard piece of platinum having a mass of one kilogram. In terms of this

$$\text{one pound} = 453.59 \text{ grams} = 0.45359 \text{ kg}$$

The inch is defined in terms of the meter as

$$\text{one inch} = \frac{1}{12} \text{ foot} = 0.025400 \text{ meters} = 2.5400 \text{ centimeters}$$

In the English system the unit of force is also called the pound. Force can be related to mass by making use of Newton's Second Law, expressed $F \sim m\,a$, or, in equational form,

$$F = \frac{m}{g_c} a \qquad (2\text{--}12)$$

Here g_c can be considered a constant of proportionality. The value of g_c could be unity if the unit of mass were so chosen that we could state:

A 1-pound force will give unit mass an acceleration of 1 foot per second. The so-called *slug* is the unit of mass that meets this requirement. However, as previously stated, the unit of mass in the English system is the pound and it happens that

$$1 \text{ slug} = 32.1740 \text{ lb}_m$$

A force of 1 lb can be defined as the gravitational pull on a 1 lb mass at a location where the acceleration of gravity is standard at 32.1740 ft per sec², or, expressed in another way, a 1 lb$_f$ is capable of giving a 1 lb mass an acceleration of 32.1740 ft per sec².

$$1 \text{ lb}_f = 1 \text{ lb}_m \times 32.1740 \text{ ft per sec}^2 \qquad (2\text{--}13)$$

$$1 \text{ lb}_f = 1 \text{ slug} \times 1 \text{ ft per sec}^2$$

Thus the numerical value of g_c in the English system is 32.1740. The units for g_c, if we refer to equation 2-12 are,

$$g_c = \frac{m}{F} a = 32.1740 \frac{\text{lb}_m}{\text{lb}_f} \frac{\text{ft}}{\text{sec}^2} = \frac{\text{slug ft}}{\text{lb}_f \text{ sec}^2} \qquad (2\text{--}14)$$

or g_c can be considered to be merely the ratio of two mass units

$$g_c = \frac{\text{slug}}{\text{lb}_m} = 32.1740$$

There is no inconsistency in g_c having a dual character since its non-dimensional aspect arises when Newton's Second Law is used as a dimensional definition for force, $F = M^1 L^1 T^{-2}$, and in this connection g_c is merely a numerical ratio.

The weight of a body is the force which the earth's gravitational pull exerts on the mass of that body at any point on the earth. However, since the acceleration of gravity on the earth's surface is not greatly variable, a one-pound mass, weighing one pound, exerts closely one pound of force almost everywhere. Since weighing is usually done by comparing the mass to be weighed with other masses (weights), calibrated in standard pounds, weighing is a precise measure of the mass of a body, except where spring balances are used.

Thermodynamics, Steam, and Gas Properties 41

In the metric system a gram force (gr_f) can be defined as

$$1 \text{ gram}_f = 1 \text{ gram}_m \times 980.665 \text{ cm/sec}^2 \qquad (2\text{--}15)$$

where 980.665 is the standard acceleration of gravity in cm per sec². The metric system does not employ a mass unit similar to the slug but sets as its conventional units of force the dyne where

$$1 \text{ dyne} = 1 \text{ gram} \times 1 \text{ cm per sec}^2 \qquad (2\text{--}16)$$

and the Newton, where

$$1 \text{ Newton} = 1 \text{ kilogram} \times 1 \text{ meter per sec}^2$$

But

$$1 \text{ gram}_f = \frac{\text{gram}_m}{g_c'} \times 1 \text{ cm per sec}^2$$

where

$$g_c' = \frac{980.665 \text{ gr}_m \text{ cm}}{\text{gr}_f \text{ sec}^2}$$

Similar to the dyne, in the English (British) system, we can define a force unit called the poundal

$$1 \text{ poundal} = \frac{\text{lb}_f}{32.174}$$

Note that 1 pound force = 4.44 Newtons = 32.174 poundals.

Example 2–1. A ton (2000 lb_m) of supplies loaded on a plane in Chicago is delivered to north-central Greenland. The supplies are reweighed on arrival in Greenland on a conventional platform scales using standard weights on its beam. The local acceleration of gravity in Chicago g_L = 32.16 ft per sec² and g_L = 32.24 ft per sec² in north-central Greenland. (a) What did the supplies weigh in Greenland compared to their prior weight in Chicago? (b) What force did these supplies exert on their loading platforms when in Chicago and when in Greenland?

Solution: (a) The supplies weighed exactly the same at both places provided they were weighed on scales using standard weights.

b) To find the force exerted, use Newton's second-law equation

$$F = \frac{w_m a}{g} = \frac{w_m g_L}{g}$$

$$F = \frac{(2000)(32.16)}{32.174} = 1999.5 \text{ lb}_f \text{ in Chicago}$$

$$F = \frac{(2000)(32.24)}{32.174} = 2004.4 \text{ lb}_f \text{ in Greenland}$$

Corresponding to the British thermal unit (Btu) of the English system we meet the calorie of the metric system which must further be clarified as the kilogram-calorie (large calorie) or the gram-calorie (small calorie). In this text when calorie is used it will usually refer to the kg cal, equivalent to 1000 gram-cal. The kg cal is equivalent to the heat required to raise 1 kilogram of water from 14.5 C to 15.5 C or 1/100 of the heat required to raise 1 kg of water from 0 C to 100 C.

$$1 \text{ Btu} = 0.251996 \text{ calorie (kg cal)} \approx 0.252 \text{ calorie} = 778.16 \text{ ft lb}_f$$

$$1 \text{ Calorie (kg cal)} = 3.968 \text{ Btu} = 3088 \text{ ft lb}_f$$

The conventional work units in the metric system are the erg where 1 erg = 1 dyne-centimeter and the kilogram-meter where

$$1 \text{ kg}_f \text{ meter} = 7.233 \text{ ft-lb} = 9.8066 \times 10^7 \text{ ergs}$$

$$1 \text{ ft lb}_f = 0.1383 \text{ kg}_f \text{ meter} = 1.355 \times 10^7 \text{ ergs}$$

Power is defined as the rate at which work is performed. The *horsepower* (hp) a common unit in engineering practice is by definition work performed at the rate of 33,000 ft-lb$_f$ per minute, 550 ft-lb$_f$ per second, 2544 Btu per hour, or 0.7068 Btu per second. The power unit in the metric system is the watt or more commonly the kilowatt (kw)

$$1 \text{ watt} = 10^7 \text{ ergs per sec} = 1 \text{ joule per sec}$$

$$1 \text{ kw} = 1.34 \text{ hp} = 3412.8 \text{ Btu per hr} = 10^{10} \text{ ergs per sec}$$

The application of power for a period of time creates energy so terms such as horsepower-hour (hp-hr) and kilowatt hour (kwhr) are no longer power units but energy units. Thus,

$$1 \text{ kwhr} = 3413 \text{ Btu} = 10^{10} \text{ ergs}$$

A final power unit used extensively in refrigeration and cooling is the *ton of refrigeration* which is energy interchange at 200 Btu per minute, 12,000 Btu per hour, or 3.51 kilowatts. Although the name is a complete misnomer, the term is used so extensively that it must be mentioned.

Common energy and power units are thus a conglomeration of the English and metric systems and flexibility in changing back and forth is essential for those working in the area. There is coming into play increased usage of the metric system in engineering technology. More precisely, the metric system should be called the International System of Units (SI System).

In sections 2-2 and 2-3 unusual attention has been given to illustrating the relationships between the types of units that are employed for energy and other physical measures. Similar care is required in all numerical solutions whether these are done in the English, metric or a mixed system of units. However, in the derivation of equations and other relationships, units are usually disregarded until the final formulation is made following which consistency in use of units again is imperative.

2-4. THE SECOND LAW AND ENTROPY

The second law of thermodynamics is an expression of man's experiences and the conclusions he has reached from his dealings with nature. One wording of the second law states: It is impossible to develop a machine that can operate continuously at the expense of energy from a single reservoir at one temperature. The significance of this statement is not at first apparent, unless we realize that if it were false, boats could travel over the ocean by merely taking energy from the sea and using this energy to power the vessel. It is indeed possible to develop work from a single reservoir at a given temperature but to do so continuously leads to an impossible situation, in terms of our experience. We find that at least two reservoirs at different temperatures are needed if an engine is to produce work continuously, with the engine taking energy from the higher temperature source, producing some work from this energy and then discharging the remainder of the energy, received by its working medium, to a sink reservoir at lower temperature. We should note that temperature is a peculiar property of matter which can indicate the direction in which heat will flow of itself between two bodies at different temperatures. In terms of temperature an alternate wording of the second law is sometimes employed: It is impossible to form a self-acting machine which can make heat flow continuously from a region of lower to one of higher temperature.

Analyses based on the second law, in addition to considerations regarding temperature, bring into play the concept of reservoirs and the concept of reversibility. Reversibility implies for a thermal process, that merely changing the direction of the driving force by an infinitesimal amount will cause a process to reverse itself and return to essentially its initial condition. Reversibility is an idealized or limiting condition that can never be fully realized because of the natural phenomena which inhibit its realization, one of the most obvious of which is friction. Heat flow to be capable of reversibility must take place under infinitesimally small temperature differences and would never be possible with the finite and often large temperature differences that naturally occur. Fluid friction and turbulence negate reversibility as is also the case for a fluid suddenly expanding into a region of lower pressure or even into a vacuum. Nevertheless, as a measure of optimum attainment or as a base from which to make comparisons, reversibility is a valuable concept.

Stemming from the idea of reversibility, we can arrive at two extremely important conclusions. First, no engine operating between two reservoirs at different temperatures can have a higher efficiency than a reversible engine, and second if this is true, all reversible engines operating between the two reservoirs will have the same efficiency. On the basis of this premise it is possible to develop the so-called absolute thermodynamic temperature scale, which was first envisaged by Carnot and later developed by Lord Kelvin.

The efficiency of any heat engine working in a cycle with the medium returning to its original state can be expressed:

$$\eta = \frac{W}{Q_a} = \frac{Q_a - Q_R}{Q_a} = 1 - \frac{Q_R}{Q_a} \qquad (2\text{-}17)$$

since by the first law, if a cycle exists, the work, **W**, must be equivalent to the difference between the heat added, Q_a, and the heat rejected Q_R. We can envisage that for each reservoir a temperature scale can be found that bears a direct relationship to the heat delivered from the reservoir to a reversible engine, Q_a, and to the heat rejected by the reversible engine, Q_R, to the sink reservoir. Such a temperature scale, which we could justifiably call the thermodynamic scale of temperature, would define efficiency in the same way as equation 2–17 and could be written:

$$\eta = \frac{W}{Q_a} = \frac{T_a - T_R}{T_a} = 1 - \frac{T_R}{T_a} \qquad (2\text{-}18)$$

It follows that

$$\frac{Q_R}{Q_a} = \frac{T_R}{T_a}$$

and

$$\frac{Q_a}{T_a} = \frac{Q_R}{T_R} \qquad (2\text{-}19)$$

This last relationship for a reversible process brings together the ratio of the heat received (or rejected) to the thermodynamic temperature at which the heat exchange occurred and tacitly indicates that each exchange occurred at a constant temperature. Since many processes occur under conditions for which the temperature is changing, let us introduce the symbol, s, and use infinitesimals for representation of varying-temperature heat addition or rejection, then:

$$ds = \frac{dQ}{T}\bigg)_{rev} \qquad (2\text{-}20)$$

$$s_2 - s_1 = \int_1^2 \frac{dQ}{T}\bigg)_{rev} \qquad (2\text{-}21)$$

The name entropy has been given to the above defined ratio, with the symbol s, and it should be noted that by its definition and derivation it is applicable only for reversible conditions and for this reason the restriction, *rev*, has been noted in equations 2–20 and 2–21. The same restriction also holds for equation 2–19 even though not written there.

Thermodynamics, Steam, and Gas Properties

Entropy has been developed at this point primarily as a mathematical concept yet it can be used in many ways. Some of these will be shown, and at the same time its physical significance will be developed. However, at this point only two comments will be made: 1. Entropy, s, is a thermodynamic property in the same sense that internal energy, u, pressure, P, temperature, T, and volume v, also are properties. 2. Entropy can be used for analyzing irreversible processes by replacing the irreversible process between two state points by reversible paths which can lead to the same end point.

Establishment of quantitative values for the thermodynamic temperature scale is possible since computations can be carried out by which the efficiency of a reversible engine working between any chosen temperature state points can be found. The most convenient reference points for such a computation are the boiling point of water at one atmosphere 212 F or 100 C and the ice point 32 F or 0 C. The result of such a computation will show an efficiency of 26.80 per cent and thus

$$\eta = 1 - \frac{T_{\text{ice point}}}{T_{\text{steam point}}} = 0.2680$$

$$T_{\text{steam point}} - T_{\text{ice point}} = 180 \text{ degrees F}$$

$$T_{\text{steam point}} - T_{\text{ice point}} = 100 \text{ degrees C}$$

Solving the above equalities with the efficiency equation we find that on the Fahrenheit scale, the absolute or thermodynamic work scale temperature is

$$T_{\text{steam point}} = 671.69 \text{ R}$$

$$T_{\text{ice point}} = 491.69 \text{ R}$$

Here R stands for the absolute temperature, expressed in degrees Rankine, and Rankine and Fahrenheit temperatures are related thus:

$$°R = °F + 459.7$$

or sufficiently closely in most cases

$$°R = °F + 460$$

Similarly on the Centigrade scale we find

$$T_{\text{steam point}} = 373.16 \text{ K}$$

$$T_{\text{ice point}} = 273.16 \text{ K}$$

where K stands for absolute temperature, expressed in degrees Kelvin, and Kelvin and Centigrade temperatures are related thus:

$$°K = °C + 273.16$$

In thermodynamic computations absolute temperatures on any chosen scale, °R or °K, should always be employed except where the computation involves merely temperature differences in which case the additive term 459.7 or 273.16 is automatically eliminated.

2-5. SPECIFIC HEAT

The amount of energy required to raise the temperature of a unit weight of a substance one degree in temperature is known as the *specific heat*. In ordinary engineering units, specific heat is customarily expressed as Btu per pound degree Fahrenheit (numerically equivalent to specific heat in metric units of calories per gram degree centigrade). The specific heat of all substances varies with temperature. However, over a relatively small temperature range, the mean or average value may be used with sufficient accuracy for many engineering problems.

There are two kinds of specific heat that are of importance when dealing with gases: that at constant volume, C_v, and that at constant pressure, C_p. If, for example, one pound of air is heated in a closed container in which no change in volume is possible, 0.171 Btu will be required to cause a temperature rise of one degree Fahrenheit. If, however, the air is heated through the same temperature range but is free to expand against the surrounding constant pressure, 0.24 Btu will be required. This extra energy in the specific-heat value really represents the work that must be done by the air as it expands against the surrounding pressure. For liquids the volume change is so small that the distinction between C_p and C_v is negligible.

To represent these specific heats in mathematical terminology, let us differentiate our defining expression for enthalpy,

$$h = u + Pv$$

then

$$dh = du + Pdv + vdP \tag{2-22}$$

For a constant pressure process this becomes

$$\partial h)_p = (\partial u + P\partial v)_p \tag{2-23}$$

Express the first law for unit mass as

$$dQ = du + Pdv \tag{2-24}$$

$$C_p = \frac{dQ}{dt}\bigg)_p \tag{2-25}$$

or in partial differentiation notation

$$C_p = \frac{\partial u + p\partial v}{\partial t}\bigg)_p = \frac{\partial h}{\partial t}\bigg)_p \tag{2-26}$$

TABLE 2-1
PROPERTIES OF GASES AND VAPORS

Gases and Vapors	Chemical Symbol	Molecular Weight	Gas Constant for $PV = WRT$		Density of Vapor at 32 F and 14.7 (psi)	Melting Point at 14.7 PSI (deg F)	Latent Heat of Fusion (Btu per lb)	Specific Heat of Liquid (Btu per lb deg F)		Boiling Point at 14.7 PSI (deg F)	Specific Heat (Btu per lb deg F Vapor at about 70 F)		$k = C_p/C_v$	Saturated Vapor Conditions		
			R for P (psf)	R, for P (psi)				At Deg F Indicated	C_p		C_p	C_v		Pressure (psi)	h_{fg} Latent (Btu per lb)	Density (lb per cu ft)
Air		28.97	53.34	0.3704	.08071						0.240	0.171	1.40	14.7	91.8	.368
Argon	Ar	39.90	38.70	0.2688	.11135	-306					0.124	0.074	1.66	14.7	67.9	.699
Helium	He	4.00	386.0	2.6805	.01114	-456					1.25	0.75	1.66	14.7	10.7	.083
Hydrogen	H_2	2.02	765.86	5.3185	.005611	-434	25.2				3.42	2.44	1.40	14.7	192	.288
Nitrogen	N_2	28.02	54.99	0.3819	.07807	-346.2	11.0				0.246	0.176	1.40	14.7	86	.296
Oxygen	O_2	32.00	48.25	0.3351	.08912	-362.4	6.0				0.217	0.155	1.40	14.7	91.8	
Vapors for Which $PV = WRT$ Can Be Used Only If Superheated																	
Acetylene	C_2H_2	26.02	59.34	0.4121	.07323	-107.8	194.4			-118.5	0.361	0.290	1.26	14.7	589.3	.0555
Ammonia	NH_3	17.03	90.50	0.6285	.04813	-107.8	194.4		1.12	-28.1	0.523	0.399	1.31	14.7	118.5	3.76
Carbon dioxide	CO_2	44.00	35.12	0.2438	.12341	-69.9	81.4	5–86	0.478	-109.3	0.206	0.156	1.30	332 (5 F)		
Carbon monoxide	CO	28.00	55.14	0.3829	.07807	-340.8	14.4	32	0.0589	-312.7	0.243	0.172	1.40	14.7	90.7	.174
Dichlorodifluoromethane (F-12)	CCl_2F_2	120.92	12.77	0.0887	.338	-247		32	0.22	21.5	0.148	0.129	1.14	14.7	72.0	
Dichloroethylene (dielene)	$C_2H_2Cl_2$	96.93	15.93	0.1106				–339–(–310)	0.268	119	0.18	0.143	1.14	6.9 (86 F)	133	.1176
Dichloromonofluoromethane (F-21)	$CHCl_2F$	102.92	15.00	0.1042				5–86	0.258	48	0.18	0.161	1.12			
Dichlorotetrafluoroethane (F-114)	CCl_2F_4	158.92	9.72	0.0672		-277.6				39						
Ethane	C_2H_6	30.05	51.38	0.3568	.08469	-272.9				-127	0.413	0.345	1.22	14.7	211	
Ethylene	C_2H_4	28.03	55.08	0.3825	.07868	-217.7		32	0.385	-154.8	0.385	0.308	1.25			
Ethyl chloride	C_2H_5Cl	64.50	23.94	0.1663		-229		32		55.5	0.27	0.23	1.17	14.6	164.5	.1845
Isobutane	C_4H_{10}	58.08	26.58	0.1846	.1669	-296.5	26.2		0.368	10.5	0.550	0.495	1.11	14.6	158.5	.174
Methane	CH_4	16.03	96.31	0.6688	.0448	-296.5	26.2		0.55	-258.5	0.515	0.390	1.32	15.0	248.4	.1548
Methyl chloride	CH_3Cl	50.48	30.59	0.2124	.14406	-144		5–86	0.35	11.0	0.24	0.20	1.29	14.7	184.2	
Methylene chloride (carrene #1)	CH_2Cl_2	84.93	18.18	0.1263		-142.0		5–86	0.37	104.6	0.154	0.128	1.20		138.5	
Methyl formate	$C_2H_4O_2$	60.03	25.72	0.1786		-147.6		55–84	0.516	89	0.20	0.18	1.12		204.6	
Nitric oxide	NO	30.01	51.40	0.3569	.08367	-268.6	33.1			243.4	0.231	0.165	1.40			
Nitrous oxide	N_2O	44.02	35.04	0.2433	.1235	-152.3				-129.6	0.221	0.171	1.26	14.7		.1292
Propane	C_3H_8	44.06	35.04	0.2433	.1261	-309.8		32	0.576	48	0.576	0.510	1.13	12.6(-50 F)	184.5	.176
Sulphur dioxide	SO_2	64.07	24.10	0.1674	.18272	-104.8	-98.9	5–86	0.346	13.6	0.154	0.123	1.25	13.4 (10 F)	168.1	.0397
Trichloroethylene	$CHCl_3$	131.38	11.75	0.0816		-123.5		68	0.223	188.5	0.12	0.105	1.14	1.72(86F)	109.5	
Trichloromonofluoromethane (F-11)	CCl_3F	137.37	11.24	0.0781				35–90	0.136	71.7	0.14	0.12	1.13	14.7	78.8	
Trichlorotrifluoroethane (F-113)	$C_2Cl_3F_3$	187.37	8.24	0.0572						119	0.16	0.15	1.08			
Water	H_2O	18.01	85.6	0.5944		+32	143.3	32–212	1.0	212	0.46	0.36	1.28	14.7	970.3	.03728

Similarly for constant-volume specific heat

$$C_v = \left(\frac{dQ}{dt}\right)_v = \left(\frac{\partial u}{\partial t}\right)_v \qquad (2\text{-}27)$$

Table 2-1 gives values of specific heat for gases and vapors, and Table 4-1 (page 136) gives values of specific heat for miscellaneous substances. It should be noticed that the value for water at moderate temperatures is unity, and also that the value for most other substances is less than unity. Specific-heat values of materials such as meats and vegetables containing varying amounts of water are variable. The specific heat of a frozen or solidified substance is different from that of the same substance in liquid form; for example, C_p for ice is 0.487, C_p for water is about 1.0. Specific-heat values for most substances increase with temperature. This is particularly true of gases, for which the specific heat shows a large increase in temperature ranges near and above 1000 F. However, it is fortunately true that in the range below 500 F, variations in the specific-heat values are not excessive.

2-6. THE PERFECT GAS

A gas which fully satisfies the equation of state represented by

$$Pv = RT \qquad (2\text{-}28)$$

is called a perfect or ideal gas. While no actual gas has an equation of state as simple as that represented by equation 2-28, nevertheless many gases in the range where their temperatures are far from their critical temperature and where their pressures are low in relation to their critical pressure closely obey this relationship. Equation 2-28 can be derived, if a derivation is considered necessary, by making use of Boyle's law which states that at constant temperature the volume of a fixed mass of gas varies inversely with its pressure and by use of Charles' law which states that at constant pressure the volume of a gas varies in direct relation to its absolute temperature. These laws readily show that

$$\frac{Pv}{T} = \frac{P_1 v_1}{T_1} = \frac{P_2 v_2}{T_2} = R \qquad (2\text{-}29)$$

R is a constant for a particular gas with v appearing as specific volume. The expression can be generalized for any volume V by introducing the mass of gas, m,

$$\frac{PV}{Tm} = \frac{PV_1}{T_1 m_1} = \frac{PV_2}{T_2 m_2} = R$$

$$PV = mRT \qquad (2\text{-}30)$$

In consistent units for the English system:

Thermodynamics, Steam, and Gas Properties

P = pressure, in pounds per square foot = 144 p, when p is in pounds per square inch;
T = temperature, in degrees Rankine
V = volume, in cubic feet;
m = pounds of gas;
v = specific volume, in cubic feet per pound;
R = gas constant in foot-pounds per pound degree Rankine.

The gas constant R is different for each gas when based on unit mass of that gas. However, for a mol (molecular weight) of a gas, the value of molal **R**, the Universal Gas Constant, is invariant and is the same for each and every gas. It has the following numerical values:

R = 1545.3 ft-lb per pound mol degree Rankine
  = 1.986 Btu per pound-mol degree Rankine
  = 1.986 kilo-calories per kg-mol degree Kelvin, calories per gram-mol degree Kelvin
  = 0.84789 kg-meter per kg-mol degree Kelvin.

The mol values (molecular weights) of many gases are tabulated in Table 2-1.

Thus, in usage we find the value of R for use in equation 2-30, with M the molecular weight, as follows

$$R = \frac{1545.3}{M} \quad \text{(2-31)}$$

If in equation 2-30 pressure is expressed in pounds per square inch and one wishes to omit use of the multiplying factor, 144, then

$$R_1 = \frac{1545.3}{144M} = \frac{10.73}{M} \quad \text{(2-32)}$$

At low pressures approaching zero, equation 2-30 is extremely precise for any gas, and the accuracy is almost independent of the temperature range. The equation is also accurate for most gases and vapors at atmospheric pressure, and even up to some 150 psi. However, when pressures exceed this range, deviations from the equation can become appreciable as other compressibility deviations take place. When gases (vapors) are close to their condensation temperature for the pressure in question, the perfect-gas equation may be far from reliable.

To illustrate the extent of deviation in the perfect-gas equation, which occurs when a vapor if close to its region of condensation, Fig. 2-2 has been prepared using water-vapor (steam). For water-vapor (steam), extensive tabulations of thermodynamic properties have been prepared. (Tables 2-2 and 2-3 show abbreviated properties.) The tabulations for

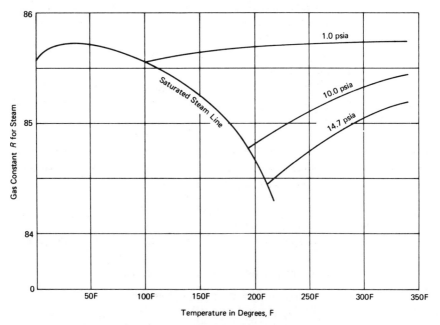

FIG. 2–2. R for steam computed from steam tables data.

all phases of water—solid, liquid, and vapor—are based on experimental measurements of water properties and on thermodynamic compilations which relate the various properties. Hence selecting a number of corresponding tabular values of P, v, and T, the values of R in equation 2–30 were computed. From these results plotted in Fig. 2–2, it can be seen that for steam, R is definitely not a constant, particularly when near its condensation (saturation) line and at pressures above 1.0 psia. R for steam (H_2O) by perfect-gas computation, equation 2–31, is

$$R = \frac{1545.3}{M} = \frac{1545.3}{18.016} = 85.77$$

where molecular weight of hydrogen is the 18.016 shown. Note, however, that for steam at low pressures such as those that exist for water-vapor (steam) in air, which are at less than 1.0 psia, the perfect-gas value of R and tabular values are in close agreement and the perfect-gas equation can be used to advantage.

2–7. PERFECT-GAS RELATIONSHIPS

The specific internal energy of a perfect gas is a function of temperature only, with the value of u not influenced by pressure. Joule demonstrated

Thermodynamics, Steam, and Gas Properties

this in 1843 when he connected two small pressure tanks together by a pipe in which a valve had been installed. When closed, the valve isolated the gaseous contents of the two tanks. For his experiment, he filled one tank with air under high pressure and he evacuated the other tank. Both tanks were then placed in a water bath surrounded by an insulated shell. The tanks were allowed to reach thermal equilibrium and the temperature of the water was precisely recorded. The valve was then opened and air rushed from the high-pressure tank into the evacuated one, quickly reaching pressure equalization. Although this highly irreversible process raised the temperature in the evacuated tank and lowered the temperature in the pressure tank, Joule was unable to detect any temperature change in the water bath after equilibrium was reached. He thus had to conclude that no net heat transferred to or from the total air mass. With such the case, $dQ = 0$, and with no work being done $dW = 0$, thus by equation 2-1, $du = 0$. Thus, there is no change in internal energy as long as the temperature is constant and the property, internal energy, is independent of pressure and volume. The conclusion that must be reached is that changes in u for a perfect gas only occur under changes in temperature and we can state that

$$u = f(T)$$

This conclusion is not exactly true for real gases and if Joule had possessed more precise instrumentation, he would have found that a slight change in temperature resulted in the bath following expansion, which would have indicated that u is not a unique function of temperature. Other evidence shows that this perfect-gas relationship is fully honored only at pressures approaching zero, under which condition the molecules of a gas are extremely far apart. However, the approximation is sufficiently close under conditions of low and moderate pressure to be of great value in making computations for real gases.

Recall equation 2-27

$$C_v = \frac{\partial u}{\partial T}\bigg)_v$$

We can rewrite this for the perfect gas as

$$C_v = \frac{du}{dT} \tag{2-33}$$

since the variations in u are independent of pressure and volume. This same independence can be extended to apply to enthalpy, and following the same reasoning, we can write for a perfect gas that

$$C_p = \frac{dh}{dT} \tag{2-34}$$

Differentiate the perfect gas equation 2-28 and obtain

$$Pdv + vdP = RdT$$

$$R = \frac{Pdv + vdP}{dT} \tag{2-35}$$

Differentiate the property enthalpy with respect to T and $h = u + Pv$ becomes

$$\frac{dh}{dT} = \frac{du}{dT} + \frac{Pdv + vdP}{dT} \tag{2-36}$$

Then

$$C_p = C_v + R$$
$$C_p - C_v = R \tag{2-37}$$

Thus, for a perfect gas or a real gas at low pressure, the difference between specific-heat values is equal to R for that gas.

Let us apply equation 2-6 to a gas involved in a reversible process, which we will further limit to being adiabatic, that is one in which no heat is transferred, thus:

$$dQ = 0 = du + Pdv$$

Refer to equations 2-33 and 2-35 and note that since

$$du = C_v \, dT \quad \text{and} \quad dT = \frac{Pdv + vdP}{R}$$

when these are substituted there results

$$0 = C_v dT + Pdv = C_v \left(\frac{Pdv + vdP}{R}\right) + Pdv$$

$$C_v \, (Pdv + vdP) + R \, Pdv = 0$$

Now substitute the value of R from equation 2-37 and find

$$C_v \, (Pdv + vdP) + (C_p - C_v) \, Pdv = 0$$

$$\frac{dP}{P} + \frac{C_p}{C_v} \frac{dv}{v} = 0$$

Call $C_p/C_v = k$ and integrate

$$\log P + k \log v = \text{constant}$$

$$Pv^k = \text{constant}$$

This can also be expressed

$$Pv^k = P_1 v_1^k = P_2 v_2^k \cdots \tag{2-38}$$

Thermodynamics, Steam, and Gas Properties

The value of k can be found from computed or experimentally-determined values of C_p and C_v. Values of k can also be found by generalizations developed from the kinetic theory of gases. These relations show that for diatomic gases, such as oxygen, nitrogen, air, and hydrogen, the value of k is in the neighborhood of 1.4. For triatomic gases such as CO_2, H_2O, and SO_2, the value of k is in the range of 1.3. Monatomic gases, of which helium is one, have a value of 1.66 for k.

Equation 2–38 is thus a relationship between pressure and volume during a reversible-adiabatic expansion or compression.

Similarly the perfect gas equation can be written

$$RT = Pv = P_1v_1 = Pv_2 \cdots$$

and

$$Pv = P_1v_1 = Pv_2 \cdots \quad (2\text{–}39)$$

is the relationship between pressure and volume for a constant temperature (isothermal) process.

2–8. PERFECT-GAS WORK PROCESSES

The work produced by a gas during a reversible-adiabatic process can be found by integration, making use of the relationship

$$d\mathbf{W} = Pdv \quad \text{and} \quad \mathbf{W}_{(1-2)} = \int_1^2 Pdv \quad (2\text{–}40)$$

Referring to Fig. 2–3, let us integrate along the reversible path 1 to 2. Here integration is the summation of the infinite number of Pdv areas which constitute the total area under the line 1–2. This line, 1–2, in turn, is the locus of all the state points of the gas during this reversible expansion. Note from equation 2–38 that

$$P = \frac{P_1v_1^k}{v^k} = \frac{P_2v_2^k}{v^k} \cdots$$

substitute in equation 2–40

$$\mathbf{W}_{(1-2)} = \int_1^2 \frac{P_1v_1^k}{v^k} dv = P_1v_1^k \int_1^2 \frac{dv}{v^k} = P_1v_1^k \left[\frac{v^{1-k}}{1-k}\right]_1^2$$

$$\mathbf{W}_{(1-2)} = \frac{P_2v_2 - P_1v_1}{1-k} = \frac{P_1v_1 - P_2v_2}{k-1} \quad (2\text{–}41)$$

or make use of equation 2–38 and eliminate v_2 in the preceding result to give

$$\mathbf{W}_{(1-2)} = \frac{P_1v_1}{k-1}\left[1 - \left(\frac{P_2}{P_1}\right)^{\frac{k-1}{k}}\right] \quad (2\text{–}42)$$

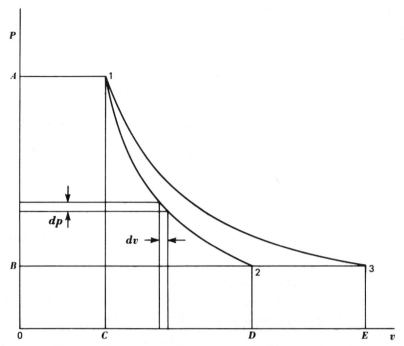

Fig. 2-3. Pv diagram to illustrate work under isentropic (1-2) and isothermal (1-3) conditions.

The area under the curve 1–2 is the work of expansion if the process moves from 1 to 2 or is the work of compression if the process goes from 2 to 1. When compression is involved, numerical answers resulting from equations 2–41 or 2–42 will be negative, representing work done on the gas.

These equations are developed for non-flow conditions, that is they measure the work delivered (or received) as the gas expands (or is compressed) but do not include the work of admission and the work of delivery from the compressor. These additional amounts of work can be found as follows. Referring to Fig. 2–3, we can readily see that for an expander engine (turbine) the work done by the gas in merely entering (flow work) is represented by the area $A1C0$ and equals P_1v_1 in magnitude. The flow work of delivery is $B2D0$ and equals P_2v_2. If we add these values to equations 2–41 and 2–42, we find that for a steady-flow machine

$$W = P_1v_1 + \frac{P_2v_2 - P_1v_1}{k-1} - P_2v_2 = \frac{k}{k-1}(P_1v_1 - P_2v_2) \quad (2\text{--}43)$$

or

$$W = \frac{k}{k-1} P_1v_1 \left[1 - \left(\frac{P_2}{P_1}\right)^{\frac{k-1}{k}} \right]$$

Thermodynamics, Steam, and Gas Properties

It is of interest to note that for a steady-flow process of the type described that the work area involved in Fig. 2-3 is $A12B$ in net magnitude. A study of the figure would show that this area could also be measured by integrating an equation of the form

$$W_{(1-2)} = \int_1^2 v\,dP \tag{2-44}$$

Substituting from equation 2-38 and integrating will yield equation 2-43.

Let us now consider the work done during an isothermal expansion. Equation 2-39 shows that during such a process

$$Pv = P_1v_1 = P_2v_2 = P_3v_3 \cdots$$

and the path of an isothermal process is illustrated by the line 1–3 in Fig. 2-3.

$$W_{(1-3)} = \int_1^3 P\,dv = \int_1^3 \frac{P_1v_1}{v}\,dv = \int_1^3 \frac{P_3v_3}{v}\,dv$$

$$W_{(1-3)} = P_1v_1 \log_e \frac{v_3}{v_1} = P_1v_1 \log_e \frac{P_1}{P_3} = RT \log_e \frac{P_1}{P_3} \tag{2-45}$$

Let us now consider an isothermal expansion in the light of equation 2-1, which we can write in integrated form as

$$Q_{(1-3)} = u_3 - u_1 + W_{(1-3)}$$

Since for an isothermal process both the temperature and internal energy are constant, it follows that

$$u_3 = u_1$$

and

$$Q_{(1-3)} = W_{(1-3)} = P_1v_1 \log_e \frac{v_3}{v_1} \tag{2-46}$$

This equation shows the obvious fact that since work is not created at the expense of internal energy, all of it must be supplied by heat addition during an expansion process. During isothermal compression conversely all of the work supplied is simultaneously delivered as heat since no additional internal energy can reside in the gas. Note also that the work during a reversible isothermal expansion, represented by the area under the line 1–3 in Fig. 2-3, is appreciably greater than the work produced under a reversible-adiabatic expansion for the same pressure range and initial conditions.

The steady-flow work under isothermal compression (or expansion) of a perfect gas by happenstance is exactly the same as non-flow work since the work of admission (P_1v_1) and of delivery (P_3v_3) are the same except for opposite sign and thus, of course, offset each other.

Example 2–2. Air at atmospheric pressure 14.7 psia and 70 F is compressed through a pressure compression ratio of 8 to 117.6 psia in a steady-flow compressor. Find the work required per pound of air delivered if (a) compressed isentropically, that is under reversible-adiabatic conditions, or (b) compressed under reversible isothermal conditions.

Solution: The specific volume of the air at entry to the compressor can be found by equation 2–30, R for air from Table 2–1 is 53.34.

$$(144)\,(14.7)\,(v) = (1)\,(53.34)\,(460 + 70)$$
$$v = 13.34 \text{ cu ft per lb}$$

a) Make use of equation 2–43 and select $k = 1.4$ from Table 2–1

$$W = \left(\frac{1.4}{1.4 - 1}\right)(144)\,(14.7)\,(13.34)\left[1 - 8^{\frac{1.4-1}{1.4}}\right]$$
$$= 98,900\,[1 - 1.811] = -80,200 \text{ ft lb}$$

The work is thus 80,200 ft lb per lb of air compressed and delivered. The minus sign merely indicates that work is done on the air during compression.

b) For the isothermal compression use equation 2–45

$$W = (144)\,(14.7)\,(13.34)\,(2.3)\,\log_{10}\frac{14.7}{117.6} = (28,380)\,(2.3)\,(-0.903)$$
$$= -58,800 \text{ ft lb}$$

The work is thus 58,800 ft lb per lb of air compressed and delivered. The minus sign merely indicates that work is done on the air during compression. Note also that 58,800 ft lb of heat per lb of air is transferred from the air during compression.

2–9. PRESSURE, VOLUME, TEMPERATURE, ENTHALPY, AND ENTROPY RELATIONSHIPS FOR THE PERFECT GAS

Let us consider a perfect gas undergoing an isentropic (reversible-adiabatic) process. Equation 2–38 gives a relationship for pressure and volume at any point during the process, similarly equation 2–28 gives the relation of temperature to pressure and volume during the process. It is then easy to use these equations, namely

$$P_1 v_1^k = P_2 v_2^k \quad \text{and} \quad P_1 v_1 = RT_1,\; P_2 v_2 = RT_2$$

and by simple algebra derive the following useful ratios.

$$\frac{T_2}{T_1} = \left(\frac{P_2}{P_1}\right)^{\frac{k-1}{k}} \tag{2-47}$$

$$\frac{T_2}{T_1} = \left(\frac{v_1}{v_2}\right)^{k-1} \tag{2-48}$$

It is thus easy to find temperature in terms of the pressure ratio or in terms of the volume ratio in which a reversible-adiabatic process has been involved.

Temperature has here been singled out because both enthalpy and

Thermodynamics, Steam, and Gas Properties

internal energy can be expressed in terms of it. To show this, refer to equation 2–34 and set it up in integral form as follows:

$$h - h_0 = \int_{T_0}^{T} C_p dT \quad (2\text{–}49)$$

When C_p is a constant this equation readily integrates into the form

$$h - h_0 = C_p (T - T_0) \quad (2\text{–}50)$$

This equation states that the change in enthalpy is directly proportional to a temperature change. The datum temperature (T_0) can be any arbitrary value such as 0 F, -40 F, or -460 F and at this temperature it is customary to give h_0 the value of zero so that the simplified expression for enthalpy becomes

$$h = C_p (T - T_0) \quad (2\text{–}51)$$

However, since tables are made for real gases and C_p is not a true constant, a step by step summation (or integration) using proper values of C_p is required if the temperature range is extensive, and equation 2–51 is more accurately written as

$$h = \int_{T_0}^{T} C_p dT \quad (2\text{–}52)$$

The air constituent in Table 3–1, for example, is based on 0 F as its datum whereas the elaborate tables of gases prepared by Keenan and Kaye use 0 R (-460 F) as their datum (see Reference 1).

The corresponding relations for internal energy would appear as

$$u - u_0 = \int_{T_0}^{T} C_v dT \quad (2\text{–}53)$$

or if u_0 is taken as zero

$$u = \int_{T_0}^{T} C_v dT \quad (2\text{–}54)$$

For small temperature changes with C_v constant

$$u - u_0 = C_v (T - T_0) \quad (2\text{–}55)$$

If the steady-flow energy equation 2–9 is applied to a compressor or expander engine operating under adiabatic conditions ($Q_{(1-2)} = 0$) and with kinetic and potential energy trivial as would usually be the case, we would find that for a compressor

$$W_{(1-2)} = h_2 - h_1 \tag{2-56}$$

and for an engine (turbine) expander

$$W_{(1-2)} = h_1 - h_2 \tag{2-57}$$

That is the work absorbed or the work produced in an adiabatic work device is exactly equal to the enthalpy change experienced by the working medium.

Example 2-3. Solve (a) of example 2-2 by employing the enthalpy function.
Solution: Use equation 2-47 to find the temperature after compression

$$\frac{T_2}{460 + 70} = \left(\frac{117.6}{14.7}\right)^{\frac{1.4-1}{1.4}} = (8)^{0.286} = 1.811$$

$$T_2 = 960 \text{ R or } 500 \text{ F}$$

C_p for air is given as 0.240 in Table 2-1 and if taken as constant over this range, we can use equation 2-50 to find the change in enthalpy

$$\Delta h = h - h_0 = 0.240 \ (960 - 530) = 103.2$$

$$W = \Delta h = 103.2 \text{ Btu per lb air or } 103.2 \times 778 = 80{,}200 \text{ ft lb per lb air}$$

Note that (b) of example 2-2 cannot be solved by use of enthalpy because the process is not adiabatic. The enthalpy change in an isothermal process is obviously zero for a gas.

A useful thermodynamic relation, involving enthalpy, for an isentropic (reversible-adiabatic) process can be found by combining the differentiated form of the defining equation for enthalpy, $h = u + Pv$, and equation 2-1

$$dh = du + Pdv + vdP$$

$$dQ = 0 = du + Pdv$$

Then

$$dh = vdP \tag{2-58}$$

and

$$h - h_0 = \int_{P_0}^{P} vdP \tag{2-59}$$

Entropy equation developments start from the mathematical definition, equation 2-20

$$dQ = Tds$$

Substitute this in the first-law expression, equation 2-1, then

$$Tds = du + Pdv \tag{2-60}$$

Divide by T and introduce equivalents from equations 2-28 and 2-33

Thermodynamics, Steam, and Gas Properties

$$ds = \frac{du}{T} + \frac{P}{T}dv$$

$$ds = \frac{C_v dT}{T} + R\frac{dv}{v} \tag{2-61}$$

Integrating

$$s_2 - s_1 = \int_{T_1}^{T_2} \frac{C_v dT}{T} + R \int_{v_1}^{v_2} \frac{dv}{v}$$

$$= \int_{T_1}^{T_2} \frac{C_v dT}{T} + R \log_e \frac{v_2}{v_1} \tag{2-62}$$

When C_v is constant (particularly over a small temperature range) we can integrate to show

$$s_2 - s_1 = C_v \log_e \frac{T_2}{T_1} + R \log_e \frac{v_2}{v_1} \tag{2-63}$$

Entropy can be expressed in terms of temperature and pressure if we make use of the enthalpy relationship and C_p for a perfect gas

$$\mathrm{h} = u + Pv \quad \text{and} \quad d\mathrm{h} = C_p dT$$

$$d\mathrm{h} = du + Pdv + vdP = C_p dT$$

Substitute for du in equation 2–60

$$Tds = C_p dT - Pdv - vdP + Pdv$$

$$ds = C_p \frac{dT}{T} - \frac{vdP}{T} = C_p \frac{dT}{T} - R\frac{dP}{P} \tag{2-64}$$

$$s_2 - s_1 = \int_{T_1}^{T_2} C_p \frac{dT}{T} - R \log_e \frac{P_2}{P_1} \tag{2-65}$$

For small temperature ranges or whenever C_p is essentially a constant value

$$s_2 - s_1 = C_p \log_e \frac{T_2}{T_1} + R \log_e \frac{P_1}{P_2} \tag{2-66}$$

Earlier in this chapter, relationships between pressure and volume during two types of expansions and/or compressions were shown to exist, namely:

    Isentropic (reversible-adiabatic) $Pv^k = \text{constant}$
    Isothermal $Pv = \text{constant}$

It is obvious that for a constant pressure process

$$Pv^0 = P_1 v_1^0 = P_2 v_2^0 = \text{constant}$$

For a constant volume process we can write

$$Pv^{\pm\infty} = P_1v_1^{\pm\infty} = Pv_2^{\pm\infty} = \text{constant}$$

As a further step we can use an exponent n and generalize the relationship for P and v for a gas into a so-called polytropic form

$$Pv^n = P_1v_1^n = P_2v_2^n = \text{constant} \qquad (2\text{-}67)$$

Values of n, descriptive of a number of reversible processes range from $+\infty$ to 0 to $-\infty$. In such processes heat or work or both types of energy interchange can occur between the gas and its energy environment. Only a few values of n are of much utility, largely in the range 0 to 3, but all values are mathematically conceivable.

It is also possible to conceive of irreversible processes with a gas for which

$$Pv^n = P_1v_1^n = P_2v_2^n = \text{constant}$$

represents a series of Pv values or points which apply for the gas as it irreversibly expands or is compressed.

Following the method used to develop equations 2–41, 2–42, 2–43, and 2–44 similar relations can be found for the work of a reversible polytropic process with n merely replacing k in the equations. For equation 2–42 we would get

$$W_{(1-2)} = \frac{P_1v_1}{n-1}\left[1 - \left(\frac{P_2}{P_1}\right)^{\frac{n-1}{n}}\right] \qquad (2\text{-}68)$$

or for steady flow, equation 2–44 becomes

$$W_{(1-2)} = \frac{n}{n-1} P_1v_1 \left[1 - \left(\frac{P_2}{P_1}\right)^{\frac{n-1}{n}}\right] \qquad (2\text{-}69)$$

Note that these expressions are not applicable for values of $n = 1$, $n = 0$, $n = \pm\infty$ but the proper forms are easily derived as was done for $n = 1$ earlier in this chapter. Work is, of course, zero for $n = \pm\infty$, where $v = \text{constant}$, since $dv = 0$.

Figure 2–4 is a plot on a Pv plane showing representative slopes that would be applicable for values of n, for a representative diatomic gas.

The pressure, volume, temperature relationships developed for the isentropic case in equations 2–47 and 2–48 are also applicable for reversible expansions or compressions under polytropic conditions with the appropriate value of n used in place of k.

Example 2–4. Air at 100 psia initially at 70 F is expanded slowly against a piston in a long cylinder with heated walls. The air in the cylinder expands as though follow-

Thermodynamics, Steam, and Gas Properties

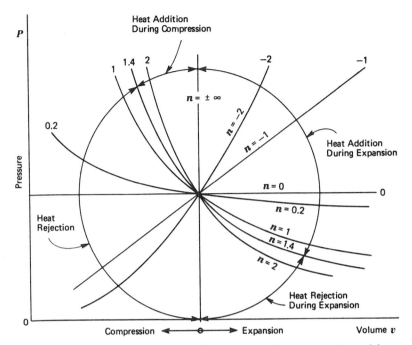

FIG. 2-4. Plot of reversible polytropic n values on Pv plane constructed for a diatomic gas.

ing a reversible polytropic with $n = 1.2$. The expansion proceeds until the pressure reaches 20 psia. For this expansion, compute: (a) the air temperature after expansion to 20 psia, (b) the work done by the expanding air, and (c) the heat added to the gas.

Solution: (a) Make use of equation 2–47 with $n = 1.2$

$$T_2 = (460 + 70)\left(\frac{20}{100}\right)^{\frac{1.2-1}{1.2}} = (530)\left(\frac{1}{5}\right)^{0.167} = \frac{530}{1.308} = 405\ R$$

b) Compute the work by using equation 2–68 after finding v_1 by use of equation 2–28, or use the equivalent RT_1 for P_1v_1. With $v_1 = 1.96$ cu ft per lb

$$W_{(1-2)} = \frac{(144)(100)(1.96)}{1.2 - 1}\left[1 - \frac{1}{1.308}\right] = 33{,}300 \text{ ft lb per lb}$$

$$= \frac{33{,}300}{778} \text{ Btu per lb} = 42.8 \text{ Btu per lb}$$

c) Use the non-flow form of the first-law equation, and express the internal energy change in terms of temperature

$$Q_{(1-2)} = C_v(T_2 - T_1) + W_{(1-2)} = 0.171(405 - 530) + 42.8$$

$$Q_{(1-2)} = 21.4 \text{ Btu per lb heat added}$$

2-10. WATER-STEAM

Earlier material in this chapter has made use of the perfect-gas relationships to illustrate thermodynamic principles. This served a useful purpose because the simplicity of the perfect-gas equation of state makes it easy to use and the results obtained from its use are sufficiently accurate for many engineering calculations. However, thermodynamic substances serve not only in their vapor (gaseous) phase but in the liquid, liquid-vapor, and solid phases. For such a spectrum, simple equations of state are not applicable and formulations based on experimental data and generalized thermodynamic relationships are required. However, these are usually so complex that for convenience it is customary to prepare tables of properties for widely-used materials.

Water in its three states, solid (ice), liquid (water), and vapor (steam) has been thoroughly investigated and elaborate tables of its complete properties are available. In this text, abbreviated tables for water-steam appear as Tables 2–2, 2–3, and 2–4 and also in connection with air as Table 3–1. Water thus serves as a good medium to illustrate the range of properties for a representative multi-phase substance.

The thermodynamic properties of state already introduced apply to the solid and liquid as well as to the gaseous phase, but further explanation is necessary in some instances. Specific volume (v) has been mentioned earlier as the volume occupied by unit mass of a medium. It has conventional units of cu ft per lb or in the metric system either cu cm per gram or cu meters per kg. Density (d) is the reciprocal of specific volume and appears in conventional units of lb per cu ft, or grams per cu cm or kg per cu meter. Tables of properties may record either specific volume or density.

The density or specific volume of a liquid is very slightly affected by changes in pressure but alters appreciably under changes in temperature. There is no simple relationship to describe the change in specific volume of a liquid with temperature and so it is customary to give tabular values. The water-steam Tables 2–2 and 2–3 employ v_f, the specific volume, instead of density. For example in Table 2–2, water at 100 F shows v_f equal to 0.01613 cu ft per lb. The corresponding density of water at 100 F is then $1/0.01613 = 61.99$ lb per cu ft.

For each liquid upon which a given pressure is being exerted there exists a certain definite temperature for that pressure at which boiling or vaporization will take place. This state at which a liquid and vapor are in temperature-pressure equilibrium is known as the *saturation state*. Tables of the properties of steam (tables 2–2, 2–3, and 2–4) all list saturation pressures and corresponding temperatures. For example, at 14.696-psia pressure (atmospheric) the saturation or boiling temperature at 212 F; at 100 psia water will not boil until its temperature reaches 327.81 F.

Thermodynamics, Steam, and Gas Properties

TABLE 2-2*

Saturated Steam
1. Temperature Table

Tempera-ture F t	Absolute Pressure Lb per sq in. p	Specific Volume		Enthalpy			Entropy	
		Sat. Liquid v_f	Sat. Vapor v_g	Sat. Liquid h_f	Evap. h_{fg}	Sat. Vapor h_g	Sat. Liquid s_f	Sat. Vapor s_g
32°	0.08854	0.01602	3306	0.00	1075.8	1075.8	0.0000	2.1877
35	0.09995	0.01602	2947	3.02	1074.1	1077.1	0.0061	2.1770
40	0.12170	0.01602	2444	8.05	1071.3	1079.3	0.0162	2.1597
45	0.14752	0.01602	2036.4	13.06	1068.4	1081.5	0.0262	2.1429
50	0.17811	0.01603	1703.2	18.07	1065.6	1083.7	0.0361	2.1264
60°	0.2563	0.01604	1206.7	28.06	1059.9	1088.0	0.0555	2.0948
70	0.3631	0.01606	867.9	38.04	1054.3	1092.3	0.0745	2.0647
80	0.5069	0.01608	633.1	48.02	1048.6	1096.6	0.0932	2.0360
90	0.6982	0.01610	468.0	57.99	1042.9	1100.9	0.1115	2.0087
100	0.9492	0.01613	350.4	67.97	1037.2	1105.2	0.1295	1.9826
110°	1.2748	0.01617	265.4	77.94	1031.6	1109.5	0.1471	1.9577
120	1.6924	0.01620	203.27	87.92	1025.8	1113.7	0.1645	1.9339
130	2.2225	0.01625	157.34	97.90	1020.0	1117.9	0.1816	1.9112
140	2.8886	0.01629	123.01	107.89	1014.1	1122.0	0.1984	1.8894
150	3.718	0.01634	97.07	117.89	1008.2	1126.1	0.2149	1.8685
160°	4.741	0.01639	77.29	127.89	1002.3	1130.2	0.2311	1.8485
170	5.992	0.01645	62.06	137.90	996.3	1134.2	0.2472	1.8293
180	7.510	0.01651	50.23	147.92	990.2	1138.1	0.2630	1.8109
190	9.339	0.01657	40.96	157.95	984.1	1142.0	0.2785	1.7932
200	11.526	0.01663	33.64	167.99	977.9	1145.9	0·2938	1.7762
210°	14.123	0.01670	27.82	178.05	971.6	1149.7	0.3090	1.7598
212	14.696	0.01672	26.80	180.07	970.3	1150.4	0.3120	1.7566
220	17.186	0.01677	23.15	188.13	965.2	1153.4	0.3239	1.7440
230	20.780	0.01684	19.382	198.23	958.8	1157.0	0.3387	1.7288
240	24.969	0.01692	16.323	208.34	952.2	1160.5	0.3531	1.7140
250°	29.825	0.01700	13.821	218.48	945.5	1164.0	0.3675	1.6998
260	35.429	0.01709	11.763	228.64	938.7	1167.3	0.3817	1.6860
270	41.858	0.01717	10.061	238.84	931.8	1170.6	0.3958	1.6727
280	49.203	0.01726	8.645	249.06	924.7	1173.8	0.4096	1.6597
290	57.556	0.01735	7.461	259.31	917.5	1176.8	0.4234	1.6472
300°	67.013	0.01745	6.466	269.59	910.1	1179.7	0.4369	1.6350
310	77.68	0.01755	5.626	279.92	902.6	1182.5	0.4504	1.6231
320	89.66	0.01765	4.914	290.28	894.9	1185.2	0.4637	1.6115
330	103.06	0.01776	4.307	300.68	887.0	1187.7	0.4769	1.6002
340	118.01	0.01787	3.788	311.13	879.0	1190.1	0.4900	1.5891
350°	134.63	0.01799	3.342	321.63	870.7	1192.3	0.5029	1.5783
360	153.04	0.01811	2.957	332.18	862.2	1194.4	0.5158	1.5677
370	173.37	0.01823	2.625	342.79	853.5	1196.3	0.5286	1.5573
380	195.77	0.01836	2.335	353.45	844.6	1198.1	0.5413	1.5471
390	220.37	0.01850	2.0836	364.17	835.4	1199.6	0.5539	1.5371
400°	247.31	0.01864	1.8633	374.97	826.0	1201.0	0.5664	1.5272
410	276.75	0.01878	1.6700	385.83	816.3	1202.1	0.5788	1.5174
420	308.83	0.01894	1.5000	396.77	806.3	1203.1	0.5912	1.5078
430	343.72	0.01910	1.3499	407.79	796.0	1203.8	0.6035	1.4982
440	381.59	0.01926	1.2171	418.90	785.4	1204.3	0.6158	1.4887
450°	422.6	0.0194	1.0993	430.1	774.5	1204.6	0.6280	1.4793
460	466.9	0.0196	0.9944	441.4	763.2	1204.6	0.6402	1.4700
470	514.7	0.0198	0.9009	452.8	751.5	1204.3	0.6523	1.4606
480	566.1	0.0200	0.8172	464.4	739.4	1203.7	0.6645	1.4513
490	621.4	0.0202	0.7423	476.0	726.8	1202.8	0.6766	1.4419
500°	680.8	0.0204	0.6749	487.8	713.9	1201.7	0.6887	1.4325
520	812.4	0.0209	0.5594	511.9	686.4	1198.2	0.7130	1.4136
540	962.5	0.0215	0.4649	536.6	656.6	1193.2	0.7374	1.3942
560	1133.1	0.0221	0.3868	562.2	624.2	1186.4	0.7621	1.3742
580	1325.8	0.0228	0.3217	588.9	588.4	1177.3	0.7872	1.3532
600°	1542.9	0.0236	0.2668	617.0	548.5	1165.5	0.8131	1.3307
620	1786.6	0.0247	0.2201	646.7	503.6	1150.3	0.8398	1.3062
640	2059.7	0.0260	0.1798	678.6	452.0	1130.5	0.8679	1.2789
660	2365.4	0.0278	0.1442	714.2	390.2	1104.4	0.8987	1.2472
680	2708.1	0.0305	0.1115	757.3	309.9	1067.2	0.9351	1.2071
700°	3093.7	0.0369	0.0761	823.3	172.1	995.4	0.9905	1.1389
705.4	3206.2	0.0503	0.0503	902.7	0	902.7	1.0580	1.0580

*Reprinted, by permission, from J. H. Keenan and F. G. Keyes, *Thermodynamic Properties of Steam* (New York: John Wiley & Sons, Inc., 1936).

TABLE 2-3*

Saturated Steam
2. Pressure Table

Absolute Press. Lb per Sq in. p	Tempera- ture F t	Specific Volume		Enthalpy			Entropy	
		Sat. Liquid v_f	Sat. Vapor v_g	Sat. Liquid h_f	Evap. h_{fg}	Sat. Vapor h_g	Sat. Liquid s_f	Sat. Vapor s_g
1.0	101.74	0.01614	333.6	69.70	1036.3	1106.0	0.1326	1.9782
2.0	126.08	0.01623	173.73	93.99	1022.2	1116.2	0.1749	1.9200
3.0	141.48	0.01630	118.71	109.37	1013.2	1122.6	0.2008	1.8863
4.0	152.97	0.01636	90.63	120.86	1006.4	1127.3	0.2198	1.8625
5.0	162.24	0.01640	73.52	130.13	1001.0	1131.1	0.2347	1.8441
6.0	170.06	0.01645	61.98	137.96	996.2	1134.2	0.2472	1.8292
7.0	176.85	0.01649	53.64	144.76	992.1	1136.9	0.2581	1.8167
8.0	182.86	0.01653	47.34	150.79	988.5	1139.3	0.2674	1.8057
9.0	188.28	0.01656	42.40	156.22	985.2	1141.4	0.2759	1.7962
10	193.21	0.01659	38.42	161.17	982.1	1143.3	0.2835	1.7876
11	197.75	0.01662	35.14	165.73	979.3	1145.0	0.2903	1.7800
12	201.96	0.01665	32.40	169.96	976.6	1146.6	0.2967	1.7730
13	205.88	0.01667	30.06	173.91	974.2	1148.1	0.3027	1.7665
14	209.56	0.01670	28.04	177.61	971.9	1149.5	0.3083	1.7605
14.696	212.00	0.01672	26.80	180.07	970.3	1150.4	0.3120	1.7566
15	213.03	0.01672	26.29	181.11	969.7	1150.8	0.3135	1.7549
16	216.32	0.01674	24.75	184.42	967.6	1152.0	0.3184	1.7497
17	219.44	0.01677	23.39	187.56	965.5	1153.1	0.3231	1.7449
18	222.41	0.01679	22.17	190.56	963.6	1154.2	0.3275	1.7403
19	225.24	0.01681	21.08	193.42	961.9	1155.3	0.3317	1.7360
20	227.96	0.01683	20.089	196.16	960.1	1156.3	0.3356	1.7319
21	230.57	0.01685	19.192	198.79	958.4	1157.2	0.3395	1.7280
22	233.07	0.01687	18.375	201.33	956.8	1158.1	0.3431	1.7242
23	235.49	0.01689	17.627	203.78	955.2	1159.0	0.3466	1.7206
24	237.82	0.01691	16.938	206.14	953.7	1159.8	0.3500	1.7172
25	240.07	0.01692	16.303	208.42	952.1	1160.6	0.3533	1.7139
26	242.25	0.01694	15.715	210.62	950.7	1161.3	0.3564	1.7108
27	244.36	0.01696	15.170	212.75	949.3	1162.0	0.3594	1.7078
28	246.41	0.01698	14.663	214.83	947.9	1162.7	0.3623	1.7048
29	248.40	0.01699	14.189	216.86	946.5	1163.4	0.3652	1.7020
30	250.33	0.01701	13.746	218.82	945.3	1164.1	0.3680	1.6993
35	259.28	0.01708	11.898	227.91	939.2	1167.1	0.3807	1.6870
40	267.25	0.01715	10.498	236.03	933.7	1169.7	0.3919	1.6763
45	274.44	0.01721	9.401	243.36	928.6	1172.0	0.4019	1.6669
50	281.01	0.01727	8.515	250.09	924.0	1174.1	0.4110	1.6585
55	287.07	0.01732	7.787	256.30	919.6	1175.9	0.4193	1.6509
60	292.71	0.01738	7.175	262.09	915.5	1177.6	0.4270	1.6438
65	297.97	0.01743	6.655	267.50	911.6	1179.1	0.4342	1.6374
70	302.92	0.01748	6.206	272.61	907.9	1180.6	0.4409	1.6315
75	307.60	0.01753	5.816	277.43	904.5	1181.9	0.4472	1.6259
80	312.03	0.01757	5.472	282.02	901.1	1183.1	0.4531	1.6207
85	316.25	0.01761	5.168	285.39	897.8	1184.2	0.4587	1.6158
90	320.27	0.01766	4.896	290.56	894.7	1185.3	0.4641	1.6112
95	324.12	0.01770	4.652	294.56	891.7	1186.2	0.4692	1.6068
100	327.81	0.01774	4.432	298.40	888.8	1187.2	0.4740	1.6026
110	334.77	0.01782	4.049	305.66	883.2	1188.9	0.4832	1.5948
120	341.25	0.01789	3.728	312.44	877.9	1190.4	0.4916	1.5878
130	347.32	0.01796	3.455	318.81	872.9	1191.7	0.4995	1.5812
140	353.02	0.01802	3.220	324.82	868.2	1193.0	0.5069	1.5751
150	358.42	0.01809	3.015	330.51	863.6	1194.1	0.5138	1.5694
160	363.53	0.01815	2.834	335.93	859.2	1195.1	0.5204	1.5640
170	368.41	0.01822	2.675	341.09	854.9	1196.0	0.5266	1.5590
180	373.06	0.01827	2.532	346.03	850.8	1196.9	0.5325	1.5542
190	377.51	0.01833	2.404	350.79	846.8	1197.6	0.5381	1.5497
200	381.79	0.01839	2.288	355.36	843.0	1198.4	0.5435	1.5453
250	400.95	0.01865	1.8438	376.00	825.1	1201.1	0.5675	1.5263
300	417.33	0.01890	1.5433	393.84	809.0	1202.8	0.5879	1.5104
400	444.59	0.0193	1.1613	424.0	780.5	1204.5	0.6214	1.4844
500	467.01	0.0197	0.9278	449.4	755.0	1204.4	0.6487	1.4634
600	486.21	0.0201	0.7698	471.6	731.6	1203.2	0.6720	1.4454
700	503.10	0.0205	0.6554	491.5	709.7	1201.2	0.6925	1.4296
800	518.23	0.0209	0.5687	509.7	688.9	1198.6	0.7108	1.4153
1000	544.61	0.0216	0.4456	542.4	649.4	1191.8	0.7430	1.3897
2000	635.82	0.0257	0.1878	671.7	463.4	1135.1	0.8619	1.2849
3000	695.36	0.0346	0.0858	802.5	217.8	1020.3	0.9731	1.1615
3206.2	705.40	0.0503	0.0503	902.7	0	902.7	1.0580	1.0580

*Reprinted, by permission, from J. H. Keenan and F. G. Keyes, *Thermodynamic Properties of Steam* (New York: John Wiley & Sons, Inc., 1936).

TABLE 2-4
Superheated Steam*

Pressure (psia) Sat. Temp.		Sat. Vapor	140 F	180 F	220 F	260 F	300 F	400 F	500 F	600 F	700 F	800 F	1000 F
1.0 101.74 F	v h s	333.6 1106.0 1.9782	356.6 1123.3 2.0081	380.6 1141.4 2.0373	404.5 1159.5 2.0647	428.4 1177.6 2.0907	452.3 1195.8 2.1153	512.0 1241.7 2.1720	571.6 1288.3 2.2233	631.2 1335.7 2.2702	690.8 1383.8 2.3137	750.4 1432.8 2.3542	869.5 1533.5 2.4283
2.0 126.08 F	v h s	173.7 1116.2 1.9200	177.96 1122.6 1.9308	190.04 1140.9 1.9603	202.1 1159.1 1.9879	214.1 1177.3 2.0140	226.0 1195.6 2.0387	255.9 1241.6 2.0955	285.7 1288.2 2.1468	315.5 1335.6 2.1938	345.4 1383.8 2.2372	375.1 1432.8 2.2778	434.7 1533.5 2.3519
5.0 162.24 F	v h s	73.52 1131.1 1.8441		75.71 1139.4 1.8574	80.59 1158.1 1.8857	85.43 1176.5 1.9121	90.25 1195.0 1.9370	102.26 1241.2 1.9942	114.22 1288.0 2.0456	126.16 1335.4 2.0927	138.10 1383.6 2.1361	150.03 1432.7 2.1767	173.87 1533.4 2.2509
10 193.21 F	v h s	38.42 1143.3 1.7876			40.09 1156.2 1.8017	42.56 1175.1 1.8341	45.00 1193.9 1.8595	51.04 1240.6 1.9172	57.05 1287.5 1.9689	63.03 1335.1 2.0160	69.01 1383.4 2.0596	83.32 1432.5 2.1118	96.58 1533.3 2.1860
14.696 212.00 F	v h s	26.80 1150.4 1.7566			27.15 1154.4 1.7624	28.85 1173.8 1.7902	30.53 1192.8 1.8160	34.68 1239.9 1.8743	38.78 1287.1 1.9261	42.86 1334.8 1.9734	46.94 1383.2 2.0170	51.00 1432.3 2.0576	59.13 1533.1 2.1319
20 227.96 F	v h s	20.09 1156.3 1.7319				21.11 1172.2 1.7545	22.36 1191.6 1.7808	25.43 1239.2 1.8396	28.46 1286.6 1.8918	31.47 1334.4 1.9392	34.47 1382.9 1.9829	37.46 1432.1 2.0235	43.44 1533.0 2.0978
60 292.71 F	v h s	7.175 1177.6 1.6438					7.259 1181.6 1.6492	8.357 1233.6 1.7135	9.403 1283.0 1.7678	10.427 1331.8 1.8162	11.441 1380.9 1.8605	12.449 1430.5 1.9015	14.454 1531.9 1.9762
100 327.81 F	v h s	4.432 1187.2 1.6026						4.937 1227.6 1.6518	5.589 1279.1 1.7085	6.218 1329.1 1.7581	6.835 1378.9 1.8029	7.446 1428.9 1.8443	8.656 1530.8 1.9193
300 417.33 F	v h s	1.5433 1202.8 1.5104							1.7675 1257.6 1.5701	2.005 1314.7 1.6268	2.227 1368.3 1.6751	2.442 1420.6 1.7184	2.859 1525.2 1.7954
500 467.01 F	v h s	0.9278 1204.4 1.4634							0.9927 1231.3 1.4919	1.1591 1298.6 1.5588	1.3044 1357.0 1.6115	1.4405 1412.1 1.6571	1.6996 1519.6 1.7363

*Abridged from J. H. Keenan and F. G. Keyes, *Thermodynamic Properties of Steam* (New York: John Wiley & Sons, Inc., 1936). Copyright 1936 by the authors.

The steam tables take water at 32 F as a datum point from which to start enthalpy values. This means that the magnitude of the **h** of water at 32 F is arbitrarily taken as zero. If 1 lb of water were heated at atmospheric pressure from 32 F to 212 F, then (since the mean specific heat of water over this range is unity), the heat added, equal to the increase in enthalpy, would be

$$WC_p(t_2 - t_1) = (1)(1)(212 - 32) = 180.0 \text{ Btu}$$

and this is the value for h_f at 212 F. The value 180.07 that appears in the tables is based on the Btu arbitrarily defined in terms of the International Steam Table Calorie.

As the specific heat of water is not unity over a large range, h_f should not be calculated but selected from the tables. For example, at 100 psia for saturated liquid, $h_f = 298.33$ Btu per lb. For water at temperatures below 212 F, at any temperature t the enthalpy is very closely

$$h_f = (1)(t - 32) \text{ Btu per lb} \tag{2-70}$$

If a pound of water at 212 F is heated at atmospheric pressure it will be found that the temperature remains constant, but that a large amount of energy from the heating source is utilized in changing this pound of water into vapor (steam). This energy required to transform one pound of water into one pound of steam is known as the *heat of vaporization*, or *latent heat*. At 212 F (14.696 psi pressure) it amounts to 970.3 Btu and is listed under the heading h_{fg} (Table 2–2). The enthalpy of each pound of dry saturated vapor referred to a 32 F datum is the sum of $h_f + h_{fg}$ and is called h_g. At 212 F (14.696 psia), $h_g = 180.07 + 970.3 = 1150.4$ Btu per lb.

During rapid boiling it often happens that some of the water is entrained in the steam delivered, in the form of fine droplets or mist. If it should happen that the steam carries 10 per cent by weight of moisture it is obvious that there is 90 per cent of dry steam present. This weight per cent of dry steam present is known as the *steam quality* and the customary symbol for it is x.

The enthalpy of wet steam at any quality x (expressed as a decimal) is

$$h_x = h_f + xh_{fg} \text{ Btu per lb} \tag{2-71}$$

The quality x can range from 0.0 to 1.0.

The specific volume of a vapor or gas is affected by both pressure and temperature. In the case of a vapor or imperfect gas, tables of properties of the vapor must be used to find all values. In the case of the steam properties given in Tables 2–2 and 2–3, the column v_g represents the specific volume in cubic feet per pound of saturated steam at the corresponding pressure-temperature indicated. The symbol v_{fg} indicates the amount by which the volume of one pound of the medium increases in

Thermodynamics, Steam, and Gas Properties

changing from liquid to dry saturated steam. The reciprocal of v_g, namely $1/v_g$, represents the density (the weight, in pounds, of one cubic foot of steam).

The specific volume of wet steam is

$$v_x = v_f + xv_{fg} = (1 - x)v_f + xv_g \text{ cu ft per lb} \qquad (2\text{-}72)$$

Or

$$v_x = xv_g \text{ cu ft per lb} \qquad (2\text{-}73)$$

is used as a close approximation.

Example 2-5. (a) Find the enthalpy **h** of steam delivered from a boiler operating at 85.3 psig pressure if its quality is 95 per cent. The barometric pressure is 14.7 psi. (b) Find also the specific volume of the steam delivered. (c) At what temperature does the steam and water mixture leave the boiler, and in what way does the 5 per cent moisture affect the temperature?

Solution: (a) Boiler pressure = 85.3 + 14.7 = 100.0 psia. At this pressure, from Table 2-3,

$$\mathbf{h}_f = 298.4 \text{ Btu} \quad \text{and} \quad \mathbf{h}_{fg} = 888.8 \text{ Btu}$$

In this case all of the 298.4 Btu has been supplied to bring the liquid up to saturation conditions above the datum of the steam tables, but only 95 per cent of the heat of vaporization has been supplied. Consequently the enthalpy of this wet steam, $\mathbf{h}_x$, must be

$$\mathbf{h}_x = \mathbf{h}_f + x\mathbf{h}_{fg} = 298.4 + (0.95)(888.8) = 1142.8 \text{ Btu per lb} \qquad Ans.$$

NOTE. Multiplying $x\mathbf{h}_g$ to find $\mathbf{h}_x$ is incorrect and must never be done.

b) Similarly, the specific volume of the wet steam v_x can be found, using $v_x = v_f + xv_{fg}$ or an equivalent form:

$$v_x = v_f(1 - x) + xv_g = 0.01774(0.05) + 0.95(4.432) = 4.21 \text{ cu ft per lb} \qquad Ans.$$

It is theoretically wrong to say $v_x = xv_g$, but for low-pressure steam, and at relatively high qualities, this approximation is often made.

c) The temperature of the saturated steam–water mixture at 100 psia is 327.81 F (col. 2) and is not affected by the quality variations. *Ans.*

If steam is heated until all traces of moisture are gone, its temperature will rise above the saturation temperature corresponding to the pressure. Such steam, at a temperature higher than saturation temperature, is called *superheated*. Superheated steam resembles a gas in its behavior and in some cases can be treated by using the simple gas laws but for most calculations, except those at very low pressures, recourse should be had to complete tables of the properties of steam in the superheated state. Table 2-4 is an abridged table of superheated-steam properties and lists for each pressure at various superheat temperatures, specific volume v, enthalpy **h,** and entropy s.

2-11. TEMPERATURE-ENTROPY (Ts) DIAGRAM FOR WATER-STEAM

It will be noticed that the last two columns of Table 2-2 and 2-3 list the entropy of saturated liquid and the entropy of saturated vapor.

Similarly in Table 2–4, the entropy of superheated steam is included. The property entropy is extremely valuable for analyzing processes involving work. It has already been shown that processes taking place under isentropic (reversible and adiabatic) conditions produce optimum work and the entropy function when plotted on a Ts plane can show many important relationships to advantage. One important usage relates to the fact that areas on a Ts plane can represent the heat added or rejected during a process.

A temperature entropy-diagram for steam appears in Fig. 2–5 and because it is similar to Ts diagrams for most fluids, it will be described in detail. For areas to be in scale, absolute temperatures are required. Temperature is usually the ordinate with entropy the abscissa. The zero of entropy for plotting such diagrams can be arbitrarily selected and in the case of water-steam, the zero value of entropy is taken at 32 F under a pressure of 0.08854 psia. The line from this datum point (A) up to the critical pressure point (C) is known as the saturated-liquid line while the line from C to F represents the saturated-vapor line. Along the saturated-liquid and saturated-vapor lines at any temperature there is a corresponding pressure of saturation. For example, at 212 F under a pressure of 14.696 psia, liquid will change into vapor (steam) and increase in volume when heat is added with the temperature remaining constant at 212 F until all of the liquid is converted to steam. If further heat is added to the water-free vapor, it will increase in temperature and become superheated to a higher temperature such as E. The saturated liquid-vapor line for 14.696 psia thus appears as BD and the area under BD, namely $BDdb$ is a measure of the amount of heat required to change water to steam under steady-flow conditions, the h_{fg} value of Table 2–3. If the steam were under a higher pressure, let us say 200 psia, vaporization of the liquid would not start until the temperature reached 381.7 F, as indicated on the chart.

It is a characteristic of most substances that as the saturation temperature and pressure are increased, the value of the heat of vaporization (latent heat) diminishes until at a point C, no latent heat is needed and the liquid and vapor phase become indistinguishable. This point is known as the critical point and represents the transition (maximum) point of the liquid-vapor line. For steam the critical pressure is 3206 psia and the temperature 705.4 F. Above its critical temperature, it is not possible to liquefy a vapor. This is true although at the usually high pressures and temperatures involved, it may not be possible to distinguish the liquid phase from the vapor phase so the statement may be of theoretical importance only. The critical temperature and pressure are most significant in the analysis of any vapor.

Thermodynamics, Steam, and Gas Properties

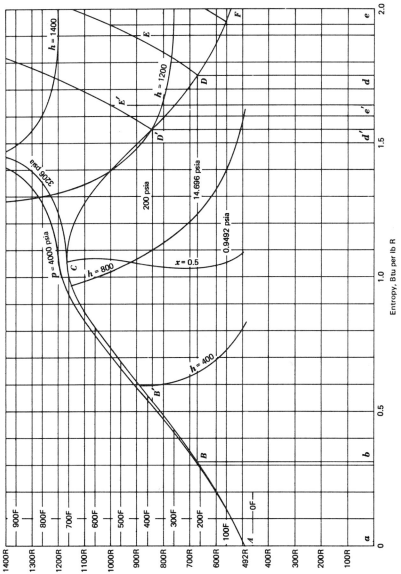

Fig. 2-5. Temperature—entropy (Ts) diagram for steam.

It is easily possible to take a saturated liquid, such as water at 212 F and 14.696 psia, and compress it to a higher pressure, such as 200 psia or even 5000 psia. Such a liquid is classified as a compressed liquid and although its specific volume is not significantly altered by the higher pressure, the state point of such a compressed liquid does not lie precisely on the saturation line but is to the left of the saturated liquid line and the region to the left of the line ABC is known as the compressed-liquid region. A representative line has been drawn in for 4000 psia in Fig. 2–5. Below the critical temperature it is customary to speak of a liquid as being compressed or saturated. Above the critical temperature, it is customary to speak of the medium as compressed vapor or superheated vapor (superheated steam in the case of water).

The lines of constant enthalpy in the superheat region swing down and to the right, tending to level off as the steam becomes far removed from its saturated-vapor state.

The area under the curve DE (i.e. $EedD$) naturally represents the heat required to superheat steam from the temperature D to the temperature E. The area under the curve AB represents the heat required to warm liquid (water) along the saturation curve from 32 F to 212 F. Similar areas are applicable for other pressures such as those under AB', $B'D'$, and $D'E'$ for a pressure of 200 psia.

2–12. GAS MIXTURES

Mixtures of gases are so common that methods for computing the properties of gaseous mixtures are most necessary. The mol concept, which has been mentioned earlier, is particularly useful. A *mol* of a chemical compound or element is a mass (weight) of that compound equal to its molecular weight expressed in pounds (grams in the metric system). Further, the volume occupied by a mol of any essentially perfect gas is the same at any specified temperature and pressure (379.6 cu ft at 60 F and 14.7 psia). Thus it follows that the density of a gas is closely proportional to the molecular weight of that particular gas. The constancy of molal volume should be recognized as only a close approximation, for except at low pressures no real gas satisfies either Avogadro's law or the perfect gas equation. It should also be noted that the gas constant, 1545.3 ft-lb per lb mol deg F, used earlier in this chapter, is a molal value and applies to a gas in any range where the gas approaches perfect gas behavior.

In gaseous mixtures at low and moderate pressures, the respective gases in the mixture frequently resemble perfect gases in their behavior and also follow a further generalization known as the Gibbs-Dalton law of partial pressures. This law states: *In a given mixture of gases or vapors each gas or vapor exerts the same pressure it would exert if it occurred alone in the same space and at the same temperature as exists in the mixture.* Partial pressures, volumetric percentages, and molal concentrations are directly related to each other in gaseous mixtures which obey the Gibbs-Dalton law.

Example 2-6. A natural gas consists of 95.6 per cent methane (CH_4), 3.8 per cent nitrogen (N_2), and 0.6 per cent water vapor (H_2O), with each constituent measured by volume. The gas is under a pressure of 15 psia and at 60 F. Compute the partial pressure of each constituent in the mixture, the molal concentrations, the weight analysis, the molecular weight, and the density.

Solution: It is most convenient for this type of problem to use a solution in tabular form as shown on the following page.

In the table, note that volumetric percentages (col. 2) and mole concentrations (col. 3) are the same. To change to a weight analysis, multiply each volumetric or molal fraction by its respective molecular weight and find what fraction each resulting product is of the total. At the bottom of column 5 is indicated a composite molecular weight of the gas, and one mole of gas can be said to weigh 16.46 lb. This value is found from the representative total of column 6. Use this molecular weight with equation 2-30 to find the density:

Constituent	Volumetric Percentages	Moles per 100 moles	Partial Pressure $p_T \times \frac{\text{Col. 2}}{100}$ (psi)	Molecular Weight	Col. 2 × Col. 5	Weight Percentage
1	2	3	4	5	6	7
Methane............	95.6	95.6	14.34	16	1529	92.89
Nitrogen	3.8	3.8	0.57	28	106	6.44
Water vapor	0.6	0.6	0.09	18	11	0.67
Total............	100.0	100.0	15.00	(16.46)	1646	100.00

$$(144)(15)(1) = m\left(\frac{1545.3}{16.46}\right)(460 + 60)$$

$$m = 0.0442 \text{ lb per cu ft} \qquad Ans.$$

Alternate Solution. An alternate solution would employ the molal volume. Thus, at 14.7 psia,

$$v = \frac{379.6}{16.46} = 23.06 \text{ cu ft per lb}$$

and

$$\text{Density} = \frac{1}{v} = \frac{1}{23.06} = 0.0435 \text{ lb per cu ft}$$

At constant temperature the density of a gas is directly proportional to pressure. Thus, at 60 F and 15 psia,

$$\text{Density} = 0.0435\left(\frac{15}{14.7}\right) = 0.0442 \text{ lb per cu ft} \qquad Ans.$$

In the event that the analysis of a gas is given with the constituents expressed in weight percentages, a transformation can be made to a volumetric (molal) concentration basis by dividing each constituent by its appropriate molecular weight. After adding the resulting quotients, the volumetric analysis can be found by using each constituent quotient as a fraction of the total.

2-13. GAS MIXING AND SEPARATION

Imagine a container containing two gases, A and B, at the same pressure but separated from each other by a thin partition. If the partition is removed it is well known that gas A will diffuse into gas B and gas B will diffuse into gas B. Diffusion is really the transport of molecular matter (mass) actuated by a concentration gradient. Diffusion is a natural process and also one which is inherently irreversible, that is a gas once diffused into another cannot of itself counter diffuse or resegregate itself. Diffusion also takes place when the molecules evaporating from a liquid surface enter a gaseous domain to diffuse there just as would the molecules of another gas. However, mass diffusivity, which is a time parameter, varies with each different species of molecule. For example, different liquids on vaporization diffuse in quiet air at widely different rates.

The mixing process with gases results in no change in internal energy, temperature, enthalpy or total pressure but there is a marked increase in entropy as each gas drops in molal concentration (partial pressure) and diffuses into the other. In the case of real gases, particularly at high pressure, there can be changes in the thermodynamic functions but except for entropy the changes are essentially trivial. Entropy change can be computed for perfect gases by use of equation 2–65 or 2–66.

Example 2–5. A chamber has two compartments of different size. In one compartment of volume 5.0 cu ft, hydrogen at 20 psia and 68 F is stored, in the other chamber, of 10.0 cu ft volume, nitrogen is stored also at 20 psia and 68 F. If the partition between the two chambers is opened and the gases diffuse into each other, compute (a) the enthalpy before and after mixing, (b) the total pressure and the partial pressures of each gas before and after mixing, and (c) the change in entropy of each gas and the total entropy change for the process.

Solution: The mass of each gas can be found by use of equation 2–30 using data from Table 2–1. R for H_2 = 765.86 ft lb per lb °R (0.961 Btu per lb °R) C_p = 3.42; R for N_2 = 54.99, and C_p = 0.246.
For the H_2
$$(144)(20)(5.0) = m(765.86)(460 + 68)$$
$$m = 0.0355 \text{ lb of } H_2$$
For the N_2
$$(144)(20)(10.0) = m(54.99)(460 + 68)$$
$$m = 0.992 \text{ lb of } N_2$$

a) Since the temperature during the process does not change, the enthalpy also is constant and can be computed for any selected temperature datum by use of equation

Thermodynamics, Steam, and Gas Properties

2-51. Choosing 0 F as a datum

$$h = mC_p (T - 460) = mC_p (68 - 0)$$

$h = (1) (3.42) (68 - 0) = 232.6$ Btu per lb of H_2 or $(0.0355) (232) = 8.3$ Btu for the m lb of H_2

$h = (1) (0.246) (68 - 0) = 16.72$ Btu per lb of N_2 or $(0.992) (16.72) = 16.6$ Btu for the m lb of N_2

Total enthalpy = 24.9 Btu, both before and after mixing

b) Each gas finally occupies the whole volume independent of the other, thus

$$\frac{P_1}{P_2} = \frac{V_2}{V_1}$$

For the hydrogen

$$\frac{20}{P_2} = \frac{15.0}{5.0} \qquad P_2 = 6.667 \text{ psia}$$

For the nitrogen

$$\frac{20}{P_2} = \frac{15.0}{10.0} \qquad P_2 = 13.333 \text{ psia}$$

c) Use equation 2-66 and recall that $T_2 = T_1$, thus

For the 0.0355 lb of H_2

$$s_2 - s_1 = 0.0355 \left[2.3 \, C_p \log_{10} \frac{T_2}{T_1} + 2.3 \, R \log_{10} \frac{P_1}{P_2} \right]$$

$$s_2 - s_1 = 0.0355 \left[0 + 2.3 \left(\frac{765.86}{778} \right) \log_{10} \frac{20}{6.667} \right]$$

$$= 0.0385 \text{ Btu per } °R \text{ for the } m \text{ lb } H_2$$

For the 0.992 lb of N_2

$$s_2 - s_1 = 0.0992 \left[0 + 2.3 \left(\frac{54.99}{778} \right) \log_{10} \frac{20}{13.333} \right]$$

$$= 0.0283 \text{ Btu per } °R \text{ for the } m \text{ lb } N_2$$

Total entropy change = 0.0283 + 0.0385 = 0.0668 Btu per °R for the 1.0276 lb of mixture

In addition to molecular diffusion it should be recognized that gravity affects the gross mixing of gases when the gases involved are of different densities. For example a stream of carbon dioxide entering an air chamber at one side would move to the lower part of the chamber because of its greater density. Ultimately if the chamber were isolated, the carbon dioxide would diffuse throughout the whole space by molecular diffusion.

Gaseous Separation by Use of Semipermeable Membranes. It is possible that a membrane might be found which when put in contact with an air mixture would permit only O_2 gas to pass through in one direction while blocking the flow of other gases. It is also possible that a similar membrane can be found which is permeable only to the passage of N_2 gas. Semipermeable membranes do exist and are used in the separation of certain gaseous and liquid mixtures and an analysis of gas separation based on the existence of such membranes can yield interesting results.

Consider the cylinder of Fig. 2–6 containing n mols of air at pressure P (atmospheric or otherwise), provided with two pistons. One of these pistons has its head constructed of a membrane permeable only to O_2 while the head of the other is permeable only to N_2. Consider the pistons initially in position I. Now assume that the pistons slowly and reversibly move inward with the total pressure P remaining constant while each gas passes through its respective membrane as shown in diagrams II and III

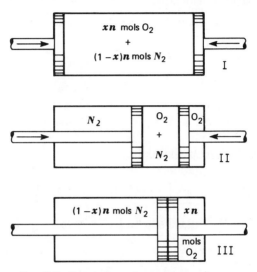

FIG. 2–6. Separation of an oxygen-nitrogen mixture by use of semi-permeable membranes in the piston heads.

until the separation is complete. The right-hand piston reacts to the O_2 as though it were a perfect sieve while it compresses the N_2 from its initial partial pressure to the final total pressure. The left-hand piston performs similarly for the N_2 as it compresses the O_2 at the right end of the piston. With slow piston movement and a constant-temperature environment work done by the piston can be computed by treating the compression as a reversible isothermal. For such compression use should be made of the partial pressure change or of molal concentration change for each component.

In a gaseous mixture containing n_a mols of gas a, n_b mols of gas b, n_c mols of gas c up to any number of gases, i, the molal concentration of each gas always appears:

$$x_a = \frac{n_a}{n_a + n_b + n_c}, \quad x_b = \frac{n_b}{n_a + n_b + n_c}$$

$$x_c = \frac{n_c}{n_T} = 1 - x_a - x_b \tag{2-74}$$

Thermodynamics, Steam, and Gas Properties

or for any number of gases

$$x_i = \frac{n_i}{n_T} = 1 - x_a - x_b \cdots - x_{i-1} \tag{2-75}$$

$$n_i = x_i n_T \tag{2-76}$$

For the n mols of air in the cylinder of Fig. 2-6 consisting of xn mols of O_2 and $(1 - x)\,n$ mols of N_2, the characteristic gas equations would appear under the initial conditions

$$PV = n\mathbf{R}T \tag{2-77}$$

$$PxV = xn\mathbf{R}T \tag{2-78}$$

$$P(1-x)V = (1-x)n\mathbf{R}T \tag{2-79}$$

Note that with n expressed in mols, $\mathbf{R}$, the Universal Gas Constant is the same for each gas 1545.3 ft-lb per lb mol degree Rankine or 1.986 Btu per lb mol degree Rankine.

The contribution of the oxygen component to the isothermal work of compression can be computed by equation 2-45 and for n mols of air undergoing separation, using a minus sign to show compression

$$\mathbf{W}_{O_2} = -xn\int_{v_1}^{v_2} P\,dv = -xn\mathbf{R}T\log_e \frac{v_2}{v_1}$$

$$= -xn\mathbf{R}T\log_e \frac{P}{Px} = -xn\mathbf{R}T\log_e \frac{1}{x} \tag{2-80}$$

For the nitrogen per n mols of air

$$W_{N_2} = -(1-x)n\mathbf{R}T\log_e \frac{P}{P(1-x)}$$

$$-W_{N_2} = (1-x)n\mathbf{R}T\log_e \frac{1}{1-x} \tag{2-81}$$

Total work for this isothermal gaseous separation of n mols of the two-component mixture in the molal ratios that these gases appear is

$$-W_T = n\mathbf{R}T\left[x\log_e \frac{1}{x} + (1-x)\log_e \frac{1}{1-x}\right]$$

$$= 2.3\,n\mathbf{R}T\left[x\log_{10}\frac{1}{x} + (1-x)\log_{10}\frac{1}{1-x}\right] \tag{2-82}$$

The total work of isothermal reversible gaseous separation of n mols of a three-component mixture having molal concentrations of x_a, x_b, x_c is

$$-W_T = 2.3\,n\mathbf{R}T\left[x_a\log_{10}\frac{1}{x_a} + x_b\log_{10}\frac{1}{x_b} + x_c\log_{10}\frac{1}{x_c}\right] \tag{2-83}$$

or in general for n mols of an i component mixture

$$-W_T = 2.3\, n\, \mathbf{R}\, T \sum_i x_i \log_{10} \frac{1}{x_i} \qquad (2\text{-}84)$$

Note in equation 2–79 that with the molal concentration of certain components known the others are determined by difference, that is:

$$x_b = 1 - x_a - x_c,\ x_a = 1 - x_b - x_c,\ x_c = 1 - x_a - x_b$$

Example 2–6. Compute the minimum theoretical work to separate one mol (28.97 lb) of air into its major gaseous constituents O_2, N_2, and A starting with air at 1.0 atm and 68 F. The molal concentration of these constituents can be considered as O_2,0.2095; N_2,0.7810; A,0.0095. Also find the work required to separate one pound of each of the constituents.

Solution: Make use of equation 2–83

$$-W_T = (2.3)(1.0)(1.986)(528)$$

$$\left[0.2095 \log \frac{1}{0.2095} + 0.781 \log \frac{1}{0.781} + 0.0095 \log \frac{1}{0.0095} \right]$$

$$= 2410\, [0.1423 + 0.0838 + 0.01923]$$

$$= 592.0 \text{ Btu per lb mol of air, isothermal work of compression}$$

$$\text{Work per lb of air separated} = \frac{592.0}{28.97} = 20.44 \text{ Btu}$$

In this one mol of air there are 0.2095 mol of O_2 thus

$$\text{Work per mol of } O_2 \text{ separated} = \frac{592.0}{0.2095} = 2823 \text{ Btu}$$

$$\text{Work per lb of } O_2 = \frac{2823}{32} = 83.3 \text{ Btu}$$

$$\text{Work per mol of } N_2 \text{ separated} = \frac{592.0}{0.781} = 758 \text{ Btu}$$

$$\text{Work per lb of } N_2 = \frac{758}{28} = 27.1 \text{ Btu}$$

$$\text{Work per mol of argon separated} = \frac{592}{0.0095} = 62{,}300 \text{ Btu}$$

$$\text{Work per lb of argon} = \frac{62{,}300}{39.94} = 1560 \text{ Btu}$$

This hypothetical method of gas separation is of interest in that it points out the minimum work of gaseous separation and also shows that a greater work is required for the gases which are at low concentration in a mixture. Gas separation is actually accomplished by liquefaction processes which will be discussed in a later chapter. Such a process shows for example that to produce one pound of gaseous oxygen from air about 175 watt hours of energy are required, this is in striking contrast to the 83.3 Btu (24.4 watt hours).

Thermodynamics, Steam, and Gas Properties   77

2–14. COMPRESSIBILITY

Real gases behave as perfect gases through only part of their vapor range. In particular, with pressure increase or with decreasing temperatures, the perfect-gas relationships become progressively less accurate. In general, it can be said that for gases, which are at temperatures far above their critical or which are at elevated temperatures relative to their saturation line and at low pressure, the deviations from the perfect-gas laws become relatively small.

Variation from the perfect or ideal gas relationship is also influenced by the complexity of the gaseous molecules. For example, air and its constituents, nitrogen and oxygen, have diatomic molecules consisting of two atoms relatively simply arranged and consequently follow the gas laws. More complex gases, such as CO_2, H_2O, and a refrigerant, such as CCl_2F_2, adhere less closely to perfect-gas relationships even under low pressures and higher temperatures. The term, higher temperature, is a relative one and primarily relates to the critical temperature of the substance and its saturation line temperatures. For example, helium has a critical temperature of 9.5R and a critical pressure of 33.2 psia. Thus, at cold temperatures in excess of 100R, the monatomic gas helium has its properties approximated by the perfect-gas laws provided its pressure is not excessive. At higher pressures even helium deviates greatly from the gas laws.

The term compressibility factor (symbol Z) is frequently used to describe the amount of deviation a gas has relative to perfect-gas conditions. By definition

$$Z = \frac{Pv}{RT} \qquad (2\text{–}85)$$

It is obvious that if the gas were perfect, the value of Z would be 1. However, real gases are more or less compressible than a perfect gas and values of Z between 0.5 to 1.5 are not unusual. Taking helium as an example, representative values show that at $300K$ ($540R$), $Z = 1.001$ at 1.2 atm, $Z = 1.043$ at 90 atm; while at a cryogenic temperature of 40 K ($72R$), $Z = 1.001$ at 1.3 atm and 1.298 at 90 atm. At 20 K and at pressures below 10 atm, helium is slightly more compressible than a perfect gas would be under the same conditions.

Figure 2–7 is a graph of compressibility data for gaseous oxygen and it can be seen that this gas is relatively more compressible than the perfect gas. Notice also by how little Z is less than unity at normal temperatures 260 K to 300 K for pressures up to 150 psia. Figure 2–8 is a similar graph for air and is applicable also for nitrogen, the major constituent of air at 77 per cent by weight. The critical point of oxygen is shown on its com-

78 Thermodynamics, Steam, and Gas Properties

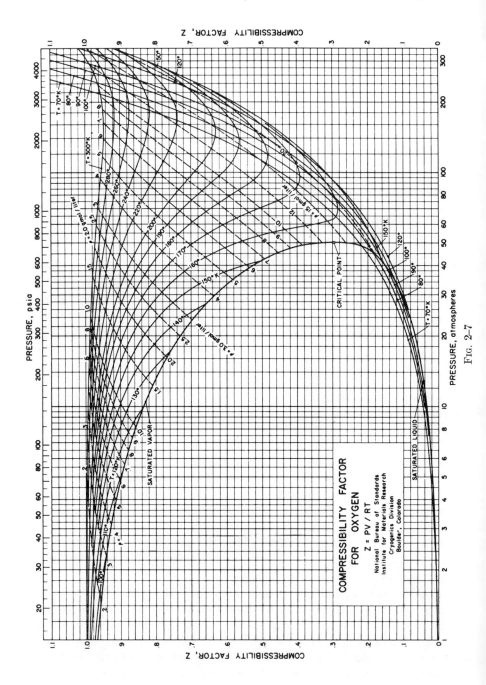

Fig. 2-7

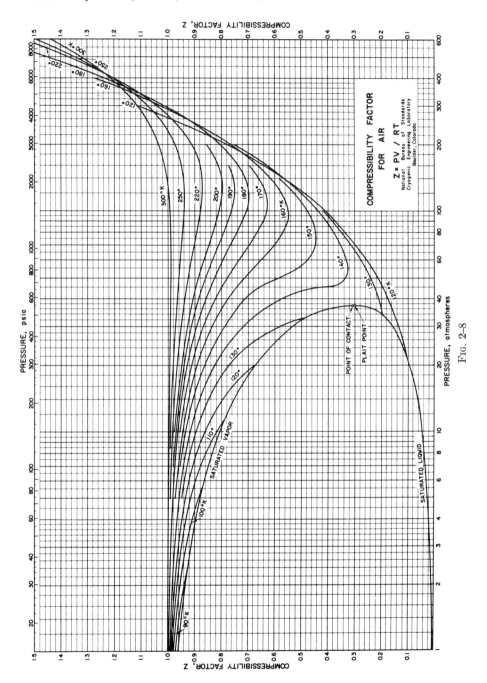

Fig. 2-8

pressibility chart. Air on the other hand being a mixture of gases, does not have a unique critical pressure. Its transition points corresponding to the critical are the *plait point* where the pressure has a maximum value in this region and the *point of contact* where the temperature is a maximum. F. Din (ref. 2) lists for the plait point 132.42 K at 37.25 atm, 88.28 cu cm per gram mol and for the point of contact 132.52 K at 37.17 atm, 90.52 cu cm per gram mol.

In this text, since most of the gases to be used have charts or other tabular data concerning them, no additional tabulations of Z will be given. It should be mentioned that to correlate the heterogeneous data relating to many different gases, usage is frequently made of the terminology and practice of reduced temperature, reduced pressure and reduced volume where these refer to the actual absolute temperature, pressure and volume expressed as ratios to the corresponding critical values of temperature, pressure and volume, since it has been found that most gases behave similarly to other gases when these are at similar reduced ratio values. (See ref. 3 and 4.)

2–15. SYMBOLS AND UNITS

These have been explained at most places where used in the chapter but further explanation may be necessary. We loosely speak of heat units and work units and although these are fundamentally precisely the same, different terminology is used and we must apply proper conversion factors where needed in an equation. For example, in equations of the type shown in 2–41, 2–42, 2–44, and 2–45, it is necessary to use absolute pressure and to express the pressure in units consistent with volume. Pounds per square inch should not be multiplied by volume in cubic feet without making prior use of a factor 144 square inches per square foot to give conventional usable units of foot pounds. Thus

P = lb per sq ft, = 144 × psi

v = cu ft or cu ft per lb

T = degrees Rankine, °F + 460

W = work in ft lb or ft lb per lb or if divided by 778 in Btu or Btu per lb

P_1/P_2 or T_1/T_2, ratios can be in any consistent absolute units

C_p, C_v specific heat values are usually expressed in Btu per lb mass degree F or in numerically equivalent units of kilo-calories per kg degree K

h enthalpy, $\mathbf{h} = u + Pv$. Since **h** is commonly in Btu per lb and Pv conventionally in ft-lb per lb_m here we should divide Pv by 778 to put this term into a consistent form for u since the latter is usually expressed in Btu per lb_m.

Thermodynamics, Steam, and Gas Properties

In the metric system pressures are usually expressed in kg per cm^2, (also known as metric atmospheres, since 1 kg per cm^2 is equal to a possible atmosphere at 14.223 psia), and v should be expressed in cubic centimeters for consistency. However, if v is in cubic meters, a multiplying factor of 10,000 would be required to yield an answer in kg-meters for a Pv product. To convert kg meters to large calories (k cal), division by 427.0 would be necessary.

Tables A-1, A-2, A-3, and A-4 in the appendix give the conversion factors for a number of different units.

PROBLEMS

2-1. A man carries a 12-lb steel ball to the top of a 50-ft building and lets it drop to the ground. (a) How much potential energy is stored in the ball when it is at the top of the building? (b) What is its velocity and kinetic energy just before it strikes the ground? *Ans.* (a) 600 ft-lb; (b) 27.8 fps

2-2. What is the ideal work-input to a pump that is delivering 200 gpm of water, if the pump delivers at 60 psig and draws in water at a vacuum of 5 psi? The water enters the pump with a velocity of 3 fps and leaves with a velocity of 10 fps. (In the solution of the problem, write the general energy equation and eliminate the terms that do not apply.) *Ans.* 7.6 hp

2-3. Assume that 80 lb of air are heated by an open furnace from 12 F to 70 F, and find the Btu required. *Ans.* 1114 Btu

2-4. A steel tank weighing 300 lb contains 3 cu ft of hydrogen at 40 F and 500 psia pressure. (a) Find the heat required to raise the tank and its contents to 77 F. (b) What pressure does the hydrogen exert when the temperature is 77 F?
Ans. (a) 1383 Btu; (b) 537 psia

2-5. In a certain ice plant, 1000 lb of brine per minute are cooled from 25 F to 14 F. The specific heat of the brine is 0.748. Find the tons of refrigeration. *Ans.* 41.1

2-6. Nitrogen gas in an ammonia synthesizing plant is compressed from a storage tank at 100.3 psig at 70 F to a pressure of 1135.3 psig. Assume that the actual compression is approximated by the conditions of a reversible polytropic for which the exponent n is 1.3. Compute (a) specific volume of the gas at start of compression, (b) temperature of gas leaving compressor, (c) work required to carry out the compression of each pound of gas. By making an energy balance between the $C_p(\Delta T)$ change and the work input, (d) find the heat rejection during the compression. Note that equations of the form of 2-47 and 2-69 are applicable with use being made of the appropriate n value.

2-7. Dry air is supplied to two expander turbines in series at 20 atm (294 psia) and expands to 2 atm (29.4 psia). Assume that the expansion is reversibly adiabatic. (a) Compute the temperature at exit from the second expander turbine if air enters the first expander at 70 F (530R). (b) Compute the power produced if 15 lb per minute flow through the expanders. (c) If 75 per cent of this power is actually delivered at the turbine shaft make use of an energy balance to find the actual temperature of the air leaving the turbine.

2-8. Dry air is compressed isothermally at 530R from 14.7 psia to 147.0 psia.

(a) Compute the work required and the heat rejected. (b) Compute the change in entropy which takes place.

Ans. (a) 83.5 Btu per lb; (b)—0.1576 Btu per lb°R

2-9. Dry air expands isothermally from 105 psia to 15 psia at 530R. (a) Compute the work produced per lb of air and the heat added during the expansion. (b) Compute the entropy change and find the heat addition by making a $T\Delta s$ computation.

2-10. (a) How much energy must be added to 10 lb of ice at 25 F to change it to water and warm it to 70 F? (b) What is the enthalpy of the 70 F water? (c) What additional amount of energy is required to warm this 70 F water and change it into dry saturated steam at 212 F and 14.7 psi?

Ans. (a) 1847.1 Btu; (b) 1150.4 Btu/lb; (c) 11.124 Btu

2-11. Find the specific volume and enthalpy of a pound of steam at a pressure of 200 psia and 95 per cent quality. *Ans.* 2.17, 1156.2

2-12. Twelve tons of ice from an indoor skating rink are disposed of by using an atmospheric-pressure steam blast. Assume that the ice is at 26 F at the start of the process and that the disposed water reaches a temperature of 40 F before disposal to the sewer drains. Dry saturated steam at 14.7 psia condenses and cools to mix with the drain water at 40 F. How many pounds of steam are required to warm, melt, and dispose of the ice as water, if 20 per cent of the useful enthalpy associated with the steam and its subcooled water is lost to the atmosphere? *Ans.* 4550 lb

2-13. Water at 300 psia in a saturated state enters a drain trap serving a steam line, and it drops to atmospheric pressure by the time it enters the trap discharge pipe. (a) Find the final temperature of the discharged water-steam and (b) compute what fraction of each original pound of water flashes to steam. Note that in a process of this type the enthalpy stays constant, so the enthalpy of the original water equals the enthalpy (eq 2-12) of the discharged mixture at 14.7 psia. *Ans.* (a) 212 F; (b) 0.22

2-14. Water at 100 psia in a saturated state enters a drain trap serving a steam line and drops to atmospheric pressure by the time it enters the trap discharge pipe. (a) Find the final temperature of the discharged water-steam and (b) compute what fraction of each original pound of water flashes to steam. Note that in a process of this type the enthalpy stays constant, so that the enthalpy of the original water equals the enthalpy (eq 2-12) of the discharged mixture at 14.7 psia. *Ans.* (a) 212 F; (b) 0.122

2-15. By volume analysis, the dry flue gases from a furnace consist, in percentages, of CO_2, 11; O_2, 8; CO, 1; and N_2, 80. Compute (a) the weight analysis of the flue gas, (b) its molecular weight, and (c) its density measured at 68 F and 14.7 psia.

Ans. (a) 14.9, 8.6, 0.9, 75.6; (b) 29.64; (c) 0.0768 lb per cu ft

2-16. The wet flue gases from the combustion of a gas show a volumetric analysis, in percentages, of CO_2, 8.5; O_2, 0.5; CO, 0.9; H_2O, 18.9; and N_2, 71.2. Compute (a) the weight analysis of the wet flue gas, (b) its molecular weight, and (c) its density at 300 F and 14.7 psia. *Ans.* (a) 13.6, 0.6, 0.9, 12.4, 72.5; (b) 27.49; (c) 0.0496

2-17. If the flue gas of problem 2-14 is cooled to 70 F, most of the water vapor condenses and the resulting dry analysis by volume is, in percentages, CO_2, 10.5; O_2, 0.6; CO, 1.1; and N_2, 87.8. Assume that this dry gas is reheated to 300 F at 14.7 psia and compute (a) the weight analysis, (b) the molecular weight, and (c) the density.

Ans. (a) 15.6, 0.6, 1.1, 82.7; (b) 29.7; (c) 0.0536

2-18. Compute the specific volume in cu ft per lb for oxygen at 300 psia and 540 R by use of the perfect gas equation and correct the value by making use of the compressibility factor chart. Carry out a similar computation for oxygen at 300 psia and 324 R (−136 F).

2-19. A chamber has two compartments of different size. In one compartment, of volume 5.0 cu ft, hydrogen at 200 psia and 68 F is stored, in the other compartment 10.0 cu ft volume, nitrogen is stored also at 200 psia and 68 F. If the partition between the two chambers is opened and the gases diffuse into each other, compute (a) the enthalpy before and after mixing, (b) the total pressure and the partial pressure of each gas before and after mixing, and (c) the change in entropy of each gas and the total entropy change for the process. Note that the mass of each gas, needed to find the enthalpy, can be found from the perfect-gas equation.

Ans. (a) $83 + 166 = 249$ Btu, same before and after; (b) 200 psia, then $H_2 = 66.7$ psia, $N_2 = 133.3$ psia; (c) $0.385 + 0.283 = 0.668$ Btu/°R for the 10.276 lb of mixture

2-20. It is proposed that a pair of semi-permeable membranes be sought that will permit the separation of oxygen and nitrogen from air. Assuming that the membranes can be found, compute the minimum work required for separating air at 1.0 atmosphere pressure and 68 F, into gaseous oxygen and nitrogen. The air can be considered to contain nitrogen and oxygen in respective mol percentages of 79 and 21. The minimum work required for separation implies the isothermal compression of each component as part of the mixture from its initial partial pressure in the mixture to 1.0 atmosphere. Express your answer in (a) Btu per mol of air, (b) Btu per mol of O_2 produced, and (c) Watt-hours per lb of O_2 produced.

Ans. (a) 540 Btu; (b) 2570 Btu; (c) 23.5 watt hr

2-21. By making use of three semi-permeable membranes compute the minimum theoretical work to separate one mol (28.97 lb) of air into its major gaseous constituents O_2, N_2 and A starting with air at 1.0 atm and 10 F. The molal concentration of these constituents can be considered as O_2, 0.2095; N_2, 0.7810; A, 0.0095. Also find the work required to separate one pound of each of the constituents.

Ans. 527 Btu per mol air, 2515 Btu per mol O_2, 675 Btu per mol N_2, 55,500 Btu per mol A

2-22. By making use of semi-permeable membranes compute the minimum work required to separate one mol of air (28.97 lb) into argon alone with the other constituents oxygen and nitrogen grouped as a composite unseparated mixture. For this note that air has a concentration of A, 0.0095 mol and other constituents 0.9905 mol. Express result in (a) Btu per lb mol of air, (b) Btu per mol A, (c) Btu per lb A.

Ans. 56.3, 5910, 148

2-23. A compressed air cylinder used as a small power source at a temperature 250 K (−10 F) contains air at a pressure of 240 atm abs. The precise gas fill space in the cylinder is 6 cu ft. Air from the cylinder is withdrawn and at a later period it is found that the cylinder pressure is 40 atm abs. and the temperature is 250 K. Compute the weight of air withdrawn taking compressibility effects into consideration.

2-24. Compute the internal tank volume required to store 1000 lb of oxygen at a pressure of 200 atm abs. and a temperature of 280 K (44 F).

2-25. Compute the internal tank volume required to store 50 lb of helium at 90 atm abs. and 300 K (540 R). Use compressibility data from section 2-14.

REFERENCES

1. J. H. Keenan and J. Kaye, *Gas Tables* (New York: John Wiley and Sons, 1948).
2. F. Din, *Thermodynamic Functions of Gases, Vol. 2 for Air* (London: Butterworth's Scientific Publications, 1956).
3. E. F. Obert and R. A. Gaggioli, *Thermodynamics*, 2d ed. (New York: McGraw-Hill Book Co., 1963), pp. 219-227.
4. B. F. Dodge, *Chemical Engineering Thermodynamics* (New York: McGraw-Hill Book Co., 1944), pp. 159-187.

3

Air and Humidity Calculations

3-1. ATMOSPHERIC AIR

Air as found in the earth's atmosphere is a mixture of constituent gases. Even when uncontaminated, its composition varies slightly at different points of the globe and at different altitudes. However, for scientific purposes it became necessary to define a composition that can be called standard air. The composition of standard air which is commonly accepted is listed in the publication entitled "The U.S. Standard Atmosphere" (ref. 1). It is reproduced here in the following tabulation with the values given in percentage composition by volume or equivalently as mols per 100 mols. In addition to the gases listed by name, there are trace amounts of other gases such as methane, sulfur dioxide, hydrogen, krypton, and xenon. These vary in composition and are simply listed in the table as other gases.

Standard Air

Nitrogen	78.084000
Oxygen	20.947600
Argon	0.934000
Carbon dioxide	0.031400
Neon	0.001818
Helium	0.000524
Other gases	0.000658
	100.000000

The apparent molecular weight of standard dry air is 28.966.

The items listed in the table are the constituents of dry air. However, most air is moist and has water vapor associated with it. Moist air is defined as a binary mixture of dry air and water vapor (steam) with the amount of steam varying from zero to the condition of saturation, that is with the water vapor in the mixture capable of coexisting in neutral equilibrium with a flat surface of liquid or solid water. Saturation denotes the maximum amount of steam that can exist in a cubic foot of space at any given temperature and is essentially independent of the weight and pressure of the air which may simultaneously exist in the same space.

For example, if water at saturation temperature were sprayed into a saturated air mixture, the water would remain in liquid phase and not evaporate. The water would either fall in the space or remain suspended as a mist or fog. Further, if the steam-saturated air were cooled in the slightest degree, it would be impossible for this existing amount of moisture to remain as steam, and the excess would condense out in the form of fog or dew. At the new, lower temperature, a different and lower saturation pressure must exist.

Saturation conditions for steam, and the corresponding maximum weight of steam that can exist in a space, are not directly predictable and must therefore be found from tabulations of the properties of steam or of the properties of steam and air mixtures. Table 3-1 in this text can be employed. It should also be mentioned that a more extensive and complete tabulation of moist air properties has been prepared by J. A. Goff and S. Gratch (refs. 6 and 7). Notice that moist air tabulations are based on a fixed total (barometric) pressure, usually that of the standard atmosphere and for other pressures, use must be made of appropriate correction factors.

3-2. MOIST-AIR TERMINOLOGY

Steam-air mixtures do not exactly follow the perfect-gas laws but for total pressures up to some three atmospheres, sufficient accuracy is possible to permit the use of these laws in engineering computations for such mixtures. This approach will first be developed, following which a more scientific analysis of the subject will be presented.

In atmospheric air-steam mixtures the Gibbs-Dalton law (sec. 2-12) is closely obeyed. This means that in any gas mixture the total pressure exerted is the summation of the partial pressures exerted independently by each of the constituent gases. Atmospheric air exists at a total pressure equal to barometric pressure (p_B), and this pressure in turn is made up of the partial pressures exerted by all the gases, mainly the nitrogen, p_{N2}; the oxygen, p_{O2}; and the water vapor (steam), p_s. Or, in mathematical terms,

$$p_B = p_{N2} + p_{O2} + p_s = p_a + p_s \qquad (3\text{-}1)$$

In this equation, since there is no need of separating the pressures of the nitrogen and oxygen, it is customary to list the total barometric pressure as the sum of the pressures of the air portion (p_a) and the steam portion (p_s). The nitrogen and oxygen together are frequently called "dry air," even though some steam is usually mixed with the air.

By referring to Table 3-1 it will be seen that at 80 F, saturated steam exerts a pressure of 1.0323 in. Hg and weighs 1/633.0, or 0.001579 lb per cu ft. These values are essentially correct whether air exists in the same

Air and Humidity Calculations

cubic foot or steam alone is present. For this low-pressure steam, even when it is saturated, the gas equation $pV = mRT$ holds closely. For steam, R is $1545/18 = 85.7$, and by using this value and the figures just given, it can be seen that the equivalence is very close. Thus,

$$(144)(0.491)(1.0323)(1) = (0.001579)(85.7)(460 + 80)$$

or
$$73.0- = 73.0+$$

It happens that over the range used for air-steam mixtures a value of $R = 85.6$ for steam gives the most consistent results, and therefore this value should be used in problems. Figure 2–2 shows values of R plotted for low-pressure steam.

The water vapor (steam) mixed with dry air in the atmosphere is known as *humidity*. The weight of water vapor, expressed in pounds or grains, occurring in each cubic foot of space is really *vapor density* (d) and in this text will be called by no other name, although in the technical literature various names are used. The weight of water vapor, expressed in pounds or grains, associated with each pound of dry-air constituents is called *humidity ratio* or *specific humidity*. In this text both names will be used and the symbol for humidity ratio is W or W_s for saturated air.

Relative humidity (ϕ) by definition is the mol fraction of water vapor (x_w) in moist air to the mol fraction (x_{ws}) of water vapor in saturated air at the same temperature (t) and total pressure (p_T)

$$\phi = \left.\frac{x_w}{x_{ws}}\right)_{t,\,p_T} \tag{3-2}$$

Under conditions where the perfect gas laws and the Gibbs-Dalton relationships hold, the above definition is equivalent to one based on partial pressures and also on densities, namely: *Relative humidity* is the ratio of the partial pressure of water vapor in the air to the pressure that saturated steam exerts at the temperature of the air. In equational form

$$\phi = \left.\left(\frac{p}{p_s}\right)\right)_{t_d} \tag{3-3}$$

or

$$\phi = \left.\left(\frac{d}{d_s}\right)\right)_{t_d} \tag{3-4}$$

where ϕ = relative humidity, expressed as a decimal;
p = partial pressure of the water vapor (steam) in the air;
p_s = pressure of saturated steam at the air temperature (dry-bulb temperature, t);
d = density of the water vapor in the air, in pounds per cubic foot;
d_s = density of saturated water vapor at the air temperature (dry-bulb), in pounds per cubic foot.

TABLE 3-1
Thermodynamic Properties of Air, Water, and Steam*
(From 32 F to 200 F and from 32 F to −120 F)

Temperature, F	Properties of Water and Steam				Properties of Dry Air at a Pressure of 29.921 in. Hg abs		Properties of Mixture of Dry Air and Sat. Steam at a Total Pressure of 29.921 in. Hg abs		
	Saturation pressure of water and steam, in. Hg	Enthalpy		Specific volume of sat. steam, cu ft/lb	True specific volume, cu ft/lb	Enthalpy, Btu/lb	Volume of mixture per lb of dry air, ft³	Enthalpy of mixture per lb of dry air, Btu	Specific humidity, grains per lb of dry air
		Saturated water, Btu/lb	Saturated steam, Btu/lb						
t	p_s	h_f	h_g	v_g	v_a	h_a	v_s	h_s	W_s
32	0.1803	0.0	1075.2	3305	12.389	7.69	12.46	11.75	26.40
33	0.1878	1.0	1075.6	3180	12.414	7.93	12.49	12.16	27.49
34	0.1955	2.0	1076.0	3062	12.439	8.17	12.52	12.57	28.63
35	0.2034	3.0	1076.5	2948	12.464	8.41	12.55	13.00	29.80
36	0.2117	4.0	1076.9	2839	12.490	8.65	12.58	13.42	31.02
37	0.2202	5.0	1077.4	2734	12.515	8.89	12.61	13.86	32.28
38	0.2290	6.0	1077.8	2634	12.540	9.13	12.64	14.30	33.58
39	0.2382	7.0	1078.2	2538	12.565	9.37	12.67	14.75	34.94
40	0.2477	8.0	1078.7	2445	12.591	9.61	12.70	15.21	36.34
41	0.2575	9.0	1079.1	2357	12.616	9.85	12.73	15.68	37.80
42	0.2676	10.1	1079.5	2272	12.641	10.09	12.76	16.16	39.30
43	0.2781	11.1	1080.0	2190	12.667	10.34	12.79	16.64	40.86
44	0.2890	12.1	1080.4	2112	12.692	10.58	12.82	17.13	42.47
45	0.3002	13.1	1080.9	2037	12.717	10.82	12.85	17.63	44.14
46	0.3119	14.1	1081.3	1965	12.742	11.06	12.88	18.13	45.86
47	0.3239	15.1	1081.7	1896	12.768	11.30	12.91	18.66	47.65
48	0.3363	16.1	1082.2	1829	12.793	11.54	12.94	19.19	49.51
49	0.3491	17.1	1082.6	1766	12.818	11.78	12.97	19.73	51.42
50	0.3624	18.1	1083.1	1704	12.844	12.02	13.00	20.28	53.40
51	0.3761	19.1	1083.5	1645	12.869	12.26	13.03	20.84	55.44
52	0.3903	20.1	1083.9	1589	12.894	12.50	13.06	21.41	57.56
53	0.4049	21.1	1084.4	1534	12.919	12.74	13.10	21.99	59.75
54	0.4200	22.1	1084.8	1482	12.945	12.98	13.13	22.59	62.01
55	0.4356	23.1	1085.2	1431	12.970	13.22	13.16	23.20	64.36

Air and Humidity Calculations

56	0.4518	24.1	1085.7	1383	13.46	13.19	23.82	66.78
57	0.4684	25.1	1086.1	1336	13.70	13.23	24.45	69.28
58	0.4856	26.1	1086.5	1292	13.94	13.26	25.10	71.86
59	0.5033	27.1	1087.0	1249	14.18	13.29	25.76	74.54
60	0.5216	28.1	1087.4	1207	14.42	13.33	26.43	77.29
61	0.5405	29.1	1087.9	1167	14.66	13.36	27.11	80.14
62	0.5599	30.1	1088.3	1129	14.90	13.40	27.82	83.09
63	0.5800	31.1	1088.7	1092	15.14	13.43	28.54	86.14
64	0.6007	32.1	1089.2	1056	15.38	13.47	29.27	89.27
65	0.6221	33.1	1089.6	1022	15.62	13.50	30.03	92.51
66	0.6441	34.1	1090.0	988.6	15.86	13.54	30.79	95.86
67	0.6668	35.1	1090.5	956.8	16.10	13.58	31.58	99.32
68	0.6902	36.1	1090.9	926.1	16.35	13.61	32.38	102.9
69	0.7143	37.1	1091.3	896.5	16.59	13.65	33.20	106.6
70	0.7392	38.1	1091.8	868.0	16.83	13.69	34.04	110.4
71	0.7648	39.1	1092.2	840.5	17.07	13.72	34.90	114.3
72	0.7911	40.1	1092.6	814.0	17.31	13.76	35.79	118.4
73	0.8183	41.1	1093.1	788.4	17.55	13.80	36.69	122.6
74	0.8463	42.1	1093.5	763.8	17.79	13.84	37.61	126.9
75	0.8751	43.1	1093.9	740.0	18.03	13.88	38.55	131.3
76	0.9047	44.1	1094.4	717.0	18.27	13.92	39.52	135.9
77	0.9352	45.1	1094.8	694.9	18.51	13.96	40.51	140.6
78	0.9667	46.1	1095.2	673.5	18.75	14.00	41.52	145.5
79	0.9990	47.1	1095.7	652.9	18.99	14.04	42.56	150.6
80	1.0323	48.1	1096.1	633.0	19.23	14.09	43.63	155.8
81	1.0665	49.1	1096.6	613.8	19.47	14.13	44.72	161.2
82	1.1017	50.1	1097.0	595.3	19.71	14.17	45.84	166.7
83	1.1380	51.1	1097.4	577.4	19.95	14.22	46.98	172.4
84	1.1752	52.1	1097.8	560.1	20.19	14.26	48.16	178.3
85	1.2136	53.1	1098.3	543.3	20.43	14.31	49.36	184.4
86	1.2530	54.0	1098.7	527.2	20.67	14.35	50.59	190.6
87	1.2935	55.0	1099.1	511.6	20.91	14.40	51.86	197.0
88	1.3351	56.0	1099.6	496.5	21.15	14.44	53.14	203.7
89	1.3779	57.0	1100.0	482.0	21.39	14.50	54.48	210.6
90	1.4219	58.0	1100.4	467.9	21.64	14.55	55.85	217.6

*Reproduced by permission of The American Society of Refrigerating Engineers from the A.S.R.E. Data Book. Compiled by C. O. Mackey, using basic Goff and Gratch data.

TABLE 3-1 (*Continued*)

Temperature, °F t	Properties of Water and Steam				Properties of Dry Air at a Pressure of 29.921 in. Hg abs		Properties of Mixture of Dry Air and Sat. Steam at a Total Pressure of 29.921 in. Hg abs			
	Saturation pressure of water and steam, in. Hg p_s	Enthalpy		Specific volume of sat. steam, cu ft/lb v_g	True specific volume, cu ft/lb v_a	Enthalpy, Btu/lb h_a	Volume of mixture per lb of dry air, ft³ v_s	Enthalpy of mixture per lb of dry air, Btu h_s	Specific humidity, grains per lb of dry air W_s
		Saturated water, Btu/lb h_f	Saturated steam, Btu/lb h_g						
91	1.4671	59.0	1100.9	454.3	13.880	21.88	14.60	57.25	224.9
92	1.5136	60.0	1101.3	441.1	13.905	22.12	14.65	58.69	232.4
93	1.5613	61.0	1101.7	428.4	13.930	22.36	14.70	60.16	240.1
94	1.6103	62.0	1102.2	416.1	13.955	22.60	14.75	61.67	248.1
95	1.6607	63.0	1102.6	404.2	13.981	22.84	14.80	63.22	256.4
96	1.7124	64.0	1103.0	392.7	14.006	23.08	14.86	64.81	264.8
97	1.7655	65.0	1103.4	381.5	14.031	23.32	14.91	66.45	273.6
98	1.8200	66.0	1103.9	370.7	14.057	23.56	14.97	68.13	282.6
99	1.8759	67.0	1104.3	360.3	14.082	23.80	15.02	69.86	291.9
100	1.9334	68.0	1104.7	350.2	14.107	24.04	15.08	71.62	301.5
101	1.9923	69.0	1105.2	340.4	14.132	24.28	15.14	73.44	311.3
102	2.0529	70.0	1105.6	331.0	14.157	24.52	15.20	75.31	321.5
103	2.1149	71.0	1106.0	321.8	14.183	24.76	15.26	77.22	332.0
104	2.1786	72.0	1106.4	313.0	14.208	25.00	15.32	79.19	342.8
105	2.2440	73.0	1106.9	304.4	14.233	25.24	15.39	81.21	353.9
106	2.3110	74.0	1107.3	296.0	14.259	25.48	15.45	83.29	365.4
107	2.3798	75.0	1107.7	288.0	14.284	25.72	15.52	85.42	377.2
108	2.4503	76.0	1108.2	280.2	14.309	25.96	15.59	87.62	389.4
109	2.5226	77.0	1108.6	272.6	14.334	26.20	15.65	89.87	402.0
110	2.5968	78.0	1109.0	265.3	14.360	26.45	15.72	92.19	414.9
111	2.6728	79.0	1109.4	258.2	14.385	26.69	15.80	94.58	428.3
112	2.7507	80.0	1109.9	251.3	14.410	26.93	15.87	97.03	442.1
113	2.8306	81.0	1110.3	244.6	14.435	27.17	15.94	99.55	456.3
114	2.9125	82.0	1110.7	238.1	14.461	27.41	16.02	102.16	471.0
115	2.9963	83.0	1111.1	231.8	14.486	27.65	16.10	104.81	486.1

Air and Humidity Calculations

116	3.0823	84.0	1111.6	225.8	14.511	27.89	16.18	107.55	501.6
117	3.1703	85.0	1112.0	219.9	14.537	28.13	16.26	110.38	517.7
118	3.2606	86.0	1112.4	214.1	14.562	28.37	16.34	113.29	534.3
119	3.3530	87.0	1112.8	208.6	14.587	28.61	16.43	116.28	551.4
120	3.4477	88.0	1113.3	203.2	14.612	28.85	16.51	119.36	569.0
121	3.5446	89.0	1113.7	197.9	14.637	29.09	16.60	122.52	587.2
122	3.6439	90.0	1114.1	192.9	14.663	29.33	16.70	125.79	606.0
123	3.7455	91.0	1114.5	188.0	14.688	29.57	16.79	129.15	625.3
124	3.8496	92.0	1114.9	183.2	14.713	29.82	16.89	132.61	645.3
125	3.9561	93.0	1115.4	178.5	14.739	30.06	16.98	136.17	665.9
126	4.0651	94.0	1115.8	174.0	14.764	30.30	17.08	139.88	687.2
127	4.1768	95.0	1116.2	169.6	14.789	30.54	17.19	143.64	709.2
128	4.2910	96.0	1116.5	165.4	14.814	30.78	17.29	147.54	731.9
129	4.4078	97.0	1117.0	161.3	14.839	31.02	17.40	151.57	755.4
130	4.5274	98.0	1117.5	157.3	14.865	31.26	17.52	155.72	779.6
131	4.6498	99.0	1117.9	153.4	14.890	31.50	17.63	160.00	804.6
132	4.7750	100.0	1118.3	149.6	14.915	31.74	17.75	164.43	830.5
133	4.9030	101.0	1118.7	145.9	14.941	31.98	17.87	168.98	857.2
134	5.0340	102.0	1119.2	142.4	14.966	32.22	17.99	173.69	884.8
135	5.1679	103.0	1119.6	138.9	14.991	32.46	18.12	178.54	913.3
136	5.3049	104.0	1120.0	135.5	15.016	32.70	18.25	183.57	942.8
137	5.4450	105.0	1120.4	132.2	15.043	32.94	18.39	188.75	973.4
138	5.5881	106.0	1120.8	129.1	15.067	33.18	18.53	194.09	1000.0
139	5.7345	107.0	1121.2	126.0	15.092	33.43	18.67	199.64	1038
140	5.8842	108.0	1121.7	123.0	15.117	33.67	18.82	205.34	1071
141	6.0371	109.0	1122.1	120.0	15.143	33.91	18.97	211.27	1106
142	6.1934	110.0	1122.5	117.2	15.168	34.15	19.13	217.39	1143
143	6.3332	111.0	1122.9	114.4	15.193	34.39	19.29	223.70	1180
144	6.5164	112.0	1123.3	111.7	15.218	34.63	19.45	230.28	1219
145	6.6832	113.0	1123.7	109.1	15.244	34.87	19.62	236.94	1259
146	6.8536	114.0	1124.1	106.6	15.269	35.11	19.81	244.06	1301
147	7.0277	115.0	1124.6	104.1	15.294	35.35	19.99	251.34	1344
148	7.2056	116.0	1125.0	101.7	15.319	35.59	20.18	258.88	1389
149	7.3872	117.0	1125.4	99.32	15.345	35.83	20.37	266.71	1436
150	7.5727	118.0	1125.8	97.04	15.370	36.07	20.58	274.84	1485

TABLE 3-1 (*Continued*)

Temperature, °F	Saturation pressure of water and steam, in. Hg	Properties of Water and Steam			Properties of Dry Air at a Pressure of 29.921 in. Hg abs		Properties of Mixture of Dry Air and Sat. Steam at a Total Pressure of 29.921 in. Hg abs		
		Enthalpy		Specific volume of sat. steam, cu ft/lb	True specific volume, cu ft/lb	Enthalpy, Btu/lb	Volume of mixture per lb of dry air, ft³	Enthalpy of mixture per lb of dry air, Btu	Specific humidity, grains per lb of dry air
		Saturated water, Btu/lb	Saturated steam, Btu/lb						
t	p_s	h_f	h_g	v_g	v_a	h_a	v_s	h_s	W_s
151	7.7622	119.0	1126.2	94.81	15.395	36.31	20.79	283.25	1535
152	7.9556	120.0	1126.6	92.65	15.420	36.56	21.01	292.00	1587
153	8.1532	121.0	1127.0	90.54	15.446	36.80	21.23	301.07	1641
154	8.3548	122.0	1127.4	88.49	15.471	37.04	21.46	310.53	1698
155	8.5607	123.0	1127.8	86.50	15.496	37.28	21.71	320.34	1757
156	8.7708	124.0	1128.3	84.55	15.521	37.52	21.96	330.57	1818
157	8.9853	125.0	1128.7	82.66	15.547	37.76	22.22	341.18	1882
158	9.2042	126.0	1129.1	80.81	15.572	38.00	22.49	352.24	1948
159	9.4276	127.0	1129.5	79.02	15.597	38.24	22.77	363.29	2018
160	9.6556	128.0	1129.9	77.27	15.622	38.48	23.07	375.81	2090
161	9.8882	129.0	1130.3	75.56	15.648	38.72	23.37	388.34	2165
162	10.126	130.0	1130.7	73.90	15.673	38.96	23.69	401.45	2244
163	10.368	131.0	1131.1	72.28	15.698	39.21	24.02	415.14	2326
164	10.615	132.0	1131.5	70.71	15.724	39.45	24.37	429.44	2412
165	10.867	133.0	1131.9	69.17	15.749	39.69	24.73	444.41	2503
166	11.124	134.0	1132.3	67.67	15.774	39.93	25.11	460.10	2597
167	11.386	135.0	1132.7	66.21	15.799	40.17	25.50	476.53	2696
168	11.653	136.0	1133.1	64.79	15.824	40.41	25.92	493.77	2880
169	11.925	137.0	1133.5	63.40	15.850	40.65	26.35	511.83	2910
170	12.203	138.0	1133.9	62.04	15.875	40.89	26.81	530.86	3024

Air and Humidity Calculations

171	12.487	139.0	1134.4	60.73	15.900	41.13	27.29	550.89	3145
172	12.775	140.0	1134.8	59.44	15.925	41.37	27.79	571.97	3273
173	13.080	141.0	1135.2	58.18	15.951	41.61	28.32	594.19	3407
174	13.370	142.0	1135.6	56.96	15.976	41.85	28.88	617.65	3549
175	13.676	143.0	1136.0	55.77	16.001	42.10	29.47	642.48	3699
176	13.987	144.0	1136.4	54.60	16.026	42.34	30.09	668.67	3858
177	14.305	145.0	1136.8	53.47	16.052	42.58	30.75	696.49	4026
178	14.629	146.0	1137.2	52.36	16.077	42.82	31.46	726.04	4205
179	14.959	147.0	1137.6	51.28	16.103	43.06	32.21	757.46	4396
180	15.295	148.0	1137.9	50.22	16.128	43.30	32.99	790.88	4598
181	15.637	149.0	1138.3	49.19	16.153	43.54	33.83	826.46	4815
182	15.986	150.0	1138.7	48.19	16.178	43.78	34.74	864.74	5046
183	16.341	151.0	1139.1	47.20	16.203	44.02	35.70	905.58	5294
184	16.703	152.0	1139.5	46.25	16.229	44.26	36.74	949.49	5560
185	17.071	153.0	1139.9	45.31	16.254	44.51	37.85	996.86	5847
186	17.446	154.0	1140.3	44.40	16.279	44.75	39.04	1047.7	6156
187	17.829	155.0	1140.7	43.51	16.304	44.99	40.34	1102.8	6491
188	18.218	156.0	1141.1	42.64	16.329	45.23	41.75	1162.6	6854
189	18.614	157.0	1141.5	41.79	16.354	45.47	43.28	1227.8	7250
190	19.017	158.0	1141.9	40.96	16.380	45.71	44.94	1298.9	7682
191	19.428	159.1	1142.3	40.14	16.405	45.95	46.78	1376.9	8156
192	19.846	160.1	1142.7	39.35	16.430	46.19	48.79	1463.0	8679
193	20.271	161.1	1143.1	38.58	16.456	46.43	51.02	1558.2	9257
194	20.704	162.1	1143.5	37.82	16.481	46.68	53.50	1664.0	9901
195	21.145	163.1	1143.8	37.09	16.506	46.92	56.27	1782.6	10622
196	21.594	164.1	1144.2	36.36	16.531	47.16	59.40	1916.2	11434
197	22.050	165.1	1144.6	35.66	16.556	47.40	62.93	2067.6	12354
198	22.515	166.1	1145.0	34.97	16.582	47.64	66.99	2240.9	13409
199	22.987	167.1	1145.4	34.30	16.607	47.88	71.66	2440.9	14624
200	23.468	168.1	1145.8	33.64	16.632	48.12	77.14	2675.6	16052

TABLE 3-1 (*Continued*)

Temperature, °F	Saturation pressure of ice and steam, in. Hg	Properties of Ice and Steam			Properties of Dry Air at a Pressure of 29.921 in. Hg abs		Properties of Mixture of Dry Air and Sat. Steam at a Total Pressure of 29.921 in. Hg abs		
		Enthalpy		Specific volume of sat. steam, cu ft/lb	True specific volume, cu ft/lb	Enthalpy, Btu/lb	Volume of mixture per lb of dry air, cu ft	Enthalpy of mixture per lb of dry air, Btu	Specific humidity, grains per lb of dry air
		Saturated ice, Btu/lb	Saturated steam, Btu/lb						
t	p_s	h_i	h_g	v_g	v_a	h_a	v_s	h_s	W_s
32	0.1803	−143.4	1075.2	3305	12.389	7.69	12.46	11.75	26.40
31	0.1723	−143.9	1074.7	3453	12.363	7.45	12.43	11.32	25.21
30	0.1645	−144.4	1074.3	3608	12.338	7.21	12.41	10.90	24.07
29	0.1571	−144.9	1073.8	3771	12.313	6.97	12.38	10.49	22.98
28	0.1500	−145.4	1073.4	3943	12.287	6.73	12.35	10.09	21.92
27	0.1431	−145.9	1073.0	4122	12.262	6.49	12.32	9.70	20.92
26	0.1366	−146.4	1072.5	4311	12.237	6.25	12.29	9.31	19.96
25	0.1303	−146.9	1072.1	4509	12.211	6.01	12.27	8.92	19.04
24	0.1243	−147.4	1071.7	4717	12.186	5.77	12.24	8.55	18.16
23	0.1186	−147.9	1071.2	4936	12.161	5.53	12.21	8.18	17.32
22	0.1130	−148.4	1070.8	5166	12.136	5.29	12.18	7.81	16.51
21	0.1078	−148.9	1070.3	5408	12.110	5.05	12.15	7.45	15.73
20	0.1027	−149.4	1069.9	5662	12.085	4.81	12.13	7.10	14.99
19	$9.789 (10)^{-2}$	−149.8	1069.5	5929	12.060	4.56	12.10	6.75	14.28
18	$9.326 (10)^{-2}$	−150.3	1069.0	6210	12.035	4.32	12.07	6.40	13.61
17	$8.884 (10)^{-2}$	−150.8	1068.6	6505	12.009	4.08	12.05	6.06	12.96
16	$8.461 (10)^{-2}$	−151.3	1068.1	6817	11.984	3.84	12.02	5.73	12.34

Air and Humidity Calculations

The pressures p and p_s must be expressed in consistent units. In the case of space saturated with steam (water vapor) at 80 F, Table 3–1 shows the specific volume of steam to be 633.0 cu ft per lb; and the density of the steam is, of course, the reciprocal of specific volume, or $d = 1/v = 1/633 = 0.001579$ lb per cu ft. The saturation pressure, from Table 3–1, is 1.0323 in. Hg. If it happened that, with the temperature 80 F at a given time, only $0.001579/2$ lb of steam existed in each cubic foot, then the characteristic gas equation would show that the pressure exerted by the steam would be only half as great as the value given in Table 3–1 for 80 F. Using $pV = mRT$, we may write:

$$(144)(0.491)(p)(1) = \frac{0.001579}{2}(85.6)(460 + 80)$$

$$p_s = 0.5162 \text{ in. Hg}$$

Thus the relative humidity is

$$\phi = \frac{p}{p_s} = \frac{d}{d_s} = \frac{0.5162}{1.0323} = \frac{0.001579/2}{0.001579} = 0.50, \text{ or } 50 \text{ per cent}$$

Example 3–1. The temperature in a certain room is 70 F and the relative humidity is 30 per cent. The barometric pressure p_B is 29.2 in. Hg. Find (a) the partial pressure of the steam in the air; (b) the weight of steam per cubic foot (vapor density); (c) the weight of steam associated with each pound of dry air (humidity ratio or specific humidity).

Solution: (a) From Table 3–1, the pressure of saturated steam at 70 F is 0.7392 in. Hg. At 30 per cent relative humidity,

$$p = \phi p_s = (0.30)(0.7392) = 0.2218 \text{ in. Hg} \qquad Ans.$$

b) The actual partial pressure of the steam in the air being known, d can be found from $pV = mRT$. Thus,

$$(144)(0.491)(0.2218)(1) = (70.7)(0.2218) = d(85.6)(460 + 70)$$

and $\qquad d = 0.000345$ lb per cu ft $\qquad Ans.$

The value of d can also be found from the steam tables for saturated steam at 70 F and a relative humidity $\phi = 0.30$. Thus,

$$d = \phi d_s = (0.30)\left(\frac{1}{868.0}\right) = 0.000345 \text{ lb per cu ft}$$

c) From Dalton's law of partial pressures, expressed in equation 3–1,

$$p_B = p_a + p$$

Therefore

$$p_a = p_B - p = 29.2 - 0.2218 = 28.98 \text{ in. Hg}$$

where 29.2 is the barometric pressure and p_a is the partial pressure of the dry air. Using $pV = mRT$, with R for air $= 53.3$, we find the volume occupied by 1 lb of the dry air ($m = 1$):

$$(144)(0.491)(28.98)(v) = (70.7)(28.98)v = (1)(53.3)(460 + 70)$$

$$v = 13.79 \text{ cu ft}$$

The weight of steam per cubic foot of space = 0.000345 lb (cf. part b). Therefore the weight of steam associated with 1 lb of dry air, that is, with 13.79 cu ft of space, is

$$(13.79)(0.000345) = 0.00476 \text{ lb} \qquad Ans.$$

Humidity ratio (specific humidity) can be calculated in one step from a simple relationship derived from $pV = mRT$. The volume occupied by one pound of air ($m = 1$) at the partial pressure of the air, $p_a = p_B - p$, is, in cubic feet,

$$v = \frac{mRT}{p_a} = \frac{(1)(53.3)(T)}{p_B - p} \qquad \text{(A)}$$

The weight of steam in one pound of dry air (v cu ft) is, from $pv = mRT$,

$$m = W = \frac{pv}{RT} = \frac{pv}{(85.6)T} \qquad \text{(B)}$$

Substitution of equation A in equation B gives

$$W = \frac{p_s(53.3)T}{85.6T(p_B - p_s)} = 0.622 \frac{p}{p_B - p} \qquad \text{(3-5)}$$

where W = humidity ratio (specific humidity), in pounds of steam per pound of dry air. (For W in grains of steam per pound of dry air the constant in equation 3-5 is 0.622×7000, or 4354);
p = partial pressure of the steam in the air;
p_B = barometric pressure.

The pressures p and p_B must be expressed in the same units.

Example 3-2. Solve part c of example 3-1 by using the relationship of equation 3-5.

Solution: The barometric pressure $p_B = 29.2$, and the partial pressure of the steam in the air is

$$p = \phi p_s = (0.3)(0.7392) = 0.2218 \text{ in. Hg}$$

Therefore, from equation 3-5, the weight of steam associated with each pound of dry air, the specific humidity or humidity ratio is

$$W = 0.622 \frac{p}{p_B - p} = 0.622 \frac{0.2218}{29.2 - 0.2218} = 0.00476 \text{ lb}$$

Example 3-3. Outside air at 10 F and 60 per cent relative humidity, after passing through a heater and humidifier, enters an auditorium at 76 F and 50 per cent relative humidity. Find how much water vapor has to be added to each pound of dry outside air to bring it to inside conditions. The barometer reads 29.8 in. Hg.

Solution: Using subscript 1 to denote outside conditions, and subscript 2 to denote inside conditions,

$$p_1 = \phi_1 p_{s1} = (0.6)(0.06286) = 0.03772 \text{ in. Hg}$$

where 0.06286 is taken from Table 3-1 at 10 F.

Air and Humidity Calculations

Since $p_B = 29.8$ in. Hg, the specific humidity of the outside air is, by equation 3-5,

$$W_1 = 0.622 \frac{0.03772}{29.8 - 0.03772} = 0.00079 \text{ lb}$$

Also, $\quad p_2 = \phi_2 p_{s2} = (0.5)(0.9047) = 0.4523$ in. Hg

where 0.9047 is taken from Table 3-1 at 76 F.

The specific humidity of the inside air is

$$W_2 = 0.622 \frac{0.4523}{29.8 - 0.4523} = 0.00958 \text{ lb}$$

The weight of water vapor (steam) added to each pound of air supplied to the auditorium is therefore

$$W_2 - W_1 = 0.00958 - 0.00079 = 0.00879 \text{ lb}$$

Since 1 lb is equivalent to 7000 grains, the number of grains of water vapor added to each pound of dry air supplied to the auditorium is $0.00879 \times 7000 = 61.53$. *Ans.*

Degree of saturation (μ) also known as *saturation ratio* is the humidity ratio, W, of the moist air to the humidity ratio that air could possess at saturation (W_s) at the same temperature and pressure. In equational form

$$\mu = \left.\frac{W}{W_s}\right)_{t,\ p_T} \quad (3\text{-}6)$$

In the technical literature, degree of saturation is called by various other names, such as percentage humidity and per cent saturation.

Relationship between the relative humidity ϕ and the degree of saturation μ can be found. By definition,

$$\phi = \frac{p}{p_s}$$

and $\quad \mu = \dfrac{W}{W_s} = \dfrac{0.622[p/(p_B - p)]}{0.622[p_s/(p_B - p_s)]} \quad (3\text{-}7)$

By transformations and substitutions, we get

$$\mu = \frac{p}{p_s}\frac{p_B - p_s}{p_B - p} = \phi\frac{1 - (p_s/p_B)}{1 - (p/p_B)(p_s/p_s)} = \phi\frac{1 - (p_s/p_B)}{1 - \phi(p_s/p_B)} \quad (3\text{-}8)$$

Similarly,

$$\phi = \mu\frac{p_B - p}{p_B - p_s} = \mu\frac{1 - \phi(p_s/p_B)}{1 - (p_s/p_B)} = \frac{\mu}{1 - (1 - \mu)(p_s/p_B)} \quad (3\text{-}9)$$

It has been mentioned that use of the Gibbs-Dalton law and the perfect-gas equation can give close approximations for determination of values in connection with air-steam mixtures. However, real gaseous molecules do not behave ideally, since there are interactions between various types of molecules, and mutual solubilities between gaseous and liquid phases. More-rigorous equations covering air-steam mixtures can be developed by the methods of statistical mechanics. These are developed in some detail in references 5 and 6, listed at the end of this chapter. In order to give some idea of the magnitude of the deviations which exist, Fig. 3–1 has been prepared. The factor f_s applies for humidity ratio when consideration is given to the more precise methods of analysis. It is employed as follows:

$$W = 0.622 \frac{f_s(p/p_B)}{1 - f_s(p/p_B)} \qquad (3\text{--}10)$$

At 14.7 psia and 212 F, 100 per cent relative humidity has no significance in terms of air, since at this condition no air can be present. However, values less than 100 per cent are significant. Relative humidities of air mixtures are sometimes used at temperatures above 212 F but are low in numerical magnitude because the denominator of the expression for relative humidity is the saturation pressure of steam at the temperature actually existing, whereas the term p_s cannot exceed the atmospheric pres-

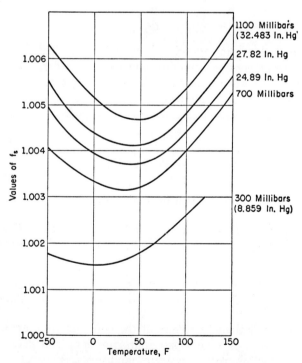

Fig. 3–1. Plot of factor f_s for use in accurate determination of humidity ratio. (From ref. 5.)

Air and Humidity Calculations

sure (or the total pressure for a compressed mixture). The degree of saturation has no significance at temperatures of 212 F and above unless the mixture is compressed.

Figure 3-2 is a graph of equation 3-8 for a barometric pressure of

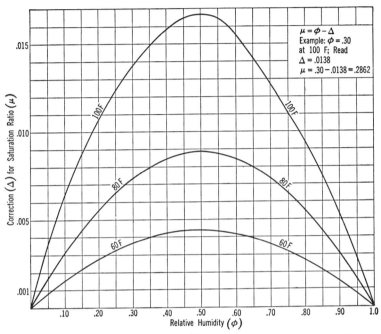

FIG. 3-2. Subtractive corrections for finding values of saturation ratio μ when relative humidity ϕ is given. Drawn for $p_B = 29.92$ in. Hg.

29.92 in. Hg and selected temperatures. At a given value of ϕ (relative humidity) it is possible to read the correction Δ to be subtracted to find the corresponding value of μ (saturation ratio); that is, $\mu = \phi - \Delta$. For example, if ϕ is 0.30 at 100 F, the value of Δ is 0.0138 and $\mu = 0.30 - 0.0138 = 0.2862$. It should be noted that ϕ and μ are numerically close at low and high values but differ appreciably in the middle range, the value of Δ increasing as the temperature becomes higher.

3-3. ENTHALPY OF AIR

The enthalpy of humid air, when the air is considered to act as a perfect gas, is found by adding the enthalpy per pound of dry air and the enthalpy of the water vapor (steam) associated with the pound of dry air.

The enthalpy h of a gas was given in equation 2-50 as $h = C_p(T - T_0)$. In most air-conditioning processes *enthalpy changes* alone are important. Thus the temperature t can be referred to any convenient datum, with 0 F customarily taken as the reference temperature, and

$$\mathbf{h}_a = C_p(t - 0) + W\mathbf{h}_v \tag{3-11}$$

where $\mathbf{h}_a$ = enthalpy of 1 lb of "dry" air, in Btu;
C_p = specific heat of air at constant pressure, usually 0.24 Btu per lb per degree Fahrenheit; 0.2403 in range 0 F to 200 F;
t = dry-bulb temperature, in degrees Fahrenheit;
W = humidity ratio or specific humidity (pounds of steam associated with 1 lb of dry air);
$\mathbf{h}_v$ = enthalpy of steam at the dry-bulb temperature, in Btu per pound (usually referred to a 32 F datum).

The value of $\mathbf{h}_v$ for steam can in some cases be taken directly from the tables compiled for superheated or saturated steam, but for the usual air-conditioning ranges of temperature and pressure the steam-table values are incomplete. It is usually more convenient to use relations derived directly from the steam-table values. For temperatures in the range from 70 F to 150 F,

$$\mathbf{h}_v = 1060.5 + 0.45t \tag{3-12}$$

For temperatures below 70 F, a more precise relation is

$$\mathbf{h}_v = 1061.7 + 0.439t \tag{3-13}$$

In equations 3-12 and 3-13, $\mathbf{h}_v$ and t have the same meanings as in equation 3-11. Where the temperature is above 70 F, equation 3-11 becomes

$$\mathbf{h}_a = C_p(t - 0) + W(1060.5 + 0.45t) \tag{3-14}$$

When greater precision in finding the enthalpy of moist air is required than can be obtained from the gas relationships, use the moist air tables and summate property values at the dry-bulb temperature, t, along with the degree of saturation, μ. In equational form

$$\mathbf{h}_a = \mathbf{h}_{a,t} + \mu \mathbf{h}_{as,t} = \mathbf{h}_{a,t} + \mu(\mathbf{h}_{s,t} - \mathbf{h}_{a,t}) \tag{3-15}$$

This equation is accurate up to 150 F but at higher temperatures shows progressively greater inaccuracy, however, to an extent that in most cases is not sufficiently great to upset engineering calculations.

3-4. AIR-HUMIDITY PROCESS, THERMODYNAMIC WET-BULB TEMPERATURE

The application of the general energy equation 2-9 to any continuous process in which air and water-vapor interactions are taking place is as follows: Consider a device, such as a humidifier, a dehumidifier, or a drier, to which air is supplied and in which water may be added or removed from the air, and in which heat Q also may be added or rejected. Refer to Fig. 3-3 and imagine that 1 lb of dry air at temperature t_1 and carrying W_1 lb of steam is supplied. In the device, steam can be condensed from the air,

Air and Humidity Calculations

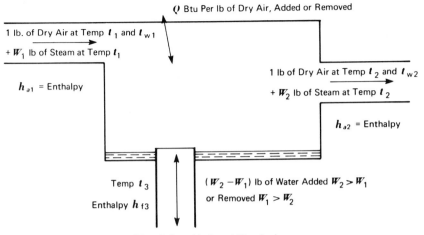

FIG. 3-3. Air-humidity device.

or water can be evaporated into the air, the action depending on the function of the device. Heat (Q Btu, based on each pound of dry air entering) may be added or removed. Finally, 1 lb of dry air, at temperature t_2, with steam content W_2, leaves. Applying the energy equation 2-9 and removing nonessential terms, we obtain:

$$h_{a1} \pm Q_{(1-2)} + (W_2 - W_1)h_{f3} = h_{a2} \qquad (3\text{-}16)$$

If heat is added, a plus sign is used with $Q_{(1-2)}$ in equation 3-16, and the Btu per pound of dry air is

$$Q_{(1-2)} = h_{a2} - h_{a1} - (W_2 - W_1)h_{f3} \qquad (3\text{-}17)$$

Substituting values from equation 3-14 for h_{a2} and h_{a1}, and algebraically rearranging, we may write:

$$Q_{(1-2)} = (0.24)(t_2 - t_1)$$
$$+ (W_2 - W_1)(1060.5 + 0.45t_2 - h_{f3}) + 0.45W_1(t_2 - t_1)$$

Since $h_{f3} = (1)(t_3 - 32)$ approximately,

$$Q_{(1-2)} = (0.24 + 0.45W_1)(t_2 - t_1)$$
$$+ (W_2 - W_1)(1092.5 - t_3 + 0.45t_2) \qquad (3\text{-}18)$$

If heat is removed, a minus sign is used for $Q_{(1-2)}$ in equation 3-16, and the Btu per pound of dry air is

$$Q_{(1-2)} = h_{a1} - h_{a2} + (W_2 - W_1)h_{f3} \qquad (3\text{-}19)$$

From this, there results:

$$Q_{(1-2)} = (0.24 + 0.45W_1)(t_1 - t_2)$$
$$+ (W_1 - W_2)(1092.5 - t_3 + 0.45t_2) \qquad (3\text{-}20)$$

Let us imagine that the device of Fig. 3–3 is provided with an insulating jacket so effective that it eliminates heat flow into or from the system to the point that it is adiabatic. Figure 3–4 shows the modified device provided with a water reservoir of large surface area. Air is shown entering

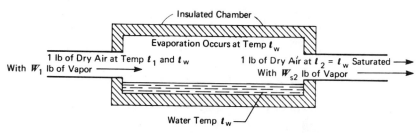

FIG. 3–4. Diagram illustrating adiabatic saturation.

it at a dry-bulb temperature t_1, and leaving in saturated condition at a dry-bulb temperature t_2. Analysis, confirmed by experimentation, will show that when such a device is provided with a continuous stream of air at a constant inlet state, the air will be brought to its "temperature of adiabatic saturation," commonly known as the "thermodynamic wet-bulb temperature." Water in the reservoir, if not at this temperature initially, will also reach the thermodynamic wet-bulb temperature. The process is one in which the air gradually cools down in contact with the water surface as water evaporates into it and the air finally becomes saturated at the thermodynamic wet-bulb temperature. This phenomenon has broad implications, one of which yields a definition namely: *Thermodynamic wet-bulb temperature* is that temperature, for any state of moist air, at which water or ice may be evaporated (sublimed) into the air to bring it to saturation at the same temperature.

Thus "thermodynamic wet-bulb temperature," usually shortened to "wet-bulb temperature" is a characteristic property for air at a given state and air can be thermodynamically and definitively described by giving its dry-bulb and wet-bulb temperatures at a given barometric (total) pressure. Air can of course also be described in several other ways, as for example, by stating dry-bulb temperature and humidity ratio and pressure.

Apply equation 3–16 to the process of adiabatic saturation for water-air. Heat $Q_{(1-2)}$ equals zero by definition; and to make adiabatic saturation a steady-flow process, enough water ($W_{s2} - W_1$) must be added at the wet-bulb temperature t_w to make up for evaporation. The resulting equation is

$$\mathbf{h}_{a1} + (W_{s2} - W_1)\mathbf{h}_{fw} = \mathbf{h}_{a2} \tag{3-21}$$

Following the pattern of equation 3–15, it is possible to rewrite equa-

Air and Humidity Calculations

tion 3-21 as follows using values from Table 3-1

$$\mathbf{h}_{a1} + \mu (\mathbf{h}_{s1} - \mathbf{h}_{a1}) + (W_{sw} - \mu W_{s1})\mathbf{h}_{fw} = \mathbf{h}_{sw} \quad (3\text{-}22)$$

Here the subscript s always means saturated moist air conditions and subscript w implies conditions at the thermodynamic wet-bulb temperature.

Refer again to equation 3-21 and computation will show that at moderate temperatures (below 100 F), the term $(W_{s2} - W_1)\mathbf{h}_{fw}$ is quite small. If this term is neglected, it can be seen that in adiabatic saturation,

$$\mathbf{h}_{a1} = \mathbf{h}_{a2} \quad (3\text{-}23)$$

That is, the enthalpy remains constant. Also, since adiabatic saturation is a process at constant wet-bulb temperature, it immediately appears that the enthalpy of air at the same wet-bulb temperature is practically constant even though the dry-bulb temperature varies. (This last statement, although not precisely true, is accurate enough for many engineering computations, and some psychrometric charts are constructed on this basis.)

Equation 3-21 may be rewritten in the form

$$\mathbf{h}_{a1} - W_1\mathbf{h}_{fw} = \mathbf{h}_{a2} - W_2\mathbf{h}_{fw}$$

Notice that inlet conditions to the process represented by this equation are restricted only in regard to the wet-bulb temperature. Therefore the inlet term can be generalized and the expression can be written as follows:

$$\Sigma = \mathbf{h}_{a2} - W_2\mathbf{h}_{fw} = \mathbf{h}_{a1} - W_1\mathbf{h}_{fw} = \mathbf{h}_{ax} - W_x\mathbf{h}_{fw} \quad (3\text{-}24)$$

This means that the *sigma function* (Σ), defined by equation 3-24, is constant for any given wet-bulb temperature.

The significance of equation 3-21 is further exemplified when it is written as follows:

$$\mathbf{h}_{a1} = \mathbf{h}_{a2} - (W_{s2} - W_1)\mathbf{h}_{fw}$$

or

$$\mathbf{h}_{a1} = \mathbf{h}_{aw} - (W_{sw} - W_1)\mathbf{h}_{fw} = \mathbf{h}_{aw} \pm D \quad (3\text{-}25)$$

This means that the enthalpy of an air-steam mixture is equal to the enthalpy of saturated air at the same wet-bulb temperature, less the small corrective term $(W_{s2} - W_1)\mathbf{h}_{fw}$. This corrective term is called the deviation and carries the symbol D.

From the adiabatic-saturation equation 3-21, it is possible to derive a valuable formula for finding the weight of water vapor in the atmosphere in terms of the wet-bulb and dry-bulb temperatures. Starting with equation 3-21, we use values for the enthalpy $\mathbf{h}_{a1}$ and $\mathbf{h}_{a2}$ from equations 3-11 and 3-14. The result is

$$C_p t_1 + W_1 \mathbf{h}_{v1} + (W_{s2} - W_1)\mathbf{h}_{fw} = C_p t_w + W_{s2}\mathbf{h}_{v2}$$

From this,

$$W_1 = \frac{C_p(t_w - t_1) + W_{s2}(\mathbf{h}_{v2} - \mathbf{h}_{fw})}{\mathbf{h}_{v1} - \mathbf{h}_{fw}} \qquad (3\text{-}26)$$

At exit condition 2, the dry-bulb temperature equals the wet-bulb temperature. Therefore $\mathbf{h}_{v2} - \mathbf{h}_{fw} = \mathbf{h}_{vw} - \mathbf{h}_{fw}$. But this latter expression equals $\mathbf{h}_{fgw}$, since $\mathbf{h}_{vw} - \mathbf{h}_{fw}$ is the change in enthalpy when liquid changes to vapor at the wet-bulb temperature, or it is the latent heat of vaporization ($\mathbf{h}_{fgw}$) at the wet-bulb temperature. By equation 3-12, the term ($\mathbf{h}_{v1} - \mathbf{h}_{fw}$) is equal to $(1060.5 + 0.45 t_1 - \mathbf{h}_{fw})$. Let us now algebraically add the identity $(0.45 t_w - 0.45 t_w)$ to the latter expression, giving

$$(1060.5 + 0.45 t_w) - \mathbf{h}_{fw} + 0.45 t_1 - 0.45 t_w$$
$$= \mathbf{h}_{vw} - \mathbf{h}_{fw} + 0.45(t_{d1} - t_w) = \mathbf{h}_{fgw} + 0.45(t_1 - t_w)$$

If these identities are applied to equation 3-24, there results

$$W_{s1} = \frac{C_p(t_w - t_1) + W_{sw}\mathbf{h}_{fgw}}{\mathbf{h}_{fgw} + 0.45(t_1 - t_w)}$$

or

$$W_s = \frac{W_{sw}\mathbf{h}_{fgw} - 0.24(t - t_w)}{\mathbf{h}_{fgw} + 0.45(t - t_w)} \qquad (3\text{-}27)$$

where W = weight of water vapor carried with each pound of dry air at dry-bulb temperature t and wet-bulb temperature t_w, in pounds, humidity ratio;

$\mathbf{h}_{fgw}$ = heat of vaporization of 1 lb of steam at the wet-bulb temperature;

W_{sw} = weight of water vapor associated with each pound of dry air when saturated at the wet-bulb temperature t_w, in pounds. The value of W_{sw} can be found by equation 3-5, or from Table 3-1, for standard barometric pressure.

The concept of wet-bulb temperature as a characteristic property of moist air has lead to the reason for calling conventional air temperature by the name *dry-bulb temperature*. In this text when unspecified air temperature is mentioned, dry-bulb temperature is meant. Its symbol is the letter t. Specifically, air temperature or dry-bulb air temperature represents air temperature in a closed or open space when the temperature is independent of radiation effects from surroundings and when air motion relative to the measuring device is not significant.

When moist air is cooled at constant pressure with its humidity ratio remaining fixed, a temperature is reached at which the air becomes saturated and further cooling results in the appearance of fog or deposition of moisture on adjacent surfaces. This saturation or condensation temperature for a given humidity ratio of air is called the *dew-point temperature*.

Air and Humidity Calculations

Instrumentation has been developed using the temperature of the dewpoint as a means of determining the exact moisture condition of humid air. The wet-bulb temperature can serve a similar purpose. Its instrument, the wet-bulb psychrometer which serves as a supplement to the saturator, is described in a subsequent section.

Example 3-3. Air leaves a well-insulated (adiabatic) saturator at 66 F, at a pressure of 29.92 in. Hg. It enters the saturator at 80 F and water is supplied as needed at 66 F. Find the humidity ratio, degree of saturation, enthalpy, and specific volume of the entering air.

Solution: Note that the saturated air at exit is at the thermodynamic wet-bulb temperature, t_w, and this is of course also the dry-bulb temperature, t_2, at that point. Read appropriate values from Table 3-1 and use with equation 3-27 or substitute values directly in equation 3-21.

At 66 F on exit $h_{sw} = h_{a2} = 30.79$ Btu per lb of air

$$W_{sw} = \frac{95.86}{7000} = 0.01369 \text{ lb per lb air}$$

By equations 3-11 and 3-12

$$h_{a1} = (0.2403)(80 - 0) + W_1[1060.5 + (0.45)(80)] = 19.23 + W_1(1096.5)$$

$$h_{fw} = 34.1 \text{ Btu per lb (Table 3-1 at 66 F)}$$

Substitute values in equation 3-21 and solve for W_1

$$19.23 + W_1(1096.5) + (0.01369 - W_1)34.1 = 30.79$$

$$W_1 = 0.01043 \text{ lb per lb air} \qquad Ans.$$

at 80 F, $W_s = 155.8$ grains $= 0.02226$ lb

$$\mu = \frac{0.01043}{0.02226} = 0.47 \qquad Ans.$$

$$h_{a1} = 19.23 + 0.01043(1096.5) = 30.68 \text{ Btu per lb at inlet} \qquad Ans.$$

Specific volume can be found by use of Table 3-1 and use of degree of saturation. At inlet temperature, 80 F,

$$v = v_a + \mu v_{as} = v_a + \mu(v_s - v_a)$$
$$= 13.60 + 0.47(14.09 - 13.60) = 13.83 \text{ cu ft per lb} \qquad Ans.$$

An alternate solution would start with equation 3-22, make use of enthalpy values as found by equation 3-15, and solve for μ. Using values from Table 3-1

$$19.23 + \mu(43.63 - 19.23) + (0.01369 - \mu 0.02226)34.1 = 30.79$$

$$\mu = 0.47 \qquad Ans.$$

$$W_1 = \mu W_{s1} = (0.47)(0.02226) = 0.1046 \text{ lb per lb air}$$

By equation 3-15

$$h_{a1} = 19.23 + 0.47(43.63 - 19.23) = 30.69 \text{ Btu per lb of dry air}$$

Specific volume is the same by either method and close agreement is observed for the other determinations.

Let us compute the middle term of equation 3-21, and compare with the values found: $(W_{sw} - W_1)\mathbf{h}_{fw}$ and we find,

$$(0.01369 - 0.01043)(34.1) = 0.11 \text{ Btu per lb air}$$

Notice that this is the amount by which the inlet enthalpy of the air $\mathbf{h}_{a1}$ is less than the enthalpy of saturated air at the thermodynamic wet-bulb temperature of the entering air.

$$\mathbf{h}_{aw} - \mathbf{h}_{a1} = 30.79 - 30.68 = 0.11 \text{ Btu per lb air}$$

Observe how small this term is in relative magnitude.

The derivations and developments in this chapter have largely been based on the perfect-gas relationships although attention has been called to precise tabular approaches to the properties of moist air. However, it is unfortunately true that each table is constructed for a single barometric pressure and the difficulty of adjusting to other barometric pressures can produce greater inaccuracy than arises from use of perfect-gas relationships. Nevertheless, mention should be made that the correction of Fig. 3-1, as expressed in equation 3-10 for humidity ratio, and the computation of enthalpy and specific volume by equations of the general form of equation 3-15 using degree of saturation, can be employed for appropriate barometric pressures. Equation 3-28 which follows, shows the interrelationships between relative humidity and degree of saturation with the corrective factor from Fig. 3-1 included.

$$\varphi = \frac{\mu}{1 - (1 - \mu)\dfrac{f_s p}{p_B}} \tag{3-28}$$

The Goff-Gratch tables noted in references 5 and 6 represent the most extensive tabular information developed on moist-air properties. The values in them differ to a slight degree from the values of Table 3-1 in this text. Although the differences are minor, slight inconsistencies arise if the different tables are used together. Internal consistency of each table leads essentially to the same final result when either is used.

Because of the problems which arise from variations in barometric pressure Table 3-2 has been included. This gives corrections to enthalpy at standard barometric pressure for a number of other pressures. To generalize for changes in barometric pressure, it should be mentioned that for a given-dry-bulb and wet-bulb temperature, the values of humidity ratio and enthalpy increase with altitude, that is for lower barometric pressures. The greatest change occurs with specific volume which for a given dry-bulb temperature and humidity ratio varies in an almost inversely proportional ratio to barometric pressure. Relative humidity changes little with changes in barometric pressure.

Air and Humidity Calculations

TABLE 3-2
Corrections to enthalpy of moist air in Btu per lb of dry air for atmospheric pressures other than 29.92 in. Hg*

Thermodynamic Wet-Bulb Temp, t_w(F)	Atmospheric Pressure, in. Hg							Thermodynamic Wet-Bulb Temp, t_w(F)	Atmospheric Pressure, in. Hg						
	24.92	25.92	26.92	27.92	28.92	30.92			24.92	25.92	26.92	27.92	28.92	30.92	
33	0.87	0.67	0.49	0.31	0.15	−0.14		62	2.67	2.05	1.49	0.95	0.46	−0.43	
34	0.91	0.70	0.51	0.32	0.16	−0.15		63	2.77	2.13	1.54	0.99	0.48	−0.45	
35	0.95	0.73	0.53	0.34	0.16	−0.15		64	2.88	2.21	1.60	1.03	0.50	−0.46	
36	0.99	0.76	0.55	0.35	0.17	−0.16		65	2.98	2.29	1.66	1.06	0.51	−0.48	
37	1.02	0.79	0.57	0.37	0.18	−0.17		66	3.09	2.38	1.71	1.10	0.53	−0.50	
38	1.06	0.82	0.59	0.38	0.18	−0.17		67	3.21	2.47	1.78	1.14	0.55	−0.51	
39	1.11	0.85	0.61	0.39	0.19	−0.18		68	3.33	2.56	1.85	1.19	0.57	−0.53	
40	1.15	0.89	0.64	0.41	0.20	−0.18		69	3.45	2.65	1.91	1.23	0.59	−0.55	
41	1.20	0.92	0.67	0.43	0.21	−0.19		70	3.58	2.75	1.98	1.27	0.62	−0.57	
42	1.25	0.97	0.70	0.45	0.22	−0.20		71	3.70	2.85	2.06	1.32	0.64	−0.59	
43	1.30	1.00	0.72	0.46	0.22	−0.21		72	3.85	2.96	2.13	1.37	0.66	−0.62	
44	1.35	1.04	0.75	0.48	0.23	−0.22		73	3.99	3.06	2.21	1.42	0.68	−0.64	
45	1.41	1.08	0.78	0.50	0.24	−0.23		74	4.14	3.18	2.29	1.47	0.71	−0.66	
46	1.46	1.12	0.81	0.52	0.24	−0.24		75	4.28	3.29	2.38	1.53	0.74	−0.68	
47	1.52	1.17	0.84	0.54	0.26	−0.24		76	4.44	3.41	2.46	1.58	0.76	−0.71	
48	1.58	1.21	0.88	0.56	0.27	−0.26		77	4.60	3.54	2.55	1.64	0.79	−0.73	
49	1.64	1.26	0.91	0.58	0.28	−0.27		78	4.77	3.66	2.64	1.70	0.82	−0.77	
50	1.68	1.31	0.95	0.61	0.29	−0.27		79	4.94	3.80	2.74	1.76	0.85	−0.79	
51	1.77	1.36	0.99	0.63	0.31	−0.29		80	5.12	3.93	2.84	1.82	0.88	−0.82	
52	1.84	1.41	1.02	0.66	0.32	−0.30		81	5.31	4.08	2.94	1.89	0.91	−0.85	
53	1.91	1.47	1.06	0.68	0.33	−0.31		82	5.50	4.22	3.04	1.95	0.94	−0.88	
54	1.98	1.52	1.10	0.71	0.34	−0.32		83	5.70	4.38	3.16	2.02	0.97	−0.91	
55	2.06	1.58	1.14	0.74	0.36	−0.33		84	5.90	4.53	3.27	2.09	1.01	−0.95	
56	2.13	1.64	1.19	0.76	0.37	−0.34		85	6.12	4.70	3.39	2.17	1.05	−0.98	
57	2.22	1.71	1.23	0.79	0.38	−0.35		86	6.35	4.86	3.50	2.25	1.08	−1.01	
58	2.30	1.77	1.28	0.82	0.40	−0.37		87	6.56	5.04	3.63	2.33	1.12	−1.05	
59	2.39	1.84	1.33	0.85	0.41	−0.38		88	6.81	5.23	3.77	2.42	1.16	−1.08	
60	2.48	1.90	1.38	0.88	0.43	−0.40		89	7.04	5.41	3.89	2.50	1.20	−1.12	
61	2.58	1.98	1.43	0.92	0.44	−0.41		90	7.30	5.60	4.04	2.59	1.25	−1.16	

*Reproduced by permission from *ASHRAE—Handbook of Fundamentals—1967*. Compilation prepared by J. C. Davis, National Bureau of Standards.

3-5. THE PSYCHROMETER

The phenomena of adiabatic saturation and thermodynamic wet-bulb temperature have been explained. However, it is obvious that an adiabatic saturator is not a workable instrument for measuring moist-air conditions and for many years an instrument known as a psychrometer has been used in its stead. The psychrometer measures the dry-bulb temperature and also a wet-bulb temperature. The wet-bulb temperature measured by this instrument is not precisely the thermodynamic wet-bulb temperature. Fortuitously however, it gives a reading which is very close to the thermodynamic wet-bulb temperature.

The *sling psychrometer* (Fig. 3-5) consists of two thermometers mounted side by side on a holder, with provision for whirling the whole device through the air. The dry-bulb thermometer is bare, and the wet bulb is covered by a wick which is kept wetted with clean water. After being whirled for a sufficient time the wet-bulb thermometer reaches its equilibrium point, and both the wet-bulb and dry-bulb thermometers are then quickly read. Rapid relative movement of the air past the wet-bulb thermometer is necessary to get dependable readings.

In the *aspiration psychrometer* (Fig. 3-6) a small fan is used to pull the air past the dry-bulb and wet-bulb thermometers to bring about wet-bulb equilibrium. If the water-supply temperature for the wick is much higher or lower than the wet-bulb temperature, readings should not be taken until it is certain that equilibrium has been reached.

To explain the phenomenon of wet-bulb temperature-depression, imagine a free surface of water existing in unsat-

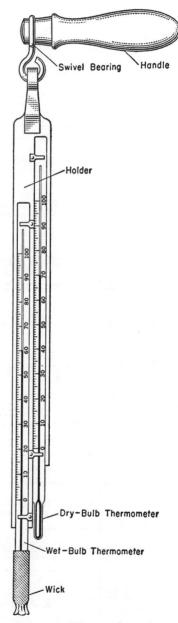

Fig. 3-5. Sling psychrometer.

Air and Humidity Calculations

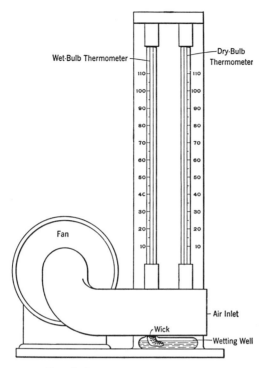

FIG. 3-6. Aspiration psychrometer.

urated air. If the temperature of the water is above the dew-point temperature, evaporation will take place from this surface into the air. The vapor leaves this surface of area, S, by diffusion through the contiguous air-vapor film. The rate of evaporation is proportional to the difference between the pressure p_w of the vapor at the liquid surface and the partial pressure p of the vapor in the air. Calling dW the weight of water evaporated in a differential of time, dT, we may write an equation as follows:

$$\frac{dW}{dT} = kS(p_w - p) \tag{A}$$

where k represents the coefficient of diffusion through a unit surface of film contiguous to the liquid.

Since the temperature t of the air is higher than the temperature t_w of the water, heat will flow from the air to the water. Calling dQ the amount of heat flowing in a differential of time dT, and calling f the coefficient of heat transfer through the film, we obtain the equation

$$\frac{dQ}{dT} = fS(t - t_w) \tag{B}$$

As heat flows from the air into the water, the temperature of the water will tend to rise and to reduce the heat inflow, but evaporation tends to increase as the temperature rises. Eventually, a condition of equilibrium will be reached when the heat dQ flowing into the water in a differential of time is exactly balanced by the heat used in evaporating the water, or $(dW/dT)(\mathbf{h}_{fg})$. Thus,

$$\frac{dQ}{dT} = \frac{dW}{dT}(\mathbf{h}_{fg}) \tag{C}$$

where $\mathbf{h}_{fg}$ = the heat of vaporization of the water;
dW/dT = the weight of water evaporated in time dT from the surface of area S.

Eliminating dW/dT from equations A and C, and using the resulting equation with equation B to eliminate dQ/dT, we get

$$p_w - p = \frac{f}{\mathbf{h}_{fg}k}(t - t_w)$$

or

$$p = p_w - \frac{f}{\mathbf{h}_{fg}k}(t - t_w) \tag{3-29}$$

Equation 3-29 gives a means of finding the water-vapor pressure in the air if the term $f/\mathbf{h}_{fg}k$ can be evaluated either experimentally or otherwise. This term can almost be considered a constant, since the value of $\mathbf{h}_{fg}$ for steam does not greatly vary over a 20- to 30-deg temperature range, and f and k depend on the gas-film thickness contiguous to the water surface. If f and k are determined for a normal set of operating conditions, as with a sling psychrometer, they, also, may be considered practically constant. Heat radiating to the wetted bulb from surrounding warmer air and surfaces tends to raise the temperature t_w slightly, and so a psychrometer should be operated with such velocity as to reduce the ratio f/k to a minimum and essentially-constant value, in order to make such radiation effects of small importance.

The value t_w in the equilibrium condition is the so-called wet-bulb temperature. To make equation 3-29 usable, the essentially constant term $f/\mathbf{h}_{fg}k$ must be evaluated, and it will be realized that the barometric pressure p_B will also affect the equation. Various forms of this equation, with the constants evaluated, are now in use. The most common of these equations are as follows:

1. The modified Apjohn equation, proposed in 1837:

$$p = p_w - \frac{p_B}{30}\frac{t - t_w}{90} \tag{3-30}$$

2. The modified Ferrel equation, proposed in 1886:

$$p = p_w - 0.000367 p_B (t - t_w)\left(1 + \frac{t_w - 32}{1571}\right) \tag{3-31}$$

Air and Humidity Calculations

3. The Carrier equation, proposed in 1911:

$$p = p_w - \frac{(p_B - p_w)(t - t_w)}{2800 - 1.3 t_w} \qquad (3\text{-}32)$$

where p = pressure of the water vapor in the atmosphere;
p_w = pressure of saturated water vapor at the wet-bulb temperature;
p_B = barometric pressure;
t = dry-bulb temperature, degrees Fahrenheit;
t_w = wet-bulb temperature, degrees Fahrenheit.

The pressures p, p_w, and p_B must be expressed in the same units, as inches of mercury or pounds per square inch.

Equations 3-31 and 3-32 have more correction terms and give results more accurate than the Apjohn equation (eq 3-30). However, in many cases equation 3-30 is sufficiently accurate, particularly when the experimental wet-bulb and dry-bulb readings have not been found to a precision closer than ±0.3 deg. The Carrier equation has been used extensively in air-conditioning work.

In actual practice, however, these equations are not used to any great extent; graphical charts made from plots of them are employed instead. Charts of this type are known as psychrometric charts, and one of them is included in this book, in three sections designated as plates I, II, and III, in a pocket attached to the inside back cover.

It must be realized that the wet-bulb temperature and the dew-point temperature are different entities. However, in the one case of saturated air, the dry-bulb, wet-bulb, and dew-point temperatures must of course all be the same. In the case of nonsaturated air, the dry-bulb temperature represents the actual temperature of the air as measured by an ordinary thermometer; the dew-point temperature is that temperature to which the air, with its existing moisture content, would have to be cooled before saturation and any condensation could occur; and the wet-bulb temperature represents that temperature which a thermometer having a bulb covered with a wetted wick would reach if whirled through the air.

Example 3-4. Air has a dry-bulb temperature of 70 F and a wet-bulb temperature of 64 F, and the barometer indicates 29.9 in. Hg. Without the use of a psychrometric chart, find (a) the relative humidity of the air; (b) the (water) vapor density in the air; (c) the humidity ratio; and (d) the dew point.

Solution: (a) Equation 3-30, 3-31, or 3-32 may be applied to find p, but equation 3-32 will be used here. Thus,

$$p = p_w - \frac{(p_B - p_w)(t - t_w)}{2800 - 1.3 t_w}$$

From Table 3-1, $p_w = 0.6007$ in. Hg for the wet-bulb temperature of 64 F. Also, $p_B = 29.9$ in. Hg; $t = 70$ F; $t_w = 64$ F. Then, the partial pressure of the vapor in the air is

$$p = 0.6007 - \frac{(29.9 - 0.6007)(70 - 64)}{2800 - (1.3)(64)} = 0.5360 \text{ in. Hg}$$

The relative humidity may now be found by equation 3-3, or $\phi = p/p_s$. From Table 3-1, $p_s = 0.7392$ in. Hg for the dry-bulb temperature of 70 F. Hence

$$\phi = \frac{0.5360}{0.7392} = 0.725, \text{ or } 72.5 \text{ per cent} \qquad Ans.$$

b) From equation 3-4,

$$d = \phi d_s$$

and from Table 3-1, the specific volume of saturated steam at $t = 70$ F is $v_g = 868.0$ cu ft per lb; and $d_s = 1/868 = 0.001152$. Therefore the weight of vapor in the air is

$$d = (0.725)(0.001152) = 0.000835 \text{ lb per cu ft} \qquad Ans.$$

c) By equation 3-5

$$W = 0.622 \frac{0.536}{29.9 - 0.536} = 0.01135 \text{ lb per lb of dry air}$$

Expressed in grains the humidity ratio is

$$(0.01135)(7000) = 79.45 \text{ grains per lb of dry air}$$

d) The dew-point temperature exists for saturated air having the above humidity ratio (specific humidity). Refer to Table 3-1 and for $W = 79.45$ grains read by interpolation the dew-point temperature as 60.8 F.

It is of interest to generalize equation 3-27 and consider the possibilities of the equation for any vapor with air. In equation 3-27, we shall replace 0.24 by the specific heat of the dry air (c_{pa}) and replace 0.45 by a term for the specific heat of the vapor (c_{pv}). Using these terms and rearranging equation 3-27, we thus obtain

$$(W_{sw} - W)\mathbf{h}_{fgw} = (c_{pa} + c_{pv}W)(t - t_w)$$

The quantities W_{sw} and W_s are the weights of vapor associated with a pound of air and, by an equation of the form of equation 3-5, they bear a proportionality to the partial pressures of the vapor in the air. By utilizing this fact, the foregoing equation can be written as

$$K(p_w - p)\mathbf{h}_{fgw} = (c_{pa} + c_{pv}W)(t - t_w) \qquad (3\text{-}33)$$

where K is a proportionality factor relating pressures and specific humidities.

Compare this form of the adiabatic-saturation equation with the dynamic type of wet-bulb equation—namely equation 3-29, which is

$$p_w - p = \frac{f}{\mathbf{h}_{fg}k}(t - t_w)$$

Notice that, for the adiabatic-saturation type of equation to agree with

Air and Humidity Calculations 113

the wet-bulb temperature for any particular vapor, the following relation must be true:

$$\frac{f}{k} = \frac{c_{pa} + c_{pv}W}{K} \qquad (3\text{-}34)$$

Fortunately, this relationship is closely true for air-steam mixtures. It is not, however, true for other mixtures, such as air-alcohol and air-gasoline.

3–6. THE PSYCHROMETRIC CHART

It has now been shown how all the important properties of air-steam mixtures are interrelated and how each of them can be calculated. To simplify the labor of making calculations and to illustrate processes, charts representing air-steam properties, drawn for a given barometric (or total) pressure, are of inestimable value. Such charts have taken many forms, and each form may have special advantages.

In this text the psychrometric chart which was developed by the American Society of Heating, Refrigerating, and Air-Conditioning Engineers was selected for inclusion. This chart uses as basic coordinates dry-bulb temperature as abscissa and humidity ratio as ordinate. However, a diagonal axis presents enthalpy as a coordinate so the chart has at least one characteristic of a Mollier chart, so-called because Richard Mollier was the first to use a psychrometric chart with enthalpy as a coordinate. An abridgement of the ASHRAE chart for normal temperature is shown in Fig. 3-7. The horizontal axis represents dry-bulb temperature by degree intervals. The dry-bulb temperature lines are straight but are not precisely parallel to each other and incline slightly from the vertical usually to the left. Humidity ratio in pounds of water per pound of dry air appears at the right side of the chart using a uniform scale with lines horizontal. The saturation curve, which includes the wet-bulb and dew-point temperature, swings upward to the right. On this curve it follows that at saturation, wet-bulb, dry-bulb and dew-point temperatures are equivalent and equal. Relative-humidity lines of similar shape are shown on 20 per cent intervals. The enthalpy lines appear drawn obliquely down the chart expressed in units of Btu per pound of dry air. The enthalpy lines are parallel to each other and close readings can be made on the scale to which the enthalpy lines are extended. The wet-bulb temperature lines are straight but are not parallel to each other and since enthalpy and wet-bulb temperature are not in a fixed equivalent ratio to each other, the enthalpy and wet-bulb temperature lines, although coincident at the saturation curve, diverge from each other in the body of the chart.

A portion of the fog region has been drawn with enthalpy and wet-bulb temperature lines extended into it. This two-phase region is a mechanical

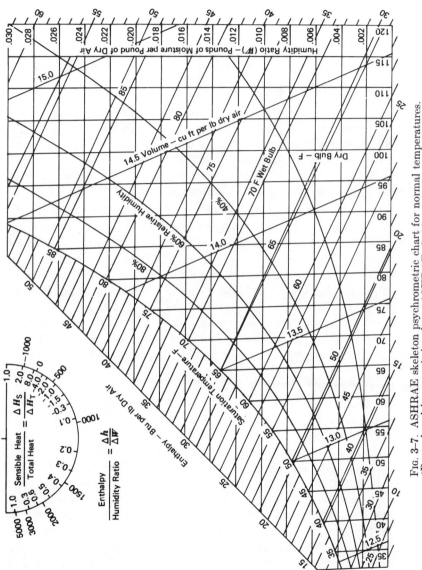

FIG. 3-7. ASHRAE skeleton psychrometric chart for normal temperatures. (Reproduced by permission from ASHRAE *Guide and Data Book 1966*.)

Air and Humidity Calculations 115

mixture of saturated moist air and liquid water in suspension. Specific volume lines are obliquely drawn at intervals of 0.5 cubic feet per pound of dry air. These lines are straight but adjacent lines are not precisely parallel to each other.

The chart has been constructed, using the thermodynamic data from the tables of Goff and Gratch (refs. 5 and 6). The basic construction of a psychrometric chart of this type is not difficult since the dry-bulb scale and the humidity-ratio scale are arbitrarily chosen, and it requires merely plotting values from the moist-air tables to locate wet-bulb and dew-point temperatures on the saturation line of the chart. Enthalpy values can be added for saturation conditions. A more difficult problem is to locate enthalpy, and wet-bulb temperatures in the body of the chart but since both enthalpy and wet-bulb temperatures are defined in terms of dry-bulb and humidity ratio each point can be uniquely located and it is possible to find the whole body of the chart. Better accuracy can be obtained if the dry-bulb lines are shifted from their orthogonal position relative to humidity ratio.

Since it is awkward to cover a large range of dry-bulb temperatures on one chart, it is customary to provide three charts for the working temperature range, Plate I for normal (medium) temperatures 32 F to 120 F, Plate II for low temperatures -40 F to 50 F and Plate III for high temperatures 60 F to 250 F. Mention has already been made that a basic psychrometric chart is applicable for one specific pressure and in this case, the three charts are for standard barometric pressure 29.92 in. Hg. If psychrometric information is needed at another pressure, either a separate chart has to be constructed for that pressure or corrections to a basic chart are required. In this text an additional chart has been provided for 5,000 ft elevation representing a standard atmospheric pressure of 24.89 in. Hg.

A protractor and nomograph appear at the left of the chart. The protractor shows two scales; one involves the ratio, enthalpy difference to humidity-ratio difference, the other, the sensible to total heat ratio. The use of the protractor nomograph will be illustrated in subsequent problems.

Example 3-5. Read from the psychrometric chart the properties of moist air at 80 F dry-bulb, 60 F wet-bulb at 29.92 in Hg barometric pressure.

Solution: Refer to the skeleton chart of Fig. 3-7 for guidance and then read from Plate I the desired results with precision.

Humidity ratio (specific humidity), W,. Move up the 80 F dry-bulb line to its intersection with the 60 F wet-bulb line. Follow the W line to the right and read $W = 0.0066$ lb of steam per lb of dry air.

Dew-point temperature. From the same intersection follow the W line to the saturation temperature curve and read $t_d = 45.9$ F.

Enthalpy. Method 1. Make use of 2 triangles and from the $t = 80$ F, $t_w = 60$ F intersection locate a line parallel to an adjacent enthalpy line and extend this to the edge scale to read $h = 26.35$ Btu per lb of dry air.

Enthalpy. Method 2. The magnitude of D in equation 3–25 shows the amount by which the enthalpy of unsaturated moist air differs from the enthalpy of saturated air at the same wet-bulb temperature. This difference is called the deviation. At $t_w = 60$ F on the saturation line, read $h_{aw} = 26.46$. To find D, read the nomograph at the upper left of Plate 1 for $W = 0.0066$ and $t_w = 60$ F as -0.11. Thus the enthalpy is $h = h_{aw} - D = 26.46 - 0.11 = 26.35$ Btu per lb dry air.

Relative humidity. Read by linear interpolation at the $t = 80$ F, $t_w = 60$ F intersection point, $\varphi = 30$ per cent.

Specific volume. At the $t = 80$ F, $t_w = 60$ F intersection point read by linear interpolation between adjacent volume lines $v = 13.72$ cu ft per lb of dry air.

3–7. AIR CONDITIONING PROCESSES USING THE PSYCHROMETRIC CHART AND MOIST-AIR TABLES

Heating of Air. If air is heated or cooled without the addition of moisture the humidity ratio (specific humidity) remains constant and this process appears as a straight horizontal line on the psychrometric chart. On the skeleton chart of Fig. 3–8 heating is shown taking place from 1 to

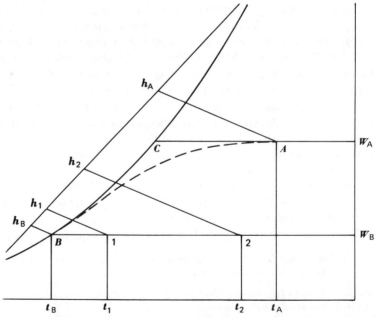

Fig. 3–8. Processes on psychrometric chart. Heating with humidity ratio constant, path 1 to 2; cooling with humidity ratio constant, path 2 to 1; cooling and dehumidification to a saturated condition, path A to B.

2 or cooling taking place from 2 to 1. In equational form, we would read

$$q_{1-2} = m_a (h_2 - h_1) \tag{3-35}$$

Air and Humidity Calculations 117

The enthalpy values can be read from the chart as shown in the diagram.

The heat added or removed during such a process which takes place at constant humidity can also be computed by use of the following equation since the process is a sensible-heat change:

$$q_{1-2} = q_s = m_a \left[C_{pa} (t_2 - t_1) + C_{ps} W (t_2 - t_1) \right]$$
$$= m_a (C_{pa} + C_{ps} W)(t_2 - t_1)$$
$$= m_a C_p (t_2 - t_1) \tag{3-36}$$

or

where q_s = heat added or removed with no moisture change, in Btu per m_a pounds of dry air;
C_{pa} = specific heat of dry air = 0.24 (approximately);
C_{ps} = specific heat of steam = 0.45 (approximately);
W = humidity ratio, in pounds per pound of dry air;
$t_2 - t_1$ = dry-bulb temperature change, in degrees Fahrenheit.

In the technical literature, the composite term $C_{pa} + C_{ps} W = C_p$ has been called *humid heat*. At the low temperatures of the air-conditioning range, W is so small that humid heat differs very little from the value of C_p for air. At higher temperatures, in particular those above 100 F, the second term of this expression for humid heat assumes much-greater significance. For general values of humid heat to use in the air-conditioning range, 0.242 to 0.245 are representative.

Example 3–6. Find the heat required to warm 5000 cfm of outside air at 39 F and 80 per cent relative humidity to 90 F without addition of moisture. Barometric pressure = 29.92 in. Hg.

Solution: Use the psychrometric chart Plate I and read that at $t_1 = 39$ F, $\varphi = 80\%$, that $h_1 = 13.68$ Btu per lb air, $W_1 = 0.0040$ lb per lb of dry air. At the same value of W_1, read at $t_2 = 90$ F that $h_2 = 26.02$ Btu per lb, $\varphi = 13.93\%$. At $t_1 = 39$ F, $\varphi = 80$ per cent, read on the chart that $v = 12.64$ cu ft per lb of dry air, then

$$m_a = \frac{5000 \text{ cfm} \times 60}{12.64} = 23{,}700 \text{ lb per hr}$$

Substituting in equation 3–35

$$q_{1-2} = 23{,}700 \, (26.02 - 13.68) = 292{,}500 \text{ Btu per hr}$$

or by use of equation 3–36

$$q_{1-2} = (23{,}700)(0.242)(90 - 39) = 292{,}500 \text{ Btu per hr}$$

Example 3–6 shows how the relative humidity decreases when air is heated with no addition of moisture. The relative-humidity decrease represents a usual winter condition. When air is cooled with no moisture change, the relative humidity, of course, increases. This condition is common in summer.

Cooling and Dehumidifying Moist Air. Whenever air is cooled to a temperature lower than its original dew point, some of the water vapor or steam in the original air necessarily condenses out. This condensation does not take place at a fixed temperature but continues over a variable path related to the surface temperature and placement of the cooling coils or the temperature and manner by which chilled water is sprayed into the air in a direct-contact cooler. The path of cooling and dehumidification to a saturated state is indicated in Fig. 3-8 by the dotted line A to B. However an actual path might be quite different perhaps moving close to C without any condensation and then dropping to A on a path much closer to the saturation curve. For computing the energy interchange the actual path is immaterial since the total energy and mass transfer from initial to final states alone are of significance, a fact which follows from a steady flow energy analysis (equation 2-9). Thus we can write an equation for the heat removal $-q_{A-B}$ as follows

$$m_a \mathbf{h}_A - q_{A-B} = m_a \mathbf{h}_B + m_a (W_A - W_B) \mathbf{h}_{fB} \qquad (3\text{-}37)$$

m_w, the water condensed from m_a lb of dry air is

$$m_a (W_A - W_B) = m_w \qquad (3\text{-}38)$$

$$q_{A-B} = m_a [(\mathbf{h}_A - \mathbf{h}_B) - (W_A - W_B)\mathbf{h}_{fB}] \qquad (3\text{-}39)$$

For m_a lb of dry air flowing per hour, q_{A-B} represents the heat removal required for the process in Btu per hr. In the equation above it is assumed that all the condensate is removed at the saturation temperature, t_B. If some is removed, at any other temperature, the enthalpy of the liquid, $\mathbf{h}_f$, must also be selected for the other temperature and used with the appropriate mass involved, to show its contribution to the total.

Example 3-7. Find the refrigeration required to cool 10,000 cfm of outside air at 90 F dry bulb, 80 F wet bulb to a condition of saturation at 56 F. Condensate is removed at 56 F and barometer is at 29.92 in. Hg.

Solution: Using Plate I, at $t_A = 90$ F, $t_{wa} = 80$ F read $\mathbf{h}_A = 43.52$ Btu per lb dry air, $\varphi = 65\%$. $W_A = 0.0195$ lb per lb dry air, $v = 14.3$ cu ft per lb dry air. Also at $t_B = 56$ F $= t_{wB} = t_{dB}$ read $\mathbf{h}_B = 23.84$ Btu per lb dry air, $W_B = 0.0096$ lb per lb dry air, $\varphi = 100\%$ at saturation. Read from Table 3-1 at 56 F that $\mathbf{h}_{fB} = 24.1$ Btu per lb water

$$m_a = \frac{10{,}000 \times 60}{14.3} = 41{,}950 \text{ lb or dry air per hour, being cooled.}$$

By use of equation 3-39 the refrigeration required is

$$q_{A-B} = 41{,}950\,[43.52 - 23.84 - (0.0195 - 0.0096)\,24.1] = 815{,}500 \text{ Btu per hour}$$

Expressed in tons of refrigeration

$$q_{A-B} = \frac{815{,}500}{12{,}000} = 67.9 \text{ tons}$$

Note that the condensate removed per hour is

$$m_a (W_A - W_B) = 41{,}950\,(0.0195 - 0.0096) = 415 \text{ lb of water per hour}$$

Air and Humidity Calculations

Air-Mixing. When two quantities of air having different enthalpies and different specific humidities are mixed, the final condition of the air mixture depends on the masses (weights) involved, and on the enthalpy and humidity ratio of each of the constituent masses which enters the mixture. If m_A pounds of air at enthalpy $\mathbf{h}_A$ and specific humidity W_A are mixed with m_B pounds of air at enthalpy $\mathbf{h}_B$ and specific humidity W_B, the following equations will obviously apply:

$$m_A \mathbf{h}_A + m_B \mathbf{h}_B = (m_A + m_B)\mathbf{h}_m \tag{3-40}$$

or
$$\mathbf{h}_m = \mathbf{h}_A + \frac{m_B}{m_A + m_B}(\mathbf{h}_B - \mathbf{h}_A) \tag{3-41}$$

and
$$m_A W_A + m_B W_B = (m_A + m_B)W_m \tag{3-42}$$

or
$$W_m = W_A + \frac{m_B}{m_A + m_B}(W_B - W_A) \tag{3-43}$$

where m_A and m_B are the weights of air mixed, in pounds or pounds per unit time; $\mathbf{h}_A$ and $\mathbf{h}_B$ are the enthalpies per pound associated with each of the weights of dry air mixed; W_A and W_B are the humidity ratios (specific humidities), associated with each pound of dry air being mixed, in grains or pounds.

If, on a psychrometric chart like the skeleton chart shown in Fig. 3–9, a straight line is drawn to connect the state points A and B of two weights of air which enter into a mixing process, the resultant mixture will be

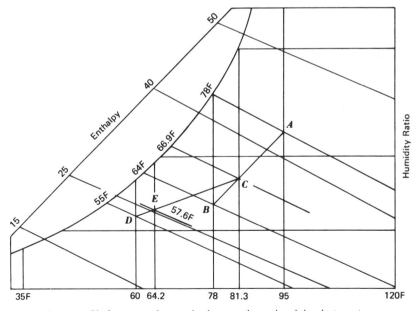

Fig. 3–9. Skeleton psychometric chart to show air mixing in two steps.

found to have a state point which falls on the line or lies very close to the line. The location of the mixing point on the line is determined by the relative weights of the materials entering into the mixing. For example, if equal weights were mixed, the point would be midway between the two end points. Similarly, if 2 parts by weight were mixed with 1 part by weight, the point would lie one-third of the distance along the line from the point corresponding to the larger quantity entering into the process. This straight-line relationship which occurs in mixing is extremely convenient, and sufficiently accurate for most types of calculation.

It is easy to rearrange equations 3-40 and 3-42 to show that

$$\frac{h_B - h_m}{h_m - h_A} = \frac{W_B - W_m}{W_m - W_A} = \frac{m_A}{m_B} \qquad (3\text{-}44)$$

In this form it is obvious that the line segments bear a direct ratio to the masses of dry air (m_A and m_B) in the mixing supply streams.

It is also possible to show, to a very close approximation, that the dry-bulb temperature resulting from two weights of air being mixed is:

$$t = t_A + \frac{m_B}{m_A + m_B} (t_B - t_A) \qquad (3\text{-}45)$$

For equations 3-41 and 3-43, a further approximation can be made—if the points are not too far apart—to the effect that the terms in the ratio $m_B/(m_A + m_B)$ may be expressed in cubic feet of air (cfm) instead of in pounds.

Example 3-9. From a conditioned space 8000 cfm of air at 78 F dry bulb and 64 F wet bulb are recirculated and mix with 2000 cfm of outside air at 95 F and 78 F wet bulb. (a) Find the condition of the resultant mixture. This mixture then enters the conditioner where 80 per cent of the mixture, by weight, is cooled and dehumidified to 60 F and 55 F wet bulb while the rest by-passes the coils and remains unchanged. The cooled and the by-passed air then mix to become the supply air for the conditioned space. (b) Find the temperature and relative humidity of this supply air.

Solution: Refer to Fig. 3-9 which illustrates this problem on a skeleton psychrometric chart. Then on Plate I locate points A (95 F, 78 F) and B (78 F, 64 F) and connect these with a straight line. Read $h_A = 41.32$ Btu per lb of dry air, $h_B = 29.21$, $v_A = 14.37$ cu ft per lb of dry air, and $v_B = 13.73$. $m_A = \frac{2000}{14.37} = 139.2$ lb per min, $m_B = \frac{8000}{13.73} = 583$.

By equation 3-41

$$h_m = 41.32 + \frac{583}{722.2} (29.21 - 41.32) = 31.52 \text{ Btu per lb dry air}$$

From this value on the diagonal enthalpy scale, use two triangles and run the enthalpy line to intersect the connecting line at point C, which is seen to be at 81.3 F, 66.9 F wb. The same result can be found by making use of the teaching of equation

Air and Humidity Calculations

3-44 which shows that lengths on the line are directly proportional to the supply mass flow rates m_A and m_B. Here

$$\frac{m_A}{m_B} = \frac{139.2}{583} = \frac{1}{4.18}$$

and the split is in the ratio of 1 to 4.18 so that the line split is $\frac{1}{5.18}$ of the length and $\frac{4.18}{5.18}$ of the length. The length can be measured in inches or millimeters and the point, C, can then be located after multiplying the total length by either ratio. Note that the shorter length lies next to the larger air-flow rate.

Of the 722.2 lb per min entering the conditioner, 80 per cent is cooled to 60 F, 55 F and the remaining 20 per cent is by-passed. Locate point D at 60 F, 55 F wb on Plate I and draw a line to the previous mixture point, C, just determined at 81.3 F, 66.9 F. Measure the length of this line and then lay off 0.20 of that length from D to locate the resultant leaving air at E having temperatures of 64.2 F and 57.6 F wb.

Humidification of Moist Air. Humidification can be accomplished when steam is supplied either from a direct source or by vaporization from heated or unheated water surfaces or from a water spray in the air supply. Equation 3-16 is applicable to a humidification process, but some modification in interpretation of the terms is necessary. In that equation, h_{f3} represents the enthalpy of water added or removed and it is tacitly assumed that the water is in a liquid state; however, this energy balance is equally applicable to water in any form (liquid, steam, or solid), provided the proper value of enthalpy is assigned. Therefore h_{f3} will be written h_3 and will indicate merely enthalpy, in Btu per pound, of the actual water, water vapor, or ice added or removed. Thus the equation takes the following form:

$$h_{a1} \pm Q_{(1-2)} + (W_2 - W_1)h_3 = h_{a2} \qquad (3\text{-}46)$$

If the process is also adiabatic, the equation becomes

$$h_{a1} + (W_2 - W_1)h_3 = h_{a2} \qquad (3\text{-}47)$$

Equation 3-47 can be written in the following form

$$\frac{h_{a2} - h_{a1}}{W_2 - W_1} = h_3 \qquad (3\text{-}48)$$

From this, for adiabatic mixing of steam and air, it can be seen that the ratio of enthalpy difference (Δh) to humidity ratio (ΔW) sets the slope of a line on the psychrometric chart, the direction of which is fixed by the enthalpy of the water (steam) supply. The line in question passes through the initial and final state points.

Example 3-10. General supply air in a certain industrial plant is available at 70 F and 60 per cent relative humidity. Air is required at 100 F for a particular process, and a higher specific humidity is not objectionable. If steam from the plant boilers at 600 psia, superheated to 1000 F (for which **h** = 1516.7 Btu per lb), is throttled into the supply air (70 F and ϕ = 60 per cent) and the final temperature is 100 F, what characteristics will the air have after mixing with the steam?

Solution: This process is adiabatic and therefore equation 3–47 is applicable. From Plate I and given values, we read

$$h_{a1} = 27.02 \text{ Btu per lb dry air}$$

$$W_1 = 0.0094 \text{ lb per dry air}$$

and

$$h_3 = 1516.7 \text{ Btu per lb steam}$$

With the values of h_{a2} and W_2 at 100 F unknown, to solve this problem locate h_{a1} at 70 F and φ = 60 per cent on Plate I and through this point draw a line parallel to a line slope of 1516.7, located on the protractor at the left of Plate I. The intersection of a line having this slope when carried to 100 F shows that h_{a2} = 52.68 Btu per lb dry air, W_2 = 0.0260 lb per lb of dry air. The parallel line can be drawn with the help of two triangles.

Fig. 3–10 is a skeleton chart showing this process for the slope 1516.7 and points 1 and 2.

If use is not made of the protractor, equation 3–47 can be written as follows:

$$27.02 + (W_2 - 0.0094)\,1516.7 = h_{a2}$$

and a solution to accurately match the two unknown values is sought.

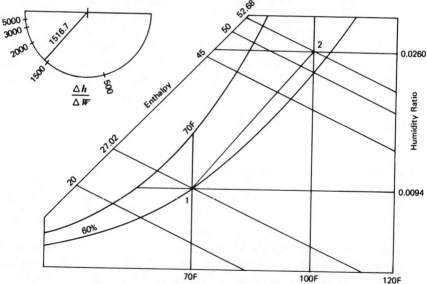

Fig. 3–10. Skeleton psychrometric chart showing use of the $\dfrac{\Delta h}{\Delta W}$ slope employed for enthalpy of steam added.

The problem involves finding the respective values of W_2 and h_{a2}

Air and Humidity Calculations

which constitute the solution to this equation. A trial-and-error solution usually can produce a close answer after about 3 or 4 substitutions of appropriate humidity ratio, enthalpy values. This approach can also be used to check the accuracy of the graphical method and this should be done since the slope of the line cannot be determined with a high degree of precision. If necessary, slightly modified values of h_{a2} and W_2 should be selected to make the equation reach the degree of accuracy desired.

It should be observed that, for the heating of air, direct addition of steam is not particularly effective, although such mixing is effective in increasing the humidity ratio. Relative humidity may increase or decrease in the process.

Evaporative Cooling. In section 3–4 it was shown that an adiabatic saturator, using recirculated water, will bring the dry-bulb temperature of the entering air to its original wet-bulb temperature whenever the air is brought to complete saturation. In the process the dry-bulb temperature is lowered toward the wet-bulb temperature because its sensible heat is absorbed in evaporating water. If the moist air leaves the saturator before complete saturation occurs, the air will be at a temperature higher than that corresponding to its wet-bulb temperature at saturation. Even so, the wet-bulb temperature of the air has not changed from its original value and the process of adiabatic evaporative cooling still takes place at the wet-bulb temperature of the entering air. On a psychrometric chart then, a line of constant wet-bulb temperature represents the process. Equation 3–21 is applicable and for the process to be truly consistent, the make-up water to the cooler would have to be provided at the wet-bulb temperature. If it is not, there is a trivial deviation from the wet-bulb line. So-called desert coolers depend on evaporative cooling to lower air temperature.

Example 3–11. On a given day, air in the southwestern United States was supplied, at 105 F dry-bulb and 20 per cent relative humidity, to an air-washer type of humidifier using recirculated water. If the humidifier had adequate spray-nozzle capacity to bring the air to within 70 per cent of the original wet-bulb depression, what were the conditions of the air leaving the humidifier?

Solution: From Plate I, original air at 105 F db and 20 per cent relative humidity has a wet-bulb temperature of 72.3 F. The original wet-bulb depression is $105 - 72.3 = 32.7$ deg. The final air temperature required to reach 70 per cent of the original wet-bulb depression is $105 - 0.7 (32.6) = 82.1$ F db.

The final condition is at 72.3 F wb and 82.1 F db and $\varphi = 62$ per cent.

This process, which is adiabatic, follows a constant wet-bulb line and therefore equation 3–21 applies. The recirculated water with a small make-up reached essentially the wet-bulb temperature of the entering air, 72.3 F, at which temperature $h_{fw} = 40.4$ Btu per lb (see Table 3–1). From Plate 1, $h_{a1} = 35.78$ Btu per lb dry air, $W_1 = 0.00945$

lb per lb dry air $h_{a2} = 36.00$ Btu per lb dry air and $W_2 = 0.0148$ lb per lb dry air when these are read at 82.1 F and 72.3 F wet bulb. Substitution in equation 3–21 shows a very close check.

$$35.78 + (0.0148 - 0.00945)(40.4) = 36.00$$

The relative humidity on leaving the humidifier has reached 62 per cent. This is not objectionable for comfort and is so preferable to a 105 F temperature that in dry atmospheres evaporative cooling can be employed to advantage, when mechanically refrigerated cooling is not available.

Spray-type equipment can also be used to dehumidify air, in which case the water supply must be externally cooled below the dew-point temperature of the supply air and the process is no longer an adiabatic one at constant wet-bulb temperature. Humidifying with externally heated hot water is also possible with spray-type equipment. This process is not an adiabatic constant wet-bulb process.

Dew Point and Relative Humidity. An important relationship holds approximately true between relative humidity and the dew-point differential. For example, for a dew-point differential 10 deg below dry-bulb temperature the relative humidity of the air is approximately 70 per cent over a moderate range of dry-bulb temperatures. Again, for a dew-point differential of 20 deg, the relative humidity is approximately 50 per cent over a moderate range of dry-bulb temperatures. Similar relative-humidity values for other dew-point differentials can be read from the psychrometric chart. This comparative constancy of relative humidity for constant dew-point differentials is utilized in some air-conditioning control systems.

3–8. AIR-SUPPLY CONDITIONS, SPACE AIR CONDITIONING

When a space is heated by warm air, the air must be supplied at a dry-bulb temperature which is sufficiently high that in cooling to the desired room temperature it offsets the heat losses from the space. The cooling load presents a similar problem, in a reverse sense. For example, the cooling load on an auditorium filled with people involves the problem of supplying air at a dry-bulb temperature low enough to keep the air in the auditorium from rising above a desired maximum temperature, and also the problem of supplying air that is sufficiently low in moisture content to keep the humidity of the air in the auditorium from rising above a desired value.

A major part of the heat load (really cooling load) in a space arises from heat transfers through boundaries and from heat sources in a space such as lights, machinery, and from a portion of the heat added by occupants in the space. This type of heat addition, if not offset, produces solely a temperature rise in the space which would be evident to the senses;

Air and Humidity Calculations

consequently it is called sensible heat (q_s). However, occupants, through their metabolic processes, also deliver moisture into the air, as do many cooking and drying processes so that steam may also be added to the air of a space. The supply air must thus be sufficiently dry to keep the moisture load from exceeding a predetermined value. The energy associated with steam additions is obviously the various masses of steam Σm_v multiplied by appropriate enthalpy values, h_v for the steam. Let us then call $\Sigma m_v h_v$ a summation term for all of the steam additions made at respective enthalpy values h_v. An energy equation for a place provided with an air stream of m_a lb of dry air per hour, with entering enthalpy h_{a1} and with sensible energy (q_s), and steam additions being added in the space would appear:

$$m_a h_{a1} + q_s + \Sigma m_v h_v = m_a h_{a2} \qquad (3\text{-}49)$$

This is a true equation since the same amount of "dry" air (m_a) must leave the space as entered and the leaving enthalpy must carry out the total entering enthalpy and added energy from the space. The moisture that leaves the space in the air stream must equal both the steam added in the space and that brought in by the original supply stream, thus

$$\Sigma m_v = m_a (W_2 - W_1) \qquad (3\text{-}50)$$

where W_2 and W_1 are the respective humidity ratios in the leaving and entering air streams in lb per lb of dry air. Combine equations 3–49 and 3–50 by division,

$$\frac{m_a (h_{a2} - h_{a1})}{m_a (W_2 - W_1)} = \frac{q_s + \Sigma m_v h_v}{\Sigma m_v}$$

$$\frac{h_{a2} - h_{a1}}{W_2 - W_1} = \frac{\Delta h_a}{\Delta W} = \frac{q_s + \Sigma m_v h_v}{\Sigma m_v} \qquad (3\text{-}51)$$

In this form it can be seen that the rate of inlet air flow has disappeared and the ratio $\dfrac{\Delta h_a}{\Delta W}$ is a slope related only to the ratio of the total space cooling load, $q_s + \Sigma m_v h_v$ and to the moisture added in the space, Σm_v. If the final conditions in and leaving the space, h_{a2} and W_2 are set, the values of h_{a1} and W_1 may be allowed to vary but they must always lie on the line set by the slope $\Delta h_a/\Delta W$.

Values of $\Delta h_a/\Delta W$ appear on the protractor at the left of Plate I.

The water-vapor (steam) produced by occupants for most calculations can be considered as though produced in the range 85 F to 95 F and for a round value $h_v = h_g = 1100$ Btu per lb can be used.

It should be recognized that in a cooling process the water vapor removed from air follows a similar pattern and the heat removal in condensing out water vapor from air, involving as it does removal of latent heat and subcooling of liquid, can be expressed

$$Q_L = m_a (W_1 - W_2) \mathbf{h}_v = \Sigma m_v \mathbf{h}_v \qquad (3\text{--}52)$$

where Q_L = latent (water-vapor) load, Btuh
m_a = lb of dry air per hour air flow through conditioner
W_1 = humidity-ratio (specific-humidity) of air into and
W_2 = air at exit from conditioner, lb per lb of dry air
$\mathbf{h}_v$ = enthalpy of water vapor, Btu per lb
$\Sigma m_v = m_a (W_1 - W_2)$ = lb per hour of water vapor condensed out, or alternatively created in a space.

Example 3-12. (*Cooling Load*). An auditorium is to be maintained at a temperature not to exceed 76 F db and 66 F wb. The air supplied to the auditorium should not be lower than 64 F. It is found that the sensible heat load on the auditorium is 300,000 Btuh and 130 lbs of water are added to the air per hour. (a) Compute the ratio of enthalpy difference to moisture added for the auditorium. (b) Using this value, determine the supply air conditions to the auditorium for minimum air flow. (c) Making use of the enthalpy of the entering and leaving air compute the air flow. (d) Find the air flow and the other data needed without making use of the line slope from the protractor.

Solution: (a) For this problem q_s = 300,000 Btu per hr and $\Sigma m_v \mathbf{h}_v$ = (130) (1100) = 143,000 Btu per hr
By equation 3-50

$$\frac{\Delta \mathbf{h}_a}{\Delta W} = \frac{300{,}000 + 143{,}000}{130} = 3408 \text{ Btu per lb of steam}$$

b) Locate this ratio on the protractor of Plate 1 to find the slope and then on the chart body of Plate 1 draw the line parallel to it which passes through the design leaving air conditions, 76 F and 66 F ($\mathbf{h}_{a2}$ = 30.76). For air supplied at 64 F and lying on this line the wet-bulb temperature is 60.0 F, and the enthalpy is $\mathbf{h}_{a1}$ = 26.42 Btu per lb of dry air.

c) The air flow rate can be found by solving equation 3-51 for m_a

$$m_a = \frac{q_s + \Sigma m_v \mathbf{h}_v}{\mathbf{h}_{a2} - \mathbf{h}_{a1}} = \frac{443{,}000}{30.76 - 26.42} = 102{,}100 \text{ lb per hr air flow.}$$

The humidity-ratio of the supply air is 0.0102 lb per lb of dry air and the dew-point temperature is 57.5 F.

d) An alternate solution to this problem would be to make use of equation 3-36 and apply it to the sensible heat load q_s, here then

$$300{,}000 = m_a (0.244) (76 - 64)$$

$$m_a = 102{,}100 \text{ lb of dry air per hour}$$

The total moisture load of 130 lb must be absorbed by the supply air and the pick up per pound of dry air is thus

$$W_2 - W_1 = \frac{130}{102{,}100} = 0.00127 \text{ lb per lb of dry air}$$

Air and Humidity Calculations

Since W_2 for the leaving air at 76 F, 66 F is 0.0114 lb per lb of dry air from Plate 1, then $W_1 = 0.0114 - 0.00127 = 0.01013$.

For $t = 64$ F and $W_1 = 0.01013$, the wet-bulb temperature is found from Plate 1 as 59.9 F. The slight difference in this answer from the slope method arises from the difficulty in reading the slope with a high degree of accuracy and the alternate method may be a preferable method under some conditions.

Sensible to Total-heat Ratio. Instead of using the water added in a space as one of the variables in establishing the condition line the ratio of Sensible to Total heat may serve equally well. Here

$$\frac{q_s}{q_T} = \frac{q_s}{q_s + \Sigma m_v \mathbf{h}_v} = \frac{\Delta H_s}{\Delta H_T} \qquad (3\text{-}53)$$

Under the symbolism used at the right of equation 3–53 the Sensible to Total-heat Ratio $\left(\dfrac{\Delta H_s}{\Delta H_T}\right)$ is plotted on the inside curve of the protractor which is provided on the psychrometric charts Plates 1 through 4. $\Delta H_s/\Delta H_T$ is a true dimensionless ratio. Its numerical values are usually decimal fractions and thus are easier to work with on the protractor than the large numerical values involved in the use of $\Delta h/\Delta W$ with its customary dimensions of Btu per lb of water. Both of these ratio-type parameters are used in the same way, to find a slope on the protractor and then this slope has to be transferred to the body of Plates 1, 2, 3, or 4.

3–9. PSYCHROMETRIC PROPERTIES UNDER VARYING BAROMETRIC PRESSURES

Attention is again called to the fact that Table 3–1 and the psychrometric charts (Plates 1, 2, and 3) have been constructed for standard barometric pressure, essentially at sea-level elevation, 29.92 inches of mercury. When these charts or Table 3–1 are used for computations involving other than standard pressure, some error in the result arises. The error is not serious for small variations of less than an inch of mercury pressure but are significant for higher deviations. Plate 4, constructed for 24.90 in. Hg. essentially 5000 feet altitude, is available for mountainous or high-plateau regions but also needs correction for sizeable deviations from its basic pressure or elevation.

A number of methods have been developed for making corrections with a correction table for enthalpy included as Table 3–2. For other properties recourse must be had to use of the basic equations of this chapter, in nearly all cases with help necessary from generalizations based on the perfect-gas laws. With barometric pressure, dry-bulb and dew-point temperatures known, it is easy to get the other properties. However, if dry-bulb and wet-bulb are the temperatures known, equation 3–27 is awkward to use and the humidity ratio, W_{sw}, for saturation at the thermodynamic wet-bulb temperature must be computed for the barometric pressure before the equation can be used.

Example 3–12. Compute the properties of moist air in a region where the barometric pressure is 26.92 in. Hg and the measured temperatures are $t = 80$ F dry bulb, $t_w = 70$ F wet-bulb.

Solution: The humidity ratio for saturated air at the wet-bulb temperature can be found by equation 3–5, or by equation 3–10 if slightly more accuracy is desired. The pressure of saturated steam at 70 F is independent of barometric pressure and thus can be read from Table 3–1 or Table 2–2 as 0.7392 in. Hg. If equation 3–10 is used $f_s = 1.0046$ from Fig. 3–1.

$$W_{sw} = 0.622 \frac{(1.0046)(0.7392/26.92)}{1 - (1.0046)(0.7392/26.92)} = 0.0176$$

lb per lb of dry air.

Comparison of this value with the saturation humidity ratio from Table 3–1 at standard barometer namely 1104 grains or 0.0158 lb shows that W_s is significantly greater at a reduced barometric pressure.

Employing equation 3–27 and reading $\mathbf{h}_{fgw}$ at the 70 F wet-bulb in Table 3–1 as $1091.8 - 38.1 = 1053.7$

$$W = \frac{(0.0176)(1053.7) - 0.24(80 - 70)}{1053.7 + 0.45(10)} = 0.01525$$

lb per lb of dry air, is the humidity ratio.

Enthalpy can be found from equation 3–14

$$\mathbf{h}_a = 0.24(80 - 0) + 0.01525(1060.5 + 0.45 \times 80)$$

$$= 35.92 \text{ Btu per lb of air}$$

Enthalpy can also be found by reading the enthalpy at standard barometer 29.92 from Plate 1 and adding the corrective term from Table 3–2, thus

$$\mathbf{h}_a = 34.00 + 1.98 = 35.98 \text{ Btu per lb dry air}$$

Enthalpy by the two methods shows reasonable agreement.

The partial pressure of the water vapor in the air can be found by solving for the partial pressure of the water vapor using equation 3–5 or equation 3–10. Using equation 3–5

$$0.01525 = 0.622 \frac{p}{26.92 - p}$$

$$p = 0.645 \text{ in. Hg}$$

The dew-point temperature corresponds to this pressure and can be read from Table 3–1, closely 66 F.

Specific volume of the moist air can also be found by using $p_a V = m R_a T$ with $p_a = p_B - p = 26.92 - 0.645 = 26.275$ in. Hg

$$(144)(0.491)(26.275)v = 1(53.34)(460 + 80)$$

$$v = 15.5 \text{ cu ft per lb dry air}$$

Comparison with values at standard barometric pressure will show that the relative humidity is little changed by pressure deviation but that great changes arise with v, $\mathbf{h}$ and W.

The labor in the computation of psychrometric properties at a pressure

Air and Humidity Calculations

varying from standard is very great if an appropriate chart is not available, so much so that it may be advisable to construct such a chart, at least in skeleton form, to minimize labor. For this purpose the layout of the standard chart can be used as a base on which to construct a new chart.

Based on the perfect gas and Gibbs-Dalton relations it is easily possible to develop the following working equations:

$$v = \frac{0.754\,(t+460)}{p_B}\left(1 + \frac{W}{0.622}\right) \quad (3\text{-}53)$$

$$p = \frac{Wp_B}{0.622 + W} \quad (3\text{-}54)$$

where v = specific volume, cu ft per lb of dry air with associated water vapor

t = dry-bulb temperature, F

p_B = barometric pressure, in. Hg

W = humidity ratio (specific humidity) in lb per lb of dry air.

PROBLEMS

3-1. Air is supplied to a certain room from the outside, where the temperature is 20 F and the relative humidity is 70 per cent. It is desired to keep the room at 70 F and 60 per cent relative humidity. How many pounds of water must be supplied to each pound of air entering the room if these conditions are to be met? The barometric pressure is 29.5 in. Hg. *Ans.* 0.00797 lb/lb

3-2. The temperature of a certain room is 72 F and the relative humidity is 50 per cent. The barometric pressure is 29.92 in. Hg. Find (a) the partial pressures of the air and water vapor, (b) the vapor density, and (c) the humidity ratio of the mixture.
Ans. (a) 14.496 psi, 0.1941 psi; (b) 0.000613 lb per cu ft; (c) 0.00833

3-3. Certain air has a dry-bulb temperature of 75 F and a relative humidity of 50 per cent, and the barometric pressure is 29.8 in. Hg. Calculate (a) the weight of 1 cu ft of the mixture of air and moisture and (b) the weight of moisture per pound of dry air.
Ans. (a) 0.0734 lb per cu ft; (b) 0.0093 lb per lb

3-4. Air with a dry-bulb temperature of 70 F and a wet-bulb temperature of 65 F is at a barometric pressure of 29.92 in. Hg. Without making use of the psychrometric chart, find (a) the relative humidity of the air; (b) the vapor density; (c) the dew-point temperature; (d) the humidity ratio; (e) the volume occupied by the mixture associated with a pound of dry air. *Ans.* (a) 76.8%; (b) 0.000883 lb per cu ft; (c) 62.4 F; (d) 0.01204 lb; (e) v = 13.62

3-5. Rework problem 3-4, but assume that the barometric pressure is 28.0 in. Hg.
Ans. (a) 77.3%; (b) 0.000891 lb per cu ft; (c) 62.5 F; (d) 0.0129 lb; (e) v = 14.56

3-6. For a barometric pressure of 29.5 in. Hg, compute the saturation ratio at 80 F for relative humidities of 30, 50, and 90 per cent. Compare these values with the 80 F line of Fig. 3-2.

3-7. Air is supplied to a room at a 72 F dry-bulb temperature and a 68 F wet-bulb temperature from outside air at 40 F db and 37 F wb. The barometric pressure is 29.92 in. Hg. Find (a) the dew-point temperatures of the inside and outside air; (b) the moisture added to each pound of dry air; (c) the enthalpy of the outside air; (d) the enthalpy of the inside air.

Ans. (a) 66.1 F, 33.6 F; (b) 0.0098 lb; (c) 13.86 Btu/lb; (d) 32.39 Btu/lb

3-8. (a) Compute the humidity ratio of air at 90 F and 60 per cent relative humidity, at 29.92 in. Hg, by use of equation 3-5. (b) Then recompute, making use of equation 3-10 and Fig. 3-1. (c) Find the percentage error involved in using equation 3-5.

Ans. (a) 0.018251; (b) 0.018346; (c) 0.5%

3-9. (a) Find from the psychrometric chart the dew point and humidity ratio of one pound of dry air at 29.92 in. Hg if the dry-bulb temperature is 80 F and the wet-bulb temperature is 70 F. (b) Find also the enthalpy and specific volume.

Ans. (a) 65.5 F, W = 0.0135; (b) h = 34.0, v = 13.9 cu ft

3-10. Air is heated to 80 F, without the addition of water, from 60 F db and 50 F wb. By use of the chart find (a) the relative humidity of the original mixture; (b) the dew-point temperature; (c) the humidity ratio; (d) the original enthalpy; (e) the final enthalpy; (f) the heat added; (g) the final relative humidity.

Ans. (a) 49%; (b) 40.7 F; (c) 0.00533 lb; (d) 20.28; (e) 25.11; (f) 4.83; (g) 24%

3-11. Air is cooled from 75 F db and 70 F wb to 55 F. Find (a) the moisture removed per pound of dry air; (b) the heat removed to condense the moisture; (c) the sensible heat removed; and (d) the total amount of heat removed.

Ans. (a) 0.0054 lb; (b) 5.8; (c) 5.0; (d) 10.8

3-12. Prove that the partial pressure of the water vapor (steam) in atmospheric air is invariant as long as the specific humidity is constant, (i.e., as long as the composition of the air is unchanged). This condition holds, of course, as long as the air temperature exceeds, or, in the limit, reaches, the dew-point temperature. HINT: Assume that the perfect-gas equations hold and write

$$p(1) = d(R_s)T$$

and $$p_B(1) = (d_m)(R_m)T$$

where p is the partial pressure of the steam in the mixture; d and R_s are the density and gas constant for steam; p_B, d_m, and R_m are the corresponding values for the total air-steam mixture. Explain the meaning of each term in the answer, which is

$$p = (p_B)\left(\frac{R_s}{R_m}\right)\left(\frac{d}{d_m}\right)$$

3-13. Air at 100 F db and 65 F wb is humidified adiabatically with steam. The steam supplied contains 20 per cent of moisture saturated at 16 psia. When sufficient steam is added to humidify the air to 60 per cent relative humidity, what is the dry-bulb temperature of the humidified air? The barometer is at 29.92 in. Hg. *Ans.* 92 F

3-14. Air at 84 F db and 60 F wb and at 29.92 in. Hg is humidified with the dry-bulb temperature remaining constant. Saturated steam is supplied for humidification at 14.7 psia in wet condition. What quality must the steam have (a) to provide saturated air and (b) to provide air at 70 per cent relative humidity?

Ans. (a) (b) 94.2%, essentially constant

3-15. In the text example 3-10, air at 70 F and 60 per cent relative humidity was adiabatically humidified by adding superheated steam at 600 psia and 1000 F (h = 1516.7 Btu per lb). (a) What is the final dry-bulb temperature if enough steam is added

Air and Humidity Calculations 131

to raise the relative humidity to 65 per cent? (b) If the final temperature is 180 F, what relative humidity can be attained by the addition of an adequate amount of this steam? *Ans.* (a) 75 F and $W = 0.0121$; (b) $W = 0.077$ and 22%

3-16. Air is humidified by the addition of dry saturated steam at 200 psia (**h** = 1198). The air is initially at 60 F db and 50 F wb and at 29.92 in. Hg. By varying the weight of steam supplied to each pound of dry air, the air temperature is raised to 63 F and 62 F respectively. For each of these final temperatures, find the corresponding values of humidity ratio, relative humidity, and enthalpy.
Ans. For 63 F: 0.00111 lb, 89.5%, 27.2

3-17. Air at 90 F db and 60 F wb and at 29.92 in. Hg is humidified to a final dew point of 70 F by use of dry saturated steam at 120 psia (**h** = 1190.4). If the process takes place under adiabatic conditions, what is the final dry-bulb temperature?
Ans. 95 F

3-18. Calculate the wet-bulb temperature of dry air at 90 F. The barometric pressure is 29.92 in. Hg. Base the solution on tables 3–1 and 2–2, not on the psychrometric chart. *Ans.* 52.7 F

3-19. Air at 100 F db and 70 F wb and 29.92 in. Hg is adiabatically mixed with water supplied at 140 F, in such proportions that the mixture has a relative humidity of 80 per cent. Find the dry-bulb temperature of the mixture. *Ans.* 75 F

3-20. Air at 40 F db and 35 F wb is mixed with warm air at 100 F db and 77 F wb in the ratio of 2 lb of cool air to 1 lb of warm air. Compute the resultant humidity ratio and enthalpy of the mixed air by equations 3–43 and 3–41. On the psychrometric chart of Plate I, connect by a straight line the points representing the two kinds of air, and locate a point on this line at a distance of one-third of its length from the cooler point. Read at this point the humidity ratio and enthalpy of the mixed air, as shown on the straight line, and compare the readings with the computed values.
Ans. $W = 0.007$ lb, **h** = 22.1

3-21. Assume that 3000 cu ft of air at 50 F and 100 per cent relative humidity are mixed with 2500 cu ft of air at 75 F and 50 per cent relative humidity. Compute the temperature, relative humidity, and humidity ratio of the resulting mixture.
Ans. 61.1 F, 72%, 0.0083 lb

3-22. In an auditorium maintained at a temperature not to exceed 75 F, and at a relative humidity not to exceed 60 per cent, a sensible-heat load of 450,000 Btuh and 1,200,000 grains of moisture per hour must be removed. Air is supplied to the auditorium at 65 F. (a) How many pounds of air per hour must be supplied? (b) What is the dew-point temperature of the entering air, and what is its relative humidity? (c) How much latent-heat load is picked up in the auditorium? (d) What is the sensible heat ratio? *Ans.* (a) 183,600 lb/hr; (b) 58.1 F, 0.78; (c) 180,000 Btuh; (d) 0.715

3-23. A meeting hall is to be maintained at 80 F db and 68 F wb. The barometric pressure is 29.92 in. Hg. The space has a load of 200,000 Btuh, sensible, and 200,000 Btuh, latent. The temperature of the supply air to the space cannot be lower than 66 F db. (a) How many pounds of air per hour must be supplied? (b) What is the required wet-bulb temperature of the supply air? (c) What is the sensible heat ratio?
Ans. (a) 58,500 lb/hr; (b) 58.4 F; (c) 0.5

3-24. A building with a heat loss of 200,000 Btuh is heated by warm air which is supplied at 135 F and which has a humidity ratio of 0.006 lb per pound of dry air. Air returns to the furnaces at 65 F, with no significant change in humidity ratio. Find (a) the number of pounds of air which must be circulated per hour for heating and (b) the number of cubic feet per minute measured at inlet conditions.

Ans. (a) 11,720 lb/hr; (b) 2930 cfm

3-25. If pipes carrying water at 50 F run through a room which has an air temperature of 70 F, what is the maximum relative humidity that can be held in the room without any water condensing on the pipes? *Ans.* 49%

3-26. Outside air at 95 F dry bulb and 80 F dew point at 27.92 in. Hg barometric pressure is cooled to saturated state at 64 F. Without using of the psychrometric chart find: (a) the humidity-ratio of the moist air before and after cooling, (b) the weight of moisture condensed, (c) the enthalpy of the moist air before and after cooling, and (d) the heat removed in Btu per hour when 300 lb per min of air pass through the air cooler.

3-27. (a) What is the relative humidity of air at 70 F if the dew-point temperature is 52 F? (b) How much heat is required to bring such air from 52 F dew point to 70 F?

Ans. (a) 0.52; (b) 4.4 Btu/lb

3-28. Outdoor air with a temperature of 40 F db and 35 F wb, and with a barometric pressure of 29 in. Hg, is heated and humidified under steady-flow conditions to a final dry-bulb temperature of 70 F and 40 per cent relative humidity. (a) Find the weight of water vapor added to each pound of dry air. (b) If the water is supplied at 50 F, how much heat is added per pound of dry air?

Ans. (a) 21.9 grains; (b) 10.5 Btu/lb

3-29. Outdoor air at 95 F db and 79 F wb, and at a barometric pressure of 28.92 in. Hg, is cooled and dehumidified under steady-flow conditions until it becomes saturated at 60 F. (a) Find the weight of water condensed per pound of dry air. (b) If the condensate is removed at 60 F, what quantity of heat is removed per pound of dry air?

Ans. (a) 0.00707 lb; (b) 15.4 Btu/lb

3-30. If the air in problem 3-29, after cooling, were reheated to 70 F, what would be its relative humidity? *Ans.* 70.4%

3-31. For a saturation ratio of 60 per cent at 160 F, what is the relative humidity? The barometric pressure is 29.92 in. Hg. *Ans.* 68.9%

3-32. The partial pressure of steam in 400 F air flowing through a certain drying oven is purported to be 10 psia. What is the relative humidity? The oven pressure is 30 in. Hg abs. *Ans.* 4.05%

REFERENCES

1. W. H. Carrier, "Rational Psychrometric Formulas," *Trans. ASME*, Vol. 23 (1911), p. 1005.

2. W. K. Lewis, "The Evaporation of a Liquid into a Gas," *Trans. ASME*, Vol. 44 (1922), p. 325.

3. J. H. Arnold, "The Theory of the Psychrometer," *Physics*, Vol. 4 (1933), pp. 255, 334.

4. B. H. Jennings and A. Torloni, "Psychrometric Charts for Use at Altitudes above Sea Level," *Refrigerating Engineering*, Vol. 62 (June, 1954), pp. 71–76, 118–26.

5. J. A. Goff and S. Gratch, "Thermodynamic Properties of Moist Air," *Trans. ASHVE*, Vol. 51 (1945), pp. 31–36.

6. E. P. Palmatier and D. D. Wile, "A New Psychrometric Chart," *Refrigerating Engineering*, Vol. 52 (July, 1946), pp. 31-36.

7. E. P. Palmatier, "Construction of the Normal Temperature Psychrometric Chart," *ASHRAE Journal*, Vol. 5 (May, 1963), pp. 55-60.

8. ASHRAE, *Handbook of Fundamentals* (1967), Chapter 6.

9. M. Costantino et al, "ASHRAE Psychrometric Chart Converted to Metric System," *ASHRAE Journal*, Vol. 8 (April, 1966), pp. 68-69.

4

Heat Transfer and Transmission Coefficients

4-1. MODES OF HEAT TRANSFER

The necessity of calculating heat flow occurs in so many engineering problems that the need of a thorough understanding of heat flow cannot be overestimated. The design of every heating or cooling system is based primarily on the heat-transfer characteristics of the building structure. Also, when both the heat flow to or from the building and the internal heat load have been calculated, the heat-transfer problem again appears in finding the size (surface) of heaters, cooling coils, or other appliances to carry the load.

Heat is gained or lost through the walls and structure of a building in two general ways: first, by transmission through the wall from the air on one side to the air on the other side, and second, by actual leakage of warmer or colder air into the building. Thus, to reduce heat transfer the insulating quality of the walls must be improved by use of building insulation or by use of insulating air spaces in walls and between roofs and ceilings. Leakage is reduced by the installation of weather strips, by use of double windows and doors, and by caulking or otherwise reducing air flow through cracks.

Transfer of heat takes place by conduction, convection, radiation, or by some combination of these processes, whenever a temperature difference exists.

Conduction is a process in which heat is transmitted from and to adjacent molecules along the path of flow, whereby some of the thermal agitation of the hotter molecules is passed on to the adjacent cooler molecules. An example of heat transfer by conduction is the passage of heat along an iron bar one end of which is being heated in a fire. In conduction of heat there is no appreciable displacement of the particles of the material.

Convection is (1) the transfer of heat between a *moving* fluid medium (liquid or gas) and a surface or (2) the transfer of heat from one point to another within a fluid by movements within the fluid, which movements

Heat Transfer and Transmission Coefficients

intermix different portions of the fluid. The final method of heat transfer in convection is eventually some form of conduction or radiation. In convection, if the fluid moves because of differences in density resulting from temperature changes, the process is called *natural convection*, or *free convection*; if the fluid is moved by mechanical means (pumps or fans) the process is called *forced convection*.

Radiation is a process in which hot bodies give off radiant energy in all directions. A colder body on which this energy falls absorbs some of the energy from the source and, as a result, evidences an increase in internal energy and usually a rise in temperature. Some media are *diathermous*; that is, they are capable of transmitting radiant energy without being affected, and capable of delivering the radiant energy undiminished to another body. Air, glass, and many gases are relatively diathermous, although both water vapor and carbon dioxide absorb radiant energy in appreciable amounts. Two bodies at different temperatures both emit radiation and absorb impinging radiation, but the hotter body emits more than it receives. The net result is a transfer of heat from the hotter to the colder body.

4-2. HEAT-TRANSFER EQUATIONS

The theory of heat conduction was first mathematically developed by the French mathematician J. B. Fourier, although Sir Isaac Newton had long before started work on the subject. Fourier's equation for unidirectional heat flow, based on experimental evidence, is

$$\frac{dQ}{d\theta} = -kA\frac{dt}{dx} \qquad (4\text{-}1)$$

where $dQ/d\theta$ = heat transfer per unit of time (θ);
A = area of the section through which heat is flowing;
dt = temperature difference causing the heat flow;
dx = length of path through the material, in the direction of the heat flow;
k = a proportionality factor called thermal conductivity.

In the terms dQ, $d\theta$, dt, and dx the differential notation is used to indicate infinitesimal changes in these quantities. It will be noted that the rate of flow of heat is inversely proportional to the thickness of the insulation; that is, less heat is transferred as the thickness of the insulation is increased. Equation 4-1, although of little more than academic interest for the purposes of this book, forms a convenient starting point for the development of more-important equations.

When the temperature t varies with both time θ and position x, as when a substance is warming or cooling, the flow is called *unsteady flow* and solutions of the differential equation 4-1 are usually very complex. When, however, equilibrium in heat transfer is reached and temperature depends only on position, the flow is called *steady flow*. For steady flow, which is a very common case, the heat transferred is constant, $dQ/d\theta = q$ in Btu per unit of time (usually Btu per hour), and thus

$$q = -kA\frac{dt}{dx} \tag{4-2}$$

The minus sign usually has no particular utility in steady flow and will be dropped. The units of k are:

$$k = \frac{q}{A} \times \frac{dx}{dt} = \frac{\text{(Btu)(ft)}}{\text{(hr)(sq ft)(deg F)}}$$

Often the units of k are also expressed in tables as

$$\frac{\text{(Btu)(in.)}}{\text{(hr)(sq ft)(deg F)}} = \frac{\text{Btu}}{\text{(hr)(sq ft)(deg F/in.)}}$$

$$\frac{\text{(Btu)(ft)}}{\text{(hr)(sq ft)(deg F)}} \times 12 = \frac{\text{(Btu)(in.)}}{\text{(hr)(sq ft)(deg F)}} \tag{4-3}$$

and $\quad\dfrac{\text{(Btu)(ft)}}{\text{(hr)(sq ft)(deg F)}} \times 0.00413 = \dfrac{\text{(gram cal)(cm)}}{\text{(sec)(sq cm)(deg C)}} \tag{4-4}$

The value of k for different materials varies over wide limits; and it even varies for the same material, with different temperatures and density of packing. Typical values of k for general substances and for building materials appear in Table 4-1.

TABLE 4-1

THERMAL CONDUCTIVITY AND OTHER PROPERTIES OF MISCELLANEOUS SUBSTANCES

Material	Specific Heat C_p (Btu per lb deg F)	Density at 68 F (lb per cu ft)	Conductivity k $\left[\dfrac{\text{(Btu)(in.)}}{\text{(hr)(sq ft)(deg F)}}\right]$	Temperature Range (F)
Air, still.................	0.24	.....	0.169–0.215	32–200
Aluminum...............	0.21	168.0	1404–1429	32–600
Ammonia				
Liquid................	1.128	38.0	3.48	5–86
Vapor.................	0.52	0.67	0.144	32
Asbestos board with cement	0.20	123	2.7	85
Asbestos, wool...........	0.20	25.0	0.62	32
Bagasse..................	0.32	13.5	0.336	68
Benzol...................	0.34	55.5	1.18	68
Brass				
Red..................	0.090	536.0	715.0	32
Yellow...............	0.088	534.0	592.0	32
Brick				
Common..............	0.22	112.0	5.0	...
Face..................	0.22	125.0	9.2	...
Fire..................	0.20	115.0	6.96	392

Heat Transfer and Transmission Coefficients

TABLE 4-1 (Continued)

Material	Specific Heat C_p (Btu per lb deg F)	Density at 68 F (lb per cu ft)	Conductivity k $\left[\dfrac{(\text{Btu})(\text{in.})}{(\text{hr})(\text{sq ft})(\text{deg F})}\right]$	Temperature Range (F)
Bronze	0.10	509.0	522.0	32
Cellulose, dry	0.37	94.0	1.66	59
Celotex	0.32	13.2	0.34	...
Cement mortar	0.19	118.0	12.0	...
Chalk	0.21	142.0	6.35	...
Cinders	0.18	40–45	1.1	...
Clay				
Dry	0.22	63.0	3.5–4.0	68–212
Wet	0.60	110.0	4.5–9.5	...
Concrete				
Cinder	0.18	97.0	4.9	75
Stone	0.19	140.0	12.0	75
Cork, granulated	0.42	8.1	0.31	90
Corkboard	0.42	8.3	0.28	60
Cornstalk, insulating board	0.32	15.0	0.33	71
Cotton	0.32	5.06	0.39	32
Foamglas	0.16–0.19	10.5	0.40	50
Gasoline	0.53	42.0	0.94	86
Glass wool	0.22	1.5	0.27	75
Glass				
Common thermometer	0.20	164.0	5.5	68–212
Flint	0.12	247.0	5.1	50–122
Pyrex	0.20	140.0	7.56	...
Gold	0.031	1205.0	2028	64–212
Granite	0.20	159.0	15.4	...
Gypsum, solid	0.26	78.0	3.0	68
Hair felt	0.33	13.0	0.26	90
Ice	0.50	57.5*	15.6	32
Iron				
Cast	0.13	442.0	326.0	129–216
Wrought	0.11	485.0	417.0	64–212
Iron oxide	0.17	306–330	3.63	68
Lampblack	...	10.0	0.45	104
Lead	0.030	710.0	240.0	64–212
Leather, sole	0.36	54.0	1.10	86
Lime				
Mortar	0.22	106.0	2.42	...
Slaked	0.13	81.0–87.0	...	...
Limestone	0.22	132.0	10.8	75
Marble	0.21	162.0	20.6	32–212
Mineral wool				
Board	0.25	15.0	0.33	75
Fill-type	0.20	9.4	0.27	103

* At 32 F.

TABLE 4-1 (Continued)

Material	Specific Heat C_p (Btu per lb deg F)	Density at 68 F (lb per cu ft)	Conductivity k $\left[\dfrac{\text{(Btu)(in.)}}{\text{(hr)(sq ft)(deg F)}}\right]$	Temperature Range (F)
Nickel	0.10	537.0	406.5	64–212
Paper	0.32	58.0	0.90	...
Paraffin	0.69	55.6	1.68	32–68
Plaster				
Cement and sand	0.20	73.8	5.00	68
Gypsum	0.20	46.2	5.60	73
Redwood bark	....	5.0	0.26	75
Rock wool	0.20	10.0	0.27	90
Rubber				
Hard	0.40	74.3	11.0	100
India	0.48	59.0	1.302	68–212
Sand, dry	0.19	94.6	2.28	68
Sandstone	0.22	143.0	12.6	68
Silver	0.056	656.0	2905.0	64–212
Soil				
Crushed quartz (4% moisture)	[0.16–0.19 dry reaching to 0.3 wet]	100 (dry)	11.5	40
Crushed quartz (4% moisture)		110 (dry)	16.0	40
Fairbanks sand				
Moisture, 4%		100 (dry)	8.5	40
Moisture, 10%		110 (dry)	15.0	40
Dakota sandy loam				
Moisture, 4%		110 (dry)	6.5	40
Moisture, 10%		110 (dry)	13.0	40
Healy clay				
Moisture, 10%		90 (dry)	5.5	40
Moisture, 20%		100 (dry)	10.0	40
Steam	0.48	0.037*	0.151	212
Steel				
1% C	0.12	487.0	310.0	64–212
Stainless	0.12	515.0	200.0	...
Tar, bituminous	0.35	75.0	....	86
Water				
Fresh	1.00	62.4	4.10	70
Sea	0.94	64.0	3.93	64
Wood				
Fir	0.65	34.0	0.80	75
Maple	....	40.0	1.2	75
Red oak	0.57	48.0	1.10	86
White pine	0.67	31.2	0.780	86
Wood fiber board	0.34	16.9	0.34	90
Wool	0.33	4.99	0.264	86

* At 212 F and 14.7 psi.

Heat Transfer and Transmission Coefficients

Conduction through a plain wall leads to the following simplification of Fourier's equation:

$$q = k\frac{A}{x}(t_1 - t_2) = k\frac{A}{x}\Delta t \qquad (4\text{-}5)$$

where q = heat transferred per unit of time (hour), in Btu;
A = area of the wall, in square feet;
x = thickness of the wall, in feet or inches (depending on the units of k);
k = thermal conductivity, in units of Btu ft (or in.) per hr sq ft deg F;
$t_1 - t_2 = \Delta t$ = the temperature difference on the two sides of the wall which causes heat flow, in degrees Fahrenheit.

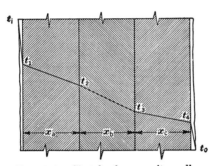

FIG. 4-1. Sketch of composite wall.

Conduction through a composite wall, shown in Fig. 4-1, can be treated by using equation 4-5 in connection with the so-called resistance concept. Writing equation 4-5 in more-general form and algebraically rearranging the terms, there results for a plain wall,

$$q = k\frac{A}{x}(t_1 - t_2) = \frac{t_1 - t_2}{x/kA} = \frac{t_1 - t_2}{R} \qquad (4\text{-}6)$$

where $R = x/kA$ is called the thermal resistance.

In the case of the composite wall shown in Fig. 4-1, the heat flow under steady conditions through each section of the wall is the same; that is, in Btu per hour through the whole wall,

$$q = q_a = q_b = q_c$$

and

$$q = \frac{t_1 - t_2}{R_a} = \frac{t_2 - t_3}{R_b} = \frac{t_3 - t_4}{R_c} = \frac{t_1 - t_4}{R_t} \qquad (4\text{-}7)$$

And in the general case,

$$q = \frac{t_1 - t_n}{R_t} = \frac{t_1 - t_n}{\dfrac{x_a}{k_a A_a} + \dfrac{x_b}{k_b A_b} + \dfrac{x_c}{k_c A_c} + \cdots + \dfrac{x_n}{k_n A_n}} \qquad (4\text{-}8)$$

In the case of flat walls per square foot of surface, in Btu per hour,

$$q = \frac{t_1 - t_n}{\dfrac{x_a}{k_a} + \dfrac{x_b}{k_b} + \dfrac{x_c}{k_c}} = \frac{t_1 - t_n}{R_t} \tag{4-9}$$

In heat flow between a fluid (gas or liquid) and a solid, there always exists a thin fluid film which tends to cling to the surface as a relatively stagnant layer and which acts as an additional resistance to heat flow. The thickness of this film is greatly influenced by the convection conditions of the system. For example, in building-construction, a wind blowing on the outside walls of a building greatly changes the film condition over that which exists on a quiet day with little wind. The films, and the temperature drops through them, are indicated in Fig. 4-1.

In the case of heat-transfer surfaces, such as those for heat exchangers, boiler tubes, and evaporators, the film conditions are affected by (1) the fluid velocity, (2) the shape and kind of surface, (3) whether natural or forced convection exists, and (4) whether boiling or condensing is occurring. The resistance of gas films to heat flow is usually many times greater than that of similar liquid films. Because of the many variables involved, exact calculation of heat transfer through such films is in many cases impossible, and recourse must be had to over-all heat-transfer coefficients based on experimental determinations and set up for use in the form of empirical equations or tables. In heat transfer through metal surfaces, the films offer the greatest resistance to heat flow and must be considered. In transfer through effective insulation, the film resistance may be relatively unimportant.

In the technical literature, f is used as a symbol for this *film coefficient*, or *surface coefficient*, although the symbol h is common in industrial heat transfer (boilers, condensers, etc.). The units of f in the engineering system are Btu per hr sq ft deg F. Dimensionally, $f = k/x$, but since x, the film thickness, is most often indeterminate, f is usually given as a value independent of k and x for each medium considered. Thus

$$\frac{1}{f} = \text{film resistance to heat flow per square foot} \tag{4-10}$$

or

$$\frac{1}{fA} = \text{film resistance to heat flow for an area } A \tag{4-11}$$

Considering films f_o and f_i on two sides of a material, equation 4-9 becomes, in units of Btu per hr sq ft,

$$q = \frac{t_i - t_o}{\dfrac{1}{f_i(1)} + \dfrac{x_a}{k_a(1)} + \dfrac{x_b}{k_b(1)} + \dfrac{x_c}{k_c(1)} + \cdots + \dfrac{1}{f_o(1)}} \tag{4-12}$$

Heat Transfer and Transmission Coefficients

The values of the film- or surface-conductance coefficient f increase with the increasing roughness of the surface involved, and they increase almost linearly with the air or wind velocity over the surface. Houghten and McDermott, using their own and Rowley's data, indicated that the following equations represent the range in values of f for building materials, when v is expressed in miles per hour:

$$f = 1.4 + 0.28v \text{ for very smooth surfaces} \quad (4\text{-}13)$$

$$f = 1.6 + 0.3v \text{ for smooth wood and plaster} \quad (4\text{-}14)$$

$$f = 2.0 + 0.4v \text{ for cast concrete and smooth brick} \quad (4\text{-}15)$$

$$f = 2.1 + 0.5v \text{ for rough stucco surfaces} \quad (4\text{-}16)$$

The values of f also increase with increasing temperature difference. Because surface, wind, and temperature conditions are often rather indeterminate, it has been customary to take a single value, $f_i = 1.65$ Btu per hr sq ft deg F, as being representative of the film coefficient for inside still-air conditions on nonreflective surfaces, while for outside surfaces a value of $f_o = 6.0$ has been most widely used for a wind velocity of 15 mph. Recently more attention has been given to the variations of the inside (still-air) film

TABLE 4-2

THERMAL CONDUCTANCES OF VERTICAL AIR SPACES AT VARIOUS MEAN TEMPERATURES*

MEAN TEMPERATURE (F)	WIDTH OF AIR SPACE (IN.)						
	0.128	0.250	0.364	0.493	0.713	1.00	1.50
	Value of Conductance a (Btu per hr sq ft deg F)						
20	2.300	1.370	1.180	1.100	1.040	1.030	1.022
30	2.385	1.425	1.234	1.148	1.080	1.070	1.065
40	2.470	1.480	1.288	1.193	1.125	1.112	1.105
50	2.560	1.535	1.340	1.242	1.168	1.152	1.149
60	2.650	1.590	1.390	1.295	1.210	1.195	1.188
70	2.730	1.648	1.440	1.340	1.250	1.240	1.228
80	2.819	1.702	1.492	1.390	1.295	1.280	1.270
90	2.908	1.757	1.547	1.433	1.340	1.320	1.310
100	2.990	1.813	1.600	1.486	1.380	1.362	1.350
110	3.078	1.870	1.650	1.534	1.425	1.402	1.392
120	3.167	1.928	1.700	1.580	1.467	1.445	1.435
130	3.250	1.980	1.750	1.630	1.510	1.485	1.475
140	3.340	2.035	1.800	1.680	1.550	1.530	1.519
150	3.425	2.090	1.852	1.728	1.592	1.569	1.559

*Trans. ASHVE, Vol. 35 (1929), p. 165.

coefficient with respect to position and character of surface. Representative values are given in Table 4–3. An examination of these values shows an extensive variation in the value of the coefficient. The values $f_i = 1.65$ and $f_o = 6.0$ have been widely used throughout the tabulated material which follows in this chapter, since it is impractical to take account of minor variations in the application of f values to thermal barriers.

Many walls contain air cells in their make-up, and for such walls it is difficult to estimate both film and air resistance to heat flow. Therefore use should be made of Table 4–2, which gives values for conductances of air spaces. In the table, a is the conductance of the air space and $1/a$ is the resistance. It should be noted that for air spaces of over $\frac{3}{4}$-in. width the value of a does not decrease to any significant extent, and that therefore a representative value of $a = 1.10$ is used for such types of spaces. The value of a does increase, however, with increasing temperature difference across the space, because of radiant heat transfer and because of more-active convection currents. Interposition of bright metallic surfaces of aluminum foil or other polished compositions greatly reduces the radiant heat transfer and improves the insulating quality of the space. Conductance values for aluminum foil-faced surfaces should be taken from Table 4–3. This table also gives data on air spaces which do not have vertical orientation. With the passage of time, reflective surfaces collect dust and undergo chemical action. Therefore emissivity values (sec. 4–10) of wall surfaces adjacent to air spaces in building construction are not always constant, and thus the rate of heat flow across an air space may change.

It is often impracticable to calculate the heat-flow conditions through the various subsections of a heat-transfer barrier, but an over-all heat-transfer coefficient can be found, either experimentally or from tabulations for surfaces built up in a similar manner. For a composite wall, equation 4–12 becomes

$$q = UA(t_i - t_o) \text{ Btuh} \qquad (4\text{–}17)$$

where U is the over-all coefficient of heat transfer expressed in units of Btu per hr sq ft deg F, found either by direct experiment or by calculation from various items in equation 4–12, or selected, when possible, from suitable tabulated values of heat-transfer data.

For composite walls of standard construction, actual tests or computations to determine values of the over-all heat-transfer coefficient U, the heat conductivity C, or the resistance R have been made by various investigators, and where these values can be found they should be used in preference to making detailed calculations. Tables 4–4 through 4–17 give some of the values, including those for film resistance on each side of a wall, with a 15-mph wind assumed on the outside wall.

TABLE 4-3

THERMAL CONDUCTIVITIES (k) AND CONDUCTANCES (C) OF BUILDING AND INSULATING MATERIALS*

$$\left[k \text{ in } \frac{(\text{Btu})(\text{in.})}{(\text{hr})(\text{sq ft})(\text{deg F})}; C \text{ in } \frac{\text{Btu}}{(\text{hr})(\text{sq ft})(\text{deg F})}, \text{ for thickness of material in question} \right]$$

Material	Type and Condition	Density (lb per cu ft)	Mean Temp (deg F)	Thermal Conductivity k	Resistance per Inch $\frac{1}{k}$	Thermal Conductance C	Resistance $\frac{1}{C}$
Air film (surface):							
Still air (f_i)	General value	....					
Heat flow, up	Horizontal	....	50–90	....	....	1.65	0.61
	Sloping, 45°	....	50–90	....	....	1.63	0.61
Heat flow, down	Horizontal	....	50–90	....	....	1.60	0.62
	Sloping, 45°	....	50–90	....	....	1.08	0.92
Heat flow, horizontal	Vertical	....	50–90	....	....	1.32	0.76
Wind, 15 mph; (f_o)	Any position	....	50–90	....	....	1.46	0.68
Wind, 7.5 mph; (f_o)	Any position	....	...	....	....	6.00	0.17
		....	...	....	....	4.00	0.25
Air spaces:	*Bounded by structural materials*						
Heat flow, up	Horizontal, ¾″ to 4″	....	50–90	....	....	1.18	0.85
	Sloping 45°, ¾″ to 4″	....	50–90	....	....	1.11	0.90
Heat flow, horizontal	Vertical, ¾″ to 4″	....	50–90	....	....	1.03	0.97
	Sloping 45°, ¾″ to 4″	....	50–90	....	....	0.97	1.03
Heat flow, down	Horizontal, ¾″	....	50–90	....	....	0.98	1.02
	Horizontal, 1½″	....	50–90	....	....	0.87	1.15
	Horizontal, 4″	....	50–90	....	....	0.81	1.23

TABLE 4-3 (*Continued*)

$$\left[k \text{ in } \frac{(\text{Btu})(\text{in.})}{(\text{hr})(\text{sq ft})(\text{deg F})}; C \text{ in } \frac{\text{Btu}}{(\text{hr})(\text{sq ft})(\text{deg F})}, \text{ for thickness of material in question}\right]$$

Material	Type and Condition	Density (lb per cu ft)	Mean Temp (deg F)	Thermal Conductivity k	Resistance per Inch $\frac{1}{k}$	Thermal Conductance C	Resistance $\frac{1}{C}$
Air spaces:	*With reflective surface*						
Space bounded by aluminum foil on one surface	Vertical or horizontal, over $\frac{3}{4}$ in. wide	......	50	......	......	0.46	2.17
Space bounded by aluminum foil on facing surfaces	Over $\frac{3}{4}$ in. wide	......	50	......	......	0.41	2.44
Space divided in two with single curtain of aluminum foil	Both sides of foil bright and each space over $\frac{3}{4}$ in.	......	50	......	......	0.23	4.35
Space bounded on one side with aluminum foil	Space only $\frac{3}{8}$ in. wide	......	50	......	......	0.62	1.612
Building boards:							
Asbestos cement	Compressed cement and asbestos	118.0	119	4.0	0.25	...	...
Asbestos	Corrugated asbestos board	20.4	110	0.48	2.08	...	...
	Sheet asbestos	48.3	110	0.29	3.44	...	...

* Data from many sources, with large number selected from 1955 and 1956 editions of American Society of Heating and Air Conditioning Engineers *Guide*.

Heat Transfer and Transmission Coefficients

Material	Density	Temp.	k	1/k	C	R
Gypsum						
Gypsum board (gypsum between layers of heavy paper)	62.8	70	1.41	0.71	...	...
3/8-in. gypsum board	...	...	...	...	3.10	0.32
1/2-in. gypsum board	53.5	90	...	...	2.25	0.45
Insulating						
Average of insulating boards without special finish: 1/2 in. thick	16.5–21.8	90	0.33–0.40	3.03–2.50	0.66–0.80	1.52–1.25
Insulating From sugar cane fibers	13.5	70	0.33	3.03	...	...
Insulating From wood fiber	15.0	70	0.33	3.03	...	...
Sheathing Fir or yellow pine	...	...	...	...	1.02	0.98
Fir and building paper, 25/32 in.	...	...	...	...	0.86	1.16
Plywood, 1/4 in.	34	...	...	...	3.21	0.31
Plywood, 5/16 in.	34	...	...	...	2.56	0.39
Plywood, 3/8 in.	34	...	...	...	2.12	0.47
Plywood, 1/2 in.	34	...	...	...	1.60	0.63
Exterior finishes (frame walls):						
Brick veneer 4 in. thick, nominal	...	...	...	...	2.27	0.44
Stucco 1 in.	...	...	12.50	0.08	...	...
Wood shingles	...	...	...	...	1.28	0.78
Yellow-pine lap siding	...	...	...	...	1.28	0.78
1-in. fir sheathing, building paper, and stucco	...	20	...	...	0.82	1.22
Flooring construction:						
Asphalt tile, 1/8 in.	120	...	...	...	24.8	0.04
California redwood, dry	28.0	75	0.70	1.43	...	...
Ceramic tile, 1 in.	...	...	...	...	12.50	0.08

TABLE 4-3 (Continued)

$$\left[k \text{ in } \frac{(\text{Btu})(\text{in.})}{(\text{hr})(\text{sq ft})(\text{deg F})}; C \text{ in } \frac{\text{Btu}}{(\text{hr})(\text{sq ft})(\text{deg F})}, \text{ for thickness of material in question}\right]$$

Material	Type and Condition	Density (lb per cu ft)	Mean Temp (deg F)	Thermal Conductivity k	Resistance per Inch $\frac{1}{k}$	Thermal Conductance C	Resistance $\frac{1}{C}$
	Maple across grain	40.0	75	1.20	0.83	...	...
	Rubber tile, ⅛ in.	110.0	...	...	...	42.4	0.02
	White pine	31.2	86	0.78	1.28	...	...
	Yellow pine, dry	36.0	75	0.91	1.10	...	...
	Linoleum, ⅛ in.	......	...	...	...	12.00	0.08
Insulation:							
Bat- or blanket-type	Made from mineral or vegetable fiber, or animal hair; closed or open	......	...	...	...	...	...
Bat-type	Corkboard, no added binder	......	...	0.27	3.70	...	...
	Mineral wool (made from rock slag or glass)	......	...	0.30	3.33	...	...
	Chemically treated wood fiber	......	...	0.27	3.70	...	...
	Fibrous material made from dolomite and silica	4.0	75	0.28	3.57	...	...
Loose-fill type	Fibrous material made from slag	1.50	75	0.27	3.70	...	...
	Glass wool fibers	9.4	103	0.27	3.70	...	...
	Redwood bark	1.50	75	0.27	3.70	...	...
		3.0	90	0.31	3.22	...	...
		5.0	75	0.26	3.84	...	...

Heat Transfer and Transmission Coefficients

Material						
Regranulated cork in particles	8.1	90	0.31	3.22	.....	.....
Rock wool	10.0	90	0.27	3.70	.....	.....
Sawdust	12.0	90	0.41	2.44	.....	.....
Vermiculite, expanded	7.0	70	0.48	2.08	.....	.....
Wood planer shavings	8.8	90	0.41	2.44	.....	.....
Interior finishes and plasters:						
Composition wallboard $\frac{3}{16}$ in. to $\frac{3}{8}$ in. thick	.....	.....	0.50	2.00	.....	.....
Gypsum board Plain or decorated ($\frac{3}{8}$ in.)	.....	.....	.....	.....	3.7	0.27
Gypsum lath, $\frac{3}{8}$ in. (and plaster) Plaster thickness assumed $\frac{1}{2}$ in.	.....	.....	.....	.....	2.4	0.42
Insulating board, $\frac{1}{2}$ in. Plain or decorated	.....	.....	.....	.....	0.66	1.52
Insulating board lath, $\frac{1}{2}$ in. (and plaster) Plaster thickness $\frac{1}{2}$ in.	.....	.....	.....	.....	0.60	1.67
Insulating board lath, 1 in. (and plaster) Plaster thickness $\frac{1}{2}$ in.	.....	.....	.....	.....	0.31	3.18
Metal lath and plaster Plaster thickness $\frac{3}{4}$ in.	.....	.....	.....	.....	4.40	0.23
Cement and sand	.....	.....	5.0	0.20	.....	.....
High cement mix	.....	.....	8.0	0.13	.....	.....
Gypsum and sand	.....	.....	5.6	0.18	.....	.....
Plaster, vermiculite-gypsum	.....	.....	1.7	0.59	.....	.....
Wood lath and plaster Total thickness $\frac{3}{4}$ in.	.....	.....	.....	.....	2.50	0.40

TABLE 4-3 (*Continued*)

$$\left[k \text{ in } \frac{(\text{Btu})(\text{in.})}{(\text{hr})(\text{sq ft})(\text{deg F})}; C \text{ in } \frac{\text{Btu}}{(\text{hr})(\text{sq ft})(\text{deg F})}, \text{ for thickness of material in question} \right]$$

Material	Type and Condition	Density (lb per cu ft)	Mean Temp (deg F)	Thermal Conductivity k	Resistance per Inch $\frac{1}{k}$	Thermal Conductance C	Resistance $\frac{1}{C}$
Masonry materials:							
Brick	Adobe, 4 in. thick	.....	...	...	...	0.89	1.12
	Common, 4 in. thick	.....	...	5.0	0.20	1.25	0.80
	Face, 4 in. thick	.....	...	9.0	0.11	2.27	0.44
	One tier common clay brick, one tier face brick, approximately 8 in. thick						
Cement mortar	....	.....	...	...	...	0.77	1.30
Clay tile, hollow sections	3 in. thick	.....	...	12.0	0.08	...	...
	4 in. thick	.....	...	...	...	1.25	0.80
	6 in. thick	.....	...	...	...	0.90	1.11
	8 in. thick	.....	...	...	...	0.66	1.52
	10 in. thick	.....	...	...	...	0.54	1.85
	12 in. thick	.....	...	...	...	0.45	2.22
	16 in. thick	.....	...	...	...	0.40	2.50
						0.31	3.23
Concrete	Light aggregate of expanded slag, clay, or pumice	.....	...	2.50	0.40	...	...
Concrete	Sand and gravel aggregate	.....	...	12.0	0.08	...	...
3-in. concrete blocks	Hollow; cinder aggregate					1.28	0.70

Heat Transfer and Transmission Coefficients

4-in. concrete blocks	Hollow; cinder aggregate			0.90	1.11
8-in. concrete blocks	Hollow; cinder aggregate			0.58	1.72
12-in. concrete blocks	Hollow; cinder aggregate			0.53	1.89
8-in. concrete blocks	Hollow; sand and gravel aggregate			0.90	1.11
12-in. concrete blocks	Hollow; sand and gravel aggregate			0.78	1.28
8-in. concrete blocks	Hollow; lightweight aggregate			0.58	1.72
12-in. concrete blocks	Hollow; lightweight aggregate			0.44	2.27
3-in. gypsum tile	3-cell partition			0.74	1.35
4-in. gypsum tile	3-cell partition			0.60	1.67
3-in. gypsum tile	Hollow			0.61	1.64
4-in. gypsum tile	Hollow			0.46	2.18
Stone, typical			12.50	0.08	
Stucco			12.50	0.08	
Roofing:					
Asbestos-cement shingles		120		4.76	0.21
Asphalt shingles		70.0	75	2.27	0.44
Built-up roofing	Assumed thickness $\tfrac{3}{8}$ in.		75	3.00	0.33
Heavy rolled roof material				1.10	0.91
Slate shingles	Assumed thickness $\tfrac{1}{2}$ in.		10.0	20.0	0.05
Wood shingles				1.06	0.94

TABLE 4-3 (*Continued*)

$$\left[k \text{ in } \frac{(Btu)(in.)}{(hr)(sq\ ft)(deg\ F)};\ C \text{ in } \frac{Btu}{(hr)(sq\ ft)(deg\ F)},\ \text{for thickness of material in question} \right]$$

Material	Type and Condition	Density (lb per cu ft)	Mean Temp (deg F)	Thermal Conductivity k	Resistance per Inch $\frac{1}{k}$	Thermal Conductance C	Resistance $\frac{1}{C}$
Cellular glass (Foamglas)	....	9.0	75	0.42	2.38	...	...
		9.0	60	0.38	2.62	...	...
		9.0	300	0.55	1.82	...	...
Corkboard	No added binder	10.6	90	0.30	3.33	...	...
		7.0	90	0.27	3.70	...	...
Corkboard	Asphaltic binder	14.5	90	0.32	3.12	...	...
Mineral (rock) wool	....	15.7	90	0.32	3.12	...	...
		15.7	30	0.29	3.44	...	...
Shredded wood and cement	....	29.8	...	0.77	1.30	...	...
Styrofoam		1.3–2.0	40	0.23–0.30	4.35–3.33	...	...
Sugar cane, fiber insulation	Encased in asphalt membrane	13.8	70	0.30	3.33	...	...
Thermoflex, refractory fiber	Felted blanket	6.0	500	0.47	2.12	...	...
		6.0	1000	0.92	1.09	...	...
Woods:	Balsa	8.8	90	0.38	2.63	...	...
	California redwood, dry	28.0	75	0.70	1.43	...	...
	Long leaf yellow pine, dry	40.0	75	0.86	1.16	...	...
	Red oak, dry	48.0	75	1.18	0.85	...	...
	Short leaf yellow pine, dry	36.0	75	0.91	1.10	...	...
	Fir or pine, average	......	...	0.80	1.25	...	...
	Maple or oak, average	......	...	1.10	0.91	...	...
Sheathing	1-in. fir-sheathing, building-paper, and pine-lap siding				......	0.50	2.00

Heat Transfer and Transmission Coefficients

Symbols Used in Heat-Transfer Equations and Tables

U = over-all coefficient of heat transfer, in units of Btu per hr sq ft deg F, existing between the air or other fluid on the two sides of a wall, floor, ceiling, roof, or heat-transfer surface under consideration.

k = thermal conductivity, in units of Btu in. per hr sq ft deg F.

f = film or surface coefficient, for computing heat transfer through the gas or liquid film by conduction, radiation, and convection to the adjacent medium, in units of Btu per hr sq ft deg F.

C = thermal conductance (the heat transmitted through a nonhomogeneous or composite material of the thickness and type for which C is given), expressed in units of Btu per hr sq ft deg F. Film effects not usually considered.

a = thermal conductance of an air space, in units of Btu per hr sq ft deg F.

R = resistance to heat flow, usually for one square foot of area. A general term for various kinds of resistances.

= $1/U$ = overall resistance, in units of hr sq ft deg F per Btu.

= $1/C$ = resistivity of a composite section. Film effects not considered.

= $1/f$ = film resistance.

= $1/a$ = air-space resistance.

For a composite wall, general expressions for resistance and heat transfer are as follows:

$$R = \frac{1}{f_i} + \frac{x_1}{k_1} + \frac{1}{a_1} + \frac{x_2}{k_2} + \frac{1}{a_2} + \ldots + \frac{1}{f_o} = \frac{1}{U} \tag{4-18}$$

$$= \frac{1}{f_i} + \frac{1}{C} + \ldots + \frac{1}{f_o} = \frac{1}{U} \tag{4-19}$$

$$q = UA(t_i - t_o) \text{ Btuh} \tag{4-20}$$

$$q = \frac{A}{R}(t_i - t_o) \text{ Btuh} \tag{4-21}$$

$$q = CA(t_i - t_2) \text{ Btuh (no films considered)} \tag{4-22}$$

$$q = \frac{k}{x}A(t_1 - t_2) \text{ Btuh through homogeneous substance of thickness } x \text{ inches (no films considered)} \tag{4-6}$$

where t_i and t_o are temperatures of the air on two sides of barrier, in degrees Fahrenheit, and A = area in square feet. [*Text continues on page 158.*]

TABLE 4-4
COEFFICIENTS OF TRANSMISSION (U) OF FRAME WALLS
[No Insulation between Studs* (See Table 4-5)]

EXTERIOR FINISH	INTERIOR FINISH	Type of Sheathing				WALL NUMBER
		Gypsum ($\frac{1}{2}$ In. Thick)	Plywood ($\frac{5}{16}$ In. Thick)	Wood¶ ($\frac{25}{32}$ In. Thick) Bldg. Paper	Insulating Board ($\frac{25}{32}$ In. Thick)	
		A	B	C	D	
Wood Siding (Clapboard)	Metal lath and plaster†	0.33	0.32	0.26	0.20	1
	Gypsum board ($\frac{3}{8}$ in.) decorated	0.32	0.32	0.26	0.20	2
	Wood lath and plaster	0.31	0.31	0.25	0.19	3
	Gypsum lath ($\frac{3}{8}$ in.) plastered‡	0.31	0.30	0.25	0.19	4
	Plywood ($\frac{3}{8}$ in.) plain or decorated	0.30	0.30	0.24	0.19	5
	Insulating board ($\frac{1}{2}$ in.) plain or decorated	0.23	0.23	0.19	0.16	6
	Insulating board lath ($\frac{1}{2}$ in.) plastered‡	0.22	0.22	0.19	0.15	7
	Insulating board lath (1 in.) plastered‡	0.17	0.17	0.15	0.12	8
Wood Shingles§	Metal lath and plaster†	0.25	0.25	0.26	0.17	9
	Gypsum board ($\frac{3}{8}$ in.) decorated	0.25	0.25	0.26	0.17	10
	Wood lath and plaster	0.24	0.24	0.25	0.16	11
	Gypsum lath ($\frac{3}{8}$ in.) plastered‡	0.24	0.24	0.25	0.16	12
	Plywood ($\frac{3}{8}$ in.) plain or decorated	0.24	0.24	0.24	0.16	13
	Insulating board ($\frac{1}{2}$ in.) plain or decorated	0.19	0.19	0.19	0.14	14
	Insulating board lath ($\frac{1}{2}$ in.) plastered‡	0.19	0.18	0.19	0.13	15
	Insulating board lath (1 in.) plastered‡	0.14	0.14	0.15	0.11	16
Stucco	Metal lath and plaster†	.43	0.42	0.32	0.23	17
	Gypsum board ($\frac{3}{8}$ in.) decorated	0.42	0.41	0.31	0.23	18
	Wood lath and plaster	0.40	0.39	0.30	0.22	19
	Gypsum lath ($\frac{3}{8}$ in.) plastered‡	0.39	0.39	0.30	0.22	20
	Plywood ($\frac{3}{8}$ in.) plain or decorated	0.39	0.38	0.29	0.22	21
	Insulating board ($\frac{1}{2}$ in.) plain or decorated	0.27	0.27	0.22	0.18	22
	Insulating board lath ($\frac{1}{2}$ in.) plastered‡	0.26	0.26	0.22	0.17	23
	Insulating board lath (1 in.) plastered‡	0.19	0.19	0.16	0.14	24
Brick Veneer‖	Metal lath and plaster†	0.37	0.36	0.28	0.21	25
	Gypsum board ($\frac{3}{8}$ in.) decorated	0.36	0.36	0.28	0.21	26
	Wood lath and plaster	0.35	0.34	0.27	0.20	27
	Gypsum lath ($\frac{3}{8}$ in.) plastered‡	0.34	0.34	0.27	0.20	28
	Plywood ($\frac{3}{8}$ in.) plain or decorated	0.34	0.33	0.27	0.20	29
	Insulating board ($\frac{1}{2}$ in.) plain or decorated	0.25	0.25	0.21	0.17	30
	Insulating board lath ($\frac{1}{2}$ in.) plastered‡	0.24	0.24	0.20	0.16	31
	Insulating board lath (1 in.) plastered‡	0.18	0.18	0.15	0.13	32

NOTE. Coefficients are expressed in Btu per (hour) (square foot) (Fahrenheit degree difference in temperature between the air on both sides), and are based on an outside wind velocity of 15 mph.
* Coefficients not weighted; effect of studding neglected.
† Plaster assumed $\frac{3}{4}$ in. thick.
‡ Plaster assumed $\frac{1}{2}$ in. thick.
§ Furring strips (1 in. nominal thickness) between wood shingles and all sheathings except wood.
‖ Small air space and mortar between building paper and brick veneer neglected.
¶ Nominal thickness, 1 in.

TABLE 4-5

Coefficients of Transmission (U) of Frame Walls and Roofs with Insulation between Framing*

Coefficient with NO Insulation between Framing	Coefficient with Insulation between Framing				Number
	Mineral Wool or Vegetable Fibers in Blanket or Bat Form†			$3\frac{5}{8}$ In. Mineral Wool between Framing‡	
	1 In. Thick	2 In. Thick	3 In. Thick		
	A	B	C	D	
0.11	0.078	0.063	0.054	0.051	33
0.13	0.088	0.070	0.058	0.055	35
0.15	0.097	0.075	0.062	0.059	37
0.17	0.10	0.080	0.066	0.062	39
0.19	0.11	0.084	0.069	0.065	41
0.21	0.12	0.088	0.072	0.067	43
0.23	0.12	0.091	0.074	0.069	45
0.25	0.13	0.094	0.076	0.071	47
0.27	0.14	0.097	0.078	0.073	49
0.29	0.14	0.10	0.080	0.075	51
0.31	0.14	0.10	0.081	0.076	53
0.33	0.15	0.10	0.083	0.077	55
0.35	0.15	0.11	0.084	0.078	57
0.37	0.16	0.11	0.085	0.080	59
0.39	0.16	0.11	0.086	0.081	61
0.41	0.16	0.11	0.087	0.082	63
0.43	0.17	0.11	0.088	0.082	65

* Coefficients corrected for 2 × 4 framing, 16 in. on centers—15 per cent of surface area.
† Based on one air space between framing.
‡ No air space.

Note. In tables 4-4 to 4-18 inclusive, computed values of U are given for composite wall structures. For outside walls and roofs, a film factor based on a 15-mph wind has been considered for the outside surface, with a still-air film factor used for all inside-wall surfaces. Table 4-18 provides a convenient basis for modifying the over-all U values for different outside wind velocities. The value of U for the over-all wall is given in units of Btu per (hr)(sq ft)(deg F temperature difference between air temperatures on the two sides).

Tables 4-4 through 4-18 are reprinted from Chapter 9 of *Heating Ventilating Air Conditioning Guide 1955*, by permission.

TABLE 4-6

Coefficients of Transmission (U) of Masonry Walls

Type of Masonry	Thickness of Masonry (in.)	Plain Walls—No Interior Finish	Plaster (½ in.) on Walls	Metal Lath and Plaster¶ Furred ††	Gypsum Board (⅜ in.) Decorated—Furred ††	Gypsum Lath (⅜ in.) Plastered**—Furred ††	Insulating Board (½ in.) Plain or Decorated—Furred ††	Insulating Board Lath (½ in.) Plastered**—Furred ††	Insulating Board Lath (1 in.) Plastered**—Furred ††	Gypsum Lath** Plastered Plus 1 in. Blanket Insulation—Furred ††	Wall Number
		A	B	C	D	E	F	G	H	I	
Solid Brick*	8	0.50	0.46	0.32	0.31	0.30	0.22	0.22	0.16	0.14	67
	12	0.36	0.34	0.25	0.25	0.24	0.19	0.19	0.14	0.13	68
	16	0.28	0.27	0.21	0.21	0.20	0.17	0.16	0.13	0.12	69
Hollow Tile† (Stucco Exterior Finish)	8	0.40	0.37	0.27	0.27	0.26	0.20	0.20	0.15	0.13	70
	10	0.39	0.37	0.27	0.27	0.26	0.20	0.19	0.15	0.13	71
	12	0.30	0.28	0.22	0.22	0.21	0.17	0.17	0.13	0.12	72
	16	0.24	0.24	0.19	0.19	0.18	0.15	0.15	0.12	0.11	73
Stone ‡	8	0.70	0.64	0.39	0.38	0.36	0.26	0.25	0.18	0.16	74
	12	0.57	0.53	0.35	0.34	0.33	0.24	0.23	0.17	0.15	75
	16	0.49	0.45	0.31	0.31	0.29	0.22	0.22	0.16	0.14	76
	24	0.37	0.35	0.26	0.26	0.25	0.19	0.19	0.15	0.13	77
Poured Concrete §	6	0.79	0.71	0.42	0.41	0.39	0.27	0.26	0.19	0.16	78
	8	0.70	0.64	0.39	0.38	0.36	0.26	0.25	0.18	0.16	79
	10	0.63	0.58	0.37	0.36	0.34	0.25	0.24	0.18	0.15	80
	12	0.57	0.53	0.35	0.34	0.33	0.24	0.23	0.17	0.15	81
Hollow Concrete Blocks	\multicolumn{11}{c}{Gravel Aggregate}	
	8	0.56	0.52	0.34	0.34	0.32	0.24	0.23	0.17	0.15	82
	12	0.49	0.46	0.32	0.31	0.30	0.22	0.22	0.16	0.14	83
	\multicolumn{10}{c}{Cinder Aggregate}	
	8	0.41	0.39	0.28	0.28	0.27	0.21	0.20	0.15	0.13	84
	12	0.38	0.36	0.26	0.26	0.25	0.20	0.19	0.15	0.13	85
	\multicolumn{10}{c}{Lightweight Aggregate ‖}	
	8	0.36	0.34	0.26	0.25	0.24	0.19	0.19	0.15	0.13	86
	12	0.34	0.33	0.25	0.24	0.24	0.19	0.18	0.14	0.13	87

* Based on 4-in. hard brick and remainder common brick.
† The 8-in. and 10-in. tile figures are based on two cells in the direction of heat flow. The 12-in. tile is based on three cells in the direction of heat flow. The 16-in. tile consists of one 10-in. and one 6-in. tile, each having two cells in the direction of heat flow.
‡ Limestone or sandstone.
§ These figures may be used with sufficient accuracy for concrete walls with stucco exterior finish.
‖ Expanded slag, burned clay or pumice.
¶ Thickness of plaster assumed ¾ in.
** Thickness of plaster assumed ½ in.
†† Based on 2-in. furring strips; one air space.

TABLE 4-7
Coefficients of Transmission (U) of Brick- and Stone-Veneer Masonry Walls

Typical Construction	Facing	Backing	Plain Walls—No Interior Finish	Plaster (½ in.) on Walls	Metal Lath and Plaster ‖—Furred **	Gypsum Board (⅜ in.) Decorated—Furred **	Gypsum Lath (⅜ in.) Plastered ¶—Furred **	Insulating Board (½ in.) Plain or Decorated—Furred **	Insulating Board Lath (½ in.) Plastered ¶—Furred **	Insulating Board Lath (1 in.) Plastered ¶—Furred **	Gypsum Lath Plastered ¶ Plus 1-in. Blanket Insulation—Furred **	Wall Number
			A	B	C	D	E	F	G	H	I	
		6-in. hollow tile†	0.35	0.34	0.25	0.25	0.24	0.19	0.18	0.14	0.13	88
		8-in. hollow tile†	0.34	0.32	0.25	0.24	0.23	0.19	0.18	0.14	0.13	89
	4-in. Brick Veneer*	6-in. concrete	0.59	0.54	0.35	0.35	0.33	0.24	0.23	0.17	0.15	90
		8-in. concrete	0.54	0.50	0.33	0.33	0.31	0.23	0.23	0.17	0.15	91
		8-in. concrete blocks‡ (gravel aggregate)	0.44	0.41	0.29	0.29	0.28	0.21	0.21	0.16	0.14	92
		8-in. concrete blocks‡ (cinder aggregate)	0.34	0.33	0.25	0.24	0.24	0.19	0.18	0.14	0.13	93
		8-in. concrete blocks‡ (lightweight aggregate)§	0.31	0.29	0.23	0.23	0.22	0.18	0.17	0.14	0.12	94
		6-in. hollow tile†	0.37	0.35	0.26	0.26	0.25	0.19	0.19	0.15	0.13	95
		8-in. hollow tile†	0.36	0.34	0.25	0.25	0.24	0.19	0.19	0.14	0.13	96
	4-in. Cut-Stone Veneer*	6-in. concrete	0.63	0.58	0.37	0.36	0.34	0.25	0.24	0.18	0.15	97
		8-in. concrete	0.57	0.53	0.35	0.34	0.33	0.24	0.23	0.17	0.15	98
		8-in. concrete blocks‡ (gravel aggregate)	0.47	0.44	0.30	0.30	0.29	0.22	0.21	0.16	0.14	99
		8-in. concrete blocks‡ (cinder aggregate)	0.36	0.34	0.25	0.25	0.24	0.19	0.19	0.15	0.13	100
		8-in. concrete blocks‡ (lightweight aggregate)§	0.32	0.30	0.23	0.23	0.22	0.18	0.17	0.14	0.12	101

* Calculation based on ½-in. cement mortar between backing and facing, except in the case of the concrete backing, which is assumed to be poured in place.
† The hollow tile figures are based on two air cells in the direction of heat flow.
‡ Hollow concrete blocks.
§ Expanded slag, burned clay or pumice.
‖ Thickness of plaster assumed ¾ in.
¶ Thickness of plaster assumed ½ in.
** Based on 2-in. furring strips; one air space.

TABLE 4-8
Coefficients of Transmission (U) of Frame Partitions or Interior Walls*

Type of Interior Finish	Single Partition (Finish on One Side Only of Studs)	Double Partition (Finish on Both Sides of Studs)		Partition Number
		No Insulation between Studs	1-in. Blanket § between Studs. One Air Space	
	A	B	C	
Metal lath and plaster†	0.69	0.39	0.16	1
Gypsum board (⅜ in.) decorated	0.67	0.37	0.16	2
Wood lath and plaster	0.62	0.34	0.15	3
Gypsum lath (⅜ in.) plastered‡	0.61	0.34	0.15	4
Plywood (⅜ in.) plain or decorated	0.59	0.33	0.15	5
Insulating board (½ in.) plain or decorated	0.36	0.19	0.11	6
Insulating board lath (½ in.) plastered‡	0.35	0.18	0.11	7
Insulating board lath (1 in.) plastered‡	0.23	0.12	0.082	8

* Coefficients not weighted; effect of studding neglected.
† Plaster assumed ¾ in. thick.
‡ Plaster assumed ½ in. thick.
§ For partitions with other insulations between studs refer to Table 4-5, using values in column B of above table, in left-hand column of Table 4-5. *Example:* What is the coefficient of transmission (U) of a partition consisting of gypsum lath and plaster on both sides of studs with 2-in. blanket between studs? *Solution:* According to above table, this partition with no insulation between studs (No. 4B) has a coefficient of 0.34. Interpolating from Table 4-5, it will be found that a wall having a coefficient of 0.34 with no insulation between studs, will have a coefficient of 0.10 with 2 in. of blanket insulation between studs (No. 55B).

TABLE 4-9
Coefficients of Transmission (U) of Masonry Partitions

Type of Partition		Thickness of Masonry (in.)	Type of Finish			Partition Number
			No Finish (Plain Walls)	Plaster One Side	Plaster Both Sides*	
			A	B	C	
Hollow clay tile		3	0.50	0.47	0.43	9
		4	0.45	0.42	0.40	10
Hollow gypsum tile		3	0.35	0.33	0.32	11
		4	0.29	0.28	0.27	12
Hollow Concrete Tile or Blocks	Cinder aggregate	3	0.50	0.47	0.43	13
		4	0.45	0.42	0.40	14
	Lightweight aggregate†	3	0.41	0.39	0.37	15
		4	0.35	0.34	0.32	16
Common brick		4	0.50	0.46	0.43	17

* 2-in. solid plaster partition, $U = 0.53$.   † Expanded slag, burned clay or pumice.

TABLE 4–10

COEFFICIENTS OF TRANSMISSION (U) OF FRAME-CONSTRUCTION CEILINGS AND FLOORS

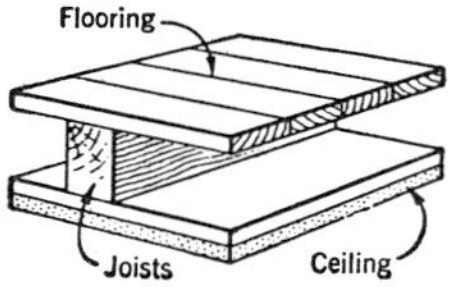

TYPE OF CEILING	Insulation between, or on Top of, Joists (No Flooring Above)												With Flooring** (On Top of Ceiling Joists)		NUMBER
	None	Insulating Board on Top of Joists		Blanket or Bat Insulation¶ between Joists*			Vermiculite Insulation between Joists*			Mineral Wool Insulation between Joists*			Single Wood Floor†	Double Wood Floor‡	
		½ In.	1 In.	1 In.	2 In.	3 In.	2 In.	3 In.	4 In.	2 In.	3 In.	4 In.			
	A	B	C	D	E	F	G	H	I	J	K	L	M	N	
No ceiling		0.37	0.24										0.45	0.34	1
Metal lath and plaster§	0.69	0.26	0.19	0.19	0.12	0.093	0.18	0.14	0.11	0.12	0.093	0.077	0.30	0.25	2
Gypsum board (⅜ in.) plain or decorated	0.67	0.26	0.18	0.19	0.12	0.092	0.18	0.13	0.10	0.12	0.092	0.077	0.30	0.24	3
Wood lath and plaster	0.62	0.25	0.18	0.19	0.12	0.091	0.17	0.13	0.10	0.12	0.091	0.076	0.28	0.24	4
Gypsum lath (⅜ in.) plastered‖	0.61	0.25	0.18	0.19	0.12	0.091	0.17	0.13	0.10	0.12	0.091	0.076	0.28	0.24	5
Plywood (⅜ in.) plain or decorated	0.59	0.24	0.18	0.19	0.12	0.091	0.17	0.13	0.10	0.12	0.091	0.076	0.28	0.23	6
Insulating board (½ in.) plain or decorated	0.36	0.19	0.15	0.16	0.10	0.082	0.14	0.12	0.097	0.10	0.082	0.069	0.22††	0.19††	7
Insulating board lath (½ in.) plastered‖	0.35	0.19	0.15	0.15	0.10	0.081	0.14	0.11	0.096	0.10	0.081	0.068	0.21	0.18	8
Insulating board lath (1 in.) plastered‖	0.23	0.15	0.12	0.12	0.089	0.072	0.12	0.097	0.084	0.089	0.072	0.061	0.16	0.14	9

* Coefficients corrected for framing on basis of 15 per cent area, 2 in. by 4 in. (nominal) framing, 16 in. on centers.

† 25/32-in. yellow pine or fir.

‡ 25/32-in. pine or fir subflooring plus 13/16-in. hardwood-finish flooring.

§ Plaster assumed ¾ in. thick.

‖ Plaster assumed ½ in. thick.

¶ Based on insulation in contact with ceiling, and consequently no air space between.

** For coefficients for constructions in columns M and N (except No. 1) with insulation between joists, refer to Table 4–5. *Example:* The coefficient for No. 3-N of Table 4–10 is 0.24. With 2-in. blanket insulation between joists, the coefficient will be 0.093. (See Table 4–5.) (Column D of Table 4–5 applicable only for 3⅝-in. joists.)

†† For 25/32-in. insulating board sheathing applied to the under side of the joists, the coefficient for *single* wood floor (col. M) is 0.18 and for *double* wood floor (col. N) is 0.16. For coefficients with insulation between joists, see Table 4–5.

TABLE 4-11

Coefficients of Transmission (U) of Concrete-Construction Floors and Ceilings

Type of Ceiling	Thickness of Concrete** (in.)	Type of Flooring					Number
		No Flooring (Concrete Bare)	Tile* or Terrazzo Flooring on Concrete	$\frac{1}{8}$-In. Asphalt Tile† Directly on Concrete	Parquet‡ Flooring in Mastic on Concrete	Double Wood Floor on Sleepers§	
		A	B	C	D	E	
No ceiling	3	0.68	0.65	0.66	0.45	0.25	1
	6	0.59	0.56	0.58	0.41	0.23	2
	10	0.50	0.48	0.49	0.36	0.22	3
$\frac{1}{2}$-in. plaster applied to underside of concrete	3	0.62	0.59	0.60	0.43	0.24	4
	6	0.54	0.52	0.53	0.39	0.22	5
	10	0.46	0.44	0.45	0.34	0.21	6
Metal lath and plaster‖—suspended or furred	3	0.38	0.37	0.37	0.30	0.19	7
	6	0.35	0.34	0.35	0.28	0.18	8
	10	0.32	0.31	0.32	0.26	0.17	9
Gypsum board ($\frac{3}{8}$ in.) and plaster¶—suspended or furred	3	0.36	0.35	0.35	0.28	0.19	10
	6	0.33	0.32	0.33	0.27	0.18	11
	10	0.30	0.29	0.30	0.24	0.17	12
Insulating board lath ($\frac{1}{2}$ in.) and plaster¶ suspended or furred	3	0.25	0.24	0.25	0.21	0.15	13
	6	0.23	0.23	0.23	0.20	0.15	14
	10	0.22	0.21	0.22	0.19	0.14	15

*Thickness of tile assumed to be 1 in.
† Conductivity of asphalt tile assumed to be 3.1.
‡ Thickness of wood assumed to be $\frac{13}{16}$ in.; thickness of mastic, $\frac{1}{8}$ in. ($k = 4.5$). Column D may also be used for concrete covered with carpet.
§ Based on $\frac{25}{32}$-in. yellow pine or fir subflooring and $\frac{25}{32}$-in. hardwood-finish flooring with an air space between subfloor and concrete.
‖ Thickness of plaster assumed to be $\frac{3}{4}$ in.
¶ Thickness of plaster assumed to be $\frac{1}{2}$ in.
** For other thickness of concrete, interpolate.

Example 4-1. Assume that the composite wall of Fig. 4-1 is made up of 4 in. of common brick against 6 in. of concrete, with $\frac{1}{2}$ in. of cement plaster on the inside (concrete) wall. Assume still air at 76 F inside, outside air at 20 F, and a 15-mph wind velocity. Find (a) the thermal resistance of the wall, (b) the over-all conductivity of the wall, (c) the heat transferred per hour, and (d) the heat transfer if the film resistances are disregarded.

Solution: (a) Referring to Table 4-3, we find the following thermal conductivities, in units of Btu in. per hr sq ft deg F:

$$k_b = 5.00 \text{ for common brick}$$

$$k_c = 12.00 \text{ for concrete}$$

$$k_p = 8.00 \text{ for cement plaster}$$

[*Text continues on page 163.*]

Heat Transfer and Transmission Coefficients

TABLE 4-12

COEFFICIENTS OF TRANSMISSION (U) OF FLAT ROOFS COVERED WITH BUILT-UP ROOFING. NO CEILING—UNDERSIDE OF ROOF EXPOSED

Type of Roof Deck	Thickness of Roof Deck (in.)	No Insulation	Insulation on Top of Deck (Covered with Built-up Roofing)							Number
			Insulating Board Thickness				Corkboard Thickness			
			½ In.	1 In.	1½ In.	2 In.	1 In.	1½ In.	2 In.	
		A	B	C	D	E	F	G	H	
Flat Metal Roof Deck*		0.94	0.39	0.24	0.18	0.14	0.23	0.17	0.13	1
Precast Cement Tile	1⅝ in.	0.84	0.3	0.24	0.17	0.14	0.22	0.16	0.13	2
Concrete	2 in.	0.82	0.36	0.24	0.17	0.14	0.22	0.16	0.13	3
	4 in.	0.72	0.34	0.23	0.17	0.13	0.21	0.16	0.12	4
	6 in.	0.65	0.33	0.22	0.16	0.13	0.21	0.15	0.12	5
Gypsum Fiber Concrete† on ½-in. Gypsum Board	2½ in.	0.38	0.24	0.18	0.14	0.12	0.17	0.13	0.11	6
	3½ in.	0.31	0.21	0.16	0.13	0.11	0.15	0.12	0.10	7
Wood‡	1 in.	0.49	0.28	0.20	0.15	0.12	0.19	0.14	0.12	8
	1½ in.	0.37	0.24	0.17	0.14	0.11	0.17	0.13	0.11	9
	2 in.	0.32	0.22	0.16	0.13	0.11	0.16	0.12	0.10	10
	3 in.	0.23	0.17	0.14	0.11	0.096	0.13	0.11	0.091	11

* Coefficient of transmission of bare corrugated iron (no roofing) is 1.50 Btu per (hr) (sq ft of projected area) (F deg difference in temperature) based on an outside wind velocity of 15 mph.

† 87½ per cent gypsum, 12½ per cent wood fiber. Thickness indicated includes ½ in. gypsum board.

‡ Nominal thicknesses specified—actual thicknesses used in calculations.

TABLE 4–13

Coefficients of Transmission (U) of Flat Roofs Covered with Built-up Roofing. With Lath and Plaster Ceilings*

Type of Roof Deck	Thickness of Roof Deck (in.)	No Insulation	Insulation on Top of Deck (Covered with Built-up Roofing)							Number
			Insulating Board Thickness				Corkboard Thickness			
			½ In.	1 In.	1½ In.	2 In.	1 In.	1½ In.	2 In.	
		A	B	C	D	E	F	G	H	
Flat Metal Roof Deck		0.46	0.27	0.19	0.15	0.12	0.18	0.14	0.11	12
Precast Cement Tile	1⅝ in.	0.43	0.26	0.19	0.15	0.12	0.18	0.14	0.11	13
Concrete	2 in.	0.42	0.26	0.19	0.14	0.12	0.18	0.14	0.11	14
	4 in.	0.40	0.25	0.18	0.14	0.12	0.17	0.13	0.11	15
	6 in.	0.37	0.24	0.18	0.14	0.11	0.17	0.13	0.11	16
Gypsum Fiber Concrete† on ½-in. Gypsum Board	2½ in.	0.27	0.19	0.15	0.12	0.10	0.14	0.12	0.097	17
	3½ in.	0.23	0.17	0.14	0.11	0.097	0.13	0.11	0.091	18
Wood‡	1 in.	0.31	0.21	0.16	0.13	0.11	0.15	0.12	0.10	19
	1½ in.	0.26	0.19	0.15	0.12	0.10	0.14	0.11	0.095	20
	2 in.	0.24	0.17	0.14	0.11	0.097	0.13	0.11	0.092	21
	3 in.	0.18	0.14	0.12	0.10	0.087	0.11	0.095	0.082	22

* Calculations based on metal lath and plaster ceilings, but coefficients may be used with sufficient accuracy for gypsum lath or wood lath and plaster ceilings. It is assumed that there is an air space between the under side of the roof deck and the upper side of the ceiling.

† 87½ per cent gypsum, 12½ per cent wood fiber. Thickness indicated includes ½ in. gypsum board.

‡ Nominal thicknesses specified—actual thicknesses used in calculations.

Heat Transfer and Transmission Coefficients

TABLE 4-14
COEFFICIENTS OF TRANSMISSION (U) OF PITCHED ROOFS

Type of Ceiling (Applied Directly to Roof Rafters)	Wood Shingles (On 1 × 4 Wood Strips* Spaced 2 In. Apart)				Asphalt Shingles or Roll Roofing (On Solid Wood Sheathing)*				Slate or Tile† (On Solid Wood Sheathing)*				Number
	None	Blanket or Bat (Thickness Below)			None	Insulation Between Rafters			None	Blanket or Bat (Thickness Below)			
		1 In.	2 In.	3 In.		Blanket or Bat (Thickness Below)				1 In.	2 In.	3 In.	
						1 In.	2 In.	3 In.					
	A	B	C‡	D‡	E	F	G‡	H‡	I	J	K‡	L‡	
No ceiling applied to rafters	0.48§	0.15	0.10	0.081	0.52§	0.15	0.11	0.084	0.55§	0.16	0.11	0.085	1
Metal lath and plaster‖	0.31	0.14	0.10	0.081	0.33	0.15	0.10	0.083	0.34	0.15	0.10	0.083	2
Gypsum board (⅜ in.) decorated	0.30	0.14	0.10	0.080	0.32	0.15	0.10	0.082	0.33	0.15	0.10	0.083	3
Wood lath and plaster	0.29	0.14	0.10	0.080	0.31	0.14	0.10	0.081	0.32	0.15	0.10	0.082	4
Gypsum lath (⅜ in.) plastered¶	0.29	0.14	0.10	0.079	0.31	0.14	0.10	0.081	0.32	0.15	0.10	0.082	5
Plywood (⅜ in.) plain or decorated	0.29	0.14	0.099	0.079	0.30	0.14	0.10	0.081	0.31	0.15	0.10	0.081	6
Insulating board (½ in.) plain or decorated	0.22	0.12	0.090	0.072	0.23	0.12	0.091	0.074	0.24	0.13	0.092	0.074	7
Insulating board lath (½ in.) plastered¶	0.22	0.12	0.088	0.072	0.22	0.12	0.090	0.073	0.23	0.12	0.091	0.074	8
Insulating board lath (1 in.) plastered¶	0.16	0.10	0.078	0.064	0.17	0.10	0.079	0.065	0.17	0.10	0.080	0.066	9

* Sheathing and wood strips assumed ⅜ in. thick.
† Figures in columns I, J, K, and L may be used with sufficient accuracy for rigid asbestos shingles on wood sheathing. Layer of slater's felt neglected.
‡ Coefficients corrected for framing on basis of 15 per cent area, 2 in. by 4 in. (nominal), 16 in. on centers.
§ No air space included; all other coefficients based on one air space.
‖ Plaster assumed ¾ in. thick.
¶ Plaster assumed ½ in. thick.

TABLE 4-15

Combined Coefficients of Transmission (U) of Unvented Pitched Roofs* and Horizontal Ceilings—Based on Ceiling Area†

Ceiling Coefficients (From Table 4-10)	Type of Roofing and Roof Sheathing						Number
	Wood Shingles on Wood Strips ‡			Asphalt Shingles § or Roll Roofing on Wood Sheathing ‖			
	No Roof Insulation (Rafters Exposed) ($U_r = 0.48$)	½-In. Insulating Board on Underside of Rafters ($U_r = 0.22$)	1-In. Insulating Board on Underside of Rafters ($U_r = 0.16$)	No Roof Insulation (Rafters Exposed) ($U_r = 0.53$)	½-In. Insulating Board on Underside of Rafters ($U_r = 0.23$)	1-In. Insulating Board on Underside of Rafters ($U_r = 0.17$)	
	A	B	C	D	E	F	
0.10	0.085	0.073	0.066	0.087	0.074	0.067	19
0.11	0.092	0.078	0.07	0.094	0.079	0.071	20
0.12	0.099	0.082	0.074	0.10	0.083	0.075	21
0.13	0.11	0.087	0.078	0.11	0.088	0.079	22
0.14	0.11	0.091	0.081	0.11	0.093	0.083	23
0.15	0.12	0.096	0.084	0.12	0.097	0.086	24
0.16	0.13	0.10	0.087	0.13	0.10	0.089	25
0.17	0.13	0.10	0.090	0.13	0.10	0.092	26
0.18	0.14	0.11	0.093	0.14	0.11	0.095	27
0.19	0.14	0.11	0.095	0.15	0.11	0.098	28
0.20	0.15	0.11	0.098	0.15	0.12	0.10	29
0.21	0.15	0.12	0.10	0.16	0.12	0.10	30
0.22	0.16	0.12	0.10	0.17	0.12	0.11	31
0.23	0.16	0.12	0.10	0.17	0.12	0.11	32
0.24	0.17	0.13	0.11	0.18	0.12	0.11	33
0.25	0.17	0.13	0.11	0.18	0.13	0.11	34
0.26	0.18	0.13	0.11	0.19	0.13	0.11	35
0.27	0.18	0.13	0.11	0.19	0.13	0.12	36
0.28	0.19	0.14	0.12	0.19	0.14	0.12	37
0.29	0.19	0.14	0.12	0.20	0.14	0.12	38
0.30	0.20	0.14	0.12	0.20	0.14	0.12	39
0.34	0.21	0.15	0.12	0.22	0.15	0.13	40
0.35	0.22	0.15	0.13	0.22	0.15	0.13	41
0.36	0.22	0.15	0.13	0.23	0.15	0.13	42
0.37	0.23	0.15	0.13	0.23	0.16	0.13	43
0.45	0.25	0.17	0.13	0.26	0.17	0.14	44
0.59	0.29	0.18	0.14	0.30	0.19	0.15	45
0.61	0.29	0.18	0.15	0.31	0.19	0.15	46
0.62	0.30	0.19	0.15	0.31	0.19	0.15	47
0.67	0.31	0.19	0.15	0.33	0.20	0.16	48
0.69	0.31	0.19	0.15	0.33	0.20	0.16	49

* Calculations based on 1:3-pitch roof ($n = 1.2$), using the following formula:

$$U = \frac{U_r \times U_c}{U_r + (U_c/n)}$$

where U = combined coefficient to be used with ceiling area; U_r = coefficient of transmission of the roof; U_c = coefficient of transmission of the ceiling; n = the ratio of the area of the roof to the area of the ceiling.

† Use ceiling area (not roof area) with these coefficients.

‡ Based on 1-in. by 4-in. strips spaced 2 in. apart.

§ Coefficients in columns D, E, and F may be used with sufficient accuracy for tile, slate, and rigid asbestos shingles on wood sheathing.

‖ Sheathing assumed ⅔ in. thick.

Heat Transfer and Transmission Coefficients

TABLE 4–16
COEFFICIENTS OF TRANSMISSION (U) OF SOLID WOOD DOORS

Nominal Thickness (in.)	Actual Thickness (in.)	U *† Exposed Door	U *† With Glass Storm Door ‡
1	$\frac{25}{32}$	0.64	0.37
$1\frac{1}{4}$	$1\frac{1}{16}$	0.55	0.34
$1\frac{1}{2}$	$1\frac{5}{16}$	0.49	0.32
$1\frac{3}{4}$	$1\frac{3}{8}$	0.48	0.31
2	$1\frac{5}{8}$	0.43	0.28
$2\frac{1}{2}$	$2\frac{1}{8}$	0.36	0.26
3	$2\frac{5}{8}$	0.31	0.23

* Computed using $k = 1.10$ (for wood); $f_i = 1.46$; $f_o = 6.0$; 1.03 for air space.
† A U value of 0.85 may be used for single exposed doors containing thin wood panels or single panes of glass, and 0.39 for the same with glass storm doors.
‡ Fifty per cent glass and thin wood panels.

We also find from Table 4–3 the following film coefficients for the inside and outside walls, in units of Btu per hr sq ft deg F:

$$f_o = 6.00 \text{ for 15-mph wind}$$
$$f_i = 1.65 \text{ for still air}$$

Therefore, from equation 4–18 or 4–19, the thermal resistance is, in units of hr sq ft deg F per Btu,

$$R = \frac{1}{f_o} + \frac{x_b}{k_b} + \frac{x_c}{k_c} + \frac{x_p}{k_p} + \frac{1}{f_i} = \frac{1}{6.00} + \frac{4.0}{5.0} + \frac{6.0}{12.0} + \frac{0.5}{8.0} + \frac{1}{1.65} = 2.136 \quad Ans.$$

b) The over-all conductivity of the wall is, then, in units of Btu per hr sq ft deg F,

$$U = \frac{1}{R} = \frac{1}{2.136} = 0.468 \quad Ans.$$

c) The heat transferred is, therefore, in units of Btu per hr sq ft,

$$q = UA(t_i - t_o) = (0.468)(1)(76 - 20) = 26.2 \quad Ans.$$

d) If, in part a, $1/f_o$ and $1/f_i$ are disregarded, R becomes 1.362 and the heat transferred is, in units of Btu per hr sq ft

$$q = \frac{1}{R} A(t_i - t_o) = \frac{1}{1.362}(1)(76 - 20) = 41.2 \quad Ans.$$

NOTE. Although the fact is not mentioned in the foregoing example, usually $\frac{1}{4}$ in. to $\frac{1}{2}$ in. of bonding plaster would be used between masonry divisions, as here between the brick and concrete, or at least an air space would exist which would also contribute to the total thermal resistance.

Example 4–2. An inside room of a house faces on an enclosed porch. The room wall adjacent to the porch is 20 ft long by 8 ft high and contains one glazed door 3 ft by 7 ft. Glass forms 80 per cent of the total area of the door, which is built of $1\frac{1}{2}$-in. wood. The wall is constructed of 2-in. by 4-in. studding, with $\frac{3}{4}$ in. of plaster and wood lath on each side. There is an air space between the studding. At a certain time the temperature on the enclosed porch is 40 F and the inside room is 72 F. Find the heat loss from the room to the porch. [*Text continued on page 166.*]

TABLE 4-17

Coefficients of Transmission (U) of Windows, Skylights, and Glass-Block Walls

Section A—Vertical Glass Sheets

Number of Sheets	One	Two			Three		
Air space (in.)	None	¼	½	1*	¼	½	1*
Outdoor exposure	1.13	0.61	0.55	0.53	0.41	0.36	0.34
Indoor partition	0.75	0.50	0.46	0.45	0.38	0.33	0.32

Section B—Horizontal Glass Sheets (Heat Flow Up)

Number of Sheets	One	Two		
Air space (in.)	None	¼	½	1*
Outdoor exposure	1.40	0.70	0.66	0.63
Indoor partition	0.96	0.59	0.56	0.56

Section C—Walls of Hollow Glass Block

Description	U	
	Outdoor Exposure	Indoor Partition
5¾ by 5¾ by 3⅞ in. thick	0.60	0.46
7¾ by 7¾ by 3⅞ in. thick	0.56	0.44
11¾ by 11¾ by 3⅞ in. thick	0.52	0.40
7¾ by 3¾ by 3⅞ in. thick with glass fiber screen dividing the cavity	0.48	0.38
11¾ by 11¾ by 3⅞ in. thick with glass fiber screen dividing the cavity	0.44	0.36

Section D—Approximate Application Factors for Windows
(Multiply Flat-Glass U Values by These Factors)

Window Description	Single Glass		Double Glass †		Windows with Storm Sash ‡	
	Per Cent § Glass	Factor	Per Cent § Glass	Factor	Per Cent § Glass	Factor
Sheets	100	1.00	100	1.00	......	....
Wood sash	80	0.90	80	0.95	80	0.90
	60	0.80	60	0.85	60	0.80
Metal sash	80	1.00	80	1.20	80	1.00
Aluminum	80	1.10	80	1.30	80	1.10‖

* For 1 in. or greater.

† Unit-type double glazing (two lights or panes in same opening).

‡ Use with U values for two sheets with 1 in. air space.

§ Based on area of exposed portion of sash; does not include frame or portions of sash concealed by frame.

‖ For metal storm sash, or metal sash with attached storm pane.

Heat Transfer and Transmission Coefficients

TABLE 4-18

Conversion Table for Wall Coefficient U for Various Wind Velocities

U for 15 MPH*	U for 0 to 30 MPH Wind Velocities					
	0	5	10	20	25	30
0.050	0.049	0.050	0.050	0.050	0.050	0.050
0.060	0.059	0.059	0.060	0.060	0.060	0.060
0.070	0.068	0.069	0.070	0.070	0.070	0.070
0.080	0.078	0.079	0.080	0.080	0.080	0.080
0.090	0.087	0.089	0.090	0.090	0.091	0.091
0.100	0.096	0.099	0.100	0.100	0.101	0.101
0.110	0.105	0.108	0.109	0.110	0.111	0.111
0.130	0.123	0.127	0.129	0.131	0.131	0.131
0.150	0.141	0.147	0.149	0.151	0.151	0.152
0.170	0.158	0.166	0.169	0.171	0.172	0.172
0.190	0.175	0.184	0.188	0.191	0.192	0.193
0.210	0.192	0.203	0.208	0.212	0.213	0.213
0.230	0.209	0.222	0.227	0.232	0.233	0.234
0.250	0.226	0.241	0.247	0.252	0.253	0.254
0.270	0.241	0.259	0.266	0.273	0.274	0.275
0.290	0.257	0.278	0.286	0.293	0.295	0.296
0.310	0.273	0.296	0.305	0.313	0.315	0.317
0.330	0.288	0.314	0.324	0.333	0.336	0.338
0.350	0.303	0.332	0.344	0.354	0.357	0.359
0.370	0.318	0.350	0.363	0.375	0.378	0.380
0.390	0.333	0.368	0.382	0.395	0.399	0.401
0.410	0.347	0.385	0.402	0.416	0.420	0.422
0.430	0.362	0.403	0.421	0.436	0.441	0.444
0.450	0.376	0.420	0.439	0.457	0.462	0.465
0.500	0.410	0.464	0.487	0.509	0.514	0.518
0.600	0.474	0.548	0.581	0.612	0.620	0.626
0.700	0.535	0.631	0.675	0.716	0.728	0.736
0.800	0.592	0.711	0.766	0.821	0.836	0.847
0.900	0.645	0.789	0.858	0.927	0.946	0.960
1.000	0.695	0.865	0.949	1.034	1.058	1.075
1.100	0.742	0.939	1.039	1.142	1.170	1.192
1.200	0.786	1.010	1.129	1.250	1.285	1.318
1.300	0.828	1.080	1.217	1.359	1.400	1.430

* In first column, U is from previous tables or as calculated for 15-mph wind velocity.

Solution: Coefficients f_i and f_o will be taken as 1.65, since there is no wind. For $\frac{3}{4}$-in. wood lath and plaster, take $C = 2.50$ (from Table 4–3). For the air space between the plaster and lath surfaces, take $a = 1.17$ [from Table 4–2, for a mean temperature of $(72 + 40)/2 = 56$ F and for the maximum width given]. (The fact that the actual air space is about 4 in. wide—greater than any width listed in the table—does not appreciably change the value of a.) For the door, which is single-glazed, $U = 0.75$ for the glass and 0.49 for the wood portion, or 0.40 corrected for no wind (from tables 4–16, 4–17, and 4–18).

For the wall, the resistance is, by equation 4–18,

$$R = \frac{1}{1.65} + \frac{1}{2.50} + \frac{1}{1.17} + \frac{1}{2.50} + \frac{1}{1.65} = 2.865$$

Therefore the overall conductivity of the wall is, in units of Btu per hr sq ft deg F,

$$U = \frac{1}{R} = \frac{1}{2.865} = 0.349$$

(Reading $U = 0.34$ directly from Table 4–8 for partition 3 may be preferable)

The net wall area $= 20 \times 8 - 3 \times 7 = 139$ sq ft. Therefore the heat loss through the wall is, by equation 4–20,

$$q_w = UA(t_i - t_o) = (0.34)(139)(72 - 40) = 155 \text{ Btuh}$$

For the door, the glass area $= 3 \times 7 \times 0.80 = 16.8$ sq ft, and the wood area $= 3 \times 7 \times 0.20 = 4.2$ sq ft. Therefore the heat loss through the door is, by equation 4–20,

$$q_D = (0.75)(16.8)(72 - 40) + (0.40)(4.2)(72 - 40) = 457 \text{ Btuh}$$

The total heat loss from room to porch by heat transfer is therefore

$$1551 + 457 = 2008 \text{ Btuh} \qquad Ans.$$

4–3. COMMENTS ON INSULATING MATERIALS, TABULAR VALUES

In general, the materials used for insulation and those used for construction purposes exhibit similar characteristics. Most nonmetallic materials have a basic structure in which there are numerous cells containing air or other gas. As the temperature increases, the thermal conductivity nearly always increases, and in many cases this increase can be attributed largely to increased molecular activity within the cells of the material. If the cells are extremely small, convection effects are not significant. Carbon black (lamp black), although difficult to use, is a good insulator; the particle size is extremely small and the resulting air cells are infinitesimally small. On the other hand, there appears to be an upper limit of cell size which if exceeded leads to a rapid decrease in insulating effectiveness. For example, carbon, charcoal, or graphite, say in pieces of $\frac{1}{2}$-in. size and packed into a space, are not effective for insulating because of greater gaseous activity in the interstices, with gas actually moving from one space to another.

The insulation problem is also related to the density characteristics of the material. A compressible substance, such as glass wool, if loosely packed, is a better insulator than is the same material if closely packed

Heat Transfer and Transmission Coefficients

(compressed) to a higher density. Thus the generalization can be made that as the density of the material decreases, its insulating effectiveness usually increases. On the other hand, this premise cannot be carried too far, since it is possible to pack a material so loosely that insulation effectiveness is poor.

In general, materials which are moist have higher thermal conductivity than dry materials. For example, soil with associated moistures of (say) 25 to 15 per cent has better thermal conductivity than the same soil when it is dry. It is also of interest to note that if a moist soil exists at a temperature low enough to freeze the entrained moisture, the conductivity exceeds that of nonfrozen moist soil.

Tables 4–4 through 4–18 give either computed or test values for composite building structures. For both the computed and test values, the effect of surface films is considered. In addition to this, attention has been given to parallel heat-flow paths within the structure. For example, in a frame structure with wood studding on 16-in. centers, a representative space would consist partly of a free air space, with or without insulation, and at 16-in. intervals a solid piece of wood. The air space has a different resistance to heat flow than does the wood studding. Consequently, if an accurately representative value is desired for such a wall section, the proportional area associated with each type of heat-flow path should be considered in connection with its respective U factor. In the case of parallel paths for heat flow, it should be realized that the conductances or over-all coefficients of each path in relation to respective areas are additive in determining the composite U value, whereas, with uniform walls involving different types of insulations in series, the respective thermal resistances of the elements constituting the thermal circuit are additive. Little inaccuracy arises if the parallel paths involve materials which do not have widely different characteristics, but it should be recognized that a wall of high thermal resistance will be greatly altered if metal parts used either for support or for fastening purposes pass through the wall.

The wall transmission-coefficient tables in this text make use of parallel-circuit corrections in instances where framing increases the U value, but not where the correction would show a decrease. In connection with framed construction, it should be recognized that lumber, when planed and finished, deviates from rough indicated sizes and that it is customary to use the following values in computations:

Indicated Size	Number of Sides Surfaced	Actual Size (Approx.)
1 in.	2	$\frac{25}{32}$ in.
2 in.	2	$1\frac{5}{8}$ in.
3 in.	2	$2\frac{5}{8}$ in.
4 in.	2	$3\frac{5}{8}$ in.

Example 4-3. A controlled-temperature test room, designed to be held at 110 F, was built in the storage space of an industrial plant where the temperature is held at 50 F in winter. Using material at hand in the plant, the wall of the room was constructed of 1-in. smooth pine boards ($\frac{25}{32}$ in. thick), 4 in. of rigid mineral-wool (rock-wool) insulation blocks, and an inside cover of pine boarding similar to the outside layer. The boards were held in place over the insulation by $\frac{3}{4}$-in. steel through-bolts, in such a way that one bolt was used for each 2 sq ft of wall area. In addition, a lattice of 2-in. by 4-in. studs and cross-ties on 6-ft centers was attached to give the walls rigidity. Disregard the thermal effect of the bracing and compute (a) the U factor of a section of the wall, not considering the presence of the steel bolts, and (b) the same, but considering the effect of the steel bolts. Consider film surface factors to apply at the outer surfaces.

Solution: (a) Reference to Table 4-3 shows that the value of f_i on inside walls is 1.65, and that k for mineral-wool block at 90 F is 0.32, and 0.86 for pine wood. Thus the wall resistance is, by equation 4-18,

$$R = \frac{1}{1.65} + \left(\frac{25}{32}\right)\frac{1}{0.86} + \frac{4}{0.32} + \left(\frac{25}{32}\right)\frac{1}{0.86} + \frac{1}{1.65}$$

$$= 0.606 + 0.908 + 12.50 + 0.908 + 0.606 = 15.528 \text{ hr sq ft deg F per Btu}$$

Therefore the U factor is, in units of Btu per hr sq ft deg F,

$$U = \frac{1}{R} = 0.0644 \qquad\qquad Ans.$$

b) The thermal conductivity k for steel is 487 (from Table 4-1), and the resistance of the metallic-bolt circuit through the 5.56 in. in effective length ($\frac{25}{32} + 4 + \frac{25}{32}$) is, allowing for films at each side,

$$R = \frac{1}{1.65} + \frac{5.56}{487} + \frac{1}{1.65} = 1.213$$

The over-all coefficient of heat transfer is therefore

$$U = \frac{1}{R} = \frac{1}{1.213} = 0.824 \text{ Btu per hr sq ft deg F}$$

The bolt area is $(0.75)^2(\pi/4)(1/144) = 0.00307$ sq ft, so that the total conductance through each bolt is

$$C_m = UA_m$$
$$= (0.824)(0.00307) = 0.0025 \text{ Btu per hr deg F}$$

Since one bolt is associated with each 2 sq ft of wall surface, the total conductance of the nonmetal basic wall section of 2 sq ft gross area is

$$C_w = UA_w = (0.0644)(2 - 0.00307) = 0.1286 \text{ Btu per hr deg F}$$

The total conductance of each 2-sq-ft section is thus

$$0.1286 + 0.0025 = 0.1311 \text{ Btu per hr deg F}$$

Thus the composite-wall U value is

$$0.1311 \div 2 = 0.0655 \text{ Btu per hr sq ft deg F} \qquad Ans.$$

and the composite wall resistance = $1/0.0655 = 15.26$. Thus if the metal bolts are disregarded because of their relatively small area, the error would be

$$\frac{15.528 - 15.26}{15.26} \times 100 = 1.76 \text{ per cent}$$

This error is not great, but it shows that whenever a low-thermal-resistance circuit is used in parallel with a high-thermal-resistance circuit an error can be created if both circuits are not considered. In particular, metal ties in insulating material can seriously reduce insulation effectiveness.

Basements of houses often rest directly on the ground, and therefore, when the basement space is heated, heat is lost into the ground. At usual basement depths, the ground temperature does not follow the swings of atmospheric temperature variations but remains essentially constant, at around 50 F. It is difficult to compute accurately the heat loss through basement floors, but a factor of 2.0 Btu per hr sq ft of basement surface can be employed to estimate the loss from the basement floor into the earth when approximately a 20-deg temperature difference exists between the ground temperature and a spot 6 in. above floor level. This corresponds to an over-all U coefficient for basement floors of 0.10 Btu per hr sq ft deg F. For basements which are very shallow, or for slabs which rest directly on the ground with a house above them, the edge ground temperature closely follows atmospheric variations in temperature and it is therefore desirable to compute the loss on the basis of the *perimeter*. In this connection a factor of 0.81 Btu per hr deg F per linear foot of floor edge has been found to give reasonably good results when used in connection with the temperature difference existing between the inside and outside of the building.

4-4. WALL-SURFACE TEMPERATURES

The temperature on the inside-wall surface or ceiling surface of a building cannot be considered to be the same as the temperature inside the building. This surface temperature depends on the convection (film) conditions in the building, the insulating ability of the wall, and the outside conditions of temperature and wind. If the surface temperature is lower than the inside dew-point temperature, moisture will condense. A wet wall or ceiling results, and the moisture may cause serious damage to plaster and woodwork, besides constituting a nuisance. In winter, if the insulation effectiveness of the wall cannot be increased, this moisture formation necessitates lowering the inside relative humidity, or decreasing the inside film resistance by increasing the air circulation over the inside surfaces. Even if the dew problem does not exist, a wall of inadequate insulating capacity may chill the occupants in winter, by radiation to the cold wall, and cause discomfort in summer, by permitting too much heat inflow.

The calculation of inside surface temperature for a given wall at certain inside and outside temperatures can most easily be made by making use of the ratio of the surface film resistance to that of the whole wall. The following example will show the method.

Example 4-4. A building wall consists of 10-in. concrete with $\frac{3}{4}$-in. plaster on metal lath on the inside surface. (a) With a 15-mph wind outside at 0 F, what is the temperature on the inside-wall surface when the room is held at 74 F db and 62 F wb? (b) Will moisture condense on this wall?

Solution: (a) From Table 4–3, $f_o = 6.00$ for 15-mph wind and $f_i = 1.65$ for inside still conditions, and k for concrete = 12.00 Btu in. per hr sq ft deg F. For $\frac{3}{4}$-in. metal lath and plaster, $C = 4.40$ Btu per hr sq ft deg F.

Using equation 4–19, the thermal resistance to heat flow is

$$R = \frac{1}{f_i} + \frac{x_1}{k_1} + \frac{1}{C} + \frac{1}{f_o}$$

or
$$R = \frac{1}{1.65} + \frac{10}{12} + \frac{1}{4.40} + \frac{1}{6.0} = 1.833 \text{ hr sq ft deg F per Btu}$$

The total temperature drop of 74 deg (74 F − 0 F) is used in sending heat through this resistance of 1.833. The temperature drop through the inside film resistance of $1/1.65 = 0.606$ can then be found by the proportion

$$\frac{R_{\text{film}}}{R_{\text{total}}} = \frac{\Delta t_{\text{film}}}{\Delta t_{\text{total}}} \tag{4-23}$$

Substituting,

$$\frac{0.606}{1.833} = \frac{\Delta t_{\text{film}}}{74}$$

Computing, the temperature drop in the film = $\Delta t_{\text{film}} = 24.4$ deg. Therefore the inside-wall temperature = 74 − 24.4 = 49.6 F. *Ans.*

b) Moisture will condense if the inside dew-point temperature is higher than the inside-wall temperature. For $t_d = 74$ F and $t_w = 62$ F, the psychrometric chart (Plate I) shows a dew point of 54.7 F. Thus moisture will form on the 49.6 F wall. *Ans.*

Example 4–5. How many layers of typical insulating fiberboard $\frac{1}{2}$ in. thick with a thermal conductivity k of 0.33 must be fastened to the inside-wall surface in example 4–4 to prevent moisture deposition? Temperatures are assumed the same.

Solution: Since the dew-point temperature is 54.7 F, the allowable temperature drop in the film cannot exceed 74 F − 54.7 F, or 19.3 deg. The film resistance will not appreciably change from its value of 0.606, but the total wall resistance must change. By equation 4–23,

$$\frac{R_{\text{film}}}{R_{\text{total}}} = \frac{\Delta t_{\text{film}}}{\Delta t_{\text{total}}}$$

or
$$\frac{0.606}{R_{\text{total}}} = \frac{19.3}{74}$$

Therefore $R_{\text{total}} = 2.32$, the minimum total wall resistance needed to prevent condensation, and $2.32 - 1.833 = 0.487$, the additional resistance needed over that for the original wall.

For fiberboard, $R = 1/k = 1/0.33 = 3.03$, the resistance per inch of thickness, and $3.03/2 = 1.515$, the resistance per $\frac{1}{2}$ in. of thickness, or for one board.

As only 0.487 additional resistance is needed, one board at $R = 1.515$ is adequate to prevent condensation. *Ans.*

In winter, condensation of humidity from the inside air on windows often presents a serious problem. The use of double-glass windows or storm windows probably furnishes the best means of alleviating such a nuisance. Lowering inside humidity to reduce condensation can seldom be recommended, since the relative humidity maintained in most homes and

Heat Transfer and Transmission Coefficients

many public buildings is in general too low. That more trouble is not experienced from moisture condensation in winter often corroborates the fact that the inside humidity is too low. In some places where high humidity is maintained it is customary to mount drip pans on the lower window sash to catch the condensation if it should become excessive. Data on transmission coefficients of glass are given in Table 4-17. A great increase in thermal resistance occurs when double glazing (two glass sheets) is used instead of single glazing. For example, with two glass sheets, one or more inches apart, the U factor reduces from 1.13 for single glass to 0.53 for double glass.

4-5. INSIDE AIR TEMPERATURES

The temperatures that should be maintained in the rooms of a building during the heating season vary over a wide range, the exact temperatures depending primarily upon the use for which the rooms are planned. Table 5-1 gives temperatures considered representative of good practice for winter. These temperatures are dry-bulb temperatures and do not necessarily imply comfort conditions, since comfort is also influenced by relative humidity, air motion, activity of the occupants, and radiation effects. For spaces which are heated by radiation methods, slightly lower dry-bulb temperatures may be acceptable.

When inside temperatures for a given type of installation are selected for design purposes the temperature at the *breathing line* of 5 feet (or less frequently at the 30-inch *comfort level*) is considered. However, in making calculations for heat-transfer losses, the breathing-line temperature may be far from the average or mean room temperature because of the tendency for air to stratify as the warm air rises to the top. For ceilings not over 20 ft high it has been found that the temperature rises approximately 2 per cent for each foot of height above the breathing line.

Example 4-6. Consider a 20-ft room in which the temperature is 70 F at the 5-ft line and find the temperature at the ceiling (15 ft above the breathing line) and at the floor (5 ft below the breathing line).

Solution: The ceiling temperature = [1.00 + (0.02)(15)]70 = 91 F, and the floor temperature = [1.00 − (0.02)(5)]70 = 63 F. The average room temperature is thus taken as (91 + 63)/2 = 77 F for heat-transfer calculations. *Ans.*

For calculation of the average room temperature (t_{avg}) in a moderately high room (height H feet), the following formula is applicable when the temperature (t_b) is given at the 5-ft breathing line and, with less accuracy, at the 30-in. comfort line:

$$t_{\text{avg}} = t_b\left[1.0 + 0.02\left(\frac{H}{2} - 5\right)\right]\deg\text{F} \qquad (4\text{-}24)$$

For the previously worked case (example 4-6) this formula becomes

$$t_{avg} = 70\left[1.0 + 0.02\left(\frac{20}{2} - 5\right)\right] = 77 \text{ F (as before)}$$

The foregoing calculation applies closely when a room is heated by direct radiation. With certain types of heating systems having positive (fan) air circulation, as when the air is distributed into the space in such a manner as to oppose the tendency of warm air to rise, the temperature difference between floor and ceiling may be quite small and the foregoing rule should not be followed. For this condition a temperature-rise factor of 1 per cent per foot of height above the breathing-line temperature should be used. Where ceiling heights exceed 15 ft, allow 1 per cent per foot of height up to 15 ft, and then allow 0.1 deg for each foot of height in excess of 15 ft. Very often a temperature at the floor 5 deg less than at the breathing line is assumed.

Inside design temperatures for summer are considered in a later chapter. These should be set slightly lower than outside temperatures but not so far below as to cause temperature shock to an occupant entering the conditioned space.

4-6. DESIGN TEMPERATURES FOR UNHEATED INSIDE SPACES

If there is an unheated or uncooled room adjacent to the room for which the heat transfer is being computed, some intermediate temperature (higher than that out of doors but lower than that of the room when heating, and lower than that out of doors but higher than that of the room when cooling) must be assumed.

Reasonable accuracy for ordinary rooms may be attained if the following rules are used in determining the design temperatures:

1. *Cooling with Unconditioned Room Adjacent.* Select for computation a temperature equal to $t_i + (t_o - t_i) \times 0.667$. In other words, add to the room temperature two-thirds of the difference between the indoor and outdoor temperatures.

2. *Adjacent Room Having Unusual Heat Sources.* Here, as would be the case with (say) a kitchen or boiler room, add 10 to 20 deg to the usually taken outside cooling-design temperature in computing the heat gain.

3. *Heating Season, with Adjacent Room Unheated.* Take $\Delta t = (t_i - t_o) \times 0.5$. That is, use one-half the temperature difference between inside and outside in computing the heat loss through the wall to the adjacent room.

Example 4-7. The outside temperature is 95 F, and that of a conditioned room is 80 F, and an unconditioned room is adjacent to the conditioned room. Find the temperature to be used for the inside room in computing the heat gain.

Heat Transfer and Transmission Coefficients

Solution: By rule 1 preceding, the inside temperature for computation = $80 + (95 - 80) \times 0.667 = 90$ F. *Ans.*

Example 4-8. Assume that in example 4-7 the adjacent room has a kitchen range. Find the temperature difference to be used for computation of the heat gain.

Solution: By rule 2 preceding, add say 15 deg to the outside temperature to get the temperature of the adjacent room: $95 + 15 = 110$ F. Then Δt for heat gain = $110 - 80 = 30$ deg.

Example 4-9. In heating, the outdoor temperature is 0 F and that of a heated room is 70 F, and an unheated room is adjacent to the heated room. Find the temperature difference to be used for finding the heat loss from the heated room.

Solution: By rule 3 preceding, $\Delta t = (70 - 0) \times 0.5 = 35$ deg. *Ans.*

4. *Ground Floors.* For floors directly on the ground, 50 F to 55 F may be assumed for the ground temperature, although some people prefer merely to assume an arbitrary temperature difference existing between the inside air and the ground. See also section 4-7.

5. *Attics.* In certain cases it is possible to compute the temperature of an adjacent space from the conductivity characteristics of the building walls and partitions. This computation is easily possible in the case of spaces which are rather simply connected to the heated (or cooled) space, as in the case of an attic adjacent only to the ceiling of the top story. The heat passing through the ceiling to the attic equals the heat passing to the outside. In equational form this becomes

$$U_r A_r (t - t_o) = U_c A_c (t_i - t) = U_c \frac{A_r}{n}(t_i - t)$$

$$t = \frac{U_c t_i + n U_r t_o}{U_c + n U_r} \quad \text{deg F} \quad (4\text{-}25)$$

where t = attic air temperature, in degrees Fahrenheit;
t_i = inside air temperature, in degrees Fahrenheit;
t_o = outside air temperature, in degrees Fahrenheit;
U_r = coefficient of heat transfer of roof;
U_c = coefficient of heat transfer of ceiling;
n = ratio of roof area A_r to ceiling area A_c.

If it is desired to find an over-all heat-transfer coefficient U for a ceiling, attic, and roof, the heat transferred through the composite structure is equated to the heat transfer through either the ceiling or the roof. When a value of U, based on the ceiling area, is desired, equate

$$U A_c (t_i - t_o) = U_c A_c (t_i - t)$$

then substitute for t from equation 4-25, which leads to

$$U = \frac{U_c U_r}{U_r + \dfrac{U_c}{n}} \quad (4\text{-}26)$$

where U = coefficient of heat transfer, in Btu per square foot of ceiling area, per degree difference between inside-room and outside temperatures, when applied to a ceiling, attic, and roof. Other symbols are as for equation 4-25.

Equations 4-25 and 4-26 may not be exact, since convection conditions in the attic affected by air circulation, may change the surface factors normally used in computing the values of U_c and U_r, and also, heat transfer by radiation within the attic has some effect. However, for most cases of closed attics in heating seasons the formulas can be used. If the attic is ventilated in winter it is more accurate to assume the attic temperature as practically that of the outside and to use only the ceiling in the heat-transfer computations.

In the case of summer air conditioning, attics should be well ventilated, since the sun, shining on the roof, will raise the inside air temperature greatly above the outside temperature and thereby impose an additional load on the conditioning system. Equations 4-25 and 4-26 should generally not be used for summer air conditioning.

In the event that none of the previously mentioned methods seems applicable, or where only a rough approximation is desired for a heating calculation, the following values may be used when outside temperatures vary between 0 and 20 F:

>Basements and unheated rooms........................ 32 F
>Vestibules, frequently opened........................... 25 F
>Attic under slate or metal roof, insulating capacity poor .. 25 F
>Attic under a roof of good insulating capacity............ 40 F

4-7. GROUND FLOOR AND BASEMENT SLABS

The heat loss through basement floors and through building-wall surface below ground level is difficult to compute with accuracy, since the ground temperature varies with depth and with the amount of heat flowing into it from the structure above. The ground temperature near the surface varies with the season of the year and the climate; in the United States, however, frost seldom penetrates more than four feet, and below three or four feet the ground temperature undergoes only moderate swings the year round. In fact, it is customary to consider the temperature of ground water as indicative of subsurface earth temperature. Groundwater temperature, even in the most northerly sections of the United States, is seldom below 40 F and in warm sections of the country seldom exceeds 60 F. Using these temperatures as a range, and considering the over-all U of a representative slab as 0.10, Table 4-19 has been constructed.

In one type of building construction which has found frequent use, the structure is placed on a concrete slab laid directly on the ground over a well-drained gravel or ash fill. For heating, in this case, the effect of loss into

Heat Transfer and Transmission Coefficients

TABLE 4-19

BELOW-GRADE HEAT LOSSES FOR BASEMENT WALLS AND FLOORS*

Ground-Water Temperature	Basement-Floor Loss † (Btuh per sq ft)	Below-Grade Wall Loss † (Btuh per sq ft)
40	3.0	6.0
50	2.0	4.0
60	1.0	2.0

* Reprinted, by permission, from *Heating Ventilating Air Conditioning Guide 1955*, Chapter 12.
† Based on basement temperature of 70 F and U of 0.10.

the ground and air near the outside edge of the slab is extremely important, and it is almost essential to provide for edge insulation. It is also desirable to put a waterproof membrane over the gravel subfill to keep water from seeping through the floor by capillary action. This seepage can happen even though the ground under the floor is well-drained.

TABLE 4-20

HEAT LOSS OF CONCRETE FLOORS AT OR NEAR GRADE LEVEL*

OUTDOOR DESIGN TEMPERATURE (F)	HEAT LOSS PER FOOT OF EXPOSED EDGE (BTUH)			
	Recommended 2-In. Edge Insulation	1-In. Edge Insulation	1-In. Edge Insulation	No Edge Insulation†
−20 to −30	50	55	60	75
−10 to −20	45	50	55	65
0 to −10	40	45	50	60

* Reprinted, by permission, from *Heating Ventilating Air Conditioning Guide 1955*, Chapter 12.
† This construction not recommended; shown for comparison only.

Table 4-20 gives representative heat-loss factors for concrete floors placed at or near grade level. This loss is based on the perimeter of the floor edge, but the values are compensated to take care of total loss from the whole slab.

4-8. HEAT-TRANSFER COMPUTATION FOR BUILDINGS

Outside temperatures for building-design purposes are based on the geographical location of the building. Table 5-2 gives winter design temperatures for many places. Since temperatures approaching the lowest ever recorded are usually rare and of short duration, the recommended outside design temperatures in Table 5–2 are higher than the recorded minimum. After selecting a design outside temperature, and the inside temperature from Table 5–1 (usually 70 F) to be maintained in a structure for a given purpose, heat-transfer losses through walls of a given type can readily be obtained by using the heat-transmission values from Tables 4–4 to 4–18. Detailed methods of calculating heating and cooling loads for whole buildings, when infiltration, ventilation, exposure, sun, and internal loads may act together, are considered in chapters 5 and 14.

Example 4–10. The top-story ceiling of a building is 27 ft by 34 ft. An unventilated attic above the ceiling is surmounted by a double-sloping roof which has an area 1.8 times greater than that of the ceiling. The ceiling is made of 2-in. by 4-in. joists ($1\frac{5}{8}$ in. by $3\frac{5}{8}$ in. finished size) on 16-in. centers. The $\frac{3}{4}$-in. wood lath and plaster is attached to the joists, and on top of the joists 1-in. yellow-pine flooring is laid ($\frac{25}{32}$ in. finished size). The roof is made of exposed 2-in. by 4-in. rafters ($1\frac{5}{8}$ in. by $3\frac{5}{8}$ in. actual size) on 2-ft centers, and the rafters are surmounted by 1-in. wood sheathing ($\frac{25}{32}$ in. thick) and covered with slate shingles. The temperature just under the ceiling is 84 F, and the outside temperature is 10 F. Find (a) the coefficients of transmission for the ceiling and the roof, (b) the probable attic temperature, (c) the composite coefficient of heat transfer per square foot of roof area for the ceiling, attic, and roof combination, and (d) the heat loss from the building through the roof.

Solution: (a) Ceiling: No wind or forced circulation. Take the value of U as equal to 0.28 (from Table 4–10) for such a ceiling. Roof: Wind on one side, still air on other. Take the value of U as equal to 0.55 (from Table 4–14) for such a roof. *Ans.*

b) Use equation 4–25 to find the probable attic temperature:

$$t = \frac{U_c t_i + n U_r t_o}{U_c + n U_r} = \frac{(0.28)(84) + (1.8)(0.55)(10)}{0.28 + 1.8(0.55)} = 26.3 \text{ F} \qquad Ans.$$

c) Use equation 4–26 to find U. Thus, in Btu per square foot of roof surface per degree Fahrenheit temperature difference between inside and outside,

$$U = \frac{U_c U_r}{U_c + n U_r} = \frac{(0.28)(0.55)}{0.28 + 1.8(0.55)} = 0.12 \qquad Ans.$$

d) Using in equation 4–20 the value of U from (c), $Q = U A_r (t_i - t_o) = 0.12(1.8 \times 27 \times 34)(84 - 10) = 14,800$ Btuh, the heat loss through the top-story ceiling.

This result can also be found by using U_c for the ceiling, along with the ceiling area and the difference between the room and attic temperatures. Thus, the heat loss through the top-story ceiling is $Q = U_c A_c (t_i - t_a) = (0.28)(27)(34)(84 - 26.3) = 14,800$ Btuh. *Ans.*

Example 4–11. An apartment-house wall 30 ft long and 12 ft high faces on a vestibule hall which is open to the outside and in which the temperature is 25 F when the inside temperature is 70 F at the breathing line. The wall is made of $\frac{3}{4}$-in. metal lath and plaster on each side of 2-in. by 4-in. studding ($1\frac{5}{8}$ in. by $3\frac{5}{8}$ in. actual size) on 18-in. centers. One side of the space between studs is faced with bright aluminum foil. Find (a) the over-all coefficient of the wall, (b) the mean room temperature, and (c) the heat loss through the wall.

Solution: (a) Since this wall is not in Table 4–8, compute U. Use 1.65 for both f_i and f_o, as there is no wind (Table 4–3). For $\frac{3}{4}$-in. metal lath and plaster, $C_p = 4.40$

Heat Transfer and Transmission Coefficients

(Table 4-3). For wood, use $k = 1.0$ (Table 4-3). For air space faced one side with bright aluminum foil, use $C_a = 0.46$ (Table 4-3).

Then the wall resistance between studding is, by equation 4-19,

$$R_{wa} = \frac{1}{f_i} + \frac{1}{C_p} + \frac{1}{C_a} + \frac{1}{C_p} + \frac{1}{f_o}$$

$$= \frac{1}{1.65} + \frac{1}{4.40} + \frac{1}{0.46} + \frac{1}{4.40} + \frac{1}{1.65} = 3.84$$

The resistance at the studding section is different and could be calculated and prorated for the space involved, but when the stud space is relatively small this procedure is often dispensed with and the resistance of the air space is used alone. Thus the overall coefficient of the wall, in units of Btu per sq ft deg F hr, is

$$U_w = \frac{1}{R_w} = \frac{1}{3.84} = 0.26 \qquad Ans.$$

b) The mean room temperature is, by equation 4-24,

$$t_{\text{avg}} = t_b\left[1.0 + 0.02\left(\frac{H}{2} - 5\right)\right]$$

$$= 70[1.0 + 0.02(\tfrac{12}{2} - 5)] = 71.4 \text{ F} \qquad Ans.$$

c) The heat loss through the wall is, by equation 4-2,

$$Q = U_w A(t - t_o) = (0.26)(30 \times 12)(71.4 - 25) = 4350 \text{ Btuh} \qquad Ans.$$

4-9. HEAT TRANSFER THROUGH PIPES, PIPE COVERINGS

In the case of pipe lagging, and similar annular coverings, the cross-section of the path through which heat must flow varies in proportion to the linear distance through the section.

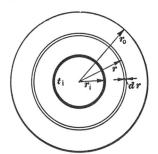

Fig. 4-2. Section through cylindrical pipe covered with lagging.

Referring to Fig. 4-2, consider the heat flow through a section of lagging, of unit length along the axis of a pipe, situated at radius r from the center of the pipe. For this case Fourier's equation (eq 4-2), which is

$$q = -kA\frac{dt}{dx}$$

becomes

$$q = -k(2\pi r \cdot 1)\frac{dt}{dr} \qquad (4\text{-}27)$$

Rearranging and integrating for the whole insulation, we see that

$$q \int_{r_i}^{r_o} \frac{dr}{r} = -2\pi k \cdot 1 \int_{t_i}^{t_o} dt$$

and

$$q \log_e \frac{r_o}{r_i} = -2\pi k(t_o - t_i)$$

The heat transferred per unit length of pipe is

$$q = \frac{2\pi k(t_i - t_o)}{\log_e (r_o/r_i)} = \frac{2\pi k(t_i - t_o)}{2.3 \log_{10} (r_o/r_i)} \quad \text{Btuh} \quad (4\text{-}28)$$

For a length L, the heat transferred becomes

$$q_L = \frac{2\pi k L(t_i - t_o)}{2.3 \log_{10} (r_o/r_i)} \quad \text{Btuh} \quad (4\text{-}29)$$

where q = Btuh transferred through the lagging;
t_i, t_o = temperatures on each side of the lagging, in degrees Fahrenheit;
L = length of the lagging, measured along the axis of the pipe, preferably in feet;
k = specific conductivity. If L is in feet, express k in units of

$$\frac{(\text{Btu})(\text{ft})}{(\text{hr})(\text{sq ft})(\text{deg F})} = \frac{1}{12} \times \frac{(\text{Btu})(\text{in.})}{(\text{hr})(\text{sq ft})(\text{deg F})}$$

r_i, r_o = radii to innermost and outermost section of lagging, in any consistent length units (inches, feet, etc.).

To find the average area of a pipe lagging, of variable cross section in the direction of heat flow, that could be accurately used in an equation of the same form as equation 4–7, equate the two expressions 4–7 and 4–28 for q:

$$k \frac{A_{\text{avg}}}{x} (t_i - t_o) = \frac{2\pi k(t_i - t_o)}{2.3 \log_{10} (r_o/r_i)}$$

$$A_{\text{avg}} = \frac{2\pi x}{2.3 \log_{10} (r_o/r_i)} = \frac{2\pi(r_o - r_i)}{2.3 \log_{10} (r_o/r_i)} \quad (4\text{-}30)$$

where area A_{avg} per unit length of pipe is either in square feet or in square inches, depending on whether r_o and r_i are in feet or inches.

This equation can also be expressed, in square feet or square inches per unit length of pipe, as

$$A_{\text{avg}} = \frac{A_o - A_i}{2.3 \log_{10} (A_o/A_i)} \quad (4\text{-}31)$$

In equations 4–30 and 4–31, A_{avg} is that average area per unit length which can be used for the lagging, as though this area were of constant cross section in the direction of heat flow. Where thin-wall tubing or insulation

Heat Transfer and Transmission Coefficients

exists, there is little difference between an arithmetic average of outside and inside area of the tubing compared with the average area as found by the accurate relation—equation 4–30 or 4–31.

Example 4–12. A 6-in. steel pipe (OD = 6.625 in.) carries saturated steam at 100 psig (337.9 F), and the surrounding air is at 80 F. Under these conditions, when the pipe was covered with 2 in. of insulation of the so-called 85 per cent magnesia type, for which $k = 0.5$ Btu in. per hr sq ft deg F, the temperature at the outer surface was found to be 100 F. Assuming that the outer surface of the steel pipe is at the same temperature as the steam, an assumption that is approximately correct, find the heat loss per hour through the insulation, per 100 ft of pipe.

Solution: By equation 4–30, the average area per unit length is expressed by the relation

$$A_{avg} = \frac{2\pi(r_o - r_i)}{2.3 \log_{10}(r_o/r_i)}$$

Therefore, since

$$r_i = \frac{6.625}{2} \text{ in., or } \frac{3.312}{12} \text{ ft}$$

and

$$r_o = \left(\frac{6.625}{2} + 2\right) \text{ in., or } \frac{5.312}{12} \text{ ft}$$

it follows that the average area per foot of pipe length is

$$A_{avg} = \frac{2\pi[(5.312 - 3.312)/12]}{2.3 \log_{10}(5.312/3.312)} = \frac{2\pi(2/12)}{2.3(0.2052)} = 2.219 \text{ sq ft}$$

Therefore the heat lost per 1-ft length of pipe is

$$q = k\frac{A_{avg}}{x}(t_i - t_o) = (0.5)\left(\frac{2.219}{2}\right)(337.9 - 100) = 131.97 \text{ Btuh}$$

And for 100 ft of length, the heat loss is

$$q_{100} = (100)(q) = 13{,}200 \text{ Btuh} \qquad Ans.$$

Alternate Solution: An alternate solution employs equation 4–29:

$$q_L = \frac{2\pi k(L)(t_i - t_o)}{2.3 \log_{10}(r_o/r_i)}$$

It is necessary to change the units of k from

$$k = 0.5 \frac{(\text{Btu})(\text{in.})}{(\text{hr})(\text{sq ft})(\text{deg F})} \quad \text{to} \quad \frac{0.5}{12} \frac{(\text{Btu})(\text{ft})}{(\text{hr})(\text{sq ft})(\text{deg F})}$$

Then the heat loss, for the length given, is

$$q_L = \frac{2\pi(0.5/12)(100)(337.9 - 100)}{2.3 \log_{10}(5.312/3.312)} = 13{,}200 \text{ Btuh} \qquad Ans.$$

Heat loss from the outside of a hot bare pipe, or even from one with some insulation, is affected not only by convection conditions around the pipe but also by radiation from the hot pipe to the cooler surroundings. To make possible the use of ordinary conduction equations for pipe, when radiation loss from the surface of the pipe is also appreciable, the so-called film factor is sometimes determined experimentally to include the effect of both convection and radiation and is called $f_c + f_r$. For this condition the factors are dependent on convection conditions and temperature difference, and also on size of pipe. Some of these values, based on the experimental work of Heilman, are found in Fig. 4-3. Although the surface

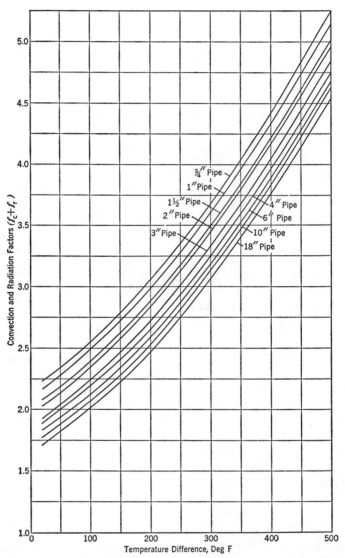

Fig. 4-3. Values of $(f_c + f_r)$ for heat loss from horizontal bare-iron pipes.

factors in Fig. 4-3 are based on bare pipe they can also be used with sufficient accuracy for the surface of insulation if the proper temperature difference between the surface of the insulation and an ambient air tem-

Heat Transfer and Transmission Coefficients 181

perature is used in connection with the extreme outside diameter of the insulation.

Example 4–13. A bare 3-in. standard pipe in a building at 70 F carries saturated steam at 100 psig (337.9 F). Find (a) the heat loss per 100 ft of pipe and (b) the weight of steam condensed as a result of this heat loss.

Solution: (a) Assuming that the temperature of the pipe surface is that of the steam (closely true), the temperature difference existing is 337.9 − 70 = 267.9 deg. Interpolating for 267.9 deg in Fig. 4–3 gives $(f_c + f_r)$ = 3.12 Btu per hr per square foot of outside pipe surface per degree temperature difference.

For 3-in. pipe the OD = 3.5 in. = 3.5/12 ft. The area of outside surface per foot of length is therefore

$$\pi \times \frac{3.5}{12} \times 1 = 0.917 \text{ sq ft}$$

The heat loss from the surface of the pipe is found as follows:

$q = (f_c + f_r)(A)(t_i - t_o)$
$q = (3.12)(0.917)(337.9 - 70) = 765.6$ Btuh per foot of length
$q_L = (100)(765.6) = 76{,}560$ Btuh per 100 ft of length   *Ans.*

b) The latent heat of steam at 100 psig (100 + 14.7 psia) is 880.7 Btu per lb (from Table 2–3). Therefore

$$\text{Steam condensed} = \frac{76{,}560}{880.7} = 86.9 \text{ lb per hr} \qquad Ans.$$

For the case of two or more different kinds of insulation in series on a pipe, use of the resistance concept of heat transfer through insulation leads to an equation of the same form as equation 4–12, but requires the use of the logarithmic mean areas of each respective insulation. Expressed in equational form,

$$q = \frac{t_i - t_o}{\dfrac{1}{f_o A_o} + \dfrac{r_o - r_1}{k_1 (A_{\text{avg}})_1} + \dfrac{r_1 - r_2}{k_2 (A_{\text{avg}})_2} + \cdots + \dfrac{1}{f_i A_i}}$$

which becomes

$$q = \frac{t_i - t_o}{\dfrac{1}{f_o 2\pi r_o} + \dfrac{r_o - r_1}{k_1 \dfrac{2\pi(r_o - r_1)}{2.3 \log_{10}(r_o/r_1)}} + \dfrac{r_1 - r_2}{k_2 \dfrac{2\pi(r_1 - r_2)}{2.3 \log_{10}(r_1/r_2)}} + \cdots + \dfrac{1}{f_i 2\pi r_i}} \qquad (4\text{--}32)$$

$$q = \frac{2\pi(t_i - t_o)}{\dfrac{1}{(f_c + f_r) r_o} + \dfrac{2.3 \log_{10}(r_o/r_1)}{k_1} + \dfrac{2.3 \log_{10}(r_1/r_2)}{k_2} + \cdots + \dfrac{1}{f_i r_i}} \qquad (4\text{--}33)$$

where q = Btuh transferred through a unit length (1 ft) of pipe having multiple insulations, per degree temperature difference between inside and outside;

$f_c + f_r$ = composite convection and radiation film factor (Fig. 4–3), in units of Btuh sq ft deg F;

k_1, k_2, etc. = conductivity of respective kinds of insulation wrapped on pipe, in units of

$$\frac{(Btu)(ft)}{(hr)(sq\ ft)(deg\ F)} = \frac{1}{12} \times \frac{(Btu)(in.)}{(hr)(sq\ ft)(deg\ F)};$$

r_o, r_i = outside and inside radii of extreme insulations, in feet;
r_1, r_2, r_3 = respective radii of individual sections of insulation;
t_o = outside air temperature, degrees Fahrenheit;
t_i = inside temperature under innermost insulation (in the case of effective insulation on pipes carrying saturated vapors, can be taken as the liquid or vapor temperature).

Example 4-14. Assume that the 3-in. pipe (3.5 OD) of example 4-13 is covered with a 1-in. layer of 85 per cent magnesia insulation, which has a value of k of 0.5 Btu in. per hr sq ft deg F. Find (a) the heat loss per 100 ft of pipe and (b) the weight of steam condensed as a result of the loss. (c) Compare results with those for bare pipe.

Solution: (a) Equation 4-33 will be used, and $f_c + f_r$ must be selected for an estimated temperature difference between the outside of the insulation and the 70 F air. Estimate 50 deg difference, giving a value of about 1.97 (from Fig. 4-3) for an outside diameter of $(3.5 + 1 + 1)$, or 5.5 in. Only one logarithmic area is involved, that for the 1-in. insulation of radii:

$$\left(\frac{3.5}{2} + 1\right) \text{in.} = \frac{2.75}{12} \text{ft} = r_o \quad \text{and} \quad \frac{3.5}{2} \text{in.} = \frac{1.75}{12} \text{ft} = r_i$$

The inside film resistance and the pipe resistance are negligible.

By equation 4-33,

$$q = \frac{2\pi(t_i - t_o)}{\dfrac{1}{(f_c + f_r)r_o} + \dfrac{2.3 \log_{10}(r_o/r_i)}{k_1}}$$

Substituting, the heat loss per 1-ft length of pipe is

$$q = \frac{2\pi(337.9 - 70)}{\dfrac{1}{1.97(2.75/12)} + \dfrac{2.3 \log_{10}(2.75/1.75)}{0.5/12}} = \frac{2\pi(337.9 - 70)}{2.21 + 10.83}$$

$$= 129.0 \text{ Btuh}$$

The heat loss per 100-ft length is therefore

$$q_{100} = (100)(129.0) = 12,900 \text{ Btuh} \qquad Ans.$$

The total temperature drop through the insulation and film occurs in direct ratio to the thermal resistances, which are proportional to the 2.21 + 10.83 above. The temperature drop through the surface film on the outer pipe insulation is thus $(337.9 - 70) \times (2.21)/(2.21 + 10.83) = 45$ deg. This result is close enough to the 50 deg assumed to make a recalculation unnecessary, inasmuch as the film resistance is such a relatively small part of the total resistance.

b) Latent heat of steam at 114.7 psia = 880.7 Btu per lb (from Table 2-3). Therefore the steam condensed = 12,900/880.7 = 14.6 lb per hr by heat loss from the 100-ft length of pipe. *Ans.*

c) This 14.6 lb of steam condensed per hour in the insulated pipe compares with 86.9 lb condensed (lost) per hour from the uninsulated pipe. *Ans.*

Heat Transfer and Transmission Coefficients 183

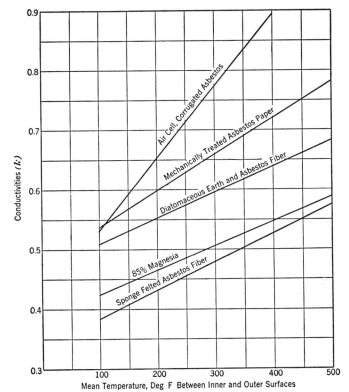

FIG. 4-4. Conductivities per inch of thickness for insulation coverings for medium- and high-temperature pipes.

Figure 4-4 shows conductivities of various types of insulating materials for covering medium- and high-temperature pipes.

4-10. RADIANT-HEAT TRANSFER

The heat transferred by radiant energy is proportional to the difference in the fourth powers of the absolute temperatures of the hot source and the more-cool receiver of the radiation. Expressed as an equation,

$$q_r = 0.172A \left[\left(\frac{T_s}{100}\right)^4 - \left(\frac{T_r}{100}\right)^4 \right] e\, F_a \qquad (4\text{-}34)$$

where q_r = Btu transferred per hour by radiation;
A = area of either the source or receiver, depending on the method of selecting F_a, in square feet;
T_s, T_r = temperatures of source and receiver, $(t_s + 460)$ and $(t_r + 460)$, in degrees Fahrenheit absolute;
e = effective absorptivity or emissivity factor, expressing the degree to which the source and receiver surfaces approach an "ideal

black body" [an ideal black body is one which could absorb (or emit) all of the radiant energy falling on it];

F_a = factor to account for the geometric configuration between the radiating surfaces. This factor must be calculated mathematically, or found from curves or tables in complete heat-transfer treatises, for various geometrical arrangements. For the case of two infinite parallel plates ($F_a = 1$), use

$$e = \frac{1}{(1/e_s) + (1/e_r) - 1}$$

This is also reasonably applicable for the panel surfaces ordinarily found in radiant-heated rooms. For the case of a completely enclosed body, small compared with enclosure ($F_a = 1$), use e for the enclosed body and also use the enclosed body area, etc.

The shape, form, and relative position of radiating bodies greatly affect the amount of radiant heat transfer and sometimes make it impossible to calculate mathematically the factor F_a.

The emissivity or absorptivity of no material is perfect ($e = 1$), although lampblack and certain oil paints may reach values of $e = 0.96$. The metals, particularly those highly polished, have low values: polished aluminum, $e = 0.039$ to 0.057; tinned sheet iron, 0.043 to 0.054; polished cast iron, 0.21; polished sheet steel, 0.55 to 0.60. Most nonmetallic substances, regardless of color, have emissivities in excess of 0.8 and the emissivities increase in value with increasing temperature. Aluminum paint has a rather low emissivity (0.3 to 0.5), but oil paints, regardless of color (including white), usually exceed 0.9. Thus painting of steam radiators or other heat-transfer surfaces has small effect on radiant heat transfer unless an aluminum or other metal suspension paint is used. Oxidized metals usually show values of e greater than 0.8. For plaster, $e = 0.91$; and for asbestos board, e approaches 0.93.

It should be mentioned that emissivity (absorptivity) factors e are different for radiation from high-temperature sources, such as the sun (or incandescent bodies), than they are for radiation from low-temperature sources, such as steam radiators and radiant-heating panels. White clothes, whitewashed surfaces, plaster, and light cream paint, for example, have e values of 0.85 to 0.95 for low-temperature (infrared) radiation, whereas for solar radiation the corresponding e values range from 0.3 to 0.5. Window glass is transparent to solar radiation ($e \equiv 0$), whereas for low-temperature radiation glass is a good absorber ($e = 0.9$ to 0.95) and such radiation is trapped. This phenomenon is used in greenhouses as a means of heating on sunny days. For black nonmetallic surfaces, there is little difference between e values for solar and other radiant sources.

Heat Transfer and Transmission Coefficients

PROBLEMS

4-1. A solid stone wall 12 in. thick is finished on the inside with $\frac{1}{2}$ in. of gypsum plaster. Find the resistance and conductance of the wall both with and without film coefficients being considered. *Ans.* $R = 1.822$ and 1.049

4-2. Find U for a wall that is made up of 10 in. of concrete, 2 in. of corkboard, and $\frac{1}{2}$ in. of plaster. Assume that there is a 15-mph wind on the concrete side of the wall.
Ans. $U = 0.119$

4-3. Determine the insulation heat loss through the four side walls of a building 40 by 60 by 20 ft. The walls are constructed of 4-in. face brick on the outside, and 6 in. of concrete, and with a $1\frac{1}{2}$- to 2-in. air space where the wall is furred for the metal lath and plaster on the inside. The building contains a total of twelve single-glazed windows $3\frac{1}{2}$ ft by 7 ft, and three 4-ft by 7-ft doors made of nominal 2-in. wood. The outside temperature $t_o = 10$ F, and the inside temperature $t_i = 80$ F. Disregard temperature variations in the room itself. What percentage of the insulation heat loss is through the doors, and what percentage is through the windows? *Ans.* Windows (23,260 Btuh), 20.3%; doors (2520 Btuh), 2.2%; net wall (88,700 Btuh), 77.5%

4-4. A building is constructed of 8-in. common-brick walls, finished with gypsum lath and plaster on furring strips (2-in. air space) on the inside surface. If the room is maintained at 70 F and the outside temperature is 20 F, what is the inside wall temperature? *Ans.* 60.9 F

4-5. The outside wall of a building is constructed of 8 in. of common brick with $\frac{1}{2}$ in. of plaster on the inside surface. The inside of the building is maintained at 75 F dry-bulb and 65 F wet-bulb temperature. How many layers of $\frac{1}{2}$-in. corkboard must be added to the wall to prevent moisture from condensing on the inside surface of the wall when the outside temperature is 0 F? *Ans.* One layer

4-6. By using values obtained from Table 4-3, check the value that appears in Table 4-7 for a wall constructed of 4-in. brick veneer, 8-in. hollow tile, and $\frac{3}{4}$-in. plaster on metal lath attached to nominal 2-in. furring strips. The intervening space contains 1-in. insulating board.

4-7. A certain room is 20 ft wide, 30 ft long, and 15 ft high. The temperature at the breathing line (5-ft level) is 70 F. In the room there is a door $3\frac{1}{2}$ ft by 7 ft and two windows 3 ft by 7 ft, with the lower edge of the windows 3 ft from the floor. What is the temperature (a) at the ceiling and (b) at the floor? (c) What average inside temperature might be used in calculating the heat loss through the walls, windows, and doors? *Ans.* (a) 84 F; (b) 63 F; (c) 73.5 F

4-8. Calculate the temperature which should be used during the heating season for estimating heat loss when there is an unheated outside room adjacent to a room which is to be held at 70 F. The outside temperature is 0 F. *Ans.* 35 F

4-9. A ranch-type home is built on a concrete floor which is laid at ground level. Waterproof insulation 1 in. thick is placed at the edge of the slab where this joins the wall and footing of the building. The insulation also extends for 6 in. down along the side of the footing. The house is rectangular in shape, 60 ft by 25 ft. Making use of Table 4-20, compute the loss at the floor slab. Assume that the house is built in a locality where the outside design temperature is -10 F. *Ans.* 7650 Btuh

4-10. The top-floor ceiling of a building 30 ft by 36 ft is constructed of 2-in. by 4-in. joists on 18-in. centers. On the underside there is metal lath and plaster, $\frac{3}{4}$ in. thick. On top of the joists there are only scattered walking planks, but the space between the joists is filled with rock wool. The temperature at the ceiling in the room below is 72 F and the attic temperature is 25 F. Find (a) the coefficient of heat transfer for the ceiling and (b) the heat loss per hour through the ceiling. *Ans.* (a) 0.077; (b) 3900 Btuh

4-11. Calculate the coefficient of heat transfer for a roof constructed of 2-in. by 4-in. rafters on 18-in. centers, with 1-in. wood sheathing and slate shingles.
Ans. 0.55 (disregarding insulating effect of rafters themselves)

4-12. Assume that a building has the ceiling of problem 4–10 and the roof of problem 4–11. (a) What is the composite coefficient of heat transfer per square foot of roof area if the ratio of roof area to ceiling area is 1.8? (b) What is the probable unheated attic temperature if the inside house temperature is 72 F and the outside temperature is 10 F? *Ans.* (a) $U = 0.0396$; (b) $t = 14.5$ F

4-13. Refer to Table 4–15. For a ceiling with a coefficient of $U_c = 0.24$ and for a roof coefficient of $U_r = 0.48$, check the value given in the table for over-all (combined) coefficient. Take n equal to 1.2. Recompute on the basis of $n = 1.8$.

4-14. One side wall of a public building is made of hollow glass blocks each $11\frac{3}{4}$ by $11\frac{3}{4}$ by $3\frac{7}{8}$ in. thick. Compute the heat loss through the 20-ft-long, 10-ft-high wall when the temperature is 70 F inside and 0 F outside. *Ans.* 7280 Btuh

4-15. In front of the glass-block wall of problem 4–14 it is usual at night to draw close-fitting drapes running from the ceiling to the floor. On the basis that the drapes produce the effect of introducing approximately a half-inch air space in additional insulating effectiveness (see Table 4–2), (a) compute the probable over-all U of the glass-block wall with its drapes and (b) find the heat loss.
Ans. (a) $U = 0.37$ (0.52 without drapes); (b) 5190 Btuh

4-16. A heated basement is under a building of 26-ft by 30-ft floor-plan size. The wall height below grade is 4.5 ft. Estimate the heat loss through the basement floor and walls in contact with ground surface when the building is located in a region of 50 F ground-water temperature. *Ans.* 3580 Btuh

4-17. Calculate the heat loss per 24 hr from 100 lin ft of standard 2-in. steel pipe carrying saturated steam at 150 psig (365.8 F). The pipe is covered with 2-in. insulation of the 85 per cent magnesia type. The temperature at the outer edge of the pipe covering is 90 F. (Neglect surface film and the thermal resistance of the pipe itself.)
Ans. 168,500 Btu per 24 hr

4-18. A bare 4-in. pipe carries steam at 150 psig through a room which is maintained at 80 F. Find (a) the heat loss per hour and (b) the weight of steam condensed per hour as a result of this heat loss, for 100 ft of pipe.
Ans. (a) 107,900 Btuh; (b) 125.8 lb per hr per 100 ft

4-19. Calculate the heat loss from 100 ft of 4-in. pipe carrying steam at 100 psig (337.9 F) and covered with 2 in. of 85 per cent magnesia insulation and 1 in. of air-cell corrugated asbestos superimposed. The room temperature is 80 F. *Ans.* 6840 Btuh

4-20. What is the heat loss from 100 ft of $\frac{1}{2}$-in. pipe carrying steam at 100 psig (a) if the pipe is bare and (b) if the pipe is covered with 1-in. insulation of 85 per cent magnesia. The room temperature is 80 F.
Ans. (a) 19,600 Btuh; (b) 4640 Btuh

REFERENCES

1. *Heating Ventilating Air Conditioning Giude*, 1955, Chapter 9.

2. *Air Conditioning Refrigerating Data Book*. [Published annually by American Society of Refrigerating Engineers to 1959.]

3. W. H. McAdams, *Heat Transmission*, 3d ed. (New York: McGraw-Hill Book Company, Inc., 1954).

4. L. B. McMillan, "Heat Transfer Through Insulation in Moderate and High Temperature Fields," *Trans. ASME*, Vol. 48 (1926), pp. 1269–1317.

5. R. H. Heilman, "Heat Losses from Bare and Covered Wrought Iron Pipe," *Trans. ASME*, Vol. 44 (1922), p. 299.

6. W. J. King, "The Basic Laws and Data of Heat Transmission," *Mechanical Engineering*, Vol. 54 (March–August, 1932).

7. *ASHRAE Handbook of Fundamentals*, 1967, Chapter 26.

5

The Heating Load

5-1. HEATING LOAD

The items entering into the heating load in a building or space are:
1. Heat loss through exposed wall area to outside, including that lost through outside walls, roofs, or ceiling to unheated attics, and through the floor to unheated spaces, but not including glass or door area.
2. Heat loss through glass surfaces and doors.
3. Heat required to warm the air entering by infiltration through outside windows and door cracks and through other points of leakage. In complete air-conditioning projects with sealed windows, the ventilation air brought in from outside may constitute this item.
4. Miscellaneous heat requirements, as for humidification of outside air, and safety factors to take care of contingencies.

Item 1. Item 1 largely constitutes the so-called insulation loss for the building. The general methods of computing this loss for various types of building construction were discussed in detail in Chapter 4. For a heat loss of this nature the following basic relationship for heat transfer applies:

$$Q_w = U_w A_w (t_i - t_o) \text{ Btu transmitted per hour} \quad (5\text{-}1)$$

In this equation, A_w is the square feet of net wall area for the space or room under consideration and equals the gross wall area less the area of any glass and doors in the space considered. In calculating gross wall area, use the length (feet) of exposed wall, measured on the inside, and use as height (feet) the distance from floor to floor when the heated space is above (otherwise use height to ceiling). The area of glass and doors, to be subtracted from the gross area, is the sum of the total sash and frame area of all windows and outside doors.

The inside temperature t_i to be used in the heated space should be modified for the height of the space or room. Values of design inside temperature should be selected from Table 5-1 for the particular type of space, although a value of 72 F can be arbitrarily taken. These temperatures are meant to apply at the breathing line, 5 ft above floor level (or at the

The Heating Load

30-in. line). With an unusually high room heated with conventional radiation, the mean wall temperature will be different from the design inside temperature, and the proper value to use can be found with equation 4-24.

The winter design temperatures of living spaces, appearing in Table 5-1, indicate a range of temperatures for each type of space. In selecting a design temperature within one of these ranges, consideration should be given to the probable relative humidity and to the amount of possible exposed cold wall and glass surface in the space. Since relative humidity in heated spaces in winter may drop appreciably below 30 per cent, a higher dry-bulb temperature may be required to give the same degree of comfort that would apply in the 45 to 50 per cent relative-humidity range. This is particularly true during cold spells, when, if proper humidification is not provided, the dry outside air entering by infiltration may reduce relative humidities in an occupied space to values as low as 10 to 25 per cent. The effect of such low humidities is to make the occupant feel as though he were in a space one or more degrees cooler. Also, an occupant in a space containing extensive cold glass area or poorly insulated exposed walls will radiate excessive heat to these areas and he will feel colder than he should in relation to the ambient dry-bulb temperature. Thus, if either of these cases is considered probable, a dry-bulb design temperature near the top of the range, or even above it, should be selected.

TABLE 5-1

INSIDE DRY-BULB TEMPERATURES USUALLY SPECIFIED FOR HEATING SEASON (WINTER)

Type of Space	Temperature Range (F)	Type of Space	Temperature Range (F)
Auditoriums	72–74	Hotel bedrooms and baths	72–76
Ballrooms	68–72	Kitchens and laundries	66
Bathrooms in general	74–80	Paint shops	80
Dining and lunch rooms	72–74	Public buildings	70–74
Factories		School classrooms	72–74
Light work	60–68	Steam baths	110
Heavy work (foundries and shops)	58–68	Stores	70–74
Gymnasiums	60–68	Swimming pools	75
Homes	72–76	Theater lounges	70
Hospitals		Toilets	70
Operating rooms	70–95		
Patients' rooms	74–76		

The design outside temperature t_o should usually be selected from Table 5-2. If the particular locality under consideration cannot be found in the table, use as a guide the United States Weather Bureau records, which give the lowest recorded temperatures for many places. Such minimum temperatures should not be used in design, since their occurrence is rare and usually of short duration; when only such minimum values are available, a figure 10 to 15 deg higher than minimum should be selected as the design temperature. One technique has been to employ an outside design temperature which is equalled or exceeded during 97.5 per cent of the time during the months of December, January, and February.* The values in Table 5-2 are partly based on this premise, but more extensively on what engineers in given communities have found to be satisfactory in making actual heat-load calculations. Wind velocities and their direction are quite variable, so that the values indicated in Table 5-2 should be considered merely indicative.

U_w should be calculated by the methods shown in Chapter 4, using the prevailing wind velocity. If the factor U_w can be found in Tables 4-4 to 4-14, no correction for wind velocity need be made unless the actual velocity is extremely different from the 15-mph velocity on which the tables were computed. Table 4-18 can be of use in case of a different wind velocity.

Item 2. For computing glass and door loss, use a relationship of the form

$$Q_g = U_g A_g (t_i - t_o) \text{ Btuh} \tag{5-2}$$

Select U_g from Table 4-17. The window area is the whole area of all the window sashes and doors in question. For very high windows, or if extreme accuracy is desired, the temperature at the mean height of the window should be used for t_i.

Item 3. The amount of air which enters through cracks and clearances around windows and doors depends mainly on the tightness of the construction and on the wind velocity. Leakage is also increased by any chimney effect the building may exert by virtue of its height and the existing temperature differences inside and outside. The crack length is the perimeter of each window sash, counting only once the crack at the meeting rail of double-hung sliding sash. Table 5-3 gives the infiltration, in cubic feet per hour per foot of crack, for different types of windows and doors, under different wind velocities.

The air entering a building by infiltration leaves at the same rate as that at which it enters. However, in the case of a building with many partitions and of tight construction, air may enter on the windward side in such quantity as to build up a slight positive pressure and thereby reduce infiltration. In general, it may be considered that the air which enters by infiltration on the windward side of a room leaves on the leeward side (or through vertical openings), so that only half of the total feet of crack of the

* During a normal winter not more than 54 of the 2160 hours in these months is less than or equal to the indicated temperature.

The Heating Load

TABLE 5-2
Outdoor Conditions For Cooling and Heating Design*

State and City	SUMMER			WINTER			Degree, Days
	Dry Bulb, °F		Wet Bulb, °F	Dry Bulb, °F		Wind Velocity, mph, and Direction	
	1%	Normal	Normal	97½%	Normal		
Alabama							
Birmingham	97	95	78	22	10	8N	2551
Mobile	95	95	80	29	15	10N	1560
Montgomery	98	95	78	26	10	7NW	2291
Alaska							
Anchorage	73	75	63	−20	−25		10,864
Fairbanks	82	80	64	−50	−50		14,279
Juneau	75	75	66	− 4	−10		9075
Arizona							
Flagstaff	84	90	65	5	−10	7SW	7152
Phoenix	108	105	76	34	25	5E	1765
Tucson	105	105	72	32	25	5NW	1800
Yuma	111	110	78	40	30	7N	974
Arkansas							
Fort Smith	101	95	76	19	10	8E	3292
Little Rock	99	95	78	23	5	10NW	3219
California							
Los Angeles	86	90	70	48	35	6NE	1391
Sacramento	100	100	72	32	30	8SE	2502
San Diego	86	85	68	44	35	5NW	1458
San Francisco	83	85	65	37	35	8N	3015
Colorado							
Colorado Springs	90	90	63	4	−25		6423
Denver	92	95	64	3	−10	8S	6283
Grand Junction	96	95	65	11	−15	5NW	5041
Connecticut							
Bridgeport	90	95	75	8	0		5617
Hartford	90	93	75	5	0	8NW	6235
New Haven	88	95	75	9	0	10N	5897
Delaware							
Wilmington	93	95	78	15	0	NW	4930
Dist. of Columbia							
Washington	94	95	78	19	0	7NW	4224
Florida							
Jacksonville	96	95	78	32	25	9	1239
Miami	92	91	79	47	35	13	214
Pensacola	92	95	78	32	20	11	1463
Tampa	92	95	78	39	30	8	683

*The data from which this table was developed came from many sources of which the most direct are the ASHRAE *Handbook of Fundamentals*, 1967, chapter 22 and the ASHRAE *Guide and Data Book*, applications Volume 1968, chapter 34. These volumes are highly recommended for the much more extensive weather data they obtain. The *Handbook of Air Conditioning Heating and Ventilating* (ref. 8) is of interest for the extensive tabulations of degree-day data and supporting information. The fundamental sources of weather information are found in references 1 through 5. Extensive data on design conditions used in many areas can be found in references 7 and 8.

TABLE 5–2 (*Continued*)

State and City	SUMMER			WINTER			
	Dry Bulb, °F		Wet Bulb, °F	Dry Bulb, °F		Wind Velocity, mph, and Direction	Degree, Days
	1%	Normal	Normal	97½%	Normal		
Georgia							
Atlanta	95	95	76	23	10	11	2961
Augusta	98	98	76	23	10	6	2397
Macon	98	95	78	27	15	7	3326
Savannah	96	95	78	27	20	9	1819
Hawaii							
Honolulu	87	90	75	62			0
Idaho							
Boise	96	95	65	10	−10	9	5809
Lewiston	98	95	65	12	− 5	7	5542
Pocatello	94	95	65	− 2	− 5	10	7033
Illinois							
Chicago	95	95	75	1	−10	11	6282
Moline	94	96	76	− 3	−10	10	6408
Peoria	94	96	76	2	−10	9	6025
Springfield	95	98	77	4	−10	14	5429
Indiana							
Evansville	96	95	78	10	0	10	4435
Fort Wayne	93	95	75	5	−10	11	6205
Indianapolis	93	95	76	4	−10	13	5699
Iowa							
Des Moines	95	95	78	− 3	−15	10NW	6588
Dubuque	92	95	78	− 7	−20	7NW	7376
Sioux City	96	95	78	− 6	−20	12NW	6951
Kansas							
Dodge City	99	95	78	7	−10	15	4986
Topeka	99	100	78	0	−10	10	5182
Wichita	102	100	75	9	−10	12	4620
Kentucky							
Lexington	94	95	78	10	0	13SW	4683
Louisville	96	95	78	12	0	10SW	4660
Louisiana							
Baton Rouge	96	95	80	30	20	8	1560
New Orleans	93	95	80	35	20		1385
Shreveport	99	100	78	26	20	9	2184
Maine							
Eastport		90	70		−10	9	8445
Portland	88	90	73	0	− 5	9	7511
Maryland							
Baltimore	94	95	78	15	0	8NW	4487
Cumberland	94	95	75	9			
Massachusetts							
Boston	91	92	75	10	0	12W	5634
Springfield	91	93	75	2	−10	9SW	

The Heating Load

TABLE 5-2 (Continued)

State and City	SUMMER			WINTER			
	Dry Bulb, °F		Wet Bulb, °F	Dry Bulb, °F		Wind Velocity, mph, and Direction	Degree, Days
	1%	Normal	Normal	97½%	Normal		
Michigan							
Detroit	92	95	75	8	−10	12SW	6232
Flint	89	95	75	3	−10	W	7377
Grand Rapids	91	95	75	6	−10	12NW	6894
Lansing	89	95	75	6	−10	10SW	6909
Marquette	88	93	73	− 4	−10	11NW	8398
Minnesota							
Duluth	85	93	73	−15	−25	13SW	10,000
Minneapolis	92	95	75	−10	−20	11NW	7966
Mississippi							
Jackson	98	95	78	24	15	8SE	2239
Meridian	97	95	79	24	10	6N	2289
Vicksburg	97	95	78	26	10	8SE	2041
Missouri							
Kansas City	100	100	76	8	−10	10NW	4711
St. Louis	98	95	78	8	0	12S	4900
Springfield	97	95	78	10	−10	11SE	4900
Montana							
Billings	94	90	66	− 6	−25	12W	7049
Butte	86	95	67	−16	−20	NW	
Helena	90	95	67	−13	−20	7SW	8129
Nebraska							
Lincoln	100	95	78	0	−10	11S	5864
North Platte	97	95	78	− 2	−10	8W	6684
Omaha	97	95	78	− 1	−10	10NW	6612
Nevada							
Las Vegas	108	110	70	26	20	S	2709
Reno	95	95	65	7	− 5	6W	6332
New Hampshire							
Concord	91	90	73	− 7	−15	6NW	7383
New Jersey							
Atlantic City	91	95	78	18	5	13SW	4812
Newark	94	95	75	15	0	11	4589
Trenton	92	95	78	16	0	9SW	4980
New Mexico							
Albuquerque	96	94	65	17	0	8SW	4348
Roswell	101	95	70	19	0	6S	3793
Santa Fe	90	90	65	11	0	6SE	6123
New York							
Albany	91	93	75	0	−10	11S	6875
Buffalo	88	93	73	6	− 5	17W	7062
New York	94	95	75	15	0	16NW	5280
Rochester	91	95	75	5	− 5	10W	6748
Syracuse	90	93	75	2	−10	11S	6756

TABLE 5-2 (*Continued*)

State and City	SUMMER			WINTER			
	Dry Bulb, °F		Wet Bulb, °F	Dry Bulb, °F		Wind Velocity, mph, and Direction	Degree, Days
	1%	Normal	Normal	97¼%	Normal		
North Carolina							
Asheville	91	93	75	17	0	9NW	4042
Charlotte	96	95	78	28	10	7SW	3191
Raleigh	95	95	78	20	10	8SW	3393
Wilmington	93	95	78	27	15	9SW	2347
Winston-Salem	94	95	78	17	10		3595
North Dakota							
Bismark	95	95	73	−19	−30	10	8851
Grand Forks	91	90	73	−23	−25		9870
Ohio							
Cincinnati	94	95	78	12	0	8SW	4410
Cleveland	91	95	75	7	0	15SW	6144
Columbus	92	95	76	7	− 5	12SW	5660
Toledo	92	95	75	5	−10	12SW	6494
Oklahoma							
Oklahoma City	100	100	77	15	0	11S	3725
Tulsa	102	100	77	16	0	N	3860
Oregon							
Eugene	91	90	68	26	−15		4726
Portland	89	90	68	24	10	7S	4635
Pennsylvania							
Bethlehem	92	95	75	5	− 5		5810
Erie	88	93	75	11	− 5	14SW	6451
Harrisburg	92	95	75	13	0	7NW	5251
Philadelphia	93	95	78	15	0	11NW	5144
Pittsburgh	90	95	75	9	0	11W	5987
Reading	92	95	75	9	0	9	4945
Scranton	89	95	75	0	− 5	8SW	6254
Rhode Island							
Providence	80	95	75	10	0	12NW	5954
South Carolina							
Charleston	95	95	78	30	15	10	2033
Columbia	98	95	75	23	10	8	2484
Greenville	95	95	75	23	10	8	2980
South Dakota							
Rapid City	96	95	70	− 6	−20	10	7345
Sioux Falls	95	95	75	−10	−20	11	7839
Tennessee							
Chattanooga	97	95	76	19	10	8	3384
Knoxville	95	95	75	17	0	7	3590
Memphis	98	95	78	21	0	12	3015
Nashville	97	95	78	16	0	10	3578

The Heating Load

TABLE 5–2 (*Continued*)

State and City	SUMMER			WINTER			Degree, Days
	Dry Bulb, °F		Wet Bulb, °F	Dry Bulb, °F		Wind Velocity, mph, and Direction	
	1%	Normal	Normal	97½%	Normal		
Texas							
Amarillo	98	100	72	12	−10	13	3985
Austin	101	100	78	29	20	10	1711
Brownsville	94	100	80	40	30	12	600
Dallas	101	100	78	24	10	10	2363
El Paso	100	100	69	25	10	10	2700
Fort Worth	102	100	78	24	10	12	2405
Galveston	91	95	80	36	20	11	1274
Houston	96	95	80	32	20	11	1396
San Antonio	99	100	78	30	20	9	1546
Utah							
Salt Lake City	97	95	65	9	−10	8	6052
Vermont							
Burlington	88	90	73	− 7	−10	12	8269
Virginia							
Lynchburg	94	95	75	19	5	8	4166
Norfolk	94	95	78	23	15	12	3421
Richmond	96	95	78	18	15	8	3865
Roanoke	94	95	76	18	0	10	4150
Washington							
Seattle	81	85	65	32	15	10	4424
Spokane	93	90	64	4	−15	8	6655
Tacoma	85	85	64	24	15	8	5145
Walla Walla	98	95	65	16	− 5	5	4805
West Virginia							
Charleston	92	95	75	14	0	9	4476
Huntington	95	95	76	14	− 5	W	4446
Parkersburg	93	95	75	12	−10	7	4754
Wheeling	91	95	75	9	− 5		5218
Wisconsin							
La Crosse	90	95	75	− 8	−25	9	7589
Madison	92	95	75	− 5	−15	11	7863
Milwaukee	90	95	75	− 2	−15	13	7635
Wyoming							
Cheyenne	89	95	65	− 2	−15	14NW	7381
Sheridan	95	95	65	− 7	−30	5NW	7680
Argentina							
Buenos Aires	91		76	34			
Bahamas							
Nassau	90		80	63			
Brazil							
São Paulo	86		74	46			

TABLE 5-2 (*Continued*)

State and City	SUMMER			WINTER			
	Dry Bulb, °F		Wet Bulb, °F	Dry Bulb, °F		Wind Velocity, mph, and Direction	Degree, Days
	1%	Normal	Normal	97½%	Normal		
Canada							
Calgary, Alta.	87	90	66	−25	−29		9703
Edmondton, Alta.	86	90	68	−26	−33		10,268
Fredericton, N.B.	89	90	75	−10	−10		8671
Halifax, N.S.	83	90	75	4	4		7361
Montreal, Que.	88	90	75	−10	−10		7899
Ottawa, Ont.	90	90	75	−13	−15		8735
Quebec, Que.	86	90	75	−13	−12		9372
Regina, Sask.	92	90	71	−29	−34		10,806
Toronto, Ont.	90	93	75	1	0		6827
Vancouver, B.C.	80	80	67	19	11		5515
Winnipeg, Man.	90	90	71	−25	−29		10,679
Yarmouth, N.S.	76	80	68	9	7		7340
Dominican Republic							
Santo Domingo	92		80	65			
Mexico							
Mexico City	83		60	39			
Monterrey	98		78	41			
Vera Cruz	91		83	62			
Panama							
Panama City	93		81	73			
Puerto Rico							
San Juan	89		80	68			
Venezuela							
Caracas	84		69	54			

openings in a room should be included in infiltration-loss calculations. In considering a single room with one exposed wall, all of the feet of crack should be used; for a room with two or more exposed walls, take the feet of crack in the wall which has the greatest amount, but in no case use less than half the total feet of crack in the room.

In the case of high buildings, chimney effect or stack effect acts, in addition to the effect of wind, in causing infiltration. That is, the warm column of air inside the building exerts less pressure at the lower part of the building than the corresponding column of cooler, denser air outside the building, and infiltration therefore occurs. This air which enters at the lower part of the building must leave at the upper part, and in the upper levels this exfiltration opposes the infiltration occurring from wind velocity. In the lower part of the building, however, wind effect and stack effect are cumulative. On the rough assumption that a neutral zone is located at the mid-height of a building, the following formulas can be used to deter-

The Heating Load

mine wind velocity in connection with Table 5-3:

$$V_e = \sqrt{V^2 - 1.75a} \qquad (5\text{-}3)$$

$$V_e = \sqrt{V^2 + 1.75b} \qquad (5\text{-}4)$$

where V_e = equivalent wind velocity, in miles per hour;
V = wind velocity upon which infiltration is based if chimney effect is neglected, in miles per hour;
a = distance from windows under consideration to mid-height of building, when windows are above mid-height, in feet;
b = distance from windows under consideration to mid-height of building, when windows are below mid-height, in feet.

To reduce chimney effect, the stair wells, elevator passages, and other vertical shafts in tall buildings should be sealed off as much as possible.

Air infiltration through walls can be neglected in most cases if the walls are lathed and plastered, or if a good application of building paper is made. With poor wood construction and no plaster or building paper, infiltration loss may be extremely high. At 10-mph wind, the air passing through each square foot of wall is 4.20 cfh for an $8\frac{1}{2}$-in. brick wall unplastered, but only 0.037 cfh for the same wall plastered.

An approximate method sometimes used for calculating air infiltration is the method of *assumed air changes*. For almost any type of room or space, a volume of air per hour equivalent to one-half to three times the volume of the room or space enters by infiltration and an equivalent volume leaves. Typical values give $1\frac{1}{2}$ changes of air per hour for a room with two sides exposed, 1 to 3 changes per hour for stores, and 1 to 2 changes for living rooms. The method is so unreliable, however, that it cannot be recommended, but at least one-half change per hour should be allowed for any space.

Expressed as an equation, the heat loss is

$$Q_{\text{inf}} = (0.244)\,(q_h)\,(d)\,(t_i - t_o)\ \text{Btuh} \qquad (5\text{-}5)$$

where Q_{inf} = heat loss by infiltration, in Btu per hour;
q_h = air entering by infiltration, in cubic feet per hour = $60\,q$;
d = density of air at conditions measured, in pounds per cubic foot;
t_i = inside temperature, in degrees Fahrenheit;
t_o = outside temperature, in degrees Fahrenheit.

For ordinary conditions a density d of air of 0.075 lb per cu ft is sufficiently accurate, so that $0.244 \times d$ in equation 5-5 reduces to 0.018 Btu per cu ft as a representative specific heat of air, on a volume basis. There-

TABLE 5–3

INFILTRATION THROUGH WINDOW AND DOOR CRACK*
(Cubic Feet per Hour per Foot of Crack)

Type of Aperture	Remarks	Wind Velocity (mph)				
		5	10	15	20	25
Double-hung wood-sash windows (unlocked)	Average; non-weather-stripped	7	21.4	39	59	80
	Average; weather-stripped	4	13	24	36	49
	Poorly fitted; non-weather-stripped	27	69	111	154	199
	Poorly fitted; weather-stripped	6	19	34	51	71
	Around window frame: masonry wall, uncalked	3	8	14	20	27
	Around window frame: masonry wall, calked	1	2	3	4	5
	Around window frame: wood frame structure	2	6	11	17	23
Double-hung metal windows	Non-weather-stripped; unlocked	20	47	74	104	137
	Non-weather-stripped; locked	20	45	70	96	125
	Weather-stripped; unlocked	6	19	32	46	60
Single-sash metal windows	Industrial; horizontally pivoted	52	108	176	244	304
	Residential casement	14	32	52	76	100
	Vertically pivoted	30	88	145	186	221
Doors	Well-fitted	27	69	110	154	199
	Poorly fitted	54	138	220	308	398

* Data based on research papers in *Trans. ASHVE*, vols. 30, 34, 36, 37, and 39.

fore, using the foregoing notation, there results

$$Q_{\text{inf}} = (0.018)(q_h)(t_i - t_o) \text{ Btuh} \tag{5-6}$$

Example 5–1. A small building has a total of twenty-eight double-hung 5-ft by 3-ft wood-sash windows of average construction, evenly distributed on all four sides. They are not weather-stripped. There are also four poorly fitting doors 6.5 ft by 3.5 ft. The wind velocity is 9 mph NW and there are no obstructions around the building. What is the heat loss by infiltration when the temperature is 70 F inside and -5 F outside?

Solution: Total window crack $= 28[2(5 + 3) + 3] = 532$ ft. Total door crack $= 4[2(6.5 + 3.5)] = 80$ ft. Consider half the total window and door crack as contributing to infiltration. Interpolation from Table 5–3 to get infiltration for 9 mph gives 18.5 cfh for the window per foot of crack, and 121.2 cfh for the door per foot of crack. Hence the air entering by window infiltration is

$$q_w = (\tfrac{1}{2} \times 532)(18.5) = 4920 \text{ cfh}$$

and the air entering by door infiltration is

$$q_d = (\tfrac{1}{2} \times 80)(121.2) = 4848 \text{ cfh}$$

Therefore the total infiltration of air is $4920 + 4848 = 9768$ cfh.

The Heating Load

Substituting in equation 5-6, the heat lost by infiltration is

$$Q_{inf} = (0.018)(9768)[70 - (-5)] = 13{,}180 \text{ Btuh} \qquad Ans.$$

Item 4. Specific computations for the miscellaneous items are often avoided, but in such cases a suitable factor for unforeseen contingencies and for unusual exposure should be added. Thus if a building is in the open sweep of wind, from 10 to 20 per cent should be added to the computed heat load. If the region where the building is located is subject to erratic and sudden temperature changes, or if the building is to be heated intermittently, the design capacity of the heating system should be increased materially.

With open fireplaces, loss from chimney effect is difficult to estimate but should be given some consideration. A figure of 2500 Btuh is sometimes arbitrarily used.

Often humidification for entering outside air is not adequately provided for, but when it is required a definite heat load results, as can be seen in the following example.

Example 5-2. If the infiltration air of example 5-1 enters at -5 F, with a relative humidity of 50 per cent, (a) how much water must be evaporated per hour to maintain the inside relative humidity at 35 per cent at 70 F, and (b) how much heat load is required for this evaporation?

Solution: (a) The humidity ratio and specific volume of the outside air at -5 F and 50 per cent relative humidity can be found from Plate II as 0.0003 lb per pound of dry air and $v = 11.42$ cu ft per lb. The humidity ratio of inside air can be read from Plate I, at 70 F and 35 per cent relative humidity, as 0.0055 lb per pound of dry air.

The water to be evaporated into each pound of infiltration air $= 0.0055 - 0.0003 = 0.0052$ pound per pound.

From example 5-1, 9768 cu ft of infiltration air, or $9768/11.42 = 855$ lb of dry air constituents, enter per hour.

The water evaporated per hour $= 0.0052 \times 855 = 4.4$ lb. *Ans.*

b) The latent heat of evaporation of water at room temperatures $= 1060$ Btu per lb (a round figure; see Table 2-2). Therefore the heat required for evaporation (humidification) $= 4.4 \times 1060 = 4664$ Btuh. *Ans.*

Many buildings, such as churches, are heated only intermittently. Heat absorption by a cold building structure when the building is warming up may amount to about one-half to twice as much as the calculated normal heat loss from the building. Consequently, where quick warming up is required, additional heating capacity should be installed. It is also true that in a building which has recently been cold, the walls for some time after heat is turned on remain much colder than the air, and the resulting excessive radiation loss from the occupants to the cold walls creates an uncomfortable sensation necessitating a higher temperature to offset this condition.

In contrast to the items constituting heating demand, there are some items which in themselves are heat producers. The heat generated by a large assemblage of people, for example, may be so excessive that the external heating supply can be reduced or even stopped. However, the full design heating capacity should be installed, since the building must be warmed before occupancy. Motors and lights also contribute largely to the heat produced in a space, particularly in manufacturing plants, and may reduce or even eliminate the need of heating capacity. Unless the work done by a motor is performed outside the building space itself, all the power supply to the motor reappears in the building. The heat output developed by a running motor can be computed as follows:

$$\frac{\text{Motor hp developed}}{\text{Motor efficiency}} \times 2545 = \text{Btuh} \qquad (5\text{-}7)$$

In the case of electric heaters or lights the heat output developed for each hour of operation can be computed thus:

$$\text{Kilowatt capacity} \times 3413 = \text{Btuh} \qquad (5\text{-}8)$$

or

$$\text{Watt capacity} \times 3.413 = \text{Btuh} \qquad (5\text{-}9)$$

Except for unusual layouts, it is not customary in designing heating systems to make deduction in capacity for power and electric lights, since these items are not usually in operation prior to occupancy. It is wise, however, to provide means of reducing the output from the heating apparatus in order to prevent overheating after these local heat supplies go into operation. In the case of well-insulated electrically-heated spaces, the lighting load constitutes a definite part of the heat supply. Also in the interior areas of large buildings the electrical loading may be so extensive that cooling is required to offset it.

5–2. HEATING-LOAD SUMMATION AND ESTIMATING

In computing the heat requirements of the rooms of a building, a systematic approach is required and it is often desirable to prepare in advance tabular forms for listing and summating the various heat losses. For an existing or proposed building, architectural drawings can furnish much of the basic information needed. Columns in a tabular form should be allowed for each room or space which is exposed to outside temperature or to the intermediate temperature of an unheated space such as an attic, cellar, or enclosed porch. The heat loss is computed by using the appropriate outside temperature or the temperature of the unheated space. It is desirable to consider together all the rooms involved on a given story before calculating the next story.

To illustrate how a tabular approach might be followed Fig. 5–1 has been worked up. The name of the room should be placed in the first column, as for example "Living Room." No space has been left to record length, width, and height, since these dimensions are misleading when one

The Heating Load

Room or Space	Gross Exposed Wall Area (sq ft)	Glass Area $U=1.13$ $\Delta t = 70$	Net Exposed Wall Area $U=0.35$ $\Delta t = 70$	Floor Area $U=0.34$ $\Delta t = 10$	Ceiling Area $U=0.28$ $\Delta t = 0$	Partitions, Fireplaces, Etc.	Crack $\left(\frac{Ft}{2}\right) q_h$ $\times (0.018)(\Delta t)$	Btuh Sum	Exposure or Other Factor	Total Room or Space Load (Btuh)	Heating Unit Size or CFM
Living Room	200	45	155	300	300		$\frac{1}{2}(57)$	11,645	0.15	13,392	177 CFM
		3560	3797	1020	0	2500	768				

FIG. 5–1. Heating data and estimate sheet filled in for one room.

is considering odd-shaped rooms and data should be obtained directly from the architectural drawings of the house. In the case of the house, used as an example in Fig. 5–1, calculations from architectural data show the exposed wall of the living room to be 200 sq ft in area, and this is so recorded in the second column.

The glass-window and frame area and the door area in the exposed wall are next calculated as 45 sq ft and are recorded in the third column, at the top of the row being filled in. The over-all coefficient of heat transfer U for the glass is 1.13, from Table 4–17, and Δt is taken as 70 deg for this problem. These values have previously been entered in the column headings, and any changes from these values for other rooms must be noted. Temperature at the mean room height or mean glass height should be used. In the bottom half of the row can now be placed the computed heat load, or $Q = UA\Delta t = (1.13)(45)(70) = 3560$ Btuh.

The value for the fourth column, which is for net exposed wall area, equals the gross area minus the window area, or $200 - 45 = 155$ sq ft, and this value is placed in the top half of the row. Since $U = 0.35$ for the wall in question and since $\Delta t = 70$, Q is found as 3797 Btuh, and this value is placed in the bottom half of the row.

The floor in this specific case is over a partly heated cellar; therefore, using $\Delta t = 10$ and $U = 0.34$, and the computed area of 300 sq ft, which has been entered in the top half of the row, under column 5, we find that $Q = 1020$ Btuh, and this value is entered in column 5, in the bottom of the row.

The ceiling area and other data are then filled in, but for these, since the bedroom above the living room is heated, it is better to use $\Delta t = 0$ and not consider slight interchanges of heat in the building except in cases where the loss to internal adjacent rooms is so great as to necessitate extra heating capacity in the space considered.

Similarly, loss to partitions is regarded as zero, but a fireplace is present, for which 2500 Btuh is allowed.

Crack calculation brings up a decision—in regard to procedure—which must be adhered to throughout. If the living room is on the windward side of the building, then cold air will be entering all the crack space and leaving on the other side of the building as warm air. The living-room radiators should be sized to handle this total infiltration load instead of half of it. If they are sized in this way, the rooms on the leeward side can be regarded as exposed only to warm air exfiltration and little or no allowance for infiltration need be made. The trouble with this method of calculation is that the wind direction is not constant, and when the wind changes direction the rooms on the leeward side may become too cool. Thus it is better as a general (but not inflexible) rule to use only one half of the crack and, for exposed rooms, add to the Btuh sum an exposure (safety) factor. For

The Heating Load

the 57 ft of crack along the double-hung non-weather-stripped windows, from Table 5-3, at 10 mph, $q_h = 21.4$ cfh. Thus

$$q = \tfrac{57}{2} (21.4)(0.018)(70) = 768 \text{ Btuh}$$

The sum of all the items is $3560 + 3797 + 1020 + 2500 + 768 = 11{,}645$ Btuh.

Because this room is considered to be on the windward side of the building, an exposure factor of 15 per cent is added.

This exposure factor, when used to modify the Btuh sum, gives the total room or space requirement as $11{,}645 \times 1.15 = 13{,}392$ Btuh.

Data on equipment and its selection are given in chapters 7, 8, 9, and 11. By adding the Btuh requirements for each room, the total design hourly heat load for the building can be found.

5-3. SPACE HEATING

With knowledge of the heat requirements of a space available, it is possible to provide sufficient steam, hot water, electric energy, or warm air to offset the heat losses of that space and maintain desired conditions in it. Later chapters treat such distribution in detail but here computational patterns for energy distribution by means of air will be presented. With space heating by use of delivered air, the process is merely one involving sensible cooling of supply air and the basic equation is

$$Q_s = 60 \, m_a \, (C_p) \, (t_2 - t_1) \text{ Btu per hr} \qquad (5\text{-}10)$$

or

$$Q_s = 60 \, \frac{q}{v} \, C_p \, (t_2 - t_1) \text{ Btu per hr} \qquad (5\text{-}11)$$

$$= 60 \, \frac{q}{v} \, (0.244) \, (t_2 - t_1)$$

$$= 14.6 \, \frac{q}{v} \, (t_2 - t_1) \text{ Btu per hr} \qquad (5\text{-}12)$$

where m_a = pounds per minute air flow
C_p = specific heat of moist air as delivered, about 0.244 Btu per pound F
q = cubic feet per minute air flow
v = specific volume of the moist air delivered, cubic feet per pound
t_2 and t_1 = temperatures of the final air in the space and as delivered degrees F
Q_s = heat load of the space, Btu per hour

For many cases it is sufficiently accurate to disregard the variation in v with temperature and using the value of $v = 13.51$ at 70 F and 50 per cent relative humidity equation 5-12 reduces to

$$Q_s = 1.08 \, q \, (t_2 - t_1) \qquad (5\text{-}13)$$

or with saturated air at 60 F,

$$v = 13.33 \text{ cu ft per lb}$$

and

$$Q_s = 1.10 \, q \, (t_2 - t_1) \tag{5-14}$$

The latter two equations suffice for many computations.

5-4. SEASONAL HEAT AND FUEL REQUIREMENTS. DEGREE-DAYS

An extensive series of tests, conducted by the American Gas Association, showed that the fuel consumption in residences and public buildings varies almost directly as the difference between the outside temperature and 65 F. Other investigations have shown that 66 F is closer than 65 F in representing a datum point, but 66 F is so close to the 65 F that the earlier datum is still used. This datum 65 F (or 66 F) indicates that when the outside temperature is 65 F or above, practically no heat is required and the fuel consumption approaches zero; also, the fuel consumption would double if an outside temperature of 55 F (10 deg difference) changed to 45 F (20 deg difference).

The difference between 65 F and the average outside temperature is important as an index of heating requirements and gives the basis for the degree-day for specifying the nominal winter heating load. A *degree-day* accrues for every degree the average outside temperature is below 65 F during a 24-hr period. Thus, in a given locality, if the outside temperature for 30 days averaged 50 F, the degree-days for the period would be (65 − 50) (30), or 450. Degree-days vary greatly from place to place, and slightly from year to year.

In large cities, airport degree-day data may be significantly different from central-city data. Degree-days at a given location also vary from year to year and consequently average values are tabulated in most instances. The data in Table 5-2 are largely based on 30-year averages but a number of shorter-period averages also appear in the table.

A normal heating season in the United States is considered to be from October 1 to May 1 (212 days, or 5088 hours). This period varies throughout the country.

By making use of the degree-day values in a given location, it is possible to estimate with a fair degree of success the probable fuel or steam consumption required by a building during a heating season. The estimations, however, may be inaccurate for such reasons as minor variations in degree-days from year to year, local wind conditions, and unusual exposures. In making estimates, therefore, it is extremely helpful to have available an accurate heat-loss analysis of the building, made under known design conditions.

The Heating Load

Many approaches to the estimation of fuel or steam consumption are in use, but only one or two will be indicated here. With the heat loss under design conditions known, it is desirable to find the heat loss per degree difference in temperature, which can be found by the relation

$$Q_D = \frac{Q_T}{t_i - t_o} \qquad (5\text{-}15)$$

where Q_T = total heat loss for a building or space, in Btu per hour, when based on:
t_i = inside design temperature, in degrees Fahrenheit;
t_o = outside design temperature (from Table 5-2), in degrees Fahrenheit;
Q_D = heat loss per degree of temperature difference, in Btu per hour.

To find the steam consumption, use

$$S = \frac{(Q_D) \times (\text{degree-days}) \times (24)}{1000} \qquad (5\text{-}16)$$

where Q_D = calculated heat loss per degree of temperature difference, in Btu per hour;
1000 = approximate number of Btu released for heating by each pound of steam;
24 = hours per day;
S = pounds of steam required for the degree-days in the period of estimation (heating season, month, etc.).

Example 5-3. For an office building in Cleveland, Ohio, having a calculated heat loss of 3,200,000 Btuh, estimate the weight of steam required during the heating season.

Solution: Degree-days in Cleveland are 6144 and the design temperatures are 0 F and 70 F. Thus, by equations 5-15 and 5-16,

$$Q_D = \frac{3,200,000}{70 - 0} = 45,700 \text{ Btuh per degree}$$

and
$$S = \frac{(45,700)(6144)(24)}{1000} = 6,740,000 \text{ lb} \qquad Ans.$$

The amount of fuel used per season depends on its heating value and on the efficiency of utilization. This latter factor is influenced by the kind of combustion equipment, and its condition, and by the manner in which it is operated. In the case of small domestic equipment, some of the otherwise wasted heat from the furnace or boiler room and chimney may be reabsorbed by the building itself. The manufacturer's published efficiency of a given unit may be used for making estimates, although such laboratory values are not usually maintained in actual practice. For rough estimation the following values can be used:

Equipment	Efficiency (per cent)
Coal-fired small boiler or furnace (heating value of coal, about 13,000 Btu per lb).................	45–70
Small oil-fired boiler or furnace (heating value of oil, about 144,000 Btu per gal)....................	60–85
Gas-fired furnace (heating value of manufactured gas, about 550 Btu per cu ft; of natural gas, about 900 Btu per cu ft*).............................	75–85

* Gas is also sold in therm units, where 1 therm = 100,000 Btu.

Example 5–4. Consider that the office building of example 5–3 is heated with its own boiler and can employ gas, oil, or coal. (a) Estimate for the heating season the cubic feet of gas, at a heating value of 800 Btu per cu ft, which would be required for this building. Also, convert the answer into therms of gas. (b) Estimate the gallons of fuel oil, at a heating value of 144,000 Btu per gal, required for the heating season. (c) Estimate the tons of coal required for the season. Assume that the coal has a heating value of 12,800 Btu per lb.

Solution: The season heat load is $(45,700)(6144)(24) = 6,740,000,000$ Btu.

a) The furnace efficiency with gas should reach 80 per cent, since gas can be burned efficiently although there is an unavoidably high loss resulting from hydrogen in the fuel (see Chap. 12). Thus

$$F_g = \frac{6,740,000,000}{(0.80)(800)} = 105,300,000 \text{ cu ft} \qquad Ans.$$

Expressed in therms, the gas required is

$$F_t = \frac{6,740,000,000}{(0.80)(100,000)} = 84,250 \text{ therms}$$

b) Considering the oil-burning efficiency to be 78 per cent,

$$F_o = \frac{6,740,000,000}{(0.78)(144,000)} = 60,000 \text{ gal} \qquad Ans.$$

c) Considering stoker firing for the coal, with good controls on the boiler, an efficiency of 70 per cent should be possible. Thus

$$F_c = \frac{6,740,000,000}{(0.70)(12,800)(2000)} = 376 \text{ tons} \qquad Ans.$$

The decision as to which fuel is most desirable depends on (1) the cost of fuel delivered, ready to burn; (2) its availability; (3) the labor costs associated with the firing of the fuel; and (4) possible ash-removal and maintenance costs as related to the different types of combustion equipment.

Table 5–4 gives data from which the steam-consumption requirements of various types of buildings can be estimated. The net volume of heated space in a building is approximately 80 per cent of the gross volume of a building. Estimates made from data such as found in this table should be recognized as furnishing only approximations of heating demands.

Example 5–5. A residence has a total volume of 40,000 cu ft and is located at a place where the degree-days are 4626. Find the probable steam consumption for a heating season.

The Heating Load

TABLE 5-4
Steam Consumption for the Heating Season for Various Classes of Buildings

Building Classification	Steam Consumption for Heating (pounds per degree-day per 1000 cu ft of net heated space)	Average Hours of Occupancy
Apartment..........................	0.962	22
Bank.............................	0.786	12
Church...........................	0.532	8
Club or hotel......................	0.990	22
Department store...................	0.385	11
Garage...........................	0.202	22
Hospital..........................	1.19	22
Lodge............................	0.390	12.5
Loft..............................	0.588	10
Manufacturing.....................	0.808	9.5
Municipal.........................	0.587	16
Office............................	0.685	12
Office and shops...................	0.617	13
Printing..........................	1.23	18
Residence........................	0.962	22
School...........................	0.592	11
Stores............................	0.624	10.5
Theatre..........................	0.482	13
Warehouse........................	0.459	9.5

Solution: The heated space in a building is usually only about 80 per cent of the total cubage, or in this case,

$$40,000 \times 0.8 = 32,000 \text{ cu ft}$$

From Table 5-4, the steam consumption per degree-day is 0.962 per 1000 cu ft of heated space. Therefore the probable steam consumption for a heating season is

$$0.962 \times \frac{32,000}{1000} \times 4626 = 142,000 \text{ lb} \qquad Ans.$$

For monthly percentage estimates of fuel consumption during the heating season it has been found that the values given in Table 5-5 apply to

TABLE 5-5
Estimated Monthly Fuel Consumption, in Per Cent of Total Consumption for Winter Season

October........... 5	February.......... 17
November........ 11	March............ 15
December........ 17	April............. 10
January.......... 20	May.............. 5

average localities in the United States (that is, not localities in the extreme north or south).

To furnish a better idea of degree-day variations throughout a season, Table 5-6 gives the monthly variation of degree-day values for several representative cities.

TABLE 5-6

AVERAGE MONTHLY DEGREE-DAY VALUES FOR REPRESENTATIVE CITIES

City	Aug., Sept.	Oct.	Nov.	Dec.	Jan.	Feb.	Mar.	Apr.	May	June, July	Total
Baltimore	33	227	526	855	921	837	637	343	95	12	4487
Chicago	95	337	712	1116	1218	1080	861	531	259	73	6282
Cleveland	107	354	684	1045	1143	1067	876	553	252	63	6144
Los Angeles	5	43	110	225	272	235	212	158	103	28	1391
Minneapolis	190	481	942	1415	1587	1372	1072	577	260	70	7966
New York	54	272	594	940	1028	953	771	465	172	31	5280

By use of a table such as Table 5-6, it is possible to compute the average temperature during a given month or during a winter season. For example, in Chicago during January, with 1218 degree-days accrued in 31 days, and using the 65 F base of the table,

$$1218 = 31(65 - t_{avg})$$

Therefore, for a typical January,

$$t_{avg} = 25.8 \text{ F}$$

To find the average temperature for a major part of the heating season in Chicago—say from October 1 to May 1, or 212 days—first subtract the degree-day values for the unwanted months from the yearly total:

$$6282 - (95 + 73 + 259) = 5855 = 212(65 - t_{avg})$$

Then compute the temperature for these seven worst months of the heating season:

$$t_{avg} = 37.4 \text{ F}$$

In the operation of heating systems, much consideration has been given to the desirability of lowering the temperature maintained in the heated space during periods when the space is not in active use or is unoccupied. Some saving usually results if the space temperature is reduced during such periods. However, if the temperature is greatly reduced, as, for example, over the night hours, the heating equipment may have to run overloaded, and inefficiently, to bring the space rapidly back to the desired

The Heating Load

temperature. If this is the case, the saving may be minimized or even disappear.

Example 5-6. Analyze the effect of lowered inside temperatures for portions of the time in relation to the building of example 5-3. Consider that the daily temperature is lowered from 70 F to 64 F during the 8 hours from 11 P.M. to 7 A.M. of any 24-hr period. Find (a) the weighted average inside temperature and (b) the possible saving in steam over a season.

Solution: (a) The weighted average inside temperature is

$$t_{avg} = \frac{(24-8)(70)+(8)(64)}{24} = 68 \text{ F} \qquad Ans.$$

b) The average outside temperature of the whole heating season for Cleveland can be found if the total days requiring heat are known. Refer to Table 5-7 and consider that heat is required during 6 days in September and during 4 days in June. When these are added to the 243 days from October 1 to May 31, the total is 253 days. To find the average outside temperature t_o for the season, use the total degree days (6144) spread over 253 days:

$$6144 = (253)(65 - t_o)$$

$$t_o = 40.7 \text{ F}$$

The 6,740,000 lb of steam required per season with a 70 F inside temperature, when referred to a 68 F inside temperature, would reduce to

$$S = 6{,}740{,}000 \,\frac{68 - 40.7}{70 - 40.7} = 6{,}280{,}000$$

The saving is thus

$$6{,}740{,}000 - 6{,}280{,}000 = 460{,}000 \text{ lb per season} \qquad Ans.$$

The saving amounts to almost 7 per cent. Although all of the saving is not realized, the possible magnitude of the saving requires giving serious consideration to inside temperature reduction at off hours.

PROBLEMS

5-1. A room in a residence is 10 ft long by 8 ft wide by 9 ft high and has two double-hung windows each 3 ft by 5 ft, and the bottom of the lower sash is 3 ft from the floor. A temperature of 70 F at the 5-ft level is maintained in the room when it is 0 F outside. (a) Compute the heat loss through the single-glazed windows, making use of the inside temperature at the mean height of the glass (5.5 ft from the floor). (b) Two of the walls of the room are exposed to the outside and have an over-all heat-transmission U value of 0.35 Btu per hr sq ft deg F. Compute the heat loss through the net exposed wall area. (c) Compute the infiltration loss through the window crackage, based on a 10-mph wind. Assume that the room faces toward the prevailing wind and therefore receives its full effect, with the infiltration then discharging through leakage cracks in other parts of the house.
Ans. (a) Glass loss, 2397 Btuh; (b) loss through two walls, 3200 Btuh; (c) 1025 Btuh

5-2. A large business office is fitted with nine poorly-fitted, double-hung wood-sash windows 3 ft wide by 6 ft high. If the outside wind is 15 mph at a temperature of -10 F, what is the saving in heat if the windows are weather-stripped? Assume an inside temperature of 70 F. Base your solution on a minimum condition in which half of the crackage undergoes infiltration and half undergoes exfiltration. *Ans.* 10,400 Btuh

5-3. Outside air at 20 F and 60 per cent relative humidity enters a small building by infiltration at a rate of 2000 cfh. (a) How much moisture must be evaporated per hour to maintain an inside relative humidity of 40 per cent at 65 F? (b) How much heat is required for humidification?   *Ans.* (a) 0.65 lb; (b) closely 700 Btuh

5-4. A room has four vertically-pivoted steel-sash windows 25 in. wide by 62 in. high. They are on the windward side of the room, which has two snugly fitting doors which open into the central part of a relatively open building. Assume that it is 10 F outside and 70 F inside, and that a prevailing 10-mph wind is blowing. Compute a) the probable infiltration with both doors open; (b) the probable infiltration with both doors closed and locked; (c) the heat loss under each of these conditions.
Ans. (a) 5100 cfh; (b) 2550 cfh; (c) 5500 and 2750 Btuh

5-5. What is the heat loss through the glass area of the room in problem 5-4 (a) when no storm windows are used and (b) when complete-coverage storm windows are employed?   *Ans.* (a) 2920 Btuh; (b) 1370 Btuh

5-6. (a) How many pounds of steam are required during a winter season for an office building located in Chicago when the heat load of the building is 3,500,000 Btuh based on an inside temperature of 65 F and an outside temperature of -10 F? Use the degree-day method. (b) If this building is heated by fuel oil burned with an over-all efficiency of 75 per cent, how many gallons of No. 6 fuel oil are required for the heating season?   *Ans.* (a) 7,040,000 lb; (b) 65,200 gal

5-7. How many pounds of steam are required each winter for an office building in Philadelphia with a heat load of 1,000,000 Btuh based on an inside temperature of 70 F and an outside temperature of 0 F? Use the degree-day method.   *Ans.* 1,630,000 lb

5-8. What is the probable fuel consumption per season for the building in problem 5-7 (a) if coal is used in the furnace and if the furnace has a 60 per cent efficiency, and (b) if oil is used and the furnace has an over-all efficiency of 70 per cent?
Ans. (a) 104 tons; (b) 16,100 gal

5-9. Estimate the district steam required per season to heat a department store having a net volume of 36,000,000 cu ft if the store is located in (a) New York; (b) St. Louis; (c) Minneapolis.   *Ans.* (a) 73,200,000 lb; (b) 63,500,000 lb; (c) 110,600,000 lb

5-10. Compute the average outside temperature throughout the Baltimore heating season, which extends from October 1 to May 1. Use data from Table 5-6.   *Ans.* 44.5 F

5-11. A certain building located in Baltimore has a heat loss of 150,000 Btuh based on 70 F inside and 0 F outside temperatures. Compute the probable fuel-oil consumption for the heating season, which extends from October 1 to May 1. Assume a heat-utilization of 77 per cent of the energy in the fuel.   *Ans.* 2010 gal

5-12. It is decided to reduce the inside temperature of the building in problem 5-8 from 70 F to 65 F from 10:30 P.M. to 7:30 A.M. Find the weighted average inside temperature for a 24-hr period. Find the probable saving in fuel per season that would accrue from this night temperature reduction.   *Ans.* 46.3 F, 8%

5-13. What is the heat loss for the first floor of the residence shown in Fig. 11-7 if one assumes that the residence is located in Pittsburgh? Consider that the prevailing winter wind at 12 mph strikes the terrace and dining-room sides of the building. There is enough waste heat from the furnace to keep the basement warm, and this prevents any significant loss downward from the first-floor rooms. Heated space is above. Make

The Heating Load 211

computations for each room and tabulate them. Allow for the outside wall surface adjacent to the steps at the right of the entry door. Consider full infiltration on the windward sides of the building and one-half the maximum on the other sides. Consider the unheated garage to be at 20 deg above the outside design temperature. Consider the fireplace structure to be 7 ft wide, and of ceiling height, with its masonry having an over-all U of 0.31 Btu per hr sq ft deg F. Also allow for excess exfiltration from the fireplace amounting to 2500 Btuh.

5-14. Compute the heat loss from the second floor of the residence of Fig. 11-8. Assume that the residence has the location and ambient conditions indicated in problem 5-13, that there is no loss to the floor beneath, and that the attic temperature is 30 deg higher than the outside design temperature. Compute each room and also the stair well, using the infiltration conditions stated in problem 5-13. Consider the chimneys adjacent to room 23 merely as conventional wall surface.

5-15. An industrial building located in the suburbs of Chicago has a plan and elevation as shown in Fig. 9-8. A temperature of 60 F is maintained at the 30-in. line, and heating is accomplished by unit heaters placed above the working space, which blow warm air to the working area. The outside winter design temperature is -10 F and the prevailing wind direction is 12 mph southwest. The walls are 12 in. thick and made of plain brick ($U = 0.36$). There are 20 steel-sash windows 6 ft by 9 ft on each side, and 8 such windows on each end. The monitor walls are of rigid asbestos shingles on solid wood sheathing and studs ($U = 0.52$). The monitor section contains 40 4.5-ft by 4.5-ft steel-frame windows in each side. The roof is asphalt shingle on solid wood sheathing, with a 1-in. insulating board on the ceiling side ($U = 0.17$). Each end of the building has a 12-ft by 12-ft upward-opening door. The floor consists of 4 in. of concrete laid directly on drained ground.

(a) Compute the temperatures at the mean height of the wall and at the wall windows, and at the monitor and monitor windows, using the temperature-increase rule in which there is assumed a one per cent increase per foot of height above the datum temperature level up to 15 ft, and a 0.1-deg rise per foot above that level. Round off the final figures for use. (b) Prepare a table similar to Fig. 5-1 for the building, first finding the gross and glass areas of the similar lower east-west walls, lower south-north walls, east-west monitor walls, and north-south monitor walls, and of the monitor roof and the lower roof. (For simplicity, disregard the slope associated with the roof areas.) Also compute the floor area. (c) Using proper temperatures and U values, compute the losses through the various walls and roofs. (d) Compute the loss through the basement floor, using data from Table 4-19 (50 F ground water). (e) On an assumed air infiltration of two complete changes of air per hour, find the infiltration loss. (f) Find the total loss for the building, with no exposure factor allowed. Since this is an industrial building, heat sources in the building should offset the need of an exposure factor.

Ans. (a) Wall and windows, 63 F; monitor wall and windows, 68 F; main roof 67.5 F. (b) Each east-west 1920 sq ft net, 1080 sq ft glass; each lower north-south wall, 1151 sq ft net and 576 sq ft glass and door; each east-west monitor wall 390 sq ft net and 810 sq ft glass; each north-south monitor wall, 195 sq ft; roof of monitor, 6000 sq ft; total lower roof 14,000 sq ft; floor area, 20,000 sq ft. (c) Total lower east-west, 279,000 Btuh; total lower north-south, 155,400; total east-west monitor, 174,400; north-south monitor, 15,800; monitor roof, 79,600; main roof, 184,400. (d) 40,000 Btuh. (e) Volume 384,600 cu ft, and 1,000,000 Btuh loss. (f) 1,888,600 Btuh.

REFERENCES

1. *Engineering Weather Data* (Army, Navy, and Air Force Manual TM 5-785) 1963.
2. *Evaluated Weather Data for Cooling Equipment Design* (Santa Rosa, Calif.: Fluor Products Co., 1958).
3. *Evaluated Weather for Cooling Equipment Design, Addendum No. 1, Summer and Winter Data*, Santa Rosa, Calif.: Fluor Products Co., 1964).
4. L. W. Crow, *Study of Weather Design Conditions*, ASHRAE Research Project No. 23, Bulletin, 1963.
5. D. W. Boyd, *Climatic Information for Building Design in Canada*, Supplement No. 1, Nat'l Building Code of Canada, Nat'l Research Council, Ottawa.
6. *Monthly Normals of Temperature, Precipation and Heating Degree Days* (Washington, D.C.: U. S. Weather Bureau, 1962).
7. Carrier Air Conditioning Co., *Handbook of Air-Conditioning System Design* (New York: McGraw Hill Book Company, Inc., 1965).
8. C. Strock and W. B. Foxhall, *Handbook of Air-Conditioning Heating and Ventilating* (New York: The Industrial Press, 1959).

6

Cooling Systems and Cooling Load

6–1. AIR-CONDITIONING SYSTEMS

The computations involved in heating load show that the heat losses occurring in a space have to be offset by heat supplied to the extent that a satisfactory thermal environment is set up for occupants in that space. In similar manner refrigeration or cooling of a space is required when environmental conditions are such that energy must be removed to provide a suitable thermal environment for occupants. Because of the complexities involved in computing cooling load, it is desirable to have some acquaintance with the means by which cooling is accomplished and some introductory material in this connection is presented before the details of cooling-load computation are presented.

Except where direct radiation for cooling a space is provided, most cooling is brought about by supplying chilled air to a space. Many methods are available for this but, in general, these modifications follow a few basic patterns. In Fig. 6–1, which involves the use of all outside air

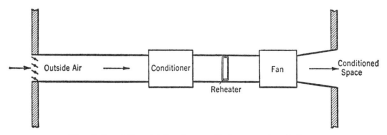

Fig. 6–1. All outside air supplied, no recirculation.

with none being recirculated, the outside air is merely chilled and dehumidified to a sufficiently low value and then delivered into the conditioned space. All of the air delivered to the space is then exhausted or wasted. A reheater may be required if the temperature of the air has been greatly reduced to drop its moisture level for control of humidity conditions in the space. Frequently a reheater is not necessary because the heat-transfer coils of the conditioner do not uniformly bring all of the air to the low dehumidification temperature and the portions, which are not so deeply cooled, mix with the coldest air to give tempered air sufficiently warm for delivery into a space. The objection or difficulty with too-cool air occurs when the occupants in the space are disturbed by the air striking them.

Figure 6-2 shows the conventional arrangement for cooling a space.

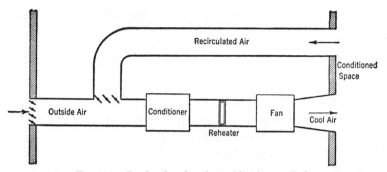

Fig. 6-2. Recirculated and outside air supplied.

In this a large amount of air from the space is returned (recirculated) with the amount of outside air used being merely that required for ventilation and in sufficient amounts to keep the odor level below objectionable limits in the space. Both the recirculated air and the outside air are cooled and dehumidified to a temperature and humidity ratio which falls on the condition line required to satisfy conditions in the space. Reheating may or may not be necessary.

Figure 6-3 shows the same arrangement of Fig. 6-2 but in this pattern

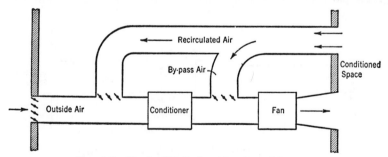

Fig. 6-3. Recirculated air with external-bypass.

some of the recirculated air is arranged to by-pass the conditioner for the purpose of tempering the conditioned air when this has to be chilled to a low temperature in the conditioner coils. This arrangement makes a reheater usually unnecessary and thus simplifies the operation. Usually the by-passing of some of the air for tempering the cold dehumidified air

Cooling Systems and Cooling Load 215

occurs entirely in the conditioner unit where two sets of dampers can direct air over the coils or permit some air to pass through without making contact with the coils, as explained in the next article.

6–2. SIMPLE AIR-CONDITIONING SYSTEM USING DAMPER CONTROL

Figure 6–4 shows a simple system for heating or cooling using face and by-pass dampers. By providing either heated or chilled water to the coil, this system can serve to heat or cool as may be required. The change-over may be a manual change with the season or can be made automatic year-round under the action of the auxiliary switch.

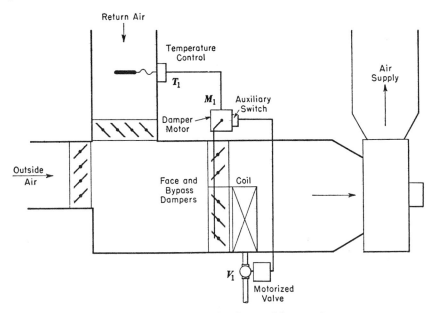

FIG. 6–4. Control system using face and bypass dampers.

In this system the temperature element T_1 is in the return air duct which leads from the space and consequently it responds directly to temperature conditions in the space. In heating, when the space temperature falls to a low value, indicated by the return air temperature, the motor damper repositions the face and bypass dampers to provide more heat, by permitting more air to flow over the heating coil and less air to flow through the bypass. When a system of this type is properly adjusted it can effectively heat over a wide range of conditions.

The same system, by adjustment of switching position, can also be operated when the coil is supplied with a refrigerant or chilled water. Cooling for a space can be accomplished under controlled conditions by operation of the face and bypass dampers. It is possible to arrange the system, when either heating or cooling, in such manner that when the coil face-dampers close completely the heating medium (or refrigerating medium) is shut off by operation of a motorized (or solenoid) valve. In cooling, a similar action can be employed to start and stop the compressor. It should be mentioned that in a simple system of this type the humidification is haphazard and thus a more complex control arrangement is required to effect humidity control.

6–3. ALL YEAR SYSTEM

Figure 6–5 shows a representative all-year system arranged for heating or cooling. The amounts of outdoor and recirculated air are controlled by the temperature elements T_{o1} and T_{o2}. These are arranged to operate so that when the outdoor temperature is either above or below a certain selected value the amount of outdoor air is changed by action of the damper motor M_o until a minimum amount, as set by the air selector control S, is reached. The thermostat T_R in the return air duct provides steam for heating through valve V_H when the return air temperature is too low, and cuts off the steam when the air is sufficiently warm. The coil is

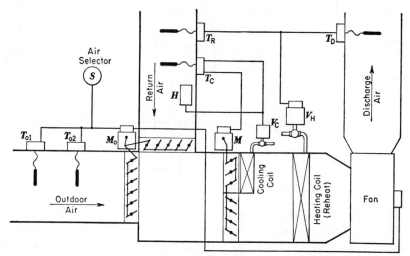

Fig. 6–5. Typical year-round system.

also responsive to the control of thermostat T_D, which can operate the heating coil to make certain that the air temperature is not less than a prescribed minimum or more than a prescribed maximum. Cooling is controlled by the cooling thermostat T_C and the humidistat H, both located in the return air duct. These operate motor M to control the face and bypass dampers of the coil. As the lower part of the temperature range is reached, the flow of cooling fluid through valve V_C is reduced or stopped. Under con-

Cooling Systems and Cooling Load 217

ditions of high humidity, H keeps valve V_C open to provide the greatest dehumidification. It is also possible to connect H to operate directly on the bypass dampers. If, under the action of the humidistat, the air temperature leaving the cooling coil is reduced to too low a value, the heating (reheat) coil, under the action of thermostat T_D, warms the air to an acceptably high level.

6-4. DUAL-DUCT SYSTEM

The system arrangements which are possible for environmental control are too numerous to describe in detail and can be hardly more than classified as to type in this text. However, because of its wide applicability, the dual-duct system will be discussed in some detail. It has been found to be very effective in structures having different zones or types of occupancy, such as office buildings, hospitals, hotels, schools, and stores. The system provides for central location of the heating and cooling plant, from which heating or cooling can be distributed to the multiple areas of a large building. Many variations in the arrangement of a dual-duct system are possible. Figure 6-6 uses more than the minimum number of elements but provides a degree of flexibility not possible with simpler dual-duct systems. All dual-duct systems use an air-supply duct carrying cool air and an air-supply duct carrying warm air. Air from these ducts enters a mixing and volume-control box which tempers and delivers the air-supply for a specific space or zone. Similar boxes serve other spaces to mix the cool and warm air to meet the desired conditions for each space.

Since both heating and cooling are available, the system can operate equally well in summer and winter and between-season shut-down of part of the system is not necessary. The system is also flexible in that it is possible to use outdoor air to its fullest extent for between-season cooling when the outdoor temperature is suitable. This is made possible by having the system arranged with two outside air inlets—the one provides for maximum outdoor air supply and the other for minimum outdoor air supply (ventilation requirement only). Under summer conditions it is undesirable to use more than minimum outdoor air and the maximum outdoor air dampers, which are interlocked with the others, are closed. Each of two supply fans carries about 50 per cent of the total air flow and at high loads the air flow in the two ducts is in this ratio. The cold air stream in summer usually provides air chilled to a 50 F to 55 F temperature range with the air almost saturated and at a low humidity ratio. When maximum cooling is required, a large fraction of this chilled air must be provided to each space and a smaller fraction of warm air is needed. Since the by-pass duct on the discharge side of the fans freely connects the two ducts, it follows that more air will flow in the cool duct. This occurs because the warm-air dampers at the mixing control boxes are largely in

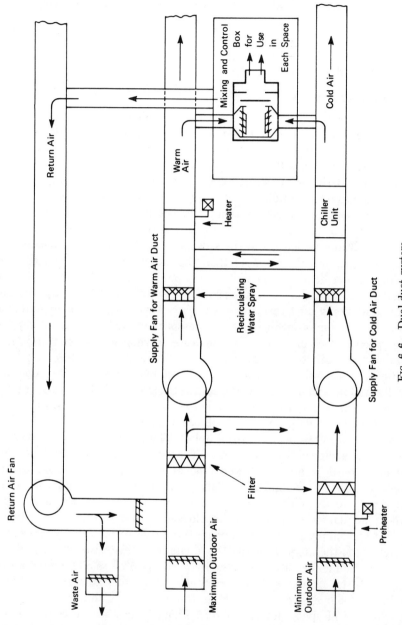

Fig. 6-6. Dual-duct system.

Cooling Systems and Cooling Load

closed position. In contrast to this, if the cooling load in a space is light, less chilled air is required and a larger fraction of warm air from the duct is needed. This type of operation delivers more high humidity warm air to the space than is justified from a moisture standpoint and the relative humidity in the space rises. To offset this happening, it is necessary to operate the heater coil in the warm-air duct. The warmer air makes it possible to provide a larger fraction of colder dryer air for the space and the relative humidity there lowers to a desirable range.

It is obvious that air cannot be stored in any space. Thus, the return or waste air system must always exhaust an amount of air equivalent to the amount of outdoor air brought into the system. Moreover, with each of the controlled spaces requiring varying amounts of cool or warm air, the flow in the two ducts can be widely different. Where minimum outdoor air is provided, the return air can move freely into either the warm air or the cold air system by means of the by-pass connection preceding the fans.

The dual-duct system is also advantageous in that for a large building in which the outside areas need heating in winter while the inside areas need cooling, it is easily possible to control and supply the needs of both types of areas. Many systems are not nearly so complex as that shown in Fig. 6–6 because in many, a single supply fan, by means of control dampers, delivers air to both of the ducts and forces the air through appropriate heaters or coolers. Similarly some systems do not employ a return air fan, although this makes control of the system less flexible. In winter the temperature of the supply air in the warm duct often ranges between 110 F and 130 F. In summer it is seldom brought above 86 F and frequently is at return-air temperature. A dual-duct system, although it cannot provide exact control of humidity in every space, particularly when the Sensible to Total-heat Ratio varies greatly in the different spaces, nevertheless does give approximate humidity control along with precise temperature control and it represents a very satisfactory method of operation.

6–5. DUAL-DUCT SYSTEM CALCULATIONS

The air flow in the system is based on the sensible-heat load and sufficient capacity must be provided in each duct to serve the maximum summer sensible cooling load (Q_{sc}) and the maximum winter sensible heating load (Q_{sw}). Making use of an equation in the form of equation 5–13

$$q_c = \frac{Q_{sc}}{1.08\,(t_r - t_c)} \tag{6-1}$$

$$q_w = \frac{Q_{sw}}{1.08\,(t_h - t_r)} \tag{6-2}$$

where q_c and q_w represent the cfm capacity required for the cold and hot ducts respectively

t_r = temperature held in a space, i.e., temperature of the recirculated air, F
t_c = temperature in cold-air duct, F
t_h = temperature in warm-air duct, F

The expressions apply to either a whole system or to a room, space, or zone of the system as may be required and the maximum flow ultimately determines the fan and duct sizes. In operation with either one or two supply fans the total flow q is

$$q = q_c + q_w \qquad (6\text{-}3)$$

It is, however, true that the solving of equations 6-1 and 6-2 does not offer an optimum solution for the summer and winter air flows because of by-passing between the two ducts and also because, for each space, air flow from both ducts occurs and the air-inlet temperature will differ both from the very low value t_c and the high value t_h. In fact the maximum flow in the warm duct usually occurs during light summer loads when the warm air temperature is held at a low level (80 F to 90 F) or during intermediate season operation. As a basis for starting the computation it is suggested that an appropriate summer cooling-air flow first be found and the warm-air flow then be expressed as an additional percentage of this flow, 30 to 40 per cent is suggested. This will give a trial air flow which can be corroborated or if necessary revised after use is made of the following equations.

For summer
$$1.08\, q_c\, (t_r - t_c) = Q_{sc} + 1.08\, q_w\, (t_h - t_r) \qquad (6\text{-}4)$$

For winter
$$1.08\, q_w\, (t_h - t_r) = Q_{sw} + 1.08\, q_c\, (t_r - t_c) \qquad (6\text{-}5)$$

Observation of equation 6-4 will show that the left-hand term, represents the cooling produced as the cold air (t_c) warms to space temperature (t_r). This cooling must serve two purposes, namely handling the sensible cooling load for summer Q_{sc} in Btuh, and also cooling the warm-duct air at t_h to the space temperature (t_r). Equation 6-5 works in similar manner but in reversed sense to offset the sensible winter heat loss, Q_{sw} Btuh and to heat the mixing air from the cold duct.

Example 6-1. For a dual-duct system find the basic air-flow patterns if the system has to serve under the following conditions: In summer a sensible cooling load of $Q_{sc} = 1{,}600{,}000$ Btuh, latent load 760 lb of water per hour, space temperature to be maintained $t_r = 78$ F (temperature of recirculated air), temperature of chilled air in cold air duct, $t_c = 52$ F, temperature of warm air in warm duct $t_h = 86$ F. In winter

Cooling Systems and Cooling Load

sensible heating load $Q_{sw} = 2{,}400{,}000$ Btuh, space temperature and temperature of recirculated air $t_r = 75$ F, temperature of chilled air in cold-air duct 55 F, temperature of hot air in warm duct $t_h = 125$ F. In summer chilled air cannot enter the space colder than 56.0 F without producing discomfort.

Solution: Make use of equation 6–1 to find a trial value of q_c

$$q_c = 1{,}600{,}000/1.08\,(78 - 52) = 57{,}000 \text{ cfm}$$

Increase this value by an assumed 35 per cent for total air flow

$$q = 57{,}000\,(1.35) = 77{,}000 \text{ cfm}$$

Use this value, 77,000 cfm with equations 6–3 and 6–4 for summer conditions

$$77{,}000 = q_c + q_w$$

$$1.08\, q_c\,(78 - 52) = 1{,}600{,}000 + 1.08\, q_w\,(86 - 78)$$

Solving these two equations simultaneously

$$q_c = 61{,}600 \text{ cfm} \qquad q_w = 15{,}400 \text{ cfm}$$

Make a rough check for the inlet temperature to the space (t_i)

$$(61{,}600)\,(t_c) + (15{,}400)\,(t_h) = 77{,}000\, t_i$$

$$t_i = 58.8 \text{ F}$$

Since this is higher than the 56 F minimum allowed on inlet, the flow ratios are suitable for summer operation.

For winter operation substitute in equations 6–3 and 6–5

$$77{,}000 = q_c + q_w$$

$$1.08\, q_w\,(125 - 75) = 2{,}400{,}000 + 1.08\, q_c\,(75 - 55)$$

$$q_w = 53{,}700 \text{ cfm} \qquad q_c = 23{,}300 \text{ cfm}$$

The problem is not complete until a check is made to determine whether the inlet air can be sufficiently dry at 52 F to absorb the 760 lb of water latent load without having the space humidity rise above an acceptable relative humidity. The situation is complicated by also having to consider the moisture brought in with the warm air stream. The solution will not be worked out here but is presented at the end of the chapter as a problem.

Notice that this total air flow of 77,000 cfm can effectively serve the sensible load with different warm and cool air flows for summer and winter. Other air flows are also possible to make a satisfactory solution and attention must also be given to part load operation.

6–6. OTHER AIR CONDITIONING SYSTEMS

So many air-conditioning-system arrangements exist that it is hardly possible to do more than classify them as to type. The basic central air-conditioning systems, which have already been described, involve the distribution of heating or cooling completely by means of convected air. One variant is the split system which involves the transport of heating or cooling by use of water or steam as well as by air. Split systems are particularly useful in the exposed or outer zones of large buildings where in winter so much heat loss occurs that it would be difficult to supply all the heating needed by air from the central system. Consequently, direct radiation is used to provide heat and the air-conditioning system provides tempered air, primarily for ventilation and humidity control.

The term high-velocity or conduit system is loosely applied to a split system arrangement in which specially-designed units in given rooms or spaces are provided with chilled water (direct refrigerant) and also with a limited amount of air, tempered and dehumidified to a proper range. This air, in passing over the chiller coil, inducts some air from the space to cool it and the mixed air entering the space is then able to control the temperature and humidity over relatively wide limits. Because air, in limited amounts only, is required, no provision is made for recirculating any air and all of the supply air is exhausted from the conditioned space. Because the units are able to provide for absorption of the air noise, the air can be supplied at high velocity and at moderate pressure, thus ducts of small size are possible with minimum demand on space usage. For year-round operation a warm water source is also provided to deliver tempered or even hot-water to the cooling coil for winter operation.

A different arrangement involves the use of units in a given space which do not receive conditioned air but are provided with fans for recirculating air from the space along with outside air for ventilation. This air is passed over coils supplied with chilled or warm water provided from a central unit. Here the central source provides no air but only a cooling or heating liquid medium. Finally, mention should be made of the complete air-conditioning units which are installed within spaces. These necessarily involve a refrigeration unit for providing the cooling, a means of dissipating the heat from the refrigeration unit either through an air circuit to outside or by means of circulating water, and, finally, a fan for recirculating air from the conditioned space along with sufficient outside air for ventilation. Frequently these are free-standing units that are placed along a wall of the space with suitable connections made for the various services required. A small version of a unit of this type is the familiar window unit which serves a similar function for an individual room or for a small space.

The equipment to heat and cool spaces will be discussed in later chapters of this book. The early pages of this chapter have merely called attention to certain means of producing desired results. The remainder of this chapter will be devoted to methods of computing the refrigeration or cooling potential required to serve a space under consideration and this is known as the cooling load.

Cooling Systems and Cooling Load

6-7. COOLING LOAD

Heat Sources for Cooling Load. In a space, the heat load which must be removed by means of cooling equipment arises from the following sources:

1. Heat transmission through barriers such as walls, doors, windows, ceilings, floors, and partitions, and caused by the different temperatures existing on the two sides of the barrier.
2. Heat from solar effects.
   a. Transmitted by radiation through glass and absorbed by inside surfaces and furnishings.
   b. Absorbed by walls or roofs exposed to rays of sun and transferred to the inside.
3. Heat and moisture introduced with infiltration air.
4. Heat load from occupants (sensible and latent).
5. Heat load from machinery, appliances, lights, and combustion equipment.

In addition, the cooling equipment must be adequate to bring ventilation air to the desired space temperature.

Item 1: Heat Transmission through Barriers. The transmission of heat through barriers is computed by the methods developed in Chapter 5 for computing the heating load. The basic equation 5-1, using appropriate inside and outside design temperatures, is applicable. In the case of a cooling load, the outside temperature is normally higher than the inside temperature, and consequently the equation appears as

$$Q = UA(t_o - t_i) \tag{6-6}$$

The inside design dry-bulb temperature can be selected from Table 10-1, and for general usage might be considered as 80 F. For most localities, the outside design temperature can be selected by use of Table 5-2.

TYPICAL OUTSIDE DAILY TEMPERATURE VARIATIONS IN SUMMER

Time	Dry Bulb (F)	Wet Bulb (F)
10 A.M.	89	74
12 M.	93	75
2 P.M.	95	76
4 P.M.	95	76
6 P.M.	94	76
8 P.M.	91	75
10 P.M.	88	74

This table gives the top temperature which experience has shown will not be exceeded more than one per cent of the time. The normal design values are based on this along with the wet-bulb temperature which might be expected to occur simultaneously with outside design dry-bulb temperature.

The dry-bulb temperature throughout a day is variable, usually reaching its maximum about 4:00 P.M., as can be seen in the accompanying tabulation.

The simple relation of equation 6–6, while perfectly accurate, is not

TABLE 6–1

INSTANTANEOUS RATES OF HEAT GAIN FROM TRANSMITTED DIRECT AND DIFFUSE OR SKY SOLAR RADIATION BY SINGLE-SHEET UNSHADED COMMON WINDOW GLASS

[For Clear Atmospheres and 18-Deg Declination, North (August 1)]

LATITUDE	SUN TIME		INSTANTANEOUS HEAT GAIN (BTU PER HR SQ FT)								
	A.M.→		N	NE	E	SE	S	SW	W	NW	Horiz.
30 Deg North	6 A.M.	6 P.M.	25	98	108	52	5	5	5	5	17
	7	5	23	155	190	110	10	10	10	10	71
	8	4	16	148	205	136	14	13	13	13	137
	9	3	16	106	180	136	21	15	15	15	195
	10	2	17	54	128	116	34	17	16	16	241
	11	1	18	20	59	78	45	19	18	18	267
	12		18	19	19	35	49	35	19	19	276
40 Deg North	5 A.M.	7 P.M.	3	7	6	2	0	0	0	0	1
	6	6	26	116	131	67	7	6	6	6	25
	7	5	16	149	195	124	11	10	10	10	77
	8	4	14	129	205	156	18	12	12	12	137
	9	3	15	79	180	162	42	14	14	14	188
	10	2	16	31	127	148	69	16	16	16	229
	11	1	17	18	58	113	90	23	17	17	252
	12		17	17	19	64	98	64	19	17	259
50 Deg North	5 A.M.	7 P.M.	20	54	54	20	3	3	3	3	6
	6	6	25	128	149	81	8	7	7	7	34
	7	5	12	139	197	136	12	10	10	10	80
	8	4	13	107	202	171	32	12	12	12	129
	9	3	14	54	176	183	72	14	14	14	173
	10	2	15	18	124	174	110	16	15	15	206
	11	1	16	16	57	143	136	42	16	16	227
	12		16	16	18	96	144	96	18	16	234
		P.M.→	N	NW	W	SW	S	SE	E	NE	Horiz.

NOTE. For total instantaneous heat gain, add these values to the Table 6–2 values.

TABLE 6-2

VALUES OF THE FACTOR X

INSTANTANEOUS RATES OF HEAT GAIN BY CONVECTION AND RADIATION FROM SINGLE-SHEET UNSHADED COMMON WINDOW GLASS

[For Clear Atmospheres and 18-Deg Declination, North (August 1); 80 F Indoor Temperature]

Sun Time	Dry-Bulb (deg F)	North Latitude Degrees	Instantaneous Heat Gain (Value X) (Btu per hr sq ft)								
			N	NE	E	SE	S	SW	W	NW	Horiz.
5 A.M.	74	↑	−6	−6	−6	−6	−6	−6	−6	−6	−6
6	74		−5	−4	−4	−5	−5	−6	−6	−6	−5
7	75		−5	−2	−2	−3	−5	−5	−5	−5	−3
8	77		−3	0	1	0	−2	−3	−3	−3	0
9	80		0	2	4	3	1	0	0	0	3
10	83		3	4	6	6	5	3	3	3	8
11	87		8	8	10	11	10	9	8	8	13
12	90	30, 40, 50	12	12	12	13	14	13	12	12	16
1 P.M.	93		15	15	15	16	17	17	17	15	20
2	94		16	16	16	16	18	19	19	17	21
3	95		17	17	17	17	19	21	21	19	21
4	94		16	16	16	16	17	20	20	19	19
5	93		15	15	15	15	15	18	19	18	17
6	91		13	13	13	13	13	14	15	15	13
7	87		8	8	8	8	8	8	8	8	8
8	85		6	6	6	6	6	6	6	6	6
9	83	↓	3	3	3	3	3	3	3	3	3

NOTE. For total instantaneous heat gain, add these values to the Table 6-1 values.

TABLE 6-3

APPLICATION FACTORS TO APPLY TO TABLES 6-1, 6-2, AND 6-4 TO OBTAIN INSTANTANEOUS RATES OF HEAT GAIN FOR VARIOUS TYPES OF SINGLE FLAT GLASS AND COMBINATIONS OF TWO SHEETS OF FLAT GLASS SPACED AT $\frac{1}{4}$ IN.

Glass *	Normal Incidence Transmittance	Factor to Apply to Table 6-1 (F_t)	Factors to Apply to Tables 6-2 and 6-3	
			C_1	C_2
Single common window..........	0.87	1.00	$1.0(X)$†	$+ 0.0(Y)$‡
Single regular plate.............	0.77	0.87	$1.0(X)$	$+ 0.25(Y)$
Single heat-absorbing plate.......	0.41	0.46§	$1.0(X)$	$+ 1.00(Y)$
Double common window.........	0 76	0.85	$0.6(X)$	$+ 0.10(Y)$
Double regular plate............	0 60	0.66§	$0.6(X)$	$+ 0.55(Y)$
Heat-absorbing plate outdoors ⎱ ... Regular plate indoors ⎰	0.35	0.37§	$0.6(X)$	$+ 0.75(Y)$

*Common window glass ⅛ in. thick. Plate glass ¼ in. thick.
†X values are Table 6-2 values.
‡Y Values are Table 6-4 values.
§For better precision, increase factors 10 per cent when glass is in the shade.

applicable in many instances because of the complicating effects of solar energy, which can so greatly increase the temperature of an outside wall surface that it greatly exceeds the outside air temperature, thus making the outside air temperature incorrect for use. Consequently, the methods described under "Item 2" (below) should be used whenever solar effects are significant in influencing the heat flow through a roof or wall.

Item 2: Heat from Solar Effects. When the sun's rays strike a pane of glass, a small amount of their energy is reflected, and the glass absorbs some energy, with consequent rise in temperature, but the greater part of the energy passes through. The solar radiation which enters depends on the angle of incidence of the sun's rays to the glass, more radiation being reflected as the angle differs from 90 deg. In the case of clear single glass perpendicular to the rays, about 87 per cent of the radiation passes through the glass undiminished.

The intensity of the sun radiation on walls or through glass varies with the time of day, with the direction in which each wall or window faces, with the day and season of the year, and with the latitude. The solar heat which passes through glass is largely absorbed by the interior furnishings, and by the inside walls and floors. Some of this heat is quickly given up and warms the inside air; but the heat which flows into heavy floors and walls may not be given up to the room air until hours later, so that the peak air temperature lags behind the peak of radiation from the sun. Glass, which is relatively diathermous to the radiant energy of the sun, acts as a screen preventing the outflow of low-temperature radiation from inside the building. Little sun energy which has entered a space can be reradiated out because of the "trap" effect of glass on low-temperature (long-wave-length) radiation. When the sun shines directly on glass, or when the glass receives diffused radiation from sun-warmed surfaces, the glass itself is heated to higher than outside temperature and thus heat is transferred from the glass.

Extensive research and experimentation have been carried out on the relationships between solar phenomena and building response. It is obvious that with the many variables involved, extremely elaborate procedures for computing solar loading have resulted. To consider all of these variables brings in complexities that become too awkward for ordinary computation unless computer programs to handle them are brought into play. Even when this is done, the end result is still dependent upon the variability of solar effects, radiation, diffusivity, altered response in relation to adjacent buildings, trees and the like. The ASHRAE Guide and Data Book (ref. 5) presents these detailed patterns most effectively and computer-program data are included. In this text a more simplified approach, which in itself is complex, is used for the solar computation. It has been found to give satisfactory results for design.

Cooling Systems and Cooling Load

The heat delivery of sun radiation through glass may be materially reduced by proper use of awnings, Venetian blinds, or shades, thereby reducing the load for the air-conditioning equipment. When awnings or other barriers to direct solar rays are employed they are most effective when placed outside the glass. The effectiveness of various window-shading devices in reducing sun radiation can be observed from the factors F_S of Table 6-5.

Tables 6-1 and 6-2, with the help of tables 6-3, 6-4, and 6-5, give a basis for finding the heat gain through glass under representative summer conditions.[1] Although the values indicated are for August, they can be used for other summer periods.

Example 6-2. Find the total instantaneous gain per square foot of single-sheet plate glass in a northwest wall at 3 P.M. in 40° north latitude location. Consider outside air at 95 F and inside conditions 80 F dry bulb.

Solution: Read, from Table 6-1, in the column labelled NW at the bottom, and opposite 40 deg North and 3 P.M., 79 Btuh per square foot for the direct-transmitted and diffuse-radiation gain. Next, correct this value by a proper factor from Table 6-3, namely 0.87, for single-plate glass: (79) (0.87) = 68.6. Read the convection and radiation gain from the glass surface, by use of Table 6-2, as 19. This reading is called the X value. For the second value, read from Table 6-4, in the NW column, and opposite 3 P.M. and 40 deg North latitude, the value 15; this is the Y value to use in the summation term $1.0X + 0.25Y$ of Table 6-3. Thus

$$(1.0)(19) + (0.25)(15) = 22.7$$

The total heat gain through the glass by direct transmittance, and by convection, and radiation from the glass surface, is thus

$$q_{glass} = (q_{tr})(F_t) + (C_1)(X) + (C_2)(Y) \tag{6-7}$$

Summing up, the total solar gain through the glass is

$$q_{glass} = (79)(0.87) + [(1.0)(19) + (0.25)(15)]$$
$$= 91.5 \text{ Btuh per sq ft} \qquad Ans.$$

This total in turn may have to be modified if shading is used, in which case an appropriate factor is selected from Table 6-5.

For walls and roofs the computations for solar heat gain are more complex, since the sun warms the surface and sets up a pulse flow of heat into the space as the solar warming rises to a peak and then falls off. The complex relationships associated with this problem can be somewhat simplified by use of the sol-air temperature concept, developed by Mackey and Wright (refs. 1, 2). The sol-air temperature is that fictitious temperature of outside air which in the absence of all radiation effects would give the same rate of heat entry into the surface as would exist under the actual combination of incident solar radiation, radiant energy exchange with the sky and outdoor surroundings, and convective heat exchange

---

[1] Tables 6-1 through 6-8 have been reprinted, by permission, from *Heating Ventilating Air Conditioning Guide 1955*, Chapter 13.

TABLE 6-4

HEAT ABSORBED IN GLASS. VALUES OF Y TO BE USED WITH FACTORS IN TABLE 6-3 IN THE DETERMINATION OF INSTANTANEOUS RATES IN HEAT GAIN DUE TO CONVECTION AND RADIATION FOR VARIOUS TYPES OF SINGLE GLASS AND COMBINATIONS OF TWO SHEETS OF GLASS SPACED AT $\frac{1}{4}$ IN.

[For Clear Atmospheres and 18-Deg Declination, North (August 1)]

Sun Time	Latitude	Values of Y (Btu per hr sq ft)*								
		N	NE	E	SE	S	SW	W	NW	Horiz.
5 A.M.	↑	0	0	1	0	0	0	0	0	0
6		4	16	18	9	1	1	1	1	3
7		2	24	30	20	2	2	2	2	11
8		2	22	33	25	2	2	2	2	21
9		2	16	30	29	8	3	3	3	32
10		3	5	25	27	14	3	3	3	37
11	40	3	3	12	21	18	3	3	3	42
12	Degrees	3	3	3	15	19	12	3	3	45
1 P.M.	North	3	3	3	3	19	22	10	3	44
2	Latitude	3	3	3	3	16	27	24	4	41
3		3	3	3	3	10	30	31	15	35
4		3	3	3	3	4	29	36	23	26
5		2	2	2	2	2	23	34	27	17
6		4	1	1	1	1	14	24	21	6
7	↓	0	0	0	0	0	2	3	3	1

Sun Time	Latitude	SE	S	SW	Sun Time	Latitude	SE	S	SW
5 A.M.	↑	0	0	0	5 A.M.	↑	2	0	0
6		7	1	1	6		13	1	1
7		18	2	2	7		22	2	2
8		22	2	2	8		28	3	2
9		24	3	3	9		30	13	3
10	30†	22	5	3	10	50†	31	20	3
11	Degrees	16	7	3	11	Degrees	27	25	5
12	North	6	9	4	12	North	20	27	17
1 P.M.	Latitude	3	9	14	1 P.M.	Latitude	9	25	26
2		3	6	21	2		3	22	32
3		3	5	27	3		3	16	33
4		3	3	26	4		2	7	31
5		2	2	21	5		2	2	26
6		1	1	11	6		1	1	17
7	↓	0	0	0	7	↓	0	0	7

* Values of Y for 8 and 9 P.M. are zero.
† For N, NE, E, W, NW, and horizontal, use 40 deg north latitude values.

TABLE 6-5
Effect of Shading upon Instantaneous Solar Heat Gain through Single Thickness of Common Window Glass (F_S)

Type of Shading	Finish on Side Exposed to Sun	Fraction of Gain through Unshaded Window	
Canvas awning			
Sides open..........................	Dark or medium	0.25	
Top and sides tight against building.......	Dark or medium	0.35	
Inside roller shade			
Fully drawn*.,.......................	White, cream	0.41	
	Medium	0.62	
	Dark	0.81	
Half-drawn*.........................	White, cream	0.71	
	Medium	0.81	
	Dark	0.91	
Inside Venetian blind, slats set at 45 deg†....	White, cream	0.56	
	Diffuse-reflecting aluminum metal	0.45	
	Medium	0.65	
	Dark	0.75	
Outside Venetian blind			
Slats set at 45 deg†.....................	White, cream	0.15	
Slats set at 45 deg,†‡ extended as awning fully covering window...............	White, cream	0.15	
Slats set at 45 deg, extended as awning covering ⅔ of window‡...............	White, cream	0.43	
Outside shading screen		Dark§	Green Tint‖
Solar altitude			
10 deg.............................	....	0.52	0.46
20 deg.............................	....	0.40	0.35
30 deg.............................	....	0.25	0.24
Above 40 deg.......................	....	0.15	0.22

* Roller shades are assumed to be opaque. Some white shades may transmit considerable solar radiation. For white translucent shades fully drawn, use 0.55; when half-drawn, use 0.77.

† Venetian blinds are fully drawn and cover window. It is assumed that the occupant will adjust slats to prevent direct rays from passing between slats. If slats are fully closed (slats set at 90 deg), use same factors as used for roller shade fully drawn.

‡ Commercial shade with wide slats. The sun may shine on window through sides of shade. Estimate the exposed portion of glass as unshaded.

§ Commercial shade, bronze. Metal slats 0.05 in. wide, 17 per inch, and set at 17-deg angle with horizontal. At solar altitudes below 40 deg, some direct solar rays are allowed to pass between slats, and this amount becomes progressively greater at low solar altitudes.

‖ Commercial aluminum shade. Slats 0.057 in. wide, 17.5 per inch, set at 17-deg angle with horizontal. At solar altitudes below 40 deg, some direct solar rays are allowed to pass between slats and this amount becomes progressively greater at low solar altitude.

with outdoor air. For example, for a dark-colored horizontal roof at 3 P.M., in full midsummer sun, the sol-air temperature might reach 132 F, in contrast with an air temperature of 95 F. The representative heat transfer through sun-activated walls and roofs is responsive to the character of the structure, to the daily range of temperature variation, to surface character (that is, dark or light), to the time of day, to latitude, and to other factors.

For solving this complex problem of solar heat gain, tables have been prepared which give the representative temperature differential to use with a building structure to find the total heat transmission resulting from the combination of air-temperature difference and solar effects. The tables are based on a 15-deg (95 F − 80 F) design datum for ordinary air-temperature difference. If the air-temperature difference is other than 15 deg, a correction must be made. Tables 6–6, 6–7, and 6–8 are to be employed for walls and roofs in finding all transmission summer space heat gains where solar effects are a factor.

Example 6–3. Find the total heat gain at 4 P.M. through a roof of 6-in. concrete and 2-in. insulation when the inside temperature is 80 F and the outside design temperature is 97 F.

Solution: Read 42 deg from Table 6–7 and, from Table 6–8, $U = 0.13$. Then the uncorrected gain from solar effect is

$$q = (0.13)(42) = 5.5 \text{ Btuh per sq ft}$$

A correction should be made because the design air-temperature difference is 97 F − 80 F = 17 deg instead of the 15 deg of the table. Use note 5 of Table 6–7 and add the 2-deg excess, so that

$$q = (0.13)(42 + 2) = 5.7 \text{ Btuh per sq ft} \qquad Ans.$$

A similar pattern applies for wall surfaces, for which Table 6–6 is used. The foregoing analysis of the solar and transmission load represents a method of accounting for the many variables which enter into the heat-flow pattern, and in connection with this pattern it should be realized that, during portions of the time, heat is flowing *into* the building from solar effects and thus a gradual *warming* of the masonry and structure is taking place, while at the other times heat is flowing *out of* the building and thus a gradual *cooling* of the masonry and structure is occurring. Often the full effect of direct sunlight falling on a surface does not occur until several hours subsequent to the exposure. Solar transmittance through glass causes a quick response, but on heavy, well-insulated walls the response can be much delayed. The tabular values in this chapter, when properly employed, partly compensate for the effect of the variables. However, in selecting the time at which the maximum load does occur from external effects, it is quite important to study the tables, particularly tables 6–1, 6–2, and 6–7, to find the time at which the maximum condition does occur.

TABLE 6-6
Total Equivalent Temperature Differentials for Calculating Heat Gain through Sunlit and Shaded Walls

NORTH LATITUDE WALL FACING:	Sun Time																SOUTH LATITUDE WALL FACING:
	A.M.						P.M.										
	8		10		12		2		4		6		8		10		12	
	Exterior Color of Wall (D = Dark, L = Light)																
	D	L	D	L	D	L	D	L	D	L	D	L	D	L	D	L	

								Frame									
NE	22	10	24	12	14	10	12	10	14	14	14	10	10	6	4	2	2	SE
E	30	14	36	18	32	16	12	12	14	14	14	10	10	6	6	2	2	E
SE	13	6	26	16	28	18	24	16	16	14	14	14	10	10	6	4	2	2	NE
S	−4	−4	4	0	22	12	30	20	26	20	16	14	10	10	6	6	2	2	N
SW	−4	−4	0	−2	6	4	26	22	40	28	42	28	24	20	6	4	2	2	NW
W	−4	−4	0	0	6	6	20	12	40	28	48	34	22	22	8	8	2	2	W
NW	−4	−4	0	−2	6	4	12	10	24	20	40	26	34	24	6	4	2	2	SW
N (shade)	−4	−4	−2	−2	4	4	10	10	14	14	12	12	8	8	4	4	0	0	S (shade)

							4-In. Brick or Stone Veneer + Frame										
NE	−2	−4	24	12	20	10	10	6	12	10	14	14	12	12	10	10	6	4	SE
E	2	0	30	14	31	17	14	14	12	12	14	14	12	12	10	8	6	6	E
SE	2	−2	20	10	28	16	26	16	18	14	14	14	12	12	10	8	6	6	NE
S	−4	−4	−2	−2	12	6	24	16	26	18	20	16	12	12	8	8	4	4	N
SW	0	−2	0	−2	2	2	12	8	32	22	36	26	34	24	10	8	6	6	NW
W	0	−2	0	0	4	2	10	8	26	18	40	28	42	28	16	14	6	6	W
NW	−4	−4	−2	−2	2	2	8	6	12	12	30	22	34	24	12	10	6	6	SW
N (shade)	−4	−4	−2	−2	0	0	6	6	10	10	12	12	12	12	8	8	4	4	S (shade)

							8-In. Hollow Tile or 8-In. Cinder Block										
NE	0	0	0	0	20	10	16	10	10	6	12	10	14	12	12	10	8	8	SE
E	4	2	12	4	24	12	26	14	20	12	12	10	14	12	14	10	10	8	E
SE	2	0	2	0	16	8	20	12	20	14	14	12	14	12	12	10	8	6	NE
S	0	0	0	0	2	0	12	6	24	14	26	16	20	14	12	10	8	6	N
SW	2	0	2	0	0	6	4	12	10	26	18	30	20	26	18	8	6	NW
W	4	2	4	2	4	2	6	4	10	8	18	14	30	22	32	22	18	14	W
NW	0	0	0	0	2	0	4	2	8	6	12	10	22	18	30	22	10	8	SW
N (shade)	−2	−2	−2	−2	−2	−2	0	0	6	6	10	10	10	10	10	10	6	6	S (shade)

							8-In. Brick or 12-In. Hollow Tile or 12-In. Cinder Block										
NE	2	2	2	2	10	2	16	8	14	8	10	6	10	8	10	10	10	8	SE
E	8	6	8	6	14	8	18	10	18	10	14	8	14	10	14	10	12	10	E
SE	8	4	6	4	6	4	14	10	18	12	16	12	12	10	12	10	12	10	NE
S	4	2	4	2	4	2	4	2	10	6	16	10	16	12	12	10	10	8	N
SW	8	4	6	4	6	4	8	4	10	6	12	8	20	12	24	16	20	14	NW
W	8	4	6	4	6	4	6	6	8	6	10	6	14	8	20	16	24	16	24	16	W
NW	2	2	2	2	2	2	4	2	6	4	8	6	10	8	16	14	18	14	SW
N (shade)	0	0	0	0	0	0	0	0	2	2	6	6	8	8	8	8	6	6	S (shade)

							12-In. Brick										
NE	8	6	8	6	8	4	8	4	10	4	12	6	12	6	10	6	10	6	SE
E	12	8	12	8	12	8	10	6	12	8	14	10	14	10	14	8	14	8	E
SE	10	6	10	6	10	6	10	6	10	6	12	8	14	10	14	10	12	8	NE
S	8	6	8	6	6	4	6	4	6	4	8	4	10	6	12	8	12	8	N
SW	10	6	10	6	10	6	10	6	10	6	10	6	8	6	10	6	12	8	14	10	NW
W	12	8	12	8	12	8	10	6	10	6	10	6	10	6	12	8	16	10	W
NW	8	6	8	6	8	4	8	4	8	4	8	4	8	4	6	10	6	10	6	SW
N (shade)	4	4	2	2	2	2	2	2	2	2	2	2	2	2	4	4	6	6	S (shade)

TABLE 6-6 (Continued)

North Latitude Wall Facing:	Sun Time																South Latitude Wall Facing:
	A.M.						P.M.										
	8		10		12		2		4		6		8		10		12	
	Exterior Color of Wall (D = Dark, L = Light)																
	D	L	D	L	D	L	D	L	D	L	D	L	D	L	D	L	D	L	

8-In. Concrete or Stone or 6-In. or 8-In. Concrete Block

	D	L	D	L	D	L	D	L	D	L	D	L	D	L	D	L	D	L	
NE	4	2	4	0	16	8	14	8	10	6	12	8	12	10	10	8	8	6	SE
E	6	4	14	8	24	12	24	12	18	10	14	10	14	10	12	10	10	8	E
SE	6	2	6	4	16	10	18	12	18	12	14	12	12	10	12	10	10	8	NE
S	2	1	2	1	4	1	12	6	16	12	18	12	14	12	10	8	8	6	N
SW	6	2	4	2	6	2	8	4	14	10	22	16	24	16	22	16	10	8	NW
W	6	4	6	4	6	4	8	6	12	8	20	14	28	18	26	18	14	10	W
NW	4	2	4	0	4	2	4	4	6	6	12	10	20	14	22	16	8	6	SW
N (shade)	0	0	0	0	0	0	2	2	4	4	6	6	8	8	6	6	4	4	S (shade)

12-In. Concrete or Stone

	D	L	D	L	D	L	D	L	D	L	D	L	D	L	D	L	D	L	
NE	6	4	6	2	6	2	14	8	14	8	10	8	10	8	12	10	10	8	SE
E	10	6	8	6	10	6	18	10	18	12	16	10	12	10	14	10	14	10	E
SE	8	4	8	4	6	4	14	8	16	10	16	10	14	10	12	10	12	10	NE
S	6	4	4	2	4	2	4	2	10	6	14	10	16	12	14	10	10	8	N
SW	8	4	8	4	6	4	6	4	8	6	10	8	18	14	20	14	18	12	NW
W	10	6	8	6	8	6	10	6	10	6	12	8	16	10	24	14	22	14	W
NW	6	4	6	2	6	2	6	4	6	4	8	6	10	8	18	12	20	14	SW
N (shade)	0	0	0	0	0	0	0	0	2	2	4	4	6	6	8	8	6	6	S (shade)

NOTES:

$$\left\{\begin{array}{l}\text{Total heat transmission from}\\ \text{solar radiation and tempera-}\\ \text{ture difference, Btu per hr sq}\\ \text{ft wall area}\end{array}\right\} = \left\{\begin{array}{l}\text{Temperature differential}\\ \text{from table}\end{array}\right\} \times \left\{\begin{array}{l}\text{Heat-transmission co-}\\ \text{efficient for wall, Btu}\\ \text{per hr sq ft deg F*}\end{array}\right\}$$

*Select coefficients from proper tables in Chapter 4.

1. *SOURCE.* Same as Table 6-7. A north wall has been assumed to be a wall in the shade; this is practically true. Dark colors on exterior surface of walls have been assumed to absorb 90 per cent of solar radiation and reflect 10 per cent; white colors, to absorb 50 per cent and reflect 50 per cent. This includes some allowance for dust and dirt, since clean, fresh white paint normally absorbs only 40 per cent of solar radiation.

2. *APPLICATION.* These values may be used for all normal air-conditioning estimates, usually without corrections, when the load is calculated for the hottest weather. Correction for latitude (note 3) is necessary only where extreme accuracy is required. There may be jobs where the indoor room temperature is considerably above or below 80 F, or where the outdoor design temperature is considerably above 95 F, in which case it may be desirable to make correction to the temperature differentials shown. The solar intensity on all walls other than east and west varies considerably with time of year.

3. *CORRECTIONS. Outdoor minus Room Temperature.* If the outdoor maximum design temperature minus room temperature is different from the base of 15 deg, correct as follows: When the difference is greater (or less) than 15 deg, add the excess to (or subtract the deficiency from) the above differentials.

Outdoor Daily Range Temperature. If the daily range of temperature is less than 20 deg, add 1 deg to every 2 deg lower daily range; if the daily range is greater than 20 deg, subtract 1 deg for every 2 deg higher daily range. For example, the daily range in Miami is 12 deg, or 8 deg less than 20 deg; therefore the correction is +4 deg.

Color of Exterior Surface of Wall. Use temperature differentials for light walls only where the permanence of the light wall is established by experience. For cream colors use the values for light walls. For medium colors interpolate halfway between the dark and light values. Medium colors are medium blue, medium green, bright red, light brown, unpainted wood, natural-color concrete, etc. Dark blue, red, brown, green, etc., are considered dark colors.

Latitudes Other Than 40 Deg North, and in Other Months. These table values will be approximately correct for the east or west wall in any latitude (0 deg to 50 deg north or south) during the hottest weather.

4. *INSULATED WALLS.* Use same temperature differentials used for uninsulated walls.

TABLE 6-7

TOTAL EQUIVALENT TEMPERATURE DIFFERENTIALS FOR CALCULATING HEAT GAIN THROUGH SUNLIT AND SHADED ROOFS

Description of Roof Construction*	Sun Time							
	A.M.			P.M.				
	8	10	12	2	4	6	8	10	12

Light-Construction Roofs—Exposed to Sun								
1″ wood,† or 1″ wood† + 1″ or 2″ insulation	12	38	54	62	50	26	10	4	0

Medium-Construction Roofs—Exposed to Sun								
2″ concrete, or 2″ concrete + 1″ or 2″ insulation, or 2″ wood†	6	30	48	58	50	32	14	6	2
2″ gypsum, or 2″ gypsum + 1″ insulation, or 1″ wood† or 2″ wood† or + 4″ rock wool 2″ concrete or in furred ceiling 2″ gypsum	0	20	40	52	54	42	20	10	6
4″ concrete, or 4″ concrete with 2″ insulation	0	20	38	50	52	40	22	12	6

Heavy-Construction Roofs—Exposed to Sun								
6″ concrete	4	6	24	38	46	44	32	18	12
6″ concrete + 2″ insulation	6	6	20	34	42	44	34	20	14

Roofs Covered with Water—Exposed to Sun								
Light-construction roof with 1″ water	0	4	16	22	18	14	10	2	0
Heavy-construction roof with 1″ water	−2	−2	−4	10	14	16	14	10	6
Any roof with 6″ water	−2	0	0	6	10	10	8	4	0

Roofs with Roof Sprays—Exposed to Sun								
Light construction	0	4	12	18	16	14	10	2	0
Heavy construction	−2	−2	2	8	12	14	12	10	6

Roofs in Shade								
Light construction	−4	0	6	12	14	12	8	2	0
Medium construction	−4	−2	2	8	12	12	10	6	2
Heavy construction	−2	−2	0	4	8	10	10	8	4

TABLE 6-7 (Continued)

NOTES:

$$\left\{ \begin{array}{l} \text{Total heat transmission from} \\ \text{solar radiation and temperature} \\ \text{difference, Btu per hr sq ft of} \\ \text{roof area} \end{array} \right\} = \left\{ \begin{array}{l} \text{Temperature differential} \\ \text{from table} \end{array} \right\} \times \left\{ \begin{array}{l} \text{Heat transmission co-} \\ \text{efficient for summer,} \\ \text{Btu per hr sq ft deg F} \end{array} \right\}$$

1. *SOURCE.* Calculated by Mackey and Wright method. Estimated for July in 40 deg north latitude. For typical design day, where the maximum outdoor temperature is 95 F and minimum temperature at night is approximately 75 F (daily range of temperature, 20 F); mean 24 hr temperature 84 F for a room temperature of 80 F. All roofs have been assumed a dark color which absorbs 90 per cent of solar radiation, and reflects only 10 per cent.

2. *APPLICATION.* These values may be used for all normal air conditioning estimates, usually without correction, in latitude 0 deg to 50 deg north or south when the load is calculated for the hottest weather. Note 5 explains how to adjust the temperature differential for other room and outdoor temperatures.

3. *PEAKED ROOFS.* If the roof is peaked and the heat gain is primarily due to solar radiation, use for the area of the roof the area projected on a horizontal plane.

4. *ATTICS.* If the ceiling is insulated and if a fan is used in the attic for positive ventilation, the total temperature differential for a roof exposed to the sun may be decreased 25 per cent.

5. *CORRECTIONS. For Temperature Difference When Outdoor Maximum Design Temperature Minus Room Temperature Is Other Than 15 Deg.* If the outdoor design temperature minus room temperature is other than the base of 15 deg, correct as follows: When the difference is greater (or less) than 15 deg add the excess to (or subtract the deficiency from) the above differentials.

For Outdoor Daily Range of Temperature Other Than 20 Deg. If the daily range of temperature is less than 20 deg, add 1 deg for every 2 deg lower daily range; if the daily range is greater than 20 deg, subtract 1 deg for every 2 deg higher daily range. For example, the daily range in Miami is 12 deg, or 8 deg less than 20 deg; therefore the correction is +4 deg at all hours of the day.

For Light Colors. Credit should not be taken for light-colored roofs except where the permanence of the light color is established by experience, as in rural areas or where there is little smoke. When the exterior surface of roof exposed to the sun is a light color, such as white or aluminum (which absorb approximately 50 per cent and reflect 50 per cent of the solar radiation), add to the temperature differential for a roof in the shade 55 per cent of the difference between the roof in the sun and the roof in the shade. When the roof exposed to the sun is a medium color such as light gray, or blue, or green, or bright red, add 80 per cent of this difference.

* Includes ⅜ in. felt roofing with or without slag. May also be used for shingle roof.
† Nominal thickness of the wood.

In the case of inside rooms, where heat flows through partitions, equation 6-6 can be employed directly, and in this equation the t_o term should, of course, be the temperature on the far side of a barrier wall.

Item 3: Heat and Moisture Introduced in Infiltration Air. Infiltration air enters by leakage through window cracks, through doors when opened, and through porous walls or other openings. Table 5-3 furnishes data on infiltration air through window cracks, and Table 6-9 gives values for determining the amount of air entering through door openings in commercial establishments.

In the case of an air-conditioning system in which there is an adequate supply of ventilating air, and in which the conditioned air is delivered into a space which has no exhaust fans, the door and infiltration loss will be greatly reduced, and may even cease, when the air supplied to the space flows outward. Expressed in other terms, whenever the fan system provides an excess pressure in the enclosure, infiltration is reduced or eliminated. Infiltration air which enters a space brings with it not only the high-temperature outside air, with its associated sensible heat as it cools to inside temperature, but also the moisture associated with the frequently humid outside air, which entails a latent heat load on the space.

Independent of infiltration air, the air-conditioning system must circulate a certain amount of fresh air required for ventilation, while it must also recirculate an amount of air sufficiently large that the total passed through

TABLE 6-8
SUMMER COEFFICIENTS OF HEAT TRANSMISSION U OF FLAT ROOFS COVERED WITH BUILT-UP ROOFING*
[Btu per (Hour) (Square Foot) (F Deg Difference between the Air on the Two Sides)]

TYPE OF ROOF DECK (CEILING NOT SHOWN)	THICKNESS OF ROOF DECK (IN.)	INSULATION ON TOP OF DECK (COVERED WITH BUILT-UP ROOFING)								
		No Ceiling—Underside of Roof Exposed					Furred Ceiling with Air Space, Metal Lath, and Plaster			
		No Insulation	Insulating Board † Thickness (in.)				No Insulation	Insulating Board † Thickness (in.)		
			$\frac{1}{2}$	1	$1\frac{1}{2}$	2		$\frac{1}{2}$	1	$1\frac{1}{2}$	2
Flat metal roof deck	4-ply felt roof	0.73	0.35	0.23	0.17	0.13	0.40	0.25	0.18	0.14	0.12
	Ditto + $\frac{1}{2}$-in. slag	0.54	0.30	0.20	0.16	0.13	0.34	0.22	0.16	0.13	0.11
Precast cement tile	4-ply felt roof $1\frac{5}{8}$	0.67	0.33	0.22	0.17	0.13	0.38	0.24	0.18	0.14	0.12
	Ditto + $\frac{1}{2}$-in. slag $1\frac{5}{8}$	0.50	0.28	0.20	0.15	0.12	0.32	0.21	0.17	0.13	0.11
Concrete	4-ply felt roof 2	0.65	0.33	0.22	0.16	0.13	0.37	0.24	0.18	0.14	0.12
	4	0.59	0.31	0.21	0.16	0.13	0.36	0.23	0.17	0.13	0.12
	6	0.54	0.30	0.20	0.16	0.13	0.33	0.22	0.17	0.13	0.11
	Ditto + $\frac{1}{2}$-in. slag 2	0.49	0.28	0.20	0.15	0.12	0.31	0.21	0.16	0.13	0.11
	4	0.46	0.27	0.19	0.15	0.12	0.30	0.21	0.16	0.13	0.11
	6	0.42	0.26	0.19	0.14	0.12	0.29	0.20	0.16	0.13	0.10
Gypsum and wood fiber‡ on $\frac{1}{2}$-in. gypsum board	4-ply felt roof $2\frac{1}{2}$	0.34	0.23	0.17	0.13	0.12	0.25	0.18	0.14	0.12	0.097
	$3\frac{1}{2}$	0.28	0.20	0.15	0.12	0.11	0.21	0.16	0.13	0.11	0.094
	Ditto + $\frac{1}{2}$-in. slag $2\frac{1}{2}$	0.29	0.20	0.16	0.13	0.11	0.22	0.16	0.13	0.11	0.093
	$3\frac{1}{2}$	0.25	0.18	0.14	0.12	0.10	0.19	0.15	0.13	0.10	0.090
Wood§	4-ply felt roof 1	0.43	0.26	0.19	0.15	0.12	0.29	0.20	0.15	0.13	0.11
	$1\frac{1}{2}$	0.33	0.22	0.17	0.13	0.11	0.24	0.18	0.14	0.12	0.097
	2	0.29	0.20	0.16	0.13	0.11	0.22	0.16	0.13	0.11	0.094
	3	0.22	0.16	0.13	0.11	0.09	0.17	0.13	0.12	0.10	0.085
	Ditto + $\frac{1}{2}$-in. slag 1	0.35	0.23	0.17	0.14	0.11	0.25	0.18	0.14	0.12	0.10
	$1\frac{1}{2}$	0.29	0.20	0.15	0.12	0.10	0.21	0.17	0.13	0.11	0.093
	2	0.26	0.19	0.14	0.12	0.10	0.20	0.15	0.13	0.10	0.090
	3	0.20	0.15	0.12	0.10	0.09	0.16	0.13	0.11	0.09	0.081

* The summer coefficients have been calculated with an outdoor wind velocity of 8 mph. For summer an inside surface conductance of 1.2 has been used instead of the regular 1.65 value. In all of these roofs a 4-ply felt roof has been assumed $\frac{3}{8}$ in. thick, thermal conductivity = 1.33. Pitch and slag have been assumed as an additional thickness of $\frac{1}{2}$ in. which has been assigned a thermal conductivity of 1.0. In both cases thermal conductivity refers to one *inch* thickness.
† If corkboard insulation is used, the coefficient U may be decreased 10 per cent.
‡ 87$\frac{1}{2}$ per cent gypsum, 12$\frac{1}{2}$ per cent wood fiber. Thickness indicated includes $\frac{1}{2}$ in. gypsum board. This is a poured roof.
§ Nominal thickness of wood is specified, but actual thickness was used in calculations.

TABLE 6-9
DOOR INFILTRATION IN SUMMER FOR COMMERCIAL ESTABLISHMENTS*

Application	Revolving and Swinging Doors Opening to Outside		Average Occupancy (Patrons and Employees) on Which Values Are Based (min.)
	Infiltration per Person in Room (cfm)		
	72-In. Revolving Door	36-In. Swinging Door	
Bank....................	7.5	10.0	20
Barber shop..............	3.5	4.5	45
Broker's office...........	5.0	6.5	30
Candy and soda store.....	5.0	6.5	30
Cigar store..............	15.0	20.0	10
Department store (small)...	5.0	6.5	30
Dress shop...............	2.0	2.5	75
Drugstore................	10.0	13.0	15
Furrier..................	2.0	2.0	90
Lunchroom...............	5.0	6.5	30
Men's shop..............	3.5	4.5	45
Office (professional)......	2.5	3.0	60
Restaurant...............	2.0	2.5	75
Shoe store...............	3.5	4.5	45

When Doors Are Left Open Continuously

72-in. revolving door (panels open)..1200 cfm
36-in. swinging door (standing open).......................................800 cfm

NOTE. The values for swinging doors and for doors left open hold only where such doors are in one wall only, or where the doors in other walls are the revolving type. If swinging doors are used for access (or if doors are left open) in more than one wall, the infiltration cannot be estimated. The values for revolving doors hold regardless of number or location.
To determine the total cfm infiltration due to opening of doors, multiply the design number of occupants by the factor from the above table for the kind of establishment in question. When there is more than one door, treat them as though there were only one, except in the case of open doors.
* Reprinted, by permission, from code of Air Conditioning and Refrigerating Machinery Association.

the conditioning equipment and delivered into the space will absorb the sensible load from all sources, and at the same time, the supply air will increase in moisture content and absorb the latent heat load.

Item 4: Heat Load Arising from Occupants in a Given Space. The heat load arising from occupants in a given space can be computed with the help of Figs. 10-3 and 10-4 or Table 10-3. The heat load should be considered in terms of the type of activity of the individuals, and it should be divided into two parts—namely that part associated with sensible cooling of the individual, and that part associated with latent cooling of the individual.

Cooling Systems and Cooling Load

TABLE 6-10
HEAT LOAD FROM EQUIPMENT

Device	Heat Dissipation During Running Time (Btuh)	
	Sensible Heat	Latent Heat
Electric lights and appliances, per installed kilowatt	3413	....
Motors, with connected load in same room,* per horsepower		
$\frac{1}{8}$–$\frac{1}{2}$ hp range	4250	....
$\frac{1}{2}$–3 hp range	3700	....
3–20 hp range	2950	....
Electric coffee urn		
3-gal	2200	1500
5-gal	3400	2300
Gas stove burner	3100	1700
Heating water	3150	3850
Domestic gas oven	8100	4000
Gas-heated coffee urn		
3-gal	2500	2500
5-gal	3900	3900
Steam-heated equipment, per square foot†		
Steam-heated surface		
Not polished	330	....
Polished	130	....
Insulated surface	80	....
Steam table	200	1000
Beauty parlor hair-driers		
Blower-type	2300	400
Helmet-type	1870	330
Restaurant, per meal served	30 [Btu]	....

* With connected load outside room, subtract 2544 Btuh per rated horsepower.
† For hooded equipment, reduce values by 50 per cent.

Item 5: Heat Load from Miscellaneous Equipment Installed in a Given Space. Data for computation of the heat load from miscellaneous equipment installed in a conditioned space are given in Table 6-10. This particular part of the heat load should always be considered in computing contributions to the cooling load, and one must carefully consider whether it is all sensible, or partly sensible and partly latent. Moreover, care must be exercised to ascertain whether or not this part of the heat load really appears within the space. For example, in a conditioned space a motor driving a shaft to which the load is attached in another room delivers only its electrical and mechanical losses into the conditioned space, and the shaft work (useful horsepower) is distributed elsewhere. Similarly, if a gas burner in a conditioned space is heating water or other material which is used elsewhere, all of the thermal energy of combustion should not be charged to the conditioned space.

The heat equivalent of the energy delivered to air by a fan in sending the air through the ductwork usually appears in the conditioned space, and becomes part of the cooling load. If the air horsepower exerted on the air is known, this heat can be easily calculated. However, a 10 per cent additional amount is often added to the sensible-heat load of a space to take care of contingencies, and the fan load is assumed in this amount.

Air in ductwork running through unconditioned space absorbs a certain amount of heat and entails an additional load on the conditioning system. With long runs this heating may be considerable and should always be calculated. In section 12-6, methods of calculation are shown.

Item 6: Ventilation Air. Air required for ventilation must be supplied in sufficient quantity to satisfy codes or ordinances when these are applicable; but even when such codes are not applicable, sufficient ventilation air must be supplied to control the odor level and to set up satisfactory conditions for occupants or for stored equipment. From a viewpoint of cooling-load design, the occupants of a space should be provided with not less than 5 to $7\frac{1}{2}$ cfm per person when no smoking is involved, and 25 to 40 cfm per person if smoking is a factor.

The ventilation load is independent of the internal space load if the ventilation air is passed through the conditioner, since under these circumstances the air is cooled before entering the space. This is in contrast to the warm infiltration air which enters directly into the space and must be removed along with the other heat loads in the space.

In analyzing the performance of cooling equipment one method is to use the so-called *bypass factor*. When using this method, it is considered that a portion of the air moves through a cooler as though unaltered by the action of the coil, while the remainder of the air is brought completely to the coil temperature. The first part of the air is known as the bypass air and, if considered unaltered, it can be regarded as entering directly into the conditioned space load in exactly the same way as does infiltration air. The mixture leaving is, of course, an average of the two types of air.

Manufacturers frequently describe their cooling equipment in terms of coil efficiency, which is merely another way of stating the bypass factor. For example, to say that a coil is 75 per cent efficient means that 25 per cent of the air bypasses the coil, while the remaining 75 per cent is brought to coil temperature. Thus if the bypass analysis for cooling load is used, the bypassed air would be considered part of the space load; whereas, using another viewpoint, the over-all action of the conditioner is considered merely to produce supply air at its mixed temperature to serve the given space. The cooling (heat-source) load of the space is specifically called either the space load or the internal cooling load.

Cooling Systems and Cooling Load

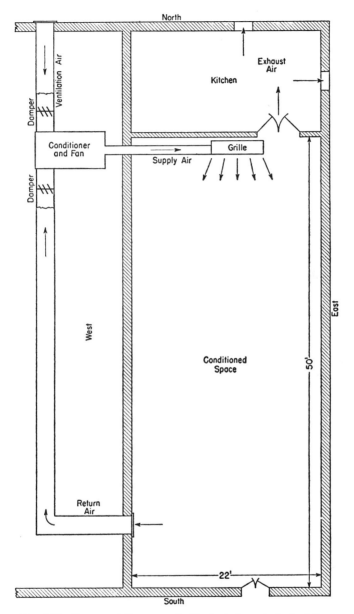

Fig. 6–7. Representative space to illustrate layout of a building for cooling load. Conditioner in adjacent space or basement.

6-8. COOLING-LOAD CALCULATIONS

In this section, examples will be worked to illustrate detailed methods of computing the cooling load of a space. Figure 6-7 shows a representative space which will be analyzed. For the conditioned space shown in the illustration, the items entering into the cooling load are:

1. Solar-type transmission loads on the east and south walls and on the roof of the building.
2. Solar-type transmission through the glazed portions of the east and south walls.
3. Wall-transmission heat gain through the north and west partition walls and through the floor.
4. Heat gain from occupants in the space.
5. Heat gain from miscellaneous sources and equipment.
6. Heat gain from infiltration air.

The specifications for the conditioned space diagrammed in Fig. 6-7 are summarized as follows:

SPECIFICATIONS OF SPACE TO BE AIR-CONDITIONED IN RESTAURANT BUILDING

Location. Philadelphia, Pa., on corner; entrance and front of building face south.

South Wall. Front 22 ft wide inside, 13 ft high; plate glass, 14 ft total width by 8 ft high; two glazed 6-ft by 7.5-ft swinging doors; remainder of wall 12-in. brick plastered inside, thin synthetic marble outside. Awning over whole front.

East Wall. Side street; 50 ft long by 13 ft high inside; 12-in. brick, plastered inside; three 3-ft by 4-ft vertically-pivoted metal windows equipped with awnings.

North-Wall Partition. Separates dining area from unconditioned kitchen, where a possible peak temperature of 109 F may occur; 22 ft wide by 13 ft high, with two 3-ft by 7-ft swinging doors to kitchen. Partition is wood frame with wood lath and plaster on both sides, and rock-wool fill between studding.

West Wall. Adjacent to unconditioned space; 12-in. brick with $\frac{3}{4}$-in. plaster on one side; 50 ft long by 13 ft high (inside dimensions).

Floor. Conventional wooden joists with wood lath and plaster below and covered with $\frac{1}{4}$-in. linoleum laid on wood flooring.

Ceiling. Under roof. Construction: 4-in. wood rafters, metal lath and plaster below, topped by 1-in. wooden roof deck, covered by roofing paper.

Occupancy. Space for total of 50 people at tables and counters; 5 employees in conditioned part of restaurant. Peak occupancy during noon hour.

Equipment. In conditioned space:
Two gas-heated coffee urns, 3-gal size.
One toaster, 2650-w rating.
(Above items hooded and vented to outside, using small exhaust fan, mounted on wall outside conditioned space)
No motors.
Electric lights: 2400 w total; never in use during peak thermal load on sunny day.

Design Conditions. From Table 5-2, for Philadelphia the outside design conditions are taken as 95 F db and 78 F wb. From Table 10-1, the inside design conditions for 95 F outside are selected, for normal application, as 80 F db and 67 F wb.

Cooling Systems and Cooling Load

Example 6–4. Calculate the transmission and solar heat gains into the restaurant at peak occupancy load about 12 noon (1:00 P.M. daylight time).

Solution: The transmission through the walls follows the general conduction formula (eq 6–6), modified, where necessary, by solar effects:

$$Q = UA(t_o - t_i) = UA\Delta t$$

South Wall

The gross area of the wall = 22 ft × 13 ft = 286 sq ft. The glass area = 14 ft × 8 ft = 112 sq ft, and the area of the door (glazed) = 6 ft × 7.5 ft = 45 sq ft. From Table 4–6, for the 12-in. brick wall with plaster, $U = 0.34$; and from Table 6–6, read 4 deg for the total equivalent temperature difference at 12 noon, sun time. Thus, for the net wall, the heat transmitted is

$$Q_w = (0.34)(286 - 112 - 45)(4) = 175 \text{ Btuh}$$

For the total glass area of 157 sq ft, read (for 12 noon at 40 deg north latitude for a south wall) the transmittance as 98 Btuh per sq ft. Correct this by the factor $F_t = 0.87$ for plate glass, from Table 6–3, and by a factor of $F_S = 0.25$ for the awning; thus the transmittance for the glass is

$$(98)(0.87)(0.25) = 21.3 \text{ Btuh per sq ft}$$

From Table 6–2, the glass convection and radiation gain at noon is 14 for the X value, and in Table 6–4 the Y value is 19. Thus the correct convection and radiation value (not greatly reduced by the awning) is, for the plate glass,

$$1.00(X) + 0.25(Y) = 14 + 0.25(19) = 18.7 \text{ Btu per sq ft}$$

The glass gain is therefore

$$(21.3 + 18.7)(157) = 6280 \text{ Btuh}$$

Thus the total gain for the south wall is

$$175 + 6280 = 6455 \text{ Btuh}$$

East Wall

The gross area of the wall = 50 ft × 13 ft = 650 sq ft. The glass area = 3 (3 ft × 4 ft) = 36 sq ft. Thus the net wall area is 614 sq ft. By Table 6–6, for the east wall (dark material) at noon, $\Delta t = 12$ deg for 12-in. brick. Thus, for the net wall, the heat transmitted is

$$Q_w = (0.34)(614)(12) = 2510 \text{ Btuh}$$

For the glass, use Table 6–1 and read 19 for the east wall at noon; by Table 6–2, read 12; and by Table 6–3, the factor $F_t = 1.0$. For the awning, by Table 6–5, read $F_S = 0.25$. Thus the glass gain per square foot is

$$(19)(F_t)(F_S) + 1.0(X) + 0.0(Y) = (19)(1.0)(0.25) + (1.0)(12) = 16.7$$

The glass gain is therefore

$$(16.7)(36) = 602 \text{ Btuh}$$

Thus the total gain for the east wall is

$$2510 + 602 = 3112 \text{ Btuh}$$

North Wall Partition

The gross area of the partition = 22 ft × 13 ft = 286 sq ft. The area of the thin panel doors = 2(3 ft × 7 ft) = 42 sq ft. Thus the over-all coefficient of heat transfer for the partition (U_w) is 0.078 (tables 4–8 and 4–5), and for the door (U_{door}), 0.85 (Table 4–16 footnote). For the partition, the heat transferred is therefore

$$Q_w = 0.078(286 - 42)(109 - 80) = 552 \text{ Btuh}$$

And, for the door,
$$Q_{door} = (0.85)(42)(109 - 80) = 1034 \text{ Btuh}$$
Thus the total north-wall heat gain is
$$552 + 1034 = 1586 \text{ Btuh}$$

West Wall

The area of the wall = 50 ft × 13 ft = 650 sq ft, and the wall is next to a building space which is perhaps 6 deg cooler than outside, say 89 F. For this wall, by equation 4–18 and Table 4–3, the thermal resistance to heat flow is
$$R = \frac{1}{U} = \frac{1}{1.65} + \frac{12}{5.0} + \frac{0.75}{5.0} + \frac{1}{1.65} = 3.76$$
Therefore the over-all coefficient of heat transfer is
$$U = \frac{1}{3.76} = 0.266 \text{ (say 0.27) Btuh per sq ft}$$
Thus the heat gain is
$$Q_w = UA(t_1 - t_2) = (0.27)(650)(89 - 80) = 1580 \text{ Btuh}$$

Floor

The area of the floor = 50 ft × 22 ft = 1100 sq ft. The over-all coefficient of heat transfer, U, is 0.24 (from Table 4–10). The cellar is cooler than outside—say 6 deg cooler. Thus the heat gain through the floor is
$$Q_f = (0.24)(1100)(89 - 80) = 2376 \text{ Btuh}$$

Ceiling and Roof

The ceiling area = 50 ft × 22 ft = 1100 sq ft. Read the summer U value as 0.29 (from Table 6–8) and the Δt value at 40 deg for medium construction at 12 noon (from Table 6–7). Therefore the heat gain through the ceiling and roof is
$$Q_{roof} = (0.29)(1100)(40) = 12{,}760 \text{ Btuh}$$

Thus the total transmission and solar gain through the walls and the roof, all in sensible form, is
$$6455 + 3112 + 1586 + 1580 + 2376 + 12{,}760 = 27{,}869 \text{ Btuh} \qquad Ans.$$

Example 6–5. Calculate the heat gain from the occupants of the restaurant discussed in the preceding example—50 diners and 5 employees. The restaurant is maintained at 80 F db and 67 F wb.

Solution: Using the values given for persons eating and for persons at moderate work, in Table 10–3, the total heat gain is found as follows:

	Sensible Heat Gain	Latent Heat Gain
50 × 250	12,500	....
50 × 250	....	12,500
5 × 305	1,525	....
5 × 545	....	2,725
Total heat gain	14,025	15,225

Example 6–6. Find the heat gain from the equipment and from other sources in the restaurant.

Solution: First, consider the two coffee urns. They are gas-heated and of 3-gal

Cooling Systems and Cooling Load

size. From Table 6–10, read 2500 Btuh as the sensible heat load and as the latent heat load. Thus

$$2 \times 2500 \text{ sensible} = 5000 \text{ Btuh}$$
$$2 \times 2500 \text{ latent} = 5000 \text{ Btuh}$$

Next, consider the single toaster at 2650 watts (2.650 kw). From Table 6–10, read 3413 Btuh as the sensible heat load. Thus the heat gain is

$$2.650 \times 3413 = 9050 \text{ Btuh, sensible}$$

At 12 noon on a sunny day there are no motors or lights in use.

Because the cooking equipment is hooded and vented, reduce the sensible gain by 50 per cent:

$$(5000 + 9050)(0.50) = 7025 \text{ Btuh}$$

Considering the meals served, at 30 Btu each, with 50 seats and 2 meals per seat per hour, there is an additional sensible heat gain of

$$30 \times 2(50) = 3000 \text{ Btuh}$$

Thus the total sensible heat gain is

$$7025 + 3000 = 10{,}025 \text{ Btuh} \qquad Ans.$$

The total latent heat gain, with the 50 per cent reduction, is

$$(5000)(0.50) = 2500 \text{ Btuh} \qquad Ans.$$

Example 6–7. For the same restaurant, find (a) the air required for ventilation, (b) the infiltration air, and (c) the cooling load for the infiltration air. (d) Sum up the total heat-gain load from all sources.

Solution: (a) For a restaurant, Table 10–4 recommends 15 cfm per person (preferred) and 12 cfm (minimum). Using 15 cfm per person, with 55 people, the outside air required for ventilation is

$$15 \text{ cfm} \times 55 = 825 \text{ cfm} \qquad Ans.$$

The restaurant volume is 50 ft × 22 ft × 13 ft = 14,300 cu ft, and 825 cfm of ventilation air would indicate that the following number of positive changes of outside air per hour take place:

$$\frac{825 \text{ cfm} \times 60 \text{ min per hr}}{14{,}300} = 3.46$$

b) Even though this air is being forced out, some infiltration can occur if wind is blowing outside. Assume a 10-mph wind and use Table 5–3 to determine the infiltration:

EAST WALL: Window crack; metal windows, vertically pivoted and center-opening. Thus

$$3 \times [(3+4)2 + 4] = 54 \text{ ft of crack}$$

or

$$88 \text{ cfh} \times 54 = 4752 \text{ cfh}$$

SOUTH WALL: Door-crack leakage only, since tight plate-glass show-windows should show no leakage. Thus

$$3(7.5) + 2(6) = 34.5 \text{ ft of crack}$$

or

$$138 \text{ cfh} \times 34.5 = 4760 \text{ cfh}$$

WEST WALL: Solid, no leakage.

NORTH WALL PARTITION: The kitchen must have adequate exhauster fans and these of course exhaust some air from the conditioned space. Take crack infiltration as one-half of the total:

$$\tfrac{1}{2}(4752 + 4760) = 4756 \text{ cfh}$$

An additional amount of infiltration occurs because of door-opening, and can be estimated by Table 6-9. Using 2.5 cfm per transient person inside the space, the door leakage is

$$2.5 \times 50 = 125 \text{ cfm}$$

or

$$125 \times 60 = 7500 \text{ cfh}$$

Thus the total infiltration is

$$4756 + 7500 = 12{,}256 \text{ cfh} \qquad Ans.$$

It is improbable that this much total infiltration will occur, considering the slight positive pressure maintained by the conditioner, but it is possible—because of exfiltration into the hood outlets and into the kitchen exhaust system. In a conservative design, this total amount should be considered.

c) The outside air is at 95 F db and 78 F wb. From Plate I, or by computation, moisture = 0.01168 lb per pound; dew point = 71.6 F; volume = 14.35 cu ft per pound of dry air.

The inside air is at 80 F db and 67 F wb. From Plate I, or by computation, the dew point = 60.1 F and 0.0112 lb per pound, and the specific volume = 13.84 cu ft per pound of dry air.

The infiltration air from outside = 12,256 cfh, or

$$\frac{12{,}256}{14.35} = 854 \text{ lb per hr}$$

The sensible heat to be removed is, by equation 3-36

$$Q_s = (854)(0.244)(95 - 80) = 3140 \text{ Btu} \qquad Ans.$$

The moisture load in the infiltration air is

$$(854)(0.0168 - 0.0112) = 4.78 \text{ pounds per hour}$$

The corresponding latent load, by equation 3-52 is

$$Q_L = 4.78 \times 1100 = 5260 \text{ Btuh}$$

d) Therefore the total space cooling load is as follows, in Btu per hour:

	Sensible	Latent	
Transmission and solar gain	27,869	....	
Human load	14,025	15,225	
Miscellaneous sources	10,025	2,500	
Outside infiltration air	3,140	5,260	
Total	55,059	22,985	Ans.

Note that this space cooling load (internal cooling load) is not the load which exists in the conditioner. The conditioner load must also include the additional load associated with cooling the ventilating air, and in some instances, any additional cooling required for extreme dehumidification followed by reheating.

Cooling Systems and Cooling Load 245

6-9. COOLING-LOAD AIR QUANTITIES

The quantity of air circulated must be adequate to handle the cooling load as the air warms up to room temperature from its supply temperature. The lower the supply temperature the less the quantity which must be circulated, but the minimum temperature is determined by the system arrangement, the necessity of avoiding drafts and cold regions, the ceiling height, and the throw required. Summer conditioning installations are usually designed to supply air at 5 to 20 deg below room temperature. Specially designed nozzles in certain suitable locations have been used with air as low as 30 deg below room temperature. For practical purposes, a reduction of 2 deg in delivered air temperature per foot of height from floor to ceiling should not be exceeded in determining an inlet temperature of the supply air. Special grilles, nozzle arrangements, room shape, or load conditions may make other temperature differentials desirable or even necessary, but the 2-deg-per-foot limit should never be exceeded for preliminary calculations. Grille and nozzle recommendations from the manufacturer's data must be followed in making final designs for temperature difference, expected air throw, and air distribution.

After the space cooling load (internal cooling load) is computed by the methods shown earlier in this chapter, the air quantity can be found by making use of the method described in section 3-8, or by the sensible heat load equation (eq 3-36) rewritten here

$$Q_S = m_a C_p (t_i - t_S) = 0.244 m_a (t_i - t_S) \qquad (6\text{-}8)$$

where Q_S = space (internal) cooling load, in Btu per hour;
m_a = air supplied to space, in pounds per hour;
C_p = specific heat of the moist air (approximately 0.244), in Btu per pound per degree Fahrenheit;
t_i = inside space temperature to be maintained, in degrees Fahrenheit dry bulb;
t_S = supply air temperature entering space, in degrees Fahrenheit dry bulb.

The $\Delta h/\Delta W$ slope, or the same slope, measured by the sensible heat ratio (SHR), is extremely important in the solution of cooling-load problems.

Refer to Fig. 6-8, which represents a skeleton psychrometric chart plot of a cooling and dehumidification process. On this chart a line has been drawn through 2, the inside space temperature, with the slope of the line set in terms of the *SHR* or $\Delta h/\Delta W$ slope. It will be noticed that this line intersects the saturation line at D. This point represents the *apparatus dew-point temperature* of the conditioner coil. Conditioner coils cannot bring all of the air passing through them to the coil surface temperature.

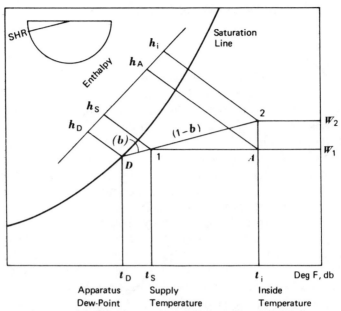

Fig. 6-8. Skeleton psychrometric chart showing use of apparatus dew point and bypass.

This fact makes the coil perform in a manner similar to what would be accomplished if a portion of the air were brought to coil temperature and the remainder bypassed. A dehumidifying coil thus produces unsaturated air at a higher temperature than the coil temperature. Again referring to Fig. 6-8, notice that in terms of the length of the whole line D to 2, the length D to 1 is proportional to the air bypassed (b), and the length 1 to 2 is proportional to the air not bypassed $(1 - b)$. Because of the chart construction, it is also closely true that in terms of dry-bulb temperatures,

$$b \equiv \frac{t_S - t_D}{t_i - t_D} \tag{6-9}$$

And in a similar manner,

$$1 - b \equiv \frac{t_i - t_S}{t_i - t_D} \tag{6-10}$$

where b is the fraction of air bypassed, or coil bypass factor, expressed as a decimal, and where the temperatures are dry-bulb values. Now, as before,

$$Q_S = m_a(0.244)(t_i - t_S)$$

and

$$Q_S = m_a(0.244)(t_i - t_D)(1 - b) \tag{6-11}$$

If air at average conditioner supply conditions is employed in the above

Cooling Systems and Cooling Load 247

equation at (say) $v = 13.5$ cu ft per pound of dry air, then

$$Q_S = \frac{(\text{cfm})(60)}{13.5}(0.244)(t_i - t_D)(1 - b) \qquad (6\text{-}12)$$

and therefore

$$\text{cfm} = \frac{Q_S}{(1.08)(t_i - t_D)(1 - b)} \qquad (6\text{-}13)$$

where cfm = volume flow capacity of supply fan delivering cooled air, in cubic feet per minute;
t_i = dry-bulb temperature of conditioned space, in degrees Fahrenheit;
t_D = apparatus dew-point temperature, in degrees Fahrenheit;
b = bypass factor of cooling coil;
Q_S = space sensible load, in Btu per hour.

It is frequently convenient to add the bypass ventilation air to the sensible and latent space loads for convenience in calculation, and in this connection the equations appear as

$$Q_{vS} = (\text{cfm})(1.08)(t_o - t_i)(b) \qquad (6\text{-}14)$$

$$Q_{vL} = \frac{(\text{cfm})(60)}{13.5}(W_o - W_i)(1100)(b) \qquad (6\text{-}15)$$

$$Q_{vL} = 4880 \, (\text{cfm})(W_o - W_i)b \qquad (6\text{-}16)$$

where Q_{vS} = sensible heat gain from ventilation air bypassed into space, in Btu per hour;
Q_{vL} = latent heat gain from ventilation air bypassed into space, in Btu per hour;
cfm = ventilation air supplied, in cubic feet per minute;
t_o, t_i = outside temperature and space temperature, in degrees Fahrenheit;
W_o, W_i = humidity ratio of outside air and space air, in pounds per pound of dry air;
b = bypass factor.

Example 6-8. For the same restaurant, and its cooling load as summarized in example 6-7, compute (a) the sensible heat ratio and (b) the enthalpy-humidity difference ratio. Assume that the minimum supply-air temperature is 64 F and find (c) the minimum supply air for the space, in pounds per hour, and (d) the humidity ratio and wet bulb of the supply air that are required in order to meet design conditions. (e) Find the amount of air delivered by the supply fan per minute. (f) When the conditioner, with its bypass condition, cools both the ventilation and recirculated air to supply conditions, what refrigeration tonnage (200 Btu per minute per ton) is absorbed by the conditioner coils?

Solution: (a) Referring to the data of example 6–7, the sensible heat ratio is

$$SHR = \frac{55{,}059}{55{,}059 + 22{,}985} = 0.705 \qquad Ans.$$

b) By equation 3–49

$$\frac{\Delta h}{\Delta W} = \frac{55{,}059 + 22{,}985}{22{,}985/1100} = 3730 \qquad Ans.$$

c) By equation 6–8,

$$55{,}059 = m_a\,(0.244)\,(80 - 64)$$
$$m_a = 14{,}000 \text{ lb per hr} \qquad Ans.$$

d) By equation 3–52, with air in the room at 80 F and 67 F, with $W_2 = 0.0112$ lb per pound of dry air, and with $h_i = 31.5$ Btu per pound of dry air,

$$22{,}985 = (14{,}000)\,(0.0112 - W_1)\,(1100)$$

Thus the humidity ratio is $W_1 = 0.0107$ lb per pound $\qquad Ans.$

From Plate I, for 0.0107 lb at 64 F db, read the wet bulb as 59.9 F. $\qquad Ans.$

The corresponding specific volume v is 13.4 cu ft per lb, and the enthalpy is 26.0 Btu per pound of dry air.

e) The amount of supply air delivered by the fan is

$$\text{cfm} = \frac{m_a v}{60} = \frac{(14{,}000)\,13.4}{60} = 3130 \text{ cfm} \qquad Ans.$$

f) Since ventilation air is supplied at 825 cfm, at 95 F and 78 F, and $v = 14.35$ cu ft per lb, then

$$m_v = \frac{(825)\,(60)}{14.35} = 3450 \text{ lb per hr}$$

Also,

$$h_o = 41.4 \text{ Btu per lb dry air}$$

and

$$W_o = 0.0168 \text{ lb per lb dry air}$$

The cooling load of the ventilating air is

$$m_v(h_o - h_S) = 3450\,(41.4 - 26.0) = 53{,}100 \text{ Btuh}$$

and the cooling load of the recirculated air is

$$m_r(h_i - h_S) = (14{,}000 - 3450)\,(31.5 - 26.0) = 58{,}030 \text{ Btuh}$$

Thus the refrigeration absorbed by the coils is

$$\frac{53{,}100 + 58{,}030}{(200)\,(60)} = 9.2 \text{ tons} \qquad Ans.$$

Example 6–9. Make use of the bypass method of solution, along with apparatus dew point, to find a solution for part f of example 6–8. Assume a bypass factor of 0.40 as a starting point.

Solution: First, for computation purposes consider that the ventilation air which bypasses the cooling coils becomes part of the space (internal) load. Then, by equations

Cooling Systems and Cooling Load

6–14 and 6–16, and for 825 cfm,

$$Q_{vS} = (825)(1.08)(95 - 80)(0.40) = 5340 \text{ Btuh}$$
$$Q_{vL} = (4880)(825)(0.0168 - 0.0112)(0.40) = 9010 \text{ Btuh}$$

Thus the sensible heat ratio is

$$SHR = \frac{Q_s}{Q_T} = \frac{55{,}059 + 5340}{78{,}044 + 5340 + 9010} = 0.65$$

Use the sensible heat ratio as shown in Fig. 6–8 and following the method described in this section, find the apparatus dew point. Run the datum line through the room condition of 80 F db and 67 F wb parallel to the SHR slope of 0.65 as shown in the protractor hemisphere. The intersection at the saturation line shows an apparatus dew point of 50.1 F for t_D. Using equation 6–9 and the by-pass factor of 0.4, given

$$b = \frac{t_S - t_o}{t_i - t_D} = \frac{t_S - 50.1}{80 - 50.1} = 0.40$$

and thus

$$t_S = 62 \text{ F}$$

If this temperature is a lower supply temperature than can be used by the occupants of the space, then a different bypass factor must be employed or else reheat must be used. For the total conditioner load we must now add the cooling load required to bring the rest of the outside air to room conditions. In this connection extend equations 6–14 and 6–16 to cover the $(1 - b)$ fraction of the outside air:

$$Q_{oS} = (\text{cfm})(1.08)(t_o - t_i)(1 - b)$$
$$= (825)(1.08)(95 - 80)(0.60) = 8010 \text{ Btuh}$$
$$Q_{oL} = (4880)(825)(0.0168 - 0.0112)(0.60) = 13{,}520 \text{ Btuh}$$

The conditioner load in sensible and latent forms is thus

$$Q_S = 55{,}059 + 5340 + 8010 = 68{,}409 \text{ Btuh}$$
$$Q_L = 22{,}985 + 9010 + 13{,}520 = 45{,}515 \text{ Btuh}$$

and the total refrigeration tonnage absorbed by the conditioner coils is therefore

$$\frac{68{,}409 + 45{,}515}{(200)(60)} = 9.5 \text{ tons} \qquad Ans.$$

The designer should set operating standards to give an economically desirable pattern of operation. As it happens, the bypass factor of 0.4 used to illustrate the foregoing problem is high; most coils operate at $b = 0.25$ or below.

6–10. SUMMARY

In this chapter the basic approaches of providing space cooling were first described. This was followed by presentation of methods for finding the magnitude of the cooling required for representative spaces—the so-called cooling load. The matching of equipment to create a system design to most effectively serve the cooling load is the problem of the engineer. Following chapters in the book introduce types of equipment which can be used both for heating and for cooling and this knowledge must be coordinated for use before endeavoring to refine designs for optimum performance.

PROBLEMS

6-1. A dual-duct system is to be used in a complex building which has a summer sensible cooling load under maximum conditions of 2,100,000 Btuh with a latent load of 810 lb of water per hour. Spaces of the system must be maintained at 78 F or lower and with wet-bulb temperature never to exceed 69 F, temperature of chilled air in the cold air duct is 54 F, temperature in the warm duct is 88 F. Outside summer design conditions are 95 F db and 69 F wb. (a) Compute a trial cooling air flow and a total air flow. (b) Use these data and find a suitable design cold-air duct flow and warm-air duct flow. (c) Find for these data an approximate value of mixed inlet air temperature to a representative space.

Ans. (a) 110,000 cfm; (b) cold 89,600 cfm, warm 20,400 cfm; (c) 60.3 F

6-2. The building of the preceding problem has a design winter heat loss of 3,700,000 Btuh. A space air temperature of 75 F is to be maintained and the warm air duct temperature can reach 130 F. Cold air duct is 55 F. The total fan capacity, sized for summer, is 110,000 cfm. Compute the probable air flow in the warm and cold air ducts at winter design operating conditions.

6-3. Refer to example 6-1 and for certain conditions assume that the total return air flow from the conditioned space, 77,000 cfm at 78 F db and not over 69 F wet bulb, is recirculated except for 15 per cent of its volume which is replaced by fresh (ventilation) air. The warm-air-duct fan delivers 15,400 cfm of the recirculated air, and at the conditions under consideration there is no flow in the by-pass duct after the fans. Air enters the conditioned spaces at 58.8 F average temperature. It leaves the conditioner at 52 F and a relative humidity of 90 per cent. The cold-air-duct air is at 52 F and a relative humidity of 90 per cent. The warm-air-duct air is at 86 F. The latent load of the conditioned spaces is 760 lb of moisture per hour. (a) In line-diagram form, sketch the system to show the air flows indicated, adding additional data as these are computed in subsequent parts of the problem. Note that of the 77,000 cfm leaving the space, 15 per cent or 11,550 cfm at 78 F dry bulb pass to waste and 65,450 cfm pass to the fans. (b) Find the weight flow rate in the cold-air duct. (c) Compute the humidity ratio of the air at exit from the conditioner. Then find how much additional moisture each lb of this air must absorb to handle the space moisture load. Note that the warm-duct air is at same humidity ratio as the space air and cannot pick up additional moisture load. (d) Find the humidity ratio, using the data of part (c) and read the space wet-bulb temperature at 78 F dry bulb from the psychrometric chart.

Ans. (b) 61,600 cfm measured at 78 F or 268,000 lb per hr; (c) $W = 0.00743$
$\Delta W = 0.00284$; (d) $W = 0.01027$, $t_{wb} = 65$ F

6-4. Refer to example 6-1 and for certain conditions assume that 15 per cent of the total air flow from the conditioned space, 77,000 cfm at 78 F, is replaced as ventilation (fresh) air with the remainder recirculated. Assume that the 78 F recirculated air should not exceed 68 F wet bulb. The latent load of the space is 760 lb of water per hour. Air enters the space at 58.8 F. Air leaves the conditioner at 52 F at $\varphi = 0.9$. Air by-passes up from the cold duct and is warmed to 86 F by the warm-air-duct heater with 15,400 cfm at this temperature, entering the warm-air duct for delivery. (a)

Cooling Systems and Cooling Load 251

Sketch the system, in line-diagram form, to show the air flows as indicated, adding temperature values as these are found in the solution. Note that of the 77,000 cfm leaving the space, 15 per cent or 11,550 cfm at 78 F pass to waste. Of the remaining 65,450 cfm, 53,900 cfm pass to the cold-duct fan and 11,550 cfm continue to the warm-duct fan. (b) Make a temperature correction to find the fresh air cfm measured at outside conditions of 95 F db and 78 F wb. (c) Making use of equation 3–45 in consistent cfm, find the temperature of the outside and recirculated air after mixing and before entry to the conditioner. (d) Find the humidity ratio of the exit air from the conditioner. (e) For the 15,400 cfm at 86 F (15,160 cfm at 78 F) warm-air supply, find how much air has to be bypassed from the cold side before the chiller, if measured at 78 F. (f) Find the resulting cfm air flow into the chiller at 78 F db and convert this to a mass flow rate in lb per hr. (g) Find the pick up per lb of chilled-air flow and the final humidity ratio of the space air at recirculation conditions.

$Ans.$ (b) 11,920 cfm; (c) 81.0 F; (d) $W = 0.00743$;

(e) 3,610 cfm at 78 F, $m = 268,200$ lb per hr; (g) $\Delta W = 0.00283$,

$W = 0.01026$ lb per lb air at 78 F and 65 F wb

6–5. Refer to problem 6–1 and check to see whether under the design conditions indicated the latent load can be properly served. All of the 78 F space air is recirculated except for 10,310 cfm of ventilation air at 95 F lb 78 F wb taken from outside. Assume that no bypassing occurs, after the fans, and that the cold and warm air duct flows are as computed for problem 6–1. (a) Make a line-diagram flow chart and indicate on this known temperatures and other pertinent data for a 2-fan dual-duct system. (b) Find the approximate mass flow rate in the cold duct and use this to compute how much water vapor in pounds is absorbed by each pound of chilled air entering the space. (c) Assuming that the cold air leaves the chiller at 54 F at 90 per cent relative humidity, compute the humidity ratio of the final space air and find its wet-bulb temperature at 78 F dry bulb.

6–6. A survey of a theater which is to be maintained at 82 F db and 68 F wb shows that at 4:00 P.M. there is a transmission and solar heat gain of 60,000 Btuh, and that on the average 500 people are then present; also, that at 8:30 P.M. there is a transmission load of 20,000 Btuh and that on the average 1000 people are then present. Determine the maximum load and the time at which it occurs. Light and equipment loads can be neglected. $Ans.$ 370,000 Btuh at 8:30

6–7. In the theater in problem 6–6, how many pounds of air at 62 F db must be supplied to the conditioned space under maximum sensible-load conditions?

$Ans.$ 54,300 lb/hr

6–8. A plate-glass window 5 ft by 7 ft is placed in the south wall of a building located near the 40-deg-latitude line. The outside design temperature is 93 F and the inside temperature is held at 80 F. (a) Compute the total heat gain through this window at 11 A.M. and 5 P.M., in Btu per hour. (b) What are the corresponding values of heat gain if an effective, open-side awning is installed which prevents all direct sunlight from falling on the glass? $Ans.$ (a) 3250 Btuh, 887 Btuh; (b) 813 Btuh, 221 Btuh

6–9. Work problem 6–8, for 5 P.M. only, but assume that the window is in a west wall in a 30-deg-latitude area. $Ans.$ (a) 1268 Btuh; (b) 317 Btuh

6–10. Find the total instantaneous gain per square foot for single-sheet plate glass in the east wall of a building located 40 deg north latitude. The time is 11 A.M. Consider the outside air at 95 F and inside conditions at 80 F db. $Ans.$ **63.4 Btuh per sq ft**

6-11. On the first floor the glass area in the south wall of a certain office building at 40 deg north latitude consists of twenty 5-ft by 8-ft steel-sash windows containing single-thickness glass. At 3 P.M. in mid-August, what is the probable heat gain by solar effects through the glazed area? *Ans.* 48,800 Btuh

6-12. Assume that the glazed area in problem 6-11 is equipped with Venetian blinds of a white to cream shade and compute the probable solar transmission under the conditions indicated in the problem. *Ans.* 27,300 Btuh

6-13. At 4 P.M. on a sunny afternoon in mid-August, what is the heat gain through the flat roof of a building 80 ft by 200 ft in size? Assume that the roof, which is approximately 6 in. thick, consists essentially of 2-in. concrete and a furred ceiling with metal lath and plaster, with no insulation. The outdoor temperature is 97 F and the indoor temperature can be considered as 80 F. *Ans.* 285,000 Btuh

6-14. For a certain conditioned space, 2000 cfm of air at 62 F db and 59 F wb are supplied. This consists in part of 800 cfm of outside air measured at 96 F db and 76 F wb. The space is maintained at 80 F db and 50 per cent relative humidity. Determine (a) the amount of air recirculated, in cubic feet per minute, and (b) the tons of refrigeration. *Ans.* (a) 1298 cfm; (b) 6.4 tons

6-15. What is the heat transmission at 5 P.M. through the south wall of problem 6-8 if the 140 sq ft of nonglass area is dark red and of brick-veneer and frame construction and has a U of 0.28 Btu per hr sq ft deg F and is not shaded in any way? *Ans.* 822

6-16. Work problem 6-15, but assume that the wall area faces southwest instead of south. *Ans.* 1252 Btuh

6-17. Determine from the following peak-load data the total occupant and equipment load in a restaurant maintained at 83 F db: average number of customers, 20; waiters, 3; one 5-gal electric coffee urn; one gas-stove burner (hooded); one 48-in. by 18-in. steam table; and a revolving display sign carrying six 60-w lights and driven by a $\frac{1}{4}$-hp motor. *Ans.* 14,358 Btuh sensible, 15,785 Btuh latent

6-18. A space to be conditioned has a load of 70,000 Btuh and a sensible heat ratio of 0.8. It is to be maintained at conditions not to exceed 78 F and 55 per cent relative humidity. The direct-expansion cooling coils have an equivalent bypass factor of 0.3. The maximum duct capacity of the system is limited to 4000 cfm. Disregard any outside air required for ventilation. (a) Make use of the SHR on the psychrometric chart (Plate I) and, by extension from room conditions, find the required temperature of the cooling coils. (b) For the bypass factor of 0.3 and room air conditions, find the air supply temperature required. (c) How much air, in pounds per minute and in cubic feet per minute, must pass through the system to carry the space load? *Ans.* (a) 58 F; (b) 64 F; (c) 3660 cfm

6-19. Find the items called for in problem 6-18, with similar specifications except that 1000 cfm of outside air at 85 F db and 75 F wb is required for ventilation. (a) First find the mixing temperature of the outside and recirculated air (assume that the total weight of air supplied to the space is the same as in problem 6-18; this condition, of course, sets the inlet temperature the same as before). (b) By using the inlet temperature and the mixing temperature, find the coil temperature. (c) What equivalent bypass factor must now hold for the equipment? If this is too high, additional bypassing external to the coil or reheat can be employed. *Ans.* (a) 79.8 F mixing; (b) 55.8 F dew point on coil; (c) 0.33 bypass ns and Cooling Load

6-20. A conditioner receives air at 80 F db and 70 F wb, and the air leaves at 63 F db. The wet coils of the conditioner are at a surface temperature of 55 F. Find (a) the bypass factor for the conditioner and (b) the amount of heat removed from each 1000 cfm supplied to the conditioner. *Ans.* (a) 0.32; (b) 31,760 Btuh

6-21. A spray-type air washer with four banks of sprays cools air at 90 F db and 75 F wb down to a water temperature of 60 F. The saturated air leaving the conditioner is reheated to 67 F db. Find (a) the weight of water removed from each pound of air entering the conditioner, (b) the heat absorbed by the spray water, and (c) the heat furnished by the reheater coils per pound of dry air supplied.
Ans. (a) 0.0042 lb; (b) 12 Btu; (c) 1.7 Btu

REFERENCES

1. C. O. Mackey and L. T. Wright, "Periodic Heat Flow—Homogeneous Walls or Roofs," *Trans. ASHVE*, Vol. 50 (1944), pp. 293-312.

2. C. O. Mackey and L. T. Wright, "Periodic Heat Flow—Composite Walls or Roofs," *Trans. ASHVE*, Vol. 52 (1946) pp. 283-96.

3. J. P. Stewart, "Solar Heat Gain Through Walls and Roofs for Cooling Load Calculations," *Trans. ASHVE*, Vol. 54 (1948) pp. 361-88.

4. G. V. Parmelee, W. W. Aubele, and R. G. Huebscher, "Measurements of Solar Heat Transmission Through Flat Glass," *Trans. ASHVE*. Vol. 54 (1948), pp. 165-86.

5. ASHRAE, *Handbook of Fundamentals*, 1967, Chapter 28.

6. ASHRAE, *Systems and Equipment Guide and Data Book*, 1967, Chapter 45, "Dual Duct Systems."

7

Steam Heating

7-1. STEAM IN HEATING SYSTEMS

When saturated steam is supplied to a radiator or convector the useful heating is caused primarily by condensation of the steam. . The heat of condensation is of course equal to the heat of vaporization (latent heat), listed as h_{fg} in tables 2-2 and 2-3. For example, at atmospheric pressure (14.7 psi) a pound of dry saturated steam, in condensing, gives up 970.3 Btu (h_{fg} from Table 2-3). During this process of condensation, just as in vaporization, the temperature remains constant (at 212 F when the pressure is 14.7 psi). Reference to Table 2-3 shows that when steam is allowed to condense at a lower pressure than atmospheric (partial vacuum) the h_{fg} values are somewhat larger than 970.3, and the saturation or condensation temperatures are lower than 212 F. In utilizing this latent heat in radiators or convectors, provision must be made for removing the condensate (water) from them. The condensate is at the same temperature as the steam until it is chilled by cool piping or radiator surface not in contact with the steam. Additional heat flow to a space, often amounting to 30 to 60 Btu per pound of steam, can occur as the liquid subcools.

Since the magnitude of h_{fg} does not greatly change over the pressure range employed in heating systems, pressure has little effect on the heat delivered by each pound of steam. However, low-pressure steam is also low-temperature steam, and since heat transfer depends primarily on the temperature difference existing between the radiator surface and the ambient air, low-pressure steam with a given radiator surface supplies less heat in a given time than high-pressure steam. The specific volume of low-pressure steam is greater than that of high-pressure steam, and so a low-pressure system, in addition to requiring somewhat greater surface in the radiators, also requires larger pipe sizes to transmit the greater volume of steam.

Radiators capable of heating a building satisfactorily in zero weather would greatly overheat the building in mild weather unless the steam were turned on and off or unless the steam temperature could be varied. With **subatmospheric** steam systems this latter arrangement is used, and the **steam** pressure (and its temperature) is reduced in mild weather to suit

Steam Heating

the heat demand. Such cooler radiators supply less heat and prevent the necessity of frequent control through operation of steam valves. With many steam-heating systems, particularly single-pipe arrangements, partly closing a valve does not give satisfactory temperature modulation.

Two brief definitions will be given at this point. A *radiator* is a heating device placed in a space so that it can direct its radiation to part or all of the heated space. It delivers heat by radiation, convection, and conduction. A *convector* is a heating device arranged to deliver heat to the air largely by convection currents. Convectors are enclosed, or concealed from the heated space.

It is always desirable to express the capacity of a radiator or convector under stated conditions in terms of Btuh (Btu per hour) or Mbh (1000 Btu per hour). However, there still exists an old unit, called by various names, such as *equivalent direct radiation* (EDR) or *square feet of radiation*, or even *feet of radiation*. Unit EDR (equivalent direct radiation) is defined as heat delivery at a rate of 240 Btuh. For rating of radiators and convectors, it is usually considered that the output is produced when steam is condensing at 215 F in 70 F ambient space. (In the case of hot-water heating an EDR is often taken as 150 Btuh.)

7-2. STEAM FLOW IN SYSTEMS

Steam flows through pipes because of pressure differences. Heat added in a boiler causes the water to change into a vapor (steam), with a resulting increase in volume. For example, at atmospheric pressure (212 F) the volume occupied by one pound of saturated steam is about 1600 times as great as the volume occupied by one pound of water; that is, $v_g/v_f = 26.80/0.01672 = 1604$ (from Table 2-2). When one pound of water is converted into steam in a heating system at a pressure above atmospheric, the steam displaces and drives out the air. As heat is transferred from the pipes and radiators the steam condenses to water and a resultant small volume. The shrinkage in volume causes a lower pressure within the system, and the rate of steam flow from the boiler increases because of the reduced pressure. Thus two forces influence the pressure difference necessary for steam flow: first, application of heat at the boiler, with evaporation and an accompanying increase in volume; second, heat removal in the radiators and pipes, with an accompanying shrinkage in volume and reduction in pressure as condensation occurs. If the heat removal is greater than the heat supply, the pressure of the system falls. When the heat supply at the boiler is diminished in airtight systems, transmission of heat in decreasing amount will continue while the steam pressure falls far into the vacuum region.

Flow of steam in a heating system is resisted by the skin friction of the various conduits, particularly in pipes, fittings, and valves. Condensation commences whenever saturated steam loses heat, so that in most pipes the fast-flowing steam rushes alongside a low-velocity stream of condensate. If the drainage slope of the pipe is such that the steam flow is counter to the water flow, friction is increased and therefore larger pipes must be used.

A boiler is an externally-fired enclosed vessel, partly filled with water. When the water boils, the steam forming above the water in the vessel will cause an increase in pressure and compress any air in the pipes of a heating system. Since the pressures throughout the system are about equal, the water line in the drop leg at the end of the steam main will be almost level with the water line in the boiler. These water surfaces will be level with each other as long as the respective pressures are equal. In the drop leg above the water line, as steam starts to flow, this water level will rise since the friction drop reduces the pressure below that in the boiler. This pressure difference also exists when steam condenses, and this will also be balanced by increased head of water in the drop leg.

When a steam coil, radiator, or convector transfers so much heat that the pressure in the drop leg is 1 psi less than boiler pressure, the water line in the drop leg will rise about 28 in. higher than the water level in the boiler. If the flow of steam is retarded, as by a valve or small-size pipe, the difference in pressure and water levels will increase. Thus in certain steam-heating systems the permissible difference between the water level in the boiler and that in the drop leg is the governing factor in choosing pipe sizes.

7-3. AIR VENTING

Water at normal temperature usually contains air some of which separates when the water is heated. Pipes and radiators contain air when steam is first developed. To heat the surfaces the air must be removed before the steam can enter, and must be vented continually thereafter.

Acceptable air valves allow the air to pass out, but close when steam starts to escape. Figure 7-1 shows a common construction. The float contains a small amount of volatile fluid which, expanding or contracting in response to temperature changes, moves the flexible bottom of the float outward, or allows it to spring inward. When the steam pressure increases above the pressure of the atmosphere, any air that is present is forced out through the escape port. After the relatively cool air escapes and the hot steam enters the valve, the confined fluid within the float chamber expands and thereby forces the flexible bottom downward. This action lifts the float and forces the valve pin into the port opening, thereby closing the valve. The float chamber will also rise if for any reason the valve

Steam Heating

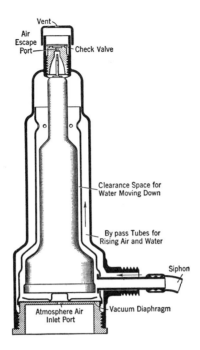

Fig. 7-1. Vacuum type of radiator valve.

becomes filled with water. Thus no water can escape through the valve. The loosely fitting siphon is used when the valve is placed on radiators, to return to the radiator any accumulation of water within the valve.

It is better to employ air valves which will retard the return of air into the system for several hours, and which thus permit steam circulation at a temperature below that of the boiling point of water at atmospheric pressures. In this connection an air valve, such as shown in Fig. 7-1, operates to discharge air and to prevent escape of water and steam, but when the steam pressure drops and the air tends to return to the system the check valve at the vent port closes to prevent such counterflow. As the steam in the system condenses, a vacuum will be formed if no air re-enters the system. The check valve, having functioned first, is then supplemented by atmospheric air pressure, which, acting through the port at the bottom of the valve, presses on the vacuum diaphragm, forces the float to rise, and thus positively seals the venting port.

The advantage of a tight air valve is its ability to keep the radiator warm for some time after the fire in the boiler has been allowed to die down. This is possible because the temperature at which water boils becomes lower as the absolute pressure decreases, and steam at 160 F or an even-lower temperature will circulate throughout an extensive piping and radiator system if air can be prevented from entering. The added cost for the vacuum-type valve is justified by the longer intervals between firing in mild weather, and by the better temperature control which can be obtained.

The air inside the steam mains cannot all escape through the air valves on the radiators. Sometimes the radiator valves are closed, shutting off communication between the radiator air valves and the main. Therefore auxiliary air valves are required on the ends of all steam mains where they enter the drop pipes and become wet. The location most likely to trap the air accumulation is on the boiler side of the last radiator and close above the water level. However, an air vent placed too close to the water line may be closed by a backsurge of water. The best location for the air vent on a steam main, all things considered, is in the top of the main just above its final drop below the boiler water line (see Fig. 7–3 and Fig. 7–4). Figure 7–2 is applicable to hot-water systems as well as to steam systems. When

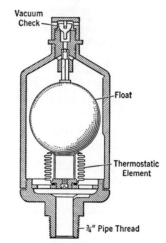

FIG. 7–2. Air eliminator. (Courtesy Sarco Company, Inc.)

water enters this type of eliminator the float rises and thereby closes the outlet, which remains closed until sufficient air collects in the chamber to drop the float, thus venting the air out at the top. If steam instead of cooled condensate enters the eliminator, the thermostatic element, which is filled with a volatile liquid, expands and forces the outlet shut, preventing escape of steam. A check valve prevents air from entering if the system is under vacuum.

7–4. TYPES OF STEAM SYSTEMS. DEFINITIONS

Numerous classifications of steam systems have been made and are in use. Many of these classifications and names, however, are merely sub-

Steam Heating

classifications of the basic types, which are summarized as follows:
  One-pipe, air-vent gravity return
  Two-pipe, air-vent gravity return
  Vapor system, simple gravity return
  Vapor system, with return trap
  Mechanical return system, nonvacuum
  Vacuum return system, with vacuum pump

Figure 7-3 illustrates a one-pipe air-vent steam system. A study of the illustration will reveal the one-pipe aspect: The supply main which

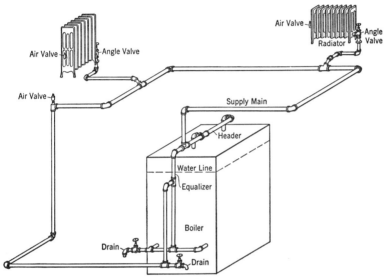

FIG. 7-3. One-pipe air-vent steam system.

starts at the boiler continues as a single pipe until it finally drops down at the end of its run to constitute the vertical drop leg and, finally, the wet return, which carries the condensate back to the boiler. Steam from the supply main passes through connecting piping to the radiators. Each of the radiators has a single angle-valve screwed into a low position on the radiator. The condensate which forms in the radiator moves counterflow to the direction of the steam entering the radiator, and after re-entering the steam main it flows in the same direction as the steam until it enters the drop leg leading down to the wet return. Systems of this type are noisy, since water and steam can surge back and forth in the radiator. It is therefore necessary to make the connecting pipes from the radiator to the main sufficiently large to serve the double purpose of steam supply and condensate return. Notice that an air vent is required on each radiator. It is also advisable to supply one on the return main in case a closure of valves on some of the radiators might make it impossible to vent the steam main and thereby prevent satisfactory operation of the system. In this and in other systems as well, steam supply mains and return pipes often run along a basement ceiling. Where possible, mains should be pitched in such a direction as to permit condensate to flow along with the steam, and where this is done the high point of the main, exclusive of risers to other floors, is normally at the boiler. A pitch of 1 in. in 20 ft is desirable.

An *overhead main* runs horizontally, or nearly so, at an elevation higher than the radiators it serves, and is supplied by a vertical main riser. Such overhead mains are frequently placed in the attics of multistoried buildings.

Supply risers are vertical pipes that pass from story to story to convey steam to the radiators or convectors on several floors. Risers are known as either upfeed or downfeed risers, the latter being those in which the steam flows downward to the radiators or convectors from an overhead main.

Return risers are those vertical pipes that take the condensate from the radiators or coils on the several stories of a building and convey it to the return main. Return risers are always of the downfeed type.

A *return main* is a nearly horizontal line of pipe that receives the condensate from the heating system and returns it to the boiler or otherwise disposes of it. The return main is usually run in the basement, but in any event it must be below all heat-transmitting surfaces, and must drain the supply mains.

A *dry-return main* is one that is run above the water line of the boiler. In some types of steam heating, a dry return conveys both water and steam, while in others which have traps at all points communicating with the steam main, a dry return carries only water and air.

A *wet-return main* is one that is run below the water line of the boiler or equivalent device and is filled with water at all times. As a rule, a wet-return main is preferable to a dry-return main, except where the main is subject to freezing temperatures. Supply mains must not be connected with other supply mains except through water seals or nonreturn traps. This type of connection is necessary because the unequal frictional resistance to steam flow, and the rate of heat dissipation in each main, produce different pressures at the respective return ends of the mains. If these ends are open through dry returns to each other, turbulence and noise will result. When the ends of supply mains in any one system are sealed separately by dropping into a wet return, the water-column level in each vertical pipe will be sufficiently different from the others to balance the pressure difference.

A *drip pipe*, or *relief*, or *bleeder*, is a pipe used to drain condensate away from the foot of supply risers or from low points, pockets, or traps in the steam main. Drip pipes usually drain into wet returns. Drip pipes in steam mains are employed at points where reduction or increase in the size of the main occurs, and where eccentric reducing fittings cannot be used. This drainage serves to prevent water hammer, by relieving the main of the water which would otherwise accumulate at such locations. The extreme end of the steam main must always be connected so as to drain the condensate into the wet-return main.

7-5. TWO-PIPE AIR-VENT SYSTEM

The two-pipe system differs from the previously described one-pipe system in that separate circuits are provided for the supply and return parts of the system. Figure 7-4 illustrates a representative two-pipe

Steam Heating 261

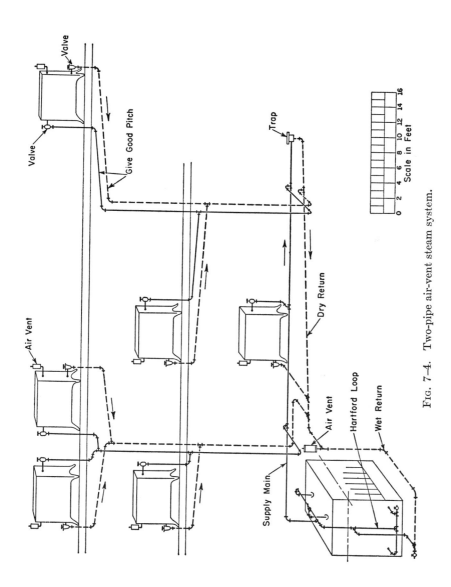

FIG. 7-4. Two-pipe air-vent steam system.

system. It will be noticed that the supply main from the boiler slopes away from the boiler and eventually connects into the return line. This connection can be made by means of a vertical downdrop water leg or, as shown, through a trap. Traps are described in detail later in this chapter, but at this point it should be mentioned that the purpose of a steam trap is to prevent the passage of steam from one part of a system to another while permitting condensate and air to pass. Thus, with the return line at lower pressure than the supply line, the trap prevents steam from passing into the return line but enables the water (condensate) and air to be removed from the supply line.

The supply main can serve risers which supply radiators on different floors, or direct radiator connections can be made into the main. With a two-pipe system, inlet to the radiator does not have to be made at a low point. However, for condensate drainage, a low point on the heater device is connected through a valve into the return piping of the system. Valves are required at both the inlet and the outlet to the radiator, and the return lines—where they are not in the form of risers—should slope toward the boiler. In the diagram, the return is indicated as a dry return; that is, it is located at a position above the water level in the boiler. It becomes a wet return when it enters the vertical drop leg from which the condensate flows into the boiler. An air vent is supplied on each radiator and, in starting, the steam from the supply main enters the radiator and expels through the air vent the air which was previously trapped in the radiator. The return line is also supplied with an air vent and this serves to vent, from both the supply main and the return main, air which is not otherwise vented at the radiators.

Where a water leg is employed in the operation of a system of this type, it is usually desirable to have the lowest radiation not less than 24 in. above the water line in the boiler. In some cases this is not possible, and where water might flood low-lying radiation a condensate return pump is used to force the water back into the boiler. The air is vented to the outside through the air vents on the radiators and on the return line. Although a two-pipe air-vent system may drop slightly into the vacuum region, it is not particularly designed to operate in this range and should normally be thought of as a positive pressure system using either gravity or mechanical return.

7-6. VAPOR SYSTEMS

The vapor system differs from the two-pipe air-vent system in that the air elimination is accomplished usually at a central point or points, and thus no air-vent valves are used on the radiators or convectors. Because it is possible for a system of this type to operate well in the vacuum region, it is customary to use packless valves, which have no opening to outside

Steam Heating

air around their stems. Traps are employed on each radiator or convector. As stated before, traps permit the passage of condensate and air and prevent the passage of steam. Thus, under ideal operating conditions, a supply of steam enters the radiator and as fast as this condenses it passes out through the trap to the return side. Thus the steam itself is not permitted to short-circuit through the radiator into the return side before it has given up its latent heat in condensing.

Figure 7-5 shows in diagrammatic fashion a layout of such a system. As before, it is desirable for the supply main to drain away from the boiler

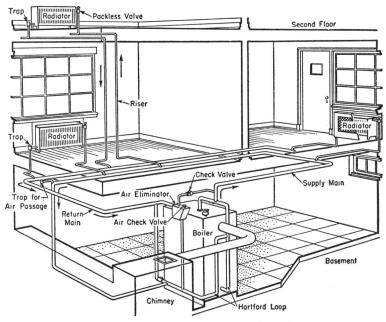

Fig. 7-5. Diagrammatic layout of vapor (steam) system.

into the return, and for the return main to connect into the boiler feed line by means of a vertical drop leg. The extreme end of the supply main also connects through a vertical drop leg into the wet return, which feeds the boiler. Water will, of course, stand at different heights in the two vertical drop legs, depending upon the pressure difference in the two parts of the system. It is desirable in a system of this type for the lowest connected radiation to be at least 24 in. above the water level in the boiler, and where this cannot be accomplished other provisions for returning the water to the boiler must be considered. The air in the supply main which does not enter the radiators and leave them through the outlet traps can pass through a connecting trap into the return main, as shown in the illustration. Normal air found in the radiators also enters the return main and moves through the dry-return part of this main to the central air eliminator of the system. Here, after some cooling, the air is eliminated to the outside, and the eliminator is so designed that when the system is under partial vacuum, outside air cannot enter through it. Because of the tightness with which vapor systems can be built, it is possible for them to operate well into the subatmospheric region. With air excluded, as long as the water-steam in the boiler is hotter than the water-steam medium in the radiators, steam can continue to flow to the radiators and maintain heating at a reduced rate even though no fuel is being burnt. It is easy to control a system of this type, because the boiler pressure can be operated over a range of positive to subatmospheric pressures and because it is possible to use a graduated type of valve at the radiator.

Vapor systems are usually satisfactory in operation because quiet circulation of steam without water hammer or air binding can be realized, and because control through variation in steam temperature is also possible. However, systems of this type usually are designed for effective operation only at low steam pressures, and consequently relatively large pipe sizes are employed. Since the condensate returns to the boiler by gravity, many systems have headroom limitations and it may be necessary to employ the return-trap arrangement described in the following section.

7-7. VAPOR SYSTEM WITH RETURN TRAP

By making use of a return-trap system, it is possible to operate the vapor system at somewhat higher steam pressures, and smaller pipe sizes are possible. However, even with a return-trap system, there are elevation limitations which must be observed in locating the return trap with respect to the return main and boiler water level. In general, slightly better control can be obtained with a return trap system than with conventional water legs.

Return traps, which are also known as alternating receivers, have, in general, a float-actuated mechanism which in one position closes a communicating port to the boiler above the water line and opens a corresponding port communicating with the dry return. In its other position, the float mechanism reverses the port openings. With the receiver open to the return system, water drains by gravity into the receiver, and as this fills the float rises. When sufficient condensate accumulates in the receiver, the float, at its high position, trips a mechanism which closes off the connection to the dry-return main and at the same time opens the return trap to the boiler. With the water in the receiver above the boiler water level, and with pressure equalized in the receiver and boiler, the water drains by gravity into the boiler. The float follows the dropping water level in the receiver until it retrips the valve mechanism to connect with the return side, and a refilling of the receiver occurs. The alternating receiver (return

Steam Heating

trap) merely makes it possible to use the vapor system more advantageously under adverse or limiting conditions; however, most new designs avoid the use of return traps.

Alternating receivers are available in a variety of forms and are designed for low-pressure operation to 15 psig, medium pressure to 45 psig, and high pressure to 100 psig. The higher-pressure units permit use of high lifts, as for example when a return, high above the heating units, is required to lead the water to a conveniently placed receiver feeding into the return trap. The receiver may need to be vented to the atmosphere to provide a low-pressure region for the lift.

7-8. VACUUM SYSTEM EMPLOYING VACUUM-PUMP RETURN

A vacuum steam-heating system differs from a vapor system in that a vacuum pump is used on the return side to maintain continuously a reduced pressure (usually subatmospheric) on the return side of the system. The vacuum-return pump removes the condensate and air from the return side, delivering the condensate to the boiler and the air to waste. With effective removal of the air, rapid circulation of steam is possible, smaller pipe sizes can be employed, and low-pressure steam or steam at partial vacuum can be fed to the radiators.

From a control viewpoint, the vacuum system is most versatile; however, the initial investment in the vacuum pump and the subsequent maintenance costs limit the use of this system to buildings larger than conventional residences. Vacuum systems are used in the majority of large steam-heated buildings, where the small pipe sizes are particularly desirable and where rapid well-balanced steam circulation is essential. Vacuum-system radiators and convectors have design and installation features similar to those found in vapor systems. However, the connections at the boiler return are so decidedly different that these are separately shown in Fig. 7-6, which illustrates the return arrangements of one type of vacuum system. Condensate and air enter the accumulator tank, which is placed at a low point to which the system can drain. If this location is below the pump suction level, lift fittings make it possible for the pump to lift water from the accumulator tank through as much as 4 ft per lift fitting. A float-operated switch on the accumulator tank starts and stops the pump motor. Since centrifugal pumps cannot effectively pump air, the pump itself is not connected to the accumulator but receives suction water from the receiving tank. The pump delivers this water at high velocity through the jets at A, where the kinetic effect of the water jets in a combining tube aspirates (pulls) water and air from the accumulator. In a diffuser tube the kinetic energy of the jets is sufficient to compress the air-water mixture to the receiving-tank pressure, which is slightly above atmospheric. The mixture in the receiving tank is separated, the air passing to the outside

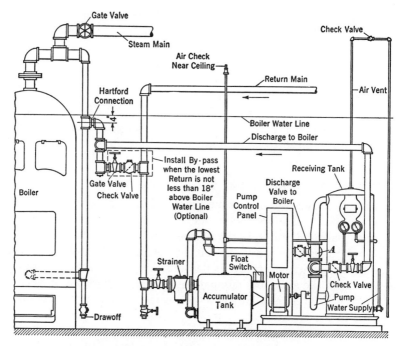

Fig. 7-6. Vacuum-system return connections and pump.
(Courtesy C. A. Dunham Company.)

and the water remaining to recirculate through the pump. Only a portion of the water discharge from the pump is required by the aspirating jets, and the remainder is sent directly into the boiler as feed water. An automatic control maintains the proper level of water in the receiving tank by closing down on the discharge valve to the boiler feed line until the level is re-established by additional water entering from the jet circuit.

Figure 7-7 shows details of a vacuum-pump-return system resembling the one just described—built as an integral unit, however, with the accumulator tank constituting the bottom left compartment and the air-separating tank the top compartment. From the top separating tank, water passes down, through a passage at the right rear of the unit, to the pump inlet. By this arrangement, water flows into the horizontal impeller of the pump, always under a positive pressure. The water delivered by the pump passes both to the boiler-feed system and to the aspirating (exhauster) jets. These jets are at sufficiently reduced pressure to pull water and air from the accumulator tank and then send it to the air-separating tank. The pump is started in response to a float-actuated switch in the condensate accumulator tank. The water flow to the boiler is controlled by a float switch in the upper tank, which restricts boiler flow until an adequate supply of

Steam Heating

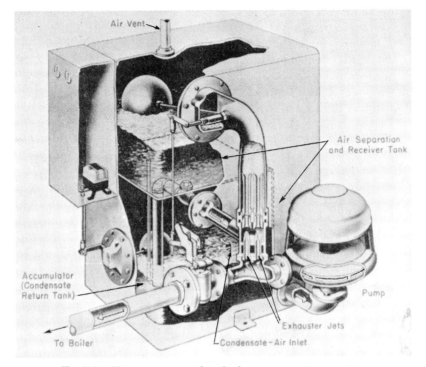

Fig. 7-7. Vacuum pump and tanks for a vacuum steam system. (Courtesy C. A. Dunham Company.)

water is available in the recirculation system to assure satisfactory operation.

In addition to the jet-aspirated types of vacuum pumps just described, mention should be made of the reciprocating vacuum pump, which resembles an air compressor in its general features, and of the rotor-type pump, which uses an elliptic housing. The rotor-type pump employs part of the condensate, which whirls rapidly inside an elliptic housing to compress and deliver the air. This pump may be driven by motor or by a low-differential-pressure steam turbine. Other designs employ separate air and water pumps on a common motor shaft. Most vacuum pumps can return the condensate directly to the boiler, provided the boiler pressure does not exceed the reasonable limits of a single-stage centrifugal pump. Usually if the boiler pressure is higher than 50 psig a separate boiler-feed pump is necessary, and in any event the pump must be designed especially for the maximum boiler pressure to be encountered.

The suction pressure for a vacuum steam-heating system may be as much as 24 in. of mercury—equivalent to about a 27-ft column of water—although a vacuum as high as this presents maintenance difficulties. High vacuums produce low steam temperatures and provide greater flexibility of the system for meeting the fluctuations of outside temperatures. Some heating systems operate at subatmospheric pressure even on the supply side, and such systems have often been found to be exceptionally economical. This economy is possible because little overheating results, since in mild weather it is possible to set the temperatures of the radiators and convectors in relation to the variations in temperature out of doors.

It should be mentioned that vacuum pumps are often made as double units, in a so-called duplex pattern. A duplex unit is able to handle the overloads that exist when a system is started, because then both units can run; or in the event that one pump fails, the other can continue in use and keep the system in operation. The capacity of the usual vacuum pump varies from some 600,000 to 24,000,000 Btuh. Representative data on the units of one manufacturer show that a 600,000 Btuh unit would handle 3.8 gpm and 1.3 cfm of air, while using a $\frac{3}{4}$-hp motor with the discharge pressure 20 psig at pump outlet. It is customary to rate pump units to maintain $5\frac{1}{2}$ in. vacuum with 160 F water. However, the usual units can operate satisfactorily at vacuums as high as 10 in. At reduced capacities or with special pumps, vacuums as high as 25 in. Hg are possible. Because pressure can be accurately controlled in the return system, and because a significant differential between inlet and outlet pressure can be maintained, it is possible to control the steam and heat input to radiators by means of variable-opening inlet valves. These restrict the flow of steam into the unit, and since steam cannot condense faster than it is supplied, the capacity of the unit is thereby limited. Some systems provide thermostatic control valves on individual radiators. In response to room temperature the control elements limit the opening of the inlet valve and thereby control the amount of steam entering a given radiator.

Although it is possible to install heat-transmitting elements below the level of the vacuum pump, this should be avoided whenever possible since such an arrangement involves lifting water which, if sufficiently hot, will partially flash to vapor when the pressure is reduced. Under such conditions pumps do not operate well, and if they fail, partial flooding of low-lying parts of the system can occur. This may stop steam circulation and can contribute to noise production. Difficulties in this connection can be reduced by employing suction lift fittings, in which alternating slugs of liquid or vapor are formed, which is equivalent to reducing the effective vertical lift through which the pump operates.

The operating pressure differential in a vacuum system is the difference between the supply pressure and the vacuum on the return side. For example, if the vacuum pump produced 6 in. of mercury vacuum and if steam were supplied at 2 psig, a simple calculation would show that the

Steam Heating

operating differential would be approximately 137 in. of water. Because of the significant differential pressure employed in vacuum systems, every connection between the supply main and the return main must be trapped by a thermostatic or mechanical device. One untrapped or leaking connection can defeat the operation of the entire system by passing enough steam to prevent the pump from maintaining a pressure below atmospheric. Vacuum traps are almost exclusively of the thermostatic type, with ports which open to pass water or air but which close tightly as steam, at its higher temperatures, starts to flow. Thus the return pipes of a vacuum system, when in proper condition, do not transport anything warmer than the condensate, which under vacuum is cooler than atmospheric-pressure steam. A typical thermostatic trap is shown in Fig. 7-8. With such traps the metallic bellows is partly filled with a volatile liquid which

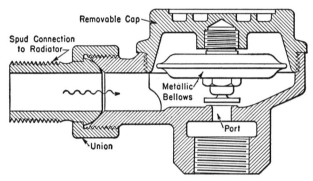

FIG. 7-8. Thermostatic trap of radiator-type design for passing water and air.

expands so as to close the port when vapor warmer than the design temperature enters through the radiator or pipe drain connection. Water at steam temperature can pass through the trap, since it flows along the bottom and does not come into contact with the bellows device to an extent sufficient to make it close the port.

Figure 7-9 shows an inverted-bucket trap. This type of trap is used on steam-heating systems but more particularly for a variety of industrial applications with process steam, such as industrial cookers, water heaters, and chemical-process vats. After a trap discharges, water still remains in the trap and, with steam or air in the inverted bucket, the bucket floats in the high (closed) position like an empty overturned can in a pond. As water, steam, and air enter the trap, the water level slowly rises inside the inverted bucket. Air and uncondensed steam pass to the top of the trap body through the vent-hole in the bucket. When insufficient air and steam remain in the bucket to hold it up, the bucket falls and the escape port

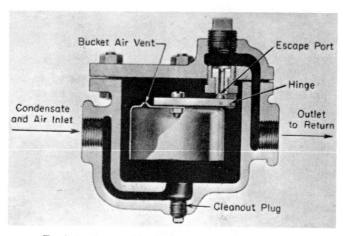

Fig. 7-9. Inverted-bucket trap. (Courtesy Hoffman Specialty Manufacturing Corporation.)

opens. The air in the top of the trap, and the water outside and inside the bucket, are blown out by the pressure difference between the supply and return systems until the air and vapor in the bucket again exert their buoyancy to close off the escape port. This design of trap is satisfactory for the operation of high-pressure as well as low-pressure systems, with pressure differentials of 5 to 200 psia. The units have capacities ranging from 900 to 11,000 lb per hour.

7-9. STEAM-PIPE ARRANGEMENTS

When water whose temperature is higher than the temperature corresponding to the pressure is released into a pipe or other container at a reduced pressure, part of the water will flash into steam. This action occurs at the suction inlet of a pump which is lifting hot water, and at the outlet of a high-pressure steam trap. If the steam for a vacuum system comes from a high-pressure source it is necessary to provide for conservation and disposal of the high-temperature condensate from high-pressure traps. The trap outlet may be connected to a flash tank, with an interior capacity at least four times that of the trap body. From the top of the flash tank there should be a valved steam connection to the low-pressure steam main, and from the bottom of the tank a thermostatic trap connection into the vacuum return. A typical piping arrangement of this type is shown in Fig. 7-10.

All vapor or temperature-modulating systems of steam heating operate on substantially the same general principle. Graduated radiator supply valves are employed, in which the area of the steam passage at the port

Steam Heating 271

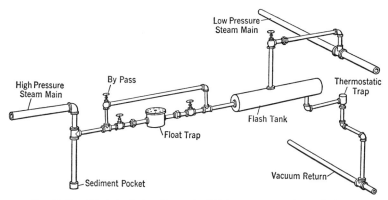

FIG. 7-10. Flash tank for disposal of high-pressure water and steam.

responds to a slight movement of the handle (Fig. 7–11). In such valves a metallic bellows expands and contracts (but does not revolve) as the valve stem rises or descends in response to turning the valve handle. The bellows forms an airtight seal. This tightness is important in any vapor

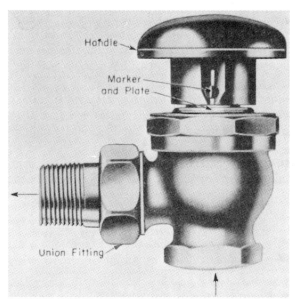

FIG. 7-11. Nonrising-stem packless radiator valve.
(Courtesy C. A. Dunham Company.)

or vacuum system which operates at subatmospheric pressure.

Most of the steam-heating-system layouts shown in this chapter use the

Hartford loop, which is designed to reduce the possibility of water leaving the boiler by any method except evaporation. In Fig. 7-4 the loop is indicated by the arrow and legend. An equalizing pipe from the steam section of the boiler leads to that point at which the feed-water return to the boiler is connected. This point is located 2 to 4 in. below the normal boiler-water level, and the downflow pipe, which is normally flooded, carries the feed water into the boiler. Under certain conditions of abnormal operation, it is possible to have a pressure much lower than boiler pressure exist in the return system and at the extreme end of the steam supply main. With such reduced pressure it would be possible for water to leave the boiler and back up in the piping, and in some instances even reach the heat-transfer equipment. Such a loss of water from the boiler might endanger the combustion boiler-heating surface. The Hartford loop can prevent such water loss, because when the water level drops below the return connection point, then steam from the downcomer equalizing line comes in direct contact with the return water, and the boiler pressure acting on both the return water and the boiler water stops the syphon action so that a further drop in the boiler water level is avoided. A similar water-loss condition might arise if, for example, a sudden breakage occurred in the return line. Again, water would not run out of the boiler by syphon action past the point where steam acts at the return inlet. Thus it is customary to provide the Hartford loop as a safety feature for steam boilers. It has no similar use in connection with hot-water systems.

In the design and erection of steam-heating plants the problem of expansion cannot be neglected. Care must be taken to have every long steam pipe free for limited movement, and so arranged that small movements of connections or branches are not blocked by building material or by structural beams or columns. A steam main should be supported from overhead with strong hangers, but the branches to the risers must be free to move to take care of expansion. At the bottom of each vertical pipe there should be two elbows. One of these elbows directly on the riser should have more than a true 90-deg angle so as to facilitate drainage. This elbow and a nipple connect with an ordinary elbow placed on its side. Thus there will always be a nipple or a piece of pipe upon which some swing or spring movement is possible for expansion and contraction of the riser.

Risers, especially, require careful design and installation to provide for expansion, since the necessary drainage pitch of runouts to radiators may easily be lost by the expansion of a tall riser.

Expansion joints must be provided in the risers of multistory buildings—with guides to control, and anchors to limit, their travel. These joints are usually placed at intervals not exceeding 40 ft. Figure 7-12

Steam Heating

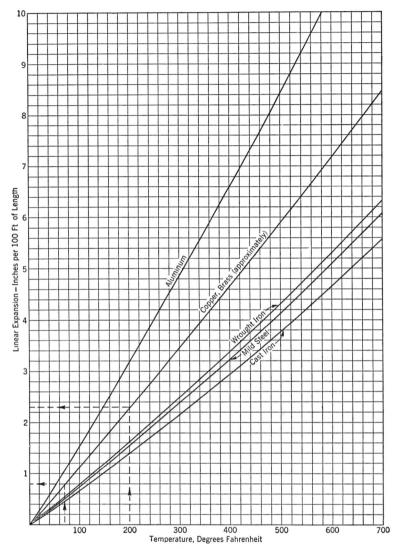

FIG. 7-12. Expansion of pipe, in inches per 100 feet, at various increases in temperature.

indicates the variation in length of metal pipes in accordance with changes in temperature.

7-10. STEEL PIPE AND TUBING CHARACTERISTICS

The most common pipe metal for industrial use is steel (wrought or cast iron is also used, but less frequently). Table 7-1 gives data for standard-weight steel pipe (Schedule 40) as well as for two heavier-wall types: extra-strong (Schedule 80) and double-extra-strong (for high-pressure service). For steam lines and hot-water heating lines, steel pipe is employed with threaded fittings, or in some cases it is connected by welding. Where threaded fittings are used, it is desirable to ream the ends of the pipe to eliminate burrs and to avoid the possibility of a partial closure of the internal bore. If this is not done, design computations for frictional loss are uncertain. Where steel pipe is used for supply water systems, it is desirable to use galvanized pipe, which is steel pipe covered with a corrosion-resistant zinc coating. Such coatings can prevent corrosion (rusting) for long periods of time.

Copper tubing is coming into increasing use for hot-water heating systems (and less frequently, for steam heating), and in many installations where water or refrigerants are being carried. Table 7-2 gives the characteristic data of conventional copper tubing, frequently known in the industry as *copper water tubing*. Copper water tubing is seamless, deoxidized copper tube supplied in three wall thicknesses. Type K, the heaviest, is supplied in coils 60 to 100 ft long, for the small bores of tubing, and in 12- to 20-ft lengths, for the larger-size tubes. It is furnished in either hard or soft temper.

Type L has a lesser wall thickness. It also is supplied in 60- to 100-ft coils up to and including one inch, and in 60-ft coils for the $1\frac{1}{4}$- and $1\frac{1}{2}$-in. sizes. Above $1\frac{1}{2}$ in. it is necessarily supplied in 12- or 20-ft lengths.

The type M tubing, which has the thinnest wall, is furnished in hard temper in 12- and 20-ft straight lengths. Whether hard or soft temper is employed depends upon the type of utilization and the appearance, which might be expected. Hard-temper tubing, which is rigid, is pleasing in appearance and is most usually used where runs of pipe are exposed. Soft-temper tubing is used for underground service or where it can be concealed in walls.

Copper water tubing, such as types K, L, and M, is not intended to be joined by threads. A most common type of connection is the solder joint, or sweated joint. In use, the pipe is cut square to size, the end mating surfaces are cleaned with steel wool, and a flux is applied to the surfaces. The pipe is inserted in the joint and the joint is then heated. Solder, usually of lead-tin or tin-antimony, melts on the hot surface and is drawn into the space between the fitting and the tubing by capillary action. After cooling and solidification a tight and permanent joint results. Where temperatures exceed the 250 F allowed for lead-tin solder, or the 312 F allowed for tin-antimony solder, it is necessary to use for the joints a brazing alloy or silver solder. Type K and type L copper water tubes are also joined by compression fittings.

Steam Heating

TABLE 7-1
Standard Steel Pipe Data

Nominal Size	External Diam (in.)	Standard-Weight Pipe (ASA Schedule 40)									Extra-Strong (ASA Schedule 80)		Double-Extra-Strong		Nominal Size
		Internal Diam (in.)	Wall Thickness (in.)	Weight per Ft Plain Ends (lb)	Threads per In.	Circumference (in.)		Transverse Area (sq in.)		Length of Pipe per Sq Ft		Wall Thickness (in.)	Weight per Ft Plain Ends (lb)	Wall Thickness (in.)	Weight per Ft Plain Ends (lb)	
						External	Internal	External	Internal	External Surface	Internal Surface					
1/8	0.405	0.269	0.068	0.244	27	1.272	0.845	0.129	0.057	9.431	14.199	0.095	0.31	......	......	1/8
1/4	0.540	0.364	0.088	0.424	18	1.696	1.144	0.229	0.104	7.073	10.493	0.119	0.54	......	......	1/4
3/8	0.675	0.493	0.091	0.567	18	2.121	1.549	0.358	0.191	5.658	7.748	0.126	0.74	......	......	3/8
1/2	0.840	0.622	0.109	0.850	14	2.639	1.954	0.554	0.304	4.547	6.141	0.147	1.09	0.294	1.71	1/2
3/4	1.050	0.824	0.113	1.130	14	3.299	2.589	0.866	0.533	3.637	4.635	0.154	1.47	0.308	2.44	3/4
1	1.315	1.049	0.133	1.678	11 1/2	4.131	3.296	1.358	0.864	2.904	3.641	0.179	2.17	0.358	3.66	1
1 1/4	1.660	1.380	0.140	2.272	11 1/2	5.215	4.335	2.164	1.495	2.301	2.768	0.191	3.00	0.382	5.21	1 1/4
1 1/2	1.900	1.610	0.145	2.717	11 1/2	5.969	5.058	2.835	2.036	2.010	2.372	0.200	3.63	0.400	6.41	1 1/2
2	2.375	2.067	0.154	3.652	11 1/2	7.461	6.494	4.430	3.355	1.608	1.847	0.218	5.02	0.436	9.03	2
2 1/2	2.875	2.469	0.203	5.793	8	9.032	7.757	6.492	4.788	1.328	1.547	0.276	7.66	0.552	13.70	2 1/2
3	3.500	3.068	0.216	7.575	8	10.996	9.638	9.621	7.393	1.091	1.245	0.300	10.25	0.600	18.58	3
3 1/2	4.000	3.548	0.226	9.109	8	12.566	11.146	12.566	9.886	0.954	1.076	0.318	12.51	0.636	22.85	3 1/2
4	4.500	4.026	0.237	10.790	8	14.137	12.648	15.904	12.730	0.848	0.948	0.337	14.98	0.674	27.54	4
5	5.563	5.047	0.258	14.617	8	17.477	15.856	24.306	20.006	0.686	0.756	0.375	20.78	0.750	38.55	5
6	6.625	6.065	0.280	18.974	8	20.813	19.054	34.472	28.891	0.576	0.629	0.432	28.57	0.864	53.16	6
8	8.625	7.981	0.322	28.554	8	27.096	25.073	58.426	50.027	0.443	0.478	0.500	43.39	0.875	72.42	8
10	10.750	10.020	0.365	40.483	8	33.772	31.479	90.763	78.855	0.355	0.381	0.593	64.3	......	......	10
12	12.750	12.000	0.375	49.562	8	40.055	37.699	127.676	113.097	0.299	0.318	0.687	88.5	......	......	12

Note. Pipe sizes 14 in. and above are designated by outside diameter, and the wall thickness is specified.

TABLE 7-2
Standard Dimensions, Weight, and Other Data for Copper Water Tubes

Nominal Size	Outside Diameter (in.)	Wall Thickness (in.)	Weight per Foot (lb)	Bore Area (sq in.)	Surface Area (sq ft per ft) Outer	Surface Area (sq ft per ft) Inside	Safe Inside Pressure (psi)
\multicolumn{8}{c}{Type K. Heavy Wall (Furnished Hard or Soft Temper)}
$\frac{1}{4}$	0.375	0.035	0.145	0.076	0.0981	0.0890	1120
$\frac{3}{8}$	0.500	0.049	0.269	0.127	0.1309	0.1052	1170
$\frac{1}{2}$	0.625	0.049	0.344	0.218	0.1636	0.1380	920
$\frac{5}{8}$	0.750	0.049	0.418	0.333	0.1963	0.1707	760
$\frac{3}{4}$	0.875	0.065	0.641	0.436	0.2291	0.1950	880
1	1.125	0.065	0.839	0.778	0.2945	0.2605	680
$1\frac{1}{4}$	1.375	0.065	1.04	1.22	0.3600	0.3259	550
$1\frac{1}{2}$	1.625	0.072	1.36	1.72	0.4254	0.3877	520
2	2.125	0.083	2.06	3.01	0.5563	0.5129	450
$2\frac{1}{2}$	2.625	0.095	2.93	4.66	0.6872	0.6375	420
3	3.125	0.109	4.00	6.64	0.8181	0.7611	410
$3\frac{1}{2}$	3.625	0.120	5.12	9.00	0.9490	0.8862	380
4	4.125	0.134	6.51	11.7	1.080	1.010	370
5	5.125	0.160	9.67	18.1	1.342	1.258	360
6	6.125	0.192	13.9	25.9	1.604	1.503	350
8	8.125	0.271	25.9	45.2	2.127	1.985	....
10	10.125	0.338	40.3	70.12	2.651	2.474	....
\multicolumn{8}{c}{Type L. Medium Wall (Furnished Hard or Soft Temper)}
$\frac{1}{4}$	0.375	0.030	0.126	0.078	0.0981	0.0903	900
$\frac{3}{8}$	0.500	0.035	0.198	0.145	0.1309	0.1126	800
$\frac{1}{2}$	0.625	0.040	0.285	0.233	0.1636	0.1427	740
$\frac{5}{8}$	0.750	0.042	0.362	0.348	0.1963	0.1744	650
$\frac{3}{4}$	0.875	0.045	0.455	0.484	0.2291	0.2055	590
1	1.125	0.050	0.655	0.825	0.2945	0.2683	510
$1\frac{1}{4}$	1.375	0.055	0.884	1.26	0.3600	0.3312	460
$1\frac{1}{2}$	1.625	0.060	1.14	1.78	0.4254	0.3940	430
2	2.125	0.070	1.75	3.09	0.5563	0.5197	370
$2\frac{1}{2}$	2.625	0.080	2.48	4.77	0.6872	0.6453	350
3	3.125	0.090	3.33	6.81	0.8181	0.7710	330
$3\frac{1}{2}$	3.625	0.100	4.29	9.21	0.9490	0.8967	320
4	4.125	0.110	5.28	12.0	1.080	1.022	300
5	5.125	0.125	7.61	18.7	1.342	1.276	280
6	6.125	0.140	10.21	26.8	1.604	1.530	260
8	8.125	0.200	19.3	46.9	2.127	2.022	....
10	10.125	0.250	30.1	72.8	2.651	2.520	....

Steam Heating

TABLE 7-2 (Continued)

Nominal Size	Outside Diameter (in.)	Wall Thickness (in.)	Weight per Foot (lb)	Bore Area (sq in.)	Surface Area (sq ft per ft) Outer	Surface Area (sq ft per ft) Inside	Safe Inside Pressure (psi)

Type M. Light Wall (Furnished Hard Temper Only)

Nominal Size	OD	Wall	Weight	Bore	Outer	Inside	PSI
3/8	0.500	0.025	0.145	0.159	0.1309	0.1243	....
1/2	0.625	0.028	0.204	0.254	0.1636	0.1562	....
3/4	0.875	0.032	0.328	0.517	0.2291	0.1944	....
1	1.125	0.035	0.465	0.874	0.2945	0.2852	....
1¼	1.375	0.042	0.682	1.31	0.3600	0.3489	....
1½	1.625	0.049	0.940	1.83	0.4254	0.4124	....
2	2.125	0.058	1.46	3.17	0.5563	0.5409	....
2½	2.625	0.065	2.03	4.89	0.6872	0.6698	280
3	3.125	0.072	2.68	6.98	0.8181	0.7988	260
3½	3.625	0.083	3.58	9.40	0.9490	0.9268	260
4	4.125	0.095	4.66	12.2	1.080	1.055	260
5	5.125	0.109	6.66	18.9	1.342	1.312	240
6	6.125	0.122	8.92	27.2	1.604	1.571	230

TABLE 7-3

Standard Dimensions, Weights, and Other Data for Threaded Standard-Pipe-Size Copper Pipe and Red-Brass Pipe

Nominal Size	Actual OD (in.)	Wall Thickness (in.)	Weight (lb per lin ft) Copper	Weight (lb per lin ft) Red Brass	Linear Feet Containing One U.S. Gal	Safe Working Pressure (psi) Copper	Safe Working Pressure (psi) Red Brass

Regular Pipe

Nominal Size	OD	Wall	Copper	Red Brass	Lin Ft/Gal	Copper PSI	Red Brass PSI
1/8	0.405	0.062	0.259	0.253	310.6	533	712
1/4	0.540	0.082	0.457	0.447	175.5	845	1128
3/8	0.675	0.090	0.641	0.627	100.2	818	1090
1/2	0.840	0.107	0.955	0.934	62.9	715	952
3/4	1.050	0.114	1.30	1.27	36.2	653	868
1	1.315	0.126	1.82	1.78	21.7	511	680
1¼	1.660	0.146	2.69	2.63	13.1	550	732
1½	1.900	0.150	3.20	3.13	9.6	505	674
2	2.375	0.156	4.22	4.12	5.8	435	580
2½	2.875	0.187	6.12	5.99	3.9	364	485
3	3.500	0.219	8.75	8.56	2.6	408	544
3½	4.000	0.250	11.4	11.2	2.0	450	600
4	4.500	0.250	12.9	12.7	1.5	400	534
5	5.563	0.250	16.2	15.8	0.96	324	432
6	6.625	0.250	19.4	19.0	0.65	272	362

7-11. EQUIVALENT PIPE LENGTH AND STEAM FLOW

The resistance to flow of steam or other fluid through a pipe is increased by the presence of valves and fittings, and the carrying capacity is thus reduced. Investigators have determined the resistance of valves and fittings in terms of lengths of straight pipe. The resistances thus expressed are added to the measured length of the pipe, and the sum is called the equivalent pipe length. Equivalent pipe length is thus the length of straight pipe of a given size that would have the same resistance as the actual pipe with its bends, fittings, valves, and other devices.

Table 8–2, which applies for smaller pipe sizes, gives the equivalent feet of run of various pipe.

Steam flowing through pipe and fittings is retarded by friction loss between the surface of the pipe and the steam, and by turbulence losses in bends and through the restricted passages which are found in valves and other devices. The design of the system should permit uniformly controlled distribution of the steam to various outlets, with minimum noise and with provision for release of air when air is in the system. Elimination and return of condensate presents an additional design problem. Furthermore, it is necessary to balance distribution in a system so that heaters remote from the steam source are not starved in relation to heaters adjacent to the steam source. This can sometimes be done by adjustment of pipe sizes, but it may also be necessary to provide orifice plates for insertion in the inlet of a radiator, to restrict or control the inflow of steam. The system and its radiators must operate satisfactorily not only under conditions of full heating load but also under conditions of partial load. Moreover, in the warming-up period the system is called upon to distribute an excessive amount of steam and to return an equivalent excess of condensate. Thus during this period the system is operating under an overload. It is obvious, therefore, that flexibility must be provided in the system, since the pressure differences causing flow may be radically different between full-load, normal-load, and partial-load conditions.

The laws governing the flow of dry or superheated steam are similar to those which apply to the flow of gases. However, additional complications exist with steam flow, as usually the steam is not 100 per cent dry, and moisture in suspension may be moving with the steam and also on the wetted walls or bottom of a steam pipe. It is also possible for the steam to be moving in one direction while condensate (water) is flowing in the opposite direction—an awkward situation which usually makes it necessary to provide larger pipe sizes to insure satisfactory flow and to reduce the possibility of impact (water-hammer) losses with accompanying noise.

Because of the complexity of the problem, pressure-loss designs based purely on theory and analytical considerations are frequently unsatisfactory. It has been found better to use semiempirical formulations, or

Steam Heating

compilations of experimental test data, for laying out designs. One of the many equations used for the flow of steam in pipes is the semiempirical Babcock formula, which is

$$m = 87 \sqrt{\frac{(\Delta P)D^5}{v\left(1 + \frac{3.6}{D}\right)L}} \tag{7-1}$$

where m = pounds of steam flowing per minute;
ΔP = loss of pressure, in pounds per square inch;
D = inside diameter of pipe, in inches;
L = equivalent length of pipe, in feet;
v = specific volume of steam, cubic foot per pound, measured at average pressure.

This equation is readily used in any case where the entire load is considered at the end of the main. When the load is distributed along the main, however, the process of sizing the main is more complicated. For ordinary heating installations, tables have been developed to provide for selection of pipe sizes.

7-12. PIPE SIZES FOR STEAM HEATING SYSTEMS

A detailed treatment of the design and pipe sizing of steam systems will not be presented here. This topic is treated completely in the ASHRAE *Guide and Data Book-Systems and Equipment Volume—1967* and in prior editions of the same book. However, to give an idea of the use of tabular pipe-sizing data, Tables 7-4, 7-5, 7-6, and 7-7 have been reproduced by permission from an earlier edition of the ASHRAE *Guide and Data Book*. Table 7-4 is satisfactory for selecting pipe sizes to use with all low-pressure steam-heating systems. In laying out a design, it is necessary to know the total pressure drop available for distributing the steam between the source and the end of the return, following which the allowable or most suitable pressure drop per 100 equivalent feet of run can be selected. The equivalent length of run of the main and return [and also the equivalent length from the steam source (boiler) to the most distant heating element] must be known, and the direction of flow of condensate—that is, whether it is with or against the direction of steam flow—must also be determined.

Table 7-4 is applicable both to one-pipe and two-pipe systems and can also be used for vapor and vacuum systems. The headings at the top of each column specifically state the conditions of use. For return piping, Table 7-5 is applicable in a similar way. It is desirable to use on the return side the same pressure drop per 100 ft that is used on the supply side of a system.

Air-vent systems of one-pipe design were formerly very common because of low initial cost. In the one-pipe system illustrated in Fig. 7-3, the single pipe from the boiler carries both the supply steam and all the

TABLE 7-4
Steam-Pipe Capacities for Low-Pressure Systems

Pipe Size (in.)	Capacities of Steam Mains and Risers								Special Capacities for One-Pipe Systems Only			
	Direction of Condensate Flow in Pipe Line								Supply Risers, Upfeed	Radiator Valves and Vertical Connections	Radiator and Riser Runouts
	With the Steam, in One-Pipe and Two-Pipe Systems						Against the Steam, Two-Pipe Only				
	$\frac{1}{32}$-PSI or $\frac{1}{2}$-Oz Drop	$\frac{1}{24}$-PSI or $\frac{2}{3}$-Oz Drop	$\frac{1}{16}$-PSI or 1-Oz Drop	$\frac{1}{8}$-PSI or 2-Oz Drop	$\frac{1}{4}$-PSI or 4-Oz Drop	$\frac{1}{2}$-PSI or 8-Oz Drop	Vertical	Horizontal			
A	B	C	D	E	F	G	H*	I†	J‡	K	L†
	Capacity Expressed in Pounds per Hour										
$\frac{3}{4}$	....	....	8	....	....	....	8	....	6	....	7
1	10	12	14	20	28	40	14	9	11	7	7
$1\frac{1}{4}$	22	25	31	43	61	87	31	19	20	16	16
$1\frac{1}{2}$	34	39	48	67	95	135	48	27	38	23	23
2	68	79	97	137	193	273	97	49	72	42	42
$2\frac{1}{2}$	112	130	159	225	318	449	159	99	116	....	65
3	206	237	291	411	581	822	282	175	200	....	119
$3\frac{1}{2}$	307	355	434	614	869	1,230	387	288	286	....	186
4	435	503	614	869	1,230	1,740	511	425	380	....	278
5	806	928	1,140	1,610	2,270	3,210	1,050	788	....	....	545
6	1,320	1,520	1,870	2,640	3,730	5,280	1,800	1,400	....	....	....
8	2,750	3,170	3,880	5,490	7,770	11,000	3,750	3,000	....	....	....
10	5,010	5,790	7,090	10,000	14,200	20,000	7,000	5,700	....	....	....
12	8,040	9,290	11,400	16,100	22,700	32,200	11,500	9,500	....	....	....
16	15,100	17,400	21,200	30,300	42,400	60,500	22,000	19,000	....	....	....
	All Horizontal Mains and Downfeed Risers						Upfeed Risers	Mains and Undripped Runouts	Upfeed Risers	Radiator Connections	Runouts Not Dripped

* Do not use column H for drops of $\frac{1}{24}$ or $\frac{1}{32}$ psi; substitute column C or column B as required.
† Pitch of horizontal runouts to risers and radiators should be not less than $\frac{1}{2}$ in. per ft. Where this pitch cannot be obtained, runouts over 8 ft in length should be one pipe size larger than called for in the table.
‡ Do not use column J for $\frac{1}{32}$ psi drop except on sizes 3 in. and over; below 3 in. substitute column B.

condensate of the system. The combined steam main and return continues until it finally drops and becomes the wet return. A water column automatically collects in the wet return at a sufficiently high level to force the water back into the boiler and counterbalance the pressure difference between the supply and return sides of the system. In two-pipe systems, as illustrated in Figs. 7–4 and 7–5, a water column is also present to force the return water into the boiler. The height of water column needed to balance the unit pressure can be selected from the following tabulation:

$$1 \text{ oz per sq in.} = 1.73 \text{ in. of water at 70 F} = \tfrac{1}{16} \text{ psi}$$
$$= 1.78 \text{ in. of water at 180 F}$$
$$= 1.80 \text{ in. of water at 200 F}$$
$$1 \text{ psi} = 28.80 \text{ in. of water at 200 F} = 2.4 \text{ ft}$$

Steam Heating

Thus for a steam-system pressure loss of $\frac{1}{4}$ psi, the water level in the return drop pipe (with 180 F water) would be 4×1.78, or 7.12 in. higher than that in the boiler.

One procedure to follow in designing low-pressure systems is to allocate not more than one-half the initial gage pressure in computing the design pressure drop.

It is desirable for the main to pitch in the direction of flow at a rate of not less than $\frac{1}{4}$ in. in 10 ft, and preferably $\frac{1}{2}$ in. The runouts to radiators and risers should pitch toward the main at not less than $\frac{1}{2}$ in. in 10 ft. If such a pitch is not practicable, or if the runout is more than 8 ft long, pipe one size larger should be used for the runout. Further, to prevent sagging between supports, it is recommended that the main should not be smaller than nominal 2-in. pipe.

In laying out two-pipe systems, the complete piping circuit should first be sketched on the architectural plan. For vapor and return-trap systems in particular, it is desirable for the supply mains to connect the convectors and other radiation with piping circuits as short and direct as possible from the boiler. With complex circuits a reversed-return system may be desirable. In such a circuit the convectors which are first supplied with steam are connected into the far end of the return main so that, although the steam gets to these convectors first, it has a long circuit to travel before returning to the boiler. By use of a reversed-circuit arrangement in a complex system, the air travel during the warm-up period is about the same for both the close and most distant radiation. Moreover, in a reversed-return arrangement both the mains and the returns pitch in the same direction which gives a much neater appearance than is the case with a direct-return arrangement. Where possible, return mains should pitch more steeply than supply mains, 1 in. in 10 ft being desirable. On full vacuum systems, return mains need not pitch more than $\frac{1}{2}$ in. in 10 ft.

Vacuum systems are normally used in large buildings because with them it is possible to obtain both close control of temperature and well-balanced steam circulation, and also, smaller pipe sizes can be used. The differential pressure in the vacuum system consists of the suction produced by the vacuum pump on the return side, and in addition, such positive (gage) pressure as may exist on the supply side. Good design practice calls for not more than $\frac{1}{8}$-psi drop per 100 ft of equivalent run and not more than 1 psi total pressure drop in a system. Pitch of main should not be less than $\frac{1}{4}$ in. in 10 ft. Supply mains smaller than 2 in. are undesirable. Connection should not be made between the steam and return sides of a vacuum system except through steam traps, as it is necessary to prevent steam in quantity from entering the return line.

TABLE 7-5. Return-Pipe Capacities for Low-Pressure Systems*

In Pounds per Hour

Capacity of Return Mains and Risers

Mains

Pipe Size (in.)	$\frac{1}{32}$-PSI or $\frac{1}{2}$-Oz Drop per 100 Ft			$\frac{1}{24}$-PSI or $\frac{2}{3}$-Oz Drop per 100 Ft			$\frac{1}{16}$-PSI or 1-Oz Drop per 100 Ft			$\frac{1}{8}$-PSI or 2-Oz Drop per 100 Ft			$\frac{1}{4}$-PSI or 4-Oz Drop per 100 Ft			$\frac{1}{2}$-PSI or 8-Oz Drop per 100 Ft		
	Wet	Dry	Vac.	Wet	Dry	Vac.	Wet	Dry	Vac.	Wet	Dry	Vac.	Wet	Dry	Vac.	Wet	Dry	Vac.
M	N	O	P	Q	R	S	T	U	V	W	X	Y	Z	AA	BB	CC	DD	EE
$\frac{3}{4}$	....	....	....	....	....	42	....	....	100	....	....	142	....	....	200	....	....	283
1	125	62	....	145	71	143	175	80	175	250	103	249	350	115	350	....	....	494
$1\frac{1}{4}$	213	130	....	248	149	244	300	168	300	425	217	426	600	241	600	....	....	848
$1\frac{1}{2}$	338	206	....	393	236	388	475	265	475	675	340	674	950	378	950	....	....	1,340
2	700	470	....	810	535	815	1,000	575	1,000	1,400	740	1,420	2,000	825	2,000	....	....	2,830
$2\frac{1}{2}$	1,180	760	....	1,580	868	1,360	1,680	950	1,680	2,350	1,230	2,380	3,350	1,360	3,350	....	....	4,730
3	1,880	1,460	....	2,130	1,560	2,180	2,680	1,750	2,680	3,750	2,250	3,800	5,350	2,500	5,350	....	....	7,560
$3\frac{1}{2}$	2,750	1,970	....	3,300	2,200	3,250	4,000	2,500	4,000	5,500	3,230	5,680	8,000	3,580	8,000	....	....	11,300
4	3,880	2,930	....	4,580	3,350	4,500	5,500	3,750	5,500	7,750	4,830	7,810	11,000	5,380	11,000	....	....	15,500
5	....	....	....	....	....	7,880	....	....	9,680	....	....	13,700	....	....	19,400	....	....	27,300
6	....	....	....	....	....	12,600	....	....	15,500	....	....	22,000	....	....	31,000	....	....	43,800

Risers

$\frac{3}{4}$	....	48	....	....	48	....	....	48	....	....	48	....	....	48	....	....	....	....
1	....	113	....	....	113	....	....	113	175	....	113	249	....	113	350	....	....	494
$1\frac{1}{4}$	....	248	....	....	248	....	....	248	300	....	248	426	....	248	600	....	....	848
$1\frac{1}{2}$	....	375	....	....	375	....	....	375	475	....	375	674	....	375	950	....	....	1,340
2	....	750	....	....	750	....	....	750	1,000	....	750	1,420	....	750	2,000	....	....	2,830
$2\frac{1}{2}$	....	....	....	....	....	1,360	....	....	1,680	....	....	2,380	....	....	3,350	....	....	4,730
3	....	....	....	....	....	2,180	....	....	2,680	....	....	3,800	....	....	5,350	....	....	7,560
$3\frac{1}{2}$	....	....	....	....	....	3,250	....	....	4,000	....	....	5,680	....	....	8,000	....	....	11,300
4	....	....	....	....	....	4,480	....	....	5,500	....	....	7,810	....	....	11,000	....	....	15,500
5	....	....	....	....	....	7,880	....	....	9,680	....	....	13,700	....	....	19,400	....	....	27,300
	....	....	....	....	....	12,600	....	....	15,500	....	....	22,000	....	....	31,000	....	....	43,800

*Table based on pipe-size data developed through research investigations of American Society of Heating and Ventilating Engineers.

Steam Heating

Example 7-1. A two-pipe vapor-vacuum system operates at 2 psig at maximum capacity and runs into the vacuum region on reduced capacity. The measured length of the longest run is 200 ft and the maximum heat delivery occurs when 800 lb of steam per hour are condensed off four risers. Making use of appropriate tabular data select pipe sizes for this system.

Solution: The total equivalent length is usually about double the measured length and thus 400 ft will be used. If half of the gage pressure namely 1 psi is allocated to the supply side, and the same to the return side the allowable pressure drop is 1 psi per 400 ft of run or $\frac{1}{4}$ psi per 100 ft of run. Reference to Table 7-4 (Column F) shows that a $3\frac{1}{2}$ in. main would be needed.

If each riser carries 200 lb of steam per hour, Column H indicates that a 2 in. riser is required.

Radiator runouts depend on the individual radiator capacity and are selected from Column H.

For the return main use the same design pressure drop, namely $\frac{1}{4}$ psi per 100 ft of equivalent run and make use of Table 7-5. Column Z here shows that a $1\frac{1}{2}$ in. wet return is satisfactory but a 2 in. pipe (Col. AA) is required for a dry-return (running above the boiler water level). For a vacuum system with a pump a $1\frac{1}{2}$ in. pipe (Col. BB) is satisfactory.

When data are taken off an architectural system layout the equivalent length should then be rechecked in terms of each elbow and fitting for the pipe sizes selected. However, this is not usually necessary. In this example it should also be noted that a $\frac{1}{4}$ psi pressure loss per 100 ft is considered a high value by some designers who prefer not to exceed $\frac{1}{8}$ psi per 100 ft. However, with pipes as large as those selected, no trouble should arise provided adequate head room between the boiler and the lowest radiation is available to allow for a high water column in the drop leg of the boiler return, when a pump is not used.

7-13. ORIFICES IN STEAM SYSTEMS

With vacuum systems, in particular, it is possible to use orifices for basic adjustment of the load and also for control. An orifice is usually supplied as a drilled hole in a plate made to be inserted in the radiator valve. The steam flow through the orifice for the small pressure ranges involved varies as the square root of the pressure differential across the orifice. In an orifice system, if the steam pressure is dropped from (say) 1 psig to 0.2 psig, with the pressure on the vacuum side remaining unchanged at -0.2 psig, the weight of steam flowing in the radiator would be reduced to

$$\frac{\sqrt{0.2 - (-0.2)}}{\sqrt{1 - (-0.2)}} = \sqrt{\frac{0.4}{1.2}} = 0.58$$

That is, only 58 per cent as much steam would flow under the changed condition as before. As the heat produced is a function of the steam flowing and condensed, it is evident immediately that the capacity can be changed through wide limits by varying the inlet pressure in an orifice system. Control for the system can be arranged to take place either by varying the inlet pressure into the steam mains feeding the various radiators, or by varying the vacuum in the return lines. For some vacuum systems, it has been observed that with the use of radiator orifice plates it is possible to eliminate traps on the discharge sides of each radiator. In long or complex systems, where there is necessarily an appreciable pressure difference between the supply and remote end of the steam main, it is possible to provide larger orifices for the more distant radiation, and smaller orifices for the radiation close to the supply source.

7–14. HIGH-PRESSURE STEAM SYSTEMS

For heating of industrial space, high-pressure steam may be used directly in suitable radiation. However, the pressure is frequently dropped through reducing valves and the low-pressure steam is used in conventional systems. Where high-pressure steam is used directly, the load-carrying capacity of steam and condensate return pipes is given in Tables 7–6 and 7–7 for 30-psig systems. With high-pressure steam, higher pressure drops can be used to force the steam throughout the system. In the 30-psig system, for example, total pressure drops from 5 to 10 psi are customary, while in a 150-psig system, 25- to 30-psi drops are employed. High-pressure mains should pitch at least $\frac{1}{4}$ in. in 10 ft, and horizontal runouts to risers and heaters should pitch at least $\frac{1}{2}$ in. per foot.

TABLE 7–6

Steam-Pipe Capacities for 30-PSIG Steam Systems*

(Lb per Hr)

(Steam and Condensate Flowing in Same Direction)

Pipe Size (in.)	Drop in Pressure (psi per 100-ft length)					
	$\frac{1}{8}$	$\frac{1}{4}$	$\frac{1}{2}$	$\frac{3}{4}$	1	2
$\frac{3}{4}$	15	22	31	38	45	63
1	31	46	63	77	89	125
$1\frac{1}{4}$	69	100	141	172	199	281
$1\frac{1}{2}$	107	154	219	267	309	437
2	217	313	444	543	627	886
$2\frac{1}{2}$	358	516	730	924	1,030	1,460
3	651	940	1,330	1,630	1,880	2,660
$3\frac{1}{2}$	979	1,410	2,000	2,450	2,830	4,000
4	1,390	2,000	2,830	3,460	4,000	5,660
6	4,210	6,030	8,590	10,400	12,100	17,200
8	8,750	12,600	17,900	21,900	25,300	35,100
10	16,300	23,500	33,200	40,600	46,900	66,400
12	25,600	36,900	52,300	64,000	74,000	104,500

Steam Heating

Unit pressure drops employed in design of high-pressure systems range from $\frac{1}{2}$ to 1 psi per 100 ft of equivalent run. With high-pressure steam, particularly when it exceeds 100 psig, it should be realized that the surface temperature of the pipes and radiator surface is in excess of 330 F. Where physical contact is made with surfaces of this temperature, bad burns can result. Thus the pipes must be so placed as to prevent physical contact. A common arrangement with high-pressure steam is the use of blast coils, in which air is warmed by being passed over the steam coils and the space is then warmed by the air. Unit heaters also can be employed. It is also possible to use finned-type pipe coil with suitable shielding. The same type of device is also satisfactory with low-pressure steam. In fact, it may be preferable to pipe the high-pressure steam to a reducing valve and drop the pressure in the reducing valve to the lower pressure and temperature ranges employed in conventional steam systems. Whenever the steam pressure is 100 psig or above, it is desirable to use two reducing valves in series to drop the pressure to, for example, 5 or 2 psig. In general, systems operating at 15 psig or less are called low-pressure systems, and those operating at pressures above 15 psig are called high-pressure systems.

TABLE 7-7
RETURN-PIPE CAPACITIES FOR 30-PSIG STEAM SYSTEMS*
(Lb per Hr)

PIPE SIZE (IN.)	DROP IN PRESSURE (PSI PER 100-FT LENGTH)				
	$\frac{1}{8}$	$\frac{1}{4}$	$\frac{1}{2}$	$\frac{3}{4}$	1
$\frac{3}{4}$	115	170	245	308	365
1	230	340	490	615	730
$1\frac{1}{4}$	485	710	1,025	1,290	1,530
$1\frac{1}{2}$	790	1,160	1,670	2,100	2,500
2	1,580	2,360	3,400	4,300	5,050
$2\frac{1}{2}$	2,650	3,900	5,600	7,100	8,400
3	4,850	7,100	10,300	12,900	15,300
$3\frac{1}{2}$	7,200	10,600	15,300	19,200	22,800
4	10,200	15,000	21,600	27,000	32,300
6	31,000	45,500	65,500	83,000	98,000

Steam heating represents one of the earliest forms of central heating and is still used extensively, particularly in public buildings. For new installations, one-pipe systems—and in fact most air-vent systems—are outmoded. For residential heating, vapor systems are still used in new designs, but more-extensive use is being made of forced-warm-air and hot-water heating systems.

PROBLEMS

7-1. Dry saturated steam at 15 psia condenses and subcools to 150 F. Making use of tables 2–2 and 2–3, compute (a) the energy released during this process and (b) the change in volume, in cubic feet per pound, during the condensation and subcooling.

Ans. (a) 1032.8 Btu/lb; (b) 26.27 cu ft

7-2. Dry steam at 12 psia condenses and subcools to 150 F. Making use of tables 2–2 and 2–3, compute (a) the energy released during this process and (b) the change in volume, in cubic feet per pound, during the condensation and subcooling.

Ans. (a) 1030.2 Btu/lb; (b) 32.38 cu ft

7-3. The boiler pressure in a one-pipe steam system, and at entry to the supply main, is 16 psia. As the supply main reaches its end and drops into a wet return feeding the boiler, the pressure is 14.8 psia. Compute the height of the water level in the drop leg with respect to the water level in the boiler. (See Fig. 7–3 for an illustration of a system of this type.) *Ans.* 1.2 psi = 34.5 in. of hot water

7-4. For a certain two-pipe system, such as the one illustrated in Fig. 7–4, the pressure in the boiler and at entry to the supply main is 16 psia. The corresponding pressure in the dry return as it turns down into the drop leg and wet return is 15 psia. Find the water-level height in the drop leg above boiler water level.

7-5. Examine Fig. 7–4 with a view to modifying the system shown there so that it becomes a vapor system. Without redrawing the figure, state exactly the minimum changes which would be required to make such a change. Assume that the mains, risers, and radiators could be used satisfactorily for either system.

7-6. Using Fig. 7–3 as a guide, diagram the system shown there and include the additions necessary to transform it into a two-pipe air-vent system. Indicate also a supply and return riser for a radiator on a higher story. Supply the valves for each end of the radiator. Notice that the return line can be satisfactorily connected to the wet return at the bottom of the drop leg, below the air valve. This system reduces the noise in single-pipe systems which occurs because the steam and water flow in opposite directions from the radiator, but the two-pipe air-vent system is being superseded by more desirable two-pipe arrangements, such as the vapor and the vacuum systems.

7-7. Repeat problem 7–6 for a two-pipe vapor system. Observe that with this system traps are used on each radiator and air valves are not required on the radiators.

7-8. A steam pipe 120 ft long is installed at 50 F. When it is filled with steam at 5.3 psig pressure, what is the increase in the length of the pipe (a) if the pipe is steel and (b) if the pipe is copper? *Ans.* (a) 1.7 in.; (b) 2.7 in.

7-9. An underground steel steam line in a tunnel is 5000 feet in length. It was laid in summer at an average ground temperature of 60 F, but in winter it will operate with steam at an average pressure of 105.3 psig. Compute the total change in length which must be absorbed by expansion joints or bends over summer and winter temperature ranges. *Ans.* 115 in. or 9.6 ft

7-10. Refer to Fig. 7–10 where it can be assumed that the float trap on the pressure

Steam Heating

main is receiving water from a 30-psia pressure system at a temperature of 240 F. The low-pressure main is in the vacuum region at 12 psia. Find (a) what percentage of each pound of water remains water in the flash tank and (b) what percentage passes into the low-pressure main as steam. (Note that the pressure of the water on the pressure side of the trap is unimportant provided it is greater than, or in the limit, reaches, the saturation pressure corresponding to the temperature. In most drain points from systems of this type the water is nearly always subcooled below saturation temperature.) *Ans.* (a) 96.1%; (b) 3.9%

7-11. Work problem 7-10 under the assumption that the water entering the trap is at 320 F and that its pressure is 100 psia (85.3 psig). The pressure in the low-pressure main is 15 psia. *Ans.* (a) 88.7%; (b) 11.3%

7-12. What size of steel pipe is required to carry 6000 lb of steam per hour through a main having an equivalent length of 450 ft if the initial pressure is 100 psig and the final pressure is 95 psig? *Ans.* 4 in.

7-13. A central steam plant provides steam to a remote building through an 8 in. welded steel pipe line of Schedule 40 pipe. An addition to the building is contemplated and question is raised as to whether the same line can serve both the original building and the addition. The linear run of pipe is 1200 ft and it is estimated the extension and equivalent length of fittings will add 400 ft additional. How many lb of saturated steam per hour can pass through this line if a pressure drop of 10 psi is allowed for the 135 psig steam at the boiler plant? *Ans.* 35,000 lb per hr

7-14. Work problem 7-2 if the allowable pressure drop is 20 psi. *Ans.* 810 lb per min.

7-15. Compute the pressure drop in a Schedule 40, 4 in. steel pipe of 400 ft equivalent length carrying 100 lb of steam per minute if the inlet steam pressure is 60 psig.
Ans. 5.4 psi

7-16. A two-pipe vapor system serving 670,000 Btuh has a total equivalent length of 200 ft, and the total pressure drop is to be 2 oz per sq in. (a) What size of main and dry return should be used? (b) To supply a radiator on the second floor delivering 18,000 Btuh, what size riser and horizontal branch should be used?
Ans. (a) 5 in., $2\frac{1}{2}$ in.; (b) $1\frac{1}{4}$ in., $1\frac{1}{4}$ in.

7-17. A vacuum system to deliver 2,160,000 Btuh has a total equivalent length of 800 ft. It is to be designed for a total pressure drop of 1 psi. What size of steam and return main should be used? *Ans.* 6 in., 3 in.

7-18. For the two-pipe, air-vent, gravity-return system of Fig. 7-4 considering the extreme right circuit as having the greatest length, compute the required size of main, return, risers, and runouts for this circuit. Each unit of radiation delivers 14,400 Btuh and requires approximately 14.4 lb of steam per hour, with a total boiler output of 86.4 lb of steam per hour.

7-19. For the two-pipe, air-vent, gravity-return system of Fig. 7-4, compute the size of the risers and runouts for the radiation which in the sketch appears to be roughly over the boiler. Refer to problem 7-18 for data regarding the system.

7-20. A central blast coil heats the air supplied to an auditorium in winter. The coil capacity at design outdoor conditions amounts to 175,000 Btuh. Steam at 5 psig is available and a vacuum return line, equipped with a pump, takes care of condensate and air venting. Find (a) the maximum steam flow to the coil, in pounds per hour, (b) the size of the feed pipe from the pressure-reducing valve to the coil if its equivalent length is 200 ft, if $\frac{1}{2}$-psi pressure drop is allowed, and, (c) the return pipe size to the vacuum pump if $\frac{1}{4}$-psi drop is allowed for the 200-ft run.

Ans. (a) 175 lb/hr; (b) 2 in.; (c) 1 in.

7-21. Rework problem 7–21 under the assumption that the blast-coil capacity is 125,000 Btuh, with no other change in data. *Ans.* (a) 125 lb/hr; (b) 2 in.; (c) $\frac{3}{4}$ in.

8

Hot Water Heating

8-1. HOT-WATER (HYDRONIC) SYSTEMS

Heating systems employing hot water as a means of conveying heat to points of utilization are called hot-water heating systems. In such systems a direct-fired boiler or steam-to-water heat exchanger delivers heat into water which then flows through pipes to radiators, convectors, or other types of heat exchangers located at points of utilization. Most present-day systems employ a pump to promote the circulation of water throughout the system, but many older systems have no pump and depend upon density differences existing in hot- and cold-water circuits to promote flow through the system. Natural-circulation systems are also called gravity-circulation systems or thermal-circulation systems.

In addition to conventional hot-water heating systems employing convectors, many of the panel, or radiant-heat, systems also use hot water as the heat source. All hot-water heating systems must make provision for changes in the volume of the water contained in the system between the hot and cold states of the system. This is accomplished by means of expansion tanks of either the open or closed variety.

8-2. PRESSURE-TEMPERATURE RELATIONS

The boiling temperature of water is governed by the pressure maintained on the water. It is possible, therefore, to use water in heating systems at a temperature slightly lower than that for saturated steam at the pressure maintained on the water. Table 2-3, which gives saturation temperatures, can be used to find limiting temperatures for hot-water heating systems operating at known pressures.

The surface of a radiator or convector will transmit heat to air at essentially the same rate from water as from steam, provided the temperature of the water is the same as that of the steam.

Because of the rapid increase in pressure which water experiences as temperatures are raised, it was formerly customary to limit the top temperatures of the system to approximately 250 F. However, there has been an increasing interest in attempts to raise the operating temperature of hot-water heating systems to values far in excess of 250 F. At the present time, systems operating in the range of 225 F to 325 F are known as medium-temperature systems, while those operating from (say) 300 F to 500 F are known as high-temperature systems. Reference to Table 2-2 would show that a pressure of some 666 psig would be required for a top temperature of 500 F. The advantages of high-temperature distribution are particularly noteworthy in the case of isolated installations where a boiler, perhaps located at a remote point, serves a widely dispersed plant. With high-temperature water, and the resultant large energy release from the cooling of this water, it necessarily follows that smaller pipe sizes can be used in the distribution system to transfer a given amount of energy than would be required with low-temperature hot water. In fact, such high-temperature systems can employ even smaller pipe sizes than are required in high-pressure steam installations.

8-3. ADVANTAGES OF HOT WATER HEATING

In hot-water heating systems the same water can be recirculated indefinitely, and thus there is minimum deposit of solids on boiler heating surfaces. Corrosion of piping is negligible, provided the oxygen-carrying fresh water for make-up is limited in amount. The temperature of the heating medium can be varied to meet weather conditions more readily than can be done with steam. For example, in mild weather the pump can circulate water to the heating surfaces at (say) 100 F to 120 F, whereas in extremely cold weather, with greater output required from the radiation, the water can be supplied at temperatures of (say) 180 F to 240 F.

Hot-water systems do not require traps or delicate appliances likely to require attention, and a water-level control is not required in a hot-water "boiler." There is little or no chance of water hammer. New hot-water systems are nearly all of the forced-circulation type, so that relatively smaller pipe sizes can be used than are required with steam. With the availability of long-life circulating pumps, precision temperature control can effectively give comfort in a heated space over a wide range of outside temperature conditions, with low maintenance cost. There is also in some instances a possibility of using the same piping and convectors for cooling in summer that are used for heating in winter.

8-4. CIRCULATION OF WATER

Any vessel which contains water and receives heat on one side will develop a circulation as the water, warmed and expanded by heat, rises, while the cooler and denser remaining water takes its place. This principle underlies the operation of all thermal-circulation (natural-circulation) hot-water heating systems. Some steam-heating systems can be changed to hot-water systems with ease; and some hot-water heating systems, with slight changes, can be made to operate with steam.

Hot Water Heating

The water in a thermal system circulates because of the difference between the pressure exerted by the hot-water column leaving the boiler on its way to the radiator and the pressure exerted by the relatively cool water column returning from the radiator. The unit pressure exerted by any water column is a function of water density and height.

Thus the pressure difference available to cause flow is expressed as

$$\Delta P = h(d_{fR} - d_{fD}) = h\left(\frac{1}{v_{fR}} - \frac{1}{v_{fD}}\right) \quad (8\text{--}1)$$

where ΔP = pressure difference for water circulation, in pounds per square foot;

h = height between boiler datum and radiator datum, in feet;

d_{fR}, d_{fD} = respective densities of the return and delivery water, in pounds per cubic foot;

v_{fR}, v_{fD} = corresponding specific volumes to be selected from Table 2–2, in cubic feet per pound.

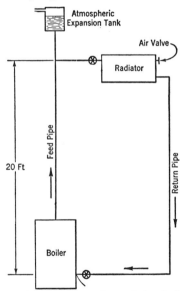

FIG. 8–1. Elements of a thermal-circulation hot-water heating system.

Equation 8–1 can be transformed so that $\Delta P'$ is expressed in milinches of water; thus

$$\Delta P' = h\left(\frac{1}{v_{fR}} - \frac{1}{v_{fD}}\right) \times 0.016 \times 12{,}000 = 193h\left(\frac{1}{v_{fR}} - \frac{1}{v_{fD}}\right) \quad (8\text{--}2)$$

Example 8-1. Refer to Fig. 8–1. If the average temperature of the water leaving the boiler is 180 F and the water cools in the radiator and piping to 160 F, what pressure difference is being used in overcoming frictional losses in the water system?

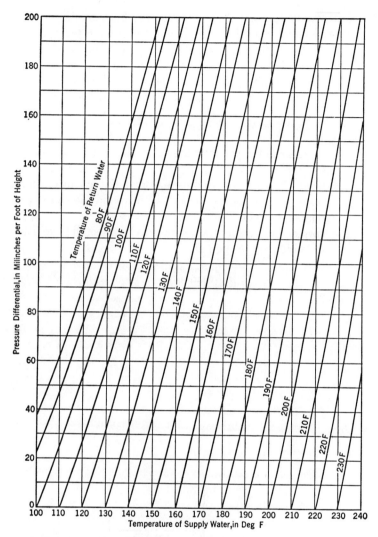

FIG. 8-2. Pressure differential per foot of height in thermal-circulation hot-water systems.

Solution: Use equation 8-1 and $h = 20$ ft. From Table 2-2, $v_{fR} = 0.01639$ cu ft per lb at 160 F, and $v_{fD} = 0.01651$ cu ft per lb at 180 F. Therefore

$$\Delta P = (20)\left(\frac{1}{0.01639} - \frac{1}{0.01651}\right) = 9.00 \text{ psf}$$

or
$$\Delta P = 9 \times 193 = 1737 \text{ milinches} \qquad Ans.$$

Example 8-1 can also be solved by making use of Fig. 8-2. For the data of example 8-1, entry is made at the supply-water temperature, 180 F,

Hot Water Heating

and the 180 F line is followed until it intersects the 160 F return temperature, at which point a pressure differential of 87 milinches per foot is read. When this is multiplied by 20 ft, the answer appears as 1740 milinches. This is in reasonable agreement with the computed answer.

If the valve at the boiler in Fig. 8–1 were suddenly closed during operation, the pressure difference of the hot and cold lines ($\Delta P = 1737$ milinches) in example 8–1 would act on the two sides of the valve, but in operation the pressure difference is entirely utilized in overcoming pipe friction (after initially accelerating the water around the circuit).

It is obvious that, with a given installation, the greater the temperature difference maintained, the more rapid the rate of circulation. However, this involves the question of the radiator cooling surface and its arrangement, and a large temperature drop may call for an undesirably large surface. In example 8–1, with a 70 F room the mean temperature difference causing heat transfer is $[(180 + 160) \div 2] - 70 = 100$ deg. With 180 F water cooled to 140 F the mean temperature difference causing the heat transfer is $[(180 + 140) \div 2] - 70 = 90$ deg. In the second case the circulation would be faster, but more surface would be required in the room to deliver the same heat than in the first case.

The warmer the water supplied by the boiler the greater the temperature difference from the radiator to the room, permitting less surface and better circulation. Atmospheric systems with open expansion tanks (like that in Fig. 8–1) are limited to water temperatures less than 212 F at the top radiators, although the lower radiators can be operated hotter in a high building because of the greater water pressure at the lower levels. Closed-expansion-tank systems can be operated at temperatures appreciably above 212 F because pressures much above atmospheric can be used.

8–5. HEATING-SYSTEM ARRANGEMENT

The water in a hot-water heating system may be heated either in fuel-fired boilers, which are essentially the same as boilers used with steam, or in transfer heaters, which use steam or some liquid warmer than the water as the source of heat. The water flows to the radiators and convectors, which are essentially the same as those used for steam. The supply pipes do not pitch down in the direction of flow. The flow and return pipes and branches are generally the same size, though one or the other can be reduced. A representative arrangement of accessories and control equipment for a forced-circulation-system hot-water boiler is shown in Fig. 8–3.

In the system of Fig. 8–3 the pump operates in response to a call for heat from the room thermostat. The operation of the pump is also under the control of the hot-water control, B, which prevents the pump from starting if the water is below a set temperature and later stops the pump after operation if this low temperature is reached. A main hot-water control, A,

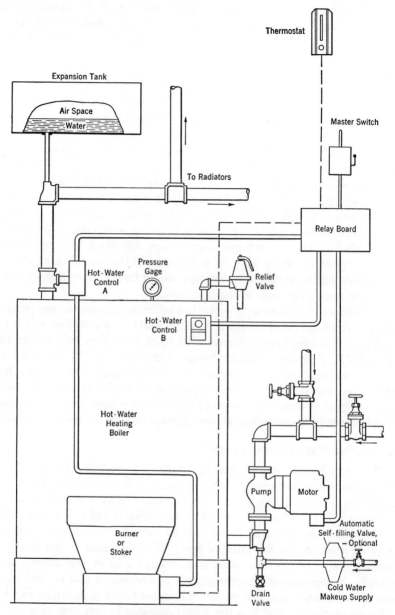

Fig. 8-3. Representative accessories and controls for a forced-circulation hot-water boiler.

Hot Water Heating

usually independent of the other controls, starts and stops the oil or gas burner or the stoker, to maintain a definite water temperature in the boiler.

The expansion tank is a very essential part of a hot-water heating system, because the difference in the total volume of water between the hot and cold conditions is great. In new installations, expansion tanks located above and near the boiler are used almost exclusively. In this type of expansion tank, air trapped in the tank is compressed as the volume of water in the system increases. If the air volume of the tank is too small, and if the pressure exceeds a predetermined set value, surplus water is released through a relief valve. The pressure-relief valve of the hot-water

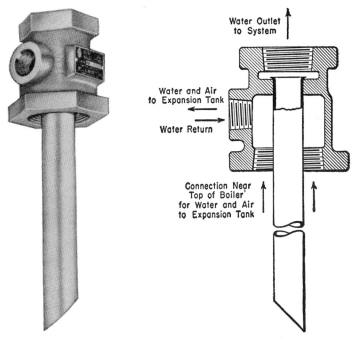

FIG. 8-4. Boiler fitting for air-elimination at heat source. (Courtesy Bell and Gossett Company.)

boiler takes the place of the safety valve of the steam boiler. With a tight system and with adequate air space in the expansion tank, little or no water need be wasted through the relief valve, and only occasional make-up of water by manual control is sufficient. Larger hot-water heating systems are provided with automatic filler or make-up devices.

A difficulty sometimes occurs in hot-water heating systems when air originally absorbed in the water moves with it and tends to separate at high points in radiators or other heat-transfer equipment. This is objectionable because it prevents water contact with part of the heat-transfer surface.

The air has to be removed through vent valves located at high points on the units. Cold water holds more gas (air) in solution than does hot water. Consequently, if the gas is driven off from the water during heating in the boiler, it can be separated and largely delivered to the expansion tank. Figure 8-4 shows a patented design which accomplishes this end. This special fitting for outflow from the boiler connects at a high point. Air is removed around the top of the fitting and is carried to the expansion tank, while the water flowing out is taken from a lower point in the fitting remote from the point of air separation. When the fresh water entering the system is properly deaerated in the boiler at the start, little difficulty occurs from air separation in radiators.

In the closed expansion tank, the air space exists above the free water surface. As the water volume in the system increases, more water flows into the bottom of the tank and compresses the air in the upper part of the tank. As the water volume in the system decreases with lowering temperature, water flows from the bottom of the tank back into the system. It is desirable that the water-flow in and out of the tank take place as far from the air cushion as possible. It is also important that the air separated in the boiler not bubble through the water in the expansion tank and thus be reabsorbed by this relatively cool water. To prevent this, a special fitting (Fig. 8-5) has been developed which sends the air from the boiler directly into the free air space of the expansion tank. Surplus tank air can also be removed through the fitting by a valve located in the bottom. The size of an expansion tank must be adequate to provide for the volume increase not only of the water in the boiler itself, but also of the water in all of the connecting piping and the heat-transfer devices.

Reference to Table 2-2 gives both the specific volume of cold water at the filling temperature of the system and also the specific volume the water will have when it reaches the highest temperature at which the system will operate. Using these values, the volume increase can be computed for the total weight of water in the system. This volume increase represents the minimum expansion-tank space which must be provided. Good practice calls for expansion-tank volume in excess of this amount. Considerable research has been carried out in an effort to find optimum expansion sizes. There are no universally accepted standards, but it is customary to use an 8-gal tank for systems of 60,000 Btuh or less, 15 gal for systems up to 85,000 Btuh, 18 gal for systems up to 120,000 Btuh, 24 gal for systems up to 240,000 Btuh, and 60 gal for systems up to 1,000,000 Btuh. These numbers apply to forced-circulation systems using smaller pipe sizes. For older, natural-circulation systems with large pipe sizes, it is desirable to increase the tank to one size higher.

In addition to the air which separates at the boiler and expansion tank, air can also be trapped in the lines and in the radiators. For air

Hot Water Heating

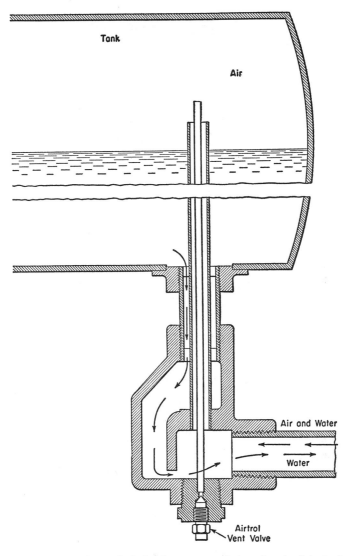

FIG. 8-5. Expansion-tank fitting for passage of air and water into tank and for return of water to system on cooling. (Courtesy Bell and Gossett Company.)

elimination at high points in the lines, units of the type illustrated in Fig. 8-6 can be used. These units are normally completely filled with water and the escape port remains closed. However, when air collects in sufficient quantity, the float drops, permitting the air to be eliminated, and as water flows in, causing the float to rise, the escape port closes. Air trapped in radiators must usually be removed through manually operated air vent

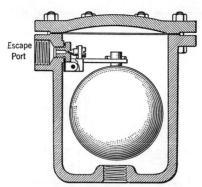

Fig. 8-6. Automatic air valve for use on hot-water heating-system line.

valves located at a high point in the radiator. In a well-adjusted system, air elimination is usually necessary only at starting, and perhaps at one other time during the heating season. Some domestic systems can function for several seasons of operation without air elimination or addition of water.

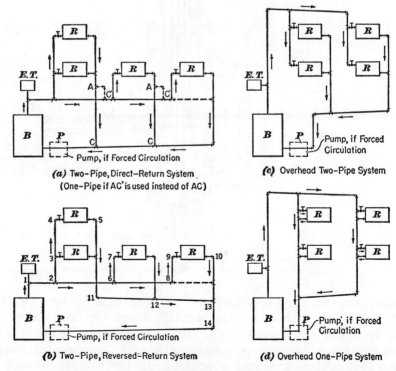

Fig. 8-7. Circulating arrangements for hot-water heating systems. B = boiler, R = radiator, P = pump, $E.T.$ = expansion tank.

Hot Water Heating

8-6. VARIETIES OF CIRCULATING SYSTEMS

There are many variations in water-circulating elements, and the nomenclature is not uniform. Several arrangements are shown in Fig. 8-7. Single-pipe systems are illustrated in Fig. 8-7d, as a variation in Fig. 8-10, and in more detail in Fig. 8-15. In single-pipe systems, a part of the water successively passes through radiator after radiator, and as the water loses heat and drops in temperature the transmitting surface in each succeeding radiator must be increased in relation to the temperature drop if the same capacity is expected from each radiator. In small systems where the total temperature drop is small, a change in radiator size may not be required.

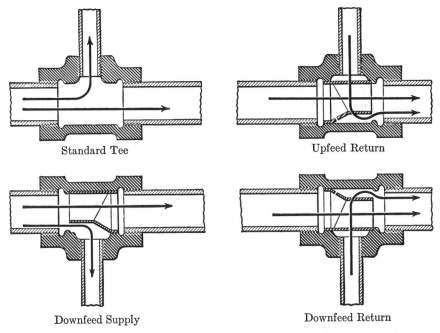

FIG. 8-8. Piping arrangements for one-pipe hot-water heating systems using diversion tees. (Courtesy Bell and Gossett Company.)

Some single-pipe systems use standard tee fittings and run the supply branch off the top of the main, with the return entering the side of the main. Others employ special tees (Fig. 8-9) having directed passages or deflectors for sending a definite portion of the water into the supply branch. The return runouts employ similar tees in which the passages are reversed to act as ejectors—with the water, flowing in the main, drawing a current from the return branch. In most cases one fitting on the radiator return side is sufficient to insure proper water flow for upfeed radiation. However, for radiation which is awkwardly placed, or remote from the main, it may be desirable to use directional tees on both inlet and outlet to the radiation. In the case of radiators below the main, where unaided natural circulation is completely ineffective, it is necessary to supply special tees for both feed and return piping. The installation and use of these tees is illustrated in Fig. 8-8, and a drawing of the representative tee of one manufacturer is shown in Fig. 8-9.

A common arrangement, particularly with two-pipe thermal-circulation systems, places the supply main above the radiator, and the return main below the radiator. Rapid water circulation is promoted by this difference in elevation between the supply and the return. With overhead feed, the water inlet valve to each radiator can be located at the top of the

FIG. 8-9. Cross section of monoflow tee as used on one-pipe systems. (Courtesy Bell and Gossett Company.)

radiator. For upfeed systems, the radiator valve should be at the bottom. If this practice is not followed, difficulty may develop with the circulation, because of air pockets forming in the upper part of the radiator and stopping circulation.

The double-pipe (two-pipe) system delivers water which has come as directly as possible from the boiler and which returns without passing through more than one radiator, as indicated in Fig. 8-7a, b, and c, and in Fig. 8-16. A modification of this double-pipe system is used in some very small installations; it has no mains but employs separate supply and return pipes all the way from the boiler to each radiator or each piping circuit.

Forced-circulation systems, because of the higher available circulating head, can be designed for higher velocities—and consequently for smaller pipe sizes than are required for gravity-circulation systems. Moreover, in forced circulation with the pump running, some heat can be delivered to a given space even before the water temperature approaches that which would be required for good circulation in a gravity-type system.

Hot Water Heating

Mechanical-circulation (forced-circulation) systems are being specified for most new installations, and where they are employed much more flexibility in design and layout is possible. For example, the supply and return mains may both be above or both be below the radiators. With double-pipe mains there are two general circuit types: direct-return and reversed-return. With the direct-return type the water from the boiler flows through the supply main to the nearest radiator, and returns as directly as possible through the return main, encountering less piping and friction loss than is met by the water going to more-distant radiators. With the reversed-return system the outgoing water to the nearest radiator makes the same journey as in the direct-return system, but the return from the radiator nearest the boiler constitutes the first section of the return main, which starts away from the boiler, turning back only after picking up the return branches from the various radiators in the order of their remoteness from the boiler (Fig. 8-16). With this scheme the round-trip travel of the water to any radiator is the same, measured along the mains, as that to any other radiator. There may be three main pipes over part of the system, increasing the cost slightly, but the reversed-return system is almost self-equalizing so far as water flow to the different radiators is concerned.

In large systems particularly, it is desirable to supply a hand-operated control valve on one end of each radiator, and a lock-shield valve on the other end. The lock-shield valve is fixed at the proper place when the heat loads are initially adjusted for the system. In small systems it is not unusual to find only one valve on each radiator. In such systems, to prevent the water in a radiator from freezing in cold weather if the valve is closed, a small hole ($\frac{1}{16}$ in. in diameter or smaller) may be drilled through the valve disk so that a slight circulation will always be maintained in the radiator even when the valve is shut. This arrangement permits enough circulation to keep the temperature of the radiator above 32 F when the outside air is below freezing temperature.

It should be realized that there are two major factors which control the distribution of water in the system; namely, the pressure created by the circulating pump (or water temperature differences in the case of a natural-circulation system) and the opposing pressure loss resulting from frictional resistance to the passage of water through the pipes and heat-transmitting elements.

8-7. HEAT TRANSMISSION WITH HOT-WATER ELEMENTS

For each radiator the feeder-pipe sizes must be adequate, to carry the required amount of water, and the heat-transfer surface in the radiator must be sufficient to absorb from the water and dissipate to the room the needed amount of heat. The heat-transmitting capacity of a radiator or convector depends on the difference between the average temperature of the water in the radiator and the temperature of the air passing over the radiator surface. Other factors, such as the proximity of colder surfaces and high or low air velocities, also affect the heat transmission, but the most significant factor is the mean temperature difference between the water and air. For ordinary radiator surface, each square foot of surface will deliver approximately 150 Btuh under conditions of 100 deg mtd (mean temperature difference). The range of approximate heat-transmission values is indicated in Table 8–1. The EDR unit of 240 Btuh, which is so common with steam heat-transfer equipment, has less utility in connection with hot-water radiation and it is therefore more common to express output in Mbh units (i.e., units of 1000 Btuh) when rating hot-water heat transmitters.

TABLE 8–1
HEAT EMISSION OF HOT-WATER RADIATORS

Mean Radiator Temperature (F)	MTD between Radiator and Room Air (F)	Approximate BTU Transmitted per Hour per Sq Ft of Surface	Approximate BTUH Transmitted per Sq Ft of Surface per Degree Difference
220	150	250	1.67
215	145	240	1.65
200	130	208	1.60
190	120	188	1.57
180	110	170	1.54
170	100	150	1.50
160	90	129	1.43
150	80	110	1.39
140	70	93	1.33
130	60	76	1.27

Tables 9–1, 9–2, and 9–3 give specific data on representative hot-water radiators and convectors. It has been mentioned before that the output of a hot-water radiator unit is determined by the weight of water passing per unit of time and by its temperature drop. This can be expressed by the following formula:

$$Q = 60W(1)(t_{in} - t_{out}) \qquad (8\text{–}3)$$

which becomes

$$Q = (60)(8.1)(\text{gpm})(t_{in} - t_{out})$$
$$= 490(\text{gpm})(t_{in} - t_{out}) \qquad (8\text{–}4)$$

where Q = output of transmitter, in Btu per hour;
W = water flowing through, in pounds per minute;
t_{in} = temperature of water in, in degrees Fahrenheit;
t_{out} = temperature of water out, in degrees Fahrenheit;
gpm = gallons per minute water flow equivalent to W lb per min.

Hot Water Heating

The factor 8.1 in equation 8–4 represents the weight per gallon of water in the range of 180 F, in contrast with the value 8.33 at 70 F. In general, it is customary to design on a 20-deg temperature drop, with 180 F entering- and 160 F leaving-water temperatures. If the water temperature drop is changed to 10 deg, then twice as much water must pass through each radiator for a given heat output. As an aid in computation, it should be mentioned that flow in pounds per hour can be transformed to flow in gallons per minute by dividing by 500 for cold water, or by 490 for hot water at about 180 F on inlet to radiators or convectors.

Compared with the standard 20-deg drop, a 30-deg drop would require only two-thirds as much water to circulate. With a direct-return piping system, the equivalent length for each radiator circuit should be computed in order to give the basis for balancing out the circuits, and in such a system it may be necessary to put restricting orifices in the shorter circuits.

In designing hydronic systems the concept of equivalent pipe length is widely used. Equivalent pipe length is the length of straight pipe of a given size that would have the same resistance as the actual pipe run with its bends, fittings, valves, and other devices. In computing equivalent length, use is often made of the *elbow equivalent*, where an elbow equivalent is equal to the resistance offered by a straight run of pipe approximately 25 times the nominal pipe diameter. Thus, for a 3-in. pipe, one elbow equivalent is $25 \times 3 = 75$ in. or 6.5 ft of linear run. Representative values of elbow equivalents show the following: elbow 1.0; open gate valve 0.5; open globe valve 12.0; open return bend 1.0; angle radiator valve 2.0; radiator 3.0; tee, 100 per cent side diversion 1.8; tee, 50 per cent side diversion 4.0; boiler 3.0; 45-degree elbow 0.7; reducer coupling 0.4. Table 8–2 presents the friction loss of fittings in length of straight pipe.

Example 8–2. In a certain hot-water system, from the boiler out to a given radiator and back to the boiler there are 12 elbows, 2 tees (50 per cent side diversion), 1 angle radiator valve, 1 radiator, 1 boiler, and 90 ft of $1\frac{1}{2}$-in. iron pipe. What is the equivalent length of the system?

Solution:

		By Table 8-2
12 elbows........	$12 \times 1 = 12$	$12 \times 3.1 = 37.2$ ft
2 tees..........	$2 \times 4 = 8$	$2 \times 12.5 = 25.0$
1 valve.........	$1 \times 2 = 2$	$1 \times 6.2 = 6.2$
1 radiator......	$1 \times 3 = 3$	$1 \times 9.4 = 9.4$
1 boiler.........	$1 \times 3 = 3$	$1 \times 9.4 = 9.4$
Total........	28	87.2 ft

Therefore, by Table 8–2, the equivalent length is $90 + 87.2 = 177.2$ ft.   *Ans.*

By elbow equivalents, 90 ft + 28 elbow equivalents $\times$ 1.5 in. $\times \frac{1}{12} \times 25 = 177.2$ ft. Solution by elbow equivalents has some convenience where the pipe size is not known at the time the layout of the circuit is started.

TABLE 8-2
FRICTION LOSS OF STANDARD PIPE FITTINGS
(In Equivalent Feet of Straight Pipe)

Fitting	Iron Pipe (in.)					Copper Tube (in.)				
	¾	1	1¼	1½	2	3	¾	1	1¼	1½	2

Fitting	¾	1	1¼	1½	2	3	¾	1	1¼	1½	2
Elbows											
90°.	1.6	2.1	2.6	3.1	4.2	6.5	1.6	2.1	2.6	3.1	4.2
45°.	1.1	1.5	1.8	2.2	2.9	4.5	1.1	1.5	1.8	2.2	2.9
90° long sweep.	0.8	1.0	1.3	1.6	2.1	3.0	0.8	1.0	1.3	1.6	2.1
Tees											
100% side diversion.	2.8	3.8	4.7	5.6	7.5	13	1.9	2.5	3.1	3.7	5.0
50% side diversion.	6.3	8.3	10.4	12.5	16.7	25	6.3	8.3	10.4	12.5	16.7
33% side diversion.	14.3	18.7	23.4	28.1	37.5	56	18.7	23.5	29.4	35.2	46.9
25% side diversion.	25.0	33.3	41.6	49.8	66.7	100	31.2	41.6	52.0	62.5	83.4
Valves											
Globe (full open).	18.7	25.0	33.8	36.8	50.0	66	26.6	35.4	44.2	53.0	70.8
Gate (full open).	0.8	1.0	1.3	1.6	2.1	3.0	1.1	1.5	1.8	2.2	2.9
Stopcock (full open).	1.6	2.1	2.6	3.1	4.2	6.5	1.6	2.1	2.6	3.1	4.2
Angle (full open).	3.6	4.2	5.2	6.2	8.3	12.5	4.7	6.3	7.8	9.4	12.5
Reducer coupling.	0.6	0.8	1.0	1.3	1.7	2.5	0.6	0.8	1.0	1.3	1.7
Boiler or radiator.	4.7	6.3	7.8	9.4	12.5	19	6.3	8.3	10.4	12.5	16.7

Hot Water Heating

8-8. HOT-WATER BASEBOARD HEATING

Hot-water baseboard heating, also called perimeter heating, is a type of radiation being used in an increasing number of installations in both public and private buildings. The baseboard elements cause minimum difficulty in regard to furniture placement, and largely prevent cold-wall downdraft. Figure 8-10 shows an arrangement of baseboard radiation as it might be employed in a residence, and Fig. 9-4 shows the product of one manufacturer.

Installation practice varies, but usually the element is placed along an outside wall of a room—along part of the wall or all of it. However, it can also be placed to cover all of the walls. Ingenious arrangements are provided for extending the baseboard panels around corners. When baseboard radiation served by copper tubing is used, it is customary to connect the unit lengths of baseboard with sweated fittings; with steel and cast-iron radiation, pipe connections with special unions are employed. Apertures for valves are also provided in the baseboard panels. Dampers are provided, with at least one design, to control the air flow over the heating surface and thereby reduce the heat output. Although the heat output is largely convective, an appreciable portion of the heat is distributed by radiation. The heated air is not hot but warm, and it moves out gently from the units. This small air motion is sufficiently great, however, to limit the usual variation in temperature between floor and ceiling to 3 deg or less.

In most installations it is common to design for a 10-deg temperature drop in buildings having a heat capacity of 50,000 Btuh or less, or a 20-deg temperature drop for buildings or zones having a loss greater than 50,000 Btuh. As the elements are usually in series, the water temperature progressively drops as it passes through the subsequent baseboard panels. For design with a 10-deg temperature drop, it is customary to split the circuit into two portions and design with a lower average temperature for the second part of the circuit. For a 20-deg temperature drop, it is customary to use three average temperatures, the second and third of which are respectively lower in sequence. Representative outputs for average water temperatures are shown in tables 9-2 and 9-3 for the products of one manufacturer. Inlet temperatures about 190 F are usual.

8-9. PIPING-SYSTEM DESIGN

In the design of a hot-water heating system, it is desirable to consider the following general procedures, which are applicable to all types of layouts:

1. The system should provide means for complete and adequate drainage at convenient low points and also provide for air venting at high points in the piping system, as well as in the radiation units.

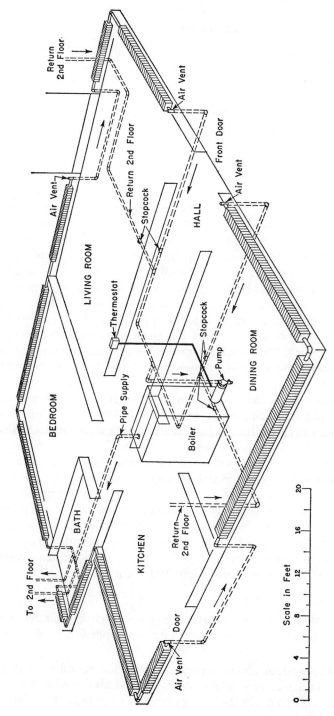

FIG. 8-10. Layout of baseboard heating system for residence.

Hot Water Heating

2. Straight runs in excess of 30 ft require provision for expansion.
3. Water temperature drops of 25 deg to 35 deg are usually assumed for gravity systems, with a 20-deg drop for forced-circulation systems. The temperature range, particularly with forced-circulation systems, can be modified where this appears to be desirable.
4. Water velocities should range from 2 fps to 4 fps where noise might be a disturbing factor. For industrial installations, and where large pipe sizes are employed, velocities up to 8 fps can be considered.
5. The design water supply temperatures for gravity systems normally range between 140 F and 200 F, while forced-circulation supply temperatures normally range from 170 F to 220 F. For high-temperature distribution of water, in systems also safe for high pressure, temperatures in excess of 300 F are employed.
6. For forced-circulation systems, the total design friction loss is set by the pressure (or head) characteristic of the available circulating pump.
7. It is customary to choose design friction values in the range 250 to 650 milinches per foot. Above 650 milinches per foot, high velocities occur; below 250 milinches, pipe sizes may be uneconomically large.
8. On large installations separate circulators for each run should be provided, with independent control.
9. Adjustable stopcocks should be provided, or provision for inserting flow-control orifices should be made, so that individual circuits can be balanced to give proper flow.
10. The specific design procedures for laying out a forced-circulation system involve the following:
    a. The radiation required for each room or heated space should be computed, and expressed preferably in Btu per hour (Btuh) or in equivalent thousands of Btu per hour (Mbh).
    b. A sketch on the building plans should be made to show pipe runs and the location of all radiators and of the boiler. On this sketch should also be shown the length of all pipe runs, and as computations are made the pipe sizes should be marked on the sketch.
    c. The gallons-per-minute (gpm) flow should be computed for each circuit by means of equation 8–4.
    d. In terms of the total gpm flow, a forced-circulation (booster) pump should be selected with the help of pump-characteristic curves (Fig. 8–11 or Fig. 8–12), picking the selection point to the right of the peak of the characteristic curve.

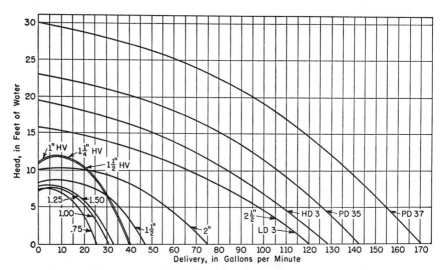

Fig. 8–11. Representative head-capacity curves for forced-circulation (booster) pumps turning at 1750 rpm, with manufacturer's data indicated. (Courtesy Bell and Gossett Company.)

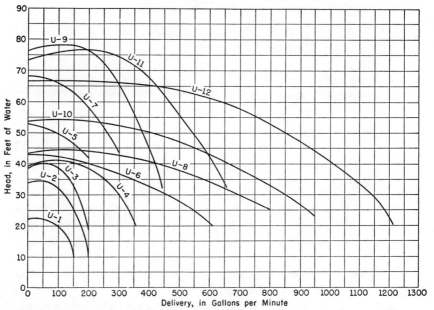

Fig. 8–12. Representative head-capacity curves for forced-circulation pumps of large capacity turning at 1750 rpm. (Courtesy Bell and Gossett Company.)

Hot Water Heating

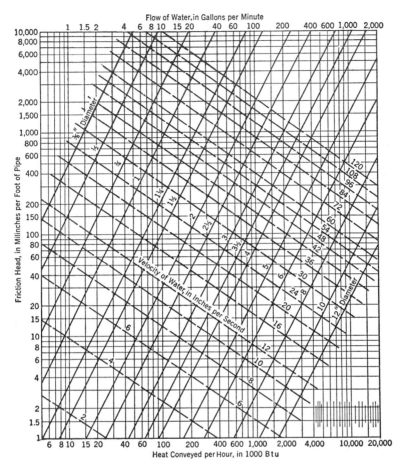

Fig. 8-13. Friction loss for hot-water flow in steel pipe. Bottom scale is based on 20-deg temperature difference between water in supply and return lines. For other temperature differences (Δt), multiply Mbh value being considered by $20/\Delta t$ before entering chart. (Reprinted, by permission, from *Heating Ventilating Air Conditioning Guide 1943*.)

    e. The linear run of the longest (or highest-friction-loss) circuit should be observed and, for a good estimate, increased by 50 or even 100 per cent to allow for fitting losses, in order to find the Total Equivalent Length. (This may later be rechecked.)

    f. Divide the pump head expressed in milinches (feet × 12,000) by the total equivalent length in feet to find the milinch friction loss per foot. Good design practice shows the range to be between 650 and 250 milinches per foot.

    g. In terms of the milinches per foot, from the preceding, and the gpm flowing, the required pipe size can be selected from Fig. 8–

13 or Fig. 8–14. If a standard 20-deg temperature drop is used the heat load at the bottom of the graphs can be used directly without correction or without referring to gpm.

h. The radiator pipe sizes can be selected with the help of Table 8–3 or Table 8–4.[1]

8–10. SINGLE-PIPE FORCED-CIRCULATION DESIGN PRACTICES

A representative system of this type is illustrated in Fig. 8–15. There are two circuits indicated, and in computation the longest (or highest-friction) circuit must be used in sizing the pump. The radiation located below the main is not desirable and in a single-pipe system can be made to work only with the help of diversion tees.

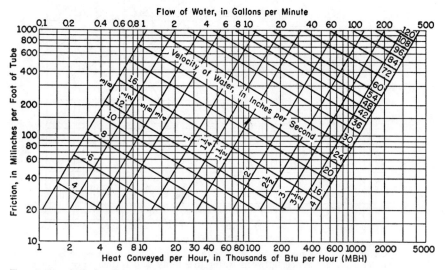

Fig. 8–14. Friction loss for hot-water flow in copper tubing. Bottom scale is based on 20-deg temperature difference between water and supply and return lines. For other temperature differences (Δt), multiply Mbh value being considered by $20/\Delta t$ before entering chart. (Reprinted, by permission, from *Heating Ventilating Air Conditioning Guide 1956*.)

Example 8–3. Work out the design features of the single-pipe system of Fig. 8–15. This system has 7 upfeed convectors, required to deliver 9200 Btuh each, and 2 downfeed convectors, required to deliver 5600 Btuh each. It is assumed that the boiler supplies water at 190 F, which cools 20 deg in each radiator. With a one-pipe system such as this, the more-remote radiators will have less capacity than those radiators close to the boiler, which are supplied with the hottest water. However, this variation in capacity will not be considered sufficiently significant to upset a simple design procedure. The drawing, when scaled, shows the left circuit as having a linear

[1] Tables 8–3 and 8–4 are reprinted, by permission, from *Heating Ventilating Air Conditioning Guide 1956*, Chapter 22.

TABLE 8-3

Heat-Carrying Capacity A (in MBH) of Standard Black Iron (or Steel) Pipes with Temperature Drop of 20 Deg*

Milinch Friction Loss per Foot of Pipe		Nominal Pipe Size (in.)													
		$\frac{3}{8}$	$\frac{1}{2}$	$\frac{3}{4}$	1	$1\frac{1}{4}$	$1\frac{1}{2}$	2	$2\frac{1}{2}$	3	$3\frac{1}{2}$	4	5	6	8	10	12
4	A	0.75	1.35	2.85	5.4	11.3	17.0	33.0	53.1	95	141	197	363	596	1250	2320	3730
	V†	1.5	1.7	2.1	2.4	2.9	3.2	3.8	4.3	5.0	5.5	6.0	7.0	7.9	9.6	11	12
6	A	0.9	1.7	3.6	6.75	14.0	21.2	41.3	66.4	119	176	248	456	748	1570	2920	4690
	V	1.8	2.1	2.6	3.0	3.6	4.0	4.7	5.3	6.2	6.9	7.5	8.8	10	12	14	16
8	A	1.05	2.0	4.2	7.9	16.4	24.8	48.4	77.9	140	207	291	535	879	1850	3440	5520
	V	2.1	2.5	3.0	3.5	4.2	4.7	5.6	6.3	7.3	8.0	8.8	10	12	14	17	19
10	A	1.2	2.2	4.7	8.9	18.6	28.0	54.7	88.1	158	234	329	605	997	2100	3910	6270
	V	2.4	2.8	3.4	4.0	4.8	5.3	6.3	7.1	8.2	9.1	9.9	12	13	16	19	27
12	A	1.35	2.45	5.2	9.8	20.5	31.0	60.4	97.4	175	259	364	671	1100	2320	4330	6950
	V	2.7	3.1	3.7	4.4	5.3	5.9	6.9	7.8	9.1	10	11	13	15	18	21	24
14	A	1.45	2.65	5.65	10.7	22.3	33.7	65.8	106	190	282	397	731	1200	2530	4730	7590
	V	2.9	3.4	4.1	4.8	5.7	6.4	7.6	8.5	9.9	11	12	14	16	20	23	26
16	A	1.55	2.85	6.05	11.5	24.0	36.3	70.8	114	205	303	428	787	1300	2730	5100	8190
	V	3.1	3.6	4.4	5.1	6.2	6.9	8.1	9.7	11	12	13	15	17	21	25	28
20	A	1.75	3.25	6.85	13.0	27.1	41.0	80.0	129	232	344	484	892	1470	3100	5790	9300
	V	3.5	4.1	4.9	5.8	7.0	7.7	9.2	10	12	13	15	17	20	24	28	32
25	A	2.0	3.65	7.75	14.7	30.6	46.3	90.5	146	263	389	548	1010	1670	3510	6570	10560
	V	4.0	4.6	5.6	6.5	7.9	8.8	10	12	14	15	17	19	22	27	32	36
30	A	2.2	4.0	8.55	16.2	33.8	51.2	100	162	290	430	607	1120	1850	3900	7280	11710
	V	4.4	5.1	6.1	7.2	8.7	9.7	11	13	15	17	18	22	25	30	35	40
35	A	2.35	4.4	9.3	17.6	36.8	55.7	109	176	316	469	661	1220	2010	4250	7940	12780
	V	4.7	5.5	6.7	7.9	9.5	11	13	14	16	18	20	23	27	33	39	44
40	A	2.55	4.7	10.0	18.9	39.6	59.9	117	189	341	505	712	1320	2170	4580	8570	13780
	V	5.1	5.9	7.2	8.4	10	11	13	15	18	20	22	25	29	35	42	47
50	A	2.85	5.3	11.3	21.4	44.7	67.7	133	214	386	572	807	1490	2460	5190	9720	15650
	V	5.7	6.7	8.1	9.5	12	13	15	17	20	22	24	29	33	40	47	54
60	A	3.15	5.85	12.4	23.6	49.4	74.9	147	238	427	633	893	1650	2730	5760	10780	17360
	V	6.3	7.4	8.9	11	13	14	17	19	22	25	27	32	36	44	52	60
70	A	3.45	6.35	13.5	25.7	53.8	81.4	160	258	465	690	973	1800	2970	6280	11760	18950
	V	6.9	8.0	9.7	11	14	15	18	21	24	27	29	35	40	48	57	65
80	A	3.7	6.8	14.5	27.6	57.9	87.6	172	278	500	743	1050	1940	3200	6770	12690	20440
	V	7.4	8.6	10	12	15	17	20	22	26	29	32	37	43	52	62	70
100	A	4.15	7.7	16.4	31.1	65.4	99.0	194	314	566	840	1190	2200	3630	7680	14400	23200
	V	8.3	9.7	12	14	17	19	22	25	30	33	36	42	48	59	70	80
150	A	5.2	9.6	20.4	38.8	81.6	124	243	393	709	1050	1490	2760	4560	9650	18120	29220
	V	10	12	15	17	21	23	28	32	37	41	45	53	61	74	88	101
200	A	6.05	11.2	23.9	45.4	95.5	145	285	461	832	1240	1750	3240	5360	11350	21320	34400
	V	12	14	17	20	25	27	33	37	43	48	53	62	71	87	104	118
300	A	7.5	13.9	29.7	56.6	119	181	356	577	1040	1550	2190	4060	6730	14270	26830	43300
	V	15	18	21	25	31	34	41	46	54	60	66	78	90	110	131	149
400	A	8.75	16.2	34.7	66.2	140	212	417	676	1220	1820	2570	4780	7910	16790	31580	51000
	V	18	21	26	30	36	40	48	54	64	71	78	92	105	129	154	175
500	A	9.85	18.3	39.2	74.8	158	239	471	765	1380	2060	2910	5410	8970	19040	35840	57880
	V	20	23	29	33	41	45	54	62	72	80	88	104	119	147	174	199
600	A	10.9	20.2	43.2	82.5	174	264	521	846	1530	2280	3220	5990	9930	21100	39740	64210
	V	22	26	32	37	45	50	60	68	80	89	97	115	132	162	193	221
800	A	12.7	23.6	50.5	96.5	204	310	610	992	1790	2670	3780	7030	11670	24820	46780	75620
	V	25	30	37	43	52	59	70	80	94	104	114	135	155	191	228	260

* For other temperature drops the pipe capacities may be changed correspondingly. For example, with a temperature drop of 30 deg the capacities shown in this table are to be multiplied by 1.5.

† V = velocity, inches per second.

TABLE 8-4
Heat-Carrying Capacity A (in MBH) of Type L Copper Tubing with Temperature Drop of 20 Deg*

Nominal Tube Size (in.)		Milinch Friction Loss per Foot of Tube										
		720	600	480	360	300	240	180	150	120	90	75	60
$\frac{3}{8}$	A	8.9	7.8	7.0	5.9	5.4	4.7	3.9	3.6	3 1	2.7	2.3	2.1
	V†	23.6	20.8	18.6	15.7	14.4	12.5	10.4	9.6	8.2	7.2	6.1	5.6
$\frac{1}{2}$	A	16.7	15.0	13.0	11.2	10.0	8.7	7.5	6.6	5.6	5.0	4.5	3.9
	V	27.6	24.8	21.5	18.5	16.5	14.4	12.4	10.9	9.3	8.3	7.4	6.4
$\frac{5}{8}$	A	29.0	26.0	22.5	19.0	17.5	15.0	13.0	11.5	10.0	8.5	7.6	6.7
	V	32.2	28.8	25.0	21.1	19.4	16.6	14.4	12.8	11.1	9.4	8.4	7.4
$\frac{3}{4}$	A	43.5	39.0	34.5	29.0	26.5	23.0	19.6	17.5	15.0	13.0	12.0	10.5
	V	34.6	31.1	27.5	23.1	21.1	18.3	15.6	13.9	12.0	10.4	9.6	8.4
1	A	93	84	74	63	57	50	42.5	38	34	28.5	26	23
	V	43	39	34	29	27	23	20	18	16	13	12	11
$1\frac{1}{4}$	A	160	145	128	107	97	85	73	65	57	48.5	44	39
	V	49	45	39	33	30	26	22	20	18	15	14	12
$1\frac{1}{2}$	A	260	240	206	175	160	140	118	106	93	79	71	62
	V	56	52	45	38	35	30	26	23	20	17	15	13
2	A	560	510	450	380	340	300	250	225	195	170	150	133
	V	70	64	56	47	42	37	31	28	24	21	19	17
$2\frac{1}{2}$	A	1100	930	820	700	630	550	470	420	370	310	280	250
	V	89	75	66	57	51	44	38	34	30	25	23	20
3	A	1650	1500	1300	1100	990	860	730	650	565	480	430	375
	V	94	85	74	62	56	49	41	37	32	27	24	21
$3\frac{1}{2}$	A	2500	2250	2000	1750	1500	1320	1100	1000	860	730	660	580
	V	105	94	84	73	63	55	46	42	36	31	28	24
4	A	3600	3200	2800	2400	2150	1900	1600	1440	1250	1150	950	840
	V	116	103	90	77	69	61	51	46	40	37	31	27

* For other temperature drops the pipe capacities may be changed correspondingly. For example, with a temperature drop of 30 deg the capacities shown in this table are to be multiplied by 1.5.

† V = velocity, inches per second.

run of 114 ft (rounded to the nearest foot) from boiler outlet to pump return, while the right circuit is 112 ft. The mains lie approximately 6 ft above the center line of the boiler.

Solution: The total load on the system is 75,600 Btuh, with 42,400 in the left circuit and 33,200 in the right. By equation 8-4, the total flow is

$$\text{gpm} = \frac{75,600}{(490)(20)} = 7.72, \text{ or } 7.7$$

Also by equation 8-4, the flow in the left circuit is

$$\text{gpm} = \frac{42,400}{(490)(20)} = 4.33, \text{ or } 4.3$$

Therefore the flow in the right circuit is

$$7.7 - 4.3 = 3.4 \text{ gpm}$$

Referring to Fig. 8-11, it can be seen that the smallest pump shown delivers **7.7** gpm at a head of **7** ft and will be used.

Hot Water Heating

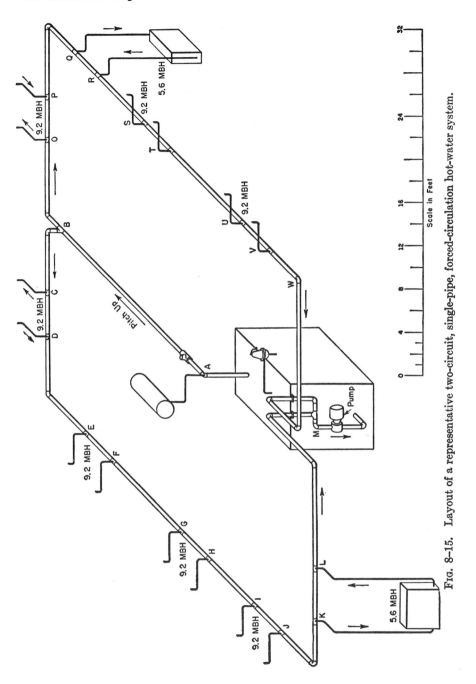

FIG. 8-15. Layout of a representative two-circuit, single-pipe, forced-circulation hot-water system.

For the longest run, 114 ft, add an additional 50 per cent to take care of equivalent length of fittings, giving a total of 171 ft ETL (equivalent total length) in this circuit. For the 7-ft head of the pump, or its equivalent, 84,000 milinches, the design friction loss per foot is

$$\frac{84,000}{171} = 490 \text{ milinches}$$

Using this friction loss with Fig. 8–13, it is seen that for a flow of 7.7 gpm a 1-in. pipe is a trifle small, so a $1\frac{1}{4}$-in. pipe should be used from A to B. For the 4.3-gpm flow from B to the pump at M, $\frac{3}{4}$-in. pipe is undersize and 1-in. pipe should therefore be used; however, because the pipe run AB is oversized, it may be preferable to specify $\frac{3}{4}$-in. pipe here. The pipe sizes can also be selected by use of Table 8–4, by interpolation for the heat loads of 75,600 Btuh and 42,400 Btuh at 490 milinches.

Consider the radiator pipes attached at C and D. Here, in the 4 ft of line from C to D, the pressure drop is approximately

$$4 \times 490 = 1960 \text{ milinches}$$

This pressure differential must be sufficient to force the required flow, namely 9200 Btuh (or 0.94 gpm by computation), through the radiator. If the linear run of this radiator circuit and its associated fittings showed, for example, an ETL of 32 ft, the friction loss for the radiator circuit would be

$$\frac{1960}{32} = 61.3 \text{ milinches per ft}$$

Choosing the pipe size from Table 8–3, at 9.2 Mbh and 61.3 milinches, it is seen that $\frac{1}{2}$ in. is too small and that a $\frac{3}{4}$-in. size is therefore required.

For the downflow radiator, two special tees arranged as in Fig. 8–8 must be employed. These cause a significant pressure drop in the line and thereby direct more flow through the radiator. They are seldom used in pairs in less than 1-in. pipe size. For the product of one manufacturer the loss in a pair of 1-in. tees at 4.3 gpm amounts to 5000 milinches, which, with the pipeline loss, could insure adequate flow through a downflow radiator. The characteristics of these tees vary greatly among manufacturers, and therefore, for precise design, the manufacturer's data should be used. For general estimation of circuit loss, consider such special tee loss to be the same as for a tee with maximum friction loss (Table 8–2 or 7–4), namely minimum side flow.

The procedure for determining the pipe size for each radiation unit will follow that indicated for the radiator at CD and will therefore not be repeated here.

For the right-hand circuit, $ABOQWM$, the pressure loss for $BQWM$ should be equal to that of $BCDKLM$. The run AB, which is common to both circuits, is 25 ft long and contains an elbow, a stopcock (flow-control valve) and a wye (or tee). For $1\frac{1}{4}$-in. pipe the equivalent length of AB is, by Table 8–2,

$$25 + 2.6 + 2.6 + 10.4 = 40.6, \text{ or } 41 \text{ ft}$$

By Fig. 8–13, for 7.7 gpm flowing in $1\frac{1}{4}$-in. pipe the friction loss is 140 milinches; thus the loss in 41 ft is

$$140 \times 41 = 5740 \text{ milinches}$$

Thus, with 5740 milinches used in AB, there remains of the 84,000 milinches pump head, $84,000 - 5740 = 78,260$ milinches for use in circuit $BQWM$ of length $(112 - 25) = 87$ ft, or of equivalent length $87 + 0.5(87) = 130.5$ ft, and the design milinch value is therefore

$$\frac{78,260}{130.5} = 600$$

Hot Water Heating

Thus it can be seen, by reading at 600 milinches in Fig. 8-13, that to pass the 3.4 gpm for circuit $BQWM$ $\frac{3}{4}$-in. pipe is required. The foregoing computation was made in detail to show a method to follow, but usually this precision is not required and certainly it should not be carried out unless the left circuit, with an exact fitting computation, is also worked out. Absolute precision can never be expected in the basic design because pipe sizes are not sufficiently close together and thus the pipes are usually too large or slightly small.

When the circuits are too large the pump will not require as much head, and more water will flow until a balance is reached at a higher flow rate and lower head on the pump characteristic curve (Fig. 8-11).

The design principles outlined in this topic are equally applicable to baseboard (perimeter) radiation, which also involves a one-pipe system. Parallel circuits from the pump header should be balanced for friction losses so as to provide the required flow in each circuit. Observe also that the main flow of each circuit passes through each unit of radiation. Sections 8-8 and 9-4 should be consulted in connection with design layout.

8-11. TWO-PIPE FORCED-CIRCULATION DESIGN PRACTICES

The design principles outlined in section 8-9 are applicable here, with particular reference to the items listed under paragraph 10. The friction loss of each circuit is essentially the same in a reversed-return circuit, so that any one of the circuits can be used for selection of the required pump head after the total gpm of the system has been found. In a direct-return system, particularly, the longest (or highest-friction) circuit must be used to set the pump conditions, and more resistance must be placed in the shorter circuits either by restriction of pipe sizes or by use of an orifice or adjustable valve.

Example 8-4. For the reversed-return system shown in Fig. 8-16, with capacities and lengths as indicated, carry out the necessary design procedures to size the pipes and select a pump. Consider a 20-deg drop in water temperature, with water supplied at a design temperature of 190 F leaving the boiler. Indicate how the design would be altered with a 10-deg drop and with a 30-deg drop.

Solution: The longest circuit, ACG to radiator 5 to $MNOP$, has a linear run of 148 ft, allowing for a 4-ft rise to the trunk main at A and a 7-ft drop to the circulator at O. Add 50 per cent for fittings to find an equivalent total length, giving $148 + 74 = 222$ ft ETL. The longest circuit has a load of 65.6 Mbh, and the shortest circuit has a load of 56 Mbh, giving a total load of 121.6 Mbh. By equation 8-4, the total pump capacity is

$$\text{gpm} = \frac{121{,}600}{(490)(20)} = 12.4$$

Also by equation 8-4, the capacity of the longest circuit is

$$\text{gpm} = \frac{65{,}600}{(490)(20)} = 6.7$$

Therefore the maximum gpm in the shortest circuit is

$$12.4 - 6.7 = 5.7$$

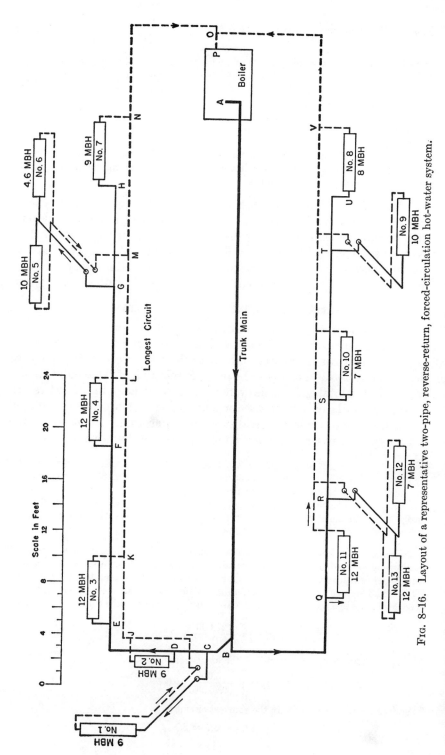

FIG. 8-16. Layout of a representative two-pipe, reverse-return, forced-circulation hot-water system.

Hot Water Heating

By reference to Fig. 8–11, it can be seen that any one of the three smallest pumps has good head characteristics and might well be investigated. Each of these pumps is equipped with a small motor of about $\frac{1}{12}$ hp. If the highest head pump of these three is selected here, it can be seen that the head at 12.4 gpm is 7.5 ft, or 7.5 × 12,000 = 90,000 milinches.

With this pump the unit design friction head for the longest circuit is thus

$$\frac{90,000}{222} = 405 \text{ milinches}$$

For the trunk main from A to the tee at B and C, Fig. 8–13 indicates that a $1\frac{1}{4}$-in. pipe size is required for 12.4 gpm, or for 121,600 Btuh. It is probably desirable to run this same size for the few feet to point C, although the flow there is only 6.7 gpm, corresponding to 65.6 Mbh. At C, 9.0 Mbh is taken off, leaving 56.6 Mbh, for which Fig. 8–13 at 405 milinches indicates that 1-in. pipe is more than adequate. Again, by Fig. 8–13,

From	At	Use
D to E	56.6 − 9.0 = 47.6 Mbh	1 in.
E to F	47.6 − 12.0 = 35.6 Mbh	$\frac{3}{4}$ in.
F to G	35.6 − 12.0 = 23.6 Mbh	$\frac{3}{4}$ in.
G to H	23.6 − 14.6 = 9.0 Mbh	$\frac{1}{2}$ in.
H to N	= 9.0 Mbh	$\frac{1}{2}$ in.

And, for the return pipe,

From	At	Use
C to J	= 9.0 Mbh	$\frac{1}{2}$ in.
D to J	= 9.0 Mbh	$\frac{1}{2}$ in.
J to K	9.0 + 9.0 = 18.0 Mbh	$\frac{3}{4}$ in.
K to L	18.0 + 12.0 = 30.0 Mbh	$\frac{3}{4}$ in.
L to M	30.0 + 12.0 = 42.0 Mbh	1 in.
M to N	42.0 + 14.6 = 56.6 Mbh	1 in.
N to O	56.6 + 9.0 = 65.6 Mbh	$1\frac{1}{4}$ in.

By the same methods, the radiator pipe size is found to be $\frac{1}{2}$ in. in every case, except that the runs from G and from M, each serving two radiators, should be $\frac{3}{4}$ in. When using steel pipe it is not desirable to use less than $\frac{1}{2}$-in. size.

In the event that this system were being sized for a 10-deg drop, the gpm flow would have to be doubled, and the pipe would be sized thus, using 24.8 gpm for the main circuit. To use the Mbh chart at the bottom of Fig. 8–13, each value should be doubled (i.e., for the first radiator at 9.0 Mbh, the pipes must carry twice as much water as at a 20-deg drop; therefore use 2 × 9.0 = 18 Mbh for the chart reading). The increase in gpm at 10-deg difference will probably indicate also that a different pump should be selected from Fig. 8–11. The same pump would deliver 24.8 gpm at 5.1 ft head, and the next-larger size would deliver 24.8 gpm at 7.6 ft head. For these two pumps the design milinches would then be

$$\frac{(5.1)(12,000)}{222} = 275 \quad \text{or} \quad \frac{(7.6)(12,000)}{222} = 411$$

Using the design milinches of either of these pumps in connection with the proper gpm, it will be found that larger pipes are required than for the 20-deg design.

If a 30-deg drop were to be used, the original gpm would be reduced by the factor $\frac{20}{30}$. If it is desired to use the Mbh values on the chart, these are applicable after multiplying by $\frac{20}{30}$ or $\frac{2}{3}$. By similar procedures it is also possible to use tables 8–3 and 8–4 with varying temperature drops.

For a precise check of results, each fitting in the design circuit should be summed in its proper size to get the true equivalent total length, and if the 50 per cent allowance for fittings is greatly in error a recalculation may be required. Such a recalculation is not usually necessary in a reversed-return design, as the balance is usually close in each circuit.

8-12. FORCED-CIRCULATION PUMPS

Circulation pumps used as boosters in hot-water systems are usually low-head, centrifugal pumps, employing horizontal shafts and turning at conventional motor speeds. Figure 8-17 shows a cross section through the pump of one manufacturer. Water flows down into the eye of the rotating impeller through which it passes while increasing in velocity and is then delivered into the volute section of the pump. In the pump volute, pressure is built up at the expense of a decrease in velocity of the water, and

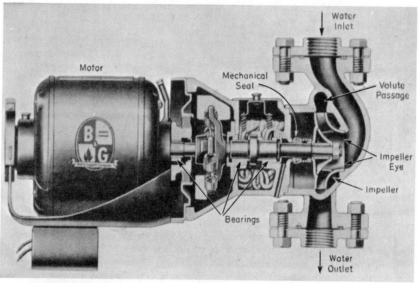

FIG. 8-17. Cross section of forced-circulation hot-water heating-system pump. (Courtesy Bell and Gossett Company.)

the water then flows downward and into the boiler. Because the water is simply being circulated in a closed circuit against resisting friction and not being lifted and delivered against external pressures, the horsepower required from the motor drive is relatively small. For example, in the representative head-capacity graphs of Fig. 8-11 the three smallest pumps have $\frac{1}{12}$-hp motors; the others, up to 2-in.-connection size, use $\frac{1}{6}$-hp motors; $2\frac{1}{2}$ in. and LD 3 use $\frac{1}{4}$ hp; HD 3 is $\frac{1}{3}$ hp; PD 35 is $\frac{1}{2}$ hp; and PD 37 is $\frac{3}{4}$ hp. For the larger higher-head pumps of Fig. 8-12, horsepowers range from 1 to 15. Motor speeds of 1750 rpm are most commonly employed. Booster

Hot Water Heating

pumps should be so made as to have long life and require a minimum of servicing. Small centrifugal pumps of this type have efficiencies which run from about 45 to 70 per cent, with performance usually improving with the size of the pump. The horsepower required can be computed as

$$\text{Hp} = \frac{\text{lb water per minute} \times \text{ft head}}{33{,}000 \times \text{pump efficiency}} \quad (8\text{-}5)$$

$$\text{Hp} = \frac{(\text{gpm})(\text{weight per gallon}) \times \text{ft head}}{33{,}000 \times \text{pump efficiency}} \quad (8\text{-}6)$$

Motor efficiencies for fractional-horsepower motors are usually low, about 50 to 60 per cent. Above 1 hp, efficiencies rapidly increase—reaching values of 75 to 90 per cent with large motors.

Example 8-5. For the pump of example 8-4, delivering 12.4 gpm against a 7.5-ft head, estimate the horsepower absorbed and the motor size which should probably be specified if the efficiency of this small pump is 52 per cent.

Solution: By equation 8-6,

$$\text{Hp} = \frac{(12.4)(8.1)(7.5)}{(33{,}000)(0.52)} = 0.044 \qquad Ans.$$

For this horsepower a $\frac{1}{12}$-hp motor (0.083 hp) is more than adequate but would probably represent a minimum size for such an installation.

With conventional forced-circulation systems the pump-head pressures employed run from 2 to 10 ft, with a common head pressure for design being 6 ft. Much higher heads are frequently employed in large heating systems, where the lower pipe investment cost can justify higher pump operating costs. Low heads are customary in small plants converted from thermal circulation. In a forced-circulation system, the ever-present thermal pressure head may assist the flow and reduce the pump power required, but is usually neglected in design computations.

8-13. THERMAL-CIRCULATION SYSTEM DESIGN

Relatively few new systems using natural circulation are being installed at the present time because forced-circulation systems use smaller pipe sizes and make possible better control and response. Nevertheless there are some currently designed systems of this type, as well as a large number of old installations which continue to give satisfactory service. Because of the smaller pressure differences which arise from thermal effects to cause flow, greater care must be exercised in the layout and design of the system. However, the methods followed are the same as those indicated for forced circulation.

Example 8-6. Design the pipe sizes for the first-story, direct-return thermal-circulation system of Fig. 8-18. Mains B and C lie 4 ft above the center of the boiler A, and radiators a, b, and c are 3 ft above the supply main. Supply water is at 190 F and return water is at 165 F.

Solution: The least favored or longest circuit is that of radiator c. This circuit has 10 elbows, 1 angle radiator valve, 1 radiator, and 1 boiler; and by Table 7–4, these are equal to a total of 18 elbow equivalents. The 4 tees in this circuit offer little resistance in the direction of main flow and therefore were not counted, but 2 each would create an appreciable loss when computing the circuits for radiator a or b and must be considered. The measured length of the circuit is 93 ft. Assume a trial equivalent length equal to double the actual length, or $2 \times 93 = 186$ ft, and later recheck. A trial length 50 per cent greater could also be used as a starting point if desired.

From Fig. 8–2, for 190 F and 165 F read 111 milinches per foot of water column. With a head of $4 + 3$, or 7 ft, the pressure head available to cause flow is $111 \times 7 = 777$ milinches. This is $777 \div 186 = 4.2$ milinches per ft friction drop, based on the trial equivalent length.

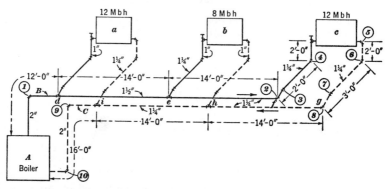

Fig. 8–18. Layout of natural-circulation hot-water system.

There is a 25-deg temperature drop, and, to use Fig. 8–13, which is constructed on a 20-deg basis, the heat loads must be reduced to $\frac{20}{25}$ of their real values. The reduced loads become: total, $\frac{20}{25}(32) = 25.6$ Mbh; for the individual radiators, $\frac{20}{25}(12) = 9.6$ Mbh and $\frac{20}{25}(8) = 6.4$ Mbh. The portion Ad of the circuit carries the complete load of 25.6 Mbh, and at 4.2 milinches Fig. 8–13 calls for a 2-in. pipe. Also, the portion de carrying $9.6 + 6.4$, or 16 Mbh, takes $1\frac{1}{2}$-in. pipe; and ef, carrying 9.6 Mbh, takes $1\frac{1}{4}$-in. pipe. Similar sizing methods apply for the return pipes from elbow 6 back to A.

To check the equivalent length of circuit it is sufficiently accurate to assume a mean size of $1\frac{1}{2}$-in. pipe, for which 18 elbow equivalents give $18 \times 25 \times 1.5 \div 12 = 56$ ft. The total equivalent length is thus $93 + 56$, or 149 ft. This value gives $777 \div 149 = 5.2$ milinches per ft as the friction pressure drop allowed in the circuit. For this friction pressure drop and the proper heat loads, Fig. 8–13 shows that the pipe sizes previously determined are still satisfactory and should be employed. However, it will be observed that each pipe is slightly oversize, and 1-in. pipe will therefore be used for the radiator risers even though this size is somewhat small for the larger radiators. As a further reduction the short run hi was dropped to $1\frac{1}{4}$-in. pipe, and iA could also be reduced to $1\frac{1}{2}$-in. pipe without unbalancing the circuit.

In a direct-return system the first radiators may receive more flow than the others and it may be necessary to insert fixed orifices or locked adjust-

Hot Water Heating

ment valves to restrict the flow to the near radiators in order to balance the flow. Refinements in piping, reductions in pipe sizes for near radiators, and increases in pipe sizes for remote radiators can also be employed.

8-14. HOT-WATER PRACTICE IN LARGE BUILDINGS

The closeness of temperature control possible in spaces heated by hot water makes this medium suitable for heating large buildings. In operation the buildings are usually zoned, with each zone representing a separate hot-water distribution system. By means of combination indoor and outdoor thermostats it is possible to establish a close temperature-control system in

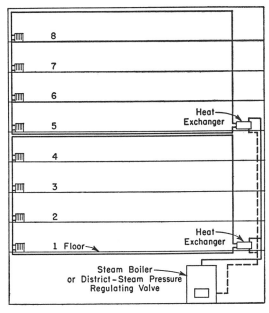

Fig. 8-19. Arrangement for hot-water heating of multistory building.

each zone. Frequently in large systems a steam boiler supplies steam to heat-exchangers in which the water is heated to the specified temperature called for by action of the thermostat. In multistory buildings, it is necessary to zone groups of floors in order to keep the static head on the heating elements within desirable limits. Figure 8-19 illustrates the type of arrangement which might be used in a multistory building. In large cities where district steam is available it is also possible to buy metered steam which in turn is furnished to the heat exchangers for warming the hot water used for heat distribution. The condensate is usually returned to the return mains of the steam supplier through condensate meters.

8-15. HIGH-TEMPERATURE WATER FOR HEATING AND INDUSTRY

Mention has previously been made of the use of high-temperature water as a means of distributing energy or directly heating a space. Figure 8-20 shows in diagrammatic form how a high-temperature distribution system might be arranged. The boiler in such a system is merely a heat exchanger between the burning fuel and the water, as the boiler operates at a pressure sufficiently high to keep the water from vaporizing. For example, if the water is heated to a temperature as high as 500 F, then the pressure of the system cannot be less than 681 psia wherever the water exists at this temperature. The heated water in turn is circulated to the points of utilization by a pump. Since the water gives up energy and cools throughout the system, it returns to the pump at appreciably lower tem-

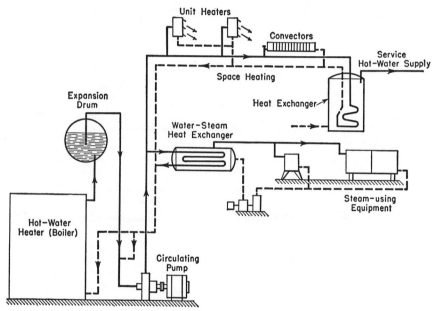

FIG. 8-20. Heating- and energy-circulation system using high-temperature water.

perature, perhaps at 150 F to 250 F. This cooler water is delivered back into the heater-boiler, but if the designer feels it to be desirable a part of it can be recirculated into the supply water to set the temperature of the supply water at any required value.

It should be noted that a hot-water supply of this type can be used as the energy source for generating steam at lower temperature and pressure where this is desirable, as might be the case where steam is required for a constant-temperature process. In process work, temperatures can be held within very precise limits merely by adjusting the pressure of the steam.

Hot Water Heating 323

PROBLEMS

8-1. In a certain thermal-circulating hot-water system the water leaves the boiler at 200 F and returns at 180 F. If the height between the boiler datum and the radiator datum is 10 ft, what is the pressure difference available to cause flow, expressed in milinches? *Ans.* 880 milinches of water measured at 190 F

8-2. A 50,000-Btuh system is reported to have a total water content of 48 gal measured at 50 F. (a) Assuming that the system is filled at 50 F and designed for operation at a maximum temperature of 220 F, compute the increase in volume of water under the temperature extremes. (b) Determine whether the suggested expansion-tank sizes in section 8–5, are adequate to serve this system, and if they are oversize suggest why.
Ans. (a) 2.2 gal; (b) tank must allow for air space in tank

8-3. A 100,000-Btuh system is reported to have a total water content of 82 gal measured at 50 F. (a) Assuming that the system is filled at 50 F and designed for operation at a maximum temperature of 220 F, compute the increase in the volume of the water in the system under the temperature extremes. (b) Determine whether the suggested expansion-tank sizes in section 8–5 are adequate to serve this system, and if they are oversize suggest why. *Ans.* (a) 3.8 gal; (b) tank must allow for air volume

8-4. The circuit from a boiler to a particular division of radiation, with a capacity of 40 Mbh, contains 40 lin ft of pipe, 12 elbows, 2 tees with 50 per cent diversion to the radiator, 1 radiator valve, 1 radiator, and 1 boiler. (a) What is the total number of elbow equivalents to be added to the measured length of run? (b) Find the size of steel pipe that should be used in this radiation circuit provided the design is to be made not to exceed 350 milinches per foot and for a 20-deg design temperature drop. (c) Assume that 20 lin ft of run for this radiator circuit, along with 14 elbow equivalents, are in the main, which is $1\frac{1}{4}$ in. in size, and find the precise equivalent length of run for the radiator circuit and its mains and compare this with an assumption of 50 per cent increase in linear length or 100 per cent increase in linear length as a means of estimating a trial equivalent length. (d) What pump head should be specified? Compute the head on the basis of actual friction loss, considering the flow in the $1\frac{1}{4}$-in. main to be twice that in the radiator run. *Ans.* (a) 28; (b) 1-in. steel pipe; (c) 106 ft, use 100%; (d) 1.4 ft

8-5. Rework problem 8–4 on the basis of using copper tubing instead of steel. Assume that the largest pipe in the circuit is 1 in. Note that copper tubing has a smaller outside diameter than steel, and also a smaller bore.
Ans. (a) 31; (b) 1 in.; (c) 105 ft; (d) 2.9 ft

8-6. A particular division of radiation with four radiators in parallel has an output of 80,000 Btuh, with each radiator furnishing 20 Mbh. The total circuit from and to the boiler, and up to the longest unit of radiation served above, consists of 40 ft of $1\frac{1}{4}$-in. steel pipe, while the radiator circuit itself consists of an additional linear run of 30 ft of smaller-size pipe. The total main and radiator circuit in question consists then of 70 ft of run and includes 14 elbows, 2 tees with 50 per cent diversion, 2 radiator valves, 1 radiator, 1 boiler, and 2 gate valves. The total boiler-radiator capacity is 120,000 Btuh. Find (a) the number of total elbow equivalents in the circuit described; (b) the size of steel pipe used for the radiator circuit if the design is not to exceed 350 milinches per foot and is based on a 20-deg design temperature drop; (c) the precise equivalent length of the run if fifteen of the elbow equivalents are in the radiator circuit and the rest are in the main circuit. (d) By what factor should the linear run of this problem have been multiplied if a trial equivalent length were required? (e) Find the pump head and the gpm capacity which should be specified.

Ans. (a) 33; (b) $\frac{3}{4}$ in.; (c) 140 ft; (d) 2 (i.e., 100% increase); (e) 2.7 ft, 12.2 gpm

8–7. Rework problem 8–6 but assume that copper tubing instead of steel is used. Note that the copper tubing, compared with steel, is smaller both in outer diameter and in bore, for the same nominal size.

Ans. (a) 37.4; (b) $\frac{3}{4}$ in.; (c) 140 ft; (d) 2 (i.e., 100% increase); (e) 3.7 ft, 12.2 gpm

8–8. A forced-circulation hot-water heating system is to be designed for outgoing water at 220 F and return water at 190 F. The total load is 1300 Mbh. The length of the longest circuit is 261 ft and this will be increased by 100 per cent to obtain a trial equivalent length. The system is to be designed on the basis of 3.0 ft of water pressure drop per 100 ft of equivalent length. Find (a) the gpm to be circulated; (b) the pump head required and the design milinches; (c) the size of pipe which should be used for all parts of the circuit carrying the full flow.

Ans. (a) 90 gpm; (b) 15.7 ft, 360 milinches per ft; (c) 3 in.

8–9. Use the same conditions as in problem 8–8, except that the supply water is at 210 F and the return water is at 200 F, and find (a) the gpm which should be circulated; (b) the pump head required and the design milinches; (c) the size of pipe required. (d) In physical size, would smaller or larger radiators be required in each of the rooms for the same heat output under the conditions of problem 8–8 and problem 8–9?

Ans. (a) 268 gpm; (b) 15.7 ft, 360 milinches; (c) $3\frac{1}{2}$ in., or reduce to 3 in. in parts
of run; (d) same size

8–10. For the system pictured in Fig. 8–16, with capacities and lengths as indicated, complete the design with steel pipe for sizing the longest circuit when the design is based on a 10-deg drop in water temperature.

Ans. Total flow 24.8 gpm, 13.4 gpm in longest circuit; for booster pump at 7.3 ft
at 24.8 gpm, use 395-milinch design basis. *A* to *BC*, 2 in.; *C* to *D*, $1\frac{1}{4}$ in.;
D to *F*, $1\frac{1}{4}$ in.; *F* to *G*, 1 in.; *G* to *N*, $\frac{3}{4}$ in.; *N* to *O*, $1\frac{1}{4}$ in.; etc.

8–11. Work problem 8–10 on the basis of copper tubing being used throughout.

8–12. For the system pictured in Fig. 8–16, with capacities and lengths as indicated, complete the design with steel pipe for sizing the longest circuit when the design is based on a 30-deg drop in water temperature.

8–13. Work problem 8–12 on the basis of copper tubing being used throughout.

8–14. A forced-circulation hot-water-heating system is designed for outgoing water at 200 F, with return water at 180 F. The total load is 1000 Mbh. The equivalent length of the longest circuit is thought to be 680 ft. The system is designed for use with a 15-ft-head pump. Find (a) the gpm circulated, (b) the total friction head, and (c) the size of pipe where maximum flow is required.

Ans. (a) 102 gpm; (b) 180,000 milinches; (c) $3\frac{1}{2}$ in.

8–15. For the perimeter heating system of Fig. 8–10, the dining-room–kitchen circuit has a linear run of 146 ft and serves a load of 36 Mbh. The living-room–bedroom circuit has a linear run of 122 ft and serves 30 Mbh. The upstairs loads taken from each of these circuits have 60 per cent of the main floor load, and each has a shorter total linear run. The system operates at an average temperature of approximately 180 F and employs a 20-deg temperature drop, with finned copper radiation. Select a

Hot Water Heating

pump of adequate capacity to serve this building, basing its output head on the longest circuit, with its known fittings, and on a reasonable friction loss per foot.

8-16. Solve problem 8-15 with no change except that a 10-deg drop is used.

8-17. Solve problem 8-15 on the basis of using cast-iron baseboard radiation of approximately the same output per foot when operated at 200 F average temperature and a 20-deg temperature drop.

8-18. Consideration is being given to heating a remote installation having a heat load of 1,000,000 Btuh, with the heat to be supplied from a central heating plant. (1) Hot water can be supplied in a low-pressure system at 240 F and returned to the heating plant at 140 F; or (2) high-temperature water can be supplied at 440 F and returned to the heating plant at 140 F; or (3) dry saturated steam can be supplied at 250 psia with condensate returning at approximately 200 F. (a) Making use of the water steam table, 2-2, compute for each of the three foregoing systems the pounds of medium flowing per hour required to heat the remote space. (b) Making use of the specific volume of water and saturated steam at maximum temperature conditions, compute the pipe cross sections which would be required to pass the hot water with the two water systems if a maximum velocity of 5 fps is employed, or to pass the steam if a maximum velocity of 50 fps is employed. From your results you will note that although steam carries the greatest energy per pound, its energy per unit of volume is smaller than that of hot water. *Ans.* (a) 9960 lb/hr, 3220 lb/hr, 968 lb/hr; (b) 1.32 sq in., 0.46 sq in., 1.43 sq in.

9

Heat-Transfer Elements and Combustion

9–1. RADIATORS AND CONVECTORS

The heating and ventilating industry makes wide use of the terms "radiator" and "convector," although in many cases the terminology is inappropriate. Any surface warmer than its surroundings gives up some heat by radiant methods (transfer through space without the intermediary action of the medium through which the heat is passing). A person some distance away from a fire feels its warming effect although the air around him may be cold and the side of his body which does not "see" (face) the fire may be quite cool. Any surface warmer than its surroundings also gives up heat to air moving past it, by convection methods. Thus in most cases the heat is transferred by both radiation and convection.

The tendency is to use the term "convector" for any device which gives up its heat largely to air currents passing by it and in which the heating elements do not "see" the space being heated. Under this definition an ordinary steam radiator still merits its name, although the larger part of its output may be delivered by convection and not by radiation.

Originally, heaters using steam and hot water were all called radiators. If the heat-transmitting device was placed in the room which was to be heated, it was called a direct radiator. If the heater was placed in the basement or in some other room, it was called an indirect radiator, though under such a condition its radiant heat did not affect the room and the hot surface warmed the air which moved around it, and thus it was really a convector. If a heater was partly enclosed it was called a semidirect radiator. If some air from outside was made to flow past its surface to mix with the room air, the device was called a direct-indirect radiator.

The early method of rating and selling these heat-transmitting elements was to measure the surface exposed to the air and to use the square feet of area of this surface as the unit of capacity and of price. Such a method was the basis for wide abuse, as the surface area in a radiator is exceedingly difficult to measure accurately, and the heat-transmitting ability of each unit of surface varies greatly.

Heat-Transfer Elements and Combustion

It is known that, per square foot of area, the end sections of cast-iron radiators accomplish much more useful heating than the interior sections of the same radiators, and that tall, wide radiators do much less heating than low, narrow ones. It is also known that radiators which deliver the warm air in a substantially horizontal direction into the room bring about more comfort in the room, with less fuel consumption than is required with radiators which deliver the air vertically. The horizontally delivered warm air mixes with the air of the room in a zone nearer to the floor and thus effects a reduction in the temperature difference between the air near the floor and the air near the ceiling.

Steam radiators are usually specified for test temperature conditions of 215 F in the radiator, with 70 F air, or a temperature difference of 145 deg. Heat transfer from a radiator occurs by conduction, convection, and radiation combined, and so is not directly proportional to the temperature difference. Tests have shown that with radiators the heat transfer is proportional to the 1.3 power of the ratio of the actual temperature difference to the standard temperature difference, and to the 1.5 power with convectors. Thus a radiator using steam at 20 psia (5.3 psig) and 228 F in a room heated to 55 F would give out $[(228 - 55)/(215 - 70)]^{1.3}$, or 1.256 times as much heat as a standard radiator. A radiator operating in the subatmospheric region at 7 psia (176.85 F) would give out only $[(176.85 - 55)/(215 - 70)]^{1.3}$, or 0.797 times as much heat as one at standard conditions. When the heat transfer with 20 psia steam is compared with that for 7 psia steam it is seen that only $(0.797/1.256) \times 100 = 63.4$ per cent as much heat will be transferred per hour in a given radiator operated at 7 psia (subatmospheric) as compared with one operated at 20 psia.

The location of the radiator in the room is also important. The most satisfactory location is alongside an exterior wall below a window, so that rising currents of warm air may counteract the dropping cool air currents caused by exposure and by leakage inward of cold outside air.

Figure 9-1 shows some of the standard types of assembled radiators. Type 1 is no longer made, though many of these radiators are in service. Type 2 is fairly standard among all manufacturers and has from one tube to seven tubes, and may be from about 12 in. to 38 in. high. Type 3 is a conventional wall radiator and may have either horizontal or vertical tubes. In order to determine the exact rating of any type of radiator, the manufacturer's catalog must usually be consulted.

To give an idea of radiator output, consider a radiator of Type 2, 4 tubes deep, having twenty sections and a height of 22 in. This radiator would have a length of 35.0 in. and a depth of $4\frac{3}{4}$ in. It would deliver 8640 Btuh when supplied with 215 F steam.

These radiators are made of cast iron, and the sections are joined to

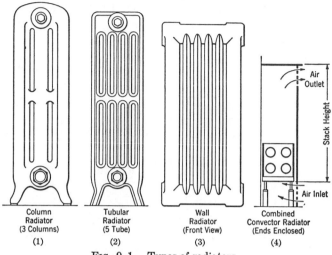

Fig. 9-1. Types of radiators.

each other by malleable slip-nipples at top and bottom. The end sections are provided with legs or not, as may be desired.

Figure 9-2 shows representative connections of a radiator for use with a vapor or vacuum steam system.

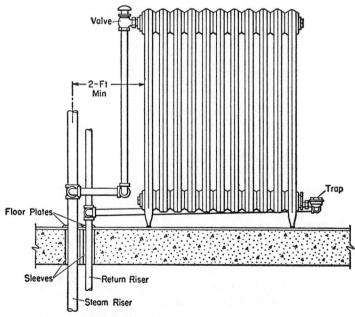

Fig. 9-2. Representative connections for steam radiator used on vapor or vacuum system.

Heat-Transfer Elements and Combustion 329

9-2. CONVECTOR DETAILS

The convector of Fig. 9-3 consists of a heater section housed by a partially recessed steel enclosure. The heater section rests near the bottom of the unit and is either attached to the back or rests on legs of adjustable height. The actual heater surface area in a convector is not the only significant factor in determining heater output, as the stack height or draft head greatly influences the velocity of the convection air currents over the fins and tubes of the heater. The free area of the airways through the heater, and the average temperature difference between the steam (or hot water) and the air, must also be considered.

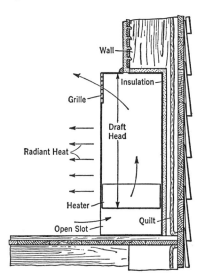

FIG. 9-3. Section through enclosed convector.

In stating output of convectors, a 65 F inlet temperature is usually taken as the basis for rating instead of the 70 F used with radiators. Convectors, like radiators, are usually placed near to and below windows; but it is sometimes necessary to locate them on interior partitions, and here they are often more effective than radiators and better in appearance. Some convectors are free-standing along a wall, but many are made partially recessed, as in Fig. 9-3, or even fully recessed, with their exposed surface flush with the wall.

Table 9-1, which is a listing of the manufacturer's data in condensed form, gives capacities of a certain hot-water convector having a 24-in. stack height. The data in this table are for a 20-deg drop in temperature, with the average water temperature indicated at the top of the appropriate columns. The lengths of coils can be changed by 4-in. increments, and the cabinets are supplied in depths of 4, 6, 8, or 10 in. For this convector

TABLE 9-1
Representative Capacities* of Floor-Mounted Convector

Cabinet Lengths (in.)	Average Water Temperature (F)					Steam at 215 F		
	170	180	190	200	210			
	Cabinet Depths (in.)							
	6	6	6	6	6	4	6	8	10
16	1.8	2.0	2.3	2.5	2.8	1.9	2.9	3.2	4.2
24	2.8	3.2	3.6	4.0	4.4	3.0	4.6	5.1	6.6
32	3.9	4.4	5.0	5.5	6.0	4.2	6.3	7.0	9.1
40	4.9	5.6	6.3	7.0	7.7	5.3	8.0	8.9	11.6
64	8.1	9.2	10.3	11.4	12.6	8.6	13.1	14.6	19.0

* Capacities in thousands of Btu per hour (Mbh), with 65 F entering air. Convector 24 in. high, with 18-in. stack height, using forced-circulation hot water at 20-deg drop in temperature.

the bottom of the coil is approximately 6 in. above the base of the cabinet; thus, for a 24-in. convector the stack height is 18 in. For comparative purposes, it might be mentioned that for a cabinet 38 in. high (essentially a 32-in. stack height) with a coil 24 in. long and 6 in. wide, at 215 F the Mbh capacity is 5.3, in contrast to the value 4.6 for the cabinet which is 24 in. high.

9-3. TUBE SHAPES AND COILS

While many tube and fin convectors are built with round water or steam passages, some types employ oval or triangular tubes. These shapes are believed to have certain advantages, such as reduced resistance to air flow through the convector, and reduced danger of bursting in case of excess internal pressure such as can occur from accidental freezing (a triangular tube with three flat sides, for example, may bulge without fracture and compensate for the expansion of water which takes place during freezing).

Practically all convectors of the general type described are as adaptable to hot-water heating as to steam heating.

Pipe Coils. Radiators and convectors may be built up of pipe and pipe fittings. Such heat-transmitting agents have a high output per unit of surface area, but are not installed as often as formerly because fabricated radiators now give better over-all performance. Pipe coils used as radiators may reach considerable lengths and should always provide allowance for expansion and contraction. This allowance is usually supplied by right-angle turns in the coils so that movement can occur at the elbows.

9-4. BASEBOARD RADIATION

Increasing use is being made of baseboard radiation both in residential and public buildings. This type of radiation has the advantage of supply-

Heat-Transfer Elements and Combustion 331

ing a continuous band of heat around the exposed perimeter of a room. This can effectively prevent cool downdrafts from exposed walls and windows. The baseboard elements are easy to install and they can blend into the room to create a pleasant appearance. Because the elements are close to the wall surface, they cause only a minimum of difficulty in regard to the placement of furniture or equipment.

Figure 9-4 shows an arrangement of the equipment of one manufacturer.

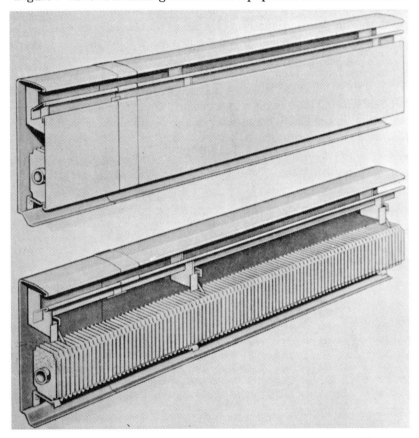

Fig. 9-4. Baseboard-type heating panels. Top view: Cover in place. Bottom view: Cover removed. (Courtesy American Radiator & Standard Sanitary Corporation.)

It should be remarked that the radiation equipment can be recessed so as to be flush with the wall, or can project out into the room a few inches. The heat-transfer element consists of pipe or tubing to which metal fins are attached. The pipe can be of steel, with steel or aluminum fins, or of copper with aluminum fins. The enclosure for the convector is usually of pressed steel formed to shape. In the unit shown in Fig. 9-4 the heat-transfer element is copper tubing with aluminum fins. This particular design is for hot-water application.

Tables 9-2 and 9-3 give the output capacity in Btu per hour for varying lengths of heater elements in baseboard enclosures, for given operating temperatures. It should also be noted that connecting lengths of bare straight pipe, and bends at corners, contribute to the heat output of the convector, and that the output of the pipe should therefore be considered in all computations. In designing with baseboard radiation, the maximum heat output in a given room is set by the available wall space along which the elements can be installed. Thus it is possible in some cases that space limitations prevent the installation of a sufficient baseboard length to offset the heat loss from a room. Where this is the case, higher hot-water temperatures supplied to the radiation may suffice, or a conventional convector can be added. With baseboard panels, hot water is more frequently used than steam, but effective operation can also be obtained with steam.

TABLE 9-2

REPRESENTATIVE CAPACITIES* OF BASEBOARD RADIATION†

Panel Length (lineal ft)	Average Water Temperature (deg F)							
	150	160	170	180	190	200	210	220
2	0.8	0.9	1.1	1.2	1.4	1.5	1.7	1.8
3	1.2	1.4	1.6	1.8	2.1	2.3	2.5	2.8
4	1.6	1.9	2.1	2.4	2.7	3.0	3.3	3.7
5	2.0	2.3	2.7	3.0	3.4	3.8	4.2	4.6
6	2.4	2.8	3.2	3.6	4.1	4.5	5.0	5.5
8	3.2	3.7	4.3	4.8	5.5	6.0	6.7	7.4
10	4.0	4.7	5.4	6.1	6.9	7.6	8.4	9.2
12	4.7	5.6	6.4	7.3	8.2	9.1	10.0	11.0
16	6.3	7.4	8.6	9.7	11.0	12.1	13.4	14.7
20	7.9	9.3	10.7	12.1	13.7	15.1	16.7	18.4
22	8.7	10.2	11.8	13.3	15.1	16.6	18.4	20.2
24	9.5	11.2	12.8	14.5	16.4	18.1	20.0	22.1
Material	Output of 1-In. Bare Pipe in Enclosure, Btuh per Lineal Foot, at Temperatures Indicated Above							
Copper	48	57	66	75	85	92	102	113
Iron	80	94	108	122	138	152	168	186

* Capacities expressed in thousands of Btu per hour (Mbh), with 65 F entering air, for copper tubing with aluminum fins.
† Table courtesy of American Radiator and Standard Sanitary Corporation (Heatrim panel).

The heat output of baseboard radiation varies greatly among manufacturers, the exact output depending on pipe and fin size, and height of baseboard, and also on whether the lighter nonferrous or heavier cast-iron

Heat-Transfer Elements and Combustion 333

type is employed. For cast iron, the heat emission per foot of panel length is usually less than for the nonferrous type because the fin area is so much less.

Many variations of this general form of radiation are made. For example, instead of the approximately 8- to 12-in. heights employed in residences, offices, and small rooms, panels of greater height having two, three, or four finned pipes placed one above the other can be used to provide much greater output. Still another type for industrial plants employs finned steel pipe located on the wall at any height above the floor, for use with high-pressure steam.

9-5. UNIT HEATERS AND COOLERS

When a fan is employed to force air over a heater coil surface, the output is increased approximately threefold over the corresponding heat delivery of a thermal-circulation convector of the same area, even with a

TABLE 9-3

REPRESENTATIVE CAPACITIES* OF BASEBOARD RADIATION FOR HOT WATER OR STEAM†

PANEL LENGTH (LINEAL FT)	AVERAGE WATER TEMPERATURE (DEG F)							STEAM AT 215 F
	150	160	170	180	190	200	220	
3	0.8	1.0	1.1	1.3	1.4	1.6	1.9	1.8
4	1.1	1.3	1.5	1.7	1.9	2.1	2.6	2.5
5	1.4	1.6	1.9	2.1	2.4	2.6	3.2	3.1
6	1.7	2.0	2.3	2.5	2.9	3.2	3.9	3.7
7	1.9	2.3	2.6	3.0	3.4	3.7	4.6	4.3
8	2.2	2.6	3.0	3.4	3.8	4.2	5.2	4.9
9	2.5	2.9	3.4	3.8	4.3	4.8	5.8	5.5
10	2.8	3.3	3.8	4.2	4.8	5.3	6.5	6.1
12	3.3	3.9	4.5	5.1	5.8	6.3	7.7	7.4
14	3.9	4.6	5.3	5.9	6.7	7.4	9.0	8.6
16	4.4	5.2	6.0	6.8	7.7	8.5	10.3	9.8
18	5.0	5.9	6.8	7.6	8.6	9.5	11.6	11.1
20	5.5	6.5	7.5	8.5	9.6	10.6	12.9	12.3
22	6.1	7.2	8.3	9.3	10.5	11.6	14.2	13.5
24	6.6	7.8	9.0	10.2	11.5	12.7	15.5	14.7

* Capacities expressed in thousands of Btu per hour (Mbh), with 65 F entering air, for cast-iron integrally-cast fin radiation and cast-iron covers.
† Table courtesy of American Radiator and Standard Sanitary Corporation (Radiantrim panel).

high draft head. An example of a forced-convection machine is the heating, ventilating, and cooling unit shown in Fig. 9-5. This recirculates part of the room air and can also take some air from outside for ventilation. A unit of this type can deliver heat into the air stream from steam or hot water in the finned coils, or can absorb heat from the air when cold water or a refrigerant is supplied to the convector. The mixing damper, responding to a thermostat in the room served by the unit ventilating machine, bypasses part of the air if necessary and sends it at the desired temperature into the room. It is always desirable with the device shown in Fig. 9–5 to install a thermostat in the air stream emerging from the unit,

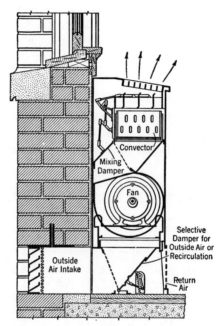

FIG. 9–5. Floor-mounted unit heater and ventilator.

so arranged as to prevent delivery of air which is too cool for comfort. The thermostat in the room might fail to give this protection when it calls for cooling. Heating and cooling units of this type are applicable to apartment buildings, and to hotels, office buildings, schools, and other public buildings.

Figure 9–6 is a photograph of a representative dual-purpose unit with its outer cover removed. Chilled water is circulated through the coil when cooling is needed, and hot water is circulated when heating is required. Only three pipe connections are required: water supply and return and a drain for condensate in summer. Electrical supply is also required for the fan motor. In this unit, which is representative of other designs as well, recirculated air and a controlled amount of outside air for ventilation are drawn through the filter section and then pass over the coils, where the air is tempered to the desired degree. A connection is usually made through the wall in the back of the unit to provide for outside air intake, the amount of which can be adjusted by a manually operated damper. The tubes of

Heat-Transfer Elements and Combustion 335

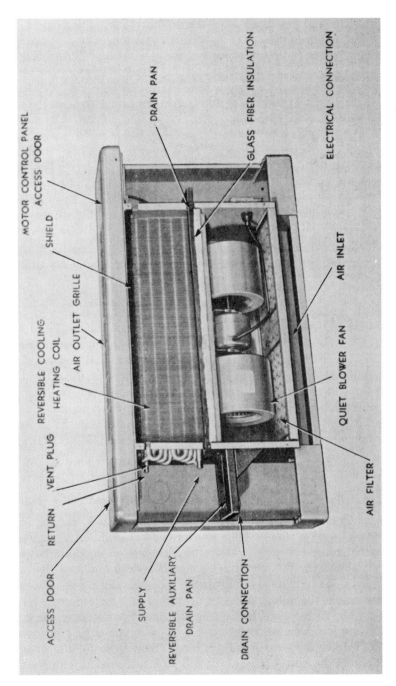

FIG. 9-6. Fan-operated convector for heating and cooling, with front and corner panels removed. (Courtesy Modine Manufacturing Co.)

this unit are of tin-coated copper of $\frac{5}{8}$-in. OD and are provided with aluminum fins. This particular unit has a four-speed motor, thus making it possible to adjust the unit to a variety of operating conditions. In summertime the coil dehumidifies as well as cools, and the condensation is removed by the pipe connection from the drain pan. Representative performance data for a large-size unit of this type would show that the $\frac{1}{10}$-hp fan handles 640 cfm; that the hot or cold water flow is 4.8 gpm; and that when the unit is supplied with 45 F entering water, with air at 85 F db and 75 F wb, the total cooling load is 24,500 Btuh; and that for winter performance, with 180 F entering water temperature and 60 F entering air, 53,800 Btuh are supplied. This unit is 64 in. long, 25 in. high, and 11 in. deep. The units can be recessed up to 5 in. Data on a smaller unit 34 in. long by 25 in. high by 11 in. deep would show capacities of 8000 Btuh for cooling and 18,000 Btuh for heating. It is customary for manufacturers to make units of this type completely recessed, and also to make them for overhead mounting where this is desirable. Figures 12-26, 12-27, and 12-28 show still another arrangement of a year-around convector supplied with conditioned air.

Unit heaters are often employed in factories and warehouses, where the units are suspended from the ceiling or mounted at a high level on a wall. Figure 9-7 shows a cutaway section through a unit heater of this type. The fan mounted in back of the unit blows room air over the heat-transfer coils, and this air, in turn, is directed outward and downward by means of adjustable louvers. Such units are satisfactory for use with either steam or hot water. The heat-transfer coils are usually made from copper tubing, and have aluminum or copper fins attached, which makes a lighter unit than is possible with a cast-iron heat-transfer surface. In addition to this arrangement of the unit, it is also possible to place the heat-transfer coils in a horizontal plane, in which case the axis of the fan is vertical and the air is directed downward. However, deflecting louvers can again direct the air partially to the sides so that it does not impinge directly downward to the floor of a space.

Unit heaters are available in a wide range of capacities, varying from about 10,000 Btuh for smaller units to over 600,000 Btuh for larger units. The capacity varies not only with the physical size of the units but also with steam pressure (temperature) or, for hot-water units, with water temperature. Many units are built with a tube-wall thickness sufficiently adequate to make the unit suitable for operation with high-pressure steam (over 100 psi). When air directly from outside is passed over the coils, the steam-condensation or water-cooling capacity of a unit is greatly increased. Motor horsepower ratings range from $\frac{1}{85}$ hp for the smallest unit to $1\frac{1}{2}$ hp in the case of large units, and air is delivered at temperatures

Heat-Transfer Elements and Combustion 337

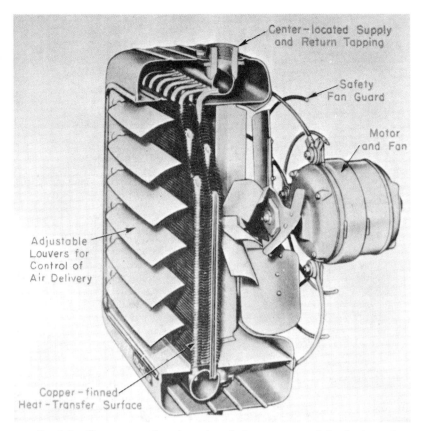

Fig. 9-7. Cutaway section through unit heater arranged for horizontal delivery of air. (Courtesy Modine Manufacturing Co.)

ranging from about 90 F to over 250 F. The higher temperatures are possible in the case of units located remote from working space.

These units are particularly applicable in heating such industrial spaces as the single-story buildings, with or without monitor roof, in use for modern factories. Such buildings usually have large glazed areas and are usually not as well insulated as the more massive multistory buildings. A plan and elevation of a representative factory building of this type appears in Fig. 9-8. This design shows one possible placement for the unit heaters. The placement of these units would be such as to direct warm air toward the corners and side walls so that a sheet of warmed air could move downward and across the floor, later rising upward for recirculation through the unit heaters.

Example 9-1. Consider the industrial building of Fig. 9-8 to be located in a suburb of Chicago. It is desired to maintain an inside temperature of 60 F at the

338 Heat-Transfer Elements and Combustion

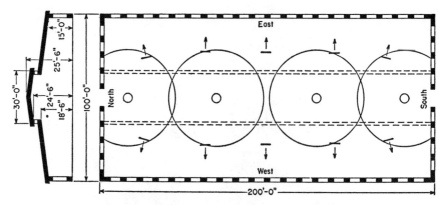

FIG. 9-8. Plan and elevation of one-story factory building.

breathing, or 30-in., line. Heating is accomplished by means of unit heaters suspended from the ceiling, and steam will be used at 215 F. The design temperature can be taken as -10 F, and the prevailing wind direction for the winter season can be considered as 12 mph southwest. The walls are plain brick 12 in. thick, and there are 20 steel-sash windows 6 ft by 9 ft on each side, with 8 windows of the same size on each end. Each end of the building has a 12-ft by 12-ft upward-opening steel door. The monitor section contains 40 $4\frac{1}{2}$-ft by $4\frac{1}{2}$-ft steel-frame windows on each side. These can be opened from below to provide ventilation. The monitor walls are of rigid asbestos shingles on solid wood sheathing mounted on studs. The roof consists of rafters on the bottom side of which is attached 1 in. of rigid insulating board. The upper surface of the roof consists of supporting sheathing covered by asphalt shingles. The floor consists of 4 in. of concrete laid on the bare ground. The design heat loss for the building was computed in a problem at the end of Chapter 5 and was found to be 1,888,600 Btuh. Plan a design pattern for the unit heaters.

Solution: This is a large building, and consequently, to give uniform coverage, it is better to use many heaters of low capacity than to use a few of large capacity. For example, 9 heaters consisting of 3 horizontal units under each roof and 3 vertical downflow units under the monitor section might serve. If each of these were of equal capacity, then each would have to deliver 1,888,600/9, or 203,200 Btuh. One manufacturer supplies such units for 60 F inlet air, 215 F steam, at 1150 fan rpm with a $\frac{1}{4}$-hp motor, to deliver 223,000 Btuh, or a 1150-rpm, $\frac{1}{2}$-hp-motor unit in vertical style, to deliver 262,000 Btuh. These could suffice. However, a better distribution, with less chance of cold spots or overheating, could be had by using 5 horizontal-discharge units on each side and 4 vertical-discharge units in the monitor section. With these, 1,888,600/14, or 85,000 Btuh per unit would be required. For this arrangement, one manufacturer's catalog lists an 1800-rpm, $\frac{1}{12}$-hp-motor unit, to deliver 88,500 Btuh, for horizontal discharge, and a vertical unit at 860 rpm with a $\frac{1}{4}$-hp motor, to deliver 85,600 Btuh. These are slightly over design size, but a slight overcapacity may be desirable to allow for door openings or other contingencies. On the basis of the units indicated, a layout has been shown in Fig. 9-8. Numerous other design possibilities exist, however, and could be used.

9-6. CENTRAL FAN SYSTEMS

In addition to the types of heating units previously described in this chapter, extensive use is made of heat-transfer elements in central heating

Heat-Transfer Elements and Combustion

and cooling systems. In winter operation with these systems, recirculated and outside air is passed over heat-transfer coils which temper the air to a desired condition for delivery into the heated space. The heated air must be sufficiently warm so that in cooling it can offset the heat losses of the space. Auditoriums, lecture halls, churches, many offices, and other building types employ this method of central heating and cooling. Steam (or sometimes hot water) is supplied to the heat-transfer surface used for tempering and warming the air.

Nonferrous heat-transfer surfaces are also used to carry chilled water or refrigerant for air-chilling under summer operating conditions. When refrigerant evaporators are used for air-chilling, it is important to distribute the refrigerant to the different fin coils of the heat transfer unit to insure a uniform temperature throughout. This is often done by having a separate distributing tube for the refrigerant to each coil. On the outside of the coil, the fins must be placed sufficiently far apart that any condensation of humidity from the air can drain off and be removed. Fin spacing is even more important if temperatures are sufficiently low to cause freezing on the coils, as a solid block of ice could prevent all throughflow of air if the fins are too close.

9-7. COOLING-UNIT LOCATION

When a room is to be cooled by a convector located inside the room, the thermal air circulation is reversed from that in effect when heating. Thus the older refrigerated storage rooms in industrial buildings have direct coolers, called bunker coils, overhead. The air, warmed by contact with the product being chilled, or by contact with the outside walls, rises above the heavier, colder air which is constantly falling down from the heat-absorbing coils.

A conventional thermal-circulation radiator or convector well placed for heating a room will not be effective when cooling the room; it will cool only a layer of air close to the floor. However, if a fan is added for use when cooling, to deliver recirculated air against the heat-absorbing surfaces, the heat-transfer rate becomes rapid and the cool-air distribution may be effective and satisfactory. In cold-storage practice, improvement in the conditioning of the product follows mechanical air circulation even though old overhead bunker coils may be retained.

Air-warmed evaporators are employed in many small air-cooling systems. They usually have distributing tubes for the liquid refrigerant so as to insure a uniform temperature throughout. The passages for the air around the surfaces must be sufficiently large not to be obstructed unduly by the water which, condensed out of the air, gathers upon them as dew and then must be drained away for ultimate disposal. The construction of air-warmed evaporators involves consideration of the air turbulence necessary for efficient heat transfer without offering too high resistance to air passage. The airways must not clog with dirt, and the liquid and gas passages must be large enough to insure ample circulation.

9-8. LIQUID-HEAT-TRANSFER ELEMENTS

Liquid-heat-transfer elements are widely employed in heating and cooling. There are two general classes of such steam-to-water devices: one in which the steam is within the tubes and the water is in a shell surrounding the tubes, and one in which these conditions are reversed. The first type may have a shell of large capacity, so as to gain a considerable storage, or may have a shell only large enough to enclose the tube bundle. The U-tube arrangement gives flexibility for expansion and contraction, and the flanged head carrying the tubes may be removed for freeing the tubes of the scale which tends to collect, especially if the water is heated above the temperature at which the dissolved solids commence to deposit. The second, or water-tube, type usually has a shell only large enough to enclose the tubes, since no storage of hot water is possible. The steam enters at the top of the shell, and condensate leaves at the bottom of the shell. In the case of liquids, cooled by a refrigerant, the evaporating refrigerant can be arranged to flow either through or around the tubes.

9-9. HEAT TRANSFER THROUGH METAL SURFACES

Heating or Cooling. Heat transfer through metal surfaces is largely limited by the film conditions which exist on the surfaces of the tubing. The transfer is affected only slightly by the conductivity of the metal itself. The film conditions depend on the fluids involved, whether these are liquid or gas, and whether vaporization or condensation may be taking place, and on velocities past or along the tubes. Convection or flow conditions in a system are classed as *free* or *forced*.

Free convection, or *natural convection,* exists when the fluid circulation is caused by changes in fluid density which occur because of temperature differences established in the fluid. *Forced convection* exists when the motion of the fluid, past the heat-transfer surfaces, is produced by some external means such as a fan or pump.

Heat transfer to or from liquids is much better than with gases, and it often happens that the capacity for heat transfer is greatly restricted by the fluid on one side of the tube, while only slightly limited on the other side. An example of this is a steam radiator in a room. Here, on the side with condensing steam, the heat transfer is comparatively good, so that the temperature of the radiator metal is almost that of the steam. The heat transfer to the air moving past the outside surfaces is so poor that heat is not transferred rapidly enough to cause an appreciable temperature gradient through the metal wall. For the same surface area, changing from cast iron to a better conductor, such as copper, would not appreciably increase the heat transfer (if the differences in polish, which affects radiation slightly, were eliminated). To increase the heat transfer, extended surfaces (fins) are often attached to the side which limits the heat transfer. This is the air or gas side in the case of those devices which use either a liquid or a vapor (undergoing a change in state) as the other medium.

The temperature difference (Δt) in heat transfer is seldom constant, and therefore it is necessary to find what mean or average temperature difference actuates heat transfer. The *logarithmic mean temperature difference* (log mtd) usually gives the most reliable value of mean temperature difference to use in heat-transfer relations:

$$\log \text{mtd (or } \Delta t) = \frac{\text{Gtd} - \text{Ltd}}{2.3 \log_{10} \frac{\text{Gtd}}{\text{Ltd}}} \qquad (9\text{-}1)$$

where log mtd = logarithmic mean temperature difference, in degrees Fahrenheit;
Gtd = greater temperature difference, in degrees Fahrenheit;
Ltd = lesser temperature difference, in degrees Fahrenheit;
$\log_{10}$ = logarithm to the base 10.

Figure 9-9 and Fig. 9-10 represent, diagrammatically, possible temperature changes occurring during counterflow or parallel flow of two fluids between which heat transfer is taking place. Counterflow is desirable, when it can be obtained with two fluids which change in temperature, since a better temperature difference can be maintained and the leaving temperatures can be brought closer to their limiting values. In case one of the fluids is a condensing or evaporating vapor, there can be no distinction between parallel flow and counterflow, as one of the fluids does not change

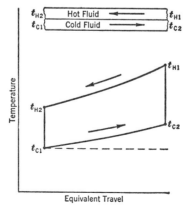

FIG. 9-9. Counterflow heat transfer.

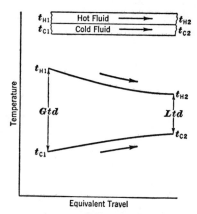

Fig. 9-10. Parallel-flow heat transfer.

in temperature. Many actual installations can neither be classed as parallel flow nor as counterflow.

Example 9-2. Water enters an ammonia condenser at 60 F and leaves at 70 F. The ammonia is condensing at 74 F. Find the log mtd.

Solution: By equation 9-1, with Gtd = 74 − 60 = 14 deg, and Ltd = 74 − 70 = 4 deg,

$$\log \text{mtd} = \frac{14 - 4}{2.3 \log_{10} 14/4} = \frac{10}{(2.3)(0.544)} = 7.98 \text{ deg} \qquad Ans.$$

The arithmetic mean temperature difference would be

$$\frac{\text{Gtd} + \text{Ltd}}{2} = \frac{14 + 4}{2} = 9 \text{ deg}$$

and is seen to be too high. There are many cases, however, when the log mtd and the arithmetic mtd are almost identical. For values of Gtd/Ltd less than 1.5, there is little difference between the two methods of computation. In this example the terms counterflow and parallel flow had no significance, since one fluid condensed at constant temperature.

Values of heat-transfer coefficients for use in the design of units must be used with caution unless it is known that the heater (or cooler) is to be used under almost exactly the same conditions for which the coefficients were determined. In installations where convection, forced or natural, is the main factor this rule is particularly cogent. Table 9-4 lists heat-transfer coefficients, gathered from many sources, which can be used for estimating the surface required. In selecting standard commercial equipment the manufacturers' tables for heat-transfer capacity and test coefficients should always be used if they are available. The effect of possible scaling or corrosion of the tubing after it has been placed in service should not be minimized in design, and adequate surface must always be supplied. Equations 4-18, 4-19, and 4-20 for heat transfer are applicable.

Heat-Transfer Elements and Combustion

Example 9-3. A cold-storage room is to be equipped with brine-pipe coils of standard $1\frac{1}{2}$-in. pipe. Fans cause a positive air circulation in the room over the coils at about 500 fpm. The air contacts the coils at 30 F room temperature and leaves the coils at 25 F. Brine enters the coils at 14 F and leaves at 20 F. Find (a) the mean temperature difference between the coils and the air, and (b) the heat transfer per square foot of pipe surface. (c) If there are 200 linear ft of pipe, what is the probable refrigeration tonnage handled in the room?

Solution: (a) As the coils probably run parallel to the ceiling and above each other it is doubtful whether the terms counterflow and parallel flow have any significance. To make the most conservative estimate, calculate as though there is parallel flow. Using equation 9-1, with Gtd = 30 − 14 = 16, and Ltd = 25 − 20 = 5,

$$\log \text{mtd} = \frac{16 - 5}{2.3 \log_{10} 16/5} = \frac{11}{2.3(0.505)} = 9.47 \text{ deg} \qquad Ans.$$

b) From Table 9-4, select the mean value for "brine pipe to moving air" as $U = 5.5$ Btu per hr sq ft deg F. *Ans.*

c) Standard $1\frac{1}{2}$-in. pipe has an outside diameter of 1.9 in. Thus

$$\text{Area per ft of pipe} = \frac{1.9 \times \pi}{12} \times 1 = 0.497 \text{ sq ft}$$

and

$$\text{Area 200 linear ft} = 200 \times 0.497 = 99.4 \text{ sq ft}$$

By equation 4-20, the heat taken up by the brine is

$$Q = UA\Delta T = (5.5)(99.4)(9.47) = 5177 \text{ Btuh}$$

Therefore the refrigeration tonnage is

$$\frac{5177}{12{,}000} = 0.431 \text{ ton} \qquad Ans.$$

Frost (ice) on the coils will seriously decrease this capacity—by 25 per cent or more as the thickness builds up to an inch.

The problem of selecting surface area for dehumidifiers is complicated by the fact that two simultaneous effects take place; namely, sensible heat is removed, and the latent heat from condensing moisture is also removed. The total heat transfer from air to coils with wetted surface is higher than to coils having dry surface. Condensation or dehumidification occurs when the coil temperature is lower than the dew-point temperature and sensible cooling and dehumidification can take place simultaneously. One method of treating coil performance is to use the equivalent-bypass approach. This considers that a coil performs just as if one part of the coil brought the air to coil surface temperature while the rest of the air passed through the coil (bypassed it) without being cooled or dried at all. When the cooled air and the bypassed air mix at outlet, the result is equivalent to what happens in a real coil. For dehumidifying coils it is desirable to use the manufacturer's data and recommended bypass factors to find performance data. However, when these are lacking and design data are required, it is possible to select a sensible surface coefficient of heat

TABLE 9-4
Heat-Transfer Coefficients* for Heaters (or Coolers) and Heat Exchangers

Type of Heat Exchanger	U, Free Convection	U, Forced Convection	Remarks
Ammonia condensers			
Atmospheric			
Gas in at top............	50–65		
Gas in at bottom.......	125–200		
Double pipe.............		150–250	Water velocity from 150 to 400 fpm
Shell and tube...........		150–300	
Submerged coil..........	30–40		Very ineffective
Baudelot coolers			
Cream..................		55	
Milk...................		70	
Oil.....................		10	
Water..................		70	
Brine coolers			
Shell-and-tube multipass		40–120	Brine velocity 100 to 400 fpm at 0 F; U decreases below, increases above, this temperature.
Submerged refrigerant coils in brine tank...	12	20	
Cooling coils			
Brine pipe			
To moving air........		3.0–8.0	Air velocity 200 to 800 fpm. Increases with temperature difference.
To still air............	1.5–3.0		
Direct expansion coils			
Finned air coils, sensible heat film factor (f_s)†	*300 400 500* [4.3 5.1 5.8]	*600 700 800* [6.5 7.2 7.7]	
Still air...............	1.0–3.0		Increases with temperature difference
Moving air............		2.0–7.0	Air velocity 200 to 800 fpm
In spray of liquid.......		80	
Refrigerant (Freon) film surface factor (f)...	[150–250]		Increases with capacity loading of tubes; also with temperature difference, metal to liquid.
Methylene chloride surface factor (f)......	[100–300]		Δt metal to liquid of 5 to 25 deg
Heat exchanger			
Double pipe...........		80	
Radiators			
Direct steam..........	1.4–2.0		
Pipe coil in room, single..	2.6		
Pipe coils above each other...............	2.6–1.6		As number increases from 1 to 8
Steam condenser			
Shell and tube type......		100–600	0.5 to 6 fps water velocity
Superheat remover			
Shell and tube..........		25	
Tubular exchanger			
Steam to oil...........		75	
Oil to oil..............	5–10	20–50	
Water to oil...........		50	
Water to water.........	25–60	150–300	
Water heater			
Steam coil in tank (brass)	30–200		Increases with temperature difference of 10 deg to 100 deg. Less corrosion with brass and copper pipe.
Steam coil in tank (steel)	10–160		

* U in Btu per hr sq ft deg F mean temperature difference, fluid to fluid, or f in Btu per hr sq ft deg F mean temperature difference, metal to fluid.

† Italic numbers above values in second and third columns are face velocities, in feet per minute.

Heat-Transfer Elements and Combustion

transfer factor (f_s) and use it in design even though dehumidification is taking place.

Forced-Convection Film Factors. Forced-convection film factors can be developed by rational analysis in terms of appropriate, dimensionless number groups, for which groups the coefficients and exponents are determined by experiment.

For liquids being heated or cooled and moving in turbulent flow outside and across a single tube,

$$f = 0.385 \frac{k}{D}\left(\frac{DV\rho}{\mu}\right)^{0.56}\left(\frac{C_p\mu}{k}\right)^{0.3} \qquad (9\text{-}2)$$

Multiply f by 1.3 if a staggered multiple-tube bank applies, and by 1.2 if the tubes are in line.

For liquids being heated inside tubes under conditions of turbulent flow,

$$f = 0.0225 \frac{k}{D}\left(\frac{DV\rho}{\mu}\right)^{0.8}\left(\frac{C_p\mu}{k}\right)^{0.4} \qquad (9\text{-}3)$$

Use an exponent 0.3 for the last term if cooling is involved.

For gases flowing over or along tube bundles, equation 9-2 can also be used. For this equation, D should be taken, by equation 12-4, equal to 4 times the hydraulic radius, where this is found by dividing the free area between the tubes by the sum of all the tube perimeters.

The film-factor values for vapors condensing on, or liquids vaporizing from, horizontal tube bundles cannot be expressed by simple equations like those just given. Such values are high, however, and the resistance to heat transfer is small through this film. For condensing or vaporizing steam, and for ammonia, values of f run between 800 and 6000 Btu per hr sq ft deg F. For making estimates, values of f between 1000 and 3000 are suggested. For the halogenated-hydrocarbon refrigerants, vaporizing and condensing, values of f range from 150 to 350 Btu per hr sq ft deg F.

The units in equations 9-2 and 9-3 may be any that are desired, provided they are dimensionally consistent. It should be observed that

$\dfrac{DV\rho}{\mu}$ is the Reynolds number (sec. 12-2)

$\dfrac{C_p\mu}{k}$ is called the Prandtl number

$\dfrac{fD}{k}$ is called the Nusselt number

All of these number groups are dimensionless. Suggested units for use in the equations are as follows:

D in feet.
f in Btu per hr sq ft deg F. See page 140.

k in Btu ft per hr sq ft deg F. (In the Prandtl number, k in Btu ft per sec sq ft deg F is suggested.)

V in fps.

ρ in lb mass (weight) per cu ft.

μ in centipoise [centipoises $\times$ 0.000672 = lb mass (weight) per ft sec]. See Figs. 12-1 and 12-2.

C_p, specific heat at constant pressure, in Btu per lb mass (weight) deg F. See tables 2-1 and 4-1.

Example 9-5. Water is being cooled in a multipass shell-and-tube evaporator, with the ammonia refrigerant in the shell boiling at 56 F. Water moves through the 1-in.-OD steel tubes (0.05-in. wall) at a velocity of 4 fps. Water enters at 66 F and leaves at 60 F. Compute the film factor f on the water side and estimate the value of U, the over-all coefficient of heat transfer.

Solution: For water at an average temperature of 63 F, $\mu = 1.09$ centipoises, from Fig. 12-1; $\rho = 1/0.01604 = 62.3$ lb per cu ft; $k = 4.10$ Btu in. per hr sq ft deg F = 4.10/12 Btu ft per hr sq ft deg F (from Table 4-1); and $C_p = 1.0$ Btu per lb mass (weight) deg F. Also, $D = (1 - 0.05 - 0.05)/12 = 0.90/12 = 0.075$ ft.

Using equation 9-3 for cooling conditions,

$$f = 0.0225 \frac{k}{D}\left(\frac{DV\rho}{\mu}\right)^{0.8}\left(\frac{C_p\mu}{k}\right)^{0.3}$$

$$= 0.0225\left[\frac{4.1}{(12)(0.075)}\right]\left[\frac{(0.075)(4)(62.3)}{(1.09)(0.000672)}\right]^{0.8}\left[\frac{(1)(1.09)(0.000672)}{4.1/[(12)(3600)]}\right]^{0.3}$$

$$= 0.0225(4.55)(25{,}600)^{0.8}(7.72)^{0.3}$$

$$= 0.0225(4.55)(3362)(1.85) = 637 \text{ Btu per hr sq ft deg F} \qquad Ans.$$

The resistance R to heat flow on the tube consists of the film resistance on the water side, the negligible resistance of the metal wall itself, and the resistance of the ammonia film. Take a value of 2000 for the ammonia film, which has good heat-transfer characteristics. Then,

$$R = \frac{1}{2000} + 0 + \frac{1}{637} = 0.0005 + 0.00157 = 0.00207$$

Therefore the over-all heat-transfer coefficient (estimated) is

$$U = \frac{1}{R} = \frac{1}{0.00207} = 483 \text{ Btu per hr sq ft deg F} \qquad Ans.$$

Because of deposition, tube scale, and the like, a value not exceeding 50 per cent of the computed result might be used in a conservative design.

9-10. BOILERS

Boilers are devices in which water can absorb heat from hot products of combustion, with or without direct radition from the burning fuel.

Steam boilers are not completely filled with water, but above the water level have a vapor space where the vapor (steam) can separate from the boiling liquid.

Hot-water boilers are not supplied with any steam-disengaging space, but are completely filled with water which circulates through the boiler by thermal action (or under the action of a pump) at a sufficiently rapid rate to prevent it from flashing into steam at the pressure under which the boiler is operating.

Boilers may be classified primarily as high-pressure or low-pressure types. Low-pressure boilers do not operate at more than 15 psig steam pressure or 30 psig water pressure. Many states and municipalities require continuous attendance of a fireman if the operating pressure exceeds these limits. Nearly all small boilers may serve hot-water heating systems as well as steam systems. Only minor changes are required for changing to either system.

Where steam is produced, the latent heat of evaporation (h_{fg}) is available at the radiators and convectors, along with a minor amount of sensible heat. The condensate is returned to the boiler through pipes smaller than those required to bring the steam to the heat transmitters.

As there is always some steam loss from a system, the water in a steam boiler eventually becomes concentrated with salts and sediment resulting from the accumulation of ingredients which do not evaporate. Therefore the water in the boiler must be changed occasionally by *blowing down*, or discharging, part of it. The interval between blowing-down periods depends upon the amount of make-up required by the system, and on the quality of the water supply.

The unevaporated material collects upon hot surfaces of the boiler and may form a hard, stonelike deposit of scale which becomes a heat insulation on the water side of the heat-transfer surfaces. Unless these deposits are removed, the metal of the boiler will eventually be overheated, as the water cannot absorb the heat with sufficient rapidity.

Steam boilers require the constant addition of new water to replace that which is being delivered in the form of steam, and there should be a gage glass, and preferably also try cocks, to indicate the water level. A safety valve must always be provided. If the metal in a steam boiler becomes overheated and if comparatively cool feed water is delivered against the hot surfaces, the metal may crack from sudden contraction. When the fuel is introduced mechanically, as in oil-burner, gas-burner, or stoker installations, automatic water-feeder and low-water-cutoff switches are highly desirable (see Fig. 9–15).

Heating boilers are of the *fire-tube type*, or *tubular type*, if, following combustion, the gases pass through tubes or comparatively small passages which are submerged in the water. They are called *water-tube boilers* if the arrangements are reversed; that is, if water is inside the tubes. Both types are effective, although, when low-pressure saturated steam is used for heating, the fire-tube construction has advantages. It is easier with this

FIG. 9-11. Boiler burner unit arranged for oil, gas, or oil-gas firing. (Courtesy Kewanee-Ross Corporation.)

Heat-Transfer Elements and Combustion 349

type to provide adequate steam-liberating area at the water line, and the larger heat-storage capacity due to a greater volume of water in the boiler gives a more uniform rate of output.

Figure 9-11 shows a steel fire-tube boiler arranged for oil or gas firing. In this H.R.T. (horizontal return tubular) boiler, the gases leaving the refractory-covered portion of the combustion chamber pass rearward through the corrugated steel combustion space to the refractory-lined back of the unit. Here the gases turn and pass forward through the fire tubes of the boiler, finally passing upward into the breeching which connects to the stack. Except for the refractory lining, this is a welded all-steel unit with the tubes rolled into the front and rear headers. These boilers are characterized by adequate steam-disengaging surface, so that relatively dry steam is delivered into the steam outlet. These boilers normally use either gas or heavy fuel oil in No. 5 or No. 6 grades. The heavy fuel oil is preheated by means of a steam connection from the boiler itself. However, provision is usually made to have a supplementary electric heater for starting a completely cold system. Most burners of this type employ a rotating device for atomizing the oil into the air stream for good admixture in the burner.

Cast-iron sectional boilers, which are made for domestic service, are also made in larger sizes for serving public buildings, apartment houses, factories, etc. They can be arranged for firing with oil or gas, or for solid fuel with stoker or hand firing. Such boilers are usually rated in accordance with standards set by the IBR (Institute of Boiler and Radiator Manufacturers). The heavy-duty, cast-iron boilers of one manufacturer, for example, range in net output capacity in single units from 650,000 Btuh to 2,100,000 Btuh.

9-11. FUELS AND COMBUSTION

The design of a boiler or of any direct-fired heater is closely related to the combustion of the fuel. Combustion is a chemical process in which the carbon and hydrogen (and sulfur) in the fuel react with the oxygen from the air, thermal energy being released. The rate at which the fuel can be burned effectively depends largely on the intimacy with which the air is brought into contact with the fuel. This intimacy is influenced mainly by the intensity of the chimney draft, by the size of the grate, by the size of the combustion chamber, by the location of the heating surfaces, and by proper control of air and fuel feed. The combustion problem in turn is influenced by the character of the fuel—whether solid, liquid, or gas.

Solid Fuels. Ordinary solid fuels are represented by the various ranks of coals, coke, and lignite. Table 9-5 shows average analyses of coals. Bituminous coals, which represent the greatest tonnage of coal mined, are

TABLE 9-5
AVERAGE ANALYSES OF TYPICAL U.S. COALS
(PER CENT)

Number of Samples	Kind of Coal	Proximate Analysis by Weight				Ultimate Analysis by Weight					Heating Value, As Received (Btu/lb)	
		Volatile Matter	Fixed Carbon	Ash	Water	Carbon	Hydrogen	Sulfur	Oxygen	Nitrogen	High	Low
5	Anthracite....	4.36	82.70	9.34	3.60	82.14	2.22	0.6	1.2	0.9	13074	12820
3	Semianthracite..	9.6	74.8	12.9	2.7	77.0	3.3	1.0	1.97	1.13	13000	12683
6	Low-volatile bituminous.	18.4	71.4	7.15	3.05	80.70	4.25	1.07	2.27	1.51	14073	13643
	High-volatile bituminous											
5	A............	35.1	53.7	7.7	3.5	74.5	5.1	1.3	6.4	1.5	13389	12852
3	B............	35.9	47.8	8.5	7.8	69.1	4.6	1.8	6.8	1.4	12126	11608
4	C............	36.9	41.9	8.5	12.7	62.6	4.4	3.5	7.2	1.1	11370	10806
2	Subbituminous .	31.9	42.8	3.90	21.4	56.70	3.80	0.35	12.70	1.15	9775	9169
2	Lignite........	28.7	30.3	6.8	34.2	42.45	2.95	0.60	12.25	0.75	7280	6614

high in volatile matter and have good calorific values. The volatile matter in coal is that portion of the coal which is driven off, in gaseous or vapor form, when the coal is heated. It consists largely of hydrocarbons and other gases resulting from decomposition and distillation. In carrying out a standard *proximate analysis* of a coal, a weighed sample is gently heated to 220 F to drive off the moisture, and the sample is again weighed. It is then heated for 7 min at 1700 F, out of contact with air, to drive off the volatile matter. The remaining sample consists of the fixed carbon plus the ash. The fixed carbon, which resembles coke in appearance, is burnt out when the sample is heated in air, the ash being left as a residue. The proximate analysis of a coal, which can be made with little trouble, gives important information about the characteristics of a fuel. However, the *ultimate analysis* gives much additional information, in that the percentage composition of the combustible matter in the fuel is furnished in terms of the chemical elements themselves.

Coke is made from the destructive distillation of bituminous coals by a heating process in which the volatile matter is driven off and the fixed carbon and ash remain.

The heating values of representative coals are indicated in Table 9-5. The higher heating value represents the thermal energy available from the fuel if the water vapor resulting from combustion of the hydrogen in the fuel is condensed and this latent heat is absorbed from the products of combustion. The lower heating value presupposes that this water vapor leaves uncondensed in gaseous form. Although it is not possible to realize the higher heating value of the fuel in combustion equipment, it is customary to compare boiler-furnace performance with the higher heating value.

The Dulong formula can be used to determine the higher heating value, in Btu per pound, of a coal or of coke when the ultimate analysis is known:

$$\text{HHV} = 14{,}540\,C + 62{,}000\left(H - \frac{O}{8}\right) + 4000\,S \qquad (9\text{-}4)$$

where C, O, H, and S are the decimal fractions by weight of the carbon, hydrogen, oxygen, and sulfur in the fuel.

Fuel Oil. Fuel oils are produced by distillation from crude petroleum and by cracking of heavy hydrocarbon products. They vary in character from light volatile distillates resembling kerosene to the heavy viscous fuel-oil No. 6, which resembles black molasses and is known as Bunker C oil. The fuel oils are hydrocarbons containing between 84 and 85 per cent carbon, between 12 and 14 per cent hydrogen, and small amounts of oxygen, nitrogen, and sulfur, with traces of moisture and solid material. Standard

CS12-48 of the U. S. Department of Commerce lists five grades of fuel oil. No. 1 fuel oil has a specific gravity in the neighborhood of 0.8, while No. 6 oils have specific gravities which sometimes are in excess of 0.98.

Specific gravity and degrees API of fuel oils are usually measured with suitable hydrometers. The relationship which relates degrees API to specific gravity is: sp gr = 141.5/(deg API + 131.5).

Table 9-6 gives the API ranges of the different grades of fuel oil, and their approximate higher heating values in Btu per gallon. The heating

TABLE 9-6

HIGHER HEATING VALUES OF FUEL OILS OF VARIOUS GRADES

Grade No.	Representative Range (deg API)	Higher Heating Value (Btu per gal)	Representative Weight (lb per gal)
1	35-40	136,000	7.0
2	30-36	138,500	7.2
4	24-29	141,000	7.5
5	18-22	148,500	7.8
6	14-16	152,000	8.0

value may also be determined in Btu per pound through use of the Bureau of Standards formula

$$\text{HHV} = 22{,}320 - (3780)(\text{sp gr at 60 F/60 F}) \tag{9-5}$$

or by use of the Sherman-Kropf formula

$$\text{HHV} = 18{,}250 + (40)(\text{deg API} - 10) \tag{9-6}$$

Values obtained by the Sherman-Kropf formula vary somewhat from those determined by the Bureau of Standards formula.

Oils with higher specific gravity have lower heating values per pound; but, as the oils with higher specific gravity contain more pounds per gallon, the heating value on a gallon basis is higher for the higher-gravity oils.

Gases. Gases are becoming of increasing importance as fuels. By virtue of the many additional pipelines being built, natural gas is being sent to sections of the country which heretofore had to depend upon manufactured gas. The price of natural gas in many cases is competitive with fuel oil, and coal.

Gas is an easy fuel to control and burn. It is sold either on the basis of cubic feet or in *therms*, where the therm represents the volume of gas

Heat-Transfer Elements and Combustion

equivalent to 100,000 Btu. Because of the wide variation in the heating value of gases per cubic foot, burners adjusted for one type of gas cannot be used with another unless readjustment of the air ratio is made. Heating value for commonly used gases in Btu per cubic foot are: carbon monoxide (CO) 322; methane (CH_4) 991; hydrogen (H_2) 315; propane (C_3H_8) 2451; coke oven gas about 575; natural gases range from 800 to 1125. These are higher heating values for the gases in Btu per cu ft, measured at 62 F at 30 in. Hg and saturated with water vapor.

9-12. COMBUSTION CALCULATIONS

The air theoretically required to burn a unit weight of the combustible elements in fuel can be readily calculated from the following chemical equations:

$$C + O_2 \to CO_2 \quad (9\text{-}7)$$
$$12 + 32 = 44$$

Thus 12 lb of carbon require 32 lb of oxygen for complete combustion, and yield 44 lb of carbon dioxide. The 12 and 16 (taken twice) are atomic weights of carbon and oxygen, respectively. For the more accurate atomic weight of carbon, 12.01, it is evident that 1 lb of C requires $32/12.01 = 2.66$ lb of O_2 and yields 3.66 lb of CO_2.

As the weight ratio of oxygen to the inert gases in air is 0.2319 to 0.7681, it follows that 1 lb of C requires $2.66/0.2319 = 11.5$ lb of air for complete combustion.

Similarly, other relations for combustion are:

$$2H_2 + O_2 \to 2H_2O \quad (9\text{-}8)$$
$$4 + 32 = 36$$

1 lb of H_2 requires $\dfrac{32}{4.03} = 7.94$ lb of O_2 and yields 8.94 lb of H_2O

1 lb of H_2 requires $\dfrac{7.94}{0.2319} = 34.3$ lb of air for complete combustion

1 lb of sulfur (S), by like calculation, requires 4.3 lb of air for complete combustion

The hydrocarbon fuels contain no oxygen, but coal and the alcohols do. This oxygen in the fuel is already chemically combined with some of the combustible matter or can readily go into combination. Consequently, allowance must be made for this oxygen in computing the weight of air required to burn a unit weight of fuel. The allowance is usually made by considering that 1 lb of this oxygen can combine with $\frac{1}{8}$ lb of hydrogen.

Let C, H, O, and S represent the proportional parts by weight of the carbon, hydrogen, oxygen, and sulfur in a fuel. The weight of air, A, theoretically required to burn a pound of the fuel is

$$A = 11.5\,C + 34.3\left(H - \frac{O}{8}\right) + 4.3\,S \qquad (9\text{–}9)$$

Combustion can hardly ever be carried on with this theoretical amount of air; a certain amount in excess of that theoretically required must be used to insure complete combustion and prevent carbon deposits. In combustion calculations the problem most usually presented is to find from an analysis of the waste gases what combustion conditions actually exist. Flue-gas and exhaust-gas analyses are usually given in volumetric percentages of the so-called dry flue gases CO_2, O_2, CO, and N_2. Generally, the most accurate method of making combustion calculations from flue-gas analyses is through the medium of a carbon balance.

The volume of any gas multiplied by its density gives the weight of that gas. Since the density of a gas is proportional to its molecular weight, the product of gas volume and molecular weight is also proportional to the weight of a gas. If the volume of each individual gas in a gas mixture is multiplied by its respective molecular weight, the sum of the resulting products is proportional to the weight of the gas mixture. Let CO_2, O_2, CO, and N_2 be the percentages or parts by volume of these respective gases in the dry waste gas. The proportional weights of these respective gases and of the mixture are $44\,CO_2$, $32\,O_2$, $28\,CO$, $28\,N_2$, and $44\,CO_2 + 32\,O_2 + 28\,CO + 28\,N_2$.

In carbon dioxide, CO_2, atomic weights show that $\frac{12}{44}$ of the total weight is carbon; that is, the proportional weight of the carbon in the CO_2 is $\frac{12}{44} \times 44\,CO_2 = 12\,CO_2$. Similarly, the weight of the carbon in carbon monoxide, CO, is $\frac{12}{28} \times 28\,CO = 12\,CO$.

The weight of the dry waste gases per unit weight of carbon existing in them must be

$$\frac{\text{Total weight of the gases}}{\text{Weight of carbon in the gases}} = \frac{44\,CO_2 + 32\,O_2 + 28\,CO + 28\,N_2}{12\,CO_2 + 12\,CO} \qquad (9\text{–}10)$$

In other words, the number of pounds of dry waste gases, G_w, per pound of carbon burned, is

$$G_w = \frac{44\,CO_2 + 32\,O_2 + 28\,CO + 28\,N_2}{12\,CO_2 + 12\,CO} = $$

$$\frac{11\,CO_2 + 8\,O_2 + 7\,CO + 7\,N_2}{3(CO_2 + CO)} \qquad (9\text{–}11)$$

Equation 9–11 should not be considered a formula. It is simply a statement of the weight of a certain gas mixture divided by the weight of carbon in the mixture. The equation is set up in this fractional form instead of in vertical tabular form in order that advantage can be taken of cancellation to reduce arithmetical work.

To find the number of pounds of dry waste gas per pound of fuel burned, the value of G_w obtained from equation 9-11 must be multiplied by C_g, where C_g is the number of pounds of carbon gasified or burnt per pound of fuel and must be determined from test data and the fuel analysis. In the case of liquid hydrocarbon fuels or gaseous fuels, essentially all of the carbon in the fuel becomes gasified or is burnt. This is not true in the case of solid fuels because some carbon nearly always appears unburned in the refuse.

The weight of air used per pound of fuel, and the combustible matter burnt per pound of fuel, appear in the waste gases either *in the dry flue gas* or *as water vapor*. Expressed in equational form, this statement becomes:

$$\text{Air used} + \text{combustible burnt} = \text{dry flue gas} + \text{water vapor} \quad (9\text{-}12)$$

or

Air used per pound of fuel + combustible burnt per pound of fuel
$$= G_w \times C_g + 9\,\text{H}$$

The combustible burnt per pound of fuel consists of all the fuel which entered into combustion. This combustible is the whole original pound of fuel, less the ash and moisture in the fuel and that portion of the combustible matter in solid fuel which remains unburnt and appears in the refuse. The term $G_w \times C_g$ has been explained previously. For 1 lb of hydrogen burnt, 9 lb of water vapor result; therefore 9 H gives an expression for the weight of water vapor per pound of fuel (H being the weight of hydrogen per pound of fuel). Thus the number of pounds of dry air, W_a, actually used per pound of fuel is

$$W_a = G_w \times C_g + 9\,\text{H} - \left(1 - \frac{\%\text{ ash} + \%\text{ moisture}}{100} - \frac{\text{combustible in refuse}}{\text{pound fuel}}\right) \quad (9\text{-}13)$$

In the case of a hydrocarbon liquid fuel, this equation usually simplifies to

$$W_a = G_w \times C_g + 9\,\text{H} - \left(1 - \frac{\%\text{ moisture}}{100}\right) \quad (9\text{-}14)$$

The following method of calculating the weight of dry air used per pound of fuel can be used *if the fuel contains a negligible amount of nitrogen*. The number of pounds of nitrogen, W_N, in the waste gas per pound of fuel is

$$W_N = \frac{28.16\,\text{N}_2}{12(\text{CO}_2 + \text{CO})} C_g = \frac{7.04\,\text{N}_2}{3(\text{CO}_2 + \text{CO})} C_g \quad (9\text{-}15)$$

All of this nitrogen must have come from the air (76.8 per cent N_2, 23.2 per cent O_2), as none was assumed in the fuel. (The value 28.16 is the molecular weight of the nitrogen-argon mixture in atmospheric air.) The number of pounds of dry air actually used per pound of fuel is

$$W_a = \frac{7.04\,N_2 \times C_g}{3(CO_2 + CO)(0.768)} \quad (9\text{-}16)$$

In the case of a fuel which contains a weight fraction of N_f parts of nitrogen, equation 9-16 can be modified into the following more-precise form:

$$W_a = \left[\frac{28.16\,N_2}{12(CO_2 + CO)}C_g - N_f\right]\frac{1}{0.768} = \left[\frac{2.35\,N_2}{CO_2 + CO}C_g - N_f\right]\frac{1}{0.768} \quad (9\text{-}17)$$

The *per cent of excess air* used is evidently found by the relation

Per cent of excess air =

$$\frac{\text{air actually used} - \text{air theoretically required}}{\text{air theoretically required}} \times 100 \quad (9\text{-}18)$$

The weight of any individual kind of dry waste gas per pound of fuel can be found by following the reasoning used to develop equation 9-11. Thus,

$$\text{Lb of } CO_2 \text{ per lb fuel} = \frac{11\,CO_2}{3(CO_2 + CO)}C_g \quad (9\text{-}19)$$

$$\text{Lb of } N_2 \text{ per lb fuel} = \frac{7.04\,N_2}{3(CO_2 + CO)}C_g \quad (9\text{-}20)$$

$$\text{Lb of } O_2 \text{ per lb fuel} = \frac{8\,O_2}{3(CO_2 + CO)}C_g \quad (9\text{-}21)$$

$$\text{Lb of } CO \text{ per lb fuel} = \frac{7\,CO}{3(CO_2 + CO)}C_g \quad (9\text{-}22)$$

In all the preceding expressions the gas items are percentages or proportional parts *by volume* and thus Orsat values can be used.

9-13. COMBUSTION

Combustion has been described as the application of the three t's: time, temperature, and turbulence. The essentials of proper combustion should be elaborated upon, however, to include (a) proper proportioning of the air to the fuel; (b) exposure of sufficient fuel surface to permit combination with the air; (c) adequate mixing of the fuel particles with the air; (d) sufficiently high temperature to assure continuing ignition; (e) arrangement of the combustion chamber so that sufficient time is allowed to complete burning before the gases leave. These principles apply to the combustion of any fuel, whether it be gaseous, liquid, or solid, but different arrangements must be made for the different fuels.

In the case of a gas, it is relatively easy to mix the gas and the air in proper proportions and to maintain ignition. Also, it is not difficult to

Heat-Transfer Elements and Combustion 357

atomize a liquid fuel into particles small enough to permit a large fuel surface to be exposed to the air. Burning of a solid fuel presents more difficulty, as the air must be brought into contact with the combustible surfaces of the fuel and must not be smothered out by the resulting ash before combustion is completed. The burning of volatile coals with a minimum of smoke presents an additional problem. These problems will be discussed further in connection with the various types of combustion equipment.

It has previously been shown how the analysis of the flue gas could be used to indicate the quantity of air used in a combustion process. Use of air greatly in excess of that theoretically required for combustion can cause appreciable loss in the operation of a boiler-furnace system. Consequently, it is desirable to adjust the air-fuel ratio to a point where good combustion is obtained and a minimum of excess air is employed. This minimum is sometimes set by other factors besides combustion, as it may be necessary to use additional air to prevent smoke or to avoid the formation of combustible products, such as carbon monoxide, in the flue gas. In addition, leaky settings permit air to enter the furnace system without contributing to the combustion process in any way.

9-14. FLUE-GAS ANALYSIS AND FURNACE LOSSES

The composition of flue gases leaving a furnace can be determined by a variety of devices, of which the Orsat analyzer is perhaps the most common. In using this apparatus a sample of flue gas is aspirated from the stack and, after being measured, passes into a solution of potassium hydroxide which absorbs the carbon dioxide from the sample. The remaining gas is then passed into a second solution, usually of pyrogallic acid, which absorbs the oxygen. The carbon monoxide in the sample is finally absorbed by a solution of cuprous chloride, and the remainder of the sample can be considered to consist of nitrogen. The fraction of each gas in the original sample is found by measuring the decrease in volume after each absorption. These data furnish a volumetric analysis of the flue gas, and by using the data it is possible to analyze combustion conditions.

Example 9-6. A heavy fuel oil has the following composition in weight fractions: C, 0.855; H, 0.115; N and other matter, 0.03. The HHV is 18,500 Btu per lb. (a) Find the volume and composition of the ideal dry products of combustion. (b) Find the air actually used if the dry-flue-gas analysis shows the following volume percentages: CO_2, 8.7; O_2, 8.5; CO, 0.1; N_2, 82.7.

Solution: (a) By equation 9-9 the air required for pound of fuel is

$$A = 11.5(0.855) + 34.3(0.115) = 13.77 \text{ lb}$$

As 76.8 per cent of air by weight is N_2, there are $13.77 \times 0.768 = 10.58$ lb of N_2 in the flue gas from 1 lb of fuel.

The weight of CO_2 per pound of fuel is

$$3.66 \times C = 3.66 \times 0.855 = 3.13 \text{ lb}$$

To put the dry flue gases, consisting of 10.58 lb of N_2 and 3.13 lb of CO_2, on a volume basis, it is convenient to make use of the molal volume of 386 cu ft at 70 F and 14.7 psi occupied by the mole weight in pounds of any gas. Thus the volume of dry gaseous products is

$$10.58 \times \frac{386}{28} + 3.13 \times \frac{386}{44} = 145.6 + 27.4 = 173.0 \text{ cu ft} \qquad Ans.$$

Therefore the percentage of carbon dioxide by volume is

$$CO_2 = \frac{27.4}{173.0} \times 100 = 15.82 \text{ per cent} \qquad Ans.$$

and of nitrogen,

$$N_2 = \frac{145.6}{173.0} \times 100 = 84.18 \text{ per cent} \qquad Ans.$$

Hence, with ideal combustion the CO_2 cannot exceed 15.8 per cent in the dry products.

b) By equation 9–11, the weight of dry flue gas per pound of carbon is

$$G_w = \frac{(11)(8.7) + (8)(8.5) + 7(0.1 + 82.7)}{3(8.7 + 0.1)} = 28.2 \text{ lb}$$

and the weight of dry flue gas per pound of fuel is

$$28.2 \times C_g = 28.2 \times 0.855 = 24.1 \text{ lb}$$

By equation 9–14, the weight of air per pound of fuel is

$$W_a = 24.1 + (9)(0.115) - (1 - 0.03) = 24.2 \text{ lb} \qquad Ans.$$

As there is little N_2 in the fuel, equation 9–16 can also be used. It represents a quicker method when it is applicable. In this case,

$$W_a = \frac{(7.04)(82.7)(0.855)}{3(8.7 + 0.1)(0.768)} = 24.5 \text{ lb}$$

This answer differs slightly from that secured by the previous method because there is some nitrogen which came from the fuel, and the nitrogen term has some significance. Use of equation 9–17 gives a more accurate answer if the nitrogen is known or assumed.

Example 9–7. An Illinois coal has 40.2 per cent volatile matter, 39.1 per cent fixed carbon, 8.6 per cent ash, and 12.1 per cent moisture. In percentages, its ultimate analysis shows C, 62.8; H, 4.6; O, 6.6; S, 4.3; and N, 1. The higher heating value is 11,480 Btu per lb and it can be shown that the maximum CO_2 by volume in the flue gas is 18 per cent. (a) Find the HHV by Dulong's formula and compare the result with the calorimeter value. (b) The dry ashpit refuse contains 25 per cent by weight of combustible, which is assumed to be carbon. Find the unburned carbon per pound of coal and the heat loss per pound of coal from this cause. (c) In percentages, the analysis of the dry flue gas by volume is: CO_2, 10; O_2, 8; CO, 0.1; N_2, 81.9. Compute the number of pounds of dry flue gas per pound of coal fired, the weight of moisture in the flue gas per pound of coal fired, and the weight of air used per pound of coal fired.

Solution: (a) The higher heating value is

$$HHV = (14{,}540)(0.628) + (62{,}000)\left(0.046 - \frac{0.066}{8}\right) + (4000)(0.043)$$
$$= 11{,}587 \text{ Btu per lb} \qquad Ans.$$

Heat-Transfer Elements and Combustion

b) The refuse contains the ash of the original coal plus unburned combustible, or 25 per cent combustible and 75 per cent ash matter. Thus 75 per cent of the refuse = 8.6 per cent of the coal, and 25 per cent of the refuse = C_R per cent of the coal unburnt. Therefore the amount of combustible in the refuse per pound of coal is

$$C_R = (25)\left(\frac{8.6}{75}\right) = 2.86 \text{ per cent, or } 0.0286 \text{ lb}$$

If this is considered carbon with a heating value of 14,540 Btu per lb, the loss per pound of coal is

$$\text{Loss} = (14{,}540)(0.0286) = 416 \text{ Btu} \qquad \textit{Ans.}$$

c) By equation 9–11, the weight of dry flue gas per pound of carbon is

$$G_w = \frac{(11)(10) + (8)(8) + 7(0.1 + 81.9)}{3(10 + 0.1)} = 24.7 \text{ lb}$$

and the weight of dry flue gas per pound of coal is

$$G_w C_g = 24.7(0.628 - 0.0286) = 14.8 \text{ lb} \qquad \textit{Ans.}$$

The moisture in the flue gas arises from the 0.121 lb of water in the original coal and from the combustion of the hydrogen, which equals 9 H. Thus

$$\text{Moisture} = 0.121 + 9(0.046) = 0.121 + 0.414 = 0.535 \text{ lb} \qquad \textit{Ans.}$$

By equation 9–13, the weight of air used per pound of coal is

$$W_a = 14.8 + 0.414 - (1 - 0.086 - 0.121 - 0.0286) = 14.98 \text{ lb} \qquad \textit{Ans.}$$

A nitrogen balance also gives a sufficiently close approximate solution. By equation 9–17, the weight of air used per pound of coal is

$$W_a = \left[\frac{(7.04)(81.9)}{3(10.1)} \, 0.5994 - 0.01\right]\frac{1}{0.768} = 14.86 \text{ lb}$$

It should be noted that the flue-gas analysis, as usually obtained, is on a dry basis; that is, the water vapor resulting from the combustion of hydrogen or from moisture in the fuel is condensed out and does not appear in the flue-gas analysis. Fuels which are high in hydrogen give dry-flue-gas analyses which are high in nitrogen and relatively low in carbon dioxide. In example 9–6, where a fuel oil was treated, it was noted that the theoretical composition of the dry flue gases showed 15.8 per cent CO_2. In contrast to this, the coal which was computed in example 9–7 would show a maximum CO_2 of 18 per cent, the fuel oil having the much higher hydrogen content. Natural gas, which is also high in hydrogen, would indicate a representative maximum CO_2, with no excess air, of around 12 per cent.

Air used in excess of that required for theoretical combustion (see Table 9–7) represents a loss in operation; the excess, with its large quantities of nitrogen, is heated to the final temperature of the stack gases and passes out the chimney. In general, the sensible heat lost up the chimney with the hot flue gases represents the biggest loss in operation and for good performance should be kept to the minimum possible.

Space does not permit a complete coverage of the items entering into a boiler-furnace heat balance, but mention will be made of the most signifi-

TABLE 9-7
AIR REQUIRED FOR COMBUSTION

Air Usage	Anthracite per Pound	Coke per Pound	Semi-bituminous per Pound	Bituminous per Pound	Lignite per Pound	Fuel Oil per Gallon
Theoretical	9.6 lb	11.2 lb	11.0 lb	10.7 lb	6.5 lb	115 lb
Converted to cubic feet at 70 F	129 cu ft	151 cu ft	148 cu ft	144 cu ft	87 cu ft	1550 cu ft
Fair practice	225 cu ft	260 cu ft	260 cu ft	250 cu ft	150 cu ft	2100 cu ft

cant items. Loss due to hot dry-flue-gas constituents is represented by the following formula:

$$Q_{dfg} = W_{dfg} C_{pm} (t_g - t_a) \qquad (9\text{-}23)$$

where W_{dfg} = pounds of dry flue gas formed during the burning of 1 lb of fuel; C_{pm} = the mean specific heat of the flue gases; t_g = temperature of the flue gases entering the stack; t_a = temperature of the air supplied for combustion; Q_{dfg} = Btu loss in dry flue gases per pound of fuel. For dry constituents of the flue gases in the range from (say) 70 F to 1000 F, C_{pm} ranges from 0.25 to 0.26.

A second loss occurs in connection with the water vapor formed in the flue gas from the combustion of the hydrogen in the fuel, and from any moisture which is carried by the fuel and which evaporates when the fuel burns. This water vapor leaves the stack in the form of steam and, as it is not usually condensed, carries away with it the latent heat associated with the steam. This loss can be represented by the following relationship, which is based on the formula which appears in the ASME Power Test Code for steam-generating units:

$$Q_{H_2O} = \left(\frac{M}{100} + \frac{9H}{100}\right)(0.46 t_g + 1089 - t_a) \qquad (9\text{-}24)$$

where Q_{H_2O} = the Btu loss from water vapor per pound of fuel fired; M = per cent of water in the fuel; H = per cent of hydrogen in the fuel; t_g and t_a = temperatures, in degrees Fahrenheit, of the flue gas and air supply, respectively.

Other losses from combustion of a fuel occur when combustible matter in the gaseous products is not completely burned out, as is the case with carbon monoxide (CO) or when hydrocarbons or actual solid combustible particles pass up the stack. Even though black smoke may be formed during combustion, this condition does not necessarily mean that large weights of combustible material are passing out of the stock; but poor combustion is definitely indicated. The loss from incomplete combustion of carbon monoxide, in Btu per pound of fuel fired, is

$$Q_{CO} = 10{,}160 C_g \left(\frac{CO}{CO_2 + CO}\right) \qquad (9\text{-}25)$$

Heat-Transfer Elements and Combustion 361

where C_g = pounds of carbon gasified (burnt) per pound of fuel fired, and where CO and CO_2 are volumetric percentages of CO and CO_2 in the flue gases.

In addition to the previously discussed losses, which apply to all types of fuels (liquid, gaseous, and solid), there is an additional loss which occurs with solid fuels—from combustible which is lost in the refuse. With solid fuels, particles of combustible consisting largely of carbon may remain unburned in the ashy refuse. The loss from this source, in Btu per pound of fuel fired, can be computed from the relation

$$Q_{CR} = 14,600 \times C_R \qquad (9\text{-}26)$$

where C_R represents the weight fraction of solid fuel remaining unburned in the refuse.

A heat balance of a boiler or furnace is an accounting of the disposition of the energy in the fuel to all points of loss and to the useful warming of the medium being heated, such as to the water-steam (or, in a warm-air furnace, to the hot air). In a carefully conducted test, the heat balance would normally show that some 95 per cent of the energy of the fuel can be accounted for in useful heating or as measurable losses. The remaining 5 per cent or so represents the so-called radiation and unaccounted-for losses. Depending upon the care with which a test is conducted, or upon the insulation effectiveness of the furnace itself, radiation and unaccounted-for losses may range from 3 to 12 per cent. Large boilers, such as those connected with power plants or large heating systems, use heat-conservation devices, such as economizers (water preheaters) in the breeching, or air preheaters for the combustion air, by means of which it is possible to reduce the furnace losses to values as low as 10 to 15 per cent. However, with small-size domestic and industrial equipment, losses in boiler-furnace operation can run from 25 to 50 per cent of the heating value of the fuel. With residential heating equipment, the computed heat losses from the furnace are not completely dissipated, since some of the lost heat passes by indirect routes into the spaces which it is desired to warm.

Example 9-8. When the fuel oil of example 9-6 burned, with the flue-gas analysis given, the stack temperature of the products leaving the boiler was 570 F. Air is supplied to the burner at 70 F. What loss occurs to the chimney (a) from the dry flue gas and (b) from the water vapor in the flue gas? (c) What is the loss from incomplete combustion? (d) Estimate the boiler efficiency.

Solution: (a) Reference to example 9-6 shows that 24.1 lb of dry flue gas are generated per pound of fuel. Use this value in equation 9-23 and take 0.255 as a representative mean specific heat for these dry-flue-gas products when the temperature is below 800 F. Thus, the loss per pound of fuel is

$$Q_{dfg} = (24.1)(0.255)(570 - 70) = 3070 \text{ Btu} \qquad Ans.$$

Thus the percentage loss to dry flue gases, based on the HHV of the fuel, is

$$\frac{3070}{18{,}500} \times 100 = 16.6 \text{ per cent}$$

b) By equation 9-24, for the dry fuel with 11.5 per cent of hydrogen, the loss from water vapor, per pound of fuel, is

$$Q_{H_2O} = \frac{(9)(11.5)}{100}[0.46(570) + 1089 - 70]$$

$$= 1315 \text{ Btu} \hspace{4em} Ans.$$

Also, expressed as a percentage of the heating value of the fuel, the loss is

$$\frac{1315}{18{,}500} \times 100 = 7.12 \text{ per cent}$$

c) By equation 9-25, the loss from incomplete combustion is, per pound of fuel,

$$Q_{CO} = 10{,}160(0.855)\left(\frac{0.1}{8.7 + 0.1}\right) = 98.5 \text{ Btu} \hspace{4em} Ans.$$

As a percentage, the loss is

$$\frac{98.5}{18{,}500} \times 100 = 0.53 \text{ per cent}$$

d) The probable radiation and unaccounted-for losses from this unit are in the neighborhood of 3 to 7 per cent. If 5 per cent is taken as representative and the other losses are added, the total is

$$5 + 16.6 + 7.12 + 0.53 = 29.25 \text{ per cent}$$

Thus, $100 - 29.25 = 70.7$ per cent of the heating value of the fuel is converted into useful heating; that is, 70.7 per cent is the probable boiler efficiency. *Ans.*

9-15. COAL COMBUSTION. STOKERS

With hand firing, coal is burned on grates; with automatic firing, it is burned on stokers which use their own fuel-bed supports (retorts, tuyères, grates). The problem associated with burning coal is that of supplying air for combustion in sufficient quantity to permit the fuel to be consumed adequately without large amounts of excess air, and yet providing enough air to prevent the occurrence of high concentrations of carbon monoxide. Smoke also raises a serious problem. In hand-firing of domestic units, it is always desirable to keep a portion of the fire bed uncovered by fresh fuel so that it presents an incandescent surface which can be used to ignite the products being distilled from the fresh coal. This procedure, called *side firing* or *alternate firing*, reduces smoke with volatile coals and presents a reasonably satisfactory method of hand firing. If frequent firing is possible, a thing feeding of green coal over the entire fuel bed is likewise satisfactory, provided the glowing coals are not completely obscured. This latter method requires the feeding of coal at short intervals, which is not usually possible in domestic installations. In firing bituminous coal, the charge

Heat-Transfer Elements and Combustion 363

may fuse into a solid piece, or bridge, and it becomes necessary to break this up with some sort of stoking bar.

As the volatile matter in the coal is being distilled off, it must come into contact with the incandescent fuel bed or with gaseous products hot enough to ignite it. After the volatile matter has been largely driven off, the coked fuel can be pushed over onto the incandescent zone of the fire bed without producing smoke. Anthracite and coke do not present a smoke hazard, but the same general principles should be adhered to. Bituminous coal burns more rapidly than anthracite and does not require so high a draft, but more attention must be given to the fuel bed when bituminous is used.

Stokers. A most common type of stoker, the underfeed, consists of a magazine or coal hopper, external to the boiler furnace, which delivers coal to a screw conveyor. The screw, rotating at slow speed, feeds coal to the stoker retort (or to grates), where combustion takes place. A separately operated fan forces air through the stoker air passages, called *tuyères*, and the coal is burned. The resulting ash from the combustion retort drops over the sides of the retort into a trough, from which it can be removed by hand or by a conveyor, or it can be stored for periods of time for later removal by any method desired. Such stokers are made in small sizes feeding as little as 15 to 60 lb of coal per hour, and in large sizes which can burn amounts of coal in excess of 1000 lb per hour.

In bituminous stokers, mechanical coal-burning devices are not limited to the underfeed type. At the front end of the boiler, there may be a coal hopper having a very small reciprocating ram which feeds the prepared-size bituminous coal to a series of rapidly revolving distributor blades. These blades are so arranged that they throw the coal uniformly over the entire firebox area. The particles of new coal, however, are so small in proportion to the mass of coke already undergoing combustion that the volatile gases are very rapidly driven off and burn without smoke and with a minimum decrease in the temperature of the firebox at that point. Many small particles of coal are burned in suspension. With this type of spreader stoker the firebox temperature is very high, and thus it is usually necessary to water-jacket the bearings of the distributor. Air for combustion purposes is furnished by a forced-draft fan.

9-16. OIL BURNERS AND COMBUSTION

Most oil burners for domestic furnaces are of the high-pressure, mechanical-atomizing type (gun type), but burners are also made in the rotary wall-flame type, and as air-atomizing burners, and finally, as simple vaporizing units. In the gun-type, or mechanical-atomizing, burner, illustrated in Fig. 9–12, a pump delivers No. 1 or No. 2 fuel oil, under a pressure of some

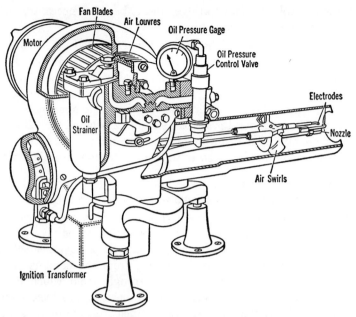

Fig. 9-12. Pressure-atomizing gun-type oil burner.

50 to 150 psi, to an atomizing nozzle, from which it leaves as a fine mist. An integrally built fan supplies the air which mixes with the oil mist. The swirling mixture of oil mist and air burns as it leaves the nozzle and passes through the combustion chamber of the furnace. Ignition of the mixture is accomplished by sparking electrodes, which receive high-tension power from a transformer. After combustion starts, the power supply to the transformer is shut off and combustion is maintained from the flame. The effectiveness of combustion in this type of burner depends on the degree of oil atomization and on the turbulence or mixing of the oil and air. Various design features are incorporated in the burner tube (gun), such as vanes or swirls for the air; and with a suitable fuel-nozzle angle and with a properly designed combustion chamber good performance can result. The design of the combustion chamber as an integral part of the burner-furnace combination is very important, as combustion is carried to completion in the chamber and proper flow patterns must exist there to complete the combustion and prevent soot formation. Combustion chambers are built from refractory bricks or furnace sections, or are frequently supplied as precast refractory units with proprietary features incorporated.

In the wall-flame type of vaporizing burner, a centrally located rotary distributor throws the oil particles out to an outer rim, where the vaporization of the oil takes place while it mixes with the combustion air. Both the oil supply and the air can be precisely regulated, and very clean com-

Heat-Transfer Elements and Combustion

bustion can be maintained. The rotor of the oil distributor, and an attached fan, constitute the only moving parts. Ignition is usually by electrical means from a suitable electrode.

9-17. BOILERS FOR BURNING GAS

Sectional boilers are employed widely for the burning of gas. It is possible to burn gas with fair efficiency in a boiler having the large combustion chamber and wide passages which are necessary for the products of combustion for coal or oil but which are undesirable with gas. If gas is to be burned, however, it is wiser to employ a heating boiler designed especially for that fuel. Such a boiler will have a relatively small combustion chamber and will have narrow passages for the products of combustion, with many baffles. These factors will promote the turbulence which aids in rapid heat transfer, by reducing the thickness of the gaseous film which clings to the heating surfaces.

In Fig. 9-13 the gas burners are seen in position beneath the cast-iron boiler sections. The flow of air to be mixed with the gas is preadjusted for

FIG. 9-13. Gas-fired boiler equipped for steam generation.
(Courtesy American Radiator & Standard Sanitary Corporation.)

the particular type of gas being used; after an adjustment is once made, only occasional readjustment is required. Boilers of the type just described are rated in accordance with specifications prepared by the AGA (American Gas Association). For this particular line of boilers, energy inputs range from 180,000 Btuh to 4,800,000 Btuh. It is customary to consider that approximately 80 per cent of the gas energy input can be realized as useful output.

The operation of gas-fired equipment presents an element of danger, in that the flame may become extinguished and the gas continue to flow. Consequently, controls for gas-fired equipment must be extremely reliable in operation. A satisfactory type of control for gas firing is pictured in Fig. 9-14. This consists of a manually operated valve which can shut

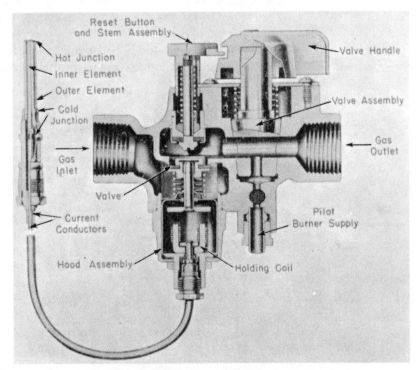

FIG. 9-14. Combination manual and automatic shutoff valve.
(Courtesy Milwaukee Gas Specialty Company.)

off the gas supply completely when the burner is not in use, and an automatic valve which completely interrupts the flow of gas in the event that the pilot light should be extinguished. The operation of the automatic valve is responsive to the action of a thermocouple. (See sec. 1-4 and Fig. 1-5.) The thermocouple is located adjacent to the pilot flame, and the

Heat-Transfer Elements and Combustion

warm junction of the couple is thus kept hot by the flame. The electromotive force produced by the couple causes sufficient current to flow through the magnetic holding coil to hold the valve open against the closing force of a spring. In the event that the pilot flame becomes extinguished, the current in the thermocouple circuit then ceases, or reduces to such a point that the holding coil can no longer keep the valve open, and the spring snaps the coil into the shut position, completely closing off the gas flow.

With this particular type of control, even should the pilot flame become reignited, it is necessary to reset the valve in the open position by means of the hand-operated reset button. This required action provides an additional safety feature. It is possible to design valves of this type with an automatic reset where such a device is desirable. In this particular valve, closure of the main valve also cuts off the gas supply to the pilot burner. Other control arrangements sometimes have a separate supply to the pilot direct from the main line.

9-18. CHIMNEYS

Chimneys, or *stacks*, are used to convey the gaseous products of combustion to an elevation sufficiently high above surrounding objects or buildings to keep the gases from being a nuisance. The height of a chimney directly influences the production of draft. As the column of hot gas inside the chimney is less dense than a similar column of cool air outside, the pressure inside the chimney near its base is less than that of the outside air. This pressure difference acting at any instant is known as the *draft* of the chimney and promotes flow of air through the furnace and up the chimney.

The theoretical draft depends on the difference between the density d_a of the outside air and the density d_{fg} of the hot flue gas; and, since both of these densities depend on temperature, it is possible to set up an expression for draft in terms of temperatures. Thus,

$$\Delta P' = H(d_a - d_{fg}) \text{ lb per sq ft} \tag{9-27}$$

By equation 2-28,

$$\Delta P' = H\left(\frac{P_B}{RT_a} - \frac{P_B}{RT_{fg}}\right)$$

which reduces to

$$\Delta P = 0.255 P_B H\left(\frac{1}{T_a} - \frac{1}{T_{fg}}\right) \text{in. of water} \tag{9-28}$$

where ΔP = theoretical draft, in inches of water;
P_B = barometric pressure, in inches of mercury;
H = height of stack, in feet;
T_a = air temperature, in degrees Fahrenheit absolute $(460 + t_a)$;
T_{fg} = average flue-gas temperature, in degrees Fahrenheit absolute $(460 + t_{fg})$.

Thus the maximum draft is directly proportional to the height of stack (H) and the atmospheric pressure (P_B), and is influenced by the air and flue-gas temperatures. This draft pressure is used in a given furnace in accelerating the air through the furnace and in overcoming frictional and impact losses in the chimney and its connections. The cross-sectional area of the chimney must be such as to permit delivering at a reasonably low velocity the volume of flue gases generated. In natural-draft chimneys, reasonable gas velocities run from less than 20 fps up to 40 fps.

The chimney diameter for the required cross-sectional area can be found from equation 9–29. In this rational formula, G is the flue-gas flow, in pounds per second; V is the velocity, in feet per second; P_B is the barometric pressure, in inches of mercury; T_{fg} is the average flue-gas temperature, in degrees Fahrenheit absolute; and D is the stack diameter, in inches:

$$D = 11.7 \sqrt{\frac{GT_{fg}}{P_B V}} \qquad (9\text{--}29)$$

Example 9–9. A fuel-oil-fired domestic unit consumes 3 gal of oil per hour. The average stack temperature is 300 F when the flue gases leave the furnace at 400 F. The barometer is at 29.9 in. Hg, the outside temperature is 40 F, and the chimney height above the combustion chamber is 30 ft. (a) What theoretical draft exists and (b) what chimney size is required to carry the flue gases?

Solution: (a) Use equation 9–28 to find the theoretical draft. Thus,

$$\Delta P = 0.255(29.9)(30)\left(\frac{1}{460 + 40} - \frac{1}{460 + 300}\right)$$

$$= 0.156 \text{ in. of water} \qquad \qquad Ans.$$

It is assumed that the flue gas is like air in its characteristics, and this assumption is closely true. Carbon dioxide tends to make the flue gas heavier than air, but water vapor from the combustion of hydrogen and from moisture in the fuel makes it lighter; therefore the flue-gas density approximates that of air.

b) Burning the fuel oil under the conditions of example 9–6 showed that 24.2 lb of air were used per pound of fuel; or 24.2 lb of air and 1 lb of fuel formed 25.2 lb of flue gas, which we can round off to 25 lb. A gallon of domestic (No. 2 grade) fuel oil weighs (say) 7.2 lb per gal (Table 9–6). Thus, for the 3 gal per hour burnt, the weight of flue gas per second is

$$G = \frac{25 \times 3 \times 7.2}{3600} = 0.15 \text{ lb}$$

Using this value in equation 9–29 and assuming a gas velocity of 20 fps, we find that the required diameter of the chimney is

$$D = 11.8 \sqrt{\frac{(0.15)(460 + 300)}{(29.9)(20)}} = 5.16 \text{ in.}$$

The chimney should be made a little larger to provide ample area for contingencies that might develop, such as in starting—when the burner fan forces a pulse of combustion

Heat-Transfer Elements and Combustion 369

gases into a cold connecting breeching and chimney. In this small size, enough excess capacity would be provided by going to 6 in., or even to 7 in., for the breeching-pipe diameter. The chimney is usually rectangular or square. The equivalent-diameter chart of Chapter 12 shows that a square chimney 6 in. by 6 in. has the same capacity as one 6.6 in. in diameter, while a chimney 7 in. by 7 in. has the capacity of one 7.7 in. in diameter. The 6-in. by 6-in. size should be a reasonable design selection. *Ans.*

Considerably less draft is required for the efficient combustion of oil or gas than is required with solid fuel, since fuel-bed resistance does not occur.

It is possible to control excessive chimney draft, or fluctuations in draft due to otherwise uncontrolled exterior conditions, by using a counter-weighted check-draft damper. With solid fuels, savings in fuel have been found where these devices have been installed. In general, no oil-burning heating boiler should be installed without such a device. In some applications the air supply for this damper control is taken directly from the boiler room, while in others, the air may be taken through a duct from out of doors. Gas stacks require an open hood or a spill vent.

9-19. BOILER SAFETY DEVICES

In the operation of boilers and other pressure vessels containing water or steam, care must be taken to see that conditions do not arise which could lead to failure of the vessel, with possible breakage or even explosion. Mention has already been made of the danger which can arise if the water level in a boiler drops below a minimum value considered to be safe. The danger arising from low water is that when hot flue gases pass over metal surfaces not in contact with water, the surfaces may become overheated to a point where the metal is weakened and permanent damage can result. Low-water cutoff devices can be provided to stop the combustion apparatus should the water level reach a dangerously low point. In Fig. 9-13 a low-water cutoff device was shown installed on a boiler.

Water-feeders are arranged to provide for a sufficiently rapid flow of water into a boiler to maintain proper water level. Figure 9-15 illustrates a combination boiler-water feeder and low-level cutoff device, similar in function to the one illustrated in Fig. 9-13. The float, shown in its chamber, is responsive to variations of water level in the boiler. The top connection to the float chamber would normally be connected into the steam space of the boiler, while the bottom connection would be below ordinary normal water level. The travel of the float is conveyed to the external control system through a bellows or packless construction so that hot boiler water is neither in contact with the electrical parts of the control system nor with the feed-water supply. The water fed into the boiler is controlled by the valve shown at the lower right in the illustration. With a low water-level in the boiler, the float, in its low position, keeps the feed line

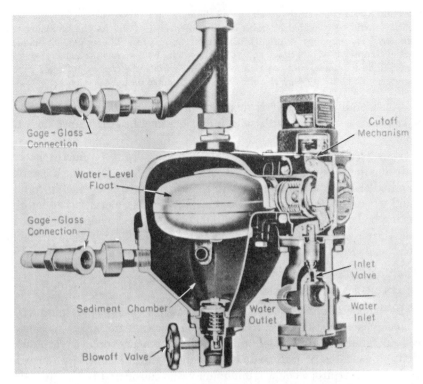

Fig. 9-15. Boiler-water feeder and feeder cutoff control. (Courtesy McDonnell & Miller, Inc.)

open, permitting water to flow into the boiler. This action continues until the level rises sufficiently to lift the float, which in turn closes off the inlet feed valve. In the upper right of the photograph the electric mechanism for cutting off the power supply to the combustion equipment is shown. This particular design is arranged to operate at a maximum steam pressure of 25 psig.

All steam boilers and water-filled pressure vessels must be provided with a pressure safety device which, when the pressure in the vessel reaches a predetermined limit, opens and permits either steam or water to leave the boiler in a quantity sufficient to bring the pressure to a safe operating condition for the system. In the case of steam boilers, the safety device is called a safety valve and is located so that steam discharges from the boiler. In the case of hot-water boilers or heaters, the safety device is called a pressure-relief valve. A pressure-relief valve of the type that would be used with a hot-water tank or heater has an adjustable spring and when the pressure in the vessel exceeds the desired pressure at which the vessel

Heat-Transfer Elements and Combustion 371

is set to operate, the valve is lifted from its seat against the restraining action of the spring, and water is eliminated from the system until the pressure falls back to a safe value.

PROBLEMS

9-1. A 16 ft × 14 ft × 10 ft high corner room with much glass area has a design heat loss of 14,800 Btuh. Explore the possibility of using either cast iron base board radiation on the outside walls or 2 recessed convectors. Hot water at an average maximum temperature of 180 F is available. Select units from tables in chapter showing Btuh excess capacity or show Btuh inadequacy for unsuitable equipment.

9-2. A room has a heat loss of 7000 Btuh. It is desired to heat this room by using baseboard radiation supplied with hot water. An average temperature of 180 F can be supplied from the boiler system. Select the lineal run of nonferrous, fin-tube baseboard radiation required to serve this room.   *Ans.* 12 ft could suffice

9-3. Obtain the data asked for in problem 9-2 but assume that representative cast-iron baseboard radiation is employed.   *Ans.* 18 ft

9-4. The manufacturer's data for a particular unit heater show that it delivers 22,600 Btuh when supplied with 60 F entering air and with 2 psig steam, using a $\frac{1}{20}$-hp motor at 1140 rpm of the fan. (a) Compute the approximate weight of steam condensed per hour in this heater when it is operating under the conditions specified. (b) When 430 cfm enter the unit, what is the approximate delivery temperature of the air? Note that C_p for air is approximately 0.245.   *Ans.* (a) 22.6 lb/hr; (b) 108 F

9-5. For the unit heater described in problem 9-4, the manufacturer's catalog states that when 40 F air enters the heater the heat transfer and condensate produced are greater than for normal design conditions by a factor 1.141. For the same air inflow, namely 430 cfm measured at 60 F, (a) compute the outlet temperature of the air from the heater under 40 F inlet conditions, and (b) estimate the pounds of steam condensed per hour.   *Ans.* (a) 94.4 F; (b) 25.8 lb/hr

9-6. Cooling water in a steam condenser enters at 65 F and leaves at 80 F. (a) If the steam is condensing at 1 psia, what is the logarithmic mean temperature difference? (b) Compare this value with the arithmetic mean temperature difference.
     Ans. (a) 28.7 log mtd; (b) 29.3 arithmetic mtd

9-7. A hot-water heater using a steam coil of the hairpin type (U type) is supplied with water at 60 F and delivers water at 140 F. The steam supply is at 1.3 psia. The maximum demand is 150 gal of water per hour. Find the total linear feet of 1-in.-OD brass tubing required in the steam coil.
     Ans. 26.3 ft (with $U = 130$)

9-8. Compute the film factor for water flowing at 0.5 fps through 2-in.-OD steel tubes (0.109-in. wall) in a heat exchanger. The water enters at 70 F and leaves at 90 F.

9-9. Water is heated in a single-pass shell-and-tube heat exchanger 12 ft long and of 14-in. OD. The shell contains twenty 2-in.-OD stainless-steel tubes with 0.093-in. wall thickness. Water is supplied at 70 F and is warmed to approximately 205 F. The water velocity in the twenty tubes is low at approximately 0.19 fps. (a) Compute the value of the film-factor (surface-transfer) coefficient on the water side of the tubes. (b) Consider the value of f for condensing steam to be 3000 Btu per hr sq ft deg F. Disregard the resistance of the clean metal itself and compute the over-all thermal resistance and U of the tube. (c) Find the logarithmic mean temperature

difference for the tubing, with steam at 115 psia (338 F) being used for heating. (d) Find the heat transferred per hour. (e) What gpm of water flow is taking place through the heater?
Ans. (a) 81.9 Btu per hr sq ft deg F; (b) 0.01255, 79.6 Btu per hr sq ft deg F; (c) 192 deg; (d) 1,740,000 Btuh; (e) 26.1 gpm

9–10. An Illinois bituminous coal as received has an analysis by weight, expressed in decimal fractions, as follows: 0.667 C, 0.0435 H, 0.0803 O, 0.0139 N, 0.0156 S, 0.0797 moisture, 0.100 ash. Compute the higher heating value of the coal, in Btu per pound, as received. Find the heating value per pound of dry coal. [HINT: Divide the as-received value by (1 − moisture).]   Ans. 11,837 and 12,862 Btu/lb

9–11. A West Virginia Pocahontas coal has a dry analysis, in per cent by weight, as follows: 85.6 C; 4.56 H; 2.85 O; 1.33 N; 0.66 S; ash, 5.0. The coal as received has 3.5 per cent moisture. (a) Compute the analysis of this coal on an as-received basis. [HINT: Multiply each dry percentage by (1 − moisture).] (b) Compute the heating value per pound of dry coal and per pound of coal as received.
Ans. (a) 82.61 C, 4.40 H, etc.; (b) 15,088 and 14,560 Btu/lb

9–12. Find the amount of air theoretically required to burn the coal of problem 9–10.   Ans. 8.89 lb per pound of coal

9–13. Find the amount of air theoretically required to burn the dry coal of problem 9–11.   Ans. 11.31 lb per pound of coal

9–14. The Illinois coal of problem 9–10, when burned in a stoker-fired furnace, showed a dry-flue-gas analysis, in per cent by volume, of 9.1 CO_2, 10.6 O_2, 0.1 CO, and 80.2 N_2. (a) Find the number of pounds of dry flue gas and the weight of air used per pound of coal, as received. (b) Find the weight of CO formed per pound of fuel burnt.
Ans. (a) 18.0, 17.6; (b) 0.017

9–15. The coal of problem 9–11 showed a dry-waste-gas analysis, in per cent by volume of 9.8 CO_2, 5.6 O_2, 0.1 CO, and 84.5 N_2. (a) Find, per pound of dry coal, the waste gases produced and the air used. (b) Find the weight of CO formed per pound of dry coal burnt. Ans. (a) 25.1 lb dry gas, 0.41 lb water vapor, 21 lb air; (b) 0.018 lb

9–16. A representative No. 6 fuel oil has the following composition, in per cent: C, 87.3; H, 10.8; S, 1.2; N, 0.2; O, moisture, and solids, 0.5. Its higher heating value is 18,500 Btu per lb. (a) Compute the weight of air required to burn theoretically 1 lb of this fuel; (b) considering its specific gravity as 0.985, compute the weight of air required to burn, theoretically, 1 gal of the fuel.
Ans. (a) 13.8 lb/lb; (b) 113.2 lb/gal

9–17. Assume that the No. 6 fuel oil in problem 9–16 showed the following dry-flue-gas analysis, in per cent; CO_2, 11; O_2, 7; and N_2, 82. Compute (a) the weight of air used per pound of fuel and (b) the percentage of excess air.
Ans. (a) 19.9 lb; (b) 44%

9–18. Under the combustion conditions of problem 9–14, the waste gases left the boiler at 610 F and air entered at 70 F. Compute the loss from (a) dry waste gases (b) water vapor in the waste gases, (c) incomplete combustion. (d) Estimate the probable boiler efficiency if 6 per cent of the loss can be considered as unaccounted for—and as due to combustible in refuse, and to radiation.
Ans. (a) 2430; (b) 612; (c) 173; (d) 66.9%

9–19. Under the combustion conditions of problem 9–15, the stack gases leave the boiler furnace at 550 F and air enters at 60 F. Compute, per pound of *dry* coal, (a)

Heat-Transfer Elements and Combustion

the loss from dry waste gases, (b) the loss from water vapor in the waste gases, and (c) the loss from incomplete combustion.

Ans. (a) 3074 Btu/lb; (b) 526 Btu/lb; (c) 183 Btu/lb

9-20. Under the combustion conditions of problem 9-15, the stack gases leave the boiler furnace at 550 F and air enters at 60 F. Compute, per pound of coal *as received*, (a) the loss from dry waste gases, (b) the loss from water vapor in the waste gases, and (c) the loss from incomplete combustion. (d) Estimate the probable boiler efficiency if 5 per cent loss is considered as due to radiation and unaccounted for.

Ans. (a) 2960 Btu/lb; (b) 552 Btu/lb; (c) 176 Btu/lb; (d) 69.7%

9-21. A poorly adjusted oil burner uses No. 2 fuel oil with the following percentage weight composition: C, 0.85; H, 0.14; inerts, 0.01. It has a higher heating value of 19,900 Btu per lb and shows the following dry-flue-gas analysis, in per cent by volume: CO_2, 8.3; O_2, 9.4; CO, 0.3; and N_2, 82.0. Deposition of soot on the boiler heating surfaces has reduced heat transfer so much that the flue gases leave at 870 F. Air enters the burner at 70 F. Compute the loss per pound of fuel from (a) dry flue gases, (b) water vapor in flue gases, and (c) incomplete combustion. Express your answers in Btu per pound of fuel and in percentages. (d) If the radiation and unaccounted-for loss is 5 per cent, estimate the boiler efficiency by difference calculation.

Ans. (a) 4900 Btu, 24.6%; (b) 1789 Btu, 9.0%; (c) 303 Btu, 1.5%; (d) 59.9%

9-22. An oil burner using No. 2 fuel oil of the composition and type indicated in problem 9-21 showed a dry-flue-gas analysis, in per cent by volume, of 11.8 CO_2, 4.5 O_2, 0.2 CO, and 83.5 N_2. Th flue gases enter the breeching at 600 F and air is supplied to the furnace at 70 F. Compute the loss per pound of fuel from (a) dry flue gases, (b) water vapor in the flue gases, (c) incomplete combustion. Express answers in Btu per pound of fuel and as percentages. If the radiation and unaccounted-for loss is 4 per cent, estimate the boiler efficiency by difference calculation.

9-23. If the flue gases from the coal of problem 9-14 have an average temperature of 540 F in the chimney when the outside temperature is 40 F, what theoretical draft is developed by the 30-ft chimney? If, under the conditions of problem 9-14, the coal is burnt at the rate of 60 lb per hr, what chimney size is required, provided the gas velocity may not exceed 25 fps?

9-24. Work problem 9-23 for a 30-ft chimney if the mean temperature in the chimney is 440 F and the outside air is at 40 F, all other conditions remaining unchanged.

9-25. Fuel prices in a particular locality were:
    No. 2 fuel oil (144,000 Btu), 12 cents per gallon
    Bituminous coal (13,000 Btu per lb), $14.00 per ton
    Gas per therm (100,000 Btu);
        21 cents for the first 10 therms
        10 cents for the next 30 therms
        7.5 cents for all over 40 therms

Compute the cost of furnishing 100,000,000 Btu by each of these fuels. Assume that the efficiency of conversion is 72 per cent for the oil, 56 per cent for the coal, and 76 per cent for the gas.

9-26. Using the data of problem 9-25, determine the cost of furnishing heat to a building for a season if the estimated requirement is 250,000,000 Btu and coal is employed.

Ans. $240.00

10
Physiological Reactions to the Environment

10-1. COMFORT AIR CONDITIONS

Air conditioning as a general term implies effective control of the physical and chemical properties of air in order to produce (1) comfort air-conditioning (the maintenance of the air surrounding human beings in such a way that it is in a condition most suitable for their comfort and health) or (2) industrial air-conditioning (the maintenance of the air surrounding a material or product in process of manufacture or storage so as best to preserve the physical stability of the material throughout its manufacturing or storage period).

So far as air environment is concerned, the factors which affect human comfort are, in order of importance: (1) temperature, (2) humidity, (3) air motion and distribution, and (4) purity (the quality of the air in regard to odor, dusts, toxic gases, and bacteria). Unless these factors are properly controlled, comfort cannot be obtained. With complete-air-conditioning systems, all four of these basic items must be considered.

Although simultaneous control of *all* these factors is required in order to produce a *fully* satisfactory environment for human comfort, many systems, as installed, do not control all of the factors and yet maintain surroundings which are pleasant and substantially conducive to comfort. This is true of many heating systems.

To understand the effect of these four factors, certain physiological and psychological responses of the human body must be taken into consideration. The objective of heating or cooling for comfort is to maintain an atmosphere of such characteristics that the people occupying the space can effectively lose enough heat to permit proper functioning of the metabolic processes in their bodies and yet not lose this heat at so rapid a rate that the body becomes chilled. The processes of combustion of food within the body produce heat in such quantity that the body temperature is normally above the temperature of the atmosphere. A complex regulating mechanism in the human body keeps the temperature of the body at or about 98.6 F. So long as an individual is capable of dissipating heat to the atmosphere at a rate equal to the rate of heat production within the body

Physiological Reactions to the Environment

it is possible to keep the body temperature constant and no difficulty is experienced. If the body temperature rises above normal, heat prostration may result, with temporary or even permanent injury.

The body dissipates heat to the ambient air moving past its surface by ordinary methods of conduction and convection. In this process, air temperature and air motion are the essential factors causing heat transfer. The body can also lose heat by radiation to colder surroundings. Some moisture is always evaporating from the surface of the skin, but when necessary the sweat glands in the body permit large quantities of water to pass through the surface of the skin. If the air in contact with the body is not saturated, this water is taken up by evaporation into the air, with the body itself supplying an appreciable portion of the latent heat. This process of body cooling is especially effective when the air humidity is low. Heating and evaporation of moisture into the air that enters the lungs also cools the body.

The processes of heat control within the body are quite complex, and although not perfectly understood, they clearly operate in two general directions: (1) to decrease or increase internal heat production (metabolism) as the body temperature rises or falls, and (2) to control the rate of heat dissipation by changing the rate of cutaneous blood circulation, and by motivating the sweat glands. When the cutaneous blood circulation is increased, more blood flows near the surface of the skin, thereby increasing the surface temperature and permitting greater heat dissipation. Similarly, under cold conditions the blood vessels near the surface of the skin contract, the blood flow at the surface is decreased, and less heat is lost. Excessive activity of the sweat glands usually comes into play following the other methods of control.

The metabolic energy in the body, if not dissipated at the same rate at which it is produced, reappears as stored energy and is evidenced by a rise in temperature of the deep body tissues. If, on the other hand, the environment is so cold that heat is lost from the body at a faster rate than that at which it is created, the deep-tissue temperature of the body slowly falls. These phenomena can be represented by the following equation:

$$M = \pm S + E \pm R \pm C \qquad (10\text{-}1)$$

where M = metabolic heat produced within the body, in Btu per hour;
S = stored energy, represented by change in body-tissue temperature, in Btu per hour;
E = evaporative heat loss, in Btu per hour;
R = heat loss or gain by radiant methods, in Btu per hour;
C = heat loss or gain by convective methods, in Btu per hour.

As previously discussed, S can increase or decrease at any particular time. This is also true of R and C, which depend upon environmental conditions.

Under nearly all conceivable conditions, E is a positive loss, with moisture evaporating from the surface of the body and absorbing some heat from that body to provide for the change of water to vapor. External heat gain to the body (minus-sign) is an unusual condition.

The use of seasonable clothing minimizes variations in the load on the human system over winter and summer conditions, but it must be remembered that the same skin area which keeps the body cool in summer remains unchanged in the winter and that added clothes alone are not sufficient insulation to maintain comfort conditions for the body. Therefore homes must be heated in winter, and it is often desirable to cool them in summer to maintain reasonable comfort.

To a certain extent the human system adapts itself to extreme atmospheric conditions. This adaptation, which is termed acclimatization, is both physiological *and* psychological. In fact, people from the tropics experience some discomfort in temperate climates until several weeks have passed. Temperatures of 85 F in winter heating are uncomfortable to most people, although in summer a temperature of 85 F is comfortable if the relative humidity is not high or if the air movement is pronounced. It is largely by changes in internal-combustion rates that the body adjusts itself to general atmospheric conditions. Of a similar nature, but requiring more-rapid internal adjustment, are "local" changes of atmospheric conditions, such as are met when an individual goes from the hot outside in summer to a conditioned interior. If the inside temperature is in a normal comfort range, yet appreciably colder than outside, an entering occupant may feel cold until readjustment of internal-body-heat controls are made. In summer months the inside air temperatures should, to a limited extent, bear some relation to the swings in outside temperature. Table 10-1 shows a set of recommendations which might be employed to allow for outside temperatures.

The variation of desired inside temperatures for the winter season is not critical. Recommended values are listed in Table 5-1.

10-2. COMFORT CHART

As far as the human body is concerned, air temperature, relative humidity, and air motion act together to produce the sensation of warmth or cold that is experienced. No simple method has as yet been devised which is completely adequate to evaluate the composite effect of all the variables involved in human response to the thermal environment. Formerly, extensive use was made of the concept of *effective temperature* (ET) as an index to express the composite effect of air temperature, relative humidity, radiation, and air motion on the human body. The numerical values of the effective temperature scale were made equal to the temperatures of calm (15 to 25 fpm) saturated air, which produces a sensation of warmth equal to that existent under the given air conditions.

TABLE 10-1

INSIDE DESIGN CONDITIONS FOR SUMMER COMFORT COOLING

A. Design Conditions Related to Outside Temperature

Outside Design Dry Bulb (deg F)	Occupancy over 40 Min			Occupancy under 40 Min		
	Dry Bulb (F)	Wet Bulb (F)	Rel Hum (%)	Dry Bulb (F)	Wet Bulb (F)	Rel Hum (%)
80	75	65	60	76	66	61
	77	63	47	78	64	47
	79	61	35	80	62	36
85	76	66	61	77	67	61
	78	64	47	79	65	48
	80	62	36	81	63	36
90	77	67	61	78	69	64
	79	65	48	80	67	52
	81	63	36	82	65	40
95	78	69	64	79	70	65
	80	67	52	81	68	52
	82	65	40	83	66	41
100	79	70	65	81	71	63
	81	68	52	83	69	50
	83	66	41	85	67	38
105	80	71	65	81	72	65
	82	69	52	83	70	54
	84	67	42	85	68	41

Tests were conducted at the ASHRAE research laboratory, and in other physiological laboratories, to determine the effect of environmental variables on the human system. In conducting the tests, normal healthy individuals were placed in rooms (or moved from room to room) in which dry- and wet-bulb temperatures (relative humidity) and air motion could be varied, and the comfort reactions of the subjects were carefully noted.

It is obvious that if the relative humidity is low, evaporation from the surface of the skin may be so rapid as to cause undue cooling as well as some drying out of the skin surface, and even if the dry-bulb temperature is high, comfort may not prevail. On the other hand, if the relative humidity is very high, evaporation from the skin surface may practically stop, and since cooling by evaporation then ceases, comfort may be obtained with a lower dry-bulb temperature. Between such limits of humidity, varying gradations of comfort will hold. From a viewpoint of body-heat dissipation alone, the range through which relative humidity is varied does not matter, but other considerations make the most desirable relative-humidity range lie between 30 and 70 per cent. For values much below 30 per cent the mucous membranes and the skin surface may become uncomfortably dry. For values above the 70 (or even 60) per cent range, there is a tendency for a clammy or sticky sensation to develop. The effect of increased air velocity is always to increase the convection and evaporative heat dissipation from the body and thereby lower the effective temperature.

A plot of effective temperatures based on responses of human subjects tested under controlled conditions (Refs. 1, 2, and 3) appears in the ASHRAE Comfort Chart of Fig. 10–1. It is known that the effective-temperature chart overemphasized the effect of relative humidity on comfort and more recent investigations at the ASHRAE laboratory and elsewhere (Refs. 4, 5, and 6) have shown that equivalent thermal response is more closely approximated by the equivalent comfort lines indicated by 3, 4, 5, and 6 in Fig. 10–1. Line 4–4 it will be noted, indicates comfort. Thermal comfort is difficult to define in direct terms so it is customary to think of thermal comfort as a condition where no thermal discomfort is apparent. It will be noted that the comfort line, 4–4, is almost straight at the lower relative humidities. This is not the case at higher relative humidities where it becomes more difficult to produce evaporative cooling.

Effective temperature indicates a striking response to the effect of relative humidity both at high and low values. The answer to the apparent discrepancy between comfort equivalence indicated by the lines of equal comfort and by effective temperature, it is believed, lies in the different manner in which the human subject testing was carried out. The earlier testing involved having subjects respond to the impressions they received immediately upon entering a controlled space and apparently such impressions were strongly influenced by relative humidity as well as by temperature. The more-recent experiments involved the responses of subjects after they remained in a space for one or more hours and adjusted to equilibrium there. Since this is a more usual condition for which comfort should be established, it is felt that the revised lines on the comfort chart should be used in preference to using effective temperature as an index of comfort. It should also be noted that in the instantaneous responses of the effective temperature approach superficial perspiration on the body and absorbed moisture in clothing were extraneous factors in the response pattern.

Reference to the comfort chart shows that thermal comfort is obtained

Physiological Reactions to the Environment

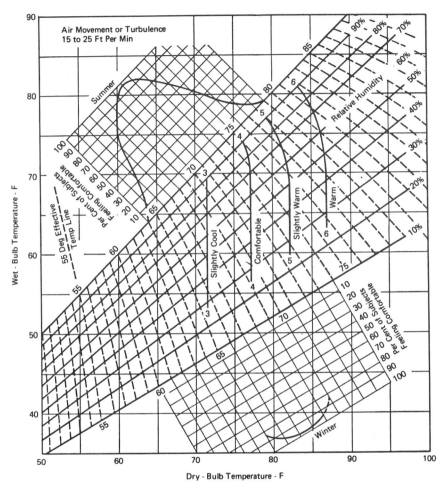

Fig. 10-1. ASHRAE comfort chart for air velocities of 15 to 25 fpm (still air). Winter and summer comfort zones apply to inhabitants of the United States. In winter the application is with heating by the usual convection methods and not by radiation methods. The summer and winter graphs show comfort as determined on the earlier effective temperature basis. (Reprinted, by permission, from ASHRAE *Handbook of Fundamentals,* Chapter 7, 1965.)

for lightly clothed individuals at a temperature of approximately 77.5 F in the mid range of relative humidities with this value dropping to approximately 76 F at high relative humidity. There is a slight shift in the comfort range between summer and winter. Because of acclimatization and the somewhat different clothing worn in winter and summer, individuals are satisfied with a slightly lower temperature in winter than in summer. The difference is small and appears to be but slightly in excess of 1°, (Ref. 5). It is customary to consider a temperature around 78 F as an upper value for pure comfort in summer with the corresponding value for winter ranging between 75 F and 77 F because of the usually heavier clothing of winter. It will be observed that these recommended values are higher than those indicated for comfort by use of effective temperature. However, it will be observed that in the mid-humidity range for summer there is good agreement between the two approaches. Comment should also be made as to the substantial variations in comfort response to thermal conditions between different individuals. Although these variations are shown in relation to effective temperature, they are equally applicable for the equivalent comfort lines.

Example 10-1. For the following conditions, state which can be considered in a comfort range for summer and winter occupancy: (a) 80 F db, 60 F wb; (b) 80 F db, 40 per cent relative humidity; (c) 70 F db, 65 F wb; (d) 70 F db, 60 F wb.
a) Warmer than desirable but acceptable in summer to some individuals
b) Warmer than desirable
c) Too cool and relative humidity is undesirably high
d) Too cool even for winter except for those with above-normal inside clothing.

The comfort chart of Fig. 10-1 does not take into account variations in comfort conditions when there are wide variations in the radiant temperature of the surroundings. Basically, the chart is applicable when the walls, floor, and ceiling are essentially at the dry-bulb temperature of the space. When properly weighted, the average surface temperature of the walls, floor, and ceiling is called the mean radiant temperature (MRT). The effect of variations in MRT on comfort has been extensively investigated. In the range of 80 F dry bulb, a rise of one degree in MRT produces approximately the same effect as a one-degree rise in dry-bulb temperature. However, as temperature increases, a greater rise in mean radiant temperature is required to produce the same physiological effect. For example, at 87 F the MRT would have to be 4 deg higher, or 91 deg MRT, to produce an equivalent effective temperature of 88 F. Thus the effect of MRT on comfort is less pronounced at high temperatures than at low temperatures, but radiant effects should always be considered if the wall surface temperature is greatly different from the room temperature. Expressed in terms of dry-bulb temperatures, a rough rule to employ is to design for a 0.3- to 1-deg increase of room temperature to offset a 1-deg drop in MRT in the conventional heating range.

10-3. HEAT LOSS FROM THE HUMAN BODY

The heat which must be dissipated by the human body is not a constant, but varies with the degree of activity, the atmospheric conditions when these are not in the comfort range, and with the individual. Table 10-2 gives some of the heat-loss rates for normal adults under different conditions

Physiological Reactions to the Environment 381

TABLE 10-2
TOTAL HEAT DISSIPATION FROM INDIVIDUALS

Type of Activity	Heat Dissipation at Room Temperature between 60 F and 90 F (Btuh)
Adults at rest	
Seated	390
Standing	440
Moderately active worker	600
Metalworker	860
Walking, 2 mph	760
Restaurant server, very busy	1000
Walking, 3 mph	1000
Walking, 4 mph; active dancing	1390
Slow run	2290
Maximum exertion	3000–4800

of activity, while Table 10-3 extends these data into working form for convenience in making heat-gain calculations.

The heat loss from the human body under varying effective temperatures is also plotted in Fig. 10-2. On curve D, which applies to men seated at rest, it can be seen that from about 66 deg ET to 82 deg ET (in the comfort zone) the heat loss is essentially constant at 400 Btuh. At lower effective temperatures the heat dissipation increases, resulting in a feeling of coolness; and at higher temperatures the ability to lose heat rapidly decreases, resulting in severe discomfort, or even prostration, at temperatures at the higher end of the chart. For curves C, B, and A, for heat loss at higher rates of activity, constant comfort-equilibrium ranges of heat dissipation can also be noticed.

At low temperatures the greater part of the heat loss is by sensible-heat methods (convection, conduction, and radiation) while at higher temperatures evaporation accounts for the greater heat loss.

Example 10-2. A certain lecture room is occupied by 100 people and is kept at a dry-bulb temperature of 74 F and a wet-bulb temperature of 62 F by the air circulated. Find (a) the relative humidity in the room, and whether the room conditions are in the comfort zone; (b) the heat given up to the air by the occupants; and (c) the portion of heat given up as latent heat by moisture evaporated into the air, and the portion given up as sensible heat.

Solution: (a) From Fig. 10-1, the effective temperature for 74 F db and 62 F wb is seen to be 69 deg ET at a relative humidity of 50 per cent. These conditions are not in the comfort zone for many individuals since they lie between the 3 and 4 lines of the chart. *Ans.*

b) From Fig. 10-2 (curve D, for people seated), the total heat loss is observed to be about 400 Btuh per person at 69 deg ET. For 100 persons, therefore, the heat loss is 400 × 100, or 40,000 Btuh. *Ans.*

c) From Fig. 10-3 (curve D, for people at rest), at a dry bulb of 74 F the heat loss by

TABLE 10-3
HEAT GAIN FROM OCCUPANTS

Type of Activity	Typical Application	Total Heat Dissipation, Adult Male (Btuh)	Total Adjusted* Heat Dissipation (Btuh)	Sensible Heat (Btuh)	Latent Heat (Btuh)
Seated at rest	Theater				
	Matinee	390	330	225	105
	Evening	390	350	245	105
Seated; very light work	Offices, hotels, apartments, restaurants	450	400	245	155
Moderately-active office work	Offices, hotels, apartments	475	450	250	200
Standing; light work; walking slowly	Dept. store, retail store	550	450	250	250
Walking; seated	Drug store	550	500	250	250
Standing; walking slowly	Bank	550	500	250	250
Sedentary work	Restaurant	490	550	275	275
Light bench work	Factory	800	750	275	475
Moderate work	Small-parts assembly	900	850	305	545
Moderate dancing	Dance hall	900	850	305	545
Walking, 3 mph; moderately heavy work	Factory	1000	1000	375	625
Bowling (participant)	Bowling alley	1500	1450	580	870
Heavy work	Factory	1500	1450	580	870

*Adjusted heat dissipation is based on the normal percentage of men, women, and children for the application listed, considering the adult female as having 85 per cent of the value for an adult male, and a child as having 75 per cent of the value for an adult male. Based on data from ASHRAE, *Handbook of Fundamentals*, Chapter 28, 1967, by permission.

evaporation is 125 Btuh per person, and 860 grains of moisture are evaporated per hour per person. Therefore the latent-heat loss from 100 people is

$$125 \times 100 = 12{,}500 \text{ Btuh} \qquad \textit{Ans.}$$

and the amount of moisture evaporated is

$$860 \times 100 = 86{,}000 \text{ grains}$$

Sensible heat loss = total loss − latent heat loss, or in this case,

$$40{,}000 - 12{,}500 = 27{,}500 \text{ Btuh}$$

Therefore the sensible heat loss in relation to the total heat loss is, the sensible heat ratio

$$\text{SHR} = \frac{27{,}500}{40{,}000} \times 100 = 68.7 \text{ per cent} \qquad \textit{Ans.}$$

Physiological Reactions to the Environment 383

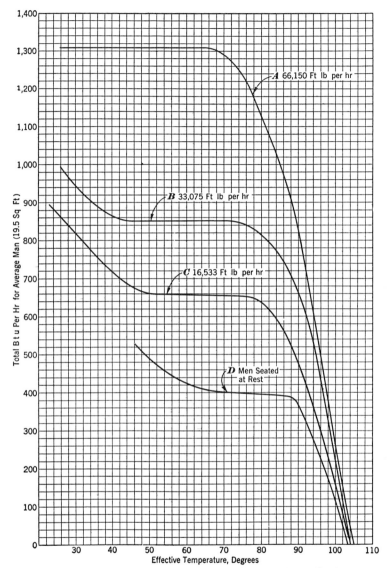

Fig. 10-2. Total heat loss from human body at various effective temperatures in still air. Curve *A*: men working 66,150 ft-lb per hr. Curve *B*: men working 33,075 ft-lb per hr. Curve *C*: men working 16,538 ft-lb per hr. Curve *D*: men seated at rest. (From *Trans. ASHVE.*)

Example 10-3. For a man doing light work when the temperature is 75 F db and 68 F wb, (a) find the heat loss in still air. (b) Separate total loss into loss by evaporation and loss by sensible heat. (c) Find also the moisture evaporated per hour.

Solution: (a) Except at temperatures outside the equilibrium range, dry bulb

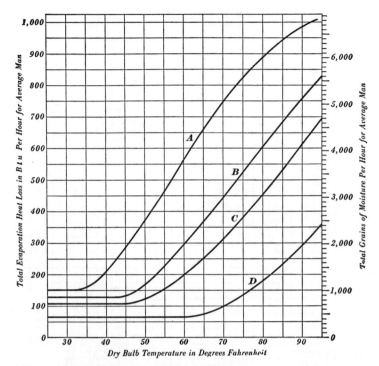

FIG. 10-3. Heat loss by evaporation, and moisture evaporated from the human body at different dry-bulb temperatures for still air conditions. Curve A: men working 66,150 ft-lb per hr. Curve B: men working 33,075 ft-lb per hr. Curve C: men working 16,538 ft-lb per hr. Curve D: men seated at rest. (From *Trans. ASHVE*.)

temperatures can be used in Fig. 10-2 and for 75 F, curve C for moderate work show 655 Btu per hr body loss in still air.

b) Fig. 10-3 shows the evaporative loss as 380 Btuh per hr at 75 F. Fig. 10-4 shows the sensible heat loss as 280 Btu per hr. This total of 660 Btuh is in reasonable agreement with that found in part (a).

c) When the evaporation heat loss is expressed in Btuh, as for example the 380 in this problem, the moisture evaporated can be found by dividing by 1050 Btu per lb (a representative h_{fg} for the evaporation of steam at temperatures around 70 F): 380/1050 = 0.362 lb, or 0.362 × 7000 = 2530 grains of moisture evaporated per hour. This result closely checks with the value of 2560 grains read directly from Fig. 10-3.

Ans.

The moisture evaporated from the surface of an individual is associated with a latent heat interchange, and a simplified formula to express this change can be developed if it is realized that, under normal atmospheric conditions, at about 70 F to 80 F the latent heat of vaporization of water is approximately 1050 Btu per lb. Thus, since 7000 grains equals one pound, the cooling provided the body by evaporation appears:

Evaporation loss = Btuh × $\frac{7000}{1050}$ = grains per hour

Physiological Reactions to the Environment

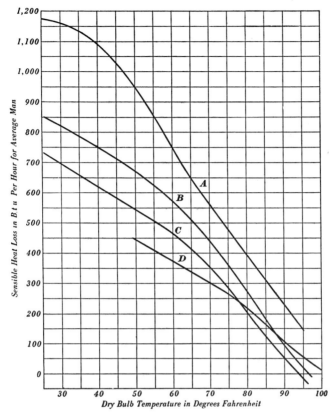

Fig. 10-4. Sensible heat loss from the human body at various dry-bulb temperatures in still air. Curve A: men working 66,150 ft-lb per hr. Curve B: men working 33,075 ft-lb per hr. Curve C: men working 16,538 ft-lb per hr. Curve D: men seated at rest. (From *Trans. ASHVE*.)

or

$$\text{Evaporation loss} = \text{Btuh} \times \tfrac{20}{3} = \text{grains per hour} \qquad (10\text{-}2)$$

Conversely,

$$\text{Grains per hour} \times \tfrac{3}{20} = \text{Btuh evaporation loss} \qquad (10\text{-}3)$$

or, since the grain is disappearing from use as a unit of mass,

$$\text{Body evaporation loss} = \text{Btuh} \times \frac{1}{1050} = \text{lb per hour} \qquad (10\text{-}4)$$

$$\text{Btu per hour, evaporative body loss} = 1050 \times \text{lb per hour} \qquad (10\text{-}5)$$

10-4. THE ENVIRONMENT UNDER STRESS CONDITIONS

Many conditions arise in industry where it is impossible to establish comfort conditions. Efforts must then be made to alleviate conditions and, in any event, to recognize the difficulties and even the changes which can arise under such conditions. For a high-temperature space the approaches that are used involve: bringing in large amounts of cooler air; insulating heat sources to the greatest possible extent; keeping humidity as low as possible by venting all sources of steam; providing minimum clothing and sometimes even insulated clothing; and, from a protective viewpoint, limiting the time of exposure that is allowed for workers in the spaces. Conditioning the men to become heat adapted to the operating conditions is also important for work in moderately hot areas where the duration period may be extended.

From the viewpoint of the individual, the time of exposure, the temperature, and the humidity in the space are important factors. It is obvious that if the space conditions are extremely hot the deep-tissue temperature of the individual rises, and exposure must be stopped before the danger point is reached. Under high temperature conditions, both the body temperature and the pulse rate rise. It is considered unsafe to let the pulse rate of an individual increase to more than 130 above the normal 75 and the rectal temperature should not be permitted to exceed 101 F when exposures exceed one hour.

High space temperatures are not unsafe if the relative humidity is low. For example, exposures of up to 0.5 hour are not dangerous even at 135 F provided the relative humidity is in the 30 per cent or lower range. On the other hand, at about 90 per cent humidity, temperatures appreciably above 100 F are not safe even for short periods. It is obvious too, that the internal metabolic rate is a factor in these decisions and, for a person performing work, the time of exposure must necessarily be much shorter than for a person at rest.

No completely satisfactory indices for measuring body performance under temperature stress conditions have been developed although a number are in use. One of these is the Effective Temperature (ET) which has already been described. It has been extended into the hot region, but exact physiological responses are not well established (refs. 8 and 9). Two other indices provide data, namely the predicted Four-Hour Sweat Rate (P4SR), and the Belding-Hatch Heat Stress Index (refs. 10 and 11).

Additional work is being carried out to obtain answers to this complex problem. Difficulty arises in carrying out experimentation with human subjects under extremely uncomfortable conditions and it is also true that danger is involved since limiting conditions could be exceeded with temporary or permanent damage to a subject.

10-4. REQUIREMENTS FOR QUALITY AND QUANTITY OF AIR

The air in an occupied space should at all times be free from toxic, unhealthful, or disagreeable fumes, and also should be relatively free from

TABLE 10-4
VENTILATION STANDARDS

APPLICATION	SMOKING	AIR PER PERSON (CFM)		MINIMUM AIR PER SQ FT OF FLOOR (CFM)
		Recommended	Minimum	
Apartment				
Average................	Some	20	15	...
De luxe................	Some	30	25	.33
Banking space............	Occasional	10	$7\frac{1}{2}$	...
Barber shops.............	Considerable	15	10	...
Beauty parlors............	Occasional	10	$7\frac{1}{2}$	...
Brokers' board rooms......	Very heavy	50	30	...
Cocktail bars.............	Heavy	30	25	...
Corridors (supply or exhaust)...	.....	...	...	.25
Department stores.........	None	$7\frac{1}{2}$	5	.05
Directors' room...........	Extreme	50	30	...
Drug stores...............	Considerable	10	$7\frac{1}{2}$	...
Factories.................	None	10	$7\frac{1}{2}$	.10
Five-and-ten-cent stores....	None	$7\frac{1}{2}$	5	...
Funeral parlors............	None	10	$7\frac{1}{2}$	...
Garage...................	.....	...	...	1.0
Hospitals				
Operating rooms *.......	None	...	...	2.0
Private rooms...........	None	30	25	.33
Wards..................	None	20	15	...
Hotel rooms..............	Heavy	30	25	.33
Kitchen				
Restaurant.............	.....	...	...	4.0
Residence..............	.....	...	...	2.0
Laboratories..............	Some	20	15	...
Meeting rooms............	Very heavy	50	30	1.25
Office				
General................	Some	15	10	...
Private................	None	25	15	.25
Private................	Considerable	30	25	.25
Restaurant				
Cafeteria..............	Considerable	12	10	...
Dining rooms..........	Considerable	15	12	...
Schoolrooms..............	None	...	...	...
Shop, retail..............	None	10	$7\frac{1}{2}$	...
Theater..................	None	$7\frac{1}{2}$	5	...
	Some	15	10	...
Toilets (exhaust)..........	.....	...	...	2.0

* Use outside air only, to overcome explosion hazard of anesthetics.

odors and dust. To obtain these conditions, enough clean outside air must always be supplied to an occupied space to counteract or adequately dilute the sources of contamination. In industrial processes, undesirable and sometimes toxic gases may arise, but in air conditioning for human occupancy alone, the problem is largely one of supplying enough air to keep odors from arising to the point of being disagreeable.

The concentration of odors in a room depends upon numerous factors, including the dietary and hygienic customs of the occupants, the type and amount of outdoor air supplied, the room volume per occupant, and the types of odor sources. In general, where smoking is not a problem, 7.5 to 30 cfm per person will take care of all conditions. Where smoking occurs, additional outside air is necessary to counteract the effect of the smoke. It is usually considered that where smoking is taking place, not less than 15 cfm per person is desirable, and 50 cfm may be necessary. There are no absolute standards as to ventilation, although certain local codes must be satisfied and in this sense these do represent standards. Table 10-4 indicates ventilation-practice standards which have been found generally satisfactory, and these can be employed advantageously provided they do not conflict with local codes. In the case of buildings or residences where infiltration is an important factor, it may be necessary to estimate the probable air changes per hour. Table 10-5 gives data on the total air change which might be expected in certain types of building construction.

The total quantity of outside air which passes through a building or space is controlled chiefly by physical considerations concerning tempera-

TABLE 10-5
PROBABLE BUILDING AIR CHANGES PER HOUR BY NATURAL EFFECTS

Type of Building	Winter		Summer	
	Min	Max	Min	Max
Ordinary factory................	$\frac{1}{2}$	2	$\frac{1}{2}$	1
Poor frame construction..........	$1\frac{1}{2}$	3	1	$1\frac{1}{2}$
Corrugated metal................	$1\frac{1}{2}$	3	1	$1\frac{1}{2}$
Stores and shops................	$\frac{1}{2}$	1	$\frac{1}{2}$	1
Commercial buildings............	$\frac{1}{2}$	$1\frac{1}{2}$	$\frac{1}{2}$	1
Office building..................	$\frac{3}{4}$	$1\frac{1}{2}$	$\frac{1}{2}$	1
Residence				
Well-constructed...............	$\frac{1}{2}$	1	$\frac{1}{4}$	$\frac{1}{2}$
Poorly constructed.............	$1\frac{1}{2}$	2	$\frac{3}{4}$	$1\frac{1}{2}$

NOTE. In commercial buildings with doors leading to nonconditioned areas, add 100 cfh for each person entering or leaving. For a 36-in. door standing open, allow 48,000 cfh. Revolving doors can be calculated at 60 cu ft per person passing through, or if the door is equipped with a brake, 50 cu ft per person.

Spaces like auditoriums, clubrooms, dance halls, and theaters usually take enough outside air to offset infiltration.

Physiological Reactions to the Environment

ture, type of air-distribution system, and air velocities. However, the type and usage of a building, the floor area, the height of room, the window area, and the type of occupancy affect the operation of the air-distribution system. In most air-conditioning systems a large amount of air is recirculated over and above the amount required to satisfy minimum ventilation conditions in regard to odor and purity.

The olfactory organs in the upper nasal cavity are extremely sensitive to very slight concentrations of odoriferous matter. However, in the presence of a given odor the sensitivity of these organs gradually lessens, but can quickly revive if the person goes to a place where the odor in question is not present and then returns. That is, breathing fresh air restores olfactory sensitivity. People on entering a place may be struck by a strong odor which the occupants do not even notice.

With recirculated air, the washing, humidifying, and dehumidifying remove a considerable proportion of body odor and may keep the outside-air requirements close to a basic minimum value of 5 cfm. For general application a minimum of 10 cfm of outside air per person, mixed with perhaps 20 cfm of recirculated air, is a good general rule.

Formerly, the carbon dioxide content in an occupied space was thought to be the ultimate index of ventilation conditions, but this index has been found to be unreliable in so many cases that its use is diminishing. A normal adult at rest, in breathing, exhales air containing about 3.5 to 4.1 per cent of carbon dioxide by volume, or about 0.5 to 0.6 cfh. Combustion processes, gas burners, open fires, and tobacco fumes also contribute to the CO_2 content of a given space.

The action of CO_2 on the human system in small quantities is harmless, but in larger quantities (over 2 per cent by volume; that is, over 200 parts in 10,000) its effect in diluting the oxygen content of the air makes breathing more rapid and causes some discomfort. With 6 per cent CO_2, breathing is very difficult; and at 10 per cent, a loss of consciousness usually results, but is not necessarily fatal. Atmospheric air contains about 3 to 4 parts of CO_2 by volume in 10,000, and the air inside a ventilated occupied space varies from about 6 to 14 parts. There are various types of apparatus on the market available for measuring these small quantities of CO_2 in the air, among which should be mentioned the Petterson-Palmquist apparatus and the Haldane apparatus.

Example 10–4. In a given occupied room 15 cfm of outside air per person is continuously circulated. If the outside air contains 300 parts CO_2 per 1,000,000 by volume, what CO_2 content exists in the room, assuming equilibrium has been established?

Solution: Each person contributes 0.5 to 0.6 (say 0.55) cu ft of CO_2 per hour to the air. During each hour, $15 \times 60 = 900$ cu ft of air is supplied, carrying

$$900 \times \frac{300}{1,000,000} = 0.27 \text{ cu ft } CO_2 \text{ per hr}$$

After equilibrium is reached, the total CO_2 associated with the air is the sum of the outside-air CO_2 plus the human CO_2, or $0.27 + 0.55 = 0.82$ cfh. This amount of CO_2 is associated with the 900 cu ft of ventilation air and therefore shows 0.82 parts of CO_2 per 900 parts of air, or

$$\frac{0.82}{900} \times \frac{1,000,000}{1,000,000} = \frac{9.1}{1,000,000} \qquad Ans.$$

That is, 910 parts of CO_2 per million exist in the room.

Example 10–5. An auditorium seating 1500 people is to be kept at 78 F db and 66 F wb when the outdoor temperature is 90 F db and 74 F wb. The heat gain through the insulation, etc., amounts to 120,000 Btuh in addition to the sensible and latent heat gain from the occupants. Find (a) the outdoor air required for ventilation; (b) the sensible heat load from the occupants; (c) the latent heat load from the occupants; (d) for the total sensible heat load, the number of pounds and cubic feet of conditioned air at 65 F db that must be circulated; (e) the wet-bulb temperature that is required in order to pick up the moisture load.

Solution: (a) From Table 10–4, considering the auditorium as a theater, $7\frac{1}{2}$ cfm per person is taken. This figure should be greatly increased if much smoking occurs. Some additional recirculation of the auditorium air is necessary, and its treatment in the conditioner washer will remove some odors, thus justifying the rather low figure of $7\frac{1}{2}$ cfm. Thus the amount of air supplied to the auditorium from the outside is

$$7\frac{1}{2} \times 1500 = 11{,}250 \text{ cfm}$$

If this air is measured at inside conditions of 78 F db and 66 F wb the specific volume is 13.8 cu ft per lb (Plate I). Therefore the outside air required is

$$11{,}250 \times \frac{1}{13.8} = 8.5 \text{ lb per min} \qquad Ans.$$

b) Using Table 10–3 read 245 Btuh sensible heat load per person. Therefore the sensible heat load from all occupants is

$$245 \times 1500 = 367{,}500 \text{ Btuh} \qquad Ans.$$

c) Using Table 10–3 read 105 Btuh latent heat loss per hour per person. Thus the latent heat load from the occupants is

$$105 \times 1500 = 157{,}500 \text{ Btuh} \qquad Ans.$$

Compute the moisture load by equation 10–4,

$$\frac{157{,}500}{1050} = 150 \text{ lb per hour}$$

d) Total sensible heat load $= Q_s = 120{,}000$ building $+ 367{,}500$ human $= 487{,}500$ Btuh. By equation 3–36,

$$Q_s = mC_p(t_i - t_c)$$

Substituting, we have

$$487{,}500 = m(0.244)(78 - 65)$$

Therefore, for the sensible heat load, the amount of air that must be circulated is

$$m = 153{,}600 \text{ lb per hr} \qquad Ans.$$

Physiological Reactions to the Environment 391

From the psychrometric chart, air at 65 F and at an estimated relative humidity of 70 per cent shows a specific volume of 13.4 cu ft per pound of "dry" air. Therefore, in cubic feet per minute, the amount of air circulated is

$$\frac{153,600}{60} \times 13.4 = 34,300 \text{ cfm} \qquad Ans.$$

e) The amount of moisture which must be picked up per pound of "dry" air circulated is

$$\frac{150}{153,600} = 0.00098 \text{ lb moisture per lb dry air}$$

Therefore, moisture per pound of air is:

Inside air at 78 F and 66 F.......... 0.01100 (Chart No. 1)
Less moisture to be absorbed........ 0.00098
Moisture content of entering air.... 0.01002 lb

From Chart No. 1, air at 65 db and 0.010 lb per pound has a 60.0 F wet bulb. *Ans.*

10–6. INDUSTRIAL CONTAMINANTS

Industrial activities and operations frequently produce substances which disseminate or become dispersed in the air, and, as such, these substances constitute contaminants or atmospheric impurities. Some of these substances are toxic in nature, others are relatively harmless. Some of them are potential fire hazards in certain concentrations, others are completely non-combustible.

It is, of course, most desirable to reduce the amount of contaminant at the source. If this cannot be done, solid particles such as dusts should be caught in separators and filters, or otherwise exhausted from the working area into the open atmosphere where dilution can take place to such an extent that the contaminant no longer represents a potential hazard. The process of exhausting is usually the only practicable method to follow with objectionable vapors and gases. Where it is not possible to remove completely the contaminant by exhaust ventilation, it then becomes necessary to supply dilution ventilation air in such quantities that the contaminant is reduced to a point where it is below a threshold danger level. Increasingly more attention has to be given to trapping vapors, fumes, and dusts at their source point to prevent the ever-increasing level of air pollution.

Highly toxic gases may be harmless to humans provided they appear in the atmosphere only in trace amounts. It is of importance then to know the amount of hazardous material which can exist in an atmosphere without causing significant harm or injury to those who must work or live in an atmosphere containing a hazardous material. It is clearly recognized that breathing air containing certain chemicals, in high concentration can be lethal in a short period of time. However, as concentrations of such chemicals are reduced, the time of exposure can be increased before lethality develops, and, if this thought is further developed, it is easy to conceive a condition where the concentration of the hazardous material may be so low as to cause no damage at all. The term *Threshold Limit* and the similar terminology *Maximum Allowable Concentration* imply that if the concentration of a hazardous material is kept below a certain level, activities in a contaminated space can be carried on without danger and often without annoyance to the occupants of that space.

It is most essential that in determining threshold limits the medical and physiological aspects of a particular contaminant be rigorously investigated, particularly in the case of toxic materials which might have cumulative effects in the human system and for which no immediate physiological reactions can be determined. The threshold limit for a given contaminant is related to a less serious response than lethality. The determination of threshold limits for a given chemical involves related experience with other chemical hazards, with animal testing, with statistical studies, and with practice over a period of years.

It is customary to express Threshold Limits for hazardous substances (Ref.12) as PPM (parts per million), a volumetric evaluation, or as mg per cu m (milligrams per cubic meter). To transform from one to the other for vapors in air it is convenient to make use of the mol volume which in the metric system is 24.0 liters at 20 C (68 F) and 760 mm Hg atmospheric pressure, or 22.4 at 0 C, 24.4 at 25 C (77 F). Using M as the molecular weight, at 760 mm Hg pressure and at 25 C, we can write

$$\frac{mg}{liter} \times \frac{1}{1000} \times \frac{24.4}{M} \times 1{,}000{,}000 = PPM$$

$$\frac{mg}{liter} = \frac{PPM}{1000} \times \frac{M}{24.4} \qquad (10\text{-}6)$$

and

$$PPM = \frac{24{,}400}{M} \frac{mg}{liter} \quad \text{at 25 C, 760 mm Hg} \qquad (10\text{-}7)$$

$$\text{Percent by volume} = \frac{PPM}{1{,}000{,}000} \times 100 \qquad (10\text{-}8)$$

Example 10–6. The threshold limit for acetone in occupied space is 1000 parts per million at 25 C (77 F). Convert this limit to milligrams per liter and milligrams per cu meter.

Solution: Acetone, C_3H_6O, has a molecular weight of 58.1. By equation 10-6

$$\frac{mg}{liter} = \frac{1000}{1000} \times \frac{58.1}{24.4} = 2.38 \qquad Ans.$$

$$\frac{mg}{cu\ meter} = 2.38 \times 1000 = 2380 \qquad Ans.$$

Physiological Reactions to the Environment

also

$$\frac{lb}{cu\ ft} = \frac{mg}{cu\ meter} \times \frac{1}{16 \times 10^6} = \frac{2380}{16 \times 10^6} = 0.000149$$

PROBLEMS

10-1. Find the effective temperatures for the following conditions: (a) 70 F db, 55 F wb; (b) 74 F db, 60 per cent rel humidity; (c) 78 F db, 68 F wet bulb. (d) Which of these are in the comfort zone for inactive conditions and 4 hour occupancy? (e) Which of these are in the comfort zone under short-term occupancy (20 minutes)?
Ans. (a) 65.1°; (b) 70.0°; (c) 72.7°; (d) cool, slightly cool, comfortable; (e) cool, comfortable, slightly warm

10-2. The air in a cinema theater, with an audience of 1000 people during an evening performance is kept at a dry-bulb temperature of 72 F and at 60 per cent relative humidity. Determine (a) whether room conditions are in the comfort zone; (b) the heat given up to the air by the occupants (use Table 10-3); (c) the portion of heat given up as latent heat by moisture evaporated into the air; and (d) the portion of heat given up as sensible heat.
Ans. (a) comfortable in winter; (b) 350,000 Btuh; (c) 105,000 Btuh; (d) 245,000 Btuh

10-3. How much air at 70 F and 45 per cent relative humidity would have to be supplied to a classroom containing 40 students if the room is to be kept at 74 F and not higher than 50 per cent relative humidity? Neglect insulation losses.
Ans. 11,000 lb/hr

10-4. Three men are doing moderate work in a room at 80 F db. (a) Using Table 10-3, how much heat is given off by them? (b) What portion of the total heat dissipation is by evaporation? (Use the sensible-latent ratio of Table 10-3); (c) How much heat is given off as sensible heat? (d) Find the moisture evaporated per hour.
Ans. (a) 2700 Btuh; (b) 730 Btuh latent; (c) 970 Btuh; (d) 1.64 lb

10-5. The threshold limit for SO_2 in air is considered to be 5 parts per million by volume while for hydrogen sulfide (H_2S) the value is 15 ppm, with both determined at 25 C and 760 mm. Compute the threshold concentrations expressed in milligrams per cu meter, atomic weight of S, 32; O, 16; H, 1.
Ans. 13, 15

10-6. How many parts of CO_2 per 10,000 by volume exist in an occupied room which is supplied with 10 cfm per person if the outside air contains 4 parts CO_2 per 10,000 by volume?
Ans. 13.2

10-7. At a certain time a residence having over-all dimensions of 45 ft by 25 ft by 18 ft has 1½ air changes per hour from infiltration. Assume that the residence is occupied by five people and compute the ventilation air from natural effects, expressed in cubic feet per minute per person.
Ans. 101 cfm

10-8. A hospital ward 130 ft by 22 ft by 10 ft occupied by a total of 30 persons (patients and staff) is to be kept at 79 F db and 68 F wb when the outdoor temperature is 95 F. The heat gain from outside amounts to 150,000 Btuh in addition to the heat gain from the occupants. Find (a) the minimum volume of outdoor air recommended for ventilation; (b) the sensible heat load from the occupants; (c) the latent-heat and moisture load from the occupants. (d) For the total sensible heat load, how many pounds per hour of conditioned air at 65 F db must be circulated? (e) What should be the wet-bulb temperature of this conditioned air in order for the air to pick up the moisture load? (f) What is the required capacity of the fan in cubic feet per minute?

10-9. A small medical center has a floor area of 80 ft by 80 ft and is three stories high, with a total height of 30 ft. (a) On the basis of $1\frac{1}{2}$ air changes per hour, what ventilation capacity in cubic feet per minute would apply for this building, and (b) if the maximum occupancy of this building at a given time is 80 adults, would this air flow satisfy the minimum ventilation standard? Smoking will probably be at a minimum throughout this building. *Ans.* (a) 4800 cfm; (b) yes

10-10. The municipal codes of certain cities specify that certain minimum amounts of air must be supplied to buildings of varying types. Assume that for the medical center discussed in problem 10-9 the municipal code specifies a minimum of 0.25 cfm per square foot of floor area. (a) For the three stories in question, compute the minimum ventilation requirements according to code. (b) Compare these with the data of the previous problem. *Ans.* (a) 4800 cfm; (b) same

10-11. A certain dance hall at peak occupancy serves 150 couples. The lighting load consists of 4000 watts. The space is not air-conditioned, but 8 fans with $\frac{1}{4}$-hp motors are used to circulate air in the space. Compute (a) the total sensible heat load and (b) the total latent heat load. Note that 1 watt = 3.41 Btuh and that 1 hp = 2545 Btuh. (The efficiency of small motors may be as low as 40 per cent.)

10-12. In winter, the hospital ward with 30 occupants mentioned in problem 10-8 has a heat loss of 220,000 Btuh at design conditions. It is presumed that natural ventilation effects cause $1\frac{1}{2}$ changes per hour if the ventilation system is not used. However, the ventilation-air-conditioning system is usually operated in winter as well as summer and offsets some of the infiltration air leakage. In winter 15 cfm of outside air per person, along with 20 cfm of recirculated air per person, are supplied at 88 F and enter the floor returns at 68 F. (a) How much heat is supplied by the ventilating-system air, and how much additional heat is required from steam radiation when the maximum winter design load occurs? (b) If the ventilation system is not in operation, what is the probable amount of outside air per person arising from natural ventilation effects?
Ans. (a) 196,800 Btuh from radiators; (b) 23.8 cfm

10-13. The threshold limit for ammonia NH_3 is reported as 35 mg per cu meter. Compute this concentration in units of parts per million measured at 25 C and 760 mm of mercury pressure.

10-14. For the ammonia of problem 10-13 compute its threshold concentration in units of parts per million at 20 C and 760 mm Hg. Also express in lb per cu ft.

REFERENCES

1. F. C. Houghten and C. P. Yaglou, "Determining Lines of Equal Comfort and Determination of the Comfort Zone," *Trans. ASHVE*, Vol. 29 (1923), pp. 163, 361.

2. C. P. Yaglou and W. E. Miller, "Effective Temperature with Clothing," *Trans. ASHVE*, Vol. 31 (1925), p. 89.

3. N. Glickman, T. Inouye, R. W. Keeton, and M. K. Fahnestock, "Physiological Examination of the Effective Temperature Index," *Trans. ASHVE*, Vol. 56 (1950), p. 51.

4. B. H. Jennings and B. Givoni, "Environment Reactions in the 80 F to 105 F Zone," *Trans. ASHRAE*, Vol. 65 (1959), p. 115.

5. W. Koch, B. H. Jennings, and C. M. Humphreys, "Environmental Study II–Sensation Responses to Temperature and Humidity under Still-Air Conditions in the Comfort Range," *Trans. ASHRAE*, Vol. 66 (1960), pp. 264-287.

6. R. G. Nevins, F. H. Robles, W. Springer, and A. M. Feyerherm, "A Temperature-humidity Chart for Thermal Comfort of Seated Persons," *ASHRAE Journal*, Vol. 8 (1966), p. 61.

7. ASHRAE *Handbook of Fundamentals*, Chapter 7, "Physiological Principles," pp. 111-126, New York, 1965.

8. C. M. Humphreys, Oscar Imalis, and Carl Gutberlet, "Physiological Response of Subjects Exposed to High Effective Temperatures and Elevated Mean Radiant Temperatures," *Trans. ASHVE*, Vol. 52 (1946), p. 153.

9. L. W. Eichna, W. F. Ashe, W. B. Bean, and W. B. Shelley, "The Upper Limits of Environmental Heat and Humidity Tolerated by Acclimatized Men Working in Hot Environments," *The Journal of Industrial Hygiene and Toxicology*, Vol. 27 (March, 1945), p. 59.

10. C. J. Wyndham, W. Bouwer, M. G. Devine, H. E. Patterson, and D. K. C. MacDonald, "Examination of Use of Heat Exchange Equations for Determining Changes in Body Temperature," *Journal of Applied Physiology*, Vol. 5 (1952), p. 299.

11. H. S. Belding and T. F. Hatch, "Index for Evaluating Heat Stress in Terms of Resulting Physiologic Strains," *Heating, Piping and Air Conditioning*, (August, 1955), p. 129.

12. Air Pollution Engineering Manual, *Public Health Service Publication*, No. 999-AP-40 (Cincinnati, Ohio, 1967), pp. 872-878.

11

Warm Air Heating

11-1. TYPES OF WARM-AIR HEATING SYSTEMS

The general term *warm-air heating* implies that heat transfer is directly from fire-heated surfaces to the air which is used in the heated space. Any stove is a warm-air heater, and any stove becomes a furnace when it is jacketed and provided with ducts so as to serve a heated space other than the one in which the furnace is placed.

The term *gravity warm-air heating*, or the synonymous term *thermal-circulation warm-air heating*, means that natural convection is the method of circulating the heated air. When air inside a duct is heated, the air expands and the weight per unit of volume becomes less than that of the surrounding cooler air. Unless the duct is horizontal, the relatively light warm air rises, and dense cool air flows in to take its place. The circulation is roughly proportional to the difference in temperature between the warm air and cool air. If a warm-air furnace is placed at the bottom of an air-duct riser, a thermally-created air movement commences.

Forced-circulation warm-air heating means that mechanical energy, rather than thermal energy, does the work of moving the air which transports the heat. The power is usually furnished by a fan. Better all-around performance is possible with mechanical circulation. Such systems are also known as *mechanical warm-air systems*.

In all warm-air heating systems, the warm air which is supplied to a space, in cooling to room conditions, counteracts the heat loss from the space.

11-2. WARM-AIR FURNACES

Warm-air furnaces can employ oil, gas, or hand- or stoker-fired solid fuels. With warm-air furnaces, it is difficult to transfer heat effectively from the firebox or combustion chamber to the air passing over the hot surfaces. Also, because radiant heat is not readily absorbed by an air stream, it is a necessity to provide secondary surface to contact and warm the air stream. Warm-air furnaces generally have metal casings, though brick encasements are sometimes employed.

Large warm-air furnaces, frequently installed in batteries, are in use

Warm Air Heating

for public buildings. These employ a fan to deliver air under pressure so that it passes upward along the combustion-chamber sides and then is deflected against secondary heat-transmitting surfaces. A battery of furnaces is usually arranged so that in mild weather some of the furnaces need not be fired.

Insofar as the heating of a building is concerned, it is essentially immaterial whether gas, oil, or solid fuel burned on a stoker, or solid fuel burned on hand grates, is employed. The availability of fuel, and the economics of the situation, determine which source of energy is to be used in a given installation.

Figure 11-1 shows an oil-fired warm-air furnace. This is a forced-circulation furnace, with the fan blower shown at the lower right. Air comes in through the passageway at the top right, passes down, and is warmed by the hot smoke pipe. It is then drawn through the filters into the fan and is delivered into the heating chamber, where it comes into con-

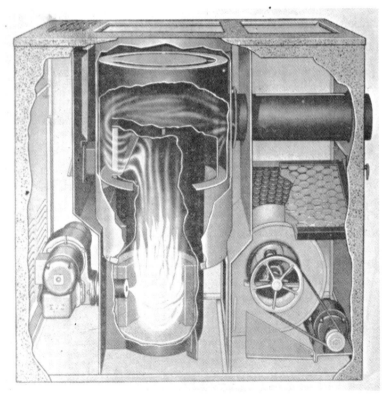

FIG. 11-1. Mechanical-circulation, oil-fired warm-air furnace. (Courtesy Ex-Cel Company.)

tact with the hot metal surfaces warmed by combustion of the oil. The passageway through the heater section is designed to bring the air into intimate contact with the furnace surfaces. The warmed air leaves through the top of the furnace into an attached duct system, and a humidifier, not shown, is usually provided near the inlet to the duct system.

Gas-burning, warm-air furnaces resemble oil-burning furnaces and are almost universally installed with fans for forcing air circulation and distribution in the rooms. They may be of one section, or have a number of sections placed side by side.

11–3. RECIRCULATION AND YEAR-ROUND OPERATION

If full recirculation for all of the furnace-warmed air throughout the heated space is employed, maximum economy will be obtained, especially in extremely cold weather. Forced-circulation warm-air systems can provide for summer operation by the addition of cooling coils so placed that chilled air can be delivered to the rooms of a building, using the same duct work employed for heating.

In thermal-circulation systems, the vertical ducts which are built in the walls and partitions for conveying the warm air are called stacks. In cross section they can be round, but most usually they are rectangular. A stack head is the duct-fitting at the room end of a duct, and this, in turn, houses the register or grille when the location is in a wall. When the outlet is in the floor, the stack terminates in a register box. In such systems, the fitting at the bottom of a stack which connects the round leader into the vertical stack is called a boot. A leader is the duct which runs almost horizontally from the furnace to the stack.

A natural-circulation system depends for its action on a modification of chimney action whereby the pressure exerted by a vertical column of warm air is less than that exerted by a similar column of cool air. A pressure difference created in this way is sufficient to promote the required circulation of air. Theoretical analyses starting from this point and making use of the space heat loss and the required air flow, and of the friction loss in ducts, could furnish a basis for design, but they would be extremely laborious and their accuracy would be dependent on the assumptions made. Consequently, it is preferable to make use of design procedures confirmed by experience and successful designs. The design procedure will not be included here but is presented in detail in publications of the National Warm Air Heating and Air Conditioning Association (Refs. 1, 2, and 5).

11–4. FORCED-CIRCULATION WARM-AIR SYSTEMS

The design of ducting for a forced-circulation system has also been developed into a simplified procedure, by the National Warm Air Heating and Air Conditioning Association (Refs. 3 to 7) and will not be reported here. For residential structures or small buildings having design heat losses not greatly in excess of 120,000 Btuh, the temperature rise occurring in the furnace should not exceed 100 deg F and the static pressure available for overcoming pressure runs about 0.2 in. of water. The furnace selected should have a bonnet capacity about 15 per cent in excess of the design Btuh heat loss (register delivery) to take care of piping loss.

When mechanical air circulation is used, rectangular trunk-line ducts for leaders are employed almost universally, since the air velocity to each branch is independent of variations in air temperature, and since such ducts need not require a particular inclination. With trunk-line air-supply ducts there is less exterior duct surface and less heat loss than with separate, thermal-circulation supply ducts.

In general, a building is heated most satisfactorily by a forced-circulation system when the blower operates for long periods in mild weather and practically continuously in weather colder than about 40 F.

The heat input should be so controlled that in mild weather the burner, or draft damper, operates frequently, but only for short periods. Moreover, the blower should operate until the furnace is cooled; otherwise, heated air may circulate by gravity through the nearest registers, or those highest above the furnace.

Frequent cycling of the source of heat, together with prolonged blower operation, insures controlled temperatures in all rooms of the building, whether in the basement, in the first story, or over an unheated garage, and temperatures are maintained at nearly constant values near the floor as well as in the living zone.

With a blower under almost continuous operation, along with intermittent operation of the heat source, the temperature of the air delivered from the registers should range from about 80 F to 150 F. The registers should be located so that the air stream is never discharged directly into space normally occupied by people at rest. Deflecting-type registers control the direction of the air stream and can reduce the air velocity quickly. The use of high sidewall registers is advantageous because there is then least danger of the air stream impinging on an occupant. They are also suitable for summer cooling and do not interfere with furniture placement.

Low sidewall and baseboard registers must be more carefully located, as the problems of obstruction by furniture and impingement of air on occupants represent possible difficulties.

A ceiling register is satisfactory, provided its outlet deflects the air sideways instead of directly downward. Floor registers usually are employed only under large glass exposures.

The design patterns provide for sizing the ducts starting at a distant room and running back through the trunk duct to the furnace and also provide for the layout of the return system.

Return-Air Intakes. Return-air intakes can be located either on inside or outside walls. Where large sources of cold air exist, the intakes should be located nearby. Intake location should be near the floor level, either in the baseboard or as a floor grille, flush with the surface.

It is desirable to put return intakes in each room except bathrooms, closets, lavatories, and kitchens. The intake should be located so as to give the return air a convenient run back to the furnace, and should be so sized that excessive amounts of return air are not drawn across the floor.

11-5. WARM-AIR PERIMETER HEATING

Warm-air perimeter heating is a relatively recent approach to heating systems which use forced warm air. It can be used with almost any type of structure, either residential or industrial, but has been found to be particularly adaptable and effective for buildings without basements. It differs from conventional warm-air heating in duct arrangement and also in that the warm air is always introduced into the heated space at or near the floor along the outside walls, preferably under windows. Moreover, the air is returned to the furnace, usually by means of high-sidewall grilles. By setting up a warm-air blanket over the cold window and wall surface, cool downdrafts are largely eliminated; the floors are characteristically warm for this reason, and also because the heat distribution system is adjacent to the floor or to part of it.

One of the simplest perimeter systems is that required for a structure without a basement, for which two representative layouts are illustrated in Figs. 11-2 and 11-3. Other variations of this are possible—as, for

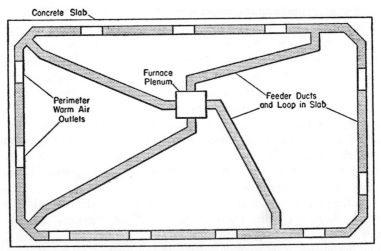

Fig. 11-2. Perimeter loop system with feeders and loop ducts in concrete slab. (After ref. 1.)

Warm Air Heating

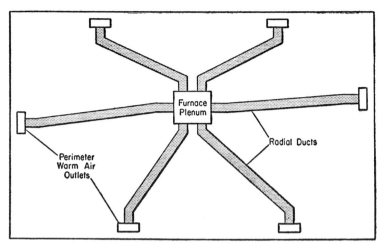

Fig. 11-3. Perimeter radial system with feeder ducts in concrete slab or crawl space. (After ref. 1.)

example, one in which the whole crawl space is enclosed and used for a plenum or pressure chamber to feed up into the outlets in the rooms. Where duct runs are made in the concrete slab around the perimeter of the building, it is desirable to provide waterproof insulation between the slab and the foundation wall. Also, it is necessary to provide a waterproof barrier near the bottom of all slabs resting on the ground, to reduce the amount of water which might be drawn up into the slab and possibly cause a damp floor.

In single-story basementless homes, the furnace is usually located in a heater closet or utility room on the first floor. The furnace delivers its heated air downward into a plenum or ductwork in the floor or crawl space. Figure 11-4 shows one arrangement diagrammatically, with feed into a crawl-space plenum and ductwork. Furnaces can also be placed in attics, in which case a downcomer duct is needed to deliver the hot air into the subfloor plenum. Furnaces should be located in heater compartments or closets, with adequate insulation or cooling air space around the furnace shell. Provision must be made for combustion air grilles into the heater closet, with or without separate air ducting. The discharge gas flue, properly encased to prevent fire hazard, is an additional requirement. In a tightly constructed building with a small air infiltration rate, a separate combustion air duct from outside is a necessity.

One simplified design system for use with residences having a crawl space or basement starts with the premise that a static pressure of 0.2 in. of static water pressure is available at the furnace, and also, that one or more 4-in. round ducts can be used to transmit the warm air to the room diffusers.

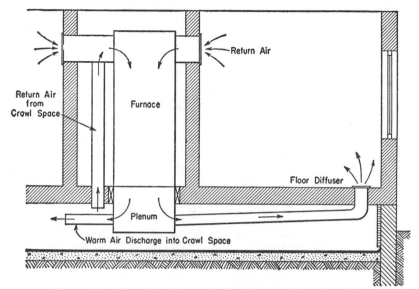

Fig. 11-4. Downflow furnace discharging into ducts located in crawl space. (After ref. 5.)

Two arrangements that can be used in this case are the so-called individual pipe arrangement and the extended-plenum arrangement. The plenum supply in both cases can be served either by downflow or upflow from the furnace. The air temperature rise in furnaces for this type of service ranges from 70 deg to 100 deg and care must be taken to see that enough 4-in. duct outlets are provided to serve the capacity of the furnace. If under certain conditions each such outlet can serve 9000 Btuh, a 75,000-Btuh furnace would require a minimum of nine outlets. More can be used at varying capacities to serve different spaces. The furnace capacity must be sufficient to provide for the heat loss from each room, the loss from the basement or crawl space, and for outside circulation air when that is required because some or all of the delivered air cannot be returned.

In making the design for a given house, the house plan should be drawn to scale, the furnace located, and then the duct runs or plenum accurately drawn to show the lengths to the diffusers and registers, recalling that these are to be located on the outside walls.

Figures 11-5 and 11-6 illustrate the layout of a perimeter system for a small one-story house with a crawl space beneath. This employs a downflow furnace and two return-air runs from the house and an additional one from the crawl space. Heat loads for each part of the house are indicated.

Warm Air Heating

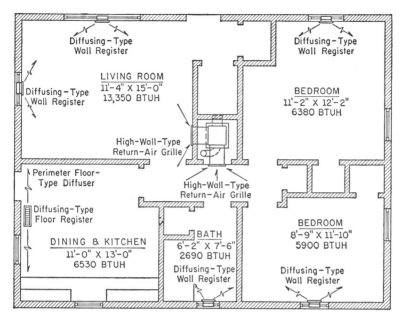

FIG. 11-5. Floor plan of one-story building over crawl space using perimeter heating. (After ref. 5.)

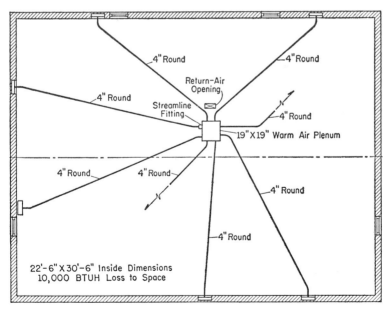

FIG. 11-6. Crawl-space piping under one-story building using perimeter heating. (After ref. 5.)

11-6. PANELS WITH WARM-AIR SYSTEMS

For certain types of installations it has been found desirable to utilize radiant heating in connection with forced-circulation warm-air systems. Most installations of this sort are arranged to operate from the ceiling of the room. The risers carry warm air through a series of passages (ducts) which are so arranged that the warm-air passages cover essentially the whole ceiling area of a room being heated. The passages are insulated on the top and sides, but directly warm the plaster or other finish of the ceiling. From the warm ceiling surface enough heat must be supplied to handle the heat loss of a room. Certain limitations in this connection are necessary, because the temperature at the ceiling is limited. Occupants are not conscious of the source of heat if the ceiling panel surface is kept at a temperature of 115 F or lower. Consequently, if ceiling temperatures in excess of this are required to handle the heat loss, it is necessary to use supplementary heat, insulate the walls, weather-strip the windows, or perhaps even provide double-pane glass.

When the system is in operation, the limit control on the furnace should be so adjusted that the air temperature does not exceed 140 F. The blower control of the furnace should start the blower in operation at 100 F and should stop the blower at about 85 F. The ability of a warm-air system to operate at controlled varying temperatures in terms of the heat load, and the relatively low heat-storage capacity of a warm-air system, indicates the possibility of more-extended use being made of warm-air panel installations. Figure 20-2 shows a typical ceiling panel for a warm-air system. An approximate idea of how much surface is required for room-heating by this method may be gained by assuming that each square foot of heated ceiling surface can supply from 50 to 100 Btuh. The higher values are associated with higher velocities in the ducts. A representative design figure of 60 Btuh per square foot is suggested.

11-7. SUMMARY

In this chapter, methods have been indicated for designing warm-air systems. The heat loss of a space heated by warm air is met by bringing into the space a sufficient quantity of warm air which, in cooling to room temperature, satisfies the loss requirements. The limit setting for air temperature at the bonnet for a low-temperature air system should not exceed 200 F, but the room-air temperature at the registers must be lower than this, particularly in small rooms where the hot air could cause discomfort to the occupants. A temperature not above 150 F, and preferably about 135 F, is suggested.

Register temperatures will be higher with gravity systems than with

Warm Air Heating

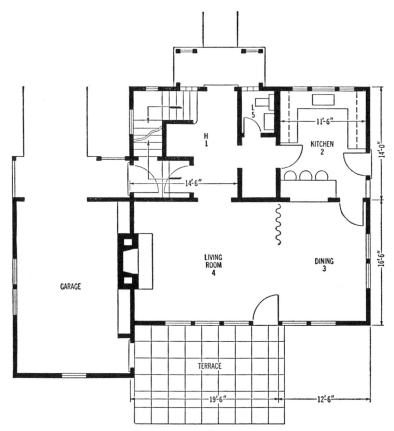

Fig. 11-7. First-floor plan of residence for design.
Data on Residence of Figs. 11-17 and 11-18

Walls—clapboard on $\frac{1}{2}$-in. insulating board; wood lath on studding. $U = 0.23$.
Living room—three metal casement windows; two-section, vertical axis, 37 in. total width of each, $38\frac{3}{8}$ in. high.
Other windows double-hung, wood frame, 30 in. by 48 in.; single thickness glass.
Outside doors—front, 7 ft by 3 ft 4 in.; back and side, 7 ft by 2 ft 8 in.
First story 9 ft high; second story 8 ft 6 in.
Ceiling to unheated attic—yellow-pine flooring on joists; metal lath and plaster with $3\frac{5}{8}$-in. rock-wool fill. $U = 0.079$.
Basement warmed by furnace. The terrace side of the building faces west.

forced-circulation systems. In fact, very satisfying operation has been obtained with a continuous-circulation system, wherein the fan runs most of the time and the air-temperature differential is kept small. The room thermostat controls the firing (combustion), and a blower control switch cuts the fan in or out as the bonnet temperature rises to the fan start-temperature or drops to the fan stop-temperature. These differentials are set close together, and the capacity of the fan is adjusted to provide the proper air flow (usually by changing fan speed through pulleys). The maximum load capacity might give an air-temperature rise of some 100 deg in passing through the furnace. Return-air temperature to a furnace is in the neighborhood of 65 F in residence heating.

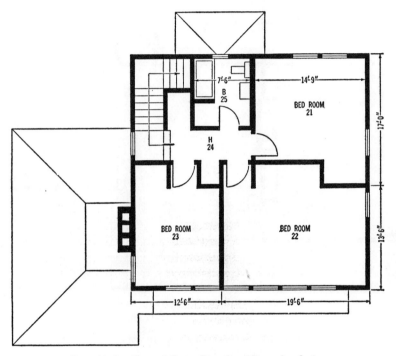

Fig. 11–8. Second-floor plan of residence for design.

The heat given up by the supply air can be expressed by a relation derived from equation 5–5. Thus,

$$Q = (0.244)(d)(60)(\text{cfm})(t_i - 65)$$

and
$$\text{cfm} = \frac{H}{14.7d(t_i - 65)} \qquad (11\text{–}1)$$

in which cfm = air supplied to space, in cubic feet per minute;
Q = heat loss, in Btu per hour;
d = air density at register temperature, in pounds per cubic foot;

Warm Air Heating

$t_i - t_o = t_i - 65$ = the supply or register temperature minus the leaving or return-air temperature, in degrees Fahrenheit.

Example 11-1. A house has a heat loss of 50,000 Btuh. (a) With a gravity system having a 135 F register temperature and a 65 F return, find how many cubic feet per minute (measured at 65 F and 30 per cent relative humidity) must flow through the system. (b) If the heat loss can be carried in a mechanical-circulation system at an average register temperature of 115 F and a 65 F return, find the flow at standard conditions, in cubic feet per minute. (c) How many tons of air per hour flow under the conditions in a and b?

Solution: (a) From the psychrometric chart (Plate I), air at 65 F and 30 per cent relative humidity at 29.92 in. Hg has a density $d = 0.075$ lb per cu ft. By equation 11-1,

$$\text{cfm} = \frac{50{,}000}{(14.7)(0.075)(135 - 65)} = 648 \qquad Ans.$$

Measured at 135 F, this would be closely

$$648 \times \frac{460 + 135}{460 + 65} = 734 \text{ cfm}$$

b) $$\text{cfm} = \frac{50{,}000}{(14.7)(0.075)(115 - 65)} = 907 \qquad Ans.$$

c) For part a,

$$\text{Tons per hour} = \frac{(648)(60)(0.075)}{2000} = 1.46 \qquad Ans.$$

For part b,

$$\text{Tons per hour} = \frac{(907)(60)(0.075)}{2000} = 2.08 \qquad Ans.$$

PROBLEMS

11-1. In a gravity warm-air heating system a room on the second floor of a certain residence has a computed heat loss of 8000 Btuh. The boot to its riser is 12 ft distant from the bonnet of the furnace. Tabular design data show that a 10 in. round leader, a 14 in. $\times$ 3¼ in. stack and a 12 in. $\times$ 8 in. sidewall register could deliver sufficient air at 135 F with a 65 F return to heat the space. Compute the necessary flow into the room in cfm, measured at 135 F.

11-2. Tabular data for the room on the first floor, corresponding to the second-floor room of problem 11-1, also has a heat loss of 8000 Btuh. Tabular design data show that a 12 in. round leader and a 13 in. $\times$ 11 in. side wall register are required. Explain why larger-sized connections are needed for the first floor as compared to the second for the same air flow and heating load.

11-3. In the residence of Figs. 11-7 and 11-8 the computed heat losses in Btuh, for certain design conditions in the different rooms are as follows, by room number: (1) 7200; (2) 6400; (3) 7710; (4) 9400; (5) 1548; (21) 7725; (22) 9373; (23) 6631; (24) 5196; (25) 2195. One of the flues at the left of the living room is the furnace chimney. Place the furnace in the basement near the easterly chimney of the fireplace and using this as a starting point graph the position a warm-air feeder trunk duct might take. Also sketch in the feeder and riser ducts from the trunk for heating the various rooms on both the first and second stories. Locate the outlets under or adjacent to windows on outside walls and more than one may be required for large rooms. Also sketch in a trunk return duct with connections to all major rooms. One central return should suffice for the hall area.

11-4. Consider that the residence of Figs. 11-7 and 11-8 is to be served by an extended-plenum (large trunk duct), with a small-pipe warm-air perimeter system. The furnace is located under the living room close to the innermost chimney, which rises alongside the fireplace. The plenum runs outward under the living room and dining room. From the plenum the small pipes for the different rooms of each floor extend to the outer walls. (a) Make a rough sketch of the house to scale, locate the furnace, the plenum, and the individual 4-in. individual supply runs for the rooms on each floor. Returns should be located on the inner walls but not necessarily one to each room.

11-5. Compute, in cubic feet per minute, the amount of air supplied for heating each first story space of the residence of Figs. 11-7 and 11-8. Use the heat-load data given in problem 11-3. Consider the air supply to be at 140 F and the return air to be at 65 F.

11-6. For a residence with a heat loss of 94,600 Btuh, compute how many cubic feet per minute of standard air and how many tons of air per hour must be circulated if a gravity system with 140 F air temperature is used. *Ans.* 1140 cfm, 2.6 tons/hr

11-7. Solve problem 11-6 but assume that the residence uses a forced-circulation system and that the air-supply temperature is 120 F. *Ans.* 1560 cfm, 3.5 tons/hr

REFERENCES
(Publications of the National Warm Air Heating and Air Conditioning Association)

1. Manual 4. *Warm-Air Perimeter Heating*, 8th ed., 1964.
2. Manual 5. *Gravity Code and Manual for the Design and Installation of Gravity Warm-Air Heating Systems*, 5th ed. 1954.
3. Manual 7. *Code and Manual for the Design and Installation of Warm-Air Winter Air Conditioning Systems.* 1953.
4. Manual 7A. *Code and Manual for Ceiling Panel Systems.* 1950.
5. Manual 10. *Small-Pipe Warm-Air Perimeter Heating.* 1953.
6. Manual 9. *Code and Manual for the Design and Installation of Warm-Air Winter Air Conditioning Systems for Large Structures*, 7th ed., 1960.
7. ASHRAE, *Guide and Data Book—Systems and Equipment*, 1967.

12

Fluid Flow, Duct Design, and Air-Distribution Systems

12-1. BASIC THEORY

In normal flow of a fluid (liquid or gas) in a restraining channel or duct, a drop of pressure occurs. The magnitude of this pressure drop depends on various factors: fluid velocity; diameter or shape of duct section, and condition of its surface; viscosity; density; temperature and pressure of the fluid; heat transfer to or from the fluid; and type of flow, viscous or turbulent. These seemingly numerous variables can be correlated into simple relationships.

When a fluid moves in a pipe or duct, there is always a thin film of the fluid which clings to the side of the pipe and does not move appreciably. In *viscous flow* or *streamline flow*, each particle of the fluid moves parallel to the motion of the other particles. No crosscurrents occur, and the velocity of the fluid particles increases as their distance from the walls of the conduit increases. The maximum velocity occurs at the center of the conduit, and the average velocity over the entire cross section is equal to one-half the maximum. In this viscous flow the pressure drop, after equilibrium of flow is established, is all used in the shearing or sliding of the various layers of fluid against each other. The magnitude of the pressure drop for viscous flow can be calculated by the Poiseuille relationship:

$$\Delta P = \frac{32 \mu L V}{g D^2} \quad (12\text{-}1)$$

The significance of the terms in the foregoing equation will be discussed later, but it should be observed that in viscous flow the pressure drop (ΔP lb per sq ft) is directly proportional to the viscosity (μ lb per ft-sec), to the equivalent length of pipe (L ft), and to the velocity (V fps); and that it is inversely proportional to $g = 32.17$ (the gravitational constant) and to the square of the equivalent diameter of the conduit (D ft).

When the fluid flow in a pipe increases above a certain critical velocity, viscous flow as just described can no longer continue and the flow becomes turbulent. In *turbulent flow* there are numerous eddies and cross currents in the stream and therefore equation 12–1 does not apply. The average velocity over the entire cross section for turbulent flow is usually about 0.8 of the maximum. Turbulent flow is the type of flow most commonly met in engineering practice, and the relationship of equation 12–2 or 12–3 is applicable. The pressure drop, in pounds per square foot, is

or

$$\Delta P = (f)\left(\frac{L}{D}\right)\rho\frac{V^2}{2g} \quad (12\text{-}2)$$

$$\Delta P = (f)\frac{L\rho V^2}{8gm} \quad (12\text{-}2a)$$

where f = friction factor, to be obtained from Fig. 12–3;
L = equivalent length of pipe, in feet;
D = diameter of the conduit, in feet;
ρ = density of the fluid, in pounds per cubic foot;
V = velocity of the fluid, in feet per second;
g = 32.17, the gravitational constant;
m = the hydraulic radius; that is, the cross-sectional area of the conduit divided by the wetted perimeter. For a circular pipe,

$$m = \frac{(\pi/4)D^2}{\pi D} = \frac{D}{4} \quad (12\text{-}3)$$

NOTE. Equations 12–2 and 12–2a often appear in the literature with a factor 4 on the right-hand side. For this condition the only difference is that the values of f are one-fourth as great.

The use of these friction-flow relationships is closely connected with the viscosity (μ) of the fluid and with the Reynolds number (sec. 12–2). The *absolute viscosity* or *coefficient of viscosity* of a fluid is defined as the tangential force required to move a plane fluid surface of unit area, at unit velocity, relative to a parallel plane surface of the fluid, a unit distance away, when the intervening space is filled with the fluid. In the CGS system the unit of viscosity is the poise. If μ (mu) is used for viscosity,

$$\mu = \frac{\text{dyne/sq cm}}{\text{cm/sec} \over \text{cm}} = \frac{\text{dyne} \times \text{sec}}{\text{sq cm}}$$

Or, in equivalent form,

$$\mu = \frac{\text{gram (mass)}}{\text{cm} \times \text{sec}} = \frac{\text{gram (force)} \times \text{sec}}{980 \times \text{sq cm}} \quad (12\text{-}4)$$

In the English system the unnamed unit of viscosity is

$$\frac{\text{Poundal} \times \text{sec}}{\text{Sq ft}} = \frac{\text{pound (mass)}}{\text{ft} \times \text{sec}} = \frac{\text{pound (force)} \times \text{sec}}{32.17 \text{ sq ft}} \quad (12\text{-}5)$$

To change units from CGS to the English system or vice versa, the con-

Fluid Flow, Duct Design, and Air-Distribution Systems

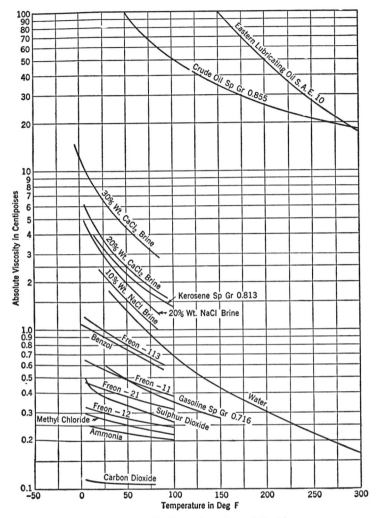

FIG. 12-1. Absolute viscosities of liquids.

stants below, following directly out of transformation from metric to English units, are applicable:

$$1 \text{ poise} = 100 \text{ centipoises} = \frac{1 \text{ dyne} \times \text{sec}}{\text{sq cm}} = \frac{\text{gram (mass)}}{\text{cm} \times \text{sec}}$$

$$= (0.0672) \frac{\text{poundal} \times \text{sec}}{\text{sq ft}} = (0.0672) \frac{\text{pounds (mass)}}{\text{ft} \times \text{sec}} \quad \textbf{(12-6)}$$

$$= (0.00209) \frac{\text{lb (force)} \times \text{sec}}{\text{sq ft}} = (0.00209) \frac{\text{slugs}}{\text{ft} \times \text{sec}}$$

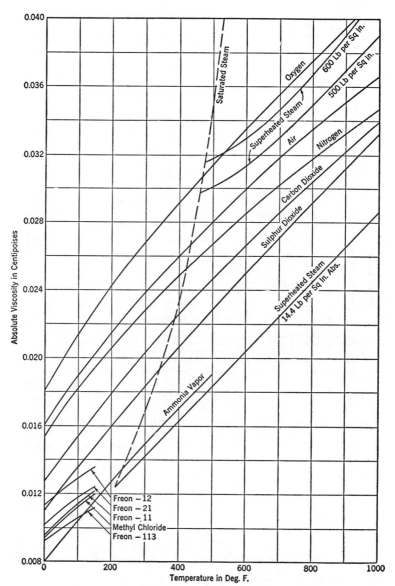

FIG. 12-2. Absolute viscosities of gases and vapors.

The viscosities of different fluids vary over wide limits and must be obtained from tabulations of experimental data (see Figs. 12-1 and 12-2). The viscosity values for a given fluid vary greatly with temperature but relatively slightly under moderate pressure changes. In the case of vapors, pressure changes may be very significant in their effect on vis-

Fluid Flow, Duct Design, and Air-Distribution Systems 413

cosity. The viscosity of pure water at 68 F is about one centipoise (exactly 1.008).

Kinematic viscosity (y or ν) is defined as absolute viscosity divided by the density of the fluid: $y = \mu/\rho$. Its units are square feet per second in the English system. Kinematic viscosity is often used instead of viscosity in computing the Reynolds number when it happens to be more convenient to do so. Many viscosimeters (instruments for measuring viscosity) determine kinematic viscosity directly. For example, the Saybolt viscosimeter, which is very commonly used, gives results in accordance with the following equation:

$$y = 0.00000237\,t - \frac{0.00194}{t} \quad (12\text{-}7)$$

where t is the time in seconds required for the sample of fluid to flow through the instrument; t is also known as the Saybolt seconds, and viscosities are often expressed in this form, which can be changed to kinematic viscosity (y) in square feet per second by the equation shown.

12-2. REYNOLDS NUMBER

The *Reynolds number* (R) is a dimensionless number of utility in correlating the friction factors for flow under varying conditions. In symbolic notation the Reynolds number appears as

$$R = \frac{DV\rho}{\mu} \quad (12\text{-}8)$$

or

$$R = \frac{DV}{y} \quad (12\text{-}9)$$

where R = Reynolds number (dimensionless);
V = velocity, in feet per second (English units) or in centimeters per second (metric units);
ρ = density, usually pounds mass (pounds weight) per cubic foot or grams mass per cubic centimeter;
μ = absolute viscosity;
y = kinematic viscosity.

To show that R is dimensionless in any consistent set of units, substitute the units of each term. In the metric system,

$$R = \frac{DV\rho}{\mu} = \frac{(\text{cm})(\text{cm/sec})(\text{grams/cm}^3)}{\text{grams}/(\text{cm} \times \text{sec})} = 1$$

In the English system,

$$R = \frac{DV\rho}{\mu} = \frac{(\text{ft})(\text{ft/sec})(\text{lb/ft}^3)}{\text{lb}/(\text{ft} \times \text{sec})} = 1$$

To calculate R numerically with customary units,

$$R = \frac{DV\rho}{\mu} = \frac{(\text{ft})(\text{ft/sec})(\text{lb/ft}^3)}{(0.0672)(\mu/100) \text{ in centipoises}}$$

$$= 1488 \frac{(\text{ft})(\text{ft/sec})(\text{lb/ft}^3)}{\mu \text{ centipoises}} \quad (12\text{-}10)$$

Equation 12-10 for the Reynolds number is convenient, as μ is usually tabulated in centipoises in standard tables and charts. The Reynolds number can also be calculated by using values for kinematic viscosity when these are available.

12-3. FRICTION FACTORS

One use of the Reynolds number R is to have it serve as a base on which to show variations in the friction factor f. Figure 12-3 covers an extended range of Reynolds numbers in the turbulent-flow region. In this region, surface roughness of the pipe or conduit has a significant influence on the values of the friction factor. The effect of surface roughness is best represented by the ratio of the surface-variation height to the diameter (or depth) of the conduit. The ordinate of Fig. 12-3 is thus relative roughness (ϵ/D). Values of ϵ for various types of pipes or conduits are given in Table 12-1; when these are used with the pipe-diameter scale at the top of Fig. 12-3, a family of lines for different kinds of pipes can be represented on the chart. It should be observed that the values of Fig. 12-3 are independent of the fluid and apply equally well to water or air, or to other gases when these are evaluated in terms of the Reynolds number. Reynolds numbers ranging between 2000 and 2500 are in a transition zone. On the low side laminar (viscous) flow occurs, while on the high side turbulent flow takes place. A value of 2300 is sometimes used as the most representative critical value.

TABLE 12-1

Recommended Values of Surface Roughness (ϵ)*

Material	ϵ (ft)
Drawn tubing, brass, lead, glass, centrifugally spun cement, bituminous lining, Transite, etc.	0.000005
Commercial steel or wrought-iron pipe	0.00015
Asphalted cast iron (asphalt-dipped)	0.0004
Galvanized iron	0.0005
Cast iron, average	0.00085
Wood stave	0.0006–0.003
Concrete	0.001–0.01 (avg 0.003)
Riveted steel	0.003–0.03 (avg 0.01)

* From L. F. Moody, *Mechanical Engineering*, Vol. 69 (1947), p. 1005.

In considering Table 12-1 it should further be noted that the roughness of a conduit not only is related to the material from which it is made, but

Fluid Flow, Duct Design, and Air-Distribution Systems

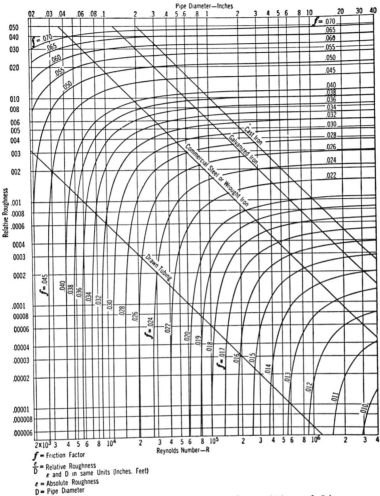

FIG. 12-3. Friction factors for fluid flow. (After ref. 3.)

also depends on the kind of workmanship used in the fabrication of the conduit and on the number and size of burrs and the amount of incrustation which may be present.

Example 12-1. Water at 200 F is flowing at 0.5 fps through a 4-in. standard-weight (ID = 4.03 in.) steel pipe of relatively smooth surface. Reference to a certain table gave the viscosity of 200 F water as 63.7×10^{-7} lb (force)-sec per sq ft. Find (a) the Reynolds number, (b) whether the flow is turbulent or viscous (streamline), and (c) the frictional loss per 100 ft of equivalent pipe length.

Solution: (a) $\mu = 63.7 \times 10^{-7}$ lb (force)-sec per sq ft becomes, in centipoises, by equation 12-6,

$$(63.7 \times 10^{-7}) \times \frac{1}{0.00209} \times 100 = 0.305 \text{ centipoises}$$

This result can also be checked from Fig. 12-1.

Density of water at 200 F = $1/v_f$ = 1/0.01663 = 60.2 lb per cu ft, from Table 2-2, and pounds (mass) per cubic foot, or pounds (weight) per cubic foot, divided by 32.17, gives slugs per cubic foot.

By equation 12-8 and use of constants from equation 12-6,

$$R = \frac{DV\rho}{\mu} = \frac{(4.03/12)(0.5)(60.2/32.17)}{63.7 \times 10^{-7}} = 49{,}300 \qquad Ans.$$

Or, by equation 12-10,

$$R = \frac{DV\rho}{\mu} = 1488\frac{(4.03/12)(0.5)(60.2)}{0.305} = 49{,}300$$

b) As R is 49,300, and thus above the top critical value for viscous flow, the flow is turbulent.
                                                                                                    Ans.

c) Use Fig. 12-3 for finding f. Read the inside pipe diameter at the top of the chart (4.03 in.), drop down to the line for commercial steel, and then move horizontally over until the value of $R = 49{,}300 = 4.9 \times 10^4$ is reached; here read $f = 0.0225$ (say 0.023). If one wished to use ϵ/D for a material not shown on the chart, the value of ϵ would be selected from Table 12-1. In this instance, $\epsilon = 0.00015$ and

$$\frac{\epsilon}{D} = \frac{0.00015}{4.03/12} = 0.000447$$

Find this value on the left axis and run horizontally to $R = 4.9 \times 10^4$, where f is read as 0.0225.

Using equation 12-2, the pressure drop in 100 ft of equivalent pipe length is found:

$$\Delta P = f\left(\frac{L}{D}\right)\rho\frac{V^2}{2g}$$

$$= (0.023)\left(\frac{100}{\frac{4.03}{12}}\right)\frac{(60.2)(0.5)^2}{64.4} = 1.6 \text{ lb per sq ft}$$

$$= \frac{1.6}{144} = 0.0111 \text{ psi} \qquad Ans.$$

For the rather unusual cases of viscous flow, the friction factor f is practically independent of the surface because of the clinging undisturbed film, and therefore does not appear in equation 12-1. However, if it is desired to calculate pressure drop in viscous flow by using equation 12-2 or 12-3, in which equations friction factors appear, a working value of f can be found by equating equation 12-1 to equation 12-2:

$$\frac{32\mu LV}{gD^2} = (f)\left(\frac{L}{D}\right)\frac{\rho V^2}{2g}$$

Or, for viscous flow only,

$$f = \frac{64\mu}{DV\rho} = \frac{64}{R} \qquad (12\text{-}11)$$

Thus, the friction factor f for viscous (streamline) flow, which can be used in equation 12-2, is found by dividing 64 by the Reynolds number. In turbulent flow, f must always be found from experimental data (using any suitable test fluid).

Fluid Flow, Duct Design, and Air-Distribution Systems

For conduits of noncircular cross section, friction factors may vary greatly from those for round sections. For example, a shallow rectangular duct offers much more resistance to fluid flow than does a square or round duct when each has the same cross-sectional area. In general, the largest area for the least perimeter of duct gives the least frictional drop. The *hydraulic radius* (m), which is defined as the cross-sectional area divided by the fluid-contact perimeter, is useful in making comparisons of odd-shaped sections.

For a circular section of diameter D the hydraulic radius is

$$m = \frac{\text{area}}{\text{perimeter}} = \frac{(\pi/4)D^2}{\pi D} = \frac{D}{4} \qquad (12\text{-}12)$$

and thus
$$D = 4m \qquad (12\text{-}13)$$

For a rectangular section of sides H and W, in consistent units of feet or inches,

$$m = \frac{H \times W}{2(H + W)} \qquad (12\text{-}14)$$

For a square conduit of side a feet or inches, the hydraulic radius is

$$m = \frac{a^2}{4a} = \frac{a}{4} \qquad (12\text{-}15)$$

and thus
$$a = 4m \qquad (12\text{-}16)$$

In the case of a round duct (conduit) it is seen that $D = 4m$. Thus if any duct is not too flattened, a Reynolds number can be calculated, using as D in equation 12-8 or 12-9 the calculated value of $4m$ for that duct. From this evaluation a friction factor can be found for use in equation 12-2 or 12-3 with the odd-shaped duct.

The preceding discussion gives a logical basis for calculating frictional losses from fluid flowing in straight runs of piping. Losses through fittings, bends, and elbows must be obtained from experimental data and are usually stated as equivalent in loss to a certain length (in diameters) of straight run of pipe. For a given type of piping or duct it is convenient to construct tables or diagrams to facilitate computations. Some of these may be found at several places in this book. In some cases it has been found that the turbulent basic-loss equations are not exactly proportional to the square of the velocity but may vary by a power somewhat less than 2 (that is, 1.8 and 1.9), but for most cases use of the square of the velocity gives the best results in making computations.

12-4. AIR-DUCT FLOW AND LOSSES

The energy equation (eq 2-3) is applicable to air that is flowing in a duct; and by eliminating from the equation irrelevant terms such as potential energy (Z) and heat transferred ($Q_{(1-2)}$), there remains the expression

$$P_1 v_1 + \frac{V_1^2}{2g} = P_2 v_2 + \frac{V_2^2}{2g} + 778(u_2 - u_1) = P_2 v_2 + \frac{V_2^2}{2g} + \text{lost head} \quad (12\text{--}17)$$

This equation states that at a given point the mechanical energy terms, namely the flow work ($P_1 v_1$) plus the kinetic energy ($V_1^2/2g$), must be equal to the flow work and kinetic energy at any other point plus any dissipation of mechanical energy to internal energy (lost head) which has meanwhile occurred. In the case of air, use is made of the term *static head*, or *static pressure*, for Pv, and *velocity head*, or *velocity pressure*, for $V^2/2g$; and the sum of static pressure and velocity pressure is the *total pressure*, or *total dynamic pressure*, or *head* (h_T). The energy units of this equation are foot-pounds per pound of air or can be expressed simply as feet of the fluid (air) considered. In the case of fans the static pressure is usually expressed in gage pressure (excess of pressure above that of the atmosphere) instead of in absolute pressure.

The static and velocity pressures of air, instead of being expressed in "feet of air" units, are more commonly changed to "inches of water" units. The transformation can be made by the use of the fundamental relation (equation 1–8),

$$h_a d_a = \frac{h}{12} d_w$$

where h_a = feet of air; d_a = density of air, pounds per cubic foot; h = inches of water (gage); d_w = density of 68 F water, 62.3 lb per cu ft. Thus, employing appropriate values at 68 F, it is seen that

$$h_a = \frac{h}{12} \frac{62.3}{d_a} = 5.19 \frac{h}{d_a} \quad (12\text{--}18)$$

For the velocity head, h_v, in inches of water,

$$\frac{V^2}{2g} = 5.19 \frac{h_v}{d_a} \quad \text{or} \quad h_v = \frac{V^2 d_a}{(5.19) 2g}$$

With V_m as fpm,

$$h_v = \frac{V_m^2 d_a}{(3600)(5.19)(2g)} = \frac{V_m^2 d_a}{1{,}202{,}000} = \left(\frac{V_m}{1096.5}\right)^2 d_a \quad (12\text{--}19)$$

$$V_m = 1096.5 \sqrt{\frac{h_v}{d_a}} \quad (12\text{--}20)$$

Standard air is air having a density of 0.075 lb per cu ft at 29.92 in. Hg. This is essentially equivalent to dry air at 70 F or to air at 30 per cent relative humidity and 65 F (actually d_a = 0.07494); and

$$V_m = 4005 \sqrt{h_v} \quad \text{or} \quad h_v = \left(\frac{V_m}{4005}\right)^2 \quad (12\text{--}21)$$

Fluid Flow, Duct Design, and Air-Distribution Systems

Call h_s the static head of the air in inches of water-gage pressure; then the total head becomes

$$h_T = h_s + h_v = h_s + \left(\frac{V_m}{1096.5}\right)^2 \times d_a = h_s + \left(\frac{V_m}{4005}\right)^2_{\text{std air}} \quad (12\text{-}22)$$

The total head h_T would remain undiminished and h_s and h_v would be mutually interchangeable from one form to the other, if it were not for frictional loss, which resists flow, and for impact losses due to sudden enlargements, contractions, and bends in the duct. These items dissipate the total head or mechanical energy, which eventually reappears as increased internal energy of the air.

For frictional loss in air ducts, equation 12-2 or 12-2a applies. In pounds per square foot,

$$\Delta P = (f)\frac{L\rho V^2}{8gm}$$

And in feet of air,

$$\Delta h_a = \frac{\Delta P}{\rho} = f\left(\frac{L}{m}\right)\frac{V^2}{8g}$$

When for a round duct the pressure loss is expressed in inches of water and the constants given in equation 12-19 are used,

$$\Delta h = f\left(\frac{L}{D}\right)\left(\frac{V_m}{1096.5}\right)^2 d_a \quad (12\text{-}23)$$

The results of tests on friction loss for air flowing in galvanized ducts have been assembled by the ASHAE and plotted in Fig. 12-4. This chart is based on standard air but can also be used, with little error, for air in the range of 50 F to 90 F. Moreover, a correction is not required for humidity variations or for small deviations in barometric pressure (not exceeding ±0.5 in. Hg). It is desirable to apply a correction factor to the chart value for unusual construction or rough ductwork. Figure 12-5 gives representative correction factors. In addition, the final design should include an over-all safety factor of (say) 10 per cent to cover possible construction contingencies.

Where the air in the duct is at a density significantly different from that for which the chart was constructed, the pressure loss should be multiplied by a factor as follows:

$$\Delta h_c = (\Delta h)\left(\frac{d_a \text{ at actual conditions}}{0.075}\right) \text{ in in. of water} \quad (12\text{-}24)$$

When the temperature is the important variable, the pressure loss should be multiplied by a temperature correction factor, giving

$$\Delta h_c = (\Delta h)\left(\frac{460 + 70}{460 + t}\right) \text{ in in. of water} \quad (12\text{-}25)$$

In the last two equations, Δh_c is the corrected value of pressure loss.

TABLE 12-2. Circular Equivalents of Rectan-

Side Rectangular Duct	4.0	4.5	5.0	5.5	6.0	6.5	7.0	7.5	8.0	8.5	9.0	9.5	10.0	10.5	11.0	11.5	12.0	12.5	13.0	13.5
3.0	3.8	4.0	4.2	4.4	4.6	4.8	4.9	5.1	5.2	5.4	5.5	5.6	5.7	5.9	6.0	6.1	6.2	6.3	6.4	6.5
3.5	4.1	4.3	4.6	4.8	5.0	5.2	5.3	5.5	5.7	5.8	6.0	6.1	6.3	6.4	6.5	6.7	6.8	6.9	7.0	7.1
4.0	4.4	4.6	4.9	5.1	5.3	5.5	5.7	5.9	6.1	6.3	6.4	6.6	6.8	6.9	7.1	7.2	7.3	7.5	7.6	7.7
4.5	4.6	4.9	5.2	5.4	5.6	5.9	6.1	6.3	6.5	6.7	6.9	7.0	7.2	7.4	7.5	7.7	7.8	8.0	8.1	8.2
5.0	4.9	5.2	5.5	5.7	6.0	6.2	6.4	6.7	6.9	7.1	7.3	7.4	7.6	7.8	8.0	8.1	8.3	8.4	8.6	8.7
5.5	5.1	5.4	5.7	6.0	6.3	6.5	6.8	7.0	7.2	7.4	7.6	7.8	8.0	8.2	8.4	8.6	8.7	8.8	9.0	9.2

Side Rectangular Duct	6	7	8	9	10	11	12	13	14	15	16	17	18	19	20	22	24	26	28	30
6	6.6																			
7	7.1	7.7																		
8	7.5	8.2	8.8																	
9	8.0	8.6	9.3	9.9																
10	8.4	9.1	9.8	10.4	10.9															
11	8.8	9.5	10.2	10.8	11.4	12.0														
12	9.1	9.9	10.7	11.3	11.9	12.5	13.1													
13	9.5	10.3	11.1	11.8	12.4	13.0	13.6	14.2												
14	9.8	10.7	11.5	12.2	12.9	13.5	14.2	14.7	15.3											
15	10.1	11.0	11.8	12.6	13.3	14.0	14.6	15.3	15.8	16.4										
16	10.4	11.4	12.2	13.0	13.7	14.4	15.1	15.7	16.3	16.9	17.5									
17	10.7	11.7	12.5	13.4	14.1	14.9	15.5	16.1	16.8	17.4	18.0	18.6								
18	11.0	11.9	12.9	13.7	14.5	15.3	16.0	16.6	17.3	17.9	18.5	19.1	19.7							
19	11.2	12.2	13.2	14.1	14.9	15.6	16.4	17.1	17.8	18.4	19.0	19.6	20.2	20.8						
20	11.5	12.5	13.5	14.4	15.2	15.9	16.8	17.5	18.2	18.8	19.5	20.1	20.7	21.3	21.9					
22	12.0	13.1	14.1	15.0	15.9	16.7	17.6	18.3	19.1	19.7	20.4	21.0	21.7	22.3	22.9	24.1				
24	12.4	13.6	14.6	15.6	16.6	17.5	18.3	19.1	19.8	20.6	21.3	21.9	22.6	23.2	23.9	25.1	26.2			
26	12.8	14.1	15.2	16.2	17.2	18.1	19.0	19.8	20.6	21.4	22.1	22.8	23.5	24.1	24.8	26.1	27.2	28.4		
28	13.2	14.5	15.6	16.7	17.7	18.7	19.6	20.5	21.3	22.1	22.9	23.6	24.4	25.0	25.7	27.1	28.2	29.5	30.6	
30	13.6	14.9	16.1	17.2	18.3	19.3	20.2	21.1	22.0	22.9	23.7	24.4	25.2	25.9	26.7	28.0	29.3	30.5	31.6	32.8
32	14.0	15.3	16.5	17.7	18.8	19.8	20.8	21.8	22.7	23.6	24.4	25.2	26.0	26.7	27.5	28.9	30.1	31.4	32.6	33.8
34	14.4	15.7	17.0	18.2	19.3	20.4	21.4	22.4	23.3	24.2	25.1	25.9	26.7	27.5	28.3	29.7	31.0	32.3	33.6	34.8
36	14.7	16.1	17.4	18.6	19.8	20.9	21.9	23.0	23.9	24.8	25.8	26.6	27.4	28.3	29.0	30.5	32.0	33.0	34.6	35.8
38	15.0	16.4	17.8	19.0	20.3	21.4	22.5	23.5	24.5	25.4	26.4	27.3	28.1	29.0	29.8	31.4	32.8	34.2	35.5	36.7
40	15.3	16.8	18.2	19.4	20.7	21.9	23.0	24.0	25.1	26.0	27.0	27.9	28.8	29.7	30.5	32.1	33.6	35.1	36.4	37.6
42	15.6	17.1	18.5	19.8	21.1	22.3	23.4	24.5	25.6	26.6	27.6	28.5	29.4	30.4	31.2	32.8	34.4	35.9	37.3	38.6
44	15.9	17.5	18.9	20.2	21.5	22.7	23.9	25.0	26.1	27.2	28.2	29.1	30.0	31.0	31.9	33.5	35.2	36.7	38.1	39.5
46	16.2	17.8	19.2	20.6	21.9	23.2	24.3	25.5	26.7	27.7	28.7	29.7	30.6	31.6	32.5	34.2	35.9	37.4	38.9	40.3
48	16.5	18.1	19.6	20.9	22.3	23.6	24.8	26.0	27.2	28.2	29.2	30.2	31.2	32.2	33.1	34.9	36.6	38.2	39.7	41.2
50	16.8	18.4	19.9	21.3	22.7	24.0	25.2	26.4	27.6	28.7	29.8	30.8	31.8	32.8	33.7	35.5	37.3	38.9	40.4	42.0
52	17.0	18.7	20.2	21.6	23.1	24.4	25.6	26.8	28.1	29.2	30.3	31.4	32.4	33.4	34.3	36.2	38.0	39.6	41.2	42.8
54	17.3	19.0	20.5	22.0	23.4	24.8	26.1	27.3	28.5	29.7	30.8	31.9	32.9	33.9	34.9	36.8	38.7	40.3	42.0	43.6
56	17.6	19.3	20.9	22.4	23.8	25.2	26.5	27.7	28.9	30.1	31.2	32.4	33.4	34.5	35.5	37.4	39.3	41.0	42.7	44.3
58	17.8	19.5	21.1	22.7	24.2	25.5	26.9	28.2	29.3	30.5	31.7	32.9	33.9	35.0	36.0	38.0	39.8	41.7	43.4	45.0
60	18.1	19.8	21.4	23.0	24.5	25.8	27.3	28.7	29.8	31.0	32.2	33.4	34.5	35.5	36.5	38.6	40.4	42.3	44.0	45.8
62	18.3	20.1	21.7	23.3	24.8	26.2	27.6	29.0	30.2	31.4	32.6	33.8	35.0	36.0	37.1	39.2	41.0	42.9	44.7	46.5
64	18.6	20.3	22.0	23.6	25.2	26.5	27.9	29.3	30.6	31.8	33.1	34.2	35.5	36.5	37.6	39.7	41.6	43.5	45.4	47.2
66	18.8	20.6	22.3	23.9	25.5	26.9	28.3	29.7	31.0	32.2	33.5	34.7	35.9	37.0	38.1	40.2	42.2	44.1	46.0	47.8
68	19.0	20.8	22.5	24.2	25.8	27.3	28.7	30.1	31.4	32.6	33.9	35.1	36.3	37.5	38.6	40.7	42.8	44.7	46.6	48.4
70	19.2	21.1	22.8	24.5	26.1	27.6	29.1	30.4	31.8	33.1	34.3	35.6	36.8	37.9	39.1	41.3	43.3	45.3	47.2	49.0
72													39 16	41.8	43.8	45.9	47.8	49.7		
74													40.0	42.3	44.4	46.4	48.4	50.3		
76													40.5	42.8	44.9	47.0	49.0	50.8		
78													40.9	43.3	45.5	47.5	49.5	51.5		
80													41.3	43.8	46.0	48.0	50.1	52.0		
82													41.8	44.2	46.4	48.6	50.6	52.6		
84													42.2	44.6	46.9	49.2	51.1	53.2		
86													42.6	45.0	47.4	49.6	51.6	53.7		
88													43.0	45.4	47.9	50.1	52.2	54.3		
90													43.4	45.9	48.3	50.6	52.8	54.8		
92													43.8	46.3	48.7	51.1	53.4	55.4		
94													44.2	46.7	49.1	51.6	53.9	55.9		
96													44.6	47.2	49.5	52.0	54.4	56.3		

* From R. G. Huebscher, *Trans. ASHVE*, Vol. 54 (1948), pp. 112-13. Reprinted by permission.

GULAR DUCTS FOR EQUAL FRICTION AND CAPACITY *

14.0	14.5	15.0	15.5	16
6.6	6.7	6.8	6.9	7.0
7.2	7.3	7.4	7.5	7.6
7.8	7.9	8.1	8.2	8.3
8.4	8.5	8.6	8.7	8.9
8.9	9.0	9.1	9.3	9.4
9.4	9.5	9.6	9.8	9.8

32	34	36	38	40	42	44	46	48	50	52	56	60	64	68	72	76	80	84	88	Side Rectangular Duct
																				6
																				7
																				8
																				9
																				10
																				11
																				12
																				13
																				14
																				15
																				16
																				17
																				18
																				19
																				20
																				22
																				24
																				26
																				28
																				30
35.0																				32
36.0	37.2																			34
37.0	38.2	39.4																		36
38.0	39.2	40.4	41.6																	38
39.0	40.2	41.4	42.6	43.8																40
39.9	41.1	42.4	43.6	44.8	45.9															42
40.8	42.0	43.4	44.6	45.8	46.9	48.1														44
41.7	43.0	44.3	45.6	46.8	47.9	49.1	50.3													46
42.6	43.9	45.2	46.5	47.8	48.9	50.2	51.3	52.6												48
43.5	44.8	46.1	47.4	48.8	49.8	51.2	52.3	53.6	54.7											50
44.3	45.7	47.1	48.3	49.7	50.8	52.2	53.3	54.6	55.8	56.9										52
45.0	46.5	48.0	49.2	50.6	51.8	53.2	54.3	55.6	56.8	57.9										54
45.8	47.3	48.8	50.1	51.5	52.7	54.1	55.3	56.5	57.8	58.9	61.3									56
46.6	48.1	49.6	51.0	52.4	53.7	55.0	56.2	57.5	58.8	60.0	62.3									58
47.3	48.9	50.4	51.8	53.3	54.6	55.9	57.1	58.5	59.8	61.0	63.3	65.7								60
48.0	49.7	51.2	52.6	54.2	55.5	56.8	58.0	59.4	60.7	62.0	64.3	66.7								62
48.7	50.4	52.0	53.4	55.0	56.4	57.7	59.0	60.3	61.6	62.9	65.3	67.7	70.0							64
49.5	51.1	52.8	54.2	55.8	57.2	58.6	59.9	61.2	62.5	63.9	66.3	68.7	71.1							66
50.2	51.8	53.5	55.0	56.6	58.0	59.5	60.8	62.1	63.4	64.8	67.3	69.7	72.1	74.4						68
50.9	52.5	54.2	55.8	57.3	58.8	60.3	61.7	63.0	64.3	65.7	68.3	70.7	73.1	75.4						70
51.5	53.2	54.9	56.5	58.0	59.6	61.1	62.6	63.9	65.2	66.6	69.2	71.7	74.1	76.4	78.8					72
52.1	53.9	55.6	57.2	58.8	60.4	61.9	63.3	64.8	66.1	67.5	70.1	72.7	75.1	77.4	79.9					74
52.7	54.6	56.3	57.9	59.5	61.2	62.7	64.1	65.6	67.0	68.4	71.0	73.6	76.1	78.4	80.9	83.2				76
53.3	55.2	57.0	58.6	60.3	62.0	63.4	64.9	66.4	67.9	69.3	71.8	74.5	77.1	79.4	81.8	84.2				78
53.9	55.8	57.6	59.3	61.0	62.7	64.1	65.7	67.2	68.7	70.1	72.7	75.4	78.1	80.4	82.8	85.2	87.5			80
54.5	56.4	58.2	60.0	61.7	63.4	64.9	66.5	68.0	69.5	71.0	73.6	76.3	79.0	81.4	83.8	86.2	88.6			82
55.1	57.0	58.9	60.7	62.4	64.1	65.7	67.3	68.8	70.3	71.8	74.5	77.2	79.9	82.4	84.8	87.2	89.6	91.9		84
55.7	57.6	59.5	61.3	63.0	64.8	66.4	68.0	69.5	71.1	72.6	75.4	78.1	80.8	83.3	85.8	88.2	90.6	92.9		86
56.3	58.2	60.1	62.0	63.7	65.4	67.0	68.7	70.3	71.8	73.4	76.3	79.0	81.6	84.2	86.8	89.2	91.6	93.9	96.3	88
56.9	58.8	60.7	62.6	64.4	66.0	67.8	69.4	71.1	72.6	74.2	77.1	79.9	82.5	85.1	87.8	90.2	92.6	94.9	97.3	90
57.4	59.4	61.3	63.2	65.0	66.8	68.5	70.1	71.8	73.3	74.9	77.8	80.8	83.4	86.0	88.7	91.2	93.6	95.9	98.3	92
57.9	60.0	61.9	63.8	65.6	67.5	69.2	70.8	72.5	74.1	75.6	78.6	81.7	84.3	86.9	89.6	92.1	94.6	96.9	99.3	94
58.4	60.5	62.4	64.4	66.2	68.2	69.8	71.5	73.2	74.8	76.3	79.4	82.6	85.2	87.8	90.5	93.0	95.6	97.9	100.3	96

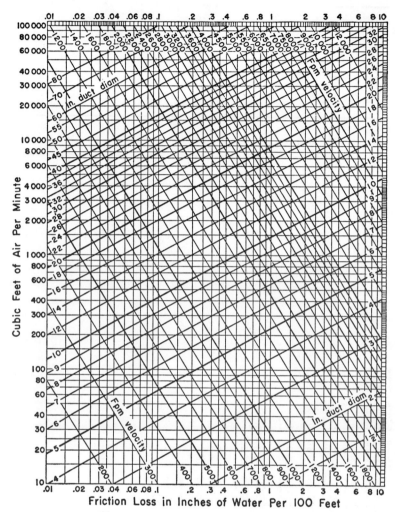

Fig. 12-4. Friction loss in air ducts. (Reproduced, by permission, from *Heating Ventilating Air Conditioning Guide 1949*.)

Example 12-2. (a) Find the friction loss in 60 ft of 30-in.-diameter duct carrying 10,000 cfm of 70 F air, by the use of the friction-loss chart in Fig. 12-4. (b) Assume that the duct is made of poor construction in the field and estimate the friction loss. (c) Use Table 12-2 to find the corresponding size of rectangular duct not more than 16 in. wide to carry this air. (d) Assume that the air is at 150 F instead of 70 F and compute the pressure loss.

Solution: (a) On Fig. 12-4, locate the 10,000-cfm line on the left axis and follow this horizontally to its intersection with the diagonal line for the 30-in. duct diameter. At this point it will be noted that the velocity is above 2000 fpm, or at (say) 2060 fpm;

Fluid Flow, Duct Design, and Air-Distribution Systems

the friction loss read at the top or bottom of the chart is 0.16 in. of water per 100 ft of run. For a 60-ft run, the loss is $0.16 \times \frac{60}{100} = 0.096$ in. of water. *Ans.*

b) Read from Fig. 12-5, at 2060 fpm for medium rough pipe, a factor of 1.35. Therefore the corrected loss = $0.096 \times 1.35 = 0.130$ in. of water. *Ans.*

c) In Table 12-2 locate, in the column for a duct of 16-in. side, a circular-duct diameter of 30 in. Interpolation between 30.3 and 29.8 shows that a 30-in. duct corresponds to a 51-in. side. Hence, employ a 16-in. by 51-in. rectangular duct for the same friction loss. *Ans.*

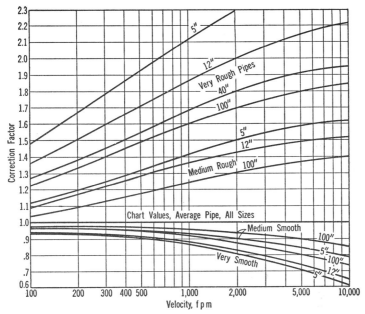

FIG. 12-5. Friction correction factors for pipe and duct roughness. (After ref. 6.)

d) To correct for temperature, use equation 12-25. Thus the pressure loss, in inches of water, is

$$\Delta h_c = (0.130)\left(\frac{460 + 70}{460 + 150}\right) = 0.113 \qquad \textit{Ans.}$$

A greater frictional loss occurs in a rectangular duct than in a circular duct of the same cross-sectional area. If the frictional loss per foot of length in a circular duct is to be made equal to that in a rectangular duct, fundamental expressions for loss for the two types of ducts are equated; and with suitable friction factors being used, there evolves

$$D = 1.3\left[\frac{(HW)^5}{(H+W)^2}\right]^{1/8} = 1.3\frac{(HW)^{0.625}}{(H+W)^{0.25}} \qquad (12\text{-}26)$$

where H, W = the two sides of a rectangular duct, respectively, in feet or inches;
D = diameter of a circular duct having the same frictional loss per foot as a rectangular duct delivering the same quantity of air, in feet or inches;
H/W = aspect ratio of a rectangular duct. In case of bends,
H = duct dimension perpendicular to plane of bend;
W = duct dimension in plane of bend.

Equation 12-26 was developed by Huebscher (ref. 3), who further concluded that for most practical purposes a rectangular duct with an aspect ratio not exceeding 8 to 1 will have the same static friction pressure loss for a certain length and mean velocity of flow as a circular duct of the same hydraulic radius. By using equation 12-26, Table 12-2 was constructed to give values of circular equivalents for rectangular and square ducts.

TABLE 12-3

PRESSURE LOSS IN DUCT FITTINGS

[Expressed in decimal parts of one velocity pressure (head), $(V_m/1096.5)^2 d_a$, or $(V_m/4005)^2$ for standard air] *

Type of Connection	Pressure Loss in Velocity Heads
Tee connection, right-angled side outlet	2.00 (V_m in branch)
Branches from duct	
15-deg angle	0.10 (V_m in branch)
30-deg angle	0.20 (V_m in branch)
45-deg angle	0.25 (V_m in branch)
Elbow, right-angled, rectangular duct aspect ratio near unity	1.15
Elbow, right-angled, round duct	0.87
Entry to pipe from chamber	0.85
Entry to pipe from chamber, coned inlet	0.20
Pipe enlargement	
Abrupt or chamber inlet	0.80–1.00
Coned, 10% slope	0.25
Pipe entering chamber, coned, 10% slope	0.50
Grilles, net area equal to duct area	1.25
Entrance loss	
Intake louvers and induction of outside air	1.50
Intake louvers, without acceleration of inlet air	0.50

* See Fig. 12-6 for velocity-head values.

Data on representative pressure loss in elbows, bends, and fittings are given in Figs. 12-7, 12-8, and 12-9, in which the loss is expressed in equivalent length of duct (L) and is measured in duct diameters (D) or duct widths (W). This method of expression is independent of the velocity of the air or gas flowing and gives the amount of straight run which will give the same loss as is occasioned by the elbow or elbows in question.

Fluid Flow, Duct Design, and Air-Distribution Systems

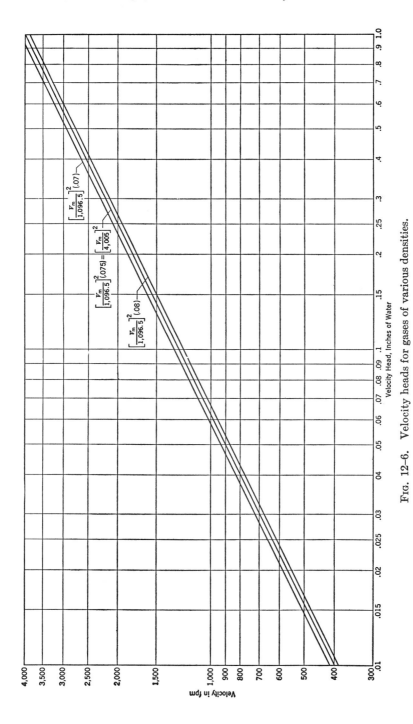

Fig. 12-6. Velocity heads for gases of various densities.

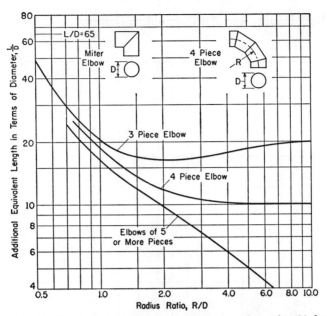

FIG. 12-7. Loss in additional elbow equivalents for 90-deg elbows of round cross section. (Reprinted, by permission, from *Heating Ventilating Air Conditioning Guide 1955*, Chapter 32.)

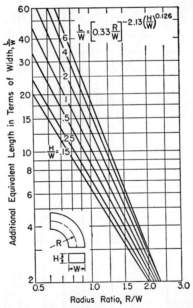

FIG. 12-8. Loss in additional equivalent widths for 90-deg elbows of rectangular cross section. To find additional equivalent length L, multiply duct width W in feet by L/W value shown. (Reprinted, by permission, from *Heating Ventilating Air Conditioning Guide 1955*, Chapter 32.)

Fluid Flow, Duct Design, and Air-Distribution Systems 427

Figures 12-7, 12-8, and 12-9, which should be used in designing duct systems, represent design values which include some allowance for actual construction. The square duct with built-in vanes, Fig. 12-9, shows how much the loss figure can vary for different arrangements. Losses expressed in velocity heads are given in Table 12-3. Figure 12-6 is a chart for converting given velocities to velocity heads.

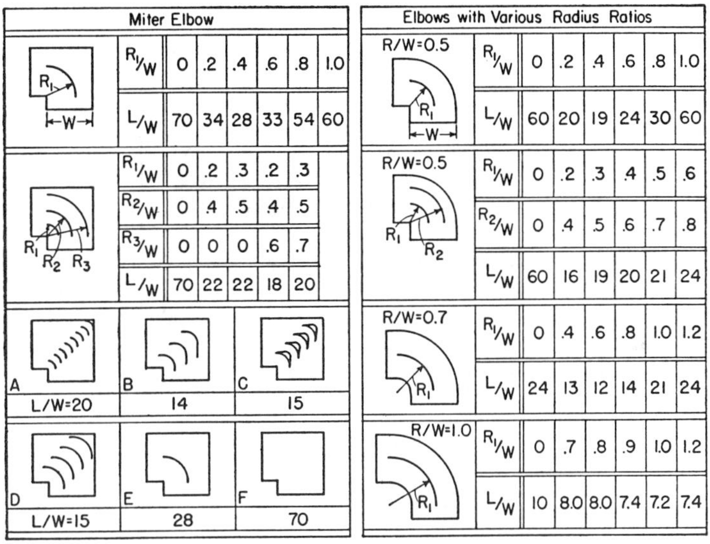

Miter Elbow						
R_1/W	0	.2	.4	.6	.8	1.0
L/W	70	34	28	33	54	60

R_1/W	0	.2	.3	.2	.3	
R_2/W	0	.4	.5	.4	.5	
R_3/W	0	0	0	.6	.7	
L/W	70	22	22	18	20	

A L/W=20	B 14	C 15
D L/W=15	E 28	F 70

Elbows with Various Radius Ratios						
R/W=0.5						
R_1/W	0	.2	.4	.6	.8	1.0
L/W	60	20	19	24	30	60

R/W=0.5						
R_1/W	0	.2	.3	.4	.5	.6
R_2/W	0	.4	.5	.6	.7	.8
L/W	60	16	19	20	21	24

R/W=0.7						
R_1/W	0	.4	.6	.8	1.0	1.2
L/W	24	13	12	14	21	24

R/W=1.0						
R_1/W	0	.7	.8	.9	1.0	1.2
L/W	10	8.0	8.0	7.4	7.2	7.4

Fig. 12-9. Pressure loss in vaned elbows of square cross section, expressed in additional equivalent duct length. Additional equivalent length L = duct width W, in feet, multiplied by L/W values shown. Vanes: A = large number of small arc vanes; B = a small number of large arc vanes; C = hollow vanes having different outside and inside curvature; D = four vanes with radius of $0.4W$; E = single splitter with radius of $0.5W$; F = no vanes or splitters. (Reprinted, by permission, from *Heating Ventilating Air Conditioning Guide 1955*, Chapter 32.)

12-5. DUCT-DESIGN PROCEDURE

In any mechanical-circulation heating, cooling, or ventilating system the fan or fans must have adequate capacity to deliver the air quantity required at a static pressure equal to or slightly greater than the total resistance offered by the duct system. The sizes of the ducts are set by the maximum air velocities which can be used without causing undue noise or without causing excessive friction loss. Large ducts will reduce frictional losses, but the space and investment requirements offset the power saving at the fan. An economic balance must be made in designing the installation. In general, however, the layout should be made as direct as possible, sharp bends should be avoided, and the duct cross-section, if rectangular, should not be too flattened. For a rectangular duct, a ratio of long to

TABLE 12-4
RECOMMENDED AND MAXIMUM DUCT VELOCITIES *

DESIGNATION	RECOMMENDED VELOCITIES (FPM)			MAXIMUM VELOCITIES (FPM)		
	Residences	Schools, Theaters, Public Buildings	Industrial Buildings	Residences	Schools, Theaters, Public Buildings	Industrial Buildings
Outside air intakes †..	700	800	1000	800	900	1200
Filters †............	250	300	350	300	350	350
Heating coils †......	450	500	600	500	600	700
Air washers.........	500	500	500	500	500	500
Suction connections...	700	800	1000	900	1000	1400
Fan outlets.........	1000–1600	1300–2000	1600–2400	1700	1500–2200	1700–2800
Main ducts.........	700–900	1000–1300	1200–1800	800–1000	1100–1600	1300–2200
Branch ducts........	600	600–900	800–1000	700–1000	800–1300	1000–1800
Branch risers........	500	600–700	800	650–800	800–1200	1000–1600

* From *Heating Ventilating Air Conditioning Guide 1949*, Chapter 41.
† These velocities are for total face area, not the net free area.

short sides of up to 6 to 1 is good practice, and the ratio should never exceed 10 to 1.

The possible resistances that a fan must overcome in delivering its air are tabulated in Table 12-5. These items are only representative; they differ for each system.

In the duct design the following procedure may be followed:

1. Lay out the most convenient system of placing the various ducts to obtain adequate distribution and to facilitate construction.

TABLE 12-5
TYPICAL FRICTIONAL LOSSES FOR DUCT SYSTEM EQUIPMENT

Item	Possible Range of Loss* (in. of water)
Air intake or fan entry.......................................	0.005 to 0.1
Air heaters or coolers, one row to several rows...................	0.1 to 0.35
Air washer...	0.2 to 0.35
Air filters..	0.2 to 0.4
Duct system (calculated for worst run)........................	0.04 to 0.4
Miscellaneous, screens, grilles, etc............................	0.1 to 0.2
Nozzle-type outlets..	0.1
Less any velocity head regain................................	0.01 up
Total static pressure loss for system (fan)..................	1.0 to 1.6 usual

* Selected from representative manufacturer's data, or calculated.

2. From the heating or cooling load, calculate the air requirements (cfm) at each duct outlet, zone, or division of the building.
3. Determine the sizes of these outlet branches, using a proper velocity or pressure drop to deliver the required quantity.
4. Calculate the size of each duct by one of the following methods:
   a. *The Assumed-Velocity Method.* The velocity in each of the various sections of the duct is assumed, in accordance with good practice, and the separate losses of each part of a definite system or circuit are added to find the total pressure loss. A modification of this approach, known as the *velocity-reduction method*, follows the general procedure indicated, but the assumed velocities are reduced progressively in the duct sections. The highest velocity is taken at the fan outlet, and velocities are lowered in the main after various branches are taken off. In general, the assumed-velocity method without refinement should be used only for relatively simple duct-system layouts. The control of flow in the various branches will be largely dependent on dampering.
   b. *The Constant-Pressure-Drop Method,* or *Equal-Friction Method.* The duct is proportioned so that the frictional loss per foot of length is constant. It is then possible for the resistances of the branches to be made essentially equal unless they are of greatly different lengths. When this method is used, it is customary to establish the constant pressure drop on the basis of the desirable velocity in the duct main beyond the fan. The branches must be dampered for control.
   c. *Balanced-Pressure-Loss Method.* This method employs pertinent procedures of method a or b, but every branch is designed to have the same pressure loss from the fan in order that minimum dependence on dampering is required. Expressed in another way, the static pressure required for the flow in any branch from its point of attachment is made equal to the static pressure of the main system at the juncture point. Theoretically, dampers for adjustment of the design flow in any branch would not be required in such a system, but it is always desirable to supply them.
   d. *Static-Regain Method.* This method endeavors to meet the objectives in paragraph 4c but employs static regain as an accessory device. The use of static regain will be discussed later.
5. Determine, from the calculations described in procedure 4 preceding, the circuit offering the greatest frictional resistance. While the circuit thus selected is frequently the longest circuit, this is not necessarily true. The maximum resistance determines the static pressure the fan must deliver to supply the air through the ducts.

Note that the fan has to supply more air than is indicated in paragraph 2, in order to make up for leakage in the duct system (often about 10 per cent) and in order to allow for heat transfer to or from the air in the duct while passing through unconditioned spaces.

In the case of factory-built conditioners, including a fan, the available static pressure of the conditioner fan determines the total allowable resistance in the duct system. In cases such as this it may be necessary to redesign the duct system for a smaller or greater resistance, to accommodate it to the characteristics of the fan supplied.

Pressure loss from air flowing in a duct occurs from frictional resistance to flow, from the shock losses of abrupt area changes, and from the shock and turbulence associated with change in direction. Pressure decreases also when the velocity of the air stream is increased. This change naturally occurs when the cross-sectional area for flow decreases (that is, in any converging section). Conversely, in a diverging or increasing section a velocity decrease can result in a pressure rise. Unfortunately, this reversed process of pressure rise when velocity decreases (known as diffusion) is a more difficult transformation. The air is necessarily flowing in the direction of increasing pressure, and turbulence loss due to separation of the stream from the walls of the passage can occur unless the divergence of the section is gradual (the included angle is less than 20°). With a very smooth passage and gradual taper, 85 to 75 per cent of the ideal pressure rise might occur, but with conventional duct arrangements 60 and 50 per cent are more representative limits. Pressure rise of this kind is called *regain*. Regain occurs also in ducts of constant cross-sectional area, as in a trunk duct beyond an outlet. For example, if a constant-diameter duct carrying 4000 cfm at 1600 fpm delivers 1500 cfm to a branch, and if the remaining 2500 cfm continue on, the velocity beyond the outlet can be only 1000 fpm, since $\frac{4000}{2500} = \frac{1600}{1000}$.

Regain Equations. By using representative coefficients for typical duct construction, the following equations for standard air can give the expected increase or decrease in pressure.

Static pressure regain (SPR), in inches of water, resulting from a lower final velocity (V_f fpm), with V_s fpm the velocity at the start of the transformation, is

$$\text{SPR} = 0.5\left[\left(\frac{V_s}{4005}\right)^2 - \left(\frac{V_f}{4005}\right)^2\right] \quad (12\text{-}27)$$

or

$$\text{SPR} = 0.5\left[\frac{(V_s - V_f)(V_s + V_f)}{16{,}040{,}000}\right] \quad (12\text{-}28)$$

A static pressure loss (SPL) occurs when the velocity increases; and, if a formula of the form of equation 12-27 or 12-28 is used, the answer is

Fluid Flow, Duct Design, and Air-Distribution Systems

minus, indicating a negative static pressure regain. The coefficient must also be modified. The relation to be used is

$$\text{SPL} = 1.05\left[\left(\frac{V_s}{4005}\right)^2 - \left(\frac{V_f}{4005}\right)^2\right] \quad (12\text{-}29)$$

or

$$\text{SPL} = 1.05\left[\frac{(V_s - V_f)(V_s + V_f)}{16{,}040{,}000}\right] \quad (12\text{-}30)$$

In the static-regain method of design, mentioned earlier in this section, the velocity in the main duct is reduced after each branch or take-off, and thus the static-pressure recovery from velocity reduction offsets or at least reduces the pressure loss in a following section. Because of this, essentially the same static pressure is available for all outlets, and remote branches are not at a disadvantage in terms of pressure required for distribution.

Example 12–3. Design a typical year-around air-conditioning duct system supplying several offices, as shown in Fig. 12–10. The maximum air capacity required in summer or winter is shown. The building limitations are such that the depth of the trunk ducts cannot exceed 16 in. and the vertical flues cannot be deeper than 10 in.

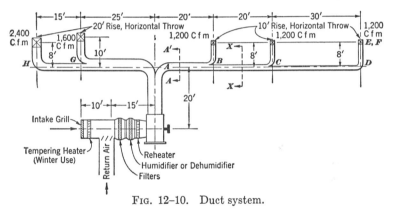

Fig. 12–10. Duct system.

Solution [by the Constant-Pressure-Drop (Equal-Friction) Method]: The total air to be delivered, neglecting duct leakage and heat gain or loss in the ducts, is indicated for the fan as $4000 + 3600 = 7600$ cfm.

The inlet duct is designed for maximum capacity at 7600 cfm, with a velocity of 1000 fpm selected from Table 12–4. The velocity in the trunk duct at the fan outlet is taken as 1200 fpm (Table 12–4), as noise should be minimized. The duct area is $A = HW/144$ sq ft. With Q in cfm and V in fpm,

$$Q = AV = \frac{HW}{144}V$$

With a maximum depth, H is 16 in. for the selected conditions, and

$$7600 = \left(\frac{16 \times W}{144}\right)(1200) \quad \text{or} \quad W = 57 \text{ in.}$$

The trunk-duct dimensions are 57 in. by 16 in. For a duct of this size, Table 12–2 shows the corresponding round size to be 31.45 (say 31.5) in. The friction chart in

432 Fluid Flow, Duct Design, and Air-Distribution Systems

Fig. 12–4 shows, for 7600 cfm in a 31.5-in. round duct, a loss of 0.076 in. of water per 100 ft of run.
The right branch of the trunk duct at AA' carries 3600 cfm; and, for 0.076 in. loss, Fig. 12–4 indicates an equivalent round size of 23.4 in. Similarly, at XX, where the required capacity is 2400 cfm, read 20.3 in.; and for run CDE at 1200 cfm, read 15.2 in. diam. Table 12–2 shows a 32-in. by 16-in. rectangular duct for the 23.4-in. round size at AA'; a 22-in. by 16-in. rectangular duct for the 20.3-in. round size at XX; a 12-in. by 16-in. rectangular duct for the 15.2-in. round size for CD and DE; and 20-in. by 10-in. rectangular size for EF. The rectangular sizes have been rounded off to the nearest inch (half-inch for small ducts).

The velocity in the riser EF is

$$V = \frac{144Q}{HW} = \frac{(144)(1200)}{(20)(10)} = 864 \text{ fpm}$$

As this is below the maximum recommended in Table 12–4, it will be considered acceptable.

TABULATION OF DATA AND RESULTS FOR EXAMPLE 12–3

Section or Elbow	Capacity (cfm)	Velocity (fpm)	Duct Diameter (in.)	Rectangular Size (in.)	Depth H Perpendicular to Plane of Bend	Width W in Plane of Bend	H/W	L/W	L (ft)
Fan	7600	1200	31.5	16 × 57	16	57	0.28	....	....
AA'	3600	1010	23.4	16 × 32	16	32	.....	....	....
A	....	....	...	....	..	..	0.5*	4.0	10.7
XX	2400	980	20.3	16 × 22	16	22	0.73	....	....
C to E	1200	900	15.2	16 × 12	16	12	.....	....	....
D	....	....	...	....	..	..	1.33*	4.7	4.7
E to F	1200	864	15.2	20 × 10	20	10	.....	....	....
E	....	....	...	....	..	..	2†	25.0	21.0
F	....	....	...	....	..	..	2†	25.0	21.0

Total of additional equivalent length for elbows..........................57.4

* R/W taken as 1.5.
† R/W taken as 0.75.

These data are expressed in tabular form in the accompanying table, for the right-hand trunk and longest branch, with Fig. 12–8 being used to find the L/W values for the appropriate elbows.

For the left branch, employ the same methods. For a 0.076-in. loss per 100 ft, the trunk to the right of G at 4000 cfm shows 26.6 in. round; and the trunk GH at 2400 cfm shows 20.3 in. round. The corresponding rectangular sizes are 39 by 16 in. before G, 22 by 16 in. from G to H, and 38 by 10 in. in the riser above H.

Calculation of Frictional Loss. The frictional loss is obviously greatest in the extreme right run, but if there should be any doubt the extreme left run and other runs should also be checked.

The right run from the fan out has a length of

20 + 20 + 20 + 30 + 8 + 10 = 108 ft
From the table, equivalent additional length for elbows = 57.4

Total equivalent length = 165.4 ft

Fluid Flow, Duct Design, and Air-Distribution Systems 433

A minimum factor of safety of 10 per cent should be added to this length. As the loss is 0.076 in. of water per 100 ft, the total duct and elbow friction loss is:

$$165.4 \times 1.10 \times \frac{0.076}{100} = 0.138 \text{ in. of water}$$

Some over-all static regain occurs in this run, and it can be found by equation 12-28. The starting and final velocities in the fan run and in the riser EF being used, we find that

$$\text{SPR} = 0.5\left[\frac{(1200 - 864)(1200 + 864)}{16,040,000}\right] = 0.022 \text{ in. of water}$$

The net static pressure loss in the duct, bends, etc. is thus

$$0.138 - 0.022 = 0.116 \text{ in. of water}$$

In addition to this net static pressure loss in the duct system, the fan is required to overcome the losses in the accessory equipment and in the return system. Computed or representative values for these losses will now be tabulated. The first item is the static pressure loss in moving the air through the intake grille and louvers, or 0.50 vel head (from Table 12-3).

$(1200/4005)^2 \times 0.5 = 0.090 \times 0.5 \ldots$ 0.045
Tempering heater (mfr.'s data)........ 0.10
Filters (mfr.'s data).................. 0.25
Dehumidifier (mfr.'s data)............ 0.22
Reheater (mfr.'s data)................ 0.10
Outlet grille (mfr.'s data)............. 0.057
Net static pressure loss, computed before 0.116

Total static pressure............ 0.888 in. of water

The fan selected must have a static pressure not less than 0.888 in. of water when delivering 7600 cfm. The total pressure of the fan, which takes into consideration the fact that the air is inducted and accelerated to the fan outlet velocity, is obviously greater than 0.888 in. of water. Fan selection methods are considered later in the chapter. In the event that the return system is also handled by this fan, pressure to cover the return-system loss should also be added.

For the other branches and the left run, similar methods of computation can be employed.

The preceding example was worked by the constant-pressure-drop method; that is, the frictional loss per foot of equivalent duct length was assumed constant. This method is generally to be preferred for solving problems in duct design, because less experience in selecting velocities is required. The branches closest to the fan have the highest static pressure, and to prevent them from getting more than their share of air they must have adequate damper control. This general method applies equally well to exhaust systems, and the procedure to be followed is the same. If it is desired to avoid much use of dampers, earlier branches can be designed for higher velocities and, therefore, for higher losses. This possibility is often limited by the maximum allowable velocity dictated by considerations of noise generation and duct vibration or bulging.

434 Fluid Flow, Duct Design, and Air-Distribution Systems

The previously mentioned *assumed-velocity method* may be used satisfactorily in simple systems where the duct losses form a relatively small part of the total loss. Even with the constant pressure drop, to start the problem a trial velocity may be assumed, as in example 12–3. A rational friction drop could have been assumed instead of a velocity, if this procedure had been desired.

Many variations in design procedure can be made in accordance with the basic rules previously outlined. A frequently used design involves delivering the air into a plenum (pressure) chamber and then employing a number of separate small ducts leading to the point of use. The plenum pressure must be maintained high enough to deliver the required quantity of air through the duct of greatest resistance.

Fundamentally the *balanced-pressure-loss method* should be most satisfactory, and the extra labor of adjusting the whole design to bring about equal total friction loss may not be difficult. However, for a short duct directly off a plenum chamber or in the main duct close to the outlet of a fan, the static pressure is close to the maximum of the system. It may thus be impossible, merely by increasing the velocity within reasonable limits, to set up the required pressure loss in this short duct; and dampers must be employed to utilize the available pressure with or without a restricted opening, such as an orifice at the supply point to supplement the dampers. In this method the basic problem is simply to design the system so that the total pressure drop from the fan to any grille is the same, and a dependence on dampers only for minor final adjustments is the desired end.

Example 12–4. For the duct system of Fig. 12–10, design the first branch on the right side to have the same total friction loss as the extreme right (longest) branch.

Solution: As determined in example 12–3, the static pressure on discharge from the fan, is 0.888 in. of water, less the pressure utilized for bringing in and carrying the air through the various devices preceding it, or $0.888 - (0.045 + 0.10 + 0.25 + 0.22 + 0.10) = 0.888 - 0.715 = 0.173$ in. of water. At leaving the fan, the total static pressure is thus 0.173 in. of water, and the velocity head is 0.09 in. for 1200 fpm (read from Fig. 12–6).

For the main duct system from the fan to point B, there is 40 ft of run plus 10.7 ft of equivalent length for elbow A; and, as this was designed for 0.076 in. of loss per 100 ft, the pressure loss is $0.076 \times 50.7/100 = 0.037$ in. of water. Thus the static pressure at B is $0.173 - 0.037 = 0.136$ in. of water. The grille absorbs 0.057 in., leaving $0.136 - 0.057 = 0.079$ in. of water; thus, disregarding velocity heads for the moment, 0.079 in. of water must be absorbed in the run of duct from B to the grille outlet—which consists of 18 ft and three elbows and in which the entry into the branch is considered as an elbow.

Consider first that the elbows in the circuit from B to its outlet grille are the same in size and type as that of branch DEF. Then the equivalent length would be $18 + 4.7 + 21 + 21 = 64.7$ ft. With the same design friction loss of 0.076, the friction would be $0.076 \times 64.7/100 = 0.049$ in. of water, which is less than the 0.079 in. of water required.

Fluid Flow, Duct Design, and Air-Distribution Systems 435

The inlet to this branch is essentially an elbow, and if so taken the circuit at B to its outlet resembles that of branch DEF, and its elbow data can be used as a first trial for equivalent length, giving $4.7 + 21 + 21 + 18$ ft of run = 64.7 ft. Thus

$$\frac{64.7}{100} \times F = 0.079$$

and $F = 0.122$ (say 0.12) in. of water per 100 ft is the trial friction loss factor to be used with Fig. 12–4 for resizing the branch starting at B. On this basis, to carry the 1200 cfm a round size 14.5 in. in diameter is indicated. For this the corresponding rectangular sizes are 16 in. by 11 in. in the horizontal run and 10 in. by 18 in. in the vertical run.

The equivalent additional lengths for the elbows, on the basis of the new sizes, and using elbow data similar to that in the table of example 12–3, are

$$18 \times \tfrac{10}{12} + 18 \times \tfrac{10}{12} + 4.7 \times \tfrac{11}{12} = 34.3 \text{ ft}$$

and $18 + 34.3 = 52.3$ ft equivalent length is sufficiently close to the assumed 64.7 ft as not to require a significant change in the indicated duct sizes of 16 in. by 11 in. and 10 in. by 18 in. for the design modified to absorb more friction. For any branch a similar

TABLE 12–6
Recommended Maximum Pressure Losses or
Gains in Branch Outlet Runs *
(With Pressures at First and Last Outlet)

Item	Average Outlet Pressure (in. of water)									
	.025	.05	.075	.10	.15	.20	.25	.30	.40	.50
Maximum allowable pressure loss in outlet run (in.) †	.01	.02	.03	.04	.06	.08	.10	.12	.16	.20
Pressure at first outlet (in.)	.03	.06	.09	.12	.18	.24	.30	.36	.48	.60
Pressure at last outlet (in.)	.02	.04	.06	.08	.12	.16	.20	.24	.32	.40

* Reprinted by permission of the Carrier Corporation.
† Consideration of the outlet selected and the individual job requirement will often indicate a selection of a loss less than the maximum value.

procedure can be followed, namely that of estimating a total equivalent length and finding the needed friction loss for 100 ft to absorb the static head available. Another procedure where leeway is available is to select a higher velocity in the branch and carry out trial computations on the basis of the assumed velocity.

In the present case, the velocity is somewhat high from a viewpoint of noise and good practice, namely

$$V = \frac{1200 \times 144}{18 \times 10} = 960 \text{ fpm}$$

However, if it is felt that this is definitely too high, little can be done to absorb the static pressure except through the use of dampers which in themselves are noisy, or through a redesign of the main duct system. The use of sound-deadening material in the riser may also be considered.

A reasonable grille velocity based on effective face area is 600 fpm. By Fig. 12–6, velocities of 600 fpm and 960 fpm show velocity pressures of 0.023 and 0.056 in. of water,

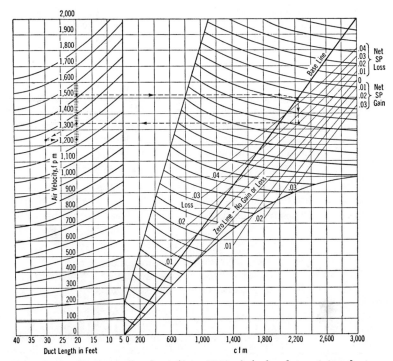

Fig. 12-11. Static-regain chart (0 to 3000 cfm), for determining duct velocity after each outlet. Chart based on 50 per cent velocity-head recovery. (Courtesy Carrier Corporation.)

EXAMPLE OF USE. Given: Velocity before outlet, 1520 fpm; air quantity after outlet, 2000 cfm; allowable net specific loss or gain, 0.02 in. (loss); length of duct section, 30 ft. Find velocity after outlet (from chart). *Solution:* (1) On left ordinate of chart, find air velocity of 1520 fpm, then move horizontally to the right until 2000-cfm line is joined; (2) run parallel to curved lines until diagonal base line is joined; (3) drop vertically to 0.02 net specific loss line; (4) go horizontally left to air-velocity base line; (5) run parallel to curved lines until 30-ft duct-length line is joined and read answer on air-velocity base line. The result is 1240 fpm (velocity after outlet).

respectively. The leaving velocity pressure of 0.023 in. is dissipated directly into the room space, and thus, of the original velocity pressure of 0.056 in. in the riser, 0.033 in. has not been accounted for. However, if the factor 0.5 in equation 12-27 is considered, it appears that only 0.033×0.5, or 0.016 in., is not accounted for. By using a nondirect take-off approaching a full right angle, the velocity head transmitted into the branch can be reduced to minor proportions.

Even with this method, which equalizes the pressure losses in each and every circuit, it is still necessary to supply dampers to make adjustments for minor inaccuracies in design and for variations in construction, as well as to permit modification of operating conditions, if desired, at some later time.

Fluid Flow, Duct Design, and Air-Distribution Systems

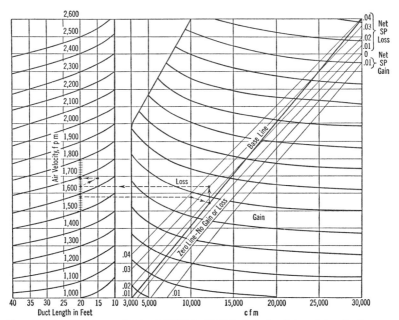

Fig. 12–12. Static-regain chart (3000 to 30,000 cfm), for determining duct velocity after each outlet. Chart based on 50 per cent velocity-head recovery. (Courtesy Carrier Corporation.)

EXAMPLE OF USE. Given: Velocity before outlet, 1580 fpm; air quantity after outlet, 10,000 cfm; allowable net specific loss or gain, 0.03 in. (loss); length of duct section, 15 ft. Find velocity after outlet (from chart). *Solution:* (1) On left ordinate of chart, find air velocity of 1580 fpm, then move horizontally to the right until 10,000-cfm line is joined; (2) run parallel to curved lines until diagonal base line is joined; (3) rise vertically to 0.03 net specific loss line; (4) go horizontally left to air-velocity base line; (5) run parallel to curved lines until 15-ft duct-length line is joined and read answer on air-velocity base line. The result is 1690 fpm (velocity after outlet).

12–6. STATIC-REGAIN METHOD OF DESIGN

The static pressure regain as the velocity is reduced in a duct system can be employed in many variations. One of the most satisfactory modifications consists of reducing the velocity after each branch by an amount sufficient to compensate for the frictional pressure loss in the succeeding section. Figures 12–11 and 12–12 have been prepared to facilitate computations of this sort, and Table 12–6 can also be used to advantage where the pressure drop between outlets is large because of the great length of the duct or because of fitting losses.

A simple example will be worked in two ways to illustrate the procedure in using static-regain methods.

Example 12–5. A duct system for a school must satisfy the conditions indicated in the diagrammatic sketch in Fig. 12–13. Size the duct system, following a reasonable

438 Fluid Flow, Duct Design, and Air-Distribution Systems

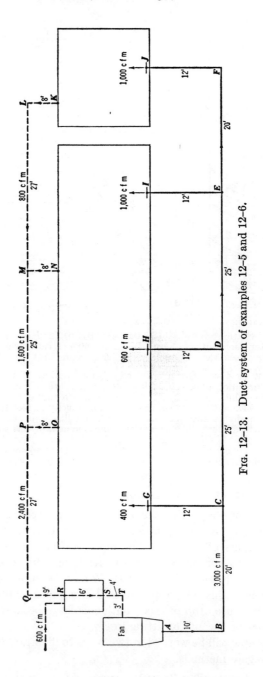

FIG. 12-13. Duct system of examples 12-5 and 12-6.

Fluid Flow, Duct Design, and Air-Distribution Systems 439

design and using a constant-pressure-drop method. The maximum depth of duct cannot exceed 10 in. on one side, and the pressure at the outlet grilles must not be less than 0.05 in. of water.

Solution: From Table 12-4 select a 1300-fpm velocity in the 3000-cfm section ABC. Find the duct size from the relation

$$A = \frac{W \times H}{144} = \frac{Q}{V}$$

Thus,

$$\frac{W \times 10}{144} = \frac{3000}{1300}$$

and

$$W = 33.2 \text{ in.}$$

Round off to 33 in. by 10 in.

From Table 12-2, a 33-in. by 10-in. duct corresponds to a 19.05-in. round duct. Figure 12-4 shows that 3000 cfm in a 19.05-in. duct produces a loss of 0.16 in. of water per 100 ft. Consider the fan to deliver into a duct-run just under a basement ceiling.

Run ABC has one bend at B, which can be assumed to be a hard bend (i. e., on the 33-in. width, in contrast to one on the 10-in. depth). By Fig. 12-8, for a 1.5 radius ratio, R/W, and for $H/W = 10/33 = 0.3$,

$$L = 3.6W = 3.6 \times \tfrac{33}{12} = 10 \text{ ft}$$

Total ELD of run $ABC = 10 + 20 + 10 = 40$ ft

Loss in $ABC = 0.16 \times \tfrac{40}{100} = 0.064$ in. of water

For flow in CD at 0.16 in. per 100 ft and 2600 cfm, Fig. 12-4 shows an 18-in. round duct; and from Table 12-2 the corresponding rectangular size is 29 in. by 10 in. Therefore

$$\text{Velocity in } CD = \frac{Q}{A} = \frac{2600 \times 144}{29 \times 10} = 1292 \text{ fpm}$$

Loss for $CD = 0.16 \times \tfrac{25}{100} = 0.04$ in. of water

For flow in DE at 0.16 in. per 100 ft and 2000 cfm, Fig. 12-4 shows a 16.4-in. round duct; and from Table 12-2 the rectangular size is 23.4 in. by 10 in. Round off to 23 in. by 10 in. Therefore

$$\text{Velocity in } DE = \frac{Q}{A} = \frac{2000 \times 144}{23 \times 10} = 1252 \text{ fpm}$$

Loss for $CD = 0.16 \times \tfrac{25}{100} = 0.04$ in. of water

For flow in EF at 0.16 in. per 100 ft and 1000 cfm, the required round size is 12.5 in. and the corresponding rectangular size is 13.2 in. by 10 in. Round off to 13 in. by 10 in. Therefore

$$\text{Velocity in } EF = \frac{1000 \times 144}{13 \times 10} = 1107 \text{ fpm}$$

Loss for $EF = 0.16 \times \tfrac{20}{100} = 0.032$ in. of water

The flow in branch FJ is 1000 cfm, and the velocity should be lowered from the 1107 fpm in EF. From Table 12-4 select 800 fpm; then the duct size is 18 in. by 10 in., since

$$\frac{W \times 10}{144} = \frac{1000}{800}$$

and

$$W = 18 \text{ in.}$$

From Table 12–2, this size corresponds to a 14.5-in. round duct; and from Fig. 12–4 the loss rate is 0.075 in. per 100 feet. Consider the bend at F to be hard (on the 18-in. width) and assume that a second bend at J just before the grille is easy (on the 10-in. dimension). From Fig. 12–8, for F at $R/W = 1.5$ and $H/W = 10/18$, read $L = 4.0W = 6$ ft; and for J leading into the outlet grille, use $R/W = 1.0$, read $H/W = 18/10$, and compute L as 12 ft. Therefore the total equivalent length from bend F to bend J is $6 + 12 + 12 = 30$ ft, and

$$\text{Loss in } FJ = 0.075 \times \tfrac{30}{100} = 0.0225 \text{ in. of water}$$

Even though a regain-type design is not considered here, there is necessarily some regain throughout the whole system; however, for this problem, regain will be computed and used only for the branch run into FJ, for which there is a large change in velocity. By equation 12–28,

$$\text{SPR} = 0.5 \left[\frac{(1107 - 800)(1107 + 800)}{16{,}040{,}000}\right] = 0.018 \text{ in. of water}$$

The results are shown in the accompanying tabulation. It can be seen that the total pressure loss from A to J is 0.181 in. of water. If 0.050 in. is allowed at the outlet grille, it is obvious that the static pressure at A (the fan outlet) must be $0.181 + 0.05$, or 0.231 in. of water. Notice that the respective pressures at branch outlets C, D, and E are higher than that at F, and thus decided dampering is required. Computations for these branches have not been included in the solution, nor has the return circuit been considered.

Example 12–6. Rework example 12–5 to utilize static regain effectively. The average outlet pressure to the branches is 0.1 in. of water.

Solution: For run ABC, with 3000 cfm at 1300 fpm,

$$A = \frac{W \times H}{144} = \frac{W \times 10}{144} = \frac{Q}{V} = \frac{3000}{1300}$$

Thus the broad dimension of the duct is

$$W = 33.2 \text{ (say 33) in.}$$

This makes V but slightly higher than 1300 fpm.

Table 12–6 shows that, for an average outlet pressure of 0.1 in. to any branch, the recommended maximum pressure loss from the first outlet to the last one should be 0.04 in. of water and thus the static pressure at the first outlet should be 0.12 in. of water.

Example 12–5 showed that, for 3000 cfm at 1300 fpm, the loss is 0.16 in. per 100 ft; and, for the run ABC, with an equivalent length of 40 ft, the pressure drop is 0.064 in. Thus the pressure at the fan (point A) must be $0.064 + 0.12 = 0.184$ in.

Using the maximum pressure loss of 0.04 in. from the first outlet C to the last outlet F, assume that this loss will occur in three equal steps; i. e., take $0.04/3 = 0.013$ in. as the drop for CD, and the same drop for DE and EF.

Consider the required regain for run CD and use the static-regain chart of Fig. 12–11. Enter the left axis of the chart at 1300 fpm and run to the intersection with the new air flow of 2600 cfm. Follow parallel to the curved guide line back to the base line. From this point, drop vertically to the line for 0.013 net specific loss (estimated). From this point move horizontally toward the left to the air velocity axis, follow parallel to the

Fluid Flow, Duct Design, and Air-Distribution Systems

Combined Tabulation of Results for Examples 12-5 and 12-6

Section	Flow (cfm)	Velocity Assumed or by Constant Pressure Drop	Duct Size W × H (in.)	Diameter of Equivalent Round (in.)	Equivalent Length (ft)	Pressure Loss (in. of water)	Static Pressure (in. of water)
A........	....	....	....	....	..	....	0.231 (A)
ABC.....	3000	1300	33 × 10	19.05	40	0.064	0.167 (C)
CD.......	2600	1292	29 × 10	18.0	25	0.040	0.127 (D)
DE.......	2000	1252	23 × 10	16.4	25	0.040	0.087 (E)
EF.......	1000	1107	13 × 10	12.5	20	0.032	0.055 (F)
FJ.......	1000	800	18 × 10	14.5	30	0.023 −.018	0.050 (J)
A–J......	....	....	....	....	..	0.181	0.050 (J)

Results for Example 12-6 by Static Regain

Section	Flow (cfm)	Velocity Assumed or by Constant Pressure Drop	Duct Size W × H (in.)	Diameter of Equivalent Round (in.)	Equivalent Length (ft)	Pressure Loss (in. of water)	Static Pressure (in. of water)
A........	....	....	....	....	..	....	0.184 (A)
ABC.....	3000	1300	33 × 10	19.05	40	0.064	0.120 (C)
CD.......	2600	1120	33.4 × 10 (33 × 10)	19.1	25	0.013	0.107 (D)
DE.......	2000	990	29 × 10	....	25	0.013	0.094 (E)
EF.......	1000	870	17 × 10	....	20	0.013	0.081 (F)
FJ.......	1000	800	10 × 18	14.5	30	0.023 −.004	0.062 (J)
A–J......	....	....	....	....	..	0.122	0.062 (J)

curved guide line to the 25-ft duct length, and then proceed horizontally back to the velocity axis. Thus, find 1120 fpm as the required velocity.

For run CD at 1120 fpm,

$$A = \frac{2600}{1120} = \frac{W \times H}{144} = \frac{W \times 10}{144}$$

Thus the broad dimension of the duct is

$$W = 33.4 \text{ in.}$$

Normally this would be rounded off to 33 in., but it will be kept at 33.4 for this example in order to illustrate the complete velocity computation and check-calculation.

For run DE at 2000 cfm, starting from 1120 fpm at net specific loss of 0.013, the velocity is found from Fig. 12-11 to be 990 fpm. Therefore

$$A = \frac{2000}{990} = \frac{W \times 10}{144}$$

and

$$W = 29.1 \text{ in.}$$

The dimensions are thus 29 in. by 10 in.

For run EF at 1000 cfm, starting from the 990 fpm just before this section, with 0.013 net specific loss, the velocity is found from Fig. 12-11, for a 20-ft run, to be 870 fpm. Therefore

$$A = \frac{1000}{870} = \frac{W \times 10}{144}$$

and
$$W = 16.6 \text{ (say 17) in.}$$

The dimensions are therefore 17 in. by 10 in.

Although not necessary in the solution, it is of interest to check how closely the solution by Fig. 12-11 meets the specific conditions. Consider section CD: The velocity changes from 1300 in run BC to 1120 in run CD. Use equation 12-27 or 12-28 to show the regain. Using equation 12-28,

$$\text{SPR} = 0.5 \left[\frac{(1300 - 1120)(1300 + 1120)}{16{,}040{,}000} \right] = 0.0136 \text{ in. of water}$$

For a 33.4-in. by 10-in. duct the equivalent round duct, from Table 12-2, is 19.1 in.; and, from Fig. 12-4, for 2600 cfm and 19.1 in. diameter, the loss is 0.11 in. per 100 ft. For 25 ft the loss becomes 0.0275 in. Since the regain partly offsets the loss, the net pressure loss is $0.0275 - 0.0136 = 0.0139$ in. of water. This shows that the design condition is met with reasonable closeness.

For FJ, which is a branch and is not considered in the regain type of analysis of the main duct, the basic calculation made in example 12-5 showed a loss of 0.023 in. However, some regain also applies here. By equation 12-28, it amounts to

$$\text{SPR} = 0.5 \left[\frac{(870 - 800)(870 + 800)}{16{,}040{,}000} \right] = 0.0036 \text{ in. of water}$$

This has been rounded off to 0.004 in the tabulation. When the tabulations for examples 12-5 and 6 are compared (see facing page), it is seen that the supply pressures to the branches C, D, E, and F are much closer in example 12-6, although even here some dampering is required. The pressure loss by the static regain method is lower, but at the same time it should be noted that the ducts are larger. On the whole, the probability of better balance and performance in the system, when the computations are based on static regain, speaks well for this method of computation, even though moderate dampering is required at J.

Return Ducts. The computations in the design of the return duct system are most usually based on the constant-pressure-drop (equal-friction) method. The net amount of air to be returned must be known, and the total pressure drop of the return system must not exceed the allowable negative suction pressure of the fan. The return system must be laid out as carefully as the supply system. Dampers should be supplied in the branches to permit adjustment of flow. If the system as originally designed has greater loss than that permissible for the fan, it must be resized (increased) to reduce the loss.

Heat Loss or Gain to Ducts. When ductwork, carrying cooled air, passes through unconditioned (warm) spaces, its temperature is raised by the heat which is transmitted into the duct. The resulting warming of

Fluid Flow, Duct Design, and Air-Distribution Systems 443

the air is a complex phenomenon, and its rate depends on the ratio of surface perimeter to cross-sectional area; the length of duct; the temperature difference; the air velocity as it affects surface convection coefficients; the composite radiation and convection effects on both sides of the duct; the type of duct surface; and the effectiveness and kind of insulation. In the case of polished aluminum or galvanized iron, the low surface emissivity to radiation may make the bare duct more effective in retarding heat transfer than a duct covering of relatively poor insulation effectiveness, such as thin asbestos sheeting. Similar conditions hold when ducts carrying warm air pass through spaces at lower ambient temperature and cooling of the duct air takes place.

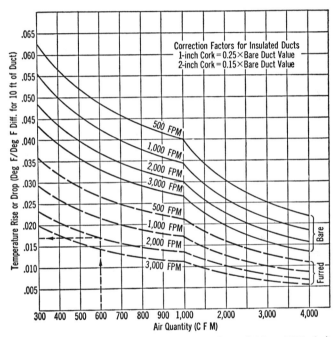

FIG. 12-14. Temperature rise or drop in ducts (300 to 4000 cfm). (Courtesy Carrier Corporation.)

Figures 12-14 and 12-15 have been prepared for ducts with aspect ratios not greater than 2 to 1. For ducts flatter than this, the following multiplying correction factors should be applied: 1.10 for a 3:1 ratio of sides; 1.18 for 4:1; 1.26 for 5:1; 1.35 for 6:1; 1.47 for 7:1; 1.50 for 8:1. In computing the values, the outside film coefficient f was taken as 1.65, with inside coefficients ranging from 1.7 to 7.2 at the higher velocities. Appropriate thermal resistances were applied for the various insulations. Furred ducts are covered with $\frac{3}{4}$ in. of metal lath and gypsum plaster.

Example 12-7. A 12-in. by 36-in. duct 80 ft long carries 3000 cfm through a space at 85 F. Chilled air enters the duct at 64 F. Find the temperature rise of the air leaving (a) if the duct is bare and (b) if the duct has a 1-in. cork covering.

Solution: (a) The temperature difference is based on conditions at start of the duct passage absorbing (or delivering) heat. In this case $\Delta t = 85 - 64 = 21$ deg. The velocity can be found from the relation

$$V = \frac{\text{cfm}}{\text{sq ft flow area}} = \frac{3000 \times 144}{12 \times 36} = 1000 \text{ fpm}$$

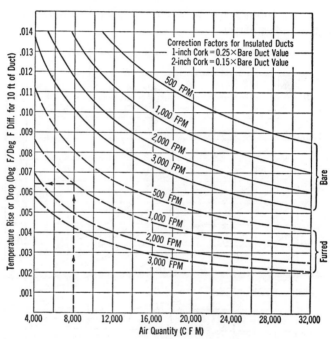

Fig. 12-15. Temperature rise or drop in ducts (4000 to 32,000 cfm). (Courtesy Carrier Corporation.)

In Fig. 12-14, start at 3000 cfm, run up to the intersection with the 1000-fpm bare-duct line, and, from this intersection, run over to the left axis. Read 0.0212 deg F as the temperature rise per degree Fahrenheit difference for each 10 ft of duct run. The temperature rise is thus

$$0.0212 \times \Delta t \times L = 0.0212 \times 21 \times \tfrac{80}{10} = 3.6 \text{ deg F}$$

Apply an aspect-ratio correction of 1.10 for 3 to 1 sides. Thus,

$$3.6 \times 1.10 = 3.96 \text{ (say 4) deg F} \qquad \qquad \textit{Ans.}$$

b) For the insulated duct apply the factor of 0.25 from Fig. 12-14. Thus the temperature rise is

$$4 \times 0.25 = 1.0 \text{ deg F} \qquad \qquad \textit{Ans.}$$

Fluid Flow, Duct Design, and Air-Distribution Systems   445

12-7. DUCT CONSTRUCTION

Some of the seams used in making ducts are illustrated in Fig. 12-16. The type of joint used is influenced somewhat by the construction facilities available and the gage of metal. The metal must be heavy enough to resist vibration and sagging between braces, and therefore the thickness is increased as the diameter or width of the duct increases. Tables 12-7 and 12-8 give recommended weights for ductwork. Large-size ducts must always be braced with transverse exterior tees (or angles), at intervals close enough to prevent sagging and possible vibration (see Fig. 12-16).

Since the end of World War II extensive use has been made of aluminum for ductwork. Aluminum is light in weight and is not difficult to fabricate; and because it is thicker than steel, for the same weight, it tends to have greater rigidity. It does have a higher coefficient of thermal expansion, which must be allowed for where there are extreme temperature changes. Recommended practices for aluminum duct construction are indicated in Table 12-7. Thickness designations of aluminum should be given in decimal fractions of inches, although gage numbers are frequently used. It is unfortunate that the gage system used for aluminum differs from that used for steel. Aluminum sheet is described by Brown and Sharp (B&S) gage numbers, and steel sheet is described by U. S. Standard gage numbers. Table 12-8 shows these gage numbers and corresponding sheet thicknesses. Aluminum gage numbers are about two numbers lower than steel gage numbers for comparative thicknesses.

It should be recognized that conventional ductwork is not leakproof and that all of the air is not delivered by the fan in the space intended. It is customary to allow 10 per cent more air flow from the fan to compensate for leakage, if the ductwork runs outside of the treated space.

In the installation of equipment, such as heaters and filters in air ducts, abrupt changes in size should be avoided. Figure 12-17 shows some arrangements which have been found satisfactory. Abrupt changes in direction and other resistance-creating conditions cause noise and reduce volume. Interior vanes should therefore be used at elbows, and obstructions should be streamlined. Some ducts must have sound-asborbing material applied on the inside. Insulation against heat transfer is usually applied on the outside, except for the reflecting types used in ducts carrying high-temperature air.

12-8. AIR DELIVERY AND DISTRIBUTION

Air must be delivered into the heated or conditioned space and distributed to the desired points at temperatures and velocities which are not objectionable to the occupants. Temperature differentials in the occupied zone of a room should not vary more than 2 deg, although in the cooling season slightly greater differentials may be permissible. An air movement

TABLE 12-7
Rectangular Steel and Aluminum Duct Construction Practice

Sheet Thickness		Maximum Side (in.)	Transverse Joint Connections	Bracing
Steel (U.S. Std gage)	Aluminum (B&S gage)			
26	24	Up to 12	S, drive, pocket or bar slips on 7' 10" centers	None
24	22	13–24	S, drive, pocket or bar slips on 7' 10" centers	None
24	22	25–30	S, drive, 1" pocket or 1" bar slips on 7' 10" centers	$1'' \times 1'' \times \frac{1}{8}''$ angles, 4 ft from joint
22	20	31–40	Drive, 1" pocket or 1" bar slips on 7' 10" centers	$1'' \times 1'' \times \frac{1}{8}''$ angles, 4 ft from joint
22	20	41–60	$1\frac{1}{2}''$ angle connections, $1\frac{1}{2}''$ pocket or $1\frac{1}{2}''$ bar slips with $1\frac{3}{8}'' \times \frac{1}{8}''$ bar reinforcing on 7' 10" centers	$1\frac{1}{2}'' \times 1\frac{1}{2}'' \times \frac{1}{8}''$ angles 4 ft from joint
20	18	61–90	$1\frac{1}{2}''$ angle connections, or $1\frac{1}{2}''$ pocket or $1\frac{1}{2}''$ bar slips on 3' 9" maximum centers with $1\frac{3}{8}'' \times \frac{1}{8}''$ bar reinforcing	$1\frac{1}{2}'' \times 1\frac{1}{2}'' \times \frac{1}{8}''$ diagonal angles, or $1\frac{1}{2}'' \times 1\frac{1}{2}'' \times \frac{1}{8}''$ angles 2 ft from joint
18	16	91 and up	2" angle connections or $1\frac{1}{2}''$ pocket or $1\frac{1}{2}''$ bar slips on 3' 9" maximum centers with $1\frac{3}{8}'' \times \frac{1}{8}''$ bar reinforcing	$1\frac{1}{2}'' \times 1\frac{1}{2}'' \times \frac{1}{8}''$ diagonal angles, or $1\frac{1}{2}'' \times 1\frac{1}{2}'' \times \frac{1}{8}''$ angles 2 ft from joint

Notes. Unless heavy insulation is used, ducts 18 in. and larger should be cross-broken. Cross-breaking may be omitted if the aluminum sheet thickness is increased by about two gage numbers.
Bracing angles may be omitted on duct sizes 25 in. to 60 in. inclusive (maximum side) if 3-ft 9-in. section lengths are used.
Round ducts should be constructed of aluminum of the same thickness as that specified for rectangular ducts having the same cross-sectional area.

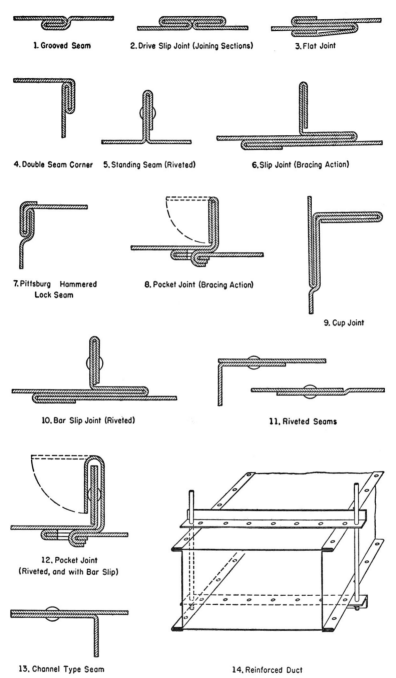

FIG. 12-16. Typical duct joints.

TABLE 12-8

COMPARATIVE GAGE NUMBERS AND WEIGHTS OF ALUMINUM AND STEEL SHEET

ALUMINUM			STEEL			ALUMINUM			STEEL	
Weight of 2s (lb/sq ft)	B&S Gage	THICKNESS (IN.)	Manufacturer's Std Gage	Weight (lb/sq ft)	Weight of 2s (lb/sq ft)	B&S Gage	THICKNESS (IN.)	Manufacturer's Std Gage	Weight (lb/sq ft)
...	...	0.2391	3	10.000	...	...	0.0673	15	2.812
3.21	3	0.2294	...	...	0.905	14	0.0641	...	...
...	...	0.2242	4	9.375	...	...	0.0598	16	2.500
...	4	0.2092	5	8.750	0.806	15	0.0571	...	...
2.88	...	0.2043	...	...	...	...	0.0538	17	2.250
...	5	0.1943	6	8.125	0.717	16	0.0508	...	...
2.57	...	0.1819	7	7.500	...	...	0.0478	18	2.000
...	6	0.1793	8	6.875	0.639	17	0.0453	...	...
...	...	0.1644	...	...	...	...	0.0418	19	1.750
2.29	...	0.1620	9	6.250	0.569	18	0.0403	...	...
...	7	0.1495	...	...	0.507	19	0.0359	20	1.500
2.04	...	0.1443	10	5.625	...	...	0.0329	21	1.375
...	8	0.1345	...	...	0.452	20	0.0320	...	...
1.81	...	0.1285	11	5.000	...	...	0.0299	22	1.250
...	9	0.1196	...	...	0.402	21	0.0285	...	...
1.61	...	0.1144	12	4.375	...	...	0.0269	23	1.125
...	10	0.1046	...	...	0.357	22	0.0253	...	...
1.44	11	0.1019	13	3.750	...	...	0.0239	24	1.000
1.28	...	0.0907	...	...	0.319	23	0.0226	...	...
...	12	0.0897	14	3.125	...	...	0.0209	25	0.875
...	...	0.0808	...	...	0.284	24	0.0201	...	...
1.14	...	0.0747	...	...	0.253	25	0.0179	26	0.749
...	13	0.0720	...	...	...	...	0.0164	27	0.686
1.01					0.224	26	0.0159	...	...

Fluid Flow, Duct Design, and Air-Distribution Systems 449

of about 25 fpm around the bodies of the people is desirable, but this velocity is usually exceeded when an appreciable volume of air has to be handled. About 50 fpm is a maximum for comfort with people seated, and a slightly higher maximum can be tolerated where people are moving. Air directed into the faces of occupants is to be preferred to air directed toward their backs or sides. Downflow with regard to occupants is preferable to upflow.

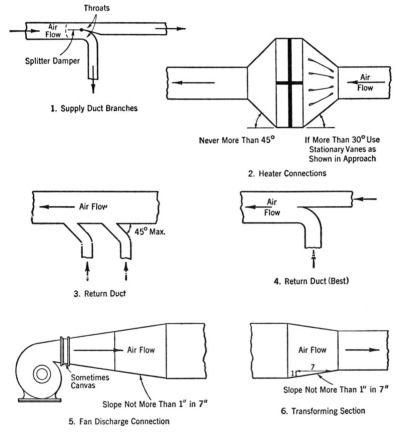

Fig. 12-17. Equipment in ducts and junction arrangements.

The *throw*, or *carry*, of an air stream is the distance, perpendicular to the face of the outlet, through which a directed air stream travels on leaving the outlet. When the stream velocity is reduced to a value of about 75 fpm, the directed energy, or throw, has largely been expended. Air will drop or rise as it leaves the outlet, the direction depending on temperature (density) differences between the entering and room air and the velocity of the entering stream. Supply air that is cooler than the air in the room falls, and warmer supply air rises. Every moving air stream mixes with and carries along some of the room air, and the momentum of the original air stream is decreased in accelerating this room air. This process of mixing, or *induction*, causes the air stream to spread out as it progresses forward. The induction of room air increases as the perimeter of the air cross-section leaving the grille is increased. This induction is thus a maximum with a flat rectangular-shaped outlet, and a minimum with a round cross-sectioned outlet of the same area.

In the case of an open outlet without vanes, the spread of the air leaving the outlet gives an included angle between the sides of the air stream of about 12 to 15 deg. The spread and induction of an air stream can be visualized by referring to Fig. 12–18. Straight vanes do not change this angle to any extent, and about 13 deg can be taken as average. Converging vanes in the outlet make the air stream converge at first, but the spread

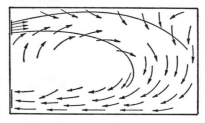

FIG. 12–18. Desirable supply and return grille locations.

which then follows is greater than that produced with straight vanes. Diverging vanes greatly increase the spread and are used to decrease the throw when this is desired. Where outlets without straightening vanes are used on the side of a duct, the air stream leaving the outlet will not flow out perpendicular to the duct but will make an angle of less than 90 deg with the direction of motion of the air in the duct. This divergence is caused by the velocity of the air in the duct tending to make the air carry forward as it leaves the grille. With high duct velocities a considerable deflection of the air stream from normal may occur. If this deflection is objectionable, a collar section with inserted vanes must be attached to the side of the duct behind the outlet attachment.

The location of outlets in a given space is often seriously affected by architectural features of the space, such as columns, beams below ceiling level, lighting fixtures, and permanent furniture, and by the location of the ducts in the walls. However, certain requirements must be met: Objectionable air currents and drafts are not permissible; stratification of warm and cool air throughout the space must be kept to a minimum; no dead spots

Fluid Flow, Duct Design, and Air-Distribution Systems 451

with stagnant air should occur; and the arrangement must be unobtrusive and must not interfere with the furnishings.

The volume of air and the allowable velocity from the standpoint of noise determine to a great extent the number of outlets or grilles to be used, but their location must be carefully considered. Figure 12-18 shows a very desirable arrangement in which a return is located on the same wall as the supply. A return on the far wall tends to create a stagnant space beneath the supply outlet. Figure 12-19 shows a *pan-type outlet* used on a ceiling. This appliance gives a rather uniform discharge around its perimeter and can be used either with the return located as shown or with a return duct built up vertically through and inside the outlet. For cooling, and even for heating, inlets at high levels are, in general, desirable. For the heating returns, the low levels are preferable. But in many cases the same system serves either function in different seasons. Satisfactory

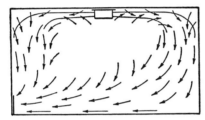

Fig. 12-19. Ceiling outlet, conditioned air.

arrangements which can be used for either heating or cooling are shown in Fig. 12-20. The air stream should be kept from impinging on walls or ceilings, to reduce formation of dirty spots and discoloration from deposited dirt.

Conventional distribution systems for theaters or auditoriums are shown in Figs. 12-21 and 12-22. The ejector system of air delivery at high velocity with a long throw can be used to advantage if the ceiling is relatively free from obstructions. Returns should, in general, be under the seats, or along the rear, or along the sides at the rear. The overhead system with downward distribution at very low velocity gives better control and usually must be employed when the ceilings are not clear of obstructions. With either system, supplementary direct radiation for winter heating may be desirable along the walls. For places of assembly, a system in which the air supply enters from beneath, as through mushroom outlets, is practicable if zoned so that the quantity and temperature of air are properly controlled for each section, but in general the downflow system is preferable.

Air for ventilating washrooms, kitchens, and other places where odors may arise should never be recirculated; it must be continually vented to

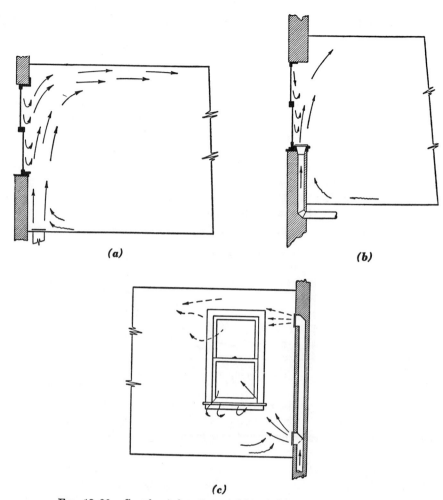

Fig. 12-20. Supply air locations. (a) and (b) winter or summer; (c) top outlet for summer, bottom outlet for winter.

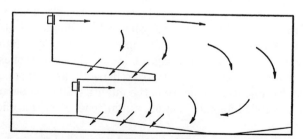

Fig. 12-21. Forward-throw distribution system for theater with plain ceiling.

Fluid Flow, Duct Design, and Air-Distribution Systems 453

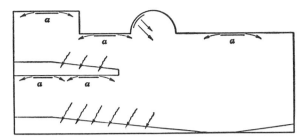

FIG. 12-22. Overhead distribution system for any type of theater.

the outside. Supply air for such rooms should generally enter from adjoining rooms, such as corridors anterooms, and dining rooms, so as to insure absence of odors in the latter rooms.

In considering grilles and registers, the term *aspect ratio* is often used. This is the ratio of the width to the height of the effective area inside the frame. A *register* consists of a grille and a base having built-in dampers or shutters. When the term *grille* is used, no base is implied—only a perforated air outlet or inlet screen.

Air Diffusers. Figure 12-23 illustrates four different designs of ceiling-type air-diffuser outlets, and Fig. 12-24 illustrates in diagrammatic form the method of operation of such diffusers. As this illustration shows, in the diffuser the primary air supply aspirates (inducts) room air into the stream and delivers tempered, mixed air into the space.

Figure 12-25 shows a diffuser arranged to act as a combination air supply and extract outlet for heating, ventilating, and cooling. These

FIG. 12-23. Various designs of air diffusers supplied for a high-velocity air system. (Courtesy Anemostat Corporation of America.)

units can be designed so that varying amounts of air—from 25 to 100 per cent of the air supplied—can be extracted through the inner rings. A dead air space between the supply-air and extract-air sections of the unit prevents short-circuiting of the supply air.

Perforated Ceilings. On many railroad cars, and in industrial buildings and some public buildings, use has been made of ceiling panels consisting of metal sheets with many uniformly distributed perforations, the space being supplied with air from a plenum chamber above the sheets. Valves for delivering air to the plenum must be arranged to provide even distribution over the conditioned space. When the valves are properly adjusted, the downward movement of air is so slow that the occupants do

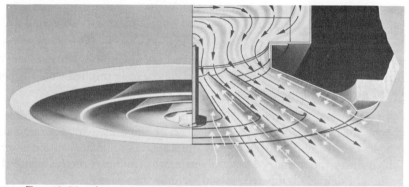

FIG. 12-24. Anemostat air diffuser in sectional view showing principle of aspiration (induction). (Courtesy Anemostat Corporation of America.)

not notice a draft. It is sometimes possible to deliver large quantities of air through such panels without annoyance to occupants, since the diffusing effect of the streams of air causes rapid mixing with the room air. In a railroad car the perforated panels are usually placed in a continuous wide strip along the center of the car ceiling.

12-9. HIGH-VELOCITY (CONDUIT) SYSTEMS

Conventional air-conditioning systems provide for supplying conditioned air to a space and for returning a portion of this air to the conditioner for recirculation. With central systems this involves the use of both supply ducts and return ducts, and these ducts require the use of valuable space in the building—perhaps for false ceilings, exposed ductwork overhead, or vertical areas which carry ductwork and services from floor to floor. To reduce the demands on available space, which become particularly critical where older buildings are being modernized, extensive use is being made of high-velocity systems, also known as conduit systems. In such systems

Fluid Flow, Duct Design, and Air-Distribution Systems   455

the air is distributed at high velocity (3000 to 6000 fpm), usually by means of round ducts called conduits. These are carefully designed to reduce friction and noise-generation. The static-regain and equal-friction methods of design are commonly used. Rolled ducts with locked seams, or commercial spiral-wound steel ducts, or light-gage welded-steel tubes,

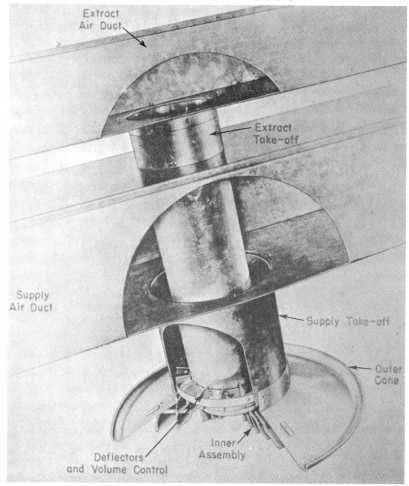

FIG. 12-25. Method of operation of a combination air supply and extract (return-air) outlet. (Courtesy Anemostat Corporation of America.)

are used. Elbows, transitions, and fittings are being made for the special needs of systems of this type.

A central air-conditioning plant located in the basement, or at some other place where space is least valuable, supplies cooled or heated air into the system. The fans supplying the air deliver at maximum static pressures of $4\frac{1}{2}$ in. to 5 in. of water, with static pressures in the system usually running from about $2\frac{1}{2}$ in. to 3 in. The air, in turn, is delivered through special diffuser outlets into the space, and no provision is made for returning the air to the conditioner. This means that frequently the amount of air delivered may not be greatly in excess of that required for ventilation. However, under special load conditions it may be desirable to supply appreciably more air. With diffusers, such as those discussed in section 12–8, air from the space mixes with the delivered air and alters its temperature to bring it close to room conditions.

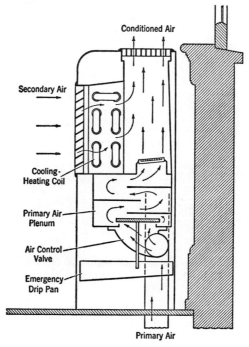

FIG. 12–26. Diagrammatic view of Carrier conduit room-air-conditioning unit.

In Fig. 12–23 an outlet box for feeding high-velocity air into a diffuser is shown. This box contains some sound-control equipment so that the noise of the high-velocity air is reduced to a level which is not objectionable. Outlet boxes of the same general design are also made for low- or high-wall mounting.

Another arrangement of a high-velocity system is shown in Fig. 12–26, where a room air-conditioning unit is illustrated in diagrammatic cross section. Here primary air from the central conditioner is supplied at high velocity (3000 to 4000 fpm) and passes around baffles for sound-absorption in the primary air plenum. The air then passes through multiejector

Fluid Flow, Duct Design, and Air-Distribution Systems

FIG. 12-27. Carrier Weathermaster conduit room-air-conditioning unit.

nozzles at about 4000 fpm (under a static pressure of 1 in. of water) to the upper part of the unit. The ejector nozzles are so located that secondary air drawn from the room is inducted into the air stream. The conditioned air delivered from the top of the unit to the room consists of about 20 per cent primary air with 80 per cent secondary, or recirculated, air. As the

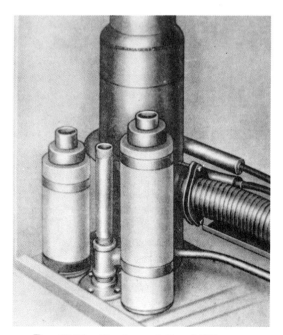

FIG. 12-28. Carrier conduit and fittings.

air quantity is essentially constant in amount, the room temperature is controlled by adjusting the quantity of cold water in summer (or hot water in winter) supplied to the cooling-heating coil. No recirculation to the central system unit is required. The primary air is all outside air from the central system and is sufficient in quantity for good year-around ventilation. Cross-circulation between rooms is thus avoided and no fans or moving parts are needed at the room unit. Figure 12–27 is an external view of the unit.

Figure 12–28 illustrates a typical conduit and its fittings. The riser conduit shown is made of 18-gage rust-proofed steel, electrically welded, and is supplied in $4\frac{1}{2}$-, $5\frac{1}{4}$-, $6\frac{1}{2}$-, and 8-in. diameters. A saddle for connecting a branch duct or feed to a room unit is shown connecting with the run of flexible tubing. The insulated pipes for supplying the cold or hot water to the coil of the unit can be seen in the foreground. The small space required for the conduit and service runs makes this system economically desirable compared with a conventional-duct system, and it may make installation in existing buildings possible with but a small amount of demand on rentable space.

PROBLEMS

12–1. Compare the viscosity of water and saturated steam, at 300 F, (a) in centipoises, (b) in pounds (force) × seconds per square foot, (c) in pounds (mass) per foot × seconds. *Ans.* (a) 0.163 and 0.01695; (b) 34.1 × 10^{-7} and 3.54 × 10^{-7}; (c) 1.096 × 10^{-4} and 1.14 × 10^{-5}

12–2. A crude oil has a density of 50.4 lb per cu ft at 250 F. Obtain from Fig. 12–1 the absolute viscosity of the oil at 250 F, in centipoises, and work out its kinematic viscosity in units of square feet per second. *Ans.* 0.00028

12–3. Gasoline (specific gravity = 0.716) at 60 F is flowing at 1 fps through a 2-in.-ID pipe with a smooth inside surface. Obtain from Fig. 12–1 the viscosity in centipoises and find the Reynolds number (a) by use of the metric system and (b) by changing the viscosity to English units. (c) Is the flow turbulent or viscous?
Ans. (a) 24,640; (b) 24,640

12–4. Determine the pressure drop for water at 100 F flowing at 1.5 fps through 100 ft of drawn brass tubing with an inside diameter of 3 in.
Ans. 0.125 psi, or 0.29 ft of water

12–5. What is the friction loss in a 120-ft length of duct 45 in. in diameter and carrying 15,000 cfm of air? The air is at 70 F and 50 per cent relative humidity and under a pressure of 14.7 psi. *Ans.* 0.05 in. of water

12–6. After computation of the duct system of example 12–3 in section 12–5, it was decided to increase the size of the last room on the right branch duct so that 2400 cfm of air were required. (a) Calculate the new duct dimensions and (b) find the fan requirements.
Ans. (a) At fan—16″ × 66″, AA—16″ × 38″, XX—16″ × 31″, 16″ × 22″, vertical—10″ × 39″; (b) 8800 cfm at 0.88 in. of water

Fluid Flow, Duct Design, and Air-Distribution Systems 459

12-7. A duct carries 6000 cfm, and at 60 ft from the fan a branch is taken off which carries 4000 cfm through three elbows and 30 ft of run for delivery through a grille. The remaining 2000 cfm continue in a 60-ft run in which there are two elbows and one grille. Size these runs to have the same pressure drop, and state the required fan capacity if a static pressure drop of 0.5 in. of water is used on the suction side of the fan, exclusive of the velocity head required to accelerate the air to fan outlet velocity. The velocity in the duct on outlet from the fan is 1400 fpm. The maximum duct depth is 12 in. What is the total pressure the fan must deliver?

12-8. Refer to example 12–5 in section 12–6 of this chapter and redesign the supply side of the system under the assumption that each outlet requires 0.8 of the quantity of air indicated in Fig. 12–13. Work by a method essentially independent of regain methods.

12-9. Work problem 12–8 by regain methods.

12-10. Standard air is flowing in a 20-in. by 40-in. duct at a velocity of 1500 fpm at a static pressure of 1 in. of water. After 200 ft of duct run, a branch receives 4000 cfm while the main duct continues with the same dimensions at reduced flow. Find (a) the drop in pressure in the first 200 ft of run; (b) the velocity in the duct after 4000 cfm are taken off; (c) the probable static regain as a result of velocity decrease; (d) the probable static pressure 400 ft down the duct from the original reference point.
Ans. (a) 0.21 in.; (b) 779 fpm; (c) 0.051 in. SPR;
(d) $1.00 - 0.21 - 0.061 + 0.051 = 0.78$ in. water

12-11. The return system of Fig. 12–13 feeds into a low-wall-type grille inlet with a static-pressure loss of 0.02 in. of water. Each branch has a bend after the grille, drops 8 ft vertically, and makes a hard bend (on greater width) into the trunk return duct. The trunk duct makes an easy bend at Q and runs for 9 ft before mixing with outside air at entry to the filters and coils. Leaving the last coil, the horizontal duct is transformed to round and, with a radius ratio of 2 (Fig. 12–7), bends into the fan inlet cone with a 10 per cent slope. From S to T, in addition to the bend, there are 7 ft of straight run. At design flow through the heaters and filters there is a total pressure drop of 0.6 in. for RS. Design the return duct system so that the static pressure loss at the fan suction for the return system does not exceed 0.80 in. of water. The static-pressure depression in accelerating the air to fan inlet velocity has been allowed for, and thus the 0.80 in. of water pressure drop need not include this item. Fix one dimension at a 16-in. maximum for the rectangular duct, and vary the other as required. Size each branch, the trunk main, and the fan inlet to satisfy design specifications.

12-12. Work problem 12–11 on the assumption that only half as much air is returned for recirculation—namely 400 cfm in each branch, with 1800 cfm supplied from outside. All other specifications are unchanged.

12-13. A 20-in. by 40-in. duct 110 ft long carries 8330 cfm of chilled air at 52 F through a space at 72 F. What is the temperature of the air on outlet (a) if the duct is bare galvanized metal and (b) if the duct is covered with 1-in. cork?
Ans. (a) 54.6 F; (b) 52.6 F

12-14. A warm-air duct, carrying air at 140 F, runs for 300 ft in a factory where the air temperature is 80 F. Its size is 30 in. by 60 in. and the velocity of the air is 2000 fpm. Find the final air temperature at exit from the duct when the duct is (a) bare galvanized iron and (b) furred and covered with $\frac{3}{4}$-in. metal lath and plaster. (c) Estimate the heat saving, in Btu per hour, resulting from covering the duct.
Ans. (a) 127.6 F; (b) 134.8 F; (c) 170,600 Btuh

12-15. A fan supplies 6000 cfm to a duct system. The system consists of a duct run of 300 ft, at the end of which there is a 90° bend through which 4000 cfm flow into a space. At 150 ft from the fan, a branch turns off the main duct at right angles and travels for 100 ft. This branch then turns at right angles again and delivers 2000 cfm of air. The duct has a maximum depth of 20 in., which is used throughout the design. Size the main and branch so that each of these has the same pressure drop from fan to outlet. Assume that the air velocity in the main duct, shortly after the air leaves the fan, is 1000 fpm, and then design the main run for the same friction drop throughout. Assume that each bend has a radius ratio of 1.5, and that the outlet grilles on each branch have equal resistances.

Ans. Main run: 20″ × 36″ and 20″ × 26″; branch: 20″ × 14″

12-16. In order to construct ductwork in a narrow passage through a heavy concrete wall, it was found necessary to reduce a 30-in. by 60-in. duct carrying 20,000 cfm of standard air to a size of 20 in. by 50 in. The static pressure in the duct, just before the duct merged into the gradual sheet-metal transformation, was 1.5 in. of water pressure. What is the velocity and probable static pressure of the air in the duct when it has reached the far side of the wall? *Ans.* 2880 fpm; 1.5 − 0.375 = 1.125 in. water

12-17. For 90 lb per min of Freon-12 at 24 F flowing in 3-in. standard pipe (3.0 in. ID) what is the pressure drop per 100 ft of length if the gas is assumed dry and saturated at 38.58 psia? Figure 12-2, Table 15-5, and Fig. 12-3 supply the required data.

Ans. 0.68 psi

12-18. Rework problem 12-17, with no other change except that copper tubing of essentially 3-in. ID is employed. *Ans.* 0.47

REFERENCES

1. M. C. Stuart, C. F. Warner, and W. C. Roberts, "Effect of Vanes in Reducing Pressure Loss in Elbows in 7-Inch Square Ventilating Duct," *Trans. ASHVE*, Vol. 48 (1942), pp. 409–24.

2. M. C. Stuart, C. F. Warner, and W. C. Roberts, "Pressure Loss Caused by Elbows in 8-Inch Round Ventilating Duct," *Trans. ASHVE*, Vol. 48 (1942), pp. 335–50.

3. R. G. Huebscher, "Frictional Equivalents for Round, Square and Rectangular Ducts," *Trans. ASHVE*, Vol. 54 (1948), pp. 111–17.

4. A. B. Stickney, "Friction in Pipes," *Refrigerating Engineering*, Vol. 53 (1947), pp. 129–31.

5. R. D. Madison, ed., *Fan Engineering*, 5th ed. (Buffalo Forge Company, 1948), pp. 131–204.

6. R. D. Madison and W. R. Elliott, "Friction Charts for Gases Including Correction for Temperature, Viscosity and Pipe Roughness," *Heating Piping and Air Conditioning*, Vol. 18, No. 10 (Oct.; 1946), pp. 107–12.

7. ASHRAE, *Guide and Data Book, Fundamentals and Equipment* (1965–66), Chapter 31, "Air-Duct Design."

8. ASHRAE, *Handbook of Fundamentals* (1967), Chapter 5, "Fluid Flow."

13

Fans and Air Distribution

13-1. FAN TYPES

Fans, which are almost universally used for the circulation of air or other gases through low-pressure systems, are made in the four general types or patterns illustrated in Fig. 13-1.

The *centrifugal fan* is widely used, since it is the most versatile type and can efficiently move large or small quantities of air over an extended range of pressures. It consists of a rotor or wheel mounted in a scroll type of housing. The wheel is turned either by direct drive or, more frequently, by motor drive employing pulleys and belts. The fan wheel, one of which can be seen in Fig. 13-2, is supplied with forward-curved, backward-curved, or radial (straight) blades. By varying the design shape of the blades, the characteristics of the fan can be changed over wide limits. The housing of the fan is usually constructed of sheet metal, although cast metal is sometimes used for smaller sizes. Where corrosive gases are handled by a fan, protective coatings of rubber, lead, asphaltum, or paint are often supplied to the housing and rotor. Details on centrifugal fan performance will be considered later in this chapter.

The *vaneaxial fan* produces an axial flow of the gases through the wheel and blades. The wheel and its blading are located in a cylindrical housing and the fan is provided with air-guide vanes before or after the wheel.

The *tubeaxial fan* is similar to the vaneaxial fan and can move air over a wide range of volumes and through medium pressure ranges.

An axial fan is shown in Fig. 13-3.

The *propeller fan* can move large quantities of air but is unable to produce a significant pressure increase on the air being moved. Its main field of usefulness is in producing air motion or in moving air into or from a space against trivial pressure differences.

13-2. STATIC AND TOTAL AIR HORSEPOWER

In section 12-4 the terms *static pressure*, *velocity pressure*, and the sum of these, *total pressure*, were discussed in connection with air. In the case

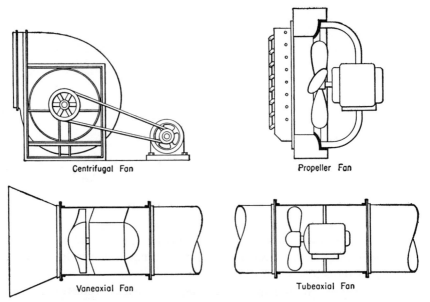

Fig. 13-1. Classification of fan types. Each type can have either belt drive or direct connection. (Courtesy National Association of Fan Manufacturers.)

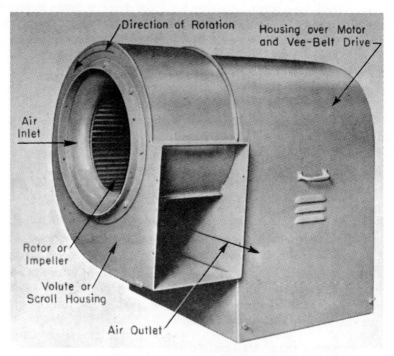

Fig. 13-2. Centrifugal fan and motor unit for ventilation service. (Courtesy Sturtevant Division, Westinghouse Electric Corporation.)

Fans and Air Distribution

of a fan, the static pressure, usually expressed in inches of water, is the pressure increase produced by the fan on the air. The velocity pressure is found from the velocity of the air stream leaving the fan outlet, by equation 12-19 or 12-21; or, in the case of a fan with ductwork to and from the fan, the velocity pressure is concerned with the net velocity change in the

FIG. 13-3. Axial-flow fan. (Courtesy American Blower Corporation.)

air passing through the fan. If a fan delivers directly into a free open space (not a plenum chamber), the discharge static pressure is considered zero, and all the energy of the air stream leaving the fan is in kinetic (velocity) form.

The useful work that a fan does consists in raising the air speed up to the discharge velocity and increasing the static pressure. This total pressure of the fan is the sum of static and velocity pressure changes, and thus the horsepower imparted to the air in static and velocity form is

$$\text{Total air hp} = \frac{[(h_T/12)(62.3)](\text{cfm})}{33{,}000} = \frac{(h_T)(\text{cfm})}{6350} \qquad (13\text{-}1)$$

It then follows that

$$\text{Total fan efficiency} = \frac{\text{total air hp}}{\text{shaft hp input to fan}} = \frac{(h_T)(\text{cfm})}{(6350)(\text{hp input})} \qquad (13\text{-}2)$$

where cfm = air delivered, in cubic feet per minute; h_T = total pressure, in inches of water; hp = horsepower. If the velocity energy of the air leaving the fan is not utilized, or if it is desired to disregard that energy, the

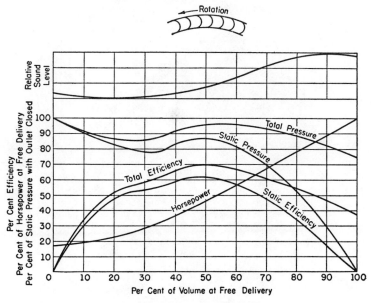

Fig. 13-4. Forward-tip-fan characteristics.

static pressure (h_s in. of water) should be used instead of the total pressure. Then

$$\text{Static air hp} = \frac{[(h_s/12)(62.3)](\text{cfm})}{33{,}000} = \frac{(h_s)(\text{cfm})}{6350} \qquad (13\text{-}3)$$

and it follows that

$$\text{Static fan efficiency} = \frac{(h_s)(\text{cfm})}{(6350)(\text{hp input})} \qquad (13\text{-}4)$$

Typical performance curves for certain types of fans are shown in Figs. 13-4, 13-5, and 13-6. In these figures the abscissas represent the range of air (gas) flow capacity, expressed as a percentage of the amount

Fans and Air Distribution

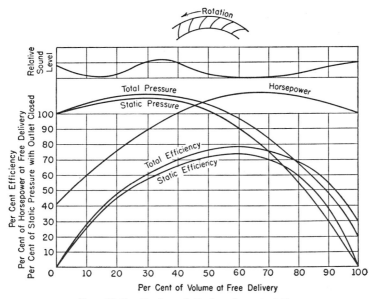

Fig. 13-5. Backward-tip-fan characteristics.

delivered when the fan is discharging freely into open space and not against the resistance of a duct system. The horsepower for this condition is usually taken as a basis for comparison, and the requirement for any other discharge can be read on the horsepower curve as a percentage of free-delivery horsepower. The static- and total-pressure curves are also expressed as relative percentages, usually in terms of the static pressure

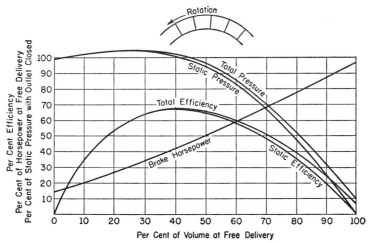

Fig. 13-6. Radial-tip-fan characteristics.

with the outlet closed. The efficiency curves are not relative but express the probable actual efficiencies for a fan of each type. Each set of curves is drawn for a constant speed (rpm) of the fan in question.

13-3. CENTRIFUGAL FANS

Centrifugal fans develop pressure largely by converting a portion of the kinetic energy imparted to the air by the impeller into a rise in pressure. In addition to the pressure rise created in this manner, a smaller pressure increase is developed in the centrifugal field created in the rotating rotor blades. However, when the blades are shallow in radial depth, the contribution to pressure rise from the centrifugal field is small, as is also the

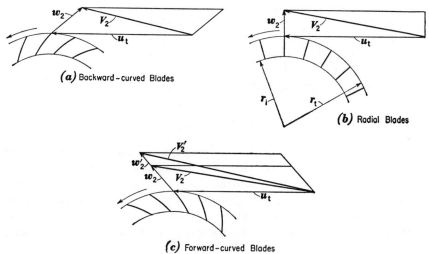

FIG. 13-7. Representative velocity diagrams for varying types of centrifugal-fan blades. Note: u_t = velocity of blade tip (fps) = $2\pi(r_t/12)(\text{rpm}/60)$; w_2 = relative velocity of air (gas) stream leaving blade passage (fps); V_2 = absolute velocity of air (gas) stream at entry to scroll (fps).

pressure rise resulting from relative velocity change in the rotor blading. The air, on leaving the impeller of the fan, enters the scroll (volute) section, which is designed to bring about a decrease in the velocity of the air. This is accomplished by fabricating the scroll as a diffusing passage (a passage with increasing cross section in the direction of flow).

Figure 13-7 shows representative impeller-exit velocity diagrams for the three conventional types of fans. The air (gas) moves outward through the rotor blade passage with a relative velocity w_2, while at the same time the rotor itself is turning and imparting to the air a tangential velocity indicated as u_t. The resultant velocity that the air possesses as it moves into the scroll (volute) is the vector sum of w_2 and u_t. In the case of backward-curved impeller blading the direction of w_2 is such that it may

Fans and Air Distribution

decrease the magnitude of V_2 compared with that of u_t. In the case of the radial (straight) blades, V_2 is increased but usually so slightly that it is not appreciably greater than u_t. On the other hand, forward-curved impeller blading produces a significant increase in the resultant velocity V_2 entering the scroll (volute). Since it is the magnitude of the velocity V_2 entering the volute, combined with effectiveness of conversion, that largely produces the pressure rise, it is obvious that forward-curved blading produces the maximum pressure rise for any fixed tip speed of an impeller. In contrast to this, backward-curved blading produces the lowest pressure rise for a given impeller tip speed.

TABLE 13-1
GOOD OPERATING VELOCITIES AND TIP SPEEDS FOR VENTILATING FANS*

Static Pressure (in. of water)	Fans with Forward-Curved Blades		Fans with Backward-Tipped and Double-Curved Blades		Tubeaxial and Vaneaxial Fans
	Outlet Velocity (fpm)	Tip Speed (fpm)	Outlet Velocity (fpm)	Tip Speed (fpm)	Wheel Velocity† (fpm)
$\frac{1}{4}$	1000–1100	1520–1700	800–1100	2600–3100	1100–1500
$\frac{3}{8}$	1000–1100	1760–1900	800–1150	3000–3500	1250–1700
$\frac{1}{2}$	1000–1200	1970–2150	900–1300	3400–4000	1400–1900
$\frac{5}{8}$	1200–1400	2225–2450	1000–1500	3800–4500	1500–2100
$\frac{3}{4}$	1300–1500	2480–2700	1100–1650	4200–5000	1650–2350
$\frac{7}{8}$	1400–1700	2660–2910	1200–1750	4500–5300	1800–2500
1	1500–1800	2820–3120	1200–1900	4800–5750	1900–2700
$1\frac{1}{4}$	1600–1900	3162–3450	1300–2100	5300–6350	2150–3000
$1\frac{1}{2}$	1800–2100	3480–3810	1400–2300	5750–6950	2350–3300
$1\frac{3}{4}$	1900–2200	3760–4205	1500–2500	6200–7550	2500–3550
2	2000–2400	4000–4500	1600–2700	6650–8050	2700–3800
$2\frac{1}{4}$	2200–2600	4250–4740	1700–2800	7050–8550	
$2\frac{1}{2}$	2300–2600	4475–4970	1800–2950	7450–9000	
3	2500–2800	4900–5365	2000–3200	8200–9850	

* Reprinted, by permission, from *Heating Ventilating Air Conditioning Guide 1955*, Chapter 33.
† Wheel velocity is the axial mean air velocity through the inside diameter of the housing cylinder at the point of wheel location.

The velocity w_2 of the air flowing through the impeller blade passage is directly related to the quantity of air being delivered by the fan. Consequently, as delivery increases under conditions of decreased system pressure the magnitude of w_2 increases. This is illustrated in Fig. 13-7c by the vectors w_2' and V_2'. The energy absorbed in a fan is a function of the quantity of flow and the pressure against which the air is delivered. In the case of a forward-tipped fan, the air flow quantity increases at a rapid rate as external (system) pressure is reduced, with the result that the horsepower required to drive the fan increases with lessened resistance on the system. Consequently, when a driving motor is selected for a fan with forward-curved blades delivering into a fixed air system, if the flow capacity of the system is underestimated and more air flows in the system for a given inlet pressure, there is an ever-present tendency to overload the motor. This can be seen in the horsepower curve of Fig. 13–4. Fans with backward-curved blades have a horsepower characteristic which reaches a peak (Fig. 13–5) and then falls off as the flow increases, while radial-bladed fans have intermediate characteristics between the two.

Table 13–1 indicates tip speeds and operating velocities of fans used in ventilating systems. It is of interest to use these figures to obtain some idea of the pressure rise created in the scroll of a fan. To do this simply, the approximate assumption is made that the air (gas) being compressed behaves as a relatively incompressible fluid. The pressure rise can be expressed, in pounds per square foot, as

$$\Delta p = K_p \frac{\rho}{2g}(V_2^2 - V_3^2) \qquad (13\text{--}5)$$

where ρ = density of the medium being compressed, in pounds per cubic foot;

g = the gravitational constant, 32.2;

V_2 = velocity of the air stream leaving the impeller, in feet per second;

V_3 = reduced velocity the air stream possesses on exit from the fan, in feet per second;

K_p = performance factor relating the conversion of velocity into pressure rise. For estimates, take K_p as ranging from 0.7 to 0.8.

Example 13–1. A fan with an impeller 21 in. in diameter is designed to turn at 600 rpm. Compute the probable pressure increase this fan might develop in its volute when passing standard air (ρ = 0.075 lb per cu ft), if the blades are forward-curved in such manner that, at a desired operating point and flow, $V_2 = 1.3u_t$. Consider K_p, the pressure-realization factor, to be 0.75, and take the outlet velocity from the fan as 1800 fpm, or 30 fps.

Solution: The blade tip speed can be found as

$$u_t = 2\pi \frac{r_t}{12}\left(\frac{\text{rpm}}{60}\right) = \frac{\pi(2r_t)(\text{rpm})}{720}$$

$$= \frac{\pi(21)(600)}{720} = 54.9 \text{ fps, or } 3290 \text{ fpm}$$

The absolute air velocity is thus

$$V_2 = 1.3u_t = (1.3)(54.9) = 71.4 \text{ fps}$$

By equation 13–5, the pressure rise is

$$\Delta p = 0.75\left(\frac{0.075}{64.4}\right)(71.4^2 - 30^2) = 3.67 \text{ psf}$$

or

$$\Delta p = (3.67)(0.192) = 0.704 \text{ in. of water}$$

Fans and Air Distribution

where 0.192 is the conversion factor for changing pounds per square foot to inches of water.

A certain manufacturer's catalog shows that a fan similar in size and operating in this range produces 1.25 in. of water, so that $1.25 - 0.704 = 0.55$ in. of water pressure is produced in the rotor of the fan by the centrifugal field and the diffusion effect.

Table 13-2 lists relative characteristics of centrifugal fans in a general way but, in addition to the three types listed, there are a number of intermediate types by means of which designers have been able to produce characteristics to serve particular purposes. For example, one design uses a backward-curved blade at the tip (outlet) edge of the blade, with forward curving at the inside (inlet) edge. Such a fan has a continuously-rising pressure characteristic, but the horsepower reaches a peak and then falls off—and thus the overloading characteristic is eliminated.

TABLE 13-2

COMPARATIVE CHARACTERISTICS OF CENTRIFUGAL-FAN TYPES

Item	Forward-curved Blades	Radial (Straight) Blades	Backward-curved Blades
Efficiency..................	Medium	Medium	High
Stability of operation........	Poor	Good	Good
Space required..............	Small	Medium	Medium
Tip speed for given pressure rise.....................	Low	Medium	High
Resistance to abrasion........	Poor	Good	Medium
Noise characteristic..........	Poor	Fair	Good

The noise emitted by a fan is of greatest importance in certain types of applications. For example, in the air supply of a lecture hall a noisy fan could not be tolerated. On the other hand, exhaust fans for industrial plants, or fans used as blowers under boilers, are not critical in regard to noise characteristics. Fan noise is frequently associated with the high tip speeds which are required to produce high pressures. For a given pressure, noise is related primarily to two quantities: the tip speed of the impeller and the air velocity leaving the wheel. Expressed in another way, noise is roughly proportional to the pressure developed, whether this is produced by a low-speed forward-curved blade or by a higher-speed backward-curved blade. For the same pressure rise, the noise in the two types will be of almost the same magnitude. In general, however, backward-curved fan blading is considered to be less sensitive to noise than is forward-curved fan blading. The graphs at the top of Figs. 13-4 and 13-5 give an indication of sound-level intensities for forward- and backward-curved fan blading. It will be noted that the forward-tip fan blading rapidly increases the intensity of the noise level as the air flow increases.

The pressure produced by the fan is naturally limited by the maximum allowable tip speed and by the design of the blading. For the production of high pressure and where noise production is not critical, rotors with straight blades, deep in a radial direction, are frequently used. These give the benefit of the more efficient type of pressure build-up which occurs in the impeller, and the absolute velocity of the medium entering the scroll is also aided by the outward-flow component (Fig. 13-7). If specially-designed high-speed fans do not supply adequate pressure, one or more fans must be used in series. Little or no difficulty need develop if fans are operated in series. However, difficulty may be encountered when fans operate in parallel, as under these circumstances each fan will deliver air into a system at fixed pressure in accordance with the characteristic static-pressure curve of the fan—and if instabilities exist in the fans, there can be some surging back and forth between the fans, particularly so if the system capacity fluctuates. Fans with forward-curved blades, in particular, cannot be operated near their point of maximum efficiency because of inherent instabilities which exist in such fans if the rate of flow through the fan reduces to values of less than about 40 per cent of wide-open delivery.

13-4. AXIAL-FLOW FANS

Axial-flow fans, while incapable of developing high pressures, are nevertheless capable of forcing air through ductwork and low-resistance equipment which may be placed in such ducts, such as light filters and also heating or cooling coils. The fans are particularly suitable for handling large volumes of air at relatively low pressures. They are low in first cost and possess good efficiency, and since they are directly in the duct system, with through air-flow, they eliminate the 90-deg change of direction which is a characteristic of the centrifugal fan.

Axial-flow fans, in general, have a large hub and blades of airfoil shape. The blades are stubby and are not close together. Figure 13-3 shows a representative fan of this type. The inlet guide vanes, which can be seen in the view, align and direct the air into the fan blades. The blades are visible in the background of the picture. The fan blades impart energy to the air and deliver it into the exhaust diffuser section of the fan.

To reduce losses, the bearing and pulley are streamlined, as shown, and a streamlined covering can also be supplied for the vee belts from the motor. In some arrangements, the fan is direct-driven and the motor itself is also mounted in the duct.

The fan blades are made in many forms, but the most effective have airfoil sections. Angle change and twist are given the blade at various positions outward from the hub to the tip. The simplest fans, however, may have little more than modified flat blades.

Fans and Air Distribution

Axial fan units have now been developed to a high degree of effectiveness. They show good efficiencies, and they can operate at high static pressures if such operation is necessary. The fan can be so designed that the horsepower characteristic is quite flat and nonoverloading. A diffusing-section outlet cone is customarily provided on the discharge side of the fan. The swirl imparted to the air by the fan blades can be eliminated by proper guide vanes on the inlet side and, in some designs, on the outlet side as well.

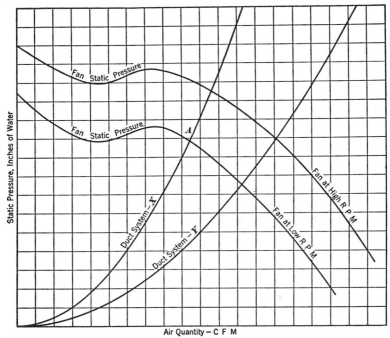

FIG. 13-8. Characteristic resistance-quantity curves for duct systems and fans.

13-5. FAN-SYSTEM CHARACTERISTIC CURVES

The importance of selecting a fan to suit accurately the characteristics of the duct system cannot be overestimated. Figure 13-8 shows the *system characteristic curves* of two different duct systems, X and Y. Both of these curves are of parabolic shape and can be represented by equations of the form

$$\text{Static pressure} \propto (\text{cfm})^2 \text{ or } (\text{cfm})^n$$

That is, the static pressure required to send air through a system is proportional to the square of the quantity (cfm) delivered. The exponent n in this equation is for systems other than those having smooth ducts. Its value is more or less than 2. For example, in sending air through a grate and furnace for forced draft, n may have a value of about 1.3.

The characteristic pressure curves for a given fan when it operates at a low speed and at a high speed are also shown in Fig. 13-8. These curves are for a fan having forward-curved blades; and, if the X-system characteristics are met by the fan for a given quantity of air at A, the choice would be satisfactory. However, if the fan were fixed to run at the higher speed shown, it would not deliver the quantity of air A without wasteful dampering because the fan pressures for quantity A become higher, as shown. For maximum efficiency, a fan having forward-curved blades should usually be selected so that the system characteristic crosses to the right of the hump. A fan having backward-curved blades is not so critical as to point of operation, but it should be selected to operate near its point of maximum efficiency. If the intersection point is selected near the steep part of the pressure curve, dampering to throttle the quantity delivered will cause a considerable pressure rise and will create high velocities and resultant noise at the dampers or registers. Notice that changing the fan speed will not change the relative point of intersection of the fan and system-characteristic curves. This relation can only be changed by selecting a different size of fan.

13-6. FAN LAWS

When a given fan is used with a given system the following fan laws hold:

1. The capacity (cfm) is directly proportional to the fan speed.
2. The pressure (static, velocity, or total) is proportional to the square of the fan speed.
3. The horsepower required is proportional to the cube of the fan speed.
4. At constant speed and capacity, the pressure and the horsepower are proportional to the density of the air.
5. At constant pressure, the speed, capacity, and horsepower are inversely proportional to the square root of the density.
6. At constant weight delivered, the capacity, speed, and pressure are inversely proportional to the density, and the horsepower is inversely proportional to the square of the density.

Example 13-2. A fan described in a manufacturer's table is rated to deliver 17,500 cfm at a static pressure (gage) of 1 in. of water when running at 256 rpm and requiring 4.54 hp. If the fan speed is changed to 300 rpm, what is the capacity, the static pressure, and the horsepower required?

Solution: By fan laws 1, 2, and 3,

$$\text{Capacity} = 17{,}500 \times (\tfrac{300}{256}) = 20{,}500 \text{ cfm} \qquad Ans.$$

$$\text{Static pressure} = 1.0 \times (\tfrac{300}{256})^2 = 1.37 \text{ in. of water} \qquad Ans.$$

$$\text{Horsepower} = 4.54 \times (\tfrac{300}{256})^3 = 7.31 \qquad Ans.$$

Fans and Air Distribution

Example 13-3. If in example 13-2, in addition to changing the speed, the air handled were at 150 F instead of standard 70 F, what capacity, static pressure, and horsepower would be required?

Solution: The density of standard dry air at 70 F and 29.92 in. Hg is 0.075 lb per cu ft, and therefore at 150 F the density is

$$0.075 \times \frac{460 + 70}{460 + 150} \times \frac{29.92}{29.92} = 0.0651 \text{ lb per cu ft}$$

The density at any temperature and any barometer is obtained by multiplying by absolute temperature and pressure ratios. The capacity, on a volume basis, is unchanged, and thus

Capacity = 20,500 cfm at 150 F   *Ans.*

By fan law 4,

$$\text{Static pressure} = (1.37) \frac{0.0651}{0.075} = 1.19 \text{ in. of water} \quad Ans.$$

and

$$\text{Horsepower} = (7.31) \frac{0.0651}{0.075} = 6.35 \quad Ans.$$

The following information is required to select properly a fan for a given installation: (a) The number of cubic feet of air per minute to be moved; (b) the static pressure that must be developed to move the air through the system; (c) the type of motive power available; (d) whether the fans are to operate singly or in parallel on any one duct; (e) the degree of noise permissible; and (f) the nature of the load, such as variable quantities or pressures of air.

When the requirements of the system are known, the main points to be considered in selecting a fan are (a) efficiency, (b) reliability of operation, (c) size and weight, (d) speed, (e) noise, and (f) cost.

To assist in choosing apparatus to the best advantage, manufacturers of fans supply tables or curves that show the following factors for each size of fan, over a wide range of static pressure:

1. The volume of standard air (0.075 lb per cu ft) handled, in cubic feet per minute.
2. The air velocity at the outlet.
3. The fan speed, in revolutions per minute.
4. The brake horsepower.
5. The peripheral speed, or speed of the blade tip, in feet per minute (if not listed, it can be computed from data shown).
6. The static pressure, in inches of water.

In the tables of fan capacities, the most efficient point of operation of the fan is sometimes indicated by printing the values in italic or boldface type.

Two especially important factors in selecting fans for ventilating systems are the efficiency, which affects the cost of operation, and the noise.

Fans and Air Distribution

Table 13–3
Pressure-Capacity Table for Two American Blower Corporation Type-ACH Fans*

CFM	Outlet Velocity	½-In. SP		⅝-In. SP		¾-In. SP		1-In. SP		1¼-In. SP		1½-In. SP	
		RPM	BHP	RPM	BHP	RPM	BHP	RPM	BHP	RPM	BHP	RPM	BHP
851	1200	848	.13	933	.16	1018	.19	....	....	....	....	....	....
922	1300	866	.15	945	.18	1019	.21	....	....	....	....	....	....
993	1400	884	.17	957	.20	1030	.23	1175	.30	....	....	....	....
1064	1500	901	.19	973	.22	1039	.26	1182	.32	....	....	....	....
1134	1600	926	.22	997	.24	1057	.29	1190	.35	1320	.43	....	....
1205	1700	954	.25	1020	.27	1078	.31	1200	.38	1325	.46	1436	.55
1276	1800	983	.28	1044	.31	1100	.34	1210	.42	1330	.50	1440	.59
1347	1900	1011	.31	1068	.35	1126	.38	1230	.46	1341	.54	1447	.63
1418	2000	1039	.35	1092	.39	1152	.42	1250	.50	1352	.59	1458	.66
1489	2100	1068	.39	1115	.43	1178	.47	1275	.54	1370	.62	1470	.72
1560	2200	1096	.44	1147	.47	1204	.51	1300	.59	1390	.67	1482	.77
1631	2300	1124	.48	1179	.52	1230	.56	1325	.64	1420	.73	1500	.83
1702	2400	1152	.53	1210	.58	1256	.62	1350	.70	1448	.78	1525	.88
5136	1200	372	.75	407	.91	444	1.06	512	1.37	....	....	....	....
5564	1300	375	.83	412	1.00	448	1.19	513	1.51	566	1.86	617	2.37
5992	1400	380	.94	417	1.12	452	1.30	517	1.69	572	2.02	619	2.41
6420	1500	385	1.05	421	1.23	458	1.46	521	1.84	575	2.21	622	2.68
6848	1600	390	1.16	426	1.35	462	1.60	526	1.98	578	2.39	626	2.80
7276	1700	395	1.30	433	1.51	466	1.74	528	2.31	580	2.59	629	3.04
7704	1800	401	1.40	438	1.64	469	1.87	530	2.47	583	2.82	633	3.29
8132	1900	409	1.56	444	1.81	475	2.04	535	2.60	590	3.02	637	3.59
8560	2000	416	1.72	449	1.98	480	2.22	540	2.72	596	3.21	641	3.90
8988	2100	423	1.98	457	2.18	488	2.43	546	2.98	601	3.53	645	4.02
9416	2200	429	2.24	465	2.35	493	2.59	550	3.12	604	3.74	648	4.28
9844	2300	435	2.31	472	2.55	500	2.78	554	3.47	610	4.02	654	4.55
10272	2400	452	2.59	477	2.76	507	2.82	562	3.65	614	4.27	658	4.90

Note. The upper data are for the model 109H fan, which has a 9-in. wheel diameter and an outlet of 0.709 sq ft. The lower data are for the model 121H fan, which has a 21-in. wheel diameter and an outlet of 4.280 sq ft.
* Courtesy American Blower Corporation.

Fans and Air Distribution

First cost, and space occupied, are secondary considerations. The fans should be selected to operate at maximum efficiency without noise. Noise may be caused not only by the fan but also by other conditions—for example, by too high a velocity of air in the ductwork, and by improper construction of ducts and airways, as well as of foundations, housings, floors, and walls. Where noise is chargeable directly to the fan, it may be caused by improper selection of fan type or by too high speed for the size. The tip speed required for a specified capacity and pressure varies with the type of blade, and a tip speed that is excessive for forward-curved blades is not necessarily so for the type curved backward or slightly backward (see Table 13-1). A fan that is operated at a point considerably beyond maximum efficiency is usually noisy.

Table 13-3 is a capacity table for two typical fans with rotors of the squirrel-cage type, and indicates the kind of data customarily supplied. The manufacturer's model 109H is a small-size fan, and model 121H is a large fan; both are types common in air-conditioning work. The fan must be selected to supply the needed cfm at the proper static pressure, and the outlet velocity must also be in the range required or suitable corrections must be made.

Example 13-4. A fan in a certain duct system must deliver 1200 cfm at a static pressure (SP) of 0.76 in. of water. The duct velocity is 1500 fpm. Ascertain if one of the fans in Table 13-3 is satisfactory.

Solution: Fan model 109H, running at 1078 rpm, delivers 1205 cfm at $\frac{3}{4}$ in. SP and requires 0.31 bhp. This should be satisfactory, particularly as the 1700-fpm outlet velocity can build up some pressure regain when air slows down in the duct with the larger cross section. By equation 12-28, the static regain is

$$\text{SPR} = 0.5 \left[\frac{(1700 - 1500)(1700 + 1500)}{16{,}040{,}000} \right] = 0.02 \text{ in.}$$

Hence the static pressure in the duct is 0.75 + 0.02 = 0.77 in. of water. If necessary, a slight change of the speed through pulley adjustment could bring this fan into the required range.

The blade-tip speed can be found by computation of the peripheral velocity from the wheel diameter and the fan speed. Thus,

$$\text{Tip speed} = \frac{\pi(D_t)(\text{rpm})}{12} = \frac{(3.14)(9)(1078)}{12} = 2540 \text{ fpm}$$

13-7. MEASUREMENT OF AIR FLOW

The quantity of air flowing in ducts can be measured by installing orifices or nozzles which develop a measurable pressure-difference. Where such a construction is not feasible, a traverse across the air stream may be made with some form of impact (pitot) tube. In Fig. 13-9 is shown a typical pitot tube inserted in a duct to measure the flow of air or other fluid. A fluid in a duct exerts its static pressure in all directions, and if the fluid is

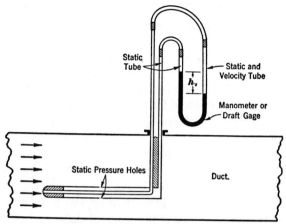

FIG. 13-9. Pitot tube in duct.

in motion it also exerts a velocity pressure because of the kinetic energy of the stream. In section 12-4 it was shown that the pressure equivalent to a given velocity could be represented for low and moderate speeds by expressions involving the velocity squared:

$$h = \frac{V^2}{2g} \tag{13-6}$$

A typical pitot tube, shown diagrammatically in Fig. 13-9, and in detail in Fig. 13-10, consists of two concentric tubes independently sealed from each other. The outer tube has small holes drilled in its side, through which the static pressure is transmitted to the manometer. The outlet of the inner tube faces the air current and consequently has two pressure effects impressed on it—that developing from the velocity of the stream,

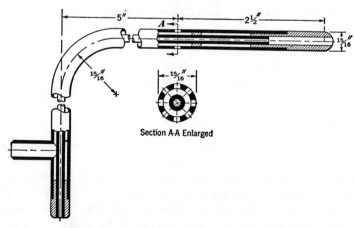

FIG. 13-10. Acceptable design of pitot tube.

Fans and Air Distribution

plus the static pressure existing in the duct. If the two tubes are connected to opposite legs of a manometer the static pressures which appear in each tube will equalize and the manometer will indicate only the pressure equivalent to velocity (h_v). The sum of the static pressure and the pressure equivalent to velocity is known as the *impact pressure*, or *total pressure*.

There are many sources of inaccuracy in using pitot tubes. The tube itself should be built in accordance with a well-tested, successful design (see Fig. 13-10), as with such a tube 100 per cent transformation of velocity to pressure will be obtained.

If excessive cross currents or eddy currents exist in a duct, the static-pressure holes may not be at right angles to the main stream flow, and impact pressures may be indicated in the static tube. If the pressure is pulsating, inaccuracies will ensue. The velocity of the air current varies between different points in the cross section of the duct, so a traverse to obtain an average velocity should always be made. This is done by dividing the cross section into a series of imaginary areas of equal size and finding the velocity pressure (h_v) in the effective center of each such division. The average of the readings thus measured will give the value from which to calculate the average velocity in the duct.

In the case of a circular duct it is customary to divide the cross section into one central area and four concentric rings each of equal area. By placing the pitot tube at the radius, corresponding to the mean area of each concentric area, the velocity pressure for that area can be found. The average velocity found from these readings, multiplied by the total area of the duct, will give the flow. The locations of these points in a duct of radius r are $0.316r$, $0.548r$, $0.707r$, $0.837r$, and $0.949r$. It is more accurate to take these readings on each side of the center line of the duct, making ten readings in all and giving rise to the descriptive name *ten-point method* for making such a traverse. For more-precise work it is desirable to make two traverses at right angles to each other, giving twenty readings.

The manometer pressure in air ducts is usually measured in inches of water. The equation $V = \sqrt{2gh}$, in which h is expressed in equivalent feet for the fluid flowing, must therefore be put in terms of inches of water. By equation 12-17 it is possible to develop the relationship

$$V = \sqrt{2g \frac{h_v d_w}{12 d_a}} = 2.31 \sqrt{h_v \frac{d_w}{d_a}} \qquad (13\text{-}7)$$

where V = fluid velocity, in feet per second;
g = 32.2, the gravitational constant;
h_v = velocity pressure from indicating-manometer, in inches;
d_w = density of measuring fluid, usually water, in pounds per cubic foot;
d_a = density of the air or fluid flowing in the duct, in pounds per cubic foot.

For water at 68 F, d_w = 62.3 lb per cu ft. With 68 F water as the measuring fluid, and with velocity expressed in feet per minute, V_m appears, from equation 13-7, as

$$V_m = 1096.5\sqrt{\frac{h_v}{d_a}} \qquad (13\text{-}8)$$

Example 13-5. A duct 4 ft in diameter carrying air at 72 F and 55 per cent relative humidity, with the barometer at 29.9 in. Hg at very low static pressure, is traversed by a pitot tube across a diameter, using the ten-point method with five equal areas. The readings, in inches of water of the velocity pressure from one side of the duct to the other, are respectively 0.210, 0.216, 0.220, 0.219, 0.220; and 0.220, 0.218, 0.219, 0.220, 0.216. Find the air flow, in cubic feet per minute and in pounds per minute.

Solution: The velocity at each point can be calculated and these results then averaged, but it is quicker, and equally accurate, to average the square roots of the velocity pressures. Thus,

$$\sqrt{h_v} = \frac{\sqrt{0.210} + \sqrt{0.216} + \sqrt{0.220} + \sqrt{0.219} + \sqrt{0.220} + \ldots + \sqrt{0.216}}{10} = 0.4668$$

The density of air at the conditions indicated can be calculated as shown in Chapter 2. Finding the weight of 1 cu ft of dry air by $PV = WRT$ and adding the weight of steam in 1 cu ft, it is seen that d_a = 0.0736 lb per cu ft and d_w = 62.3 lb per cu ft. By equation 13-7,

$$V = 2.31\sqrt{h_v \frac{d_w}{d_a}} = 2.31(0.4668)\sqrt{\frac{62.3}{0.0736}} = 31.37 \text{ fps}$$

The quantity flowing is

$$q = (\text{sq ft})_{\text{duct area}} \times \frac{\text{ft}}{\text{sec}} \times 60 = \text{cfm}$$

$$= \frac{\pi}{4} 4^2 \times 31.37 \times 60 = 23{,}650 \text{ cfm} \qquad Ans.$$

And

Lb per min = cfm × air density = 23,650 × 0.0736 = 1740.6 *Ans.*

Anemometer. An anemometer, illustrated in Fig. 13-11, can be used for finding the air velocity in ducts or at outlet grilles. An anemometer contains a miniature wind wheel which revolves on bearings having slight friction and thus turns a pointer in front of a dial calibrated to indicate air travel in linear feet. It must be used in connection with a timepiece, preferably a stop watch. An anemometer is a useful tool for comparative readings, but it is not adapted for very high velocities or extreme accuracy. An anemometer should be calibrated frequently, since the condition of the bearings affects the indications and since the instrument is easily damaged by rough handling.

For measuring the air volume through supply grilles the surface of the grille should be apportioned, by eye or measurement, into equal areas about 6 in. square. A 4-in. anemometer should be used and it should be held very close to (or even against) the grille, with the dial facing the operator.

Fans and Air Distribution

The average of the readings taken at each of the area divisions should be used. In each square, generally not less than half-minute, and preferably one-minute, readings should be taken. Then

$$\text{cfm} = CV\left(\frac{A+a}{2}\right) \qquad (13\text{-}9)$$

where V = average velocity from corrected anemometer readings, in feet per minute;
A = gross area of the grille, in square feet;
a = net free area of the grille, in square feet;
C = a flow factor determined by experiment (0.97 for velocities of 150 to 600 fpm, and 1.0 for velocities over 600 fpm).

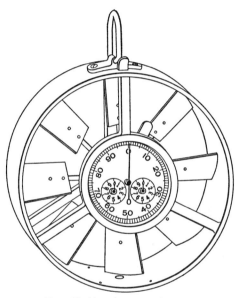

Fig. 13-11. Anemometer.

For measuring the air volume through exhaust grilles the same subdividing of the area should be done, and the same average velocities should be obtained, but the dial of the instrument should face the grille. Then

$$\text{cfm} = YVA \qquad (13\text{-}10)$$

where V = average velocity, in feet per minute;
A = gross area of grille, in square feet;
Y = a flow factor determined by experiment (0.8).

Anemometers are almost universally used in measuring the air flow through grilles and registers in ordinary ventilating work, where velocities usually are comparatively low (not over 800 fpm).

Kata Thermometer. The Kata thermometer is essentially an alcohol thermometer developed for determining very low air velocities. The bulb is heated in water until the alcohol expands and rises to a reservoir above the graduated tube. The time required for the liquid to cool by 5 degrees Fahrenheit is observed by use of a stop watch, and this time is a measure of the air movement.

Velometer. The velometer consists of a delicately poised vane within a substantial housing. The vane actuates a pointer calibrated to read directly in feet per minute the speed of the air flow, as with a pitot tube, without the necessity for timing. The velometer may be placed directly in the air stream, or it may be connected through a flexible tube to special jets which permit taking accurate velocity readings in locations inaccessible for an anemometer or pitot tube. The ordinary accuracy is within 3 per cent.

TABLE 13-4
FLOW COEFFICIENTS K FOR SQUARE-EDGED ORIFICE PLATES, WITH VELOCITY OF APPROACH INCLUDED

REYNOLDS NUMBER	DIAMETER RATIO OF ORIFICE TO INSIDE OF APPROACH DUCT D_o/D_1						
	0.30	0.40	0.50	0.55	0.60	0.65	0.70
FLANGE TAPS							
5×10^4	0.61	0.62	0.64	0.66	0.67	0.71	0.75
10^5	0.60	0.62	0.63	0.65	0.66	0.68	0.72
5×10^5	0.60	0.61	0.63	0.63	0.65	0.67	0.69
10^6	0.60	0.61	0.62	0.63	0.65	0.67	0.69
10^7	0.60	0.61	0.62	0.63	0.65	0.67	0.68
RADIUS TAPS							
10^5	0.60	0.62	0.63	0.65	0.67	0.68	0.72
5×10^5	0.60	0.61	0.62	0.64	0.65	0.67	0.71
10^6	0.60	0.61	0.62	0.64	0.65	0.67	0.70
10^7	0.60	0.61	0.62	0.63	0.65	0.67	0.70

Electrical Anemometers. Electrical anemometers are useful in measuring air currents. They operate on the principle of the variation in electrical resistance of a hot wire with temperature. The rate of cooling of this wire is responsive to the velocity of the air passing around the wire.

Orifices for Gas Flow. Orifices for measuring the flow of air, steam, and other gases are in extensive use for industrial flow-measurement. The orifice is usually a carefully fabricated opening made in a metal plate, with the plate designed for mounting between two mating flanges in a pipe or duct. Most orifices are made with a sharp (square) edge facing upstream

Fans and Air Distribution

to receive the flow, and the trailing orifice edge is chamfered away. If properly manufactured and used, the orifices can give reliable flow data. Coefficients for orifices are most frequently given for flange taps or for radius taps. For flange taps, the orifice pressure taps are located close to the orifice plate and the connections reach into the stream through the sides of the flanges. For radius taps, the upstream pressure tap is located one pipe diameter (2 pipe radii) upstream, and the downstream tap is one pipe radius downstream from the orifice. Values of the flow coefficient K are given in Table 13-4, where this coefficient as given includes the velocity-of-approach correction favor. For conditions where the density of the gas does not appreciably change in passing through the orifice, the following equations apply:

$$Q = KA_o\sqrt{2gh_g} \qquad (13\text{-}11)$$

$$Q = 8.02 KA_o\sqrt{\frac{h_w d_w}{12 d_g}} \qquad (13\text{-}12)$$

where Q = gas flow, in cubic feet per second;
A_o = orifice area, in square feet;
K = flow coefficient, including velocity of approach;
h_g = differential pressure across orifice, measured in feet of gas flowing;
h_w = differential pressure across the orifice, measured in inches of measuring medium (usually water or mercury);
d_w = density of the measuring fluid, in pounds per cubic foot (for water, 62.3 lb per cu ft; for mercury, 62.3 × 13.6 lb per cu ft);
d_g = density of the gas under measurement, in pounds per cubic foot;
D_o = orifice diameter, in inches.

When a water manometer is used and the orifice diameter (D_o) is in inches,

$$Q = K \frac{\pi}{4} \frac{D_o^2}{144} (8.02) \sqrt{\frac{h_w(62.3)}{12 d_g}}$$

which becomes

$$Q = 0.0996 K D_o^2 \sqrt{h_w/d_g} \quad \text{cfs} \qquad (13\text{-}13)$$

Example 13-5. The air flow into a certain high-velocity duct system is measured on discharge from the outlet fan by a square-edge orifice 8.0 in. in diameter, located in a pipe of closely 12-in. inside diameter. The pressure drop across the orifice is 10 in. of water and flange taps are used. What is the flow, in cubic feet per minute, entering the duct system if the air density is 0.078 lb per cu ft? Also express the flow in pounds per minute.

Solution: Use equation 13-13 and read $K = 0.68$ from Table 13-4 for a diameter ratio of $8.0/12.0 = 0.667$ at a representative Reynolds number of 5×10^5. Then

$$Q = (0.0996)(0.68)(8.0)^2 \sqrt{10/0.078}$$
$$= 49.2 \text{ cfs}$$

Flow in cfm = (49.2)(60) = 2952 *Ans.*
Flow in lb per min = $(2952)(d_g)$ = (2952)(0.078) = 230 *Ans.*

Rounded-entry orifices, or flow nozzles, are also used in measurement of gas flow. The flow coefficients K for these also vary with the diameter ratio and the Reynolds number but are so close to unity over a broad range of flow conditions that for approximate computation $K = 1.0$ can be used.

PROBLEMS

13–1. A fan delivers 3000 cfm of air under a static pressure of 1.0 in. of water. (a) What is the static air-horsepower, and (b) how much shaft horsepower is required if the static fan efficiency is 60 per cent at this load?   *Ans.* (a) 0.47; (b) 0.787

13–2. (a) If the air leaves the fan of problem 13–1 at a velocity of 1050 fpm, what is the fan total pressure in inches of water? The air density is 0.075 lb per cu ft. (b) What is the total air horsepower and the total fan efficiency if the shaft input is the same as in problem 13–1?   *Ans.* (a) 1.0687 in. water; (b) 0.505 air hp, 64.2%

13–3. Air flows in a certain duct at 80 F under a static pressure of 3 in. of water (gage), with the barometer at 29.4 in. Hg. The average velocity in the duct is 1500 fpm. Find the total pressure (head) in the duct.   *Ans.* 3.14 in. water

13–4. In a certain duct, for a flow of 6000 cfm a static pressure of 0.86 in. is needed. (a) If the duct velocity at the fan is 1400 fpm, will one of the fans for which data is given in Table 13–3 be satisfactory? If so, at how many revolutions per minute will it run?
Ans. (a) Yes; (b) closely 483

13–5. By use of the 10-point method, pitot-tube readings across a duct 2 ft in diameter and carrying standard air (0.075 lb per cu ft) were 0.15, 0.15, 0.14, 0.15, 0.14, 0.15, 0.14, 0.16, 0.15, and 0.14 in. of water. Compute (a) the average velocity and (b) the air flow in cubic feet per minute.

13–6. Air at 80 F and 60 per cent relative humidity is flowing through a duct of 2-ft diameter. The barometric pressure is 29.92 in. Hg. A pitot tube inserted in the duct indicates a velocity pressure of 0.28 in. of water. Assuming that the value for this one point can be considered a representative value, (a) compute the air velocity and (b) find the flow through the duct in cubic feet per minute and in pounds per minute.

13–7. Solve problem 13–6 on the assumption that dry carbon dioxide is flowing through the duct and that the same readings are observed.

13–8. Standard air of density 0.075 lb per cu ft is flowing through a duct and a 10-point traverse is made of the duct, which is 6 ft in diameter. The following readings, in inches of water, are observed: 0.080, 0.095, 0.086, 0.084, 0.087, 0.083, 0.090, 0.088, 0.086, and 0.084. (a) Compute the flow through the duct in cubic feet per minute and in pounds per minute. (b) Compute the flow through the third annulus, for which the readings are 0.086 and 0.088. (c) Assuming that the average of these two values can be considered to represent the flow through the whole duct, multiply this value by 5, and compare the answer with the answer obtained by the longer computation.

13–9. Standard air of density 0.075 lb per cu ft is flowing through an 8-in. by 12-in. duct. The duct cross section is split into four imaginary rectangles, and a pitot reading is taken at the center of each rectangle, yielding the following values in inches of water: 0.162, 0.168, 0.164, and 0.166. Find (a) the average velocity in feet per minute and (b) the flow in cubic feet per minute.

Fans and Air Distribution 483

13–10. A certain fan has an impeller (rotor) 21 in. in diameter, turns at 580 rpm, delivers 7276 cfm at a static pressure of 1.25 in. of water, and requires 2.59 hp. The outlet velocity is 1700 fpm. Assume that the fan speed is increased to 610 rpm and estimate for this new speed (a) the probable capacity, (b) the horsepower, and (c) the static pressure. Explain why your answers differ from the data given in Table 13–3 for this fan. A study of Fig. 13–8 should suggest an answer.

Ans. (a) 7650 cfm; (b) 3.0 hp; (c) 1.38 in. water

13–11. The fan in problem 13–10 delivers 7276 cfm at a static pressure of 1.25 in. of water, and the velocity in the outlet is 1700 fpm. If 2.59 brake horsepower are supplied by the motor, what is (a) the static air horsepower, (b) the total air horsepower, and (c) the fan efficiency, based on total pressure?

Ans. (a) 1.43 hp; (b) 1.64 hp; (c) 63.3%

13–12. A 4-in. anemometer is used in making a 12-reading traverse over the face of a 12-in. by 20-in. rectangular exhaust grille. The average of the anemometer readings, with each reading taken for one minute, is 1265, and the resultant units are necessarily in feet per minute. Assume that the factor Y for a grille of this type can be considered 0.8. Compute the probable exit flow from the grille.   *Ans.* 1686 cfm

13–13. A square-edged orifice of 15-in. diameter is mounted in a round duct which is 25 in. in diameter and which carries a gas of density 0.065 lb per cu ft. The pressure across the orifice is 8 in. of water, measured at radius-tap positions. Compute the gas flow in cubic feet per second and in pounds per minute.

13–14. Air flowing in a duct is under a static pressure of 20 in. of water, and the barometer is at 29.45 in. Hg. A square-edged orifice with flange taps shows a differential pressure of 9.4 in. of water. The diameter of the orifice is 8 in., and that of the duct is 16 in. (a) Compute the density of the air, considering it to be dry and at a temperature of 60 F. (b) Find the air flow in cubic feet per minute and in pounds per minute.

13–15. Dry saturated steam at 50 psia is flowing in a 4-in. steam line (4.03 in. ID). A 3-in. orifice with flange taps shows a differential pressure of 5 in. Hg. Compute the steam flow in the line in pounds per hour. Refer to Table 2–3 for necessary data on the steam.

BIBLIOGRAPHY

AMERICAN SOCIETY OF MECHANICAL ENGINEERS. *Flow Measurement, Power Test Codes.* New York, 1949. [Part 5, Chapter 4]

AMERICAN SOCIETY OF MECHANICAL ENGINEERS. *Fluid Meters: Their Theory and Application*, 4th ed. New York, 1937.

AMERICAN SOCIETY OF MECHANICAL ENGINEERS. *Test Code for Fans.* New York, 1946.

NATIONAL ASSOCIATION OF FAN MANUFACTURERS. *Standard Definitions, Terms and Test Codes for Centrifugal, Axial and Propeller Fans.* NAFM Bulletin No. 110. Detroit, 1952.

AIR MOVING AND CONDITIONING ASSOCIATION. *Standards and Recommended Practices for Air-Moving Devices.* 1961

14

Principles of Refrigeration

14-1. REFRIGERATION WITH ICE

Refrigeration may be considered the development, in a given space, of a temperature lower than that which exists in some other, or adjacent, space.

The melting of ice or snow was one of the earliest methods of refrigeration and is still employed. Ice (or snow) melts at 32 F[1]; so when ice is placed in a given space warmer than 32 F, heat flows into the ice and the space is cooled or refrigerated. The capacity of the ice to absorb heat arises from the fact that when ice changes state from solid to liquid, the latent heat of fusion (143.3 Btu per lb) must be supplied from the surroundings.

In the simple case of ice melting, enough heat-transfer surface must be provided to permit a desired rate of heat flow into the ice from air or water moving past it, or from metal surfaces in contact with it.

Another medium of refrigeration is solid carbon dioxide (dry ice). At atmospheric pressure carbon dioxide (CO_2) cannot exist in a liquid state and consequently, when solid CO_2 absorbs heat, it sublimes or goes directly from the solid into the vapor state. At 14.7 psia, sublimation occurs at about -109.3 F, and the heat of sublimation is approximately 246 Btu per lb at this temperature. Thus dry ice is well suited for low-temperature refrigeration. It also is used widely for the refrigeration of small packages, where the absence of any resulting liquid adds to its convenience.

Water ice was formerly employed, to a limited extent, as a cooling medium in small-size air-conditioning installations (less than 10 to 20 tons of refrigeration). In most such installations water, cooled by spraying and trickling over the ice, served the conditioning equipment. Ice systems performed well under fluctuating loads, since the rate of ice melting could be made to follow closely the cooling-load variations, and the efficiency was not greatly reduced under light loads. Although the cost of ice was

[1] Under super pressures ice does not melt at 32 F; for example, at 14,200 psi, ice melts at 16 F.

Principles of Refrigeration

relatively high and presented some inconvenience in handling, the low investment cost, compared with that for mechanical refrigeration equipment, justified the use of ice for some cooling installations. Ice and ice-salt mixtures are still used on refrigerated railway cars to maintain the quality of perishable foods in transit.

14-2. MECHANICAL VAPOR REFRIGERATION

Just as a solid, in changing its state into liquid (or gaseous) form, absorbs heat from its surroundings or other sources, so a liquid, in vaporizing, must absorb heat. For example, ammonia, at atmospheric pressure, boils at -28 F and possesses a latent heat of about 589.3 Btu per lb. Ammonia, if placed in surroundings warmer than -28 F, would cool the surrounding space as it evaporated. If the pressure on the ammonia were increased to 30.4 psia, it would boil to 0 F and thus cooling could be

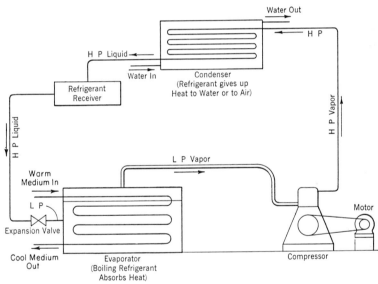

FIG. 14-1. Elements of a mechanical (compression) refrigerating system with water-cooled condenser.

accomplished, but in a higher temperature range. Mechanical refrigeration makes possible the control of pressure and temperature of the boiling refrigerant and also makes possible the use of the same refrigerant over and over again with little or no loss of the refrigerant. Theoretically, almost any stable, noncorrosive liquid can operate as a refrigerant if its pressure-temperature relations are suitable for the conditions desired. However, certain features, discussed in detail in the next chapter, determine the desirability of a given refrigerant. The temperature-pressure relationship, which is an important characteristic, can be quickly observed for many refrigerants in Fig. 15-1.

The elements of a compression refrigeration system are shown in Fig. 14-1. In the evaporator the liquid refrigerant, in vaporizing, absorbs heat from brine (or from water, or directly from the air of the space to be cooled). The low-pressure (l-p) refrigerant vapor from the evaporator is inducted into the compressor, which raises the vapor in pressure and temperature for delivery to the condenser. The refrigerant must be compressed sufficiently to have a saturation temperature higher than the temperature of the cooling medium employed so that heat can be dissipated in the condenser. After heat removal and condensation in the condenser, the liquid refrigerant may pass to a receiver, or storage tank. The high-pressure (h-p) liquid refrigerant next passes through the expansion valve, where the refrigerant throttles (drops) to the evaporator pressure of the system. In passing through the expansion valve the liquid refrigerant cools itself at the expense of evaporating a portion of the liquid. In a refrigeration system the low pressure in the evaporator is determined by the temperature which it is desired to maintain in the cooled space. The high pressure in the condenser is determined ultimately by the temperature of the available cooling medium, i.e., the circulating water or the atmosphere (air temperature). The process is one in which the refrigerant absorbs heat at a low temperature and then, by the action of mechanical work, the refrigerant is raised to a sufficiently high temperature to allow rejection of this heat. Note that the mechanical work or energy supplied to the compressor is the means used to raise the temperature of the system. Compressors can be powered by any convenient means; electric motors, steam engines, internal combustion engines, etc.

14-3. THE IDEAL (CARNOT) CRITERIA OF REFGRIGERATION

In thermodynamic analyses, wide use is made of the Carnot criterion (Carnot cycle) as a standard against which the performance of a prime mover (turbine or engine) can be compared. The reasoning underlying the Carnot criterion is also applicable to an idealized refrigeration system, constituting, as it does, a reversed heat engine or heat pump. The criterion, in its conventional form, presupposes two constant temperature levels of heat exchange and two reversible-adiabatic processes, none of which is completely possible of realization. Nevertheless the criterion will be presented here because of its importance as a measure of maximum performance.

The efficiency of a Carnot heat engine is expressed as

$$\text{Efficiency} = \frac{Q_C - Q_R}{Q_C} = \frac{T_C - T_R}{T_C} \quad (14\text{-}1)$$

Principles of Refrigeration

where T_C = high temperature of the system;
Q_C = heat interchange at T_C;
T_R = low temperature of the system;
Q_R = heat interchange at T_R.

The values Q_C and Q_R can be expressed in any consistent energy units, and T_C and T_R are expressed in absolute degrees, i.e., degrees Fahrenheit + 460 = degrees Rankine.

The cycle can advantageously be shown on the temperature-entropy plane. This plane is convenient for representation, since in this plane, for reversible processes, areas can show magnitudes of heat interchange. Thus in Fig. 14–2, for a heat engine, the heat added (Q_C) from *2* to *3* is

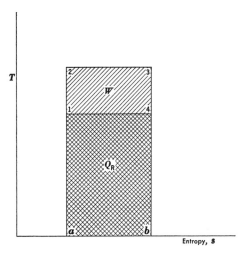

FIG. 14–2. Ideal Carnot cycle for power or refrigeration.

represented by the area *a 2 3 b*, and the path of reversible-adiabatic expansion, with temperature dropping from T_C to T_R, is shown by *3* to *4* (no area). The heat rejected (Q_R) is represented by the area *4 b a 1*. The path *1* to *2* represents reversible-adiabatic compression from temperature T_R to T_C. In a power cycle the work produced (*W*) is equal to the difference between the heat added and the heat rejected, or

$$W = Q_C - Q_R \tag{14-2}$$

Consequently the work area is *1 2 3 4*.

For refrigeration the process is reversed, with heat added at the low temperature T_R, in amount Q_R, and represented by the area *1 4 b a*. The temperature of the cycle medium is raised by compression, following the path *4* to *3*. Heat is rejected at the high temperature T_C, in amount

Q_C, and represented by the area *3 2 a b*. The work, which must be provided from an external source, is $-(Q_R - Q_C)$ and is represented by the area *3 2 1 4*.

For the refrigeration (heat pump) type of operation, conventional efficiency has little significance and it is desirable to introduce the term coefficient of performance (C P). The coefficient of performance for any refrigeration system, either ideal or actual, is expressed as the refrigeration created divided by the work required to produce it. Thus

$$\mathrm{C\,P} = \frac{Q_R}{W} \qquad (14\text{--}3)$$

For the Carnot (ideal) basis, Q_R/W appears as

$$\mathrm{C\,P} = \frac{Q_R}{Q_C - Q_R} \qquad (14\text{--}4)$$

$$= \frac{T_R}{T_C - T_R} \qquad (14\text{--}5)$$

Any consistent energy units for Q_C, Q_R, and W can be employed with Btu, or Btu per lb most usual.

Example 14–1. An ideal (Carnot) refrigeration system operates between temperature limits of -20 and 80 F. Find (a) the ideal coefficient of performance and (b) the horsepower required from an external source to absorb 12,000 Btuh at the low temperature.

Solution: (a) By equation 14–5,

$$\mathrm{C\,P} = \frac{-20 + 460}{(80 + 460) - (-20 + 460)} = \frac{440}{540 - 440} = 4.4. \qquad Ans.$$

That is, 4.4 times as much refrigeration is produced (heat is absorbed) as is required as work.

b) A horsepower is 33,000 ft-lb per min. Thus

33,000 ft-lb/min ≡ 33,000/778 or 42.4 Btu/min
≡ 2545 Btuh

With a C P of 1, the horsepower required would be 12,000/2545. However, since the ideal C P is 4.4,

$$\mathrm{Hp} = \frac{12{,}000}{(\mathrm{C\,P})(2545)} = \frac{12{,}000}{(4.4)(2545)} = 1.07 \qquad Ans.$$

14–4. REFRIGERANT DIAGRAMS, TEMPERATURE-ENTROPY

It has been mentioned that a refrigerant can be any stable, noncorrosive substance having suitable liquid-vapor phase characteristics to serve satisfactorily in the evaporator-condenser temperature range required. The mediums actually used must satisfy a number of other criteria, such as availability, cost, low-toxicity, pressure range, behavior

Principles of Refrigeration

with lubricants, and the like. These criteria are discussed in detail in Chapter 15, whereas at this point the basic thermodynamics of the refrigerating cycle as related to any refrigerant will be primarily considered. However, in this chapter reference to tabular properties of refrigerants given in Chapter 15 will be made, with frequent discussions related to the refrigerants ammonia and dichlorodifluoromethane. The latter is also known as Freon-12, genetron-12, and refrigerant 12.

The properties of a refrigerant or of any thermodynamic substance can be represented to advantage by plotting property values on coordinate

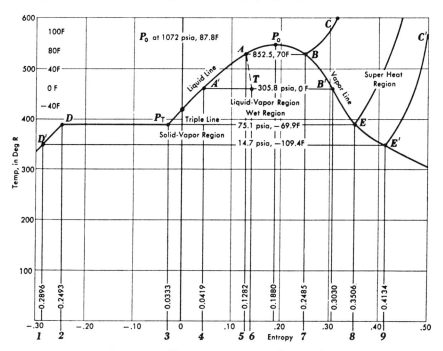

Fig. 14-3. Temperature-entropy diagram for carbon dioxide (CO_2).

diagrams. In common use are the temperature-entropy plane (Ts), the pressure-enthalpy plane (ph), and of less frequent use, the pressure-volume plane and the enthalpy-entropy plane. In refrigeration work the pressure-enthalpy diagram is by far the most useful type of plot, although the temperature-entropy plot, on which areas can represent the heat interchange, is also useful.

Figure 14-3 is a temperature-entropy plot of carbon dioxide. This substance was chosen for discussion because it illustrates a medium used, to a limited extent, as a refrigerant in all three phases (solid, liquid, and vapor). The critical point for CO_2, indicated at P_o on the diagram, has a value of 87.8 F, with a corresponding saturation pressure of 1072 psia. At temperatures higher than its critical temperature it is impossible to liquefy a gas, no matter what pressure is exerted on it. Below the critical point a definite pressure-temperature relationship applies; that is, for each pressure there is a corresponding temperature at which vaporization (or condensation) occurs. For example, liquid CO_2 at a pressure of 852.5 psia (point A on Fig. 14–3) will boil (vaporize) at 70 F, and such a liquid when heated would change completely from liquid to vapor, as shown at point B. If further heat is added to the dry saturated vapor, with pressure remaining constant at B, the temperature of the vapor will rise as it moves toward point C, and we speak of this vapor as being superheated. The region to the right of the vapor line is known as the superheated (or dry-gas) region. The region between the liquid and vapor lines consists of saturated mixtures of liquid and vapor.

It is possible for a liquid, such as the saturated liquid at A, to be cooled with the imposed pressure remaining constant, as for example, to the temperature A', for which the saturation pressure is lower than for point A. Such a liquid existing at a higher pressure than the saturation pressure corresponding to its temperature is known as *compressed liquid* or as *subcooled liquid*. The entropy of compressed liquid is trivially lower than the entropy of saturated liquid at the same temperature. However, the difference cannot be shown on a Ts diagram of this type; thus the liquid line represents saturated and compressed-liquid conditions. In most cases the properties of a liquid, particularly enthalpy and specific volume, are found in tables under the temperature listing and are independent of pressure. If a saturated (or compressed) liquid is cooled sufficiently, it will ultimately reach a point at which solidification or fusion takes place. The particular temperature and pressure at which saturated liquid, solid, and vapor are in equilibrium is known as the *triple point*, marked P_T in Fig. 14–3. If heat is removed from liquid at the triple point, solidification starts and continues from the liquid-solid mixture until all of the liquid has fused. This point is represented by D on the diagram, and the line D to P_T is thus a solid-liquid line. The triple point conditions for CO_2 are at $-$ 69.9 F and 75.1 psia. An interesting consequence of this fact is that liquid CO_2 cannot exist at atmospheric pressure (14.7 psia), since this is lower than 75 psia. Solid CO_2 (dry ice) is very familiar at atmospheric conditions, but an open container of liquid CO_2, no matter what its temperature, cannot exist. In contrast to CO_2, water (H_2O) has a triple point of 32 F at 0.0885 psia, and thus we are quite familiar with both water and ice at atmospheric pressure. Melting of solid CO_2 to liquid can take place only at pressures equal to or higher than 75.1 psia.

The line DD' in Fig. 14–3 represents loci of solid CO_2 saturation con-

Principles of Refrigeration

ditions, because, as is true for the liquid phase, there are similar solid phase saturation-temperature-pressure relationships. For each pressure there is a corresponding temperature at which sublimation takes place and the solid becomes vapor, as at E or E'. The vapor at E is unique, in that it is at the triple point and can be in equilibrium either with solid or liquid. It should also be mentioned that the triple point on the Ts diagram is the line DE.

Water ice can also sublime and go directly from the solid to the vapor (steam) phase. For example, wet articles hung outside to dry, in 0 F temperature weather, first freeze and then slowly dry as the water sublimes. Snow and ice on the ground, under subfreezing conditions, also slowly disappear by sublimation without ever reaching the liquid phase.

The thermodynamic properties of all the commonly used refrigerants have been computed and usually can be found in tabular form. The datum of reference for the tabular values is usually taken at -40 F and the corresponding saturation pressure. At this reference condition, enthalpy and entropy are usually arbitrarily given values of zero. This represents a satisfactory procedure, since our interest rests in changes in these properties and not in absolute values of them.

Areas on the Ts diagram, when properly interpreted, can represent heat energy added or withdrawn during a process. For such areas to be to scale, it is necessary to draw the diagram using temperature expressed in absolute degrees. The entropy, however, can be from an arbitrary datum, but of course in consistent units (usually Btu per lb degree if the area and result are desired in Btu). For example, in Fig. 14-3 the area under $A'A$, namely $A'A54$, represents the heat addition required to warm saturated liquid CO_2 from conditions at A' to those at A; correspondingly, the area $AB75$ represents the heat addition required to change liquid to vapor or the enthalpy of vaporization at 852.5 psia, and the area $D'E'91$ represents the enthalpy of sublimation at 14.7 psia. As most refrigeration is concerned only with the refrigerant in the liquid and vapor phases, diagrams are not usually drawn for conditions below the triple point, nor are they extended up to the critical point (P_o).

14-5. REFRIGERANT DIAGRAMS, PRESSURE-ENTHALPY

A skeleton, pressure-enthalpy (ph) diagram for dichlorodifluoromethane (refrigerant-12 popularly known as Freon-12 or genetron-12) appears in Fig. 14-4. As in the Ts diagram, the key lines are those for saturated liquid and saturated vapor. If these lines were extended higher than shown in the figure, they would meet at the critical point (P_o), which for this refrigerant happens to be at 596.9 psia and 233.6 F. The region enclosed between the saturated liquid and vapor lines is the wet-vapor region. To the right of the vapor line is the superheated-vapor region. To the left

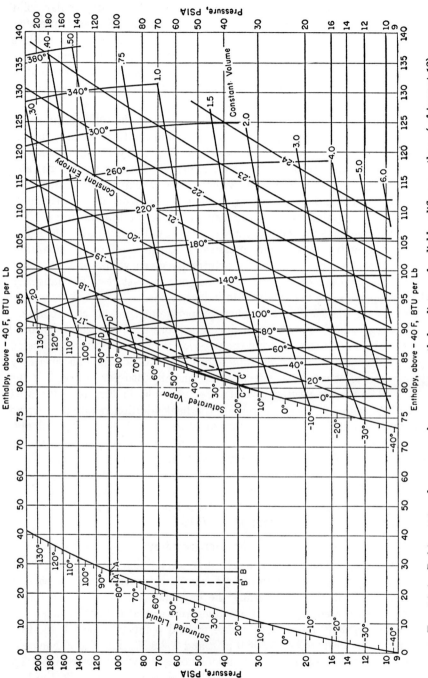

Fig. 14-4. Refrigerant cycle superposed on pressure-enthalpy diagram for dichlorodifluoromethane (refrigerant-12).

Principles of Refrigeration

of the liquid (saturated) line lies the compressed-liquid region. The conditions of compressed liquid can advantageously be shown on this type diagram at A and A'. If the liquid temperature is constant, no significant change in the enthalpy of compressed liquid occurs as the pressure on the liquid is varied. There is, however, a slight increase in compressed-liquid enthalpy with pressure increase, and although this is significant in power-plant operations (as feed-pump work for the extensive pressure ranges), in refrigeration computations the trivial enthalpy change can be disregarded.

The pressure-enthalpy diagram is important in refrigeration computations because on this diagram, enthalpy changes can be read with ease, and such changes can represent heat added or withdrawn, or work received (in adiabatic compression). Moreover, the refrigeration cycle largely employs two pressures, namely, the pressure in the evaporator associated with the useful refrigeration, and that in the condenser associated with the heat dissipation phase of the cycle. Both of these pressures are readily shown on the ph diagram.

14–6. REFRIGERATION-CYCLE, EXPANSION AND EVAPORATION

Expansion valve (throttling). Liquid refrigerant, after it leaves the condenser (or the receiver), enters the expansion valve. This valve serves the purposes of controlling the refrigerant flow and of dropping the refrigerant, both in temperature and pressure, from condenser to evaporator conditions. A flow process which takes place adiabatically without work production is a throttling process and in such it is easy to show, from the steady-flow energy equation 2–5, after eliminating irrelevant terms, that

$$\frac{V_1^2}{2gJ} + h_1 = \frac{V_2^2}{2gJ} + h_2 \qquad (14\text{–}6)$$

Because the kinetic energy at entry to and exit from an expansion valve seldom differs greatly, it is customary to disregard these terms and write the expansion valve equation as

$$h_1 = h_2 \qquad (14\text{–}7)$$

This equation states that the enthalpy remains constant in the throttling process of an expansion valve.

A certain amount of the liquid, however, always flashes into gas, because a pound of liquid at the low pressure and temperature has less enthalpy than a pound at higher temperature. Thus some vaporization must take place. The process may also be considered in this light: the bringing of warm liquid to cold evaporator temperature occurs at the expense of evaporating a portion of the liquid, with the loss of some refrigerating effect. This process results in the production of flash vapor which can accomplish no useful cooling in the evaporator. Thus the expansion-valve equation is

$$h_{f1} = h_{fR} + xh_{fgR} \qquad (14\text{–}8)$$

where h_{f1} = enthalpy of liquid refrigerant at the temperature at which it enters the expansion valve, in Btu per pound;
h_{fR} = enthalpy of liquid at evaporator pressure, in Btu per pound;
h_{fgR} = latent heat of refrigerant at evaporator pressure, in Btu per pound;
x = quality, expressed as a decimal, of the refrigerant after passing through the expansion valve, also the pound (weight) of flash gas formed per pound of refrigerant.

The expansion valve (throttling process) is represented on the ph chart (Fig. 14–4) by the straight line AB (or by $A'B$ if the liquid is originally subcooled).

On the Ts chart, a line of constant enthalpy swings down and to the right, and the area representing liquid enthalpy at the beginning of the process must equal the area for wet vapor at the end of the process. Using the Ts chart of Fig. 14–3 for illustration, the liquid condition is shown at A, before throttling, and the wet-vapor condition is indicated at T, after throttling. Realizing that the entropy datum is not significant, let us select $A'4$ as a convenient datum; then for the throttling process we can state that the area $A'A54$, before throttling, must equal the area $A'T64$, after throttling.

Example 14–2. Saturated liquid dichlorodifluoromethane (refrigerant-12) at a temperature of 86 F at 108.04 psia enters an expansion valve and expands into an evaporator in which the refrigerant is boiling at 20 F. Find (a) the enthalpy of the refrigerant and entry to and exit from the expansion valve and (b) the weight of flash gas formed per pound of refrigerant entering the valve.

Solution: (a) In Table 15–5 the enthalpy of liquid refrigerant-12 at 86 F is shown to be $h_f = 27.769$ Btu per lb. The enthalpy at exit from the valve is likewise 27.769 Btu per lb. *Ans.*

b) At 20 F evaporator temperature, from Table 15–5, the saturation pressure is 35.736 psia, $h_{fR} = 12.863$, and $h_{fgR} = 66.522$. By equation 14–8

$$27.769 = 12.863 + x(66.522)$$

$$x = 0.224 \text{ lb of flash gas formed per lb of refrigerant entering the expansion valve} \qquad Ans.$$

Example 14–3 If the liquid refrigerant of example 14–2 is subcooled to 70 F before entering the expansion valve, find the items asked for in example 14–2.

Solution: Table 15–5 shows that liquid refrigerant-12 has an enthalpy at 70 F of 24.050 Btu per lb. This is true for 70 F compressed liquid at 108.04 psia just as it is true for saturated 70 F liquid at 84.888 psia; that is, the enthalpy of liquid is a function of temperature. Thus for part (a) the enthalpy is 24.050 Btu per lb before entry and at exit from the expansion valve. *Ans.*

Principles of Refrigeration

b) At 20 F,
$$24.050 = 12.863 + x(66.522)$$
$$x = 0.168 \text{ lb of flash gas formed per lb of refrigerant entering the expansion valve} \quad Ans.$$

Note that subcooled (compressed) liquid produces less flash gas than liquid at the higher temperature corresponding to saturation.

Evaporator. In the evaporator the liquid from the expansion valve changes into vapor as it absorbs heat from the space being cooled. The heat absorbed appears as increased enthalpy of the refrigerant. The vapor leaving the evaporator may be dry saturated as at C in Fig. 14–4, superheated as at C', or slightly wet. Thus

$$Q_R = h_R - h_{f1} \quad (14\text{–}9)$$

where Q_R = Btu absorbed per pound of refrigerant in evaporator;
h_R = enthalpy of vapor leaving the evaporator, in Btu per pound;
h_{f1} = enthalpy of liquid refrigerant at temperature supplied to expansion valve, in Btu per pound.

Refrigerant flow is often expressed in pounds per minute per ton of refrigeration. As indicated before, the ton of refrigeration is a rate of heat exchange (absorption); by definition, 200 Btu per min or 12,000 Btu per hr. Thus

$$w_R = \frac{200}{Q_R} = \frac{200}{h_R - h_{f1}} \quad (14\text{–}10)$$

where w_R represents pounds of refrigerant circulated per minute per ton. The values Q_R, h_R, and h_{f1} are as they appear in equation 14–9.

Example 14–4. Refrigerant-12 is vaporizing in an evaporator at 20 F. Assuming that dry saturated vapor leaves the evaporator, and that liquid is supplied to the expansion valve at 86 F, find the rate of refrigerant flow required per ton of refrigeration and also per 10,000 Btu per hr.

Solution: Reference to Table 15–5 shows that dry saturated vapor (h_g) at 20 F and 35.736 psia has an enthalpy of 79.385 Btu per lb. By equation 14–9,

$$Q_R = h_R - h_{f1} = 79.385 - 27.769 = 51.616 \text{ Btu per lb}$$

By equation 14–10,

$$w_R = \frac{200}{51.616} = 3.87 \text{ lb per min per ton} \quad Ans.$$

For 10,000 Btu per hr,

$$\text{Flow} = \frac{10{,}000}{(60)(51.616)} = 3.23 \text{ lb per min} \quad Ans.$$

14-7. REFRIGERATION CYCLE, COMPRESSION

To maintain a given evaporator pressure the compressor must remove the vapor as fast as it is formed. If the evaporator load is small, that is, if the temperature difference between the medium cooled and the evaporator is slight, little refrigerant can evaporate and the compressor suction causes a reduction in the evaporator pressure. This pressure decrease will continue until the temperature difference between the refrigerated space and the colder evaporator becomes just sufficient to generate enough vapor to supply the effective piston displacement of the compressor. If, on the other hand, the temperature difference is great (excessive load), vapor will be generated rapidly at relatively high evaporator temperature and the compressor may be overloaded. The heat-transfer load characteristics of the evaporator must be such as not to overload the compressor.

More work is required to compress a pound of refrigerant from a low pressure (temperature) to a given condenser pressure (temperature) than from a higher pressure to the same condenser pressure. Yet, in the actual machine, the greater weight of refrigerant handled at a higher suction pressure (because of lower specific volume and compressor-clearance behavior) usually causes a greater power requirement on the driving motor as the evaporator pressure increases.

Compression of a vapor, under adiabatic conditions, requires the least work when performed *isentropically* (at constant entropy). Isentropic compression cannot be attained in actual equipment, but it forms a basis for computing ideal work under different operating conditions. From this basis the actual work can be estimated. The entropy of a given vapor is always listed in complete tables of properties of the refrigerant and is also plotted on pressure-enthalpy charts.

The theoretical (or isentropic) work of compression, using the symbols of Fig. 14–4, is found by

$$W_T = (h_D - h_C)_s \quad \text{or} \quad (h_{D'} - h_{C'})_s \quad \text{Btu per lb}$$

or in general symbolism

$$W_T = (h_D - h_R)_s \quad \text{Btu per lb of refrigerant} \quad (14\text{--}11)$$

where h_D = enthalpy at discharge pressure and conditions;
h_R = enthalpy of vapor entering the compressor, with entropy s assumed constant during compression to h_D.

The theoretical horsepower per ton of refrigeration is

$$\text{Hp} = \frac{\text{lb refrigerant}}{\text{min} \times \text{ton}} \times W_T \times \frac{1}{42.4} \quad (14\text{--}12)$$

To find the actual work or power required by a compressor from the theoretical work of compression, a factor must be employed. For rough approximations it may be considered that the isentropic work, increased by 30 to 50 per cent, approximates the shaft work that is required to carry out the compression. This factor must account for friction in the packing glands of the compressor, in the bearings, and from piston rings,

Principles of Refrigeration

heat exchanges, irreversibilities in the compressor, and the like. The factor is higher with small, inefficient compressors, and lower with larger-size, well-designed units. Centrifugal compressors, because of higher internal loss and turbulence, might be expected to lie at the higher end of the range of factor values, but, since these machines are usually large and well-designed, the factor values for centrifugal compressors usually lie at about 30 per cent.

Example 14–5. A refrigeration compressor takes refrigerant-12, dry-saturated vapor at 20 F (at 35.736 psia), and compresses and delivers it to a condenser operating at a condensing temperature of 86 F. Compute (a) the ideal work of compression, in Btu per pound, (b) the theoretical horsepower required per ton of refrigeration if this compressor serves an evaporator operating under the conditions of example 14–4, and (c) the probable actual input shaft horsepower required for a 1-ton-capacity reciprocating compressor.

Solution: (a) Refer to the ph chart for refrigerant-12 vapor (Fig. 15–4) and find the isentropic work of compression such as would be indicated by the line CD of Fig. 14–4. For the given compressor inlet condition, namely dry saturated vapor at 20 F, the enthalpy h_R or h_C on the diagram is 79.4 Btu per lb. Follow a path parallel to the lines of constant entropy until the pressure corresponding to 86 F condensation (namely 108.04 psia) is reached at which point h_D = 87.9 Btu per lb. Then by equation 14–11,

$$W_T = (h_D - h_R)_s = 87.9 - 79.4 = 8.5 \text{ Btu per lb} \qquad Ans.$$

b) From example 14–4, the refrigerant flow per ton is 3.87 lb per min, and the theoretical power is

$$\text{Hp} = w_R W_T = (3.87)(8.5) = 32.9 \text{ Btu per min}$$

and

$$\text{Theoretical hp per ton} = \frac{32.9}{42.4} = 0.78 \qquad Ans.$$

where $2545/60 = 42.4$ Btu per min $\equiv 1$ hp.

c) If we take a representative shaft work factor of 1.4, the probable shaft horsepower is

$$\text{Compressor shaft hp} = (1.4)(0.78) = 1.09 \qquad Ans.$$

The capacity of a reciprocating compressor for handling refrigerant depends on the piston displacement per minute and on the volumetric efficiency. Volumetric efficiency (charge efficiency) of a compressor is defined as the amount of vapor handled, measured in cubic feet per minute at suction pressure and temperature, divided by the piston displacement per minute. This efficiency is always less than unity because of (1) superheating of the entering vapor as it enters the warm cylinder; (2) throttling or friction loss through ports and valves, which reduces cylinder pressure; and (3) re-expansion of compressed vapor from the clearance volume, which reduces the effective volume for fresh vapor charge at suction pressure. Values of volumetric efficiency will thus vary with the type of compressor; with the operating conditions, as affected by the condition of the suction vapor and by the pressure ratio of compression; and with the refrigerant. Volumetric efficiencies η_v of reciprocating compressors range from about 76 to 90 per cent. A representative value of 86 per cent is suggested for general usage with current compressor designs. If liquid enters the compressor along with vapor, a fictitious volumetric efficiency in excess of 100 per cent may arise. A further discussion of volumetric efficiency is presented at the end of this chapter.

The effective piston displacement (PD) of a single-acting compressor can be expressed, in cubic feet per minute, as

$$\text{PD} = C \frac{\pi d^2 S N}{4 \times 1728} (\eta_v) \qquad (14\text{--}13)$$

where C = number of cylinders in compressor;
d = compressor bore, in inches;
S = compressor stroke, in inches;
N = revolutions per minute;
η_v = volumetric efficiency, expressed as a decimal;
PD = effective piston displacement, for induction of vapor at low-pressure (suction) conditions, in cubic feet per minute.

14-8. REFRIGERATION CYCLE, CONDENSATION

The condenser, whether water-cooled or air-cooled, must remove heat from the refrigerant to change the superheated gas leaving the compressor into saturated or subcooled liquid. Again referring to Fig. 14-4, the condenser must remove heat ideally to change the gas with enthalpy h_D (or $h_{D'}$) to liquid with enthalpy h_{fA} (or if subcooled to $h_{fA'}$). That is,

$$Q_C = h_D - h_A \quad \text{Btu per lb} \qquad (14\text{--}14)$$

or for other operating conditions

$$Q_C = h_{D'} - h_{A'} \quad \text{Btu per lb} \qquad (14\text{--}15)$$

Expressed another way, the heat removed in the condenser is equal to the heat absorbed at low temperature in the evaporator plus the heat equivalent of the work performed on the refrigerant in the compressor or, symbolically,

$$Q_C = Q_R + W \qquad (14\text{--}16)$$

where Q_C = heat removed from refrigerant in condenser, in Btu per pound;
Q_R = refrigerating effect (heat absorbed) in evaporator, in Btu per pound of refrigerant;
W = the work energy added to the refrigerant, in Btu per pound.

Example 14-6. A two-cylinder, 3-in. by 3-in. refrigerant-12 compressor running at 1140 rpm operates on a system with condensation occurring at 86 F and evapora-

Principles of Refrigeration

tion at 20 F. Assume that dry saturated vapor enters the compressor and the condensed refrigerant is not subcooled. Compute (a) the piston displacement; (b) the compressor capacity, in pounds of refrigerant per minute, if the volumetric efficiency is 82 per cent; (c) the tons of refrigeration produced; (d) the idealized condenser heat load, in Btu per pound of refrigerant and (e) in Btu per hour; and (f) the actual coefficient of performance (CP).

Solution: (a) Use equation 14–13 or equivalent reasoning to find the true piston displacement D.

$$D = 2\frac{(\pi)\,(3^2)\,(3)\,(1140)}{(4)\,(1728)} = 27.95 \text{ cfm} \qquad Ans.$$

The true piston displacement is independent of volumetric efficiency ($\eta_v = 1.0$).

b)   PD = (27.95) (0.82) = 22.9 cfm effective piston displacement

Because the vapor supplied to the compressor is dry and saturated and at 20 F, its specific volume can be read directly from Table 15–5 as $v_g = 1.0988$ cu ft per lb,

$$w = \frac{PD}{v_g} = \frac{(\text{cu ft}) \times (\text{lb})}{(\text{min}) \times (\text{cu ft})} = \frac{22.9}{1.0988} = 20.8 \text{ lb per min refrigerant flow rate}$$

Ans.

c) From example 14–4 the refrigeration effect per pound of refrigerant is 51.616 Btu per lb. Thus

Refrigeration produced = (51.616)20.8 = 1074 Btu per min

$$= \frac{1074}{200} = 5.37 \text{ tons} \qquad Ans.$$

d) For the given operating conditions, it was shown in example 14–5 that the isentropic work per pound of refrigerant was 8.5 and the actual work was taken 40 per cent higher, or

$$W = (8.5)\,(1.4) = 11.9 \text{ Btu per lb}$$

From part (c) the refrigeration effect is $Q_R = 51.616$ Btu per lb and by equation 14–16,

$$Q_C = 51.616 + 11.9 = 63.5 \text{ Btu per lb} \qquad Ans.$$

e) $Q = (63.5)\,(20.8)\,(60) = 79{,}200$ Btu per hr   *Ans.*

f) By equation 14–3,

$$CP = \frac{51.616}{11.9} = 4.3 \qquad Ans.$$

14-9. COMPRESSOR WORK BY THE GAS EQUATION

The work required for compression of a refrigerant should always be found from tabulations of properties (or charts) of the refrigerant when these are available, and for the commonly used refrigerants complete tables have been prepared. However, for gases and vapors used as refrigerants and when complete property tabulations are not available, use can be made of relationships developed from perfect-gas laws. The general form of the expression for isentropic work of compression, as derived in standard thermodynamics textbooks, appears as

$$W_T = \frac{144}{778} \frac{k}{k-1} p_R v_R \left[\left(\frac{p_D}{p_R}\right)^{k-1/k} - 1\right] \quad \text{Btu per lb} \quad (14\text{--}17)$$

$$W_T = \frac{k}{(778)(k-1)} RT_R \left[\left(\frac{p_D}{p_R}\right)^{k-1/k} - 1\right] \quad \text{Btu per lb} \quad (14\text{--}18)$$

$$\text{Hp} = \frac{144}{33{,}000} \frac{k}{k-1} p_1 V_1 \left[\left(\frac{p_2}{p_1}\right)^{k-1/k} - 1\right] \quad (14\text{--}19)$$

where $k = C_p/C_v$, ratio of specific heats of the refrigerant (tables 2–1 and 15–1);

p_R or p_1 = pressure at inlet to compressor, in psia;
v_R or v_1 = specific volume at compressor inlet, in cubic feet per pound;
T_R or T_1 = temperature at compressor inlet, in degrees Rankine;
R = gas constant, 1545.3/m, in foot-pounds per pound degree Rankine (see sec. 2–8);
V_R or V_1 = volume, in cubic feet per minute at inlet conditions, entering compressor.

Moreover, in terms of temperature it is closely true that

$$\frac{T_2}{T_1} = \left(\frac{p_2}{p_1}\right)^{k-1/k} \quad (14\text{--}20)$$

where the subscripts 1 and 2 refer to conditions before and after compression with temperature and pressure in same units as previous equations. Thus when a gas is compressed it may increase greatly in temperature, provided the compression is carried out so quickly that the gas has little chance to cool. In fact, gas compression in reciprocating compressors is closely adiabatic, although not truly isentropic. Moreover, when a gas expands in an expander engine or gas turbine and produces work, the gas temperature decreases and equation 14–20 (when inverted) is equally applicable to rapid (adiabatic) expansion. The work and power equations 14–17 through 14–19 are also applicable to the expansion process if the inlet terms and ratios are appropriately modified.

14–10. AIR-CYCLE REFRIGERATION

Before the development of the halogenated-hydrocarbon refrigerants, with their low toxicity or even nontoxic characteristics, air-cycle systems were often used where absolute assurance was required against possible escape of toxic refrigerant. In comparison to mechanical vapor-liquid systems, air systems were bulky and heavy; the horsepower per ton was many times greater than with vapor systems; and the cooling medium, instead of acting at a fixed temperature, warmed over a temperature range. Most of the early air systems were closed, so-called dense-air systems. A closed system made possible a plant of smaller physical size, because the air was compressed at all stages of the cycle, and humidity,

Principles of Refrigeration

as a problem, was largely eliminated. A diagrammatic layout of a closed system is shown in Fig. 14–5. In this system, air at temperature T_1, several degrees lower than the space temperature, and at pressure p_1 enters the compressor. During compression, the air increases in pressure to p_2 and in temperature to T_2. The compressed air then enters the water-cooled heat exchanger, which corresponds to the condenser of a vapor system, and is cooled to temperature T_3, which is a few degrees warmer than the inlet water temperature. The cooled compressed air then enters the expander engine, where it drops to the low pressure of the system, at

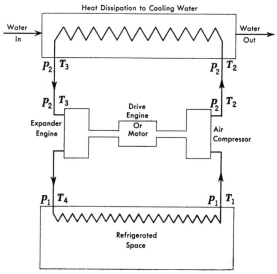

FIG. 14–5. Closed air-cycle refrigeration system.

the same time decreasing to temperature T_4. The useful refrigeration produced is

$$Q_R = w C_p (T_1 - T_4) \quad \text{Btu per min} \quad (14\text{--}21)$$

The heat dissipation is

$$Q_C = w C_p (T_2 - T_3) \quad \text{Btu per min} \quad (14\text{--}22)$$

The isentropic work input to the air compressor is

$$W_C = w C_p (T_2 - T_1) = w C_p T_1 \left[\left(\frac{p_2}{p_1}\right)^{k-1/k} - 1 \right] \quad (14\text{--}23)$$

The isentropic work delivered by the expander engine is

$$W_E = w C_p (T_3 - T_4) = w C_p T_3 \left[1 - \left(\frac{p_4}{p_3}\right)^{k-1/k} \right] \quad (14\text{--}24)$$

The net isentropic work required from the motor, in Btu per min, is

$$W_N = w C_p [T_2 - T_1 - (T_3 - T_4)] \quad (14\text{-}25)$$

$$= w C_p \left\{ T_1 \left[\left(\frac{p_2}{p_1}\right)^{k-1/k} - 1 \right] - T_3 \left[1 - \left(\frac{p_4}{p_3}\right)^{k-1/k} \right] \right\} \quad (14\text{-}26)$$

In an idealized system, in which there is no pressure loss, the compression pressure ratio p_2/p_1 equals the expansion ratio p_3/p_4. Letting r represent each of these ratios, the expression for net work simplifies to

$$W_N = w C_p \left\{ T_1 \left[r^{k-1/k} - 1 \right] - T_3 \left[1 - \frac{1}{r^{k-1/k}} \right] \right\} \quad (14\text{-}27)$$

In the preceding expressions
w = air flow in system, in pounds per minute;
C_p = specific heat of air, approximately 0.24;
p_1, p_2, p_3, p_4 = pressures in any consistent units, generally, in psia;
T_1, T_2, T_3, T_4 = temperatures in degrees Rankine, except where temperature difference only is involved, degrees Fahrenheit may be used;
W_C, W_E, W_N = values of work, in Btu per minute.

Figure 14-6 illustrates the diagrammatic layout of an air-cycle refrigeration system as employed for air conditioning the cockpit and cabin space of an airplane. The system illustrated is primarily applicable to turbojet and turbopropellor airplanes, where it is possible to bleed (draw) a supply of compressed air from the compressor. This compressed air would be at a pressure of perhaps 60 to 80 psia and at a temperature of 300 to 400 F. Thus it is necessary to cool the compressed air, and, as shown, both a precooler and a heat exchanger, which employ ram air from the atmosphere, are used for this purpose. The fan employed for drawing air from the atmosphere through the heat exchanger is driven by the expander turbine that produces the chilled air. The fan constitutes the load for the expander turbine. The temperature of the supply air entering the turbine depends on outside air temperature, and with 60 F outside air, at sea level, might be at a temperature of about 90 F. At high-altitude temperatures, which range from 20 to -80 F, the compressed air would be cooled appreciably more.

It has been mentioned that the performance of air refrigeration is poor, in that some 4 to 5 hp are required per ton on refrigeration. In contrast, a representative 1 hp per ton is required for vapor refrigeration in the air-conditioning range. However, the light weight of the expander turbine and its fan running at some 30,000 to 40,000 rpm does give a weight advantage to the air system.

For comparison, in Fig. 14-7, an open air system, such as might be used on an airplane, and a vapor system are shown together on a Ts dia-

Principles of Refrigeration

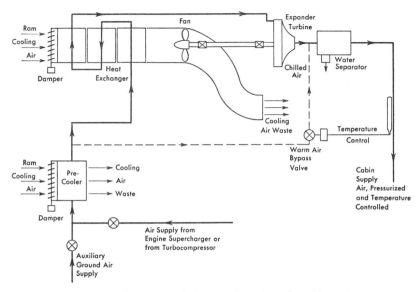

Fig. 14–6. Diagrammatic layout of an air-cycle refrigeration system for an airplane.

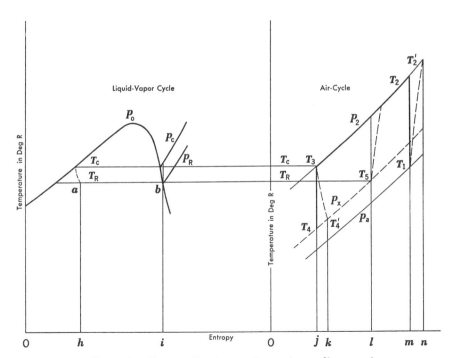

Fig. 14–7. Comparative temperature-entropy diagrams for liquid-vapor and air cycles.

gram. The air cycle indicates compression taking place isentropically from ambient air pressure (p_a) and ambient air temperature (T_1) to p_2 and T_2. (In the actual compressor, because of internal irreversibilities, the final temperature is higher than T_2 at a value T'_2.) The hot air at the high pressure of the system cools in the heat exchanger to the temperature T_3, and this value is shown as corresponding to the condensing temperature (T_c) of the vapor system. In the expander engine, the air drops to the pressure corresponding to the variable cabin pressure p_x, ideally reaching temperature T_4 (actually T'_4). The cool air warms to final temperature T_5, producing useful refrigeration, which is measured by the area T_4jlT_5 for ideal conditions (or actually T'_4klT_5). The ideal heat dissipation load is the area T_2T_3jm (or actually T'_2T_3jn).

The conventional vapor-liquid system operates between a condensing temperature T_c and an evaporator temperature T_R. Note that the useful refrigeration per pound of refrigerant is the area $ahib$, and note also that the refrigeration is all produced at a fixed temperature T_R. This area should be contrasted with the much smaller ideal air-cycle area T_4jlT_5, which represents useful refrigeration per pound of air. Notice also that the air-cycle refrigeration has the disadvantage of being produced at a variable temperature. The turbines and centrifugal compressors in the small sizes, used in the range of 10 to 40 hp, are not highly efficient. A general value for efficiency of 65 per cent relative to isentropic conditions can be used in preliminary design layout.

Example 14–7. An air-cycle refrigeration system for an airplane is supplied with compressed air bled-off from the main compressor at 66 psia and 400 F. This air, after passing through four heat-exchanger elements, arrives at the expander turbine at 60 psia and 90 F. The turbine delivers its output air at 12 psia and on test shows an isentropic performance efficiency of 65 per cent. The flow rate to the turbine under maximum conditions is 60 lb of air per min. Compute (a) the minimum possible exit temperature from the air turbine under the conditions indicated, (b) the probable exit temperature attained, and (c) the horsepower delivered to a fan coupled to the turbine. (d) The cold air is tempered by mixing with recirculated or other warm air and is then supplied to the cabin-cockpit areas, where it warms to 78 F before exhausting from the plane. Compute the total tons of refrigeration equivalent to warming the chilled air from its low temperature to 78 F.

Solution: (a) By equation 14–20,

$$\frac{T_4}{T_3} = \frac{T_4}{460+90} = \left(\frac{12}{60}\right)^{1.4-1/1.4} = \left(\frac{1}{5}\right)^{0.286} = \frac{1}{1.584}$$

$$T_4 = 347 \text{ R}$$

$$t_4 = 460 - 347 = -113 \text{ F} \hspace{4em} Ans.$$

b) This low temperature is not reached because of inefficiencies in the small, high-speed turbine, and the presence of water vapor or humidity in the air also alters the

Principles of Refrigeration

performance. Considering merely the turbine efficiency here, the actual temperature drop in the turbine is

$$\Delta t = 0.65\,(460 + 90 - 347) = 132 \text{ deg}$$

The probable exit temperature is

$$T'_4 = 550 - 132 = 418 \text{ R}$$
$$t'_4 = 460 - 418 = -42 \text{ F} \qquad Ans.$$

c) Use equation 14–24 with $C_p = 0.24$ for air. The theoretical work of the expander engine is

$$W_E = (60)\,(0.24)\,(550 - 347) = 1626 \text{ Btu per min}$$

Actual work of expander engine is

$$W'_E = (60)\,(0.24)\,[0.65\,(550 - 347)] = 1058 \text{ Btu per min}$$

$$\text{Expander hp} = \frac{1058}{42.4} = 24.9 \qquad Ans.$$

d) Refrigeration produced is

$$Q_R = wC_p\,(T_1 - T'_4)$$
$$= (60)(0.24)\,[78 - (-42)] = 1728 \text{ Btu per min}$$

$$\text{Tons} = \frac{1728}{200} = 8.6 \qquad Ans.$$

The horsepower required to compress the air was provided by the main compressor drive and was not computed in this example. However, a simple computation will show that 60 lb of air compressed per min requires a substantial amount of power, far in excess of that produced by the expander engine.

14–11. VOLUMETRIC EFFICIENCY OF RECIPROCATING COMPRESSORS

In a reciprocating compressor, when the piston reaches the end of its stroke (dead-center position), a portion of gas always remains in and is not discharged from the cylinder. The space in the cylinder which this gas occupies is known as the clearance volume, and the relative magnitude of this volume varies greatly with compressor design and valve arrangement. *Clearance* is usually expressed as a percentage of the stroke-displacement volume (i.e., the volume swept through by the piston in one stroke). Representative clearance values lie in the range of 8 to 2 per cent with 6 to 4 per cent being a common range for most refrigerant compressors now being built. Figure 14–8 is a pressure-volume diagram of events in a compressor cylinder, with V_P representing the stroke-displacement volume, and V_{C1} representing the clearance volume. The percentage clearance (m) is therefore

$$m = \frac{V_{C1}}{V_P}\,100 \qquad (14\text{–}28)$$

In internal combustion engines, clearance is expressed in another way as the volume ratio of compression (R); namely, the total gas volume in the cylinder at the start of a stroke divided by the gas-space volume present with the piston on dead center at the end of its stroke. Thus in Fig. 14-8

$$R = \frac{V_P + V_{C1}}{V_{C1}} \qquad (14\text{-}29)$$

A relationship between m and R can easily be found.

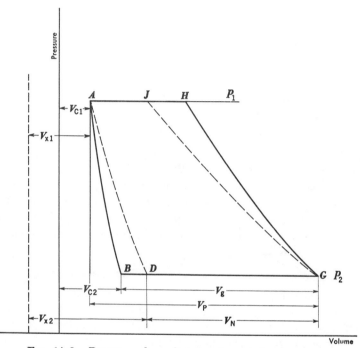

Fig. 14-8. Pressure-volume diagram of events in a compressor cylinder showing the effect of clearance.

Some clearance is necessary and desirable to provide cushioning and prevent piston slap and also to provide passage space up to and adjacent to the valve seats. However, clearance is objectionable because the capacity of two similar compressors each having the same piston displacement will be smaller for the machine having the greater clearance. In fact, some large compressors are provided with variable clearance pockets in the cylinder head, so arranged that a valve can be turned to open a pocket to increase the clearance space and thereby reduce the capacity of constant-speed compressors.

The volume (V_{C1}) of gas in the clearance space of a compressor is at

Principles of Refrigeration

discharge pressure p_1, and as the piston moves out on its suction stroke, this gas re-expands, finally reaching volume V_{C2} when the cylinder pressure has dropped to the system suction pressure p_2. The valves on the compressor operate by differential pressure and therefore cannot open until the pressure in the cylinder is less than the pressure in the suction line from the evaporator. Thus a fresh charge of gas cannot enter until the piston has swept out slightly more than the volume $(V_{C2} - V_{C1})$. The effective piston-stroke displacement remaining for new charge is thus only V_g instead of the full displacement V_P, which would be available if no clearance existed.

A gas expanding quickly does so almost adiabatically, and the re-expansion stroke is closely approximated by the equation

$$p_1 V_{C1}{}^k = p_2 V_{C2}{}^k \tag{14-30}$$

where k = ratio of specific heats, or the polytropic coefficient. Refer to Fig. 14-8, by algebra, note that

$$V_{C2} = V_{C1}\left(\frac{p_1}{p_2}\right)^{1/k} \tag{14-31}$$

and

$$V_g = V_P - (V_{C2} - V_{C1}) = V_P - V_{C1}\left[\left(\frac{p_1}{p_2}\right)^{1/k} - 1\right] \tag{14-32}$$

The volumetric efficiency based on clearance alone (η_{vC}) is thus

$$\eta_{vC} = \frac{V_g}{V_P} = \frac{V_P - V_{C1}\left[\left(\frac{p_1}{p_2}\right)^{1/k} - 1\right]}{V_P} \tag{14-33}$$

But by equation 14-28,

$$V_{C1} = m\frac{V_P}{100}$$

therefore

$$\eta_{vC} = 1 - \frac{m}{100}\left[\left(\frac{p_1}{p_2}\right)^{1/k} - 1\right] \tag{14-34}$$

or

$$\eta_{vC} = 1 + \frac{m}{100} - \frac{m}{100}\left(\frac{p_1}{p_2}\right)^{1/k} \tag{14-35}$$

where η_{vC} = volumetric efficiency related to clearance alone, expressed as a decimal fraction;
m = clearance, expressed as a percentage and based on piston displacement;
p_1 = compressor discharge pressure, usually in psia;
p_2 = compressor suction pressure, usually in psia.

Consider what might happen if a cylinder with stroke-displacement volume V_P had its clearance volume increased to a value V_{x1} (formerly V_{C1}). This might be accomplished by installing a new cylinder head or by using a clearance pocket. With the greater volume of gas in the clearance, the re-expansion line would take the position $A\ D$ instead of $A B$, and the new charge would only be inducted during the stroke in amount V_N. Thus capacity would be decreased closely in the ratio of $1 - V_N/V_g$.

Clearance is only one of the factors affecting volumetric efficiency. In section 14–7 it was mentioned that superheating of the gas to a higher temperature as it entered the cylinder, and throttling or friction loss to gas flow through ports and valves, thereby reducing the pressure, also contributed to lowering volumetric efficiency. Thus mainly three items act to produce volumetric efficiencies of less than unit value. If these two final factors are incorporated, a relation for probable actual volumetric efficiency can be written as

$$\eta_v = (\eta_{vc}) \times \frac{p_{\text{cyl}}}{p_e} \times \frac{T_e}{T_{\text{cyl}}}$$

$$= \left[1 + \frac{m}{100} - \frac{m}{100}\left(\frac{p_1}{p_2}\right)^{1/k}\right] \times \frac{p_{\text{cyl}}}{p_e} \times \frac{T_e}{T_{\text{cyl}}} \qquad (14\text{–}36)$$

where p_{cyl} = pressure of the vapor in the cylinder at the start of compression, in psia, essentially equal to p_2;

p_e = pressure of the vapor in the evaporator suction pipe at compressor inlet, in psia;

T_e = temperature of the vapor in the evaporator suction pipe at entry to the compressor, in degrees Rankine;

T_{cyl} = temperature of the vapor in the cylinder at the start of compression, in degrees Rankine.

Example 14–8. A $2\frac{1}{2}$-in.-bore, $1\frac{3}{4}$-in.-stroke compressor runs at 800 rpm. The compressor has a clearance of 4.2 per cent. Refrigerant-12 is used, the discharge pressure is 108 psia at the compressor, and the evaporator operates at 20 F and 35.7 psia. It is thought that the gas in the cylinder warms to 40 F, and there is a pressure drop of 1.5 psi on entry to the cylinder. The value of k for the re-expansion line is closely 1.14. Compute (a) the volumetric efficiency based on clearance volume (η_{vc}) and (b) the probable actual volumetric efficiency.

Solution: (a) The true pressures in the cylinder control re-expansion and must be used in equation 14–35. Thus

$$\eta_{vc} = 1 + \frac{4.2}{100} - \frac{4.2}{100}\left(\frac{108}{34.2}\right)^{1/1.14}$$

$$= 1.042 - 0.042(3.16)^{1/1.14} = 1.042 - 0.042(2.74) = 0.927 \qquad Ans.$$

b) By equation 14–36,

$$\eta_v = 0.927 \times \frac{34.2}{35.7} \times \frac{460 + 20}{460 + 40} = 0.852 \qquad Ans.$$

The effect of volumetric efficiencies less than 100 per cent is naturally to lower compressor capacity. This also contributes indirectly to an in-

Principles of Refrigeration

crease in the power consumed by the compressor per pound of refrigerant compressed. This additional power is required because at reduced capacity the compressor has to make more strokes for a given output, and the total frictional power is therefore increased per unit of regriferant handled. Moreover, the lowering in volumetric efficiency contributed by pressure loss (throttling) adds to the power consumed. The re-expansion process produces power and is not inherently a power-absorbing process.

Mention has been made of the fact that, if the isentropic work of compression is multiplied by a factor ranging from 1.3 to 1.5, a close measure of the input shaft work required for compression can be obtained. The basis on which this factor is obtained involves, among many items, turbulence of gas flow into and from the compressor, heat-transfer effects, and mechanical efficiency, with the latter in turn partly affected by volumetric efficiency. In general, about 20 per cent added to the isentropic work of compression gives a reasonable estimation figure for the indicated work of the compressor pistons. Mechanical-efficiency values for compressors vary over wide limits, but for commercial machines it should be higher than 80 per cent, say 85 per cent for a representative value. Using these values, if the isentropic factor of 1.20 has a representative mechanical efficiency of 0.85 superposed, the overall factor becomes 1.41, which can be seen to lie midway in the range of indicated factors for shaft work input. To arrive at electrical input to the driver, note that motor efficiency varies from over 90 per cent for large-horsepower motors to some 85 per cent for motors of about 5 hp, and falls to very low values, say 50 per cent, for fractional-horsepower motors.

PROBLEMS

14-1. What are the ideal Carnot CP and the horsepower requirement per ton of refrigeration for a refrigeration system operating between the standard temperature limits of 5 and 86 F? *Ans.* 5.74, 0.82

14-2. A building with a heating load of 150,000 Btu per hr is to be heated, with a reversed refrigeration system, to 68 F when the outside temperature is 0 F. Find (a) the ideal coefficient of performance (CP) and (b) the ideal horsepower required at the compressor to furnish this heat load to the building. Assume that a 10 deg temperature difference is needed. *Ans.* (a) 5.12, (b) 9.6

14-3. A refrigerant-12 system (Table 15-5) is operating between 20 and 100F. If the liquid is subcooled to 80 F, find the lb of flash vapor formed in the expansion valve per lb of refrigerant. *Ans.* 0.203

14-4. Compare the refrigerating effect obtained from ammonia operating between 0 F evaporating temperature and 100 F condensing temperature, with (a) no subcooling and dry compression from saturated vapor, and (b) subcooling to 75 F and dry compression. Refer to Table 15-3.
Ans. (a) 456.6, (b) 485.6 Btu/lb

14-5. Ammonia is used for a refrigeration system between 26.92 and 169.2 psia. Find (a) the lb of refrigerant circulated per min per ton, (b) the isentropic work of compression, and (c) the coefficient of performance (CP). Dry saturated vapor enters the compressor. *Ans.* (a) 0.424, (b) 116, (c) 4.0

14-6. Find the bore and stroke of a two-cylinder, 200-rpm, single-acting Freon-12 compressor to handle 8 tons of refrigeration when operating between 0 F and 90 F condensing, with liquid subcooling to 80 F. Make bore and stroke equal.
Ans. 7 × 7 at 0.82 vol eff

14-7. A 100-ton Freon-12 (Table 15-5, Fig. 15-4) compressor runs with a discharge pressure of 108.0 psia and a suction pressure of 26.5 psia. If the liquid is subcooled to 70 F before expansion and if the vapor entering the compressor is superheated by 15 deg, find (a) the quality of Freon leaving the expansion valve, (b) the heat absorbed in the evaporator per lb of Freon circulated, (c) the weight of Freon per min per ton of refrigeration, (d) the theoretical piston displacement per min per ton, (e) the theoretical horsepower, (f) the probable actual horsepower, and (g) the probable actual piston displacement per min per ton.
Ans. (a) 0.21, (b) 56.4, (c) 3.5, (d) 5.4 cfm per ton, (e) 102, (f) 143, (g) 6.1

14-8. If a four-cylinder, 250-rpm, 75-ton ammonia compressor is operating between 90 F condensation and 30 F evaporator conditions without subcooling and with dry compression, find (a) quality of ammonia leaving expansion valve, (b) heat absorbed in evaporator per lb of ammonia circulated, (c) weight of ammonia per min per ton of refrigeration, (d) piston displacement, (e) size of compressor, if it is single acting and the bore-stroke ratio = 1, and (f) theoretical horsepower.
Ans. (a) 0.124, (b) 477.0, (c) 0.42, (d) 152.3, 173, (e) 7.25 × 7.25, (f) 52.5

14-9. (a) A cylinder of refrigerant-12, partly filled with liquid, lies in a room at a temperature of 90 F. What pressure exists in the cylinder? (b) What pressure exists in similar cylinders of ammonia, sulfur dioxide, and trichloromonofluoromethane located in the same room? See tables 15-5 and 15-7.
Ans. (a) 114.3 psia

14-10. Using data for monochlorodifluoromethane (refrigerant-22), construct a ph chart to approximate scale on a sheet of graph paper. Select enough points to indicate the liquid line and the saturated vapor line, and show two constant-temperature lines in the superheat region. Use Table 15-6 and superheat data from Plate IV.

14-11. Repeat problem 10 for refrigerant-12, excluding data in the superheat region.

14-12. A certain new refrigerant is known to have a $C_p/C_v = k$ ratio of 1.16, and its specific volume at 10 F is computed to be 1.13 cu ft per lb at a pressure of 48 psia. (a) Compute the ideal horsepower required to compress 5 lb per min of this refrigerant to 144 psia. (b) Compute the temperature of the refrigerant on discharge from the compressor. *Ans.* (a) 1.4, (b) 547 R

14-13. Repeat problem 14-12 for air if $k = 1.4$ and the pressures are the same. Use equation 2-15 to find the specific volume of air. *Ans.* (a) 5.4, (b) 644 R

14-14. In the expander turbine of an air-cycle machine, air at 90 F and 72 psia expands to 12 psia. (a) Find the temperature of the air after ideal expansion. (b) Compute the actual temperature after expansion if the adiabatic turbine efficiency is 70 per cent. (c) Compute the actual turbine horsepower produced for 2 lb per sec air flow. (d) Find the useful refrigeration possible if the cool turbine exhaust air warms to 70 F. *Ans.* (a) 329.5 R, (b) 395.7 R, (c) 104, (d) 3870 Btu/min

Principles of Refrigeration 511

14-15. A reciprocating compressor has a clearance volume of 5 per cent. Compute the volumetric efficiency based on clearance alone for a compressor with compression ratio of (a) 2.5 and (b) 4.0. Assume that the refrigerant has a k value of 1.16. *Ans.* (a) 94%, (b) 87%

14-16. Repeat problem 14-15 for a refrigerant with a k value of 1.31.
Ans. (a) 95%, (b) 90.6%

14-17. If the refrigerant of problem 14-16 warms from -10 to 5 F in entering the compressor and the pressure drops from 23.7 to 23.1 psia after entry to the cylinder, estimate the actual volumetric efficiency. *Ans.* (a) 89.6%, (b) 85.4%

REFERENCES

1. American Society of Heating, Refrigerating, and Air-Conditioning Engineers, *ASHRAE Guide and Data Book* (1967), Chapters 11 and 14.

2. G. E. Gregg, "Air Cycle Air Conditioning," *Refrigerating Engineering*, Vol. 65. (November, 1957), pp. 35-38.

15

Refrigerants and Refrigeration Systems

15-1. REFRIGERANT CHARACTERISTICS

An ideal refrigerant as such does not exist, and even if an almost ideal chemical could be found, as a refrigerant it could never cover the complete spectrum of ranges man would need to have it serve. Thus any refrigerant selected will be a compromise, but for any particular service it should possess as many as practicable of the following qualities:

1. *Condensing pressures that are not excessive* so that extra-heavy construction will be unnecessary.

2. *Low boiling temperatures* at atmospheric pressure, so that the system does not require vacuum operation with attendant possibility of leakage of damp air into the system.

3. *High critical temperature.* It is impossible to liquefy (condense) a vapor at a temperature higher than the critical temperature, no matter how high the pressure is raised. With air-cooled equipment this fact makes it desirable to have critical temperatures higher than 130 F. With the exception of CO_2 and ethane, which have critical temperatures of 87.8 and 89.8 F, respectively, all the common refrigerants have critical temperatures higher than 200 F.

4. *High latent heat of vaporization.* The higher the latent heat, the less the weight of refrigerant which must be circulated per minute per unit of capacity.

5. *Low specific heat of liquid* is a desirable quality, since the expansion valve throttles the liquid, and the liquid refrigerant must be cooled at the expense of partial evaporation.

6. *Low specific volume of vapor* is essential with reciprocating machinery, but may not be an important item with centrifugal machines.

7. *Absence of corrosive action on metals used* is necessary.

8. *Chemical stability* of the compound is essential.

9. The refrigerant should be *nonflammable* and *nonexplosive*.

10. That the refrigerant be *nontoxic* to lungs, eyes, and general health is an important item in air-conditioning installations.

11. *Ease of locating leaks* by odor or suitable indicator is an important consideration.

Refrigerants and Refrigeration Systems

12. *Availability, low cost,* and *ease of handling* are obviously desirable features.

13. *Action of the refrigerant on lubricants* must be such as not to ruin their lubricating value.

14. *Satisfactory heat transfer* and *viscosity* coefficients are required.

15. The *freezing temperature of the liquid* should be appreciably below any temperature at which the evaporator might operate.

16. For the ratios of the compression to be used, *low, compressor-discharge temperatures* are desirable to prevent possible breakdown or deterioration of refrigerant and lubricant in the system.

Thermodynamically, on the ideal cycle the performance of all refrigerants is exactly the same between the same temperature limits, but in actual cases the irreversible action through the expansion valve and the relative amount of compression in the superheat region cause deviations from the ideal. The American Society of Heating, Refrigerating, and Air-Conditioning Engineers (ASHRAE) established 5 and 86 F as standard conditions under which to make comparisons, and Table 15-1 shows some of these values computed. These temperatures were selected during the period when ice manufacture was an important use of refrigeration, and the values are still used as reference datum points.

Figure 15-1 is a plot of the pressure-temperature relations of a large number of refrigerants. Freezing-point, triple-point, and critical-point conditions are also marked on this chart. Complete and thermodynamically consistent tabulations of properties are now available for almost all of the refrigerants which are used to any extent. Properties of the most extensively used refrigerants are given in this text, but by far the most extensive and complete coverage of all refrigerant properties will be found in the Handbook of Fundamentals of the American Society of Heating, Refrigerating and Air-Conditioning Engineers (ref. 13).

15-2. THE HALOCARBON (HALOGENATED-HYDROCARBON) REFRIGERANTS

Toward the end of the 1920 decade a team of engineers and scientists working with Dr. Thomas Midgley, Jr. successfully developed a new family of refrigerants having exceptionally desirable characteristics. The most outstanding feature of the new group was its extremely low toxicity, and this characteristic contributed to early acceptance of one of these refrigerants for widespread use in air-conditioning installations. Following the preliminary development work, the Kinetic Chemicals Division of E. I. du Pont de Nemours & Co. put the refrigerants into production under the proprietary name of "Freon." Starting about 1950, a number of other companies went into production of this refrigerant group under a variety of proprietary names such as "genetron," "isotron," "frigen," etc.

TABLE 15-1
COMPARATIVE CHARACTERISTICS OF BASIC REFRIGERANTS
(Performance based on 5 F Evaporator Temperature and 86 F Condenser Temperature)

	11 Trichloro-monofluoro-methane	12 Dichloro-difluoro-methane	22 Monochloro-difluoro-methane	113 Trichloro-trifluoro-ethane	40 Methyl Chloride	717 Ammonia
Chemical formula	CCl_3F	CCl_2F_2	$CHClF_2$	$C_2Cl_3F_3$	CH_3Cl	NH_3
Molecular weight	137.4	120.9	86.5	187.4	50.5	17.0
Boiling point (F) at 14.7 psi	74.7	−21.6	−41.4	117.6	−10.8	−28.0
Evaporator pressure at 5 F (psi)	2.9	26.5	43.0	0.98	21.1	34.3
Condensing pressure at 86 F (psi)	18.3	108.0	174.5	7.86	94.7	169.2
Freezing point (F) at 14.7 psi	−168	−252	−256	−31	−144	−108
Critical temperature (F)	388	234	205	417	289	271
Critical pressure (psi)	635	597	716	495	969	1657
Compressor discharge temperature (F)	112	100	131	86	172	210
Compression ratio (86 F/5 F)	6.24	4.07	4.06	8.02	4.48	4.94
Saturated liquid viscosity at 5 F (centipoises)	0.650	0.328	0.286	1.200	0.293	0.250
Saturated liquid viscosity at 86 F (centipoises)	0.405	0.251	0.229	0.619	0.234	0.207
Vapor viscosity at 5 F and 14.7 psi (centipoises)	0.0096	0.0114	0.0114	0.0093*	0.0095	0.0085
Vapor viscosity at 86 F and 14.7 psi (centipoises)	0.0111	0.0127	0.0131	0.0105*	0.0109	0.0102
Specific volume of saturated vapor at 5 F (cu ft/lb)	12.27	1.49	1.25	27.04	4.47	8.15
Latent heat of vaporization at 5 F (Btu/lb)	84.0	69.5	93.6	70.6	180.7	565.0
Specific heat of liquid at 86 F (Btu/lb F)	0.21	0.24	0.34	0.22	0.39	1.14
Specific heat of vapor at constant pressure of 14.7 psi and 86 F (Btu/lb F)	0.13	0.15	0.15	0.15	0.24	0.51

Property						
Specific heat ratio at 86 F and 14.7 psi $k = C_p/C_v$	1.14	1.14	1.18	1.09	1.20	1.32
Horsepower/ton refrigeration ideal	0.927	1.002	1.011	0.960	0.962	0.989
Refrigerant circulated/ton refrig. (lb/min)	2.96	3.92	2.89	3.73	1.33	0.422
Compressor displacement/ton refrig. (cfm) ideal	36.32	5.81	3.60	100.76	5.95	3.44
Thermal conductivity of saturated liquid at 32 F (Btu ft/sq ft F)	0.0680	0.0559	0.0704	0.0576	0.103	0.29
Thermal conductivity of saturated liquid at 86 F (Btu ft/sq ft F)	0.0609	0.0492	0.0595	0.0521	0.089	0.29
Thermal conductivity of vapor at 32 F and 14.7 psi (Btu ft/sq ft F)	0.0045	0.0048	0.0060	0.0038†	0.0053	0.0128
Thermal conductivity of vapor at 86 F and 14.7 psi (Btu ft/sq ft F)	0.0048	0.0056	0.0068	0.0045†	0.0065	0.0145
Stability (toxic decomposition products)	Yes	Yes	Yes	Yes	Yes	No
Toxicity (Underwriters' Laboratories Group No.)	5A	6	5A	4-5	4	2
Flammability	None	None	None	None	Yes	Yes
Odor	Ethereal	Ethereal	Ethereal	Ethereal	Ethereal	Acrid
Type of compressor in which usually used	Centr.	All	Recip.-Rotary	Centr.	All	All
Evaporator temperature range, F	−20 to 50	−100 to 50	−125 to 50	−25 to 50	−80 to 50	−90 to 20

* At 0.1 atmosphere
† At 0.5 atmosphere

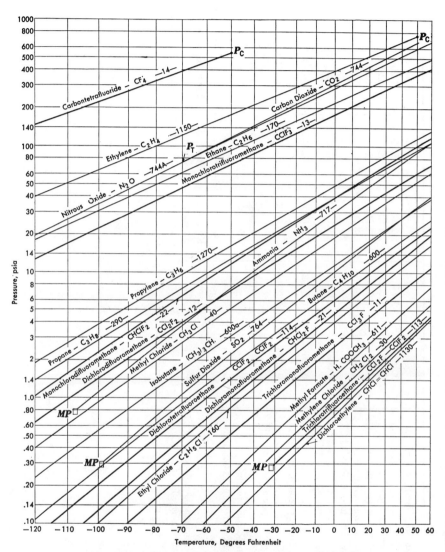

Fig. 15-1. Saturation pressure-temperature relationships for common refrigerants listed by name and number.

Refrigerants and Refrigeration Systems   517

Refrigerants of this group are substitution refrigerants in that halogen atoms, mainly chlorine and fluorine, are substituted in a hydrocarbon structure for hydrogen atoms. The hydrocarbon methane, with a formula CH_4, has been most widely featured in this pattern. For example, assume that 2 chlorine atoms and 2 fluorine atoms are used to replace the hydrogen atoms in methane. The resulting formula becomes CCl_2F_2. This can properly be called dichloro-difluoro-methane, where the prefix *di* (or *bi*) represents 2 and similarly, *mono* 1, *tri* 3, *tetra* 4, etc. Another refrigerant is $CHClF_2$, which can be named monochlorodifluoromethane; also, CCl_3F is called trichloromonofluoromethane.

Such names, although logical, are long and difficult to remember and so proprietary names were largely used instead, with dichlorodifluoromethane being called Freon-12, genetron-12, isotron-12, etc., and monochlorodifluoromethane being called Freon-22, genetron-22, frigen-22, etc. A larger number of proprietary names coming into use began to confuse the issue further, and ASHRAE decided to adopt a standard system of designation and call these refrigerants merely by number, with CCl_2F_2 being designated as refrigerant-12, $CHClF_2$ as 22, CHF_3 as 23, CH_2F_2 as 32, etc. Table 15–2 lists a large number of refrigerants tabulated in accordance with the ASHRAE standard system.

The numbers employed in the standard for the halocarbon refrigerants follow a definite pattern originally developed in connection with the Freon refrigerants. The halogen substitutions are based on two hydrocarbons, methane, CH_4, (for which two-digit numbers are used to indicate the refrigerants) and ethane, C_2H_6, (for which three-digit numbers are used to indicate the different refrigerants). For the methane derivatives, the first number is always one greater than the number of hydrogen atoms appearing in the refrigerant molecule, whereas the second digit gives exactly the number of fluorine atoms appearing in the refrigerant molecule. For example, in refrigerant-12, the number 1, being one greater than the number of hydrogen atoms, indicates no hydrogen to be present, while the 2 indicates that two fluorine atoms exist in the molecule so that the formula is CCl_2F_2. It is also possible to name this refrigerant in terms of its constituents, as dichlorodifluoromethane. For the 110-series, the first digit merely indicates that ethane is the basic hydrocarbon. The second 1 indicates no hydrogen atoms in the formula. The third digit indicates, as before, the number of fluorine atoms appearing in the molecule. The respective boiling points at atmospheric pressure of each of the refrigerants are also given in Table 15–2. It will be noticed that the boiling point decreases with the increasing number of fluorine atoms in any particular grouping. No specific formulation can be made as to toxicity, but of all the refrigerants listed, refrigerant-12 (Freon-12, genetron-12) is least toxic. The other refrigerants in the table (except for the straight hydro-

TABLE 15-2
AMERICAN SOCIETY OF HEATING, REFRIGERATING AND AIR-CONDITIONING
ENGINEERS (ASHRAE) REFRIGERANT NUMBERING SYSTEM

ASRE Standard Designation	Chemical Name	Chemical Formula	Molecular Weight	Boiling Point (deg F at 14.7 psi)
\multicolumn{5}{c}{Halocarbon and Hydrocarbon Compounds}			
10	Carbontetrachloride	CCl_4	153.8	170.2
11	Trichloromonofluoromethane	CCl_3F	137.4	74.8
12	Dichlorodifluoromethane	CCl_2F_2	120.9	− 21.6
13	Monochlorotrifluoromethane	$CClF_3$	104.5	−114.6
13B1	Monobromotrifluoromethane	$CBrF_3$	148.9	− 72.0
14	Carbontetrafluoride	CF_4	88.0	−198.4
20	Chloroform	$CHCl_3$	119.4	142
21	Dichloromonofluoromethane	$CHCl_2F$	102.9	48.1
22	Monochlorodifluoromethane	$CHClF_2$	86.5	− 41.4
23	Trifluoromethane	CHF_3	70.0	−119.9
30	Methylene chloride	CH_2Cl_2	84.9	105.2
31	Monochloromonofluoromethane	CH_2ClF	68.5	48.0
32	Methylene fluoride	CH_2F_2	52.0	− 61.4
40	Methyl chloride	CH_3Cl	50.5	− 10.8
41	Methyl fluoride	CH_3F	34.0	−109
50	Methane	CH_4	16.0	−259
110	Hexachloroethane	CCl_3CCl_3	236.8	365
111	Pentachloromonofluoroethane	CCl_3CCl_2F	220.3	279
112	Tetrachlorodifluoroethane	CCl_2FCCl_2F	203.8	199.0
112a	Tetrachlorodifluoroethane	CCl_3CClF_2	203.8	195.8
113	Trichlorotrifluoroethane	CCl_2FCClF_2	187.4	117.6
113a	Trichlorotrifluoroethane	CCl_3CF_3	187.4	114.2
114	Dichlorotetrafluoroethane	$CClF_2CClF_2$	170.9	38.4
114a	Dichlorotetrafluoroethane	CCl_2FCF_3	170.9	38.5
114B2	Dibromotetrafluoroethane	$CBrF_2CBrF_2$	259.9	117.5
115	Monochloropentafluoroethane	$CClF_2CF_3$	154.5	− 37.7
116	Hexafluoroethane	CF_3CF_3	138.0	−108.8
120	Pentachloroethane	$CHCl_2CCl_3$	202.3	324
123	Dichlorotrifluoroethane	$CHCl_2CF_3$	153	83.7
124	Monochlorotetrafluoroethane	$CHClFCF_3$	136.5	10.4
124a	Monochlorotetrafluoroethane	CHF_2CClF_2	136.5	14
125	Pentafluoroethane	CHF_2CF_3	120	− 55
133a	Monochlorotrifluoroethane	CH_2ClCF_3	118.5	43.0
140a	Trichloroethane	CH_3CCl_3	133.4	165
142b	Monochlorodifluoroethane	CH_3CClF_2	100.5	12.2
143a	Trifluoroethane	CH_3CF_3	84	− 53.5
150a	Dichloroethane	CH_3CHCl_2	98.9	140
152a	Difluoroethane	CH_3CHF_2	66	− 12.4
160	Ethyl chloride	CH_3CH_2Cl	64.5	54.0
170	Ethane	CH_3CH_3	30	−127.5
218	Octafluoropropane	$CF_3CF_2CF_3$	188	− 36.4
290	Propane	$CH_3CH_2CH_3$	44	− 44.2
600	Butane	$CH_3CH_2CH_2CH_3$	58.1	31.3
600a	Isobutane	$CH(CH_3)_3$	58.1	14
\multicolumn{5}{c}{Inorganic Compounds}			
717	Ammonia	NH_3	17	− 28.0
718	Water	H_2O	18	212
727	Air		29	−318
744	Carbon dioxide	CO_2	44	−109(subl.)
744A	Nitrous oxide	N_2O	44	−127
764	Sulfur dioxide	SO_2	64	14.0

Refrigerants and Refrigeration Systems

carbons which essentially fit the halocarbon numbering pattern) are provided with distinctive numbers. The distinctive first digit 7 in a three-figure number applies to the inorganic refrigerants.

15-3. AMMONIA (NH$_3$), REFRIGERANT 717

Ammonia is the most extensively used refrigerant, particularly in industrial and commercial refrigeration. When water-free it is known as *anhydrous ammonia*, and when mixed with water (as used in absorption refrigeration systems) it is known as *aqua ammonia*. By reference to tables 15-3 and 15-4 and Figs. 15-2 and 15-3 it can be seen that the operating pressure range is moderate, sub-atmospheric only for temperatures

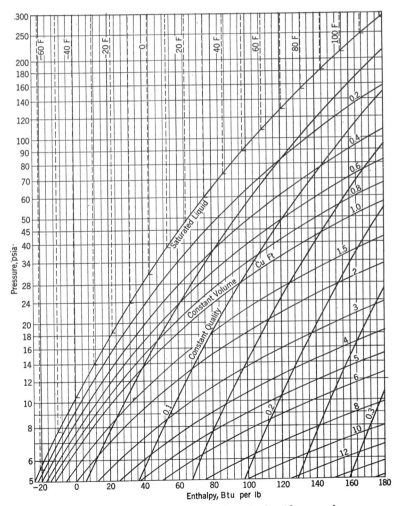

Fig. 15-2. Pressure-enthalpy chart for liquid ammonia.

below − 28 F and usually not exceeding 200 psia in the condenser. Its latent heat is exceptionally high, over 500 Btu per lb. No lubrication difficulties exist with ammonia, provided a proper mineral oil is selected. Water mixed with ammonia will not freeze at expansion valves; freezing can happen with most other refrigerants. The thermodynamic performance in a refrigerating cycle is high (Table 15–1). Ammonia also has a high critical temperature, 271.4 F at 1657 psia.

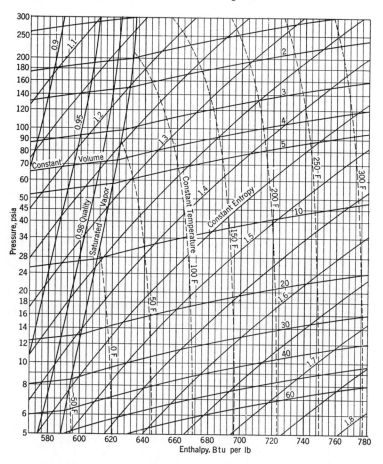

Fig. 15-3. Pressure-enthalpy chart for ammonia, vapor region.

Ammonia is noncorrosive to iron and steel materials, but rapidly corrodes copper and copper or zinc alloys (brass, bronze), so care should be exercised that these metals are not used in contact with ammonia.

Ammonia gas is irritating particularly to eyes and mucous membranes. Small amounts of ammonia in air appear to be more annoying than harm-

Refrigerants and Refrigeration Systems

TABLE 15-3
TEMPERATURE TABLE FOR SATURATED AMMONIA

Temp F t	Pressure PSIA p	Sp Vol of Liquid cu ft/lb v_f	Sp Vol of Vapor cu ft/lb v_g	Density of Vapor lb/cu ft $\frac{1}{v_g}$	Enthalpy Btu/lb above -40 F			Entropy		Temp F t
					Liquid h_f	Vaporization h_{fg}	Vapor h_g	Liquid s_f	Vapor s_g	
-40	10.41	0.02322	24.86	0.04022	0.0	597.6	597.6	0.0000	1.4242	-40
-30	13.90	.02345	18.97	.05271	10.7	590.7	601.4	.0250	1.4001	-30
-20	18.30	.02369	14.68	.06813	21.4	583.6	605.0	.0497	1.3774	-20
-10	23.74	0.02393	11.50	0.08695	32.1	576.4	608.5	0.0738	1.3558	-10
-9	24.35		11.23	.08904	33.2	575.6	608.8	.0762	.3537	-9
-8	24.97		10.97	.09117	34.3	574.9	609.2	.0786	.3516	-8
-7	25.61		10.71	.09334	35.4	574.1	609.5	.0809	.3495	-7
-6	26.26		10.47	.09555	36.4	573.4	609.8	.0833	.3474	-6
-5	26.92	0.02406	10.23	0.09780	37.5	572.6	610.1	0.0857	1.3454	-5
-4	27.59		9.991	.1001	38.6	571.9	610.5	.0880	.3433	-4
-3	28.28		9.763	.1024	39.7	571.1	610.8	.0904	.3413	-3
-2	28.98		9.541	.1048	40.7	570.4	611.1	.0928	.3393	-2
-1	29.69		9.326	.1072	41.8	569.6	611.4	.0951	.3372	-1
0	30.42	0.02419	9.116	0.1097	42.9	568.9	611.8	0.0975	1.3352	0
2	31.92		8.714	.1148	45.1	567.3	612.4	.1022	.3312	2
4	33.47		8.333	.1200	47.2	565.8	613.0	.1069	.3273	4
5	34.27	0.02432	8.150	0.1227	48.3	565.0	613.3	0.1092	1.3253	5
6	35.09		7.971	.1254	49.4	564.2	613.6	.1115	.3234	6
8	36.77		7.629	.1311	51.6	562.7	614.3	.1162	.3195	8
10	38.51	0.02446	7.304	0.1369	53.8	561.1	614.9	0.1208	1.3157	10
11	39.40		7.148	.1399	54.9	560.3	615.2	.1231	.3137	11
12	40.31		6.996	.1429	56.0	559.5	615.5	.1254	.3118	12
13	41.24		6.847	.1460	57.1	558.7	615.8	.1277	.3099	13
14	42.18		6.703	.1492	58.2	557.9	616.1	.1300	.3081	14
15	43.14	0.02460	6.562	0.1524	59.2	557.1	616.3	0.1323	1.3062	15
16	44.12		6.425	.1556	60.3	556.3	616.6	.1346	.3043	16
17	45.12		6.291	.1590	61.4	555.5	616.9	.1369	.3025	17
18	46.13		6.161	.1623	62.5	554.7	617.2	.1392	.3006	18
19	47.16		6.034	.1657	63.6	553.9	617.5	.1415	.2988	19
20	48.21	0.02474	5.910	0.1692	64.7	553.1	617.8	0.1437	1.2969	20
21	49.28		5.789	.1728	65.8	552.2	618.0	.1460	.2951	21
22	50.36		5.671	.1763	66.9	551.4	618.3	.1483	.2933	22
23	51.47		5.556	.1800	68.0	550.6	618.6	.1505	.2915	23
24	52.59		5.443	.1837	69.1	549.8	618.9	.1528	.2897	24
25	53.73	0.02488	5.334	0.1875	70.2	548.9	619.1	0.1551	1.2879	25
26	54.90		5.227	.1913	71.3	548.1	619.4	.1573	.2861	26
27	56.08		5.123	.1952	72.4	547.3	619.7	.1596	.2843	27
28	57.28		5.021	.1992	73.5	546.4	619.9	.1618	.2825	28
29	58.50		4.922	.2032	74.6	545.6	620.2	.1641	.2808	29
30	59.74	0.02503	4.825	0.2073	75.7	544.8	620.5	0.1663	1.2790	30
31	61.00		4.730	.2114	76.8	543.9	620.7	.1686	.2773	31
32	62.29		4.637	.2156	77.9	543.1	621.0	.1708	.2755	32
33	63.59		4.547	.2199	79.0	542.2	621.2	.1730	.2738	33
34	64.91		4.459	.2243	80.1	541.4	621.5	.1753	.2721	34
35	66.26	0.02518	4.373	0.2287	81.2	540.5	621.7	0.1775	1.2704	35
36	67.63		4.289	.2332	82.3	539.7	622.0	.1797	.2686	36
37	69.02		4.207	.2377	83.4	538.8	622.2	.1819	.2669	37
38	70.43		4.126	.2423	84.6	537.9	622.5	.1841	.2652	38
39	71.87		4.048	.2470	85.7	537.0	622.7	.1863	.2635	39
40	73.32	0.02533	3.971	0.2518	86.8	536.2	623.0	0.1885	1.2618	40
41	74.80		3.897	.2566	87.9	535.3	623.2	.1908	.2602	41
42	76.31		3.823	.2616	89.0	534.4	623.4	.1930	.2585	42
43	77.83		3.752	.2665	90.1	533.6	623.7	.1952	.2568	43
44	79.38		3.682	.2716	91.2	532.7	623.9	.1974	.2552	44
45	80.96	0.02548	3.614	0.2767	92.3	531.8	624.1	0.1996	1.2535	45
46	82.55		3.547	.2819	93.5	530.9	624.4	.2018	.2519	46
47	84.18		3.481	.2872	94.6	530.0	624.6	.2040	.2502	47
48	85.82		3.418	.2926	95.7	529.1	624.8	.2062	.2486	48
49	87.49		3.355	.2981	96.8	528.2	625.0	.2083	.2469	49
50	89.19	0.02564	3.294	0.3036	97.9	527.3	625.2	0.2105	1.2453	50

TABLE 15-3—(Continued)

Temp F t	Pressure psia p	Sp Vol of Liquid cu ft/lb v_f	Sp Vol of Vapor cu ft/lb v_g	Density of Vapor lb/cu ft $\frac{1}{v_g}$	Enthalpy Btu/lb above −40 F			Entropy		Temp F t
					Liquid h_f	Vaporization h_{fg}	Vapor h_g	Liquid s_f	Vapor s_g	
51	90.91		3.234	.3092	99.1	526.4	625.5	.2127	.2437	51
52	92.66		3.176	.3149	100.2	525.5	625.7	.2149	.2421	52
53	94.43		3.119	.3207	101.3	524.6	625.9	.2171	.2405	53
54	96.23		3.063	.3265	102.4	523.7	626.1	.2192	.2389	54
55	98.06	0.02581	3.008	0.3325	103.5	522.8	626.3	0.2214	1.2373	55
56	99.91		2.954	.3385	104.7	521.8	626.5	.2236	.2357	56
57	101.8		2.902	.3446	105.8	520.9	626.7	.2257	.2341	57
58	103.7		2.851	.3508	106.9	520.0	626.9	.2279	.2325	58
59	105.6		2.800	.3571	108.1	519.0	627.1	.2301	.2310	59
60	107.6	0.02597	2.751	0.3635	109.2	518.1	627.3	0.2322	1.2294	60
61	109.6		2.703	.3700	110.3	517.2	627.5	.2344	.2278	61
62	111.6		2.656	.3765	111.5	516.2	627.7	.2365	.2262	62
63	113.6		2.610	.3832	112.6	515.3	627.9	.2387	.2247	63
64	115.7		2.565	.3899	113.7	514.3	628.0	.2408	.2231	64
65	117.8	0.02614	2.520	0.3968	114.8	513.4	628.2	0.2430	1.2216	65
66	120.0		2.477	.4037	116.0	512.4	628.4	.2451	.2201	66
67	122.1		2.435	.4108	117.1	511.5	628.6	.2473	.2186	67
68	124.3		2.393	.4179	118.3	510.5	628.8	.2494	.2170	68
69	126.5		2.352	.4251	119.4	509.5	628.9	.2515	.2155	69
70	128.8	0.02632	2.312	0.4325	120.5	508.6	629.1	0.2537	1.2140	70
71	131.1		2.273	.4399	121.7	507.6	629.3	.2558	.2125	71
72	133.4		2.235	.4474	122.8	506.6	629.4	.2579	.2110	72
73	135.7		2.197	.4551	124.0	505.6	629.6	.2601	.2095	73
74	138.1		2.161	.4628	125.1	504.7	629.8	.2622	.2080	74
75	140.5	0.02650	2.125	0.4707	126.2	503.7	629.9	0.2643	1.2065	75
76	143.0		2.089	.4786	127.4	502.7	630.1	.2664	.2050	76
77	145.4		2.055	.4867	128.5	501.7	630.2	.2685	.2035	77
78	147.9		2.021	.4949	129.7	500.7	630.4	.2706	.2020	78
79	150.5		1.988	.5031	130.8	499.7	630.5	.2728	.2006	79
80	153.0	0.02668	1.955	0.5115	132.0	498.7	630.7	0.2749	1.1991	80
81	155.6		1.923	.5200	133.1	497.7	630.8	.2769	.1976	81
82	158.3		1.892	.5287	134.3	496.7	631.0	.2791	.1962	82
83	161.0		1.861	.5374	135.4	495.7	631.1	.2812	.1947	83
84	163.7		1.831	.5462	136.6	494.7	631.3	.2833	.1933	84
85	166.4	0.02687	1.801	0.5552	137.8	493.6	631.4	0.2854	1.1918	85
86	169.2		1.772	.5643	138.9	492.6	631.5	.2875	.1904	86
87	172.0		1.744	.5735	140.1	491.6	631.7	.2895	.1889	87
88	174.8		1.716	.5828	141.2	490.6	631.8	.2917	.1875	88
89	177.7		1.688	.5923	142.4	489.5	631.9	.2937	.1860	89
90	180.6	0.02707	1.661	0.6019	143.5	488.5	632.0	0.2958	1.1846	90
91	183.6		1.635	.6116	144.7	487.4	632.1	.2979	.1832	91
92	186.6		1.609	.6214	145.8	486.4	632.2	.3000	.1818	92
93	189.6		1.584	.6314	147.0	485.3	632.3	.3021	.1804	93
94	192.7		1.559	.6415	148.2	484.3	632.5	.3041	.1789	94
95	195.8	0.02727	1.534	0.6517	149.4	483.2	632.6	0.3062	1.1775	95
96	198.9		1.510	.6620	150.5	482.1	632.6	.3083	.1761	96
97	202.1		1.487	.6725	151.7	481.1	632.8	.3104	.1747	97
98	205.3		1.464	.6832	152.9	480.0	632.9	.3125	.1733	98
99	208.6		1.441	.6939	154.0	478.9	632.9	.3145	.1719	99
100	211.9	0.02747	1.419	0.7048	155.2	477.8	633.0	0.3166	1.1705	100
102	218.6		1.375	.7270	157.6	475.6	633.2	.3207	.1677	102
104	225.4		1.334	.7498	159.9	473.5	633.4	.3248	.1649	104
105	228.9	0.02769	1.313	0.7615	161.1	472.3	633.4	0.3269	1.1635	105
106	232.5		1.293	.7732	162.3	471.2	633.5	.3289	.1621	106
108	239.7		1.254	.7972	164.6	469.0	633.6	.3330	.1593	108
110	247.0	0.02790	1.217	0.8219	167.0	466.7	633.7	0.3372	1.1566	110
115	266.2	.02813	1.128	.8862	173.0	460.9	633.9	.3474	.1497	115
120	286.4	.02836	1.047	.9549	179.0	455.0	634.0	.3576	.1427	120
125	307.8	.02860	0.973	1.028	185.1	448.9	634.0	.3679	.1358	125

Refrigerants and Refrigeration Systems

TABLE 15-4
PROPERTIES OF SUPERHEATED AMMONIA VAPOR
PRESSURE, IN PSIA (SATURATION TEMPERATURE IN ITALICS)

Temp F	Sp Vol Vapor Cu Ft/Lb v	Enthalpy Btu/Lb h 34 4.66 F	Entropy s	Temp F	Sp Vol Vapor Cu Ft/Lb v	Enthalpy Btu/Lb h 48 19.80 F	Entropy s	Temp F	Sp Vol Vapor Cu Ft/Lb v	Enthalpy Btu/Lb h 60 30.21 F	Entropy s
Sat	8.211	613.2	1.3260	Sat	5.934	617.7	1.2973	Sat	4.805	620.5	1.2787
10	8.328	616.4	1.3328								
20	8.542	622.3	.3452	20	5.937	617.8	1.2976				
30	8.753	628.0	.3570	30	6.096	624.0	.3103				
40	8.960	633.6	.3684	40	6.251	630.0	.3225	40	4.933	626.8	1.2913
50	9.166	639.2	1.3793	50	6.404	635.9	1.3341	50	5.060	632.9	1.3035
60	9.369	644.7	.3900	60	6.554	641.6	.3453	60	5.184	639.0	.3152
70	9.570	650.1	.4004	70	6.702	647.3	.3561	70	5.307	644.9	.3265
80	9.770	655.5	.4105	80	6.848	652.9	.3666	80	5.428	650.7	.3373
90	9.969	660.9	.4204	90	6.993	658.5	.3768	90	5.547	656.4	.3479
100	10.17	666.3	1.4301	100	7.137	664.0	1.3868	100	5.665	662.1	1.3581
110	10.36	671.6	.4396	110	7.280	669.5	.3965	110	5.781	667.7	.3681
120	10.56	677.0	.4489	120	7.421	675.0	.4061	120	5.897	673.3	.3778
130	10.75	682.3	.4581	130	7.562	680.5	.4154	130	6.012	678.9	.3873
140	10.95	687.7	.4671	140	7.702	685.9	.4246	140	6.126	684.4	.3966
150	11.14	693.0	1.4759	150	7.842	691.4	1.4336	150	6.239	689.9	1.4058
160	11.33	698.4	.4846	160	7.981	696.8	.4425	160	6.352	695.5	.4148
170	11.53	703.8	.4932	170	8.119	702.3	.4512	170	6.464	701.0	.4236
180	11.72	709.2	.5017	180	8.257	707.7	.4598	180	6.576	706.5	.4323
190	11.91	714.5	.5101	190	8.395	713.2	.4683	190	6.687	712.0	.4409
200	12.10	720.0	1.5183	200	8.532	718.7	1.4766	200	6.798	717.5	1.4493
220	12.48	730.8	.5346	220	8.805	729.6	.4930	210	6.909	723.1	.4576
240	12.86	741.7	.5504	240	9.077	740.6	.5090	220	7.019	728.6	.4658
260	13.24	752.7	.5659	260	9.348	751.7	.5246	230	7.129	734.1	.4739
280	13.62	763.8	.5811	280	9.619	762.9	.5399	240	7.238	739.7	.4819
				300	9.888	774.1	1.5548	260	7.457	750.9	1.4976
								280	7.675	762.1	.5130
								300	7.892	773.3	.5281

TABLE 15-4—(Continued)
PROPERTIES OF SUPERHEATED AMMONIA VAPOR

Temp F		120 66.02 F		Temp F		140 74.79 F		Temp F		170 86.29 F	
Sat	2.476	628.4	1.2201	Sat	2.132	629.9	1.2068	Sat	1.764	631.6	1.1900
70	2.505	631.3	1.2255	80	2.166	633.8	1.2140	90	1.784	634.4	1.1952
80	2.576	638.3	.2386	90	2.228	640.9	.2272				
90	2.645	645.0	.2510								
100	2.712	651.6	1.2628	100	2.288	647.8	1.2396	100	1.837	641.9	1.2087
110	2.778	658.0	.2741	110	2.347	654.5	.2515	110	1.889	649.1	.2215
120	2.842	664.2	.2850	120	2.404	661.1	.2628	120	1.939	656.1	.2336
130	2.905	670.4	.2956	130	2.460	667.4	.2738	130	1.988	662.8	.2452
140	2.967	676.5	.3058	140	2.515	673.7	.2843	140	2.035	669.4	.2563
150	3.029	682.5	1.3157	150	2.569	679.9	1.2945	150	2.081	675.9	1.2669
160	3.089	688.4	.3254	160	2.622	686.0	.3045	160	2.127	682.3	.2773
170	3.149	694.3	.3348	170	2.675	692.0	.3141	170	2.172	688.5	.2873
180	3.209	700.2	.3441	180	2.727	698.0	.3236	180	2.216	694.7	.2971
190	3.268	706.0	.3531	190	2.779	704.0	.3328	190	2.260	700.8	.3066
200	3.326	711.8	1.3620	200	2.830	709.9	1.3418	200	2.303	706.9	1.3159
210	3.385	717.6	.3707	210	2.880	715.8	.3507	210	2.346	713.0	.3249
220	3.442	723.4	.3793	220	2.931	721.6	.3594	220	2.389	719.0	.3338
230	3.500	729.2	.3877	230	2.981	727.5	.3679	230	2.431	724.9	.3426
240	3.557	734.9	.3960	240	3.030	733.3	.3763	240	2.473	730.9	.3512
250	3.614	740.7	1.4042	250	3.080	739.2	1.3846	250	2.514	736.8	1.3596
260	3.671	746.5	.4123	260	3.129	745.0	.3928	260	2.555	742.8	.3679
270	3.727	752.2	.4202	270	3.179	750.8	.4008	270	2.596	748.7	.3761
280	3.783	758.0	.4281	280	3.227	756.7	.4088	280	2.637	754.6	.3841
290	3.839	763.8	.4359	290	3.275	762.5	.4166	290	2.678	760.5	.3921
300	3.895	769.6	1.4435	300	3.323	768.3	1.4243	300	2.718	766.4	1.3999
				320	3.420	780.0	.4395	320	2.798	778.3	.4153
								340	2.878	790.1	.4303

TABLE 15-5
Saturation Properties of Dichlorodifluoromethane (Refrigerant-12)*

Temp F	Pressure psia	Volume cu ft/lb		Enthalpy Btu/lb			Entropy Btu/(lb)(R)		Temp F
		Liquid v_f	Vapor v_g	Liquid h_f	Latent h_{fg}	Vapor h_g	Liquid s_f	Vapor s_g	
−140	0.2562	0.0096579	110.46	−20.652	82.548	61.896	−0.056123	0.20208	−140
−130	0.4122	0.0097359	70.730	−18.609	81.577	62.968	−0.049830	0.19760	−130
−120	0.6419	0.0098163	46.741	−16.565	80.617	64.052	−0.043723	0.19359	−120
−110	0.9703	0.0098992	31.777	−14.518	79.663	65.145	−0.037786	0.19002	−110
−100	1.4280	0.0099847	22.164	−12.466	78.714	66.248	−0.032005	0.18683	−100
−90	2.0509	0.010073	15.821	−10.409	77.764	67.355	−0.026367	0.18398	−90
−80	2.8807	0.010164	11.533	−8.3451	76.812	68.467	−0.020862	0.18143	−80
−70	3.9651	0.010259	8.5687	−6.2730	75.853	69.580	−0.015481	0.17916	−70
−60	5.3575	0.010357	6.4774	−4.1919	74.885	70.693	−0.010214	0.17714	−60
−50	7.1168	0.010459	4.9742	−2.1011	73.906	71.805	−0.005056	0.17533	−50
−45	8.1540	0.010511	4.3828	−1.0519	73.411	72.359	−0.002516	0.17451	−45
−40	9.3076	0.010564	3.8750	0	72.913	72.913	0	0.17373	−40
−39	9.5530	0.010575	3.7823	0.2107	72.812	73.023	0.000500	0.17357	−39
−38	9.8035	0.010586	3.6922	0.4215	72.712	73.134	0.001000	0.17343	−38
−37	10.059	0.010596	3.6047	0.6324	72.611	73.243	0.001498	0.17328	−37
−36	10.320	0.010607	3.5198	0.8434	72.511	73.354	0.001995	0.17313	−36
−35	10.586	0.010618	3.4373	1.0546	72.409	73.464	0.002492	0.17299	−35
−34	10.858	0.010629	3.3571	1.2659	72.309	73.575	0.002988	0.17285	−34
−33	11.135	0.010640	3.2792	1.4772	72.208	73.685	0.003482	0.17271	−33
−32	11.417	0.010651	3.2035	1.6887	72.106	73.795	0.003976	0.17257	−32
−31	11.706	0.010662	3.1300	1.9003	72.004	73.904	0.004469	0.17243	−31
−30	11.999	0.010674	3.0585	2.1120	71.903	74.015	0.004961	0.17229	−30
−29	12.299	0.010685	2.9890	2.3239	71.801	74.125	0.005452	0.17216	−29
−28	12.604	0.010696	2.9214	2.5358	71.698	74.234	0.005942	0.17203	−28
−27	12.916	0.010707	2.8556	2.7479	71.596	74.344	0.006431	0.17189	−27
−26	13.233	0.010719	2.7917	2.9601	71.494	74.454	0.006919	0.17177	−26
−25	13.556	0.010730	2.7295	3.1724	71.391	74.563	0.007407	0.17164	−25
−24	13.886	0.010741	2.6691	3.3848	71.288	74.673	0.007894	0.17151	−24
−23	14.222	0.010753	2.6102	3.5973	71.185	74.782	0.008379	0.17139	−23
−22	14.564	0.010764	2.5529	3.8100	71.081	74.891	0.008864	0.17126	−22
−21	14.912	0.010776	2.4972	4.0228	70.978	75.001	0.009348	0.17114	−21
−20	15.267	0.010788	2.4429	4.2357	70.874	75.110	0.009831	0.17102	−20
−19	15.628	0.010799	2.3901	4.4487	70.770	75.219	0.010314	0.17090	−19
−18	15.996	0.010811	2.3387	4.6618	70.666	75.328	0.010795	0.17078	−18
−17	16.371	0.010823	2.2886	4.8751	70.561	75.436	0.011276	0.17066	−17
−16	16.753	0.010834	2.2399	5.0885	70.456	75.545	0.011755	0.17055	−16
−15	17.141	0.010846	2.1924	5.3020	70.352	75.654	0.012234	0.17043	−15
−14	17.536	0.010858	2.1461	5.5157	70.246	75.762	0.012712	0.17032	−14
−13	17.939	0.010870	2.1011	5.7295	70.141	75.871	0.013190	0.17021	−13
−12	18.348	0.010882	2.0572	5.9434	70.036	75.979	0.013666	0.17010	−12
−11	18.765	0.010894	2.0144	6.1574	69.930	76.087	0.014142	0.16999	−11
−10	19.189	0.010906	1.9727	6.3716	69.824	76.196	0.014617	0.16989	−10
−9	19.621	0.010919	1.9320	6.5859	69.718	76.304	0.015091	0.16978	−9
−8	20.059	0.010931	1.8924	6.8003	69.611	76.411	0.015564	0.16967	−8
−7	20.506	0.010943	1.8538	7.0149	69.505	76.520	0.016037	0.16957	−7
−6	20.960	0.010955	1.8161	7.2296	69.397	76.627	0.016508	0.16947	−6
−5	21.422	0.010968	1.7794	7.4444	69.291	76.735	0.016979	0.16937	−5
−4	21.891	0.010980	1.7436	7.6594	69.183	76.842	0.017449	0.16927	−4
−3	22.369	0.010993	1.7086	7.8745	69.075	76.950	0.017919	0.16917	−3
−2	22.854	0.011005	1.6745	8.0898	68.967	77.057	0.018388	0.16907	−2
−1	23.348	0.011018	1.6413	8.3052	68.859	77.164	0.018855	0.16897	−1
0	23.849	0.011030	1.6089	8.5207	68.750	77.271	0.019323	0.16888	0
1	24.359	0.011043	1.5772	8.7364	68.642	77.378	0.019789	0.16878	1
2	24.878	0.011056	1.5463	8.9522	68.533	77.485	0.020255	0.16869	2
3	25.404	0.011069	1.5161	9.1682	68.424	77.592	0.020719	0.16860	3
4	25.939	0.011082	1.4867	9.3843	68.314	77.698	0.021184	0.16851	4
5	26.483	0.011094	1.4580	9.6005	68.204	77.805	0.021647	0.16842	5
6	27.036	0.011107	1.4299	9.8169	68.094	77.911	0.022110	0.16833	6
7	27.597	0.011121	1.4025	10.033	67.984	78.017	0.022572	0.16824	7
8	28.167	0.011134	1.3758	10.250	67.873	78.123	0.023033	0.16815	8
9	28.747	0.011147	1.3496	10.467	67.762	78.229	0.023494	0.16807	9

* Refrigerant-12 ("Freon-12") Copyright 1956, E. I. du Pont de Nemours & Co., Inc. Reprinted by permission.

TABLE 15-5—(Continued)

Temp F	Pressure psia	Volume cu ft/lb		Enthalpy Btu/lb			Entropy Btu/(lb)(R)		Temp F
		Liquid v_f	Vapor v_g	Liquid h_f	Latent h_{fg}	Vapor h_g	Liquid s	Vapor s_g	
10	29.335	0.011160	1.3241	10.684	67.651	78.335	0.023954	0.16798	10
11	29.932	0.011173	1.2992	10.901	67.539	78.440	0.024413	0.16790	11
12	30.539	0.011187	1.2748	11.118	67.428	78.546	0.024871	0.16782	12
13	31.155	0.011200	1.2510	11.336	67.315	78.651	0.025329	0.16774	13
14	31.780	0.011214	1.2778	11.554	67.203	78.757	0.025786	0.16765	14
15	32.415	0.011227	1.2050	11.771	67.090	78.861	0.026243	0.16758	15
16	33.060	0.011241	1.1828	11.989	66.977	78.966	0.026699	0.16750	16
17	33.714	0.011254	1.1611	12.207	66.864	79.071	0.027154	0.16742	17
18	34.378	0.011268	1.1399	12.426	66.750	79.176	0.027608	0.16734	18
19	35.052	0.011282	1.1191	12.644	66.636	79.280	0.028062	0.16727	19
20	35.736	0.011296	1.0988	12.863	66.522	79.385	0.028515	0.16719	20
21	36.430	0.011310	1.0790	13.081	66.407	79.488	0.028968	0.16712	21
22	37.135	0.011324	1.0596	13.300	66.293	79.593	0.029420	0.16704	22
23	37.849	0.011388	1.0406	13.520	66.177	79.697	0.029871	0.16697	23
24	38.574	0.011352	1.0220	13.739	66.061	79.800	0.030322	0.16690	24
25	39.310	0.011366	1.0039	13.958	65.946	79.904	0.030772	0.16683	25
26	40.056	0.011380	0.98612	14.178	65.829	80.007	0.031221	0.16676	26
27	40.813	0.011395	0.96874	14.398	65.713	80.111	0.031670	0.16669	27
28	41.580	0.011409	0.95173	14.618	65.596	80.214	0.032118	0.16662	28
29	42.359	0.011424	0.93509	14.838	65.478	80.316	0.032566	0.16655	29
30	43.148	0.011438	0.91880	15.058	65.361	80.419	0.033013	0.16648	30
31	43.948	0.011453	0.90286	15.279	65.243	80.522	0.033460	0.16642	31
32	44.760	0.011468	0.88725	15.500	65.124	80.624	0.033905	0.16635	32
33	45.583	0.011482	0.87197	15.720	65.006	80.726	0.034351	0.16629	33
34	46.417	0.011497	0.85702	15.942	64.886	80.828	0.034796	0.16622	34
35	47.263	0.011512	0.84237	16.163	64.767	80.930	0.035240	0.16616	35
36	48.120	0.011527	0.82803	16.384	64.647	81.031	0.035683	0.16610	36
37	48.989	0.011542	0.81399	16.606	64.527	81.133	0.036126	0.16604	37
38	49.870	0.011557	0.80023	16.828	64.406	81.234	0.036569	0.16598	38
39	50.763	0.011573	0.78676	17.050	64.285	81.335	0.037011	0.16592	39
40	51.667	0.011588	0.77357	17.273	64.163	81.436	0.037453	0.16586	40
41	52.584	0.011603	0.76064	17.495	64.042	81.537	0.037893	0.16580	41
42	53.513	0.011619	0.74798	17.718	63.919	81.637	0.038334	0.16574	42
43	54.454	0.011635	0.73557	17.941	63.796	81.737	0.038774	0.16568	43
44	55.407	0.011650	0.72341	18.164	63.673	81.837	0.039213	0.16562	44
45	56.373	0.011666	0.71149	18.387	63.550	81.937	0.039652	0.16557	45
46	57.352	0.011682	0.69982	18.611	63.426	82.037	0.040091	0.16551	46
47	58.343	0.011698	0.68837	18.835	63.301	82.136	0.040529	0.16546	47
48	59.347	0.011714	0.67715	19.059	63.177	82.236	0.040966	0.16540	48
49	60.364	0.011730	0.66616	19.283	63.051	82.334	0.041403	0.16535	49
50	61.394	0.011746	0.65537	19.507	62.926	82.433	0.041839	0.16530	50
51	62.437	0.011762	0.64480	19.732	62.800	82.532	0.042276	0.16524	51
52	63.494	0.011779	0.63444	19.957	62.673	82.630	0.042711	0.16519	52
53	64.563	0.011795	0.62428	20.182	62.546	82.728	0.043146	0.16514	53
54	65.646	0.011811	0.61431	20.408	62.418	82.826	0.043581	0.16509	54
55	66.743	0.011828	0.60453	20.634	62.290	82.924	0.044015	0.16504	55
56	67.853	0.011845	0.59495	20.859	62.162	83.021	0.044449	0.16499	56
57	68.977	0.011862	0.58554	21.086	62.033	83.119	0.044883	0.16494	57
58	70.115	0.011879	0.57632	21.312	61.903	83.215	0.045316	0.16489	58
59	71.267	0.011896	0.56727	21.539	61.773	83.312	0.045748	0.16484	59
60	72.433	0.011913	0.55839	21.766	61.643	83.409	0.046180	0.16479	60
61	73.613	0.011930	0.54967	21.993	61.512	83.505	0.046612	0.16474	61
62	74.807	0.011947	0.54112	22.221	61.380	83.601	0.047044	0.16470	62
63	76.016	0.011965	0.53273	22.448	61.248	83.696	0.047475	0.16465	63
64	77.239	0.011982	0.52450	22.676	61.116	83.792	0.047905	0.16460	64
65	78.477	0.012000	0.51642	22.905	60.982	83.887	0.048336	0.16456	65
66	79.729	0.012017	0.50848	23.133	60.849	83.982	0.048765	0.16451	66
67	80.996	0.012035	0.50070	23.362	60.715	84.077	0.049195	0.16447	67
68	82.279	0.012053	0.49305	23.591	60.580	84.171	0.049624	0.16442	68
69	83.576	0.012071	0.48555	23.821	60.445	84.266	0.050053	0.16438	69

Refrigerants and Refrigeration Systems

TABLE 15-5—(Continued)

Temp F	Pressure psia	Volume cu ft/lb		Enthalpy Btu/lb			Entropy Btu/(lb)(R)		Temp F
		Liquid v_f	Vapor v_g	Liquid h_f	Latent h_{fg}	Vapor h_g	Liquid s_f	Vapor s_g	
70	84.888	0.012089	0.47818	24.050	60.309	84.359	0.050482	0.16434	70
71	86.216	0.012108	0.47094	24.281	60.172	84.453	0.050910	0.16429	71
72	87.559	0.012126	0.46383	24.511	60.035	84.546	0.051338	0.16425	72
73	88.918	0.012145	0.45686	24.741	59.898	84.639	0.051766	0.16421	73
74	90.292	0.012163	0.45000	24.973	59.759	84.732	0.052193	0.16417	74
75	91.682	0.012182	0.44327	25.204	59.621	84.825	0.052620	0.16412	75
76	93.087	0.012201	0.43666	25.435	59.481	84.916	0.053047	0.16408	76
77	94.509	0.012220	0.43016	25.667	59.341	85.008	0.053473	0.16404	77
78	95.946	0.012239	0.42378	25.899	59.201	85.100	0.053900	0.16400	78
79	97.400	0.012258	0.41751	26.132	59.059	85.191	0.054326	0.16396	79
80	98.870	0.012277	0.41135	26.365	58.917	85.282	0.054751	0.16392	80
81	100.36	0.012297	0.40530	26.598	58.775	85.373	0.055177	0.16388	81
82	101.86	0.012316	0.39935	26.832	58.631	85.463	0.055602	0.16384	82
83	103.38	0.012336	0.39351	27.065	58.488	85.553	0.056027	0.16380	83
84	104.92	0.012356	0.38776	27.300	58.343	85.643	0.056452	0.16376	84
85	106.47	0.012376	0.38212	27.534	58.198	85.732	0.056877	0.16372	85
86	108.04	0.012396	0.37657	27.769	58.052	85.821	0.057301	0.16368	86
87	109.63	0.012416	0.37111	28.005	57.905	85.910	0.057725	0.16364	87
88	111.23	0.012437	0.36575	28.241	57.757	85.998	0.058149	0.16360	88
89	112.85	0.012457	0.36047	28.477	57.609	86.086	0.058573	0.16357	89
90	114.49	0.012478	0.35529	28.713	57.461	86.174	0.058997	0.16353	90
91	116.15	0.012499	0.35019	28.950	57.311	86.261	0.059420	0.16349	91
92	117.82	0.012520	0.34518	29.187	57.161	86.348	0.059844	0.16345	92
93	119.51	0.012541	0.34025	29.425	57.009	86.434	0.060267	0.16341	93
94	121.22	0.012562	0.33540	29.663	56.858	86.521	0.060690	0.16338	94
95	122.95	0.012583	0.33063	29.901	56.705	86.606	0.061113	0.16334	95
96	124.70	0.012605	0.32594	30.140	56.551	86.691	0.061536	0.16330	96
97	126.46	0.012627	0.32133	30.380	56.397	86.777	0.061959	0.16326	97
98	128.24	0.012649	0.31679	30.619	56.242	86.861	0.062381	0.16323	98
99	130.04	0.012671	0.31233	30.859	56.086	86.945	0.062804	0.16319	99
100	131.86	0.012693	0.30794	31.100	55.929	87.029	0.063227	0.16315	100
101	133.70	0.012715	0.30362	31.341	55.772	87.113	0.063649	0.16312	101
102	135.56	0.012738	0.29937	31.583	55.613	87.196	0.064072	0.16308	102
103	137.44	0.012760	0.29518	31.824	55.454	87.278	0.064494	0.16304	103
104	139.33	0.012783	0.29106	32.067	55.293	87.360	0.064916	0.16301	104
105	141.25	0.012806	0.28701	32.310	55.132	87.442	0.065339	0.16297	105
106	143.18	0.102829	0.28303	32.553	54.970	87.523	0.065761	0.16293	106
107	145.13	0.012853	0.27910	32.797	54.807	87.604	0.066184	0.16290	107
108	147.11	0.012876	0.27524	33.041	54.643	87.684	0.066606	0.16286	108
109	149.10	0.012900	0.27143	33.286	54.478	87.764	0.067028	0.16282	109
110	151.11	0.012924	0.26769	33.531	54.313	87.844	0.067451	0.16279	110
112	155.19	0.012972	0.26037	34.023	53.978	88.001	0.068296	0.16271	112
114	159.36	0.013022	0.25328	34.517	53.639	88.156	0.069141	0.16264	114
116	163.61	0.013072	0.24641	35.014	53.296	88.310	0.069987	0.16256	116
118	167.94	0.013123	0.23974	35.512	52.949	88.461	0.070833	0.16249	118
120	172.35	0.013174	0.23326	36.013	52.597	88.610	0.071680	0.16241	120
122	176.85	0.013227	0.22698	36.516	52.241	88.757	0.072528	0.16234	122
124	181.43	0.013280	0.22089	37.021	51.881	88.902	0.073376	0.16226	124
125	183.76	0.013308	0.21791	37.275	51.698	88.973	0.073800	0.16222	125
130	195.71	0.013447	0.20364	38.553	50.768	89.321	0.075927	0.16202	130
135	208.22	0.013593	0.19036	39.848	49.805	89.653	0.078061	0.16181	135
140	221.32	0.013746	0.17799	41.162	48.805	89.967	0.080205	0.16159	140
145	235.00	0.013907	0.16644	42.495	47.766	90.261	0.082361	0.16135	145
150	249.31	0.014078	0.15564	43.850	46.684	90.534	0.084531	0.16110	150
170	313.00	0.014871	0.11873	49.529	41.830	91.359	0.093418	0.15985	170
190	387.98	0.015942	0.089418	55.769	35.792	91.561	0.10284	0.15793	190
210	475.52	0.017601	0.064843	62.959	27.599	90.558	0.11332	0.15453	210
230	577.03	0.021854	0.039435	72.893	12.229	85.122	0.12739	0.14512	230
233.6 (Critical)	596.9	0.02870	0.02870	78.86	0	78.86	0.1359	0.1359	233.6 (Critical)

TABLE 15-6
Saturation Properties of Monochlorodifluoromethane (Refrigerant-22)*

Temp F	Pressure psia	Spec Volume cu ft/lb		Enthalpy Btu/lb			Entropy Btu/(lb)(R)		Temp F
		Liquid v_f	Vapor v_g	Liquid h_f	Latent h_{fg}	Vapor h_g	Liquid s_f	Vapor s_g	
−150	0.2777	0.01027	137.00	−27.70	111.68	86.98	−0.06794	0.29267	−150
−140	0.4552	0.01035	86.25	−22.56	110.71	88.15	−0.06114	0.28516	−140
−130	0.7213	0.01043	56.10	−20.42	109.75	89.33	−0.05454	0.27836	−130
−120	1.108	0.01051	37.57	−18.26	108.79	90.53	−0.04809	0.27218	−120
−110	1.657	0.01060	25.83	−16.09	107.83	91.74	−0.04179	0.26657	−110
−100	2.417	0.01069	18.20	−13.89	106.85	92.96	−0.03561	0.26146	−100
−90	3.443	0.01079	13.10	−11.67	105.85	94.19	−0.02953	0.25682	−90
−80	4.805	0.01089	9.616	−9.42	104.84	95.41	−0.02353	0.25259	−80
−70	6.576	0.01099	7.188	−7.13	103.78	96.65	−0.01759	0.24873	−70
−60	8.844	0.01111	5.461	−4.80	102.68	97.88	−0.01170	0.24520	−60
−50	11.70	0.01123	4.212	−2.43	101.53	99.10	−0.00599	0.24198	−50
−48	12.35	0.01125	4.004	−1.95	101.29	99.34	−0.00468	0.24137	−48
−46	13.04	0.01127	3.810	−1.46	101.05	99.58	−0.00350	0.24077	−46
−44	13.74	0.01130	3.626	−0.98	100.81	99.83	−0.00233	0.24018	−44
−42	14.48	0.01132	3.454	−0.49	100.57	100.08	−0.00117	0.23960	−42
−40	15.25	0.01135	3.291	0.00	100.31	100.32	0.00000	0.23903	−40
−39	15.64	0.01136	3.213	0.24	100.19	100.44	0.00058	0.23875	−39
−38	16.06	0.01138	3.138	0.49	100.06	100.56	0.00117	0.23847	−38
−37	16.47	0.01139	3.064	0.74	99.94	100.68	0.00175	0.23819	−37
−36	16.89	0.01140	2.993	0.99	99.81	100.80	0.00233	0.23792	−36
−35	17.33	0.01142	2.923	1.23	99.68	100.92	0.00292	0.23765	−35
−34	17.77	0.01143	2.855	1.48	99.56	101.04	0.00350	0.23738	−34
−33	18.22	0.01144	2.790	1.73	99.43	101.17	0.00408	0.23711	−33
−32	18.67	0.01145	2.726	1.98	99.30	101.29	0.00466	0.23685	−32
−31	19.14	0.01147	2.664	2.23	99.17	101.41	0.00525	0.23659	−31
−30	19.62	0.01148	2.604	2.48	99.04	101.53	0.00583	0.23633	−30
−29	20.10	0.01149	2.545	2.73	98.91	101.65	0.00641	0.23607	−29
−28	20.60	0.01151	2.488	2.99	98.78	101.77	0.00700	0.23581	−28
−27	21.10	0.01152	2.432	3.24	98.64	101.89	0.00758	0.23556	−27
−26	21.61	0.01154	2.379	3.50	98.51	102.01	0.00816	0.23531	−26
−25	22.14	0.01155	2.326	3.75	98.37	102.12	0.00875	0.23506	−25
−24	22.67	0.01156	2.275	4.00	98.24	102.24	0.00933	0.23481	−24
−23	23.21	0.01158	2.225	4.26	98.10	102.36	0.00991	0.23456	−23
−22	23.77	0.01159	2.177	4.52	97.96	102.48	0.01050	0.23432	−22
−21	24.33	0.01161	2.129	4.77	97.82	102.60	0.01108	0.23408	−21
−20	24.90	0.01162	2.083	5.03	97.68	102.72	0.01166	0.23384	−20
−19	25.49	0.01163	2.039	5.29	97.55	102.84	0.01225	0.23360	−19
−18	26.08	0.01165	1.995	5.54	97.40	102.95	0.01283	0.23336	−18
−17	26.69	0.01166	1.953	5.80	97.26	103.07	0.01342	0.23313	−17
−16	27.30	0.01168	1.911	6.06	97.12	103.19	0.01400	0.23289	−16
−15	27.93	0.01169	1.871	6.33	96.98	103.31	0.01459	0.23266	−15
−14	28.58	0.01170	1.832	6.59	96.83	103.42	0.01517	0.23244	−14
−13	29.23	0.01172	1.793	6.85	96.69	103.54	0.01575	0.23221	−13
−12	29.89	0.01173	1.756	7.11	96.54	103.66	0.01634	0.23198	−12
−11	30.56	0.01175	1.720	7.38	96.39	103.77	0.01692	0.23176	−11
−10	31.25	0.01176	1.684	7.64	96.25	103.89	0.01751	0.23154	−10
−9	31.95	0.01178	1.650	7.91	96.10	104.01	0.01809	0.23132	−9
−8	32.65	0.01179	1.616	8.17	95.95	104.12	0.01868	0.23110	−8
−7	33.38	0.01181	1.583	8.44	95.80	104.24	0.01827	0.23088	−7
−6	34.11	0.01182	1.551	8.71	95.64	104.35	0.01985	0.23067	−6
−5	34.86	0.01184	1.520	8.98	95.49	104.47	0.02044	0.23045	−5
−4	35.61	0.01185	1.489	9.25	95.34	104.58	0.02103	0.23024	−4
−3	36.39	0.01187	1.459	9.52	95.18	104.70	0.02161	0.23003	−3
−2	37.17	0.01188	1.430	9.79	95.03	104.81	0.02220	0.22982	−2
−1	37.97	0.01190	1.402	10.06	94.87	104.92	0.02279	0.22961	−1
0	38.78	0.01192	1.374	10.32	94.71	105.04	0.02337	0.22941	0
1	39.60	0.01193	1.347	10.60	94.55	105.15	0.02396	0.22920	1
2	40.43	0.01195	1.321	10.87	94.39	105.26	0.02455	0.22900	2
3	41.29	0.01196	1.295	11.15	94.23	105.38	0.02514	0.22880	3
4	42.15	0.01198	1.270	11.42	94.07	105.49	0.02573	0.22860	4

* Refrigerant-22 (genetron-22) copyright 1958, General Chemical Division Allied Chemical and Dye Division. Reprinted by permission.

Refrigerants and Refrigeration Systems

TABLE 15-6—(Continued)

Temp F	Pressure psia	Spec Volume cu ft/lb		Enthalpy Btu/lb			Entropy Btu/(lb)(R)		Temp F
		Liquid v_f	Vapor v_g	Liquid h_f	Latent h_{fg}	Vapor h_g	Liquid s_f	Vapor s_g	
5	43.03	0.01199	1.245	11.69	93.90	105.60	0.02632	0.22840	5
6	43.93	0.01201	1.221	11.97	93.74	105.71	0.02691	0.22820	6
7	44.83	0.01203	1.198	12.25	93.58	105.82	0.02749	0.22800	7
8	45.75	0.01204	1.175	12.53	93.41	105.94	0.02808	0.22781	8
9	46.69	0.01206	1.152	12.81	93.24	106.04	0.02867	0.22761	9
10	47.64	0.01208	1.130	13.08	93.07	106.16	0.02927	0.22742	10
11	48.60	0.01209	1.109	13.36	92.90	106.27	0.02985	0.22723	11
12	49.58	0.01211	1.088	13.64	92.73	106.38	0.03044	0.22704	12
13	50.58	0.01213	1.068	13.93	92.56	106.49	0.03104	0.22685	13
14	51.59	0.01214	1.048	14.21	92.39	106.59	0.03163	0.22666	14
15	52.61	0.01216	1.028	14.49	92.21	106.70	0.03222	0.22648	15
16	53.65	0.01218	1.009	14.77	92.04	106.81	0.03281	0.22629	16
17	54.71	0.01219	0.9904	15.06	91.86	106.92	0.03340	0.22611	17
18	55.78	0.01221	0.9722	15.34	91.68	107.03	0.03400	0.22592	18
19	56.87	0.01223	0.9544	15.63	91.50	107.13	0.03459	0.22574	19
20	57.97	0.01225	0.9369	15.91	91.32	107.24	0.03518	0.22556	20
21	59.10	0.01226	0.9198	16.21	91.14	107.35	0.03578	0.22538	21
22	60.23	0.01228	0.9031	16.49	90.95	107.45	0.03637	0.22520	22
23	61.39	0.01230	0.8868	16.78	90.77	107.56	0.03697	0.22502	23
24	62.56	0.01232	0.8708	17.07	90.59	107.66	0.03756	0.22485	24
25	63.74	0.01234	0.8552	17.36	90.40	107.76	0.03816	0.22467	25
26	64.95	0.01235	0.8399	17.66	90.21	107.87	0.03875	0.22449	26
27	66.17	0.01237	0.8249	17.95	90.02	107.97	0.03935	0.22432	27
28	67.41	0.01239	0.8103	18.24	89.83	108.07	0.03994	0.22415	28
29	68.66	0.01241	0.7959	18.54	89.64	108.18	0.04054	0.22397	29
30	69.94	0.01243	0.7819	18.83	89.45	108.28	0.04114	0.22380	30
31	71.23	0.01244	0.7681	19.13	89.24	108.38	0.04174	0.22363	31
32	72.54	0.01246	0.7546	19.42	89.06	108.48	0.04233	0.22346	32
33	73.87	0.01248	0.7415	19.72	88.86	108.58	0.04293	0.22329	33
34	75.21	0.01250	0.7285	20.02	88.66	108.68	0.04353	0.22312	34
35	76.58	0.01252	0.7159	20.32	88.46	108.78	0.04413	0.22295	35
36	77.96	0.01254	0.7035	20.62	88.26	108.88	0.04473	0.22279	36
37	79.36	0.01256	0.6914	20.92	88.06	108.98	0.04533	0.22262	37
38	80.78	0.01258	0.6795	21.22	87.85	109.07	0.04593	0.22245	38
39	82.22	0.01260	0.6678	21.52	87.65	109.17	0.04653	0.22229	39
40	83.68	0.01262	0.6564	21.82	87.44	109.27	0.04713	0.22212	40
41	85.16	0.01264	0.6452	22.13	87.23	109.36	0.04773	0.22195	41
42	86.65	0.01266	0.6343	22.44	87.02	109.46	0.04833	0.22179	42
43	88.17	0.01268	0.6236	22.74	86.81	109.55	0.04893	0.22163	43
44	89.71	0.01269	0.6130	23.05	86.60	109.65	0.04954	0.22147	44
45	91.27	0.01271	0.6027	23.36	86.39	109.74	0.05014	0.22130	45
46	92.84	0.01273	0.5926	23.66	86.17	109.83	0.05074	0.22114	46
47	94.44	0.01275	0.5827	23.97	85.95	109.93	0.05134	0.22098	47
48	96.06	0.01277	0.5730	24.28	85.73	110.02	0.05195	0.22082	48
49	97.70	0.01279	0.5634	24.59	85.51	110.11	0.05255	0.22066	49
50	99.36	0.01281	0.5541	24.90	85.29	110.20	0.05316	0.22050	50
51	101.0	0.01284	0.5449	25.22	85.07	110.29	0.05376	0.22034	51
52	102.7	0.01286	0.5359	25.55	84.83	110.38	0.05436	0.22018	52
53	104.5	0.01288	0.5272	25.85	84.61	110.46	0.05496	0.22002	53
54	106.2	0.01290	0.5185	26.16	84.39	110.55	0.05558	0.21986	54
55	107.9	0.01292	0.5100	26.48	84.16	110.64	0.05618	0.21970	55
56	109.8	0.01294	0.5017	26.79	83.93	110.72	0.05679	0.21954	56
57	111.6	0.01296	0.4935	27.10	83.70	110.81	0.05739	0.21938	57
58	113.4	0.01298	0.4855	27.43	83.46	110.89	0.05800	0.21923	58
59	115.2	0.01300	0.4776	27.75	83.23	110.97	0.05861	0.21907	59
60	117.1	0.01302	0.4699	28.07	82.99	111.05	0.05921	0.21891	60
61	119.0	0.01304	0.4623	28.39	82.75	111.14	0.05982	0.21875	61
62	120.9	0.01307	0.4549	28.71	82.51	111.22	0.06043	0.21860	62
63	122.9	0.01309	0.4476	29.03	82.27	111.30	0.06104	0.21844	63
64	124.9	0.01311	0.4404	29.35	82.07	111.38	0.06165	0.21828	64

TABLE 15-6—(Continued)

Temp F	Pressure psia	Spec Volume cu ft/lb		Enthalpy Btu/lb			Entropy Btu/(lb)(R)		Temp F
		Liquid v_f	Vapor v_g	Liquid h_f	Latent h_{fg}	Vapor h_g	Liquid s_f	Vapor s_g	
65	126.9	0.01313	0.4334	29.68	81.78	111.46	0.06225	0.21812	65
66	128.9	0.01316	0.4265	30.00	81.53	111.53	0.06286	0.21796	66
67	130.9	0.01318	0.4197	30.33	81.29	111.61	0.06347	0.21780	67
68	132.9	0.01320	0.4131	30.65	81.04	111.69	0.06408	0.21765	68
69	135.0	0.01322	0.4065	30.98	80.78	111.76	0.06469	0.21749	69
70	137.1	0.01325	0.4001	31.31	80.53	111.84	0.06530	0.21733	70
71	139.3	0.01327	0.3939	31.63	80.27	111.91	0.06590	0.21717	71
72	141.5	0.01329	0.3876	31.96	80.02	111.98	0.06651	0.21701	72
73	143.7	0.01332	0.3816	32.29	79.76	112.05	0.06712	0.21685	73
74	145.0	0.01334	0.3756	32.62	79.50	112.12	0.06773	0.21669	74
75	148.1	0.01336	0.3697	32.95	79.24	112.19	0.06834	0.21653	75
76	150.4	0.01339	0.3640	33.29	78.97	112.26	0.06895	0.21637	76
77	152.6	0.01341	0.3583	33.62	78.71	112.32	0.06956	0.21621	77
78	155.0	0.01343	0.3528	33.95	78.44	112.39	0.07017	0.21605	78
79	157.3	0.01346	0.3473	34.28	78.17	112.45	0.07078	0.21589	79
80	159.7	0.01348	0.3419	34.62	77.90	112.52	0.07139	0.21573	80
81	162.0	0.01350	0.3366	34.95	77.63	112.58	0.07200	0.21557	81
82	164.5	0.01353	0.3315	35.29	77.35	112.64	0.07261	0.21541	82
83	166.9	0.01355	0.3264	35.62	77.08	112.70	0.07321	0.21524	83
84	169.4	0.01358	0.3214	35.96	76.80	112.76	0.07382	0.21508	84
85	171.9	0.01360	0.3164	36.30	76.52	112.82	0.07443	0.21491	85
86	174.4	0.01363	0.3116	36.64	76.24	112.87	0.07504	0.21475	86
87	176.9	0.01365	0.3068	36.97	75.95	112.93	0.07565	0.21458	87
88	179.5	0.01368	0.3022	37.31	75.67	112.98	0.07625	0.21441	88
89	182.1	0.01370	0.2976	37.65	75.38	113.03	0.07686	0.21425	89
90	184.7	0.01373	0.2930	37.99	75.09	113.09	0.07747	0.21408	90
91	187.4	0.01376	0.2886	38.33	74.80	113.14	0.07807	0.21391	91
92	190.1	0.01378	0.2842	38.67	74.51	113.18	0.07868	0.21374	92
93	192.8	0.01381	0.2799	39.01	74.22	113.23	0.07928	0.21357	93
94	195.5	0.01384	0.2757	39.35	73.92	113.28	0.07989	0.21339	94
95	198.3	0.01386	0.2716	39.70	73.63	113.32	0.08049	0.21322	95
96	201.1	0.01389	0.2675	40.04	73.33	113.36	0.08109	0.21305	96
97	203.9	0.01392	0.2634	40.38	73.03	113.40	0.08169	0.21287	97
98	206.8	0.01395	0.2595	40.72	72.72	113.44	0.08230	0.21270	98
99	209.7	0.01398	0.2556	41.07	72.42	113.48	0.08290	0.21252	99
100	212.6	0.01401	0.2518	41.41	72.11	113.52	0.08349	0.21234	100
101	215.5	0.01404	0.2480	41.75	71.81	113.56	0.08410	0.21216	101
102	218.4	0.01407	0.2443	42.08	71.51	113.59	0.08469	0.21198	102
103	221.4	0.01410	0.2407	42.42	71.20	113.62	0.08529	0.21180	103
104	224.5	0.01413	0.2371	42.78	70.87	113.65	0.08588	0.21162	104
105	227.5	0.01416	0.2336	43.12	70.56	113.68	0.08648	0.21143	105
106	230.6	0.01419	0.2301	43.47	70.25	113.71	0.08707	0.21125	106
107	233.7	0.01423	0.2267	43.81	69.93	113.74	0.08766	0.21106	107
108	236.9	0.01426	0.2234	44.15	69.61	113.76	0.08825	0.21087	108
109	240.1	0.01429	0.2201	44.49	69.29	113.78	0.08884	0.21068	109
110	243.3	0.01433	0.2168	44.84	68.97	113.81	0.08943	0.21049	110
112	249.9	0.01439	0.2105	45.52	68.32	113.84	0.09059	0.21010	112
114	256.5	0.01446	0.2044	46.20	67.67	113.87	0.09175	0.20971	114
116	263.2	0.01453	0.1985	46.88	67.02	113.89	0.09290	0.20931	116
118	270.1	0.01460	0.1927	47.55	66.36	113.91	0.09404	0.20890	118
120	277.2	0.01468	0.1872	48.22	65.69	113.92	0.09516	0.20849	120
122	284.4	0.01476	0.1818	48.89	65.03	113.92	0.09628	0.20806	122
124	291.7	0.01483	0.1766	49.55	64.36	113.91	0.09738	0.20764	124
125	295.4	0.01487	0.1741	49.87	64.02	113.90	0.09792	0.20742	125
130	314.4	0.01508	0.1621	51.49	62.35	113.84	0.10057	0.20631	130
135	334.3	0.01529	0.1510	53.10	60.69	113.74	0.10310	0.20516	135
140	355.0	0.01553	0.1410	54.52	59.07	113.59	0.10548	0.20398	140
145	376.7	0.01577	0.1319	55.91	57.51	113.41	0.10767	0.20278	145
150	399.2	0.01603	0.1237	57.18	56.04	113.21	0.10966	0.20157	150

TABLE 15-7
Saturated Liquid and Vapor Tables for Various Refrigerants

Temp F t	Pressure PSIA p	Sp Vol of Liquid cu ft/lb v_f	Sp Vol of Vapor cu ft/lb v_g	Density of Vapor lb/cu ft $\frac{1}{v_g}$	Enthalpy Btu/lb above -40 F			Entropy		Temp F t
					Liquid h_f	Vaporization h_{fg}	Vapor h_g	Liquid s_f	Vapor s_g	
Carbon Dioxide, CO_2 (Refrigerant-744)										
-40	145.8	0.01437	0.6113	1.64	0.00	137.8	137.8	1.0000	1.3285	-40
-30	177.8	0.01466	0.5029	1.99	4.5	133.7	138.2	1.0107	1.3218	-30
-20	214.9	0.01498	0.4168	2.40	9.1	129.4	138.5	1.0212	1.3154	-20
-10	257.3	0.01532	0.3472	2.88	13.9	124.8	138.7	1.0314	1.3091	-10
0	305.5	0.01570	0.2904	3.44	18.8	120.1	138.9	1.0418	1.3029	0
4	326.5	0.01588	0.2707	3.69	20.8	118.0	138.8	1.0460	1.3006	4
6	337.4	0.01596	0.2614	3.83	21.8	116.9	138.7	1.0481	1.2994	6
10	360.2	0.01614	0.2437	4.10	24.0	114.7	138.7	1.0536	1.2980	10
20	421.8	0.01663	0.2049	4.88	29.4	108.9	138.3	1.0648	1.2919	20
30	490.8	0.01719	0.1722	5.81	35.4	102.4	137.8	1.0768	1.2859	30
40	567.8	0.01787	0.1444	6.93	41.7	95.0	136.7	1.0884	1.2786	40
50	653.6	0.01868	0.1205	8.30	48.4	86.6	135.0	1.1010	1.2709	50
60	748.6	0.01970	0.0994	10.06	55.5	76.6	132.1	1.1145	1.2618	60
70	853.4	0.02112	0.08040	12.44	63.7	63.8	127.5	1.1292	1.2497	70
80	968.7	0.02370	0.06064	16.49	73.9	44.8	118.7	1.1486	1.2314	80
86	1043.0	0.02686	0.04789	20.88	83.3	27.1	110.4	1.1646	1.2143	86
87.8	1066.2	0.03454	0.03454	28.95	97.0	0.0	97.0	1.1890	1.1890	87.8
Trichloromonofluoromethane (Refrigerant-11)										
-80	0.157	0.00961	189	0.0053	-7.89	90.68	82.79	-0.0197	0.1995	-80
-60	0.356	0.00974	87.5	0.0114	-3.94	89.06	85.12	-0.0096	0.2037	-60
-40	0.739	0.00988	44.2	0.0226	0.00	87.48	87.48	0.0000	0.2085	-40
-20	1.420	0.01002	24.06	0.0415	3.94	85.93	89.87	0.0091	0.2046	-20
-10	1.920	0.01010	18.17	0.0550	5.91	85.16	91.07	0.0136	0.2030	-10
0	2.555	0.01018	13.94	0.0718	7.89	84.38	92.27	0.0179	0.2015	0
5	2.931	0.01022	12.27	0.0815	8.88	84.00	92.88	0.0201	0.2009	5
10	3.352	0.01026	10.83	0.0923	9.88	83.60	93.48	0.0222	0.2003	10
20	4.342	0.01034	8.519	0.1174	11.87	82.82	94.69	0.0264	0.1991	20
30	5.557	0.01042	6.776	0.1476	13.88	82.03	95.91	0.0306	0.1981	30
40	7.032	0.01051	5.447	0.1836	15.89	81.22	97.11	0.0346	0.1972	40
50	8.804	0.01060	4.421	0.2262	17.92	80.40	98.32	0.0386	0.1964	50
60	10.90	0.01069	3.626	0.2758	19.96	79.57	99.53	0.0426	0.1958	60
70	13.40	0.01079	2.993	0.3342	22.02	78.71	100.73	0.0465	0.1951	70
80	16.31	0.01088	2.492	0.4012	24.09	77.84	101.93	0.0504	0.1947	80
86	18.28	0.01094	2.242	0.4461	25.34	77.31	102.65	0.0527	0.1944	86
90	19.69	0.01098	2.091	0.4783	26.18	76.95	103.12	0.0542	0.1942	90
100	23.60	0.01109	1.765	0.5666	28.27	76.03	104.30	0.0580	0.1938	100
110	28.09	0.01119	1.499	0.6671	30.40	75.08	105.47	0.0617	0.1935	110
120	33.20	0.01130	1.281	0.7808	32.53	74.10	106.63	0.0654	0.1933	120
Sulfur Dioxide, SO_2 (Refrigerant-764)										
-40	3.136	0.01044	22.42	0.04460	0.00	178.61	178.61	0.00000	0.42562	-40
-20	5.883	0.01063	12.42	0.08052	5.98	175.09	181.07	0.01366	0.41192	-20
-10	7.863	0.01072	9.44	0.10593	9.16	172.97	182.13	0.02075	0.40544	-10
0	10.35	0.01082	7.280	0.13736	12.44	170.63	183.07	0.02795	0.39917	0
5	11.81	0.01087	6.421	0.15574	14.11	169.38	183.49	0.03155	0.39609	5
10	13.42	0.01092	5.682	0.17599	15.80	168.07	183.87	0.03519	0.39306	10
20	17.18	0.01102	4.487	0.22287	19.20	165.32	184.52	0.04241	0.38707	20
30	21.70	0.01114	3.581	0.27925	22.64	162.38	185.02	0.04956	0.38119	30
40	27.10	0.01126	2.887	0.34638	26.12	159.25	185.37	0.05668	0.37541	40
50	33.45	0.01138	2.334	0.42589	29.61	155.95	185.56	0.06370	0.36969	50
60	40.93	0.01150	1.926	0.51921	33.10	152.49	185.59	0.07060	0.36405	60
70	49.62	0.01163	1.590	0.62893	36.58	148.88	185.46	0.07736	0.35846	70
80	59.68	0.01176	1.321	0.75700	40.05	145.12	185.17	0.08399	0.35291	80
86	66.45	0.01184	1.185	0.84388	42.12	142.80	184.92	0.08783	0.34954	86
90	71.25	0.01190	1.104	0.90580	43.50	141.22	184.72	0.09038	0.34731	90
100	84.52	0.01204	0.9262	1.07968	46.90	137.20	184.10	0.09657	0.34173	100
110	99.76	0.01219	0.7804	1.28139	50.26	133.05	183.31	0.10254	0.33611	110
120	120.93	0.01236	0.6598	1.51561	53.58	128.78	182.36	0.10829	0.33046	120

TABLE 15-8
METHYL CHLORIDE (REFRIGERANT-40), SATURATED LIQUID AND VAPOR (Ref. 1)

Temp F t	Pressure PSIA p	Sp Vol of Liquid cu ft/lb v_f	Sp Vol of Vapor cu ft/lb v_g	Density of Vapor lb/cu ft $\frac{1}{v_g}$	Enthalpy Btu/Lb above -40 F			Entropy		Temp F t
					Liquid h_f	Vaporization h_{fg}	Vapor h_g	Liquid s_f	Vapor s_g	
−80	1.953	0.01493	41.08	0.02434	−13.888	198.64	184.75	−0.0351	0.4882	−80
−60	3.799	.01523	22.09	.04527	− 7.039	194.78	187.74	−0.0172	.4703	−60
−40	6.878	.01553	12.72	.07861	0.000	190.66	190.66	0.0000	.4544	−40
−30	9.036	.01568	9.873	.1013	3.562	188.52	192.08	0.0084	.4472	−30
−20	11.71	.01583	7.761	.1289	7.146	186.34	193.49	.0166	.4405	−20
−10	14.96	.01598	6.176	.1619	10.75	184.11	194.87	.0247	.4343	−10
− 8	15.69	.01601	5.908	.1693	11.48	183.66	195.14	.0263	.4331	− 8
− 6	16.45	.01604	5.654	.1769	12.20	183.21	195.42	.0279	.4319	− 6
− 4	17.24	.01607	5.413	.1847	12.93	182.76	195.69	.0295	.4307	− 4
− 2	18.05	.01610	5.185	.1929	13.66	182.30	195.96	.0311	.4296	− 2
0	18.90	0.01613	4.969	0.2013	14.39	181.85	196.23	0.0327	0.4284	0
2	19.77	.01616	4.763	.2100	15.12	181.39	196.51	.0343	.4273	2
4	20.68	.01619	4.568	.2189	15.85	180.93	196.78	.0359	.4262	4
5	21.15	.01622	4.471	.2237	16.21	180.70	196.92	.0367	.4257	5
6	21.62	.01625	4.379	.2284	16.58	180.47	197.05	.0375	.4251	6
8	22.59	.01628	4.206	.2378	17.31	180.01	197.31	.0390	.4240	8
10	23.60	.01631	4.038	.2477	18.04	179.53	197.58	.0406	.4229	10
12	24.64	.01634	3.878	.2579	18.77	179.06	197.83	.0422	.4218	12
14	25.72	.01637	3.726	.2684	19.51	178.58	198.09	.0437	.4208	14
16	26.83	.01640	3.581	.2792	20.25	178.10	198.34	.0453	.4198	16
18	27.97	.01644	3.443	.2904	20.98	177.61	198.59	.0468	.4187	18
20	29.16	.01647	3.312	.3019	21.73	177.11	198.84	.0484	.4177	20
22	30.38	0.01650	3.186	0.3138	22.47	176.61	199.08	0.0499	0.4166	22
24	31.64	.01654	3.067	.3261	23.21	176.11	199.32	.0514	.4156	24
26	32.95	.01658	2.952	.3388	23.95	175.61	199.56	.0530	.4146	26
28	34.29	.01662	2.843	.3517	24.70	175.10	199.79	.0545	.4136	28
30	35.68	.01665	2.739	.3650	25.44	174.59	200.03	.0560	.4126	30
32	37.11	.01669	2.640	.3787	26.18	174.08	200.26	.0575	.4117	32
34	38.58	.01673	2.546	.3928	26.93	173.56	200.49	.0590	.4107	34
36	40.09	.01677	2.455	.4073	27.67	173.05	200.72	.0605	.4098	36
38	41.65	.01681	2.369	.4222	28.42	172.53	200.95	.0621	.4088	38
40	43.25	.01684	2.286	.4375	29.17	172.00	201.17	.0636	.4079	40
42	44.91	0.01688	2.206	0.4532	29.92	171.48	201.40	0.0651	0.4070	42
44	46.61	.01692	2.130	.4694	30.67	170.95	201.62	.0665	.4061	44
46	48.35	.01696	2.057	.4861	31.42	170.42	201.84	.0680	.4052	46
48	50.15	.01700	1.987	.5033	32.17	169.89	202.06	.0695	.4043	48
50	51.99	.01704	1.920	.5208	32.93	169.35	202.28	.0710	.4034	50
52	53.88	.01708	1.856	.5388	33.68	168.81	202.49	.0725	.4025	52
54	55.83	.01712	1.794	.5573	34.44	168.27	202.71	.0740	.4017	54
56	57.83	.01716	1.735	.5763	35.19	167.72	202.91	.0754	.4008	56
58	59.88	.01720	1.679	.5958	35.95	167.18	203.13	.0769	.3999	58
60	62.00	.01724	1.624	.6158	36.71	166.62	203.33	.0784	.3991	60
64	66.39	0.01732	1.522	0.6572	38.23	165.51	203.74	0.0813	0.3974	64
68	71.01	.01740	1.427	.7008	39.76	164.39	204.15	.0842	.3958	68
72	75.86	.01748	1.339	.7467	41.29	163.24	204.53	.0870	.3941	72
76	80.94	.01756	1.258	.7948	42.82	162.08	204.90	.0899	.3925	76
80	86.26	.01764	1.183	.8451	44.36	160.91	205.27	.0928	.3910	80
84	91.82	.01773	1.114	.8979	45.90	159.72	205.62	.0956	.3894	84
86	94.70	.01778	1.081	.9253	46.67	159.13	205.80	.0970	.3887	86
88	97.64	.01782	1.049	0.9531	47.44	158.52	205.96	.0984	.3879	88
92	103.7	.01791	.9889	1.011	48.99	157.31	206.30	.1012	.3865	92
96	110.1	.01800	.9333	1.072	50.54	156.08	206.62	.1041	.3850	96
100	116.7	.01808	.8814	1.135	52.09	154.85	206.94	.1069	.3836	100
104	123.6	0.01818	.8331	1.200	53.65	153.60	207.25	0.1096	0.3822	104
108	130.8	.01828	.7884	1.268	55.22	152.33	207.55	.1124	.3808	108
112	138.3	.01838	.7466	1.339	56.78	151.06	207.84	.1151	.3794	112
120	154.2	.01859	.6710	1.490	59.93	148.46	208.39	.1206	.3768	120
130	175.9	.01887	.5889	1.698	63.89	145.13	209.02	.1274	.3736	130
140	199.6	.01915	.5189	1.927	67.87	141.71	209.58	.1341	.3705	140
150	225.4	.01945	.4586	2.181	71.87	138.23	210.10	.1407	.3674	150
160	253.5	.01978	.4070	2.457	75.90	134.66	210.56	.1473	.3646	160
170	283.9	.02015	.3613	2.768	79.97	130.96	210.93	.1538	.3618	170

Refrigerants and Refrigeration Systems

ful, but in quantities approaching 0.5 per cent by volume in air, serious effects may result if exposure is prolonged beyond a few minutes. Ammonia is rarely used with direct-expansion evaporators in the air ducts for comfort air conditioning and its use in this manner is usually prohibited by law.

Ammonia burns with difficulty, but can form explosive mixtures with air between mixture ratios of 16 to 25 per cent by volume. Such mixtures do not ignite easily, however. No harmful decomposition products are formed.

TABLE 15-9

RELATIVE TOXICITY AND SAFETY OF REFRIGERANTS
(References 9, 10, 11, and 12 at end of Chapter)

REFRIGERANT	TOXICITY NUMBER OF NATL. FIRE UNDERWRITERS	LETHIALITY KILLS OR SERIOUSLY INJURES		FLAMMABLE OR EXPLOSIVE % BY VOL	ASA-B9 CODE GROUP
		Duration of Exposure in Hours	Per Cent by Volume		
Ammonia (717)	2	0.5	0.5–0.6	16–25	2
Butane (600)	5	2	37.5	1.6–6.5	3
Carbon dioxide (744)	5	0.5 to 1	5–7*	Not	...
Ethane (170)	5	2	37	3.3–10.6	3
Methyl chloride (40)	4	2	2–2.5	8.1–17.2	2
Refrigerant-11	5	2	10	Not	1
Refrigerant-12	6	2	28.5–30.4	Not	1
Refrigerant-13	6	...	...	Not	1
Refrigerant-21	...	0.5	10.2	Not	1
Refrigerant-22	5A	...	...	Slightly	1
Refrigerant-113	4	1	4.8	Not	1
Refrigerant-114A	6	2	20.1	Not	1
Refrigerant-500	5A	2	19.4	Not	1
Sulfur dioxide (764)	1	0.08	0.7	Not	...

* Not toxic, but heavy concentrations upset the regulatory mechanisms of breathing and cause suffocation.

15-4. DICHLORODIFLUOROMETHANE (CCl_2F_2), REFRIGERANT-12

Refrigerant-12, commonly known as Freon-12 or genetron-12, is used extensively in comfort air-conditioning systems. Refrigerant-12 is one of the so-called halogenated-hydrocarbon refrigerants (halocarbons), and its chemical formula can be written as CCl_2F_2.

The pressure range is moderate, as can be seen in Table 15-5 and Fig. 15-4. The latent heat is low, 50 to 85 Btu per lb, so that the weight of "12" circulated per minute per ton of refrigeration is very much greater than with ammonia, although the volume handled is but little greater. This is not a serious disadvantage but necessitates designing for the proper piston displacement and valve capacity. In general, a system de-

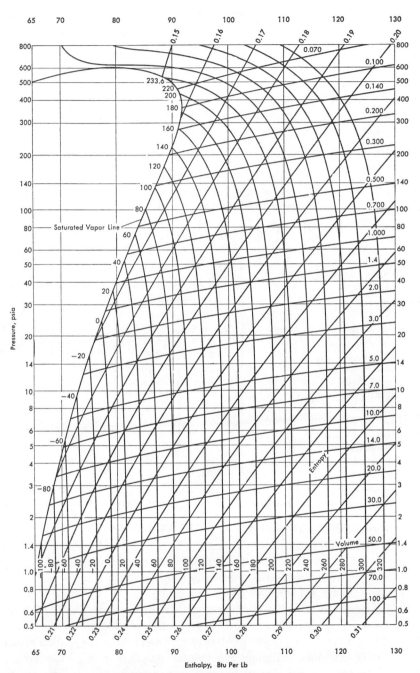

Fig. 15-4. Pressure-enthalpy chart for dichlorodifluoromethane (refrigerant-12) in vapor region.

Refrigerants and Refrigeration Systems 535

signed for one refrigerant will not work well with another without modification.

Refrigerant-12 is chemically stable and has practically no corrosive effect on the ordinary metals unless contaminated by impurities of which water is one. This refrigerant is noncombustible, although in the presence of open flames or very hot surfaces it breaks down and forms toxic gases. The vapor itself is almost perfectly nontoxic, even in concentrations above 20 per cent by volume in air; apparently the only difficulty is that of reducing the amount of oxygen. It has only a very slight odor.

Pipe joints for all of the halocarbons must be carefully made to prevent leakage of gas. Copper tubing with sweated joints is customary in small installations. Mineral oil of selected grade and free of water is used. Some difficulty may occur in the lubrication systems, since oil and refrigerant are mutually soluble in each other. Care must be exercised to see that too much oil from the compressor is not carried over into the evaporator, endangering compressor lubrication as well as reducing heat transfer in the evaporator. Rubber gasket material is inadvisable with most of the halogenated-hydrocarbon refrigerants, but various synthetic gasket compositions (neoprene) are satisfactory in many cases.

Water must be completely removed from refrigerant systems, with the exception of those using ammonia, and the systems must be kept continuously dry to prevent hydrolysis, as well as freezing and clogging at the expansion valves. Various types of dehydrators, using such materials as silica gel, activated alumina, drierite (calcium sulfate), etc., can be attached in systems either temporarily or permanently for removing water. In many cases, before starting, a high vacuum is held on the warm system to evaporate (dry out) any moisture in the piping.

15–5. MONOCHLORODIFLUOROMETHANE ($CCIHF_2$), REFRIGERANT-22

Refrigerant-22, also known as Freon-22 and as genetron-22 (formerly genetron-141), is in extensive use for reciprocating compressors. Like refrigerant-12, it is chemically stable, almost odorless and non-irritating, and exhibits no permanent deleterious effects in concentrations up to 18 per cent by volume for exposures of less than 2-hour duration (Table 15–9).

A study of the thermodynamic properties of refrigerant-22 in Table 15–6 and Plate V indicates that "22" is particularly suitable for use in the low-temperature field (-40 to -100 F) because its pressure is higher than that of ammonia, refrigerant-12, and, in fact, of most refrigerants with the exception of CO_2 and some of the hydrocarbons. The specific volume is moderate at low pressures, and, with the relatively good latent heat which this refrigerant possesses, it can be seen that reasonable piston displacements per ton are possible even at low-suction temperatures.

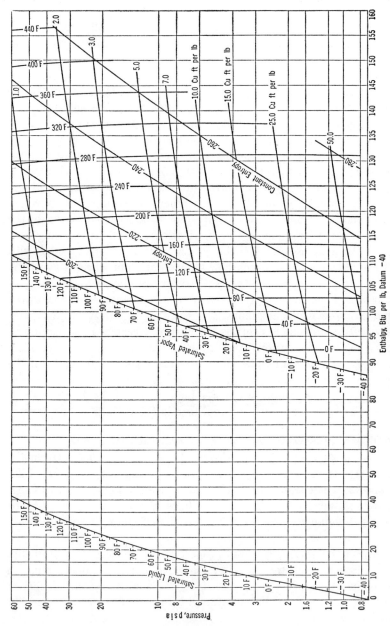

Fig. 15-5. Pressure-enthalpy chart for refrigerant-11, trichloromonofluoromethane (Freon-11). (Data source, Kinetic Chemicals, Inc.)

Refrigerants and Refrigeration Systems   537

This refrigerant is also being used extensively in the moderate temperature (air-conditioning) range because its low specific volume makes it possible, with a compressor of given physical size (piston displacement), to serve a greater tonnage load than is possible with refrigerant-12. It is practically noninflammable and nonexplosive. The specific heat of the vapor C_p, in Btu per pound degree Fahrenheit, is 0.138 at -30 F, 0.143 at -10 F, 0.152 at 40 F, and 0.160 at 100 F. The value of C_p/C_v is 1.16 at 86 F. It has a critical temperature of 205 F at 716 psia.

15-6. TRICHLOROMONOFLUOROMETHANE (CCl_3F), REFRIGERANT-11

Refrigerant-11, also known as Freon-11 and Carrene No. 2, is a so-called vacuum refrigerant; that is, refrigeration temperatures in the evaporator can be obtained only at sub-atmospheric pressures (Table 15-7 and Fig. 15-5).

Refrigerant-11 is practically odorless and relatively nontoxic, even in high concentrations, provided the exposure is not continued over very long periods. It is nonexplosive and practically noninflammable, but in the presence of flames and very hot surfaces, forms toxic decomposition products. There is no corrosive action on the common metals of construction, but rubber gaskets are attacked.

The specific volume of a vacuum refrigerant is usually so great that the large volume of gas can be handled effectively only by high-speed centrifugal compressors. For very large vapor volumes the physical dimensions required for reciprocating compressors become prohibitive. With the centrifugal compressors ordinarily used, the range of pressure is small, usually less than 30 psi.

15-7. MISCELLANEOUS HALOCARBON REFRIGERANTS

Methyl chloride, (CH_3Cl) refrigerant-40, formerly used extensively in small air-conditioning units, now has its greatest application in small commercial (storage-compartment) units. The pressure range for methyl chloride (Table 15-8 and Fig. 15-6) is moderate. Latent heat of evaporation is about 176 Btu per pound. The refrigerant is stable and relatively noncorrosive to the common metals of construction except for aluminum.

Methyl chloride is moderately inflammable and can be explosive between limits of 8.1 and 17.2 per cent by volume in air. As a halocarbon it forms toxic decomposition products in the presence of open flames and very hot surfaces.

Methyl chloride is a sweet-smelling vapor that is not seriously offensive, and so should have odorous agents added to it to give active warning of its presence. It is mildly anesthetic. It is somewhat toxic and, in concentrations over about 2 per cent, may be dangerous if the exposure

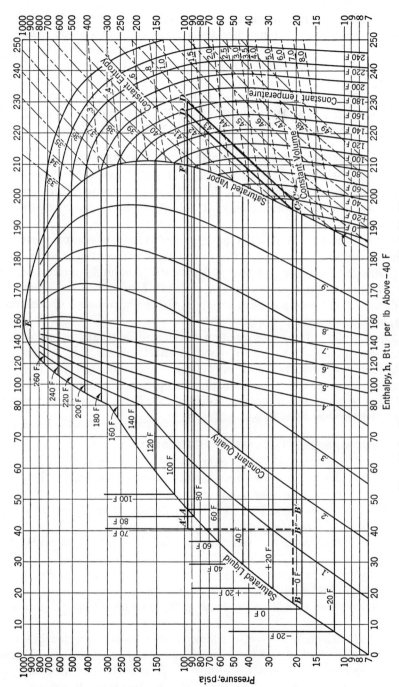

FIG. 15-6. Pressure-enthalpy diagram for methyl chloride.

Refrigerants and Refrigeration Systems 539

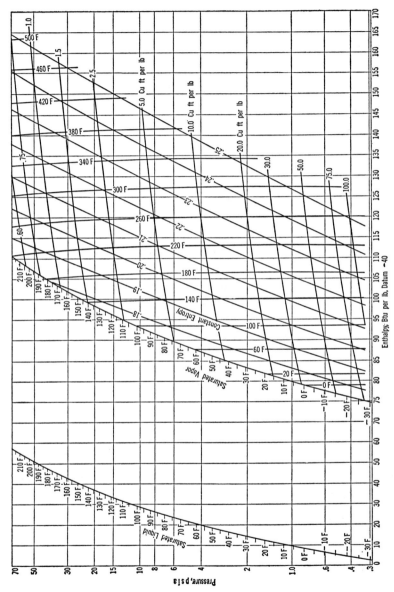

Fig. 15-7. Pressure-enthalpy chart for refrigerant-113 (Freon-113). (Data source, Kinetic Chemicals, Inc.)

is long in duration. For short periods, apparently no serious consequences arise unless the concentrations are high. Acrolein is often added as a warning agent.

Dichloromonofluoromethane ($CHCl_2F$), refrigerant-21, also called Freon-21, is a vacuum-type refrigerant. It exerts 5.24 psia pressure at 5 F and 31.23 psia at 86 F. It is somewhat less toxic than methyl chloride and is practically noninflammable and nonexplosive.

Dichlorotetrafluoroethane ($C_2Cl_2F_4$), refrigerant-114a and -114, is a low-pressure refrigerant of the halogenated-hydrocarbon class. The particular isomer of dichlorotetrafluoroethane, called 114a, has a molecular arrangement CCl_2FCF_3, and for this the boiling point at 14.7 psia is 37.6 F; at 5 and 86 F the respective saturation pressures are 7.03 psia and 22.7 psia. The specific volume of saturated vapor at 5 F is 4.04 cu ft per lb, and the ratio of specific heats (k) is 1.01. Refrigerant-114a is characterized by low toxicity, approaching that of refrigerant-12. Moreover, for a vacuum refrigerant, the required compressor displacement per ton is very moderate, being only 18.8 cfm at 5 F evaporator temperature.

Trichlorotrifluoroethane ($C_2Cl_3F_3$), refrigerant-113, also known as Freon-113, is a vacuum-type refrigerant. For its general characteristics see Fig. 15-7.

Methylene chloride (CH_2Cl_2), refrigerant-30, also known as Carrene No. 1 or dichloromethane, is another vacuum refrigerant (Fig. 15-1) suitable for centrifugal compressors. It has a mild odor similar to that of chloroform. It is nonexplosive and relatively noninflammable, although, like the other halogenated hydrocarbons, it breaks down in the presence of flames and very hot surfaces to form toxic products. It is toxic in concentrations of over 5 per cent by volume.

15-8. MISCELLANEOUS REFRIGERANTS

Carbon dioxide (CO_2), refrigerant-744, although formerly used in some air-conditioning installations, has now been almost completely supplanted by lower pressure refrigerants. Carbon dioxide is noncorrosive and inert, as well as relatively nontoxic, odorless, and nonirritating. However, in high concentrations, over 6 per cent by volume, much discomfort is experienced and loss of consciousness and ultimately death can result if the person exposed is not moved into fresh air. Carbon dioxide is a very high-pressure refrigerant (Fig. 14-3 and Table 15-7). Thus all equipment must be made extra heavy. Lubrication is simple and CO_2 is very suitable for operation at temperatures down to about -60 F. Solidification occurs at -69.88 F.

Sulfur dioxide (SO_2), refrigerant-764, which has a very irritating acrid vapor, was used formerly in domestic refrigerators. It is a moderately

Refrigerants and Refrigeration Systems

low-pressure refrigerant (Table 15-7) and must be kept dry in use because it forms an acid in the presence of water.

Hydrocarbon refrigerants are all very inflammable and explosive. They are in general not very toxic. They are somewhat soluble in lubricating oil. Different hydrocarbons can be selected to work in desired pressure and temperature ranges. Among those used are:

Butane (C_4H_{10}), refrigerant-600, has a moderately low pressure and exerts 13.1 psia pressure at 5 F and 59.5 psia at 86 F.

Propane (C_3H_8), refrigerant-290, is in an intermediate pressure range, 42.1 psia at 5 F and 185.3 psia at 86 F.

Ethane (C_2H_6), refrigerant-170, lies in a high-pressure range, about 240 psia at 5 F and 676 psia at 86 F.

Ethylene (C_2H_4), refrigerant-1150, has been used to some extent for very low-temperature work as in gas liquefaction and separation. It boils at -154.7 at 14.7 psia and has a critical temperature of 49.3 F at 743 psia.

Water-steam can be used as a refrigerant in air-conditioning applications. Low pressures (very high vacuums) are required (see Table 2-2), but these can be rather easily obtained with centrifugal or steam-jet (thermo-jet) compressors. Water may be considered the safest refrigerant of all.

Azeotrope refrigerants. An azeotrope is a mixture of two or more chemicals which maintain the same ratio of constituent chemicals in both the liquid and vapor phase. For example, the constituents of an azeotropic mixture cannot be separated by distillation. Several refrigerant azeotropes have been proposed, but only one, refrigerant-500, has been used extensively. Refrigerant-500, also known as Carrene-7, is an azeotropic mixture of 73.8 per cent of refrigerant-12 (CCl_2F_2) and 26.2 per cent of refrigerant-152a (CH_3CHF_2). This refrigerant has a saturation pressure of 31.1 psia at 5 F and a specific volume of 1.501 cu ft per lb for the vapor. At 86 F the saturation pressure is 127.6 psia. Although other azeotropes have been proposed, none of them has found wide commercial acceptance.

15-9. TOXICITY OF REFRIGERANTS

All gaseous substances, with the possible exception of air, are to a certain extent toxic. There are various degrees or levels of toxicity, since some substances produce toxic effects and danger to life merely because they exclude or reduce the amount of oxygen necessary for living processes, while others are truly poisonous. Carbon dioxide, for example, is sometimes considered harmless (since it is exhaled every time a person breathes) in amounts ranging from 3.5 to 4 per cent by volume. Nevertheless, if a person enters an atmosphere containing 6 per cent CO_2, breathing becomes difficult, and with 10 per cent CO_2 in air, loss of consciousness can quickly result. The danger in this case does not result from the fact that oxygen is excluded, but from the fact that the regulating mechanism in the respiratory passages becomes disturbed and may cease to function. However, refrigerant-12 (Freon-12), which is remarkably nontoxic, can be breathed safely in concentrations up to 20 per cent by volume without seriously upsetting normal life processes. At the toxic end of the scale are gases, such as sulphur dioxide, which are so irritating to the mucosa that, even in concentrations for less than 1 per cent by volume, they cause surface damage. This condition is also true with ammonia if its concentration exceeds the amount which can readily be diluted by surface moisture. Still other gases, when breathed, are absorbed into the system and can cause damage to internal organs. A case in point is the manner in which protracted exposures to carbon tetrachloride cause kidney damage.

Toxicity is related (1) to the nature of the material, (2) to the relative amount of the chemical in the air, and (3) to the length of time during which exposure (breathing of the vapor) takes place. For many years The National Board of Fire Underwriters has conducted a program, using animals for test, to measure the relative toxicity of refrigerants. The Board has set up a classification scheme with numbers from 1 to 6, with the refrigerants in classification 1 being extremely toxic or dangerous and ranging up to refrigerant classification 6, where toxicity is essentially absent. In Table 15–9 the common refrigerants are tabulated and their toxicity noted.

In addition to possible danger from the refrigerant itself, a second type of hazard arises in case of fire, because certain refrigerants can burn or are explosive, and other refrigerants, in the presence of an open fire or of a hot incandescent surface, break down and can form poisonous by-products. The halogenated hydrocarbons are particularly dangerous; they readily form toxic combustion products in the presence of flames because of the chlorine and fluorine they contain. The B-9.1 Safety Code of ASHRAE and the American Standards Association classifies the hazards of refrigerants under three numbers:

1. Flammable vapor-air mixtures, only dangerous within narrow limits of concentration.
2. The refrigerant vapor is under sufficiently high pressure that it tends to extinguish flame.
3. Because of its offensive odor or because of performance failure, the loss of refrigerant is usually known before a combustible concentration occurs.

Refrigerants and Refrigeration Systems

15-10. REFRIGERANT LEAK DETECTION AND GAS PURGERS

In spite of the care with which refrigerant systems are fabricated, there is always some possibility that leakage may occur. Leakage of refrigerant is expensive and can also be objectionable, particularly in the case of refrigerants with unpleasant odor and those toxic in nature. Various methods have been developed to determine points of leakage. Ammonia leaks can often be found by smell or by burning sulfur candles or wicks, which generate a dense cloud of white smoke in the presence of ammonia vapor. The halogenated-hydrocarbon refrigerants can be detected by the halide torch. In this torch an alcohol flame burns in the presence of copper and, when air with traces of these gases passes through the flame, the flame takes on a decidedly greenish color. There is a search tube, or hose, which can be moved around the suspected leak points; the combustion process aspirates air through the tube and into the flame, the appearance of which can then be observed. An aqua-ammonia swab gives a dense white smoke in the presence of SO_2 gas. In high concentrations of any refrigerant gas, except refrigerant-12, an appropriate gas mask should always be worn.

The reverse problem also exists, since it is possible for extraneous noncondensable gases to intermix with the refrigerant inside a refrigeration system. In the case of vacuum systems, whenever there are leaks in the piping, air, which contains moisture, enters the system to cause trouble. Noncondensable gases also appear in any closed system whenever there is breakdown of the refrigerant or lubricant as a result of high compression temperatures or from chemical reaction. Chemical reaction can take place when unwarranted moisture in a system promotes hydrolysis and subsequent gas formation. Air is also frequently present because of improper removal when the system is started. Noncondensable gases most frequently collect in the condenser, and their presence is indicated whenever the condenser pressure is found to be far out of agreement with the pressure which, the temperature of the cooling water or circulating air shows, should be indicated. When this is the case, the noncondensable gases must be purged out, usually with some loss of refrigerant. However, excessive condenser pressures are not always caused by noncondensable gases, since dirty tubes, which interfere with effective cooling, cause high pressures, and an overloaded condenser, fed with more refrigerant than that for which it was designed, can also show high pressures.

A purger of proprietary design for exhausting noncondensable gases with minimum loss of refrigerant is illustrated in Fig. 15-8. This uses refrigerant from the main system, fed into a coil at A and withdrawn at B, to cool the gas-laden vapor in chamber C. This cooling condenses out refrigerant which drains back to the receiver, while the chamber C holds the gases at high pressure. These gases can be discharged to waste. In operation, refrigerant feeding through the coil in chamber C chills this chamber, and liquid refrigerant is drawn into the chamber through E. After chamber C is filled with liquid, valve F is opened slightly and gas and refrigerant vapor pass into the space above the liquid refrigerant. If noncondensable gas is present, it collects in chamber C above the liquid, the level of which falls as the gas flows in. If no gas is present the refrigerant vapors merely condense in space C and the liquid level remains constant. When C does fill with gas, the waste valve D is cracked and the gas is discharged to waste. At the same time, the liquid level should

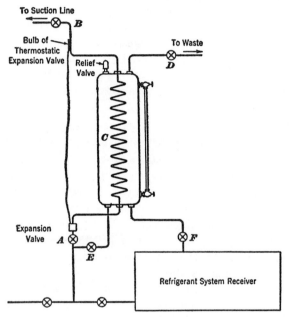

Fig. 15-8. Noncondensable gas purger for refrigeration system.

be observed through the gage glass to prevent loss of refrigerant. This level will rise as the gas is forced out. When no more gas collects in C, the refrigerant feed at A is closed. Chamber C then warms and the refrigerant drains gack through E. All valves on the purger can then be closed completely, isolating it from the system. This type of purger must always be placed higher than the receiver or condenser.

15-11. LOW-TEMPERATURE MULTISTAGE SYSTEMS

When the required temperature range between evaporator and condenser is very great, much over 100 degrees, it becomes desirable to employ multistage operation. There are many reasons why this is true, one of which is the fact that for excessive temperature lifts, the pressure ratio of compression becomes too large to handle effectively in one stage. For example, an ammonia system operating between 5 and 86 F has a ratio of $169.2/35.09 = 4.8$, whereas if the operation is from -40 to 86 F, the ratio is $169.2/10.41 = 16.2$. The latter ratio is not suitable for single-stage operation since the temperature of the gas after compression would exceed 300 F and be subject to breakdown, the volumetric efficiency (with reciprocating machinery) would be greatly reduced, and the flash gas formed in the evaporator would be excessive. If two-stage operation is considered, the pressure ratio of compression need not necessarily exceed the square root of the over-all pressure ratio. For example, for the -40 to 86 F range, the ratio value of ammonia might be $\sqrt{16.2} = 4.0$ per stage. Of course, the ratio need not be the same per stage, and it will vary in relation to the comparative sizes of the low- and high-pressure compressors. For very low temperatures (as low as -160 to -220 F), such as may be met in industrial refrigeration, three or more stages and even multiple refrigerants (in cascade) may be required. In general, pressure-ratio lifts of 6 to 8 and higher would indicate the desirability of multistaging.

Figure 15–9 is a diagrammatic layout of a two-stage system. Here, vapor from the evaporator is drawn into the low- or first-stage compressor; passes through an intercooler, using water if the compressed gas is adequately hot to be subject to water cooling; and then enters the interstage chamber. In the latter it is cooled essentially to saturation temperature at the interstage (intermediate) pressure on coming into contact with liquid refrigerant vaporizing at interstage pressure. Vapor from the interstage chamber is drawn in by the high- or second-stage compressor which raises the refrigerant to the condenser (head) pressure of the system. After condensing, refrigerant is stored in the receiver from which it may pass through a cooling coil and thence to the expansion valve E feeding the evaporator. A cooling-coil arrangement is here shown with a float-feed expansion valve F bringing a portion of the high-pressure liquid to interstage (intermediate) pressure, where in vaporizing, it cools the evaporator liquid supply and also cools the vapor entering the high-pressure compressor. By closing valve 1 and opening valve 2, an alternate arrangement is shown in which all the high-pressure liquid drops to intermediate pressure in float valve F and passes into the chamber where, cooled to saturation temperature at this pressure, the remaining liquid can then pass out through valve 2. In either arrangement, the flash vapor resulting from cooling the supply refrigerant to its temperature at E is removed at the intermediate pressure and does not have to expand down to the low pressure of the system. Both arrangements are employed and both are astisfactory. With the coil arrangement in the

546 Refrigerants and Refrigeration Systems

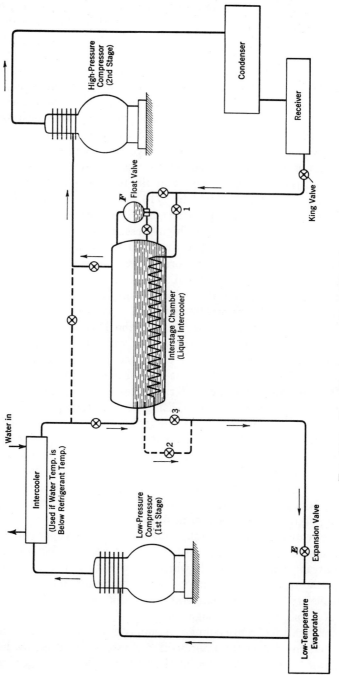

Fig. 15–9. Two-stage compression refrigeration system.

Refrigerants and Refrigeration Systems

chamber, heat-transfer temperature gradients prevent the refrigerant from quite reaching saturation temperature at interstage pressure and only one main expansion valve, E, is required, which operates at essentially fixed condenser pressure. An automatic feed, F, may also be required by both systems. The open-flash chamber system is particularly applicable for systems operating under steady load (temperature) conditions.

For given suction and condenser pressures, the interstage pressure is fixed by the volumetric capacities of the low- and high-pressure compressors and will automatically adjust to a value fixed by these capacities. Although it might seem desirable to adjust the loading equally between stages, computation will show that intermediate pressures slightly above and below the optimum do not seriously detract from the over-all efficiency of a system. The pressure compression ratio should not be excessive for either stage. By observing the amount of the flash and intercooling vapor from the interstage chamber, it can be noted that the high-pressure compressor handles more weight of refrigerant than the low-stage compressor.

Example 15–1. A two-stage refrigeration system employing refrigerant-12 (Freon-12) operates with an evaporator temperature of -70 F and delivers dry saturated vapor to the low-stage compressor which raises the pressure to the interstage pressure of 19.2 psia. A water-cooled intercooler is not feasible, but in the interstage chamber, the compressed gas is cooled to within 5 deg of saturation. The second-stage compressor delivers the vapor to the condenser, which operates at a condensing pressure of 93.00 psia, corresponding to a condensing temperature of 76 F. The liquid from the receiver, with no subcooling, enters the interstage chamber, and in passing through a cooling coil, drops to within 10 deg of the liquid refrigerant temperature and then passes to the expansion valve. Assume isentropic work of compression in both compressors and volumetric efficiencies of 86 per cent in the low-pressure compressor and 84 per cent in the high-pressure compressor, and neglect heat losses. Find (a) the refrigerating effect of each pound of refrigerant supplied to the evaporator and (b) the pounds of refrigerant circulated per minute in the low-pressure evaporator per ton of refrigeration. (c) Find the total amount of cooling required in the interstage chamber and the pounds of refrigerant used there per pound of low-pressure refrigerant.

Find the total weight of refrigerant circulated in the high-pressure compressor (d) per pound of refrigerant in the evaporator and (e) per ton of refrigeration in the evaporator. (f) Find the isentropic work in the second stage per ton of refrigeration in the evaporator. (g) Find the total isentropic work in both stages per ton. (h) Find the actual piston displacement in cfm required in the low and high stages per ton of refrigeration. (i) For a three-cylinder, single-acting, 4-in. by 4-in. stroke, high-pressure compressor running at 400 rpm, compute the low-temperature tonnage developed when the low-temperature compressor has its required displacement and compute this displacement. (j) Estimate the horsepower of the unit for this tonnage. (k) Draw a ph diagram for the two-stage system.

Solution: (a) From the evaporator, vapor that is assumed dry and saturated at -70 F enters the low-stage compressor with an h of 69.58 Btu per lb (point C on Fig. 15–10), an entropy of 0.1792 (say 0.180), and a specific volume of 8.57 cu ft per lb at a pressure of 3.96 psia. The evaporator is supplied with liquid subcooled to 0 F, i.e., within 10 deg of the saturation temperature of -10 F in the interstage cooler. For 0 F, $h_f = 8.52$ Btu per lb (A on Fig. 15–10).

$$q_R = (69.58 - 8.52) = 61.06 \text{ Btu per lb of refrigerant supplied to low-pressure evaporator} \quad Ans.$$

b) $\dfrac{200}{61.1} = 3.27$ lb of refrigerant per min circulated in low-pressure evaporator per ton of refrigeration *Ans.*

c) Work of compression, isentropic, in low stage, starts at 3.96 psia with an h of 69.58 Btu per lb and $s = 0.180$. Using this value of entropy, Fig. 15–4 at 19.2 psia shows that the h after compression is 80.9 Btu per lb (D on Fig. 15–10) and its temperature is 19 F. In the interstage chamber this gas, on coming into contact with the

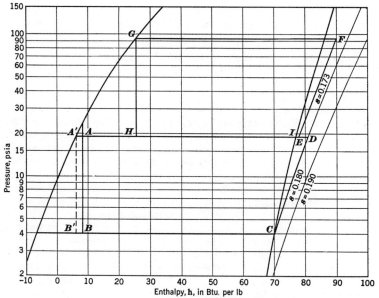

FIG. 15–10. A ph chart with refrigerant-12 (Freon-12) for two-stage compression system.

liquid at 19.2 psia, is brought to -5 F, i.e., within 5 deg of the saturation temperature of -10 F, and its enthalpy is 78.3 Btu per lb read on Fig. 15–4 (point E on Fig. 15–10).

$$80.9 - 78.3 = 2.6 \text{ Btu required to cool compressed vapor in interstage chamber}$$

Liquid from the receiver at 76 F with $h_f = 25.44$ (point G on Fig. 15–10) is subcooled to 0 F (h = 8.52), and for each pound cooled $25.44 - 8.52 = 16.92$ Btu act to vaporize liquid in the chamber. For each pound of refrigerant supplied to the evaporator, $2.6 + 16.92 = 19.5$ Btu must be removed by vaporizing (and superheating) interstage refrigerant. *Ans.*

Refrigerants and Refrigeration Systems

This heat removal requires the vaporization of

$$\frac{19.5}{h_E - h_f} = \frac{19.5}{78.3 - 25.44} = \frac{19.5}{52.89} = 0.37 \text{ lb of refrigerant in the interstage chamber} \qquad Ans.$$

Here 52.89 is the refrigerating effect of each pound of interstage refrigerant as the interstage refrigerant changes from point H to E on Fig. 15–10.

d) The total weight of refrigerant in the high-pressure compressor for each pound of refrigerant in the evaporator is $1 + 0.37 = 1.37$ lb. *Ans.*

e) Per ton of low-temperature (-70 F) refrigeration, $(3.27)(1.37) = 4.48$ lb of refrigerant are used in the high-pressure compressor. *Ans.*

f) The isentropic work of the second stage, at a starting entropy $s = 0.173$ at 19.2 psi and -5 F with $h = 78.3$ (point E on Fig. 15–10), is $89.9 - 78.3 = 11.6$ Btu per lb, where 89.9 is the enthalpy after isentropic compression, read from Fig. 15–4 (point F on Fig. 15–10).

Second-stage work = $(11.6)(1.37) = 15.9$ Btu per lb of refrigerant in evaporator

Second-stage work = $(11.6)(4.48) = 52.0$ Btu per min per ton of low-temperature refrigeration *Ans.*

g) First-stage work from part (c) is 80.9 Btu per lb less 69.58, or 11.32 Btu per lb; and per ton by part (b)

$(3.27)(80.9 - 69.58) = 37.0$
Second-stage work from part (f) = 52.0
Total work = 89.0 Btu per min per ton of refrigeration in evaporator *Ans.*

h) Specific volume of low-pressure refrigerant at -70 F, dry and saturated, is 8.57 cu ft per lb (Table 15–5). With a volumetric efficiency of 86 per cent and 3.27 lb per min per ton

$$\frac{3.27 \times 8.57}{0.86} = 32.6 \text{ cfm actual piston displacement of low-stage compressor per ton} \qquad Ans.$$

Specific volume of intermediate-pressure refrigerant at 19.2 psia is 1.973 cu ft per lb (estimated from Table 15–5 and Fig. 15–4). With 84 per cent volumetric efficiency and 4.48 lb of refrigerant per ton from part (e)

$$\frac{4.48 \times 1.973}{0.84} = 10.53 \text{ cfm actual required piston displacement of high-stage compressor per ton of refrigeration in low-temperature evaporator} \qquad Ans.$$

i) Actual piston displacement of high-pressure compressor is

$$\frac{3\pi(4)^2 (4.25)(400)}{4(1728)} = 37.1 \text{ cfm}$$

Find the tonnage of the unit by dividing the 37.1 cfm by the cfm required by high-stage unit per ton from (h). Therefore,

$$\frac{37.1}{10.53} = 3.53 \text{ tonnage of two-stage unit} \qquad Ans.$$

To develop this tonnage the necessary low-pressure displacement, using 32.6 cfm per ton from part (h), is

$$(32.6)(3.53) = 115.1 \text{ cfm} \qquad Ans.$$

j) Assume that a 20 per cent increase in isentropic hp gives the ihp, and assume a mechanical efficiency of 85 per cent. Using data from part (g) and 3.53 tons from part (i),

$$\frac{1.20}{0.85}(3.53)(37.0)\left(\frac{1}{42.4}\right) = 4.35 \text{ hp, low-stage motor}$$

$$\frac{1.20}{0.85}(3.53)(52.0)\left(\frac{1}{42.4}\right) = 6.11 \text{ hp, high-stage motor}$$

The 42.4 is the Btu per minute in 1 hp.

k) Figure 15-10 is the ph chart for the two-stage system. It is drawn on the basis of 1 lb of refrigerant throughout, but it must be realized that for each pound of refrigerant in the low-pressure system $ABCDE$, there are 1.37 lb of refrigerant in the cycle $EFGH$, with only 0.37 lb vaporizing from H to I and superheating to E on mixing with the vapor from D. One pound of refrigerant moves from H to A. Had the refrigerant from G been flashed directly into the interstage chamber without the use of a coil, the refrigerant supply to the low stage would have been at condition A'.

PROBLEMS

15-1. A certain refrigeration system using dichlorodifluoromethane, refrigerant-12 (Freon-12), condenses refrigerant at 100 F (131.7 psia), and the evaporator refrigerant temperature is 36 F (48.12 psia). The liquid at the expansion valve is not subcooled, and dry saturated vapor enters the compressor. Find (a) the enthalpy of the liquid at the expansion-valve inlet; (b) the useful refrigeration in the evaporator, in Btu per pound; (c) the isentropic work of compression, in Btu per pound; (d) the probable shaft horsepower required by this compressor per ton of refrigeration; and (e) the probable condenser heat load per ton of refrigeration.

Ans. (a) 31.1 Btu/lb; (b) 49.9 Btu/lb; (c) 7.9 Btu/lb; (d) 1.05 shp; (e) 14,700 Btuh

15-2. What size of four-cylinder, 1140 rpm, single-acting compressor is required in order to develop 12 tons of refrigeration when the compressor is operating under the conditions of problem 15-1? Consider the bore-stroke ratio as unity, and the volumetric efficiency as 87 per cent. Round off the answer to the nearest larger sixteenth inch.

Ans. $2\frac{13}{16}''$ by $2\frac{13}{16}''$

15-3. Work problem 15-1, but assume that the operating conditions are 86 F condensing temperature and 0 F evaporating temperature.

Ans. (a) 27.8 Btu/lb; (b) 49.5 Btu/lb; (c) 11.5 Btu/lb; (d) 1.53 shp; (e) 15,900 Btuh

15-4. Rework problem 15-2 under the operating conditions of problem 15-3.

Ans. $3\frac{1}{2}''$ by $3\frac{1}{2}''$

15-5. A manufacturer makes a four-cylinder, single-acting compressor with 3-in. bore and $4\frac{1}{4}$-in. stroke, which runs at 800 rpm. Under certain operating conditions this unit runs with a 40 F evaporating temperature and a 98 F condensing temperature. Using refrigerant-12 as a refrigerant, compute (a) for dry compression and no liquid subcooling, the pounds of refrigerant circulated per minute per ton of refrigeration; (b) the piston displacement; (c) the refrigerating capacity of the machine in tons, and also in units of 1000 Btu (Mbh), using a volumetric efficiency

Refrigerants and Refrigeration Systems

of 87 per cent; (d) the isentropic horsepower required to drive the compressor under full capacity. (e) Specify a motor size for this unit.

Ans. (a) 3.93 lb/ton-min; (b) 55.6 cfm; (c) 15.9 tons, 190.9 Mbh; (d) 10.4 hp; (e) 15 hp, next standard

15–6. Work problem 15–5, but assume that the evaporating temperature is 20 F instead of 40 F.

Ans. (a) 4.1 lb/ton-min; (b) 55.6 cfm; (c) 10.7 tons, 129 Mbh; (d) 10.2 hp, 14.3 shp; (e) 15 hp, standard

15–7. A refrigeration system using monochlorodifluoromethane (refrigerant-22) condenses refrigerant at 80 F (159.7 psia), and the evaporator refrigerant temperature is 30 F. The liquid at the expansion valve is not subcooled, and dry saturated vapor enters the compressor. Find (a) the enthalpy of the liquid at the expansion-valve inlet; (b) the useful refrigeration in the evaporator, in Btu per pound; (c) the isentropic work of compression, in Btu per pound; (d) the probable shaft horsepower required by the compressor per ton of refrigeration; and (e) the probable condenser heat load per ton of refrigeration.

Ans. (a) 34.6 Btu/lb; (b) 73.6 Btu/lb; (c) 9.0 Btu/lb; (d) 0.81 shp; (e) 14,000 Btuh

15–8. What size of four-cylinder, 1140-rpm, single-acting compressor is required in order to develop 12 tons of refrigeration when the compressor is operating under the conditions of problem 15–7? Consider the bore-stroke ratio as unity and the volumetric efficiency as 87 per cent. Round off the answer to the nearest larger sixteenth inch. *Ans.* $2\frac{7}{16}''$

15–9. Rework example 15–1 assuming a -60 F evaporator and no other change in the data.

Ans. (a) 62.17; (b) 3.22; (c) 19.5, 0.379; (d) 1.379; (e) 4.44; (f) 56.4 Btu/min; (g) 84.7 Btu/min; (h) 24.3 cfm, 10.4 cfm; (i) 3.57 tons, 86.7 cfm; (j) 3.3, 6.6 hp

15–10. A two-stage refrigeration system using monochlorodifluoromethane (refrigerant-22) operates with a -90 F evaporator with a condensing temperature of 80 F. The intermediate (interstage) pressure is 24.9 psia. The gas from the low-pressure compressor is cooled to -10 F in the interstage chamber. The liquid from the receiver, with no subcooling, flashes directly into the interstage chamber and cools to -20 F and then passes through the main expansion valve. Assume no heat losses, isentropic work of compression, and volumetric efficiencies of 85 and 82 per cent in the low- and high-stage compressors, respectively. Find (a) the refrigerating effect of each pound of refrigerant supplied to the evaporator and (b) the pound of refrigerant circulated per minute in the low-pressure evaporator per ton of refrigeration. (c) Find the total amount of cooling required in the interstage chamber and the pound of refrigerant used there per pound of low-pressure refrigerant. Find the total weight of refrigerant circulated in the high-pressure compressor (d) per pound of refrigerant in the evaporator and (e) per ton of refrigeration in the evaporator. Find (f) the isentropic work in the second stage per ton of refrigeration in the evaporator, (g) the total isentropic work in both stages per ton, and (h) the actual piston displacement in cfm required in the low and high stages per ton of refrigeration. (i) Compute the theoretical horsepower per ton of refrigeration.

REFERENCES

1. R. & H. Chemicals Department, E. I. du Pont de Nemours & Co., *Methyl Chloride*, 5th ed.
2. U. S. Bureau of Standards, *Tables of Thermodynamic Properties of Ammonia*, Circular No. 142.

3. American Society of Refrigerating Engineers, *Air Conditioning Refrigerating Data Book*, 10th ed., 1957–58 (1957).

4. Kinetic Chemicals, Division of E. I. du Pont de Nemours & Co., *Thermodynamic Properties of Dichlorodifluoromethane "Freon-12,"* (1955–56).

5. Kinetic Chemicals, Division of E. I. du Pont de Nemours & Co., *Thermodynamic Properties of Trichloromonofluoromethane "Freon-11" At Low Pressures,* (1942).

6. Allied Chemicals, Division of General Chemicals Co., *Thermodynamic Properties of Monochlorodifluoromethane "genetron-22,"* (1958).

7. C. H. Meyers and M. S. Van Dusen, *The Vapor Pressure of Liquid and Solid Carbon Dioxide,* Bureau of Standards Research Paper No. 538 (March, 1933).

8. Jennings and Shannon, *Tables of Properties of Aqua-Ammonia Solutions,* Lehigh University Studies, No. 1 (July, 1938).

9. National Board of Fire Underwriters' Laboratories, *The Comparative Life, Fire and Explosion Hazards of Common Refrigerants,* (1933).

10. National Board of Fire Underwriters' Laboratories, *The Comparative Life, Fire and Explosion Hazards of Monochlorodifluoromethane "Freon-22,"* (1940).

11. R. J. Thompson, "Properties and Characteristics of Refrigerants," *Refrigerating Engineering,* Vol. 44 (November, 1942), pp. 311–18.

12. American Standard Association, "B9.1-1964 Safety Code for Mechanical Refrigeration." Also known as ASHRAE Standard 15-63.

13. American Society of Heating, Refrigerating, and Air-Conditioning Engineers, *Handbook of Fundamentals* (1967), Chapters 15 and 20.

16

Refrigeration Equipment and Arrangement

16-1. EVAPORATORS

Refrigeration evaporators are classified in regard to the way they are used—direct-expansion or indirect-expansion. A direct-expansion evaporator is one in which the boiling refrigerant, in the evaporator coils, cools the air, or substance being refrigerated, by direct contact. In an indirect-expansion evaporator, water, brine, or some other medium is cooled by the refrigerant, and this medium is pumped or delivered to take up the heat load from the air or product. Both systems have advantages. In low-temperature work, a central refrigerating point delivering brine to many points may be desirable. The extra equipment for indirect expansion is often a disadvantage, and lower temperatures are required in the evaporator, to maintain a certain product or air temperature, than are necessary if direct expansion is used.

For air conditioning, where temperatures below 32 F are rare, water is well adapted as a carrier in indirect expansion. For low-temperature refrigeration, brine is almost universally used. The usual brine is made from calcium chloride ($CaCl_2$) of proper concentration to prevent freezing. Figure 16-1 gives the specific heat, specific gravity, and other data about $CaCl_2$ brines, over a range of temperatures and concentrations.

In addition to salt brines several of the glycols are used as carriers, because they can remain liquid at temperatures down to − 40 F and lower. Of the glycols, ethylene and propylene are in most extensive use. Both salt and glycol brines must be provided with inhibitors to prevent internal corrosion of the metal.

Shell and tube evaporators can be constructed by welding ends (tube sheets) on to a length of pipe (cylinder) and rolling or welding small tubes into the previously drilled and reamed tube sheets. Figure 16-2 is a cutaway detail photograph of the end of an evaporator showing the thick tube sheet into which the tubes are locked in place by the rolling operation. The particular tubes in this type are gilled (finned), along their run in the evaporator, to give more heat transfer surface for the boiling refrigerant, with a short end length left intact to permit the rolling

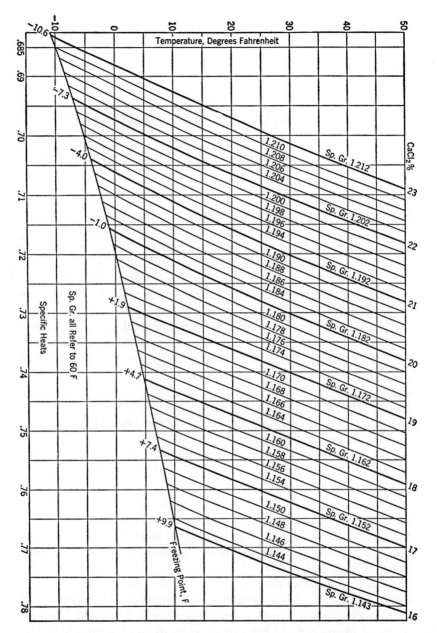

Fig. 16-1. Calcium chloride brine (chemically pure) specific heats and freezing points, at varying weight concentrations.

Refrigeration Equipment and Arrangement

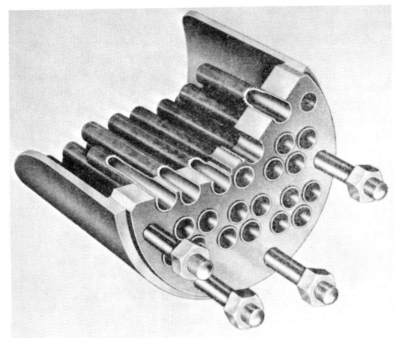

FIG. 16-2. Cut-away photograph of end of shell and tube evaporator. (Courtesy Carrier Corp.)

operation into the tube sheet. The outside diameter of the low fins does not exceed that of tube ends, and replacement is not difficult. The studs in the picture hold the external header in place and this is pulled up tight against the tube-sheet gasket by the nuts shown. In the lower left portion of Fig. 16-15 can be seen an evaporator with a two-pass external header in place. The tube sheet and tubes can be clearly seen occupying the lower half of the evaporator. The external headers, bolted on, lead the liquid to be cooled through the tubes—once for single-pass types and back and forth several times in multipass types. In Fig. 16-15 the inlet- and outlet-liquid connections for the water or brine being cooled can be seen at the lower left of the picture. Evaporators of this sort are usually operated *flooded*, that is, with liquid refrigerant surrounding the tubes. Vapor generated at the tube surfaces causes rapid ebullition in the liquid as the vapor rises to the surface, and this action promotes rapid heat transfer. The liquid level must be kept at the proper height in the evaporator; otherwise liquid carry-over may take place into the compressor. Above the tubes in Fig. 16-15 are shown eliminator plates which serve the purpose of separating out liquid particles and sending them back to the evaporator, instead of letting them pass over into the centrifugal compressor.

With certain designs of evaporators, it may be necessary to use *surge drums*, or *accumulators*, to permit separation of the liquid particles which are carried with the refrigerant vapor. Figure 16-3 shows an accumulator as used in connection with a pipe-coil evaporator. The liquid refrigerant is fed into the accumulator, and the liquid carry-over from the evaporator coils is separated out. The vapor, free of liquid, then passes to the compressor.

Figure 16-4 shows an outside view of a shell and tube cooler (evaporator). The external headers are at the ends for sending and directing the liquid to be cooled. Inlet and outlet of the water (or brine) are shown at the left along with a protective thermostat which goes into action to

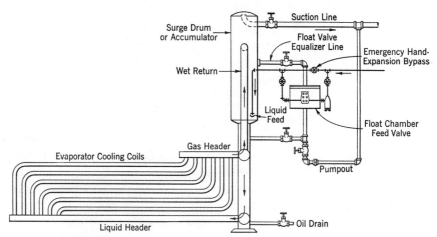

FIG. 16-3 Early design evaporator arranged for flooded operation, float-feed expansion valve.

shut off the compressor if the water or brine reaches its freezing point. High-pressure liquid refrigerant enters the vertical strainer (shown above and to the right of the left support), passes through the piston-operated automatic expansion valve, and then flows past the hand-controlled shut-off valve into the cooler. The vertical tube near the middle houses the pilot float which controls the operation of the expansion valve in terms of the liquid level desired in the evaporator. The vaporized refrigerant passes through the eliminators in the shell and leaves for the compressor via the welded suction loop at the center of the shell. At the base of the pilot-float tube is an oil bleeder drain connection, with a hand valve, solenoid valve, and sight glass in series leading to the bottom of the suc-

Refrigeration Equipment and Arrangement

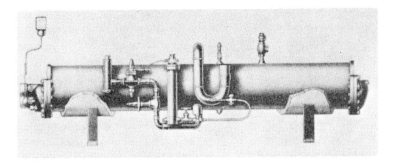

FIG. 16-4. Carrier shell and tube evaporator.

tion loop. Under switch action through the solenoid or by hand-valve action, accumulations of oil and oil-loaded refrigerant can be sent back to the compressor intermittently or continuously. At the top right can be seen the safety valve, and to the left of this appears a combination gage connection and purge valve for delivery of air and noncondensable gases.

Finned evaporators are frequently used for direct air cooling. The outside of the pipe for such evaporators is supplied with extended flanges to give an increased surface in contact with the header and may be arranged with one or more hairpin bends so as to lengthen the individual circuits.

Finned, as well as bare-pipe, evaporators may be operated flooded with accumulators, may be partly flooded with no accumulator, or may be dry. In dry evaporators the refrigerant liquid supply is controlled at the expansion valve in such amount that, in passing through the coil, it is completely vaporized and is usually somewhat superheated.

Figures 22-2 and 22-3 show, as a photograph and as a diagram, views of finned-type evaporator coils. In both figures these appear about one-third of the way down from the top. The expansion valve feeds the refrigerant into the top of the coils, and in passing down through the coils the refrigerant vaporizes and is drawn by the suction line into the compressor. The fan at the top of the cabinet draws warm air up through the coil and delivers the cooled air into the conditioned space.

Good evaporator design calls for relatively short lengths of coil through which the vapor has to travel, for adequate space to permit prompt and adequate separation of the vapor from the liquid, and for sufficient velocity in the tube to sweep the vapor from the tube surface as formed. Low-temperature work also calls for limitation of high static head (deep liquid depth) in the evaporator, as this causes a higher pressure in the lower part of the evaporator and, consequently, a higher boiling temperature.

16-2. EXPANSION VALVES

Manual expansion valves are simply specially built valves with needle-pointed or cone-pointed stems which are opened by hand in amount to feed adequate refrigerant.

Automatic high-side float valves are built along the lines of a float-operated steam trap and deliver all the liquid coming from the condenser to the evaporator. The refrigerant charge in a system with high-side float valve must be such that the liquid can largely be stored in the evaporator without danger of sending liquid slugs over to the compressor.

Automatic low-side float valves operate to maintain a definite level in an evaporator of the flooded type. The valve chamber is equalized to the evaporator with pipes on top and bottom and is placed at about the liquid level to be held in the evaporator. As refrigerant in the evaporator vaporizes, the float level drops to permit liquid to enter the evaporator until the proper level is regained.

Thermal-expansion valves are of several types, in all of which a thermostatic bulb, clamped to the side of the suction pipe or actually mounted inside this pipe, reacts to the temperature of the suction gas leaving the evaporator. If the rate of refrigerant flow to the evaporator is inadequate, it will be indicated by a leaving refrigerant temperature much higher than the saturation refrigerant temperature in the evaporator. This relatively high temperature reacts on the fluid in the thermostatic bulb to increase its pressure. This increased pressure is transmitted back to a bellows, or diaphragm chamber, and then operates against a spring resistance and push rod to open the refrigerant needle valve a greater amount. These controls can be adjusted after installation

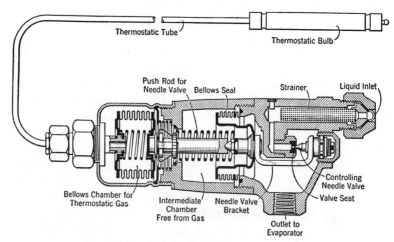

FIG. 16-5. A thermostatic expansion valve.

Refrigeration Equipment and Arrangement

on a given job or may be set for a definite temperature before installation. From 3 to 20 deg of superheat may be necessary to give desired rates of flow. The thermostatic fluid used in the bulb is frequently the same refrigerant used in the system (thus ammonia liquid is loaded into the expansion-valve bulb of an ammonia system). Figure 16–5 is a section through such a valve.

Automatic diaphragm expansion valves (constant-pressure type) have spring-loaded diaphragms which are acted on by the evaporator pressure (Fig. 16–6). As this pressure lowers, indicating insufficient refrigerant in the evaporator, the pressure on the diaphragm cannot prevent the

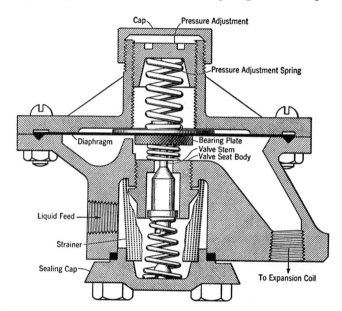

Fig. 16–6. Constant-pressure type expansion valve (diagrammatic).

spring on the far side from moving the diaphragm and this motion in turn is used to increase the reducing valve opening; increased refrigerant flow is thus permitted. This type of valve operates to keep the suction pressure essentially constant and is called a constant-back-pressure valve.

Automatic electrically operated valves connected from thermostatically controlled relay circuits are also used as expansion valves. In these a solenoid arrangement usually holds the valve open against a spring which closes the valve when the electric circuit is broken.

16–3. RECIPROCATING COMPRESSORS

Both reciprocating- and centrifugal-type compressors are in use. Centrifugal compressors use low-pressure (vacuum type) refrigerants and usually are of large capacity, 75 tons of refrigeration or over. Recipro-. cating machines are more common and more widely distributed and run in capacity from a fraction of a ton to more than 100 tons per unit. These machines are built as vertical or horizontal units, with modifications of radial-cylinder units and Y or X cylinder arrangements. Most units

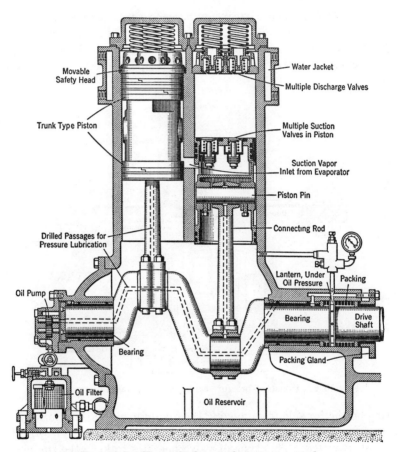

FIG. 16-7. Two-cylinder, single-acting, vertical refrigeration compressor of early type.

are single-acting, with double-acting units (refrigerant compressed by both sides of pistons) usually of the horizontal type. Compressor piston speeds for any type seldom exceed 700 fpm and usually are much lower.

Figure 16-7 shows a two-cylinder, single-acting refrigerant compressor of a type widely used with ammonia before the halocarbons reached the extensive use they now enjoy. The suction and discharge valves in this compressor are spring-loaded poppet valves which are operated in re-

Refrigeration Equipment and Arrangement 561

sponse to the different pressures on the two sides of the valve. The suction valves are located in the head (top) of the piston, and the crankcase and under side of the piston are open to the suction vapor pressure. When a piston moves down, the discharge valve will be shut and the pressure above the piston will be reduced below suction-vapor (crankcase) pressure. When this reduction occurs, the suction valves are lifted from their seats by the pressure difference and vapor passes into the cylinder above the piston. When the piston is on its upstroke, the vapor being compressed forces the suction valves tightly against their seats. The pressure increases as the piston rises until it exceeds condenser pressure, when the discharge valves in the head of the cylinder will lift from their seats and

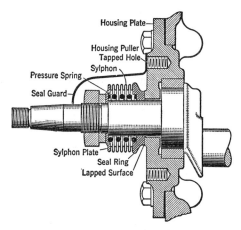

FIG. 16-8. Revolving shaft seal for a small compressor.

the compressed vapor will pass to the condenser. The internal head of the cylinder is held by heavy springs so that, if a slug of liquid should get above the piston, the head will yield enough to prevent serious damage. This permits the close clearances between the pistons and the cylinder head, which are necessary for efficient operation.

The shaft packing on this compressor (Fig. 16-7) is metallic or semimetallic and is held in place by the adjustable packing gland. An opening for inspection, called a *lantern*, is used with this packing and furnishes a means of lubricating and oil discharge. All packed glands leak to some extent and require constant attention.

Figure 16-8 shows one type of revolving shaft seal widely used to eliminate the packed joint on modern-size compressors. In this, a spring holds a revolving, hardened-steel seal ring tightly against a similar stationary surface fastened to the compressor frame. This surface is lubricated and the refrigerant or air will not leak past the rubbing surfaces.

Fig. 16-9. Double unit, 14-cylinder radial compressor (condensing unit). (Courtesy Airtemp Division, Chrysler Corp.)

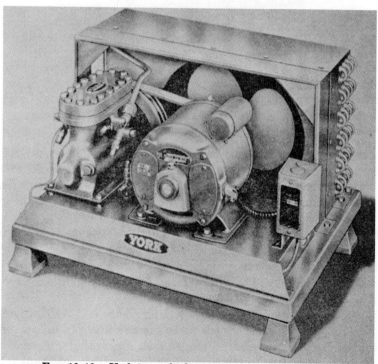

Fig. 16-10. York two-cylinder, air-cooled, condensing unit. (Courtesy York Corp.)

Refrigeration Equipment and Arrangement

FIG. 16-11. York six-cylinder, 3¾ in. by 3 in., condensing unit with V/W compressor (Courtesy York Corp.)

FIG. 16-12. Cut-away view of Chrysler Airtemp radial compressor.

While the sylphon bellows, illustrated in Fig. 16–8, turns with the shaft, this flexible element in other makes of seal may be stationary. Some very small domestic refrigerators have the motor and compressor hermetically sealed within the refrigerant-containing vessel and thus require no shaft seal. Even larger units are now being constructed as hermetic units.

The refrigerant valves of reciprocating compressors can be of the spring-loaded *poppet type*. These operate by differential pressure, often assisted by gravity. It is desirable to mount all large valves in protective cages so that the compressor may be saved from damage in case a valve should break. Some compressors employ thin ribbons of steel, forming flap-like closures. These are called *feather valves*. *Ring-plate valves* are also used.

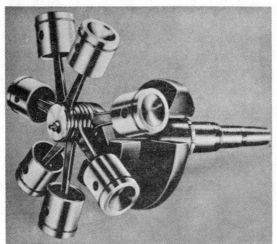

FIG. 16–13. Pistons and crankshaft of Chrysler Airtemp radial compressor.

Figure 16–9 illustrates the radial-type compressor. Here two radial compressors are shown under direct drive from a common motor. Above the compressors is mounted a water-cooled condenser. Such an integral arrangement of compressor, drive (motor), condenser, and accessories constituting the "high" pressure side of a refrigerating system is known as a *condensing unit*. Figure 16–10 shows a small condensing unit with an air-cooled condenser, and Fig. 16–11 is a six-cylinder unit with three pairs of compressor cylinders arranged in the form of the letter W, with a water-cooled condenser above.

Details of a radial compressor similar to that of Fig. 16–9 are shown in Figs. 16–12, 16–13, and 16–14. The seven pistons of the compressor, as can be seen in Fig. 16–13, do not all lie in the same vertical plane, but all are actuated by the same crank and are located on a long overhung crank pin. In Fig. 16–12, the pistons, crank, crankshaft, and crankshaft counterbalances are shown in place, with the shaft passing outside the compressor through the rotary seal to the flexible coupling at the motor. The piston (Fig. 16–14) does not have an integral valve in its head, as this compressor uses ring-plate valves on both suction and discharge.

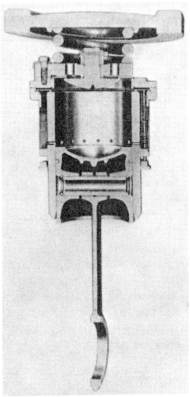

FIG. 16–14. Piston and cylinder details of Chrysler Airtemp radial compressor.

These ring-plate valves resemble large washers. The larger diameter suction valve is shown at the top end of the suction strainer, and the smaller diameter discharge plate valve rests below and almost immediately in line with the heavy spring coils which hold the inner cylinder head in place. This inner cylinder head remains fixed unless a slug of liquid, oil or refrigerant, forces this to be displaced to relieve extreme pressures. Small conical springs (not shown), placed at intervals above the ring plate valves, cause these to seat quickly when the piston reverses its motion.

The piston itself uses a single *ring*, since oil passing by the piston lubricates the top of the cylinder and the valves and is not objectionable. The oil discharged is separated out in the separator shown near the center of Fig. 16–12. By float feed, the oil returns to a storage reservoir and to the suction of the gear pump, which is driven off the compressor shaft and is shown at the lower right. The oil is delivered under pressure for lubrication and to operate controls. The suction gas enters at the pipe at the right of Fig. 16–12 and goes through strainers shown and into the cylinders.

A special feature of these compressors is the unloader device for starting and for load control. The multiple tubing for this appears at the right of Fig. 16–12, and the suction valve unloading device itself can be seen in Fig. 16–14, with the actuating collar ring shown near the middle of the cylinder. When this collar ring is forced toward the head of the cylinder, the rod forces the suction valve off its seat. Thus, gas merely surges back and forth through the open suction valve and the compressor is completely unloaded with zero capacity. At starting, with oil pressure not yet acting, a spring forces the rod, shown above the oil pump in Fig. 16–12, to the right; and this rod pulls into place a roller mechanism which forces the collar ring of the cylinder to move to the unloading position. When oil pressure, after starting, has built up, the unloader is pushed to the left and cylinder gas delivery can start. In addition to unloading at starting, a bellows device actuated by the suction pressure in the compressor can act to cut off the oil control supply, as before described, to one or more of the cylinders of the compressor. This unloading of individual cylinders can be regulated, for given operating conditions, from the outside (extreme right of Fig. 16–12). Falling suction pressure indicates reduced external load on the refrigeration system.

These radial units can operate at motor speed without reduction. The double seven-cylinder unit of Fig. 16–9 runs at 1150 rpm, has 3-in. bore, $2\frac{3}{4}$-in. stroke, single-acting cylinders, and uses a 50-hp motor. A similar unit is also built to run at 1750 rpm and uses a 75-hp motor. This latter unit is rated at 73.8 tons, with a 40 F suction evaporator and 97 F condensing temperature.

Reciprocating compressors are usually equipped with two hand-operated shut-off valves placed close to the compressor, one in the suction line and one in the discharge line. In addition, the larger machines have, inside of the main stop valves, a bypass connection, with a manually-operated valve connecting the suction and discharge lines. The bypass valve sometimes has to be opened to reduce the starting load. When a

Refrigeration Equipment and Arrangement 567

compressor is started against the suction of a warm evaporator, it may be necessary to prevent overloading by throttling the suction valve, although this sometimes causes crankcase oil to froth badly. Nearly all compressors have a safety valve which is set above normal discharge pressure and can discharge into the suction side of the system in case excessive pressures are developed in the cylinder.

16-4. RECIPROCATING COMPRESSOR CONTROL

Reciprocating compressors are frequently motor-driven by direct connection or by multiple vee belts. Direct-connected synchronous motors often drive large compressors. Many commercial installations use diesel- or steam-engine drive. Constant-speed operation makes regulation of output difficult under varying load conditions. Intermittent operation can be used with small compressors, but frequent starting and stopping of big machines is undesirable. Clearance pockets are built into some compressors by which, through opening a valve, the volume of the cylinder clearance can be increased. This operation in turn reduces the capacity by reducing the amount of fresh vapor drawn in per stroke (see also sec. 14–11).

The capacity of a refrigeration machine in relation to the load on a system can be varied in the following ways:

1. Bypassing gas from the high-pressure to low-pressure sides of the compressor
2. Bypassing internally, in the compressor, by holding open a suction valve or valves
3. By throttling the amount of suction gas entering the compressor
4. By variable speed motors

Bypassing is in most extensive use at the present time. One bypass system which is perhaps least desirable consists of a bypass line connecting the discharge side of the compressor to the suction side. Under the action of a solenoid valve the bypass opens and gases from the compressor discharge return to the suction side of the compressor so that refrigerant delivery from a bypassed cylinder of the compressor is stopped and compressor capacity is reduced. Power requirements of the compressor may be almost as high when the bypass is open as when it is closed. With the bypass open, power requirements may range from 65 to 90 per cent of normal.

A second, more desirable, method involves an arrangement, whereby the suction valve of a cylinder is held open so that the charge of gas simply surges back and forth in the cylinder. This makes the cylinder of the compressor inoperative as far as gas delivery is concerned. The most usual arrangement in this connection consists of a solenoid, enclosed in a hermetic housing, responsive to the action of a thermostat and controlling two valves. When the thermostat indicates that capacity should be reduced, the solenoid opens the valve which permits high-pressure gas to move to the operating side of an operating piston. This piston, in its small cylinder, is mounted on the compressor cylinder and carries an operating finger which pushes against, and holds open, the suction valve when it is desired to reduce capacity. When the thermostat shows that additional capacity is required, the solenoid valve closes the condenser-pressure connection and opens the suction-pressure of the system to the operating cylinder line. This releases the holding finger and the compressor suction valve resumes normal operation.

In the case of a four-cylinder compressor, it is possible to arrange for three cylinders to be inactivated, by holding open their respective suction valves. With one cylinder inoperative, the capacity is reduced by 25 per cent; with two, by 50 per cent; with three, by 75 per cent. In a six-cylinder compressor, the reduction steps are $33\frac{1}{3}$, 50, and 66 per cent with two cylinders, three cylinders, and four cylinders rendered inoperative in progressive steps. Under no circumstances should all of the cylinders in a compressor be unloaded, since it is necessary to have some cool gas passing through the compressor to prevent over-heating.

Throttling of the suction is an undesirable method of control, for reduction of suction pressure can cause the crankcase oil to froth and also cause it to be pumped from the compressor. Throttling, when used with a bypass line from the high- to low-pressure sides of the compressor, is a less objectionable arrangement.

Varying speed for capacity control is not usually applicable with a-c motors. However, it is possible to use special two-speed motors and produce a 50 per cent reduction in speed. With d-c motors, variable speeds are easily accomplished, but d-c power is not available in most communities. In connection with this type of variable speed control, it should be mentioned that very small compressors use a stop and start cycle to serve when the load falls off and comes on. This represents a two-step, speed control arrangement with a range varying from 0 to 100 per cent speed.

An illustrative control system which has been successfully used operates in the following manner. Consider a direct, air-cooling system or one using a water chiller. In either of these systems, thermostats might be used. The first thermostat is set to operate a cylinder unloader when the air (water-chiller) temperature reaches a certain minimum. The thermostat, in its operation, sets the solenoid to cause unloading of one or more compressor cylinders. If the temperature of the air (chilled-water) continues to drop, the second thermostat, through a similar solenoid, actuates the controller unloader to set a second cylinder (or pair of cyl-

Refrigeration Equipment and Arrangement

Fig. 16-15. Carrier centrifugal-compression unit. (Courtesy of Carrier Corp.)

inders) out of operation. The second thermostat may have a final step possible, such that, if the temperature reaches an absolute minimum, the main power switch is opened to stop the unit. In reverse sense, as the air (chilled-water) temperature rises, the thermostats act to put the bypassed cylinders back into operation. Provision can also be made that when the compressor is started again, a time-delay mechanism holds some suction valve controls open and thus permits the unit to start under light load.

An additional control which must be provided on a machine of this type is the high-low pressure controller, which acts to stop the compressor in the event of either excessively high pressure or excessively low pressure in the refrigeration system. There is also usually a protective device on the lubrication system such that if the pressure of the lubricant is too low, the compressor is automatically stopped.

16-5. CENTRIFUGAL COMPRESSORS

Centrifugal compressors are more limited in the pressure range that they can cover than are reciprocating compressors, but they can handle large volumes of gases effectively and can use the so-called vacuum refrigerants, including water-steam. Figure 16-15 shows a centrifugal compressor unit designed for a vacuum refrigerant such as refrigerant-11. The refrigerant is vaporized in the evaporator, shown at the bottom left of the picture, as it absorbs heat from the brine or water flowing through the tubes. The vapor passes through the eliminator plates, which throw back any entrained liquid, to the suction side of the first stage of the centrifugal compressor where, in its passage through the rotating impeller, the velocity of the vapor is greatly increased. In the flared diffusion passages of the casing, the velocity of the moving vapor decreases and a resulting increase in pressure takes place. Additional pressure increase is imparted similarly through the second stage. Two or more stages can be employed, depending on the pressure range required, or multiple units can be arranged in series. The compressed vapor enters the condenser at the top left, where, in passing over the water-cooled tubes, it is condensed and drains to the liquid outlet, at the far end of the condenser. The high-pressure liquid goes through a two-stage expansion; first a float valve (similar in design to a float steam trap) drops the pressure down to that of the suction pressure of the second stage, and then the remaining liquid is expanded through a second float feeding the liquid down to evaporator pressure. The economizer action of this two-step expansion saves power and is well justified, since the flash gas from the first expansion has to be raised only from the intermediate pressure and not from evaporator pressure (see Fig. 15-9 and sec. 15-11 for a further discussion of this problem).

Refrigeration Equipment and Arrangement

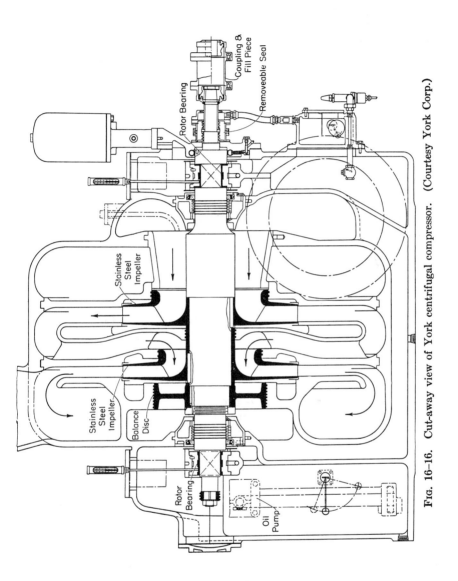

Fig. 16-16. Cut-away view of York centrifugal compressor. (Courtesy York Corp.)

The layout of the centrifugal compressor system is essentially the same as that of the reciprocating system, consisting of evaporator, compressor, condenser, expansion valve (two-step in this case), and connection back to evaporator. This machine also includes a *purge device*, whereby vapor and any noncondensable gas which may be in the system are drawn from a cool portion of the condenser and compressed in a small (0.5 hp) reciprocating compressor; the vapor is then condensed in the purge unit. The liquid refrigerant is returned to the main system, while the noncondensables are discharged to atmosphere and waste. Because the system operates under vacuum, the shaft seal of the main compressor is enclosed in a bellows and supplied with oil under pressure. Metal-to-metal contact occurs only when the machine is stopped, and then the whole seal is submerged under a head of oil in the shaft-seal reservoir. Machines similar to this are now being made of hermetic construction, which makes the seal unnecessary. Either motor drive, as shown, or turbine drive is used for these machines, which run at high speeds of 3500 to 7600 rpm and vary in capacity from 50 to about 1000 tons of refrigeration per unit. Evaporator temperatures from the air-conditioning range of 45 to 30 F down to very low temperatures of even − 150 F are possible with these units if appropriate sub-atmospheric refrigerants are used.

Figure 16–16 is a detail drawing showing the centrifugal compressor design of another manufacturer. The suction gas moves to the left, into the first-stage impeller, then passes through the diffusion section and the connecting passageway into the second stage, and finally passes out to the condenser (not shown).

The centrifugal compressor is a dynamic device and as such, it can only impart energy by means of the impeller, which accelerates the fluid passing through. In the impeller the pressure is increased, and further pressure rise takes place in the fixed diffusing vanes (or in the volute casing of the compressor). Figure 16–17 is a diagrammatic front view of an impeller, such as either of the impellers shown from the side in Fig. 16–16. The inlet section of the impeller is known as the eye, and here the gas enters the impeller in a direction essentially parallel to the axis of the machine. Thus the inlet velocity triangle shown in Fig. 16–17 would really only be completely visible if viewed at 90 deg from the plane of the figure. This figure shows inducer guide vanes into the impeller. Such vanes are optional in use, depending upon the design. However, whether or not they are used, the absolute velocity (V_1) of the gas on inlet into the eye is essentially parallel to the axis (shaft) of the compressor and consequently lies essentially perpendicular to the impeller velocity u_1. The basic equation for power interchange between an impeller and the gas passing

Refrigeration Equipment and Arrangement

through it, derived from fundamental momentum and energy relationships, is

$$P = \frac{G}{g}(V_{u2}u_2 - V_{u1}u_1) \qquad (16\text{-}1)$$

where P = power interchange, in foot-pounds per second;
 G = fluid (gas) flow through impeller, in pounds per second;
 g = 32.17;
 u_1, u_2 = velocities of points on the impeller at varying radii, in feet per second, with u_1 the velocity value at the eye section, and u_2 the velocity at the impeller tip (u_t);
 V_{u2}, V_{u1} = components of the absolute gas velocity in the plane and direction of u, in feet per second.

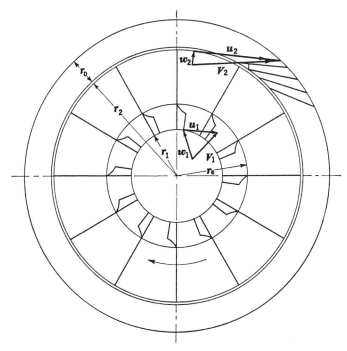

FIG. 16-17. Inlet and exit velocity conditions for a centrifugal compressor impeller.

In words, this equation states that the change in product terms multiplied by the fluid flow rate (G) is a measure of the power imparted to the fluid by the impeller. Again refer to Fig. 16-17 and note that on exit the relative velocity w_2 is usually almost radial and consequently the component of absolute velocity (V_{u2}) in the direction of u_2 is essentially equal to u_2 or $V_{u2} = \mu u_2$, where μ is a factor to account for any deviation in the exit direction of w_2 from being radial. With a radial-tipped impeller, μ can usually be considered as being 0.9 or higher. Figure 13–7 shows possible deviations as they exist for similar fan impellers. On inlet, however, with the fluid flowing into the impeller in an almost axial direction, V_{u1} approaches zero in value. For these conditions equation 16–1 becomes

$$P = \frac{G}{g}(V_{u2}u_2 - V_{u1}u_1) = \frac{G}{g}[\mu u_2 u_2 - (0)u_1]$$

$$= \frac{G}{g}(\mu u_2^2) \tag{16-2}$$

Since u_2 is the tip velocity of the impeller, it is desirable to write it thus, and equation 16–2 becomes

$$P = \frac{G}{g}(\mu u_t^2) \tag{16-3}$$

This equation is a basic measure of the power in foot-pounds per second absorbed by an impeller with a tip speed of u_t feet per second for a fluid (gas) flow of G pounds per second. However, the power imparted by the impeller is also measured by the increase in enthalpy (h) of the fluid plus the increase in kinetic energy in the fluid, and in equational form this appears as

$$P = G\left[778\,(h_3 - h_1) + \frac{V_3^2 - V_1^2}{2g}\right] \tag{16-4}$$

This becomes

$$P = G\left[778\,C_p\,(T_3 - T_1) + \frac{V_3^2 - V_1^2}{2g}\right] \tag{16-5}$$

since, for a gas, enthalpy change can be measured by temperature change (see eq. 2–50). The velocity V_3 represents the velocity at outlet from the fixed diffuser section following the impeller, at the same point of measurement as T_3. The exit passage area and the passage area on inlet are usually so designed that V_3 and V_1 are essentially equal so that the kinetic energy term approaches zero and can be dropped. Thus equation 16–5 becomes

$$P = G(778)C_p(T_3 - T_1) = 778GC_pT_1\left(\frac{T_3}{T_1} - 1\right) \tag{16-6}$$

For a gas undergoing isentropic compression, equation 14–20 applies:

$$\frac{T_3}{T_1} = \left(\frac{p_3}{p_1}\right)^{k-1/k} \tag{14-20}$$

Refrigeration Equipment and Arrangement

and substituting

$$P = 778GC_pT_1\left[\left(\frac{p_3}{p_1}\right)^{k-1/k} - 1\right] \tag{16-7}$$

Equations 16-3 and 16-7 represent the same process and are equalities, thus

$$\frac{G}{g}(\mu u_t^2) = 778GC_pT_1\left[\left(\frac{p_3}{p_1}\right)^{k-1/k} - 1\right]$$

$$\frac{p_3}{p_1} = \left(\frac{\mu u_t^2}{778C_pgT_1} + 1\right)^{k/k-1} \tag{16-8}$$

However, the compression process is not truly isentropic and if an efficiency (η_{ic}) is introduced to account for this fact, equation 16-8 then becomes usable. For general conditions, η_{ic} ranges from 0.70 to 0.80. Thus the performance of a centrifugal compressor can be expressed

$$\frac{p_3}{p_1} = \left(\frac{\eta_{ic}\mu u_t^2}{778C_pgT_1} + 1\right)^{k/k-1} \tag{16-9}$$

This equation is applicable to centrifugal air compressors, centrifugal steam compressors, or refrigerant compressors, provided the proper value of the polytropic coefficient k can be found, and the mean specific heat of the vapor during compression C_p is known. For design estimation, k can be taken as the ratio of specific heats.

Example 16-1. A two-stage centrifugal compressor has two impellers in series, each with a diameter of 34 in. The impellers are direct-connected to a motor turning at 3550 rpm and completely enclosed in a hermetic housing. Refrigerant-11 is used, and the suction temperature is to be 30 F in the evaporator. Compute the probable maximum pressure rise possible in this machine, and find a top condensing temperature.

Solution: The impeller tip speed is

$$u_t = \pi\frac{D}{12}\left(\frac{\text{rpm}}{60}\right) = \frac{\pi D}{720}\,(\text{rpm}) = \frac{(3.14)(34)}{720}(3550)$$
$$= 526 \text{ fps}$$

Using equation 16-9, assume $\eta_{ic} = 0.80$ and $\mu = 0.9$, and using $C_p = 0.136$ and $k = 1.13$ for refrigerant-11 (from Table 2-1),

$$\frac{p_3}{p_1} = \left[\frac{(0.8)(0.9)(526)^2}{(778)(0.136)(32.17)(460+30)} + 1\right]^{1.13/0.13} = 2.7$$

The pressure ratio for each stage is 2.7, and for two stages is thus

$$(2.7) \times (2.7) = 7.3$$

However, because of pressure loss between stages, perhaps allowance should be made for say 95 per cent realization. From Table 15-7 the evaporator pressure is 5.56 psia; therefore the maximum condenser pressure would be

$$5.56\,(7.3 \times 0.95) = 38.5 \text{ psia}$$
$$\text{Pressure rise} = 38.5 - 5.56 = 32.94 \text{ psia} \qquad\qquad Ans.$$

This is beyond the range of values of Table 15–8, but more extended tables would show the staturation pressure to be 129 F.   *Ans.*

Thus there is adequate pressure-ratio as well as pressure-range capacity in the machine running at this speed. The capacity design-pressure ratio would be set at a lower value as shown in Fig. 16–18.

Figure 16–18 shows the general characteristic pressure-ratio capacity curves for a centrifugal compressor. At any fixed speed, such as 4 M in the figure, it will be observed that as the flow rate decreases, the pressure ratio which can be produced, slowly increases, but soon reaches a maximum value followed by instability. The region of instability is indicated as the region to the left of the dashed line. The design operating

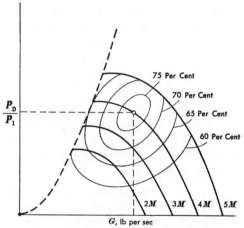

FIG. 16–18. Characteristic pressure-ratio, capacity curves at various speeds for a centrifugal compressor.

point should always be selected safely to the right of the instability region. It should be mentioned that the pressure ratio as computed by use of equation 16–9 represents essentially the maximum pressure ratio that could be produced by that machine.

A significant development in recent years has been to make centrifugal machines for small as well as large capacities, and one manufacturer has brought out such a machine (the Centravac), one model of which operates effectively at 50 tons of refrigeration. Another development has been to build hermetic units, in which the motor is completely enclosed in a shell inside the system with no shaft and packing gland leading to outside. The fact that centrifugal machines can employ low-pressure refrigerants with a total pressure range in the neighborhood of two atmospheres means that extremely heavy shells do not have to be

Refrigeration Equipment and Arrangement 577

used, and the hermetic casing for the motor is not difficult to construct. The hermetic design reduces the possibility of leakage either in or out and eliminates the need of purging. Usually two impellers are used, which permit the maximum pressure-ratio rise to be produced in two steps.

As was seen in Fig. 16–18, a characteristic of centrifugal machinery is the fact that the capacity falls off as the pressure ratio increases. In the case of a centrifugal machine running at constant speed and stable condensing pressure, as the external load falls off, the evaporator temperature and pressure start to drop. This in turn increases pressure ratio, and therefore decreases rate of refrigerant flow so that, before long, the machine stabilizes at a lower capacity point. Moreover, when it is necessary to decrease capacity at fixed evaporator temperature, a similar result can be produced by increasing the condenser temperature. This is readily accomplished merely by diminishing the amount of water flowing through the condenser. Most centrifugal machines being built at the present time have internal, hydraulically-operated regulating vanes. These are installed at the inlet eye section to the low-pressure compressor and when operated to restrict this opening, the amount of gas entering the low-pressure impeller is reduced as closure takes place.

16–6. CONDENSERS

Condensers for refrigeration units are made in many forms and designs. The trend is toward horizontal or vertical *shell and tube* types. These condensers are similar in appearance to shell and tube evaporators, and the circulation within the tubes may be for multipass or single pass. The condenser heads should be easily removable for cleaning the sludge and debris which collect on the water side. A condenser can be seen at the top of Fig. 16–15.

Double-pipe condensers, in which the water passes through the inside pipe and the refrigerant vapor condenses in the annular space between the two pipes, were formerly used very extensively. They are very effective but require elaborate joints at the ends of each pipe for connecting the two tubes.

The so-called *atmospheric-type condenser,* in which water trickled down over condenser tubes placed outdoors, usually on the roof of a building, required much surface for a given capacity (see Table 9–4), and installations are now seldom made.

The *evaporative condenser,* using finned surface, regulated water spray, and forced circulation of the water and air, is very effective. Figure 16–19 shows one type of an evaporative condenser. In this condenser the (hot) vapor from the compressor is supplied to a bank of finned tubes enclosed in a metal cabinet. A circulating pump draws water from a basin in the bottom of the cabinet and sprays it over the tubes. Fans draw large quantities of air into the lower part of the cabinet. This air passes up around the wetted condenser tubes and a portion of the water evaporates into the air. The wetted surface and the rapidly moving air produce very effective heat transfer, permitting a large condensing capacity in a very compact space. In a cooling tower, with water flowing down and air rising up through it, the water cooling may approach the wet-bulb temperature of the entering air. In an evaporative condenser the recirculated spray water leaving the bottom tubes will have a temperature which lies between the original wet-bulb temperature of the air and the temperature at which the refrigerant condenses in the tubes. Tests on a certain

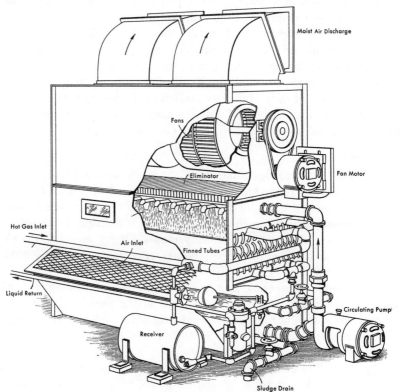

FIG. 16-19. Evaporative condenser.

evaporative condenser showed representative data as follows: temperature of refrigerant condensing, 100 F; temperature of spray water falling into recirculating basin, 89 F; wet-bulb temperature of air supply, 80 F; cfm of air circulated per ton of refrigeration, 265. Increasing the quantity of air circulated brings the water temperature closer to the wet-bulb temperature of the air, and increasing air quantities in the range from 110

Refrigeration Equipment and Arrangement 579

to 300 cfm per ton of capacity shows justifiable gains; but above this point the rate of gain is so small that the fan power costs begin to offset gains from decreased water temperatures.

16-7. EVAPORATIVE-CONDENSER CALCULATIONS

A basis for finding the load on an evaporative condenser can be had by starting back with the evaporator, where, by equation 14–9,

$$Q_R = h_R - h_{f1} \qquad [14\text{-}9]$$

and in an evaporator producing T tons of refrigeration, the refrigerant flow (w'_R), in pounds per minute, is

$$w'_R = \frac{200\,T}{Q_R} = \frac{200\,T}{h_R - h_{f1}} \qquad (16\text{-}10)$$

The heat output of the condenser is equivalent to the heat absorbed in the evaporator plus the work absorbed by the refrigerant in passing through the compressor. If we call W_k the Btu absorbed per pound as work, then the total work addition is $w'_R W_k$ and the condenser load, in Btu per minute, becomes

$$Q'_c = w'_R (h_R - h_{f1} + W_k)$$

$$= \frac{200T}{h_R - h_{f1}} (h_R - h_{f1} + W_k)$$

$$= 200T \left(1 + \frac{W_k}{h_R - h_{f1}}\right) \qquad (16\text{-}11)$$

The ratio $W_k/(h_R - h_{f1})$ varies with the effectiveness of the compressor and with the pressure ratio through which the refrigerant is compressed. The value of W_k exceeds the isentropic work, and previously in this book it has been indicated that W_k is approximately 40 per cent greater than the isentropic work, $(\Delta h)_s$. Equation 16–11 can thus be expressed

$$Q'_c = 200T \left(1 + \frac{1.4(\Delta h)_s}{h_R - h_{f1}}\right) \quad \text{Btu per min} \qquad (16\text{-}12)$$

This can be written

$$Q'_c = 200\,T\,K \quad \text{Btu per min} \qquad (16\text{-}13)$$

where K represents the composite term of equation 16–12. For representative systems the condensing load factor K will vary from 1.1 to 1.5 with a value of 1.25 being a value that might be used for estimation computations.

Refrigeration Equipment and Arrangement

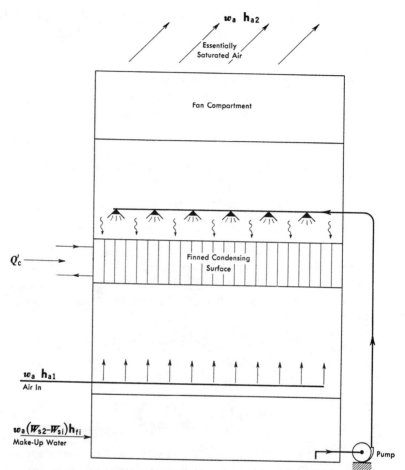

FIG. 16-20. Diagrammatic arrangement of an evaporative condenser.

Thus an over-all expression for the heat delivered per minute to the condenser in Btu per minute, for **T** tons produced, appears as

$$Q'_c = 200\text{T}\,(1.25) \tag{16-14}$$

Consider now the evaporative-condenser aspects of this problem. Referring to Fig. 16-20, the following equation can be written in terms of the pounds of air flowing through the evaporator (w_a pounds per minute) and specific humidities (W_{s2} and W_{s1})

$$w_a \text{h}_{a1} + w_a(W_{s2} - W_{s1})\text{h}_{f1} + Q'_c = w_a \text{h}_{a2}$$
$$w_a \text{h}_{a1} + w_a(W_{s2} - W_{s1})\text{h}_{f1} + 200(\text{T})(1.25) = w_a \text{h}_{a2}$$
$$w_a = \frac{200(\text{T})(1.25)}{(\text{h}_{a2} - \text{h}_{a1}) + (W_{s1} - W_{s2})\text{h}_{f1}} \tag{16-15}$$

Refrigeration Equipment and Arrangement 581

In the above expression w_a represents the pounds of dry air flowing through the evaporative condenser per minute, h_{a1} and h_{a2} are the enthalpies in Btu per lb of dry air respectively on inlet and outlet from the evaporator, W_{s2} and W_{s1} are specific humidities, and h_{f1} is the enthalpy of the replacement water supply to the condenser sump. In general, it will be found that the airflow through an evaporative condenser, in pounds per minute, approximates 20 T or expressed in volume terms, about 260 cfm per ton of refrigeration. However, this value ranges from perhaps 120 to 300 cfm per ton.

In operation, an evaporative condenser is not an adiabatic device, since heat is continuously being added from the refrigerant. Consequently, the final temperature of the water leaving the condenser coils will not reach the entering wet-bulb temperature as closely as would be the case were this a simple cooling tower. Expressed in equational form,

$$t_{\text{condensation}} > t_{\text{water in sump}} > t_{\text{wet bulb in}}$$

and $t_{\text{water in sump}}$ is $4°$ to $25° > t_{\text{wet bulb in}}$

and $t_{\text{condensation}}$ is $5°$ to $30° > t_{\text{wet bulb in}}$

The rate of pumping must be sufficiently high to wet the coils at all times, and it has been found good practice to use pumping rates which range from 1 to 2 gpm per ton of refrigeration handled.

16-8. WATER COOLING FOR CONDENSERS

Recirculated water must frequently be employed for cooling condensers, since large quantities of water from city mains may be costly and the resultant overloading of mains and sewers makes it necessary to reduce, as much as possible, the quantity of condenser water required. By using evaporative condensers, cooling towers, and the like, only about 4 to 8 per cent as much make-up water is required as when water is used directly. In some cases, water after passing through the condensers can be used for building service, but this arrangement seldom balances out perfectly. Heat dissipation from recirculated water is usually accomplished by *spray ponds, atmospheric cooling towers, mechanical-draft cooling towers,* or *evaporative condensers.*

In the case of a spray pond, the warm condenser water, under considerable pressure, is forced into the air by spray nozzles and allowed to fall into an open pond. This system requires a considerable area and is not usually adaptable to installations in cities.

In the case of an atmospheric cooling tower, the warm water is distributed by sprays or troughs at the top of the tower, allowed to trickle down over air-swept baffles, and then collected in a basin at the bottom. The sides of the tower are inclined louvres, to facilitate the circulation of

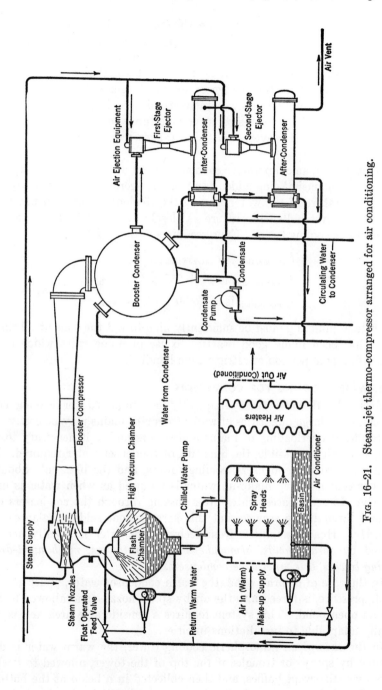

Fig. 16-21. Steam-jet thermo-compressor arranged for air conditioning.

Refrigeration Equipment and Arrangement 583

air through the tower and to prevent the water from being blown out. The degree of cooling depends on the wet-bulb temperature of the air, the velocity of wind movement, and the intimacy of contact between the water and the air.

A mechanical-draft cooling tower is similar in construction to the atmospheric type, except that it is provided with large fans for moving the air through the tower and bringing it in contact with the water. The operation of a mechanical-draft tower is independent of the wind. Such a tower generally uses interior sprays and dispenses with the elaborate baffle system necessary with an atmospheric cooling tower.

With any of the foregoing methods the amount of make-up water required is only that sufficient to replace the moisture evaporated, plus loss by windage, and the small amount required for periodical flushing and cleaning. About 0.03 to 0.06 gpm of water per ton of refrigeration capacity is lost through evaporation into the air and carry-out as mist. Mechanical-draft cooling towers bring water to within 2 to 6 deg of the wet-bulb temperature of the air; with economy in operation (fan power, atomization, etc.) dictating 4 to 6 deg above the wet-bulb as an economical range to meet.

16-9. STEAM-JET COMPRESSORS

In steam-jet compressors, also known as steam-jet ejectors or thermocompressors, steam is supplied to one or more nozzles, where it is permitted to expand. The steam (Fig. 16-21), leaving the nozzles at high velocity, draws vapor from the flash chamber, and in the booster compressor the velocity energy of this mixture is reduced and a resultant increase in pressure occurs. The final pressure must be sufficiently high for the steam to condense at the temperature maintained in the condenser. Reference to Table 2-2 shows that for a 40 F evaporation temperature in the flash chamber a pressure of 0.1217 psia is necessary, and for condensation at 90 F the pressure is 0.6982 psia. Thus, for this case, the pressure rise is 0.5765 psi and the pressure ratio of compression is 5.7. The steam jet, in drawing water vapor from the flash chamber, loses velocity in moving some of the flash vapor from the evaporator, and the mixture then is compressed under further reduction in velocity in the flared booster compressor tube.

In the booster condenser, any air or other noncondensable gases will collect and increase the total pressure if not continuously removed. This removal of air, saturated with vapor, is accomplished in the air-ejection equipment. This equipment in Fig. 16-21 is represented as a two-stage steam-jet compressor which discharges the undesired air at atmospheric pressure to the outside. The air load is relatively small compared to the booster compressor load but the compression is so great in the example mentioned, from 0.6982 to about 14.7 psi atmospheric pressure, that it is more economical to do this in two stages than in one. Notice that the flash-chamber steam needs to be compressed only to condenser pressure.

The chilled water from the flash chamber must be pumped out and then delivered to the air conditioner at whatever pressure above atmospheric may be required. The cool water may be used in a spray dehumidifier or in convectors as desired. The water, in spraying, picks up more air than when convectors are used and the necessity for removing

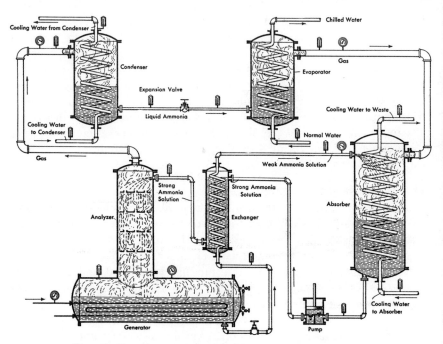

FIG. 16-22. Conventional absorption refrigeration system.

this air causes an appreciable additional load on the air-ejection equipment. The water, after absorbing the heat load, returns to the flash chamber and on being sprayed therein, drops in temperature as part of it evaporates. The cycle then repeats.

The thermo-compressor fits well into installations where steam is low in cost and where a large water-cooling capacity is available, either from city mains or from a cooling tower. The system is adequate for comfort air conditioning but is not particularly feasible for water temperatures below 40 F. The condenser load (water required) per ton of refrigeration is 3 to 4 times as great with thermo-compressors as with mechanical compression. About 30 to 40 lb of steam at 100 psig are required per hour per ton of refrigeration. High steam pressures are preferable, but

Refrigeration Equipment and Arrangement 585

operation is practicable at 50 psig pressure. About 10 per cent of the steam is used for the air-ejection equipment. The cost of installation is low and the system can operate effectively under light loads.

16-10. ABSORPTION REFRIGERATION

In Fig. 16–22 is shown a diagrammatic arrangement of a conventional absorption system using ammonia and aqua ammonia. Water will absorb large quantities of ammonia vapor, the amount absorbed increasing with the external pressure and decreasing with rising temperature. The absorber, which operates at about evaporator pressure, is supplied with a cooled solution of water ammonia not saturated with ammonia. This so-called weak aqua absorbs ammonia gas from the evaporator suction line until the liquid becomes saturated at the evaporator (absorber) pressure. Heat is generated during the absorption process and is removed by cooling water. The saturated aqua is then pumped through a heat exchanger into the generator. The generator, which operates at about condenser pressure, is supplied with steam, or other heat supply, and ammonia is boiled off from the mixture until the aqua is reduced to a saturated condition at generator pressure and temperature. The ammonia vapor from the generator eventually passes to the condenser, where it is condensed, and as liquid, passes through the expansion valve into the evaporator. The weak hot aqua from the generator passes through the heat exchanger where it is cooled in warming the strong liquid, and is throttled as it passes into the absorber again to absorb a charge of ammonia vapor from the evaporator.

The absorption system resembles the compression system in several ways: The condenser, expansion valve, and evaporator are interchangeable with either system—the compressor is paralleled by the absorber for the suction stroke of the piston, by the aqua pump for the compression stroke, and by the generator for the delivery of the ammonia. The energy input to the system consists of a small amount of power consumed by the aqua pump and a large amount of thermal energy supplied by the heating medium to the generator. There are three major circuits of fluid throughout the system: The ammonia circuit from the generator eventually to the absorber; the strong aqua circuit from the absorber eventually to the generator; the weak aqua circuit from the generator finally to the absorber; and, in addition, steam and circulating water are required.

The vapor which rises from the solution in the generator consists of ammonia vapor along with small quantities of steam. As this vapor is cooled, the steam (saturated with ammonia) condenses out first. The analyzer performs this function of dehydration by bringing the vapor into contact with the aqua richest in ammonia and by cooling the vapor with this aqua. If the dehydration is not complete enough in the analyzer, an added water-cooled vessel called a rectifier may be used to complete the process for sending anhydrous (dry) ammonia to the condenser. Traces of moisture in the ammonia are not serious, although this moisture collects in the evaporator (if of flooded or partly flooded type) and must periodically be purged back to the absorber.

Tables of properties of aqua solutions[1] show the following values for a given pressure and liquid concentration: saturation temperature, enthalpy of the liquid, enthalpy of the vapor rising from such liquid, and the ammonia concentration in the vapor in equilibrium with the liquid. Typical values for saturated liquid leaving an absorber maintained at 93 F at 40 psia are

$x_f = 42$ per cent ammonia by weight in liquid,

$t = 93.0$ F,

$h_f = -45.3$ Btu per lb of liquid (32 F datum for tables),

$h_v = 587.9$ Btu per lb of vapor (32 F datum for tables),

$x_v = 99.2$ per cent of ammonia in vapor.

For aqua leaving the generator saturated at 180 psia and at 239.5 F, values are $x_f = 28$ per cent, $h_f = 135.9$ Btu per lb, $h_v = 710.2$ Btu per lb, $x_v = 88.8$ per cent.

The ammonia-water absorption system has not been used very extensively in recent years except where steam has been cheap or a by-product and where low refrigerant temperatures are required. However, the direct application of heat for energy supply has certain advantages, and absorption systems using media other than ammonia-water are being developed for air-conditioning work. One of these systems has employed dichloromonofluoromethane as the refrigerant and dimethyl ether of tetraethyleneglycol as the absorbent.

The Electrolux refrigerator is gas-heated, uses aqua ammonia as the absorbent refrigerant, and dispenses with the aqua pump by using an inert gas (hydrogen) in the evaporator and absorber. The sum of the partial pressure of the hydrogen and the partial pressure of the ammonia in the evaporator is thus made equal to the condenser pressure of the system. Circulation of the various media in the cycle occurs from thermal (density) differences which develop in the unit.

16-11. SALT-SOLUTION-CYCLE ABSORPTION REFRIGERATION

For air conditioning in which refrigeration temperatures below 32 F are not needed, an absorption refrigeration system using water as the refrigerant and lithium-bromide solution as the absorbent has been suc-

[1] See reference 8 of Chapter 15.

Refrigeration Equipment and Arrangement

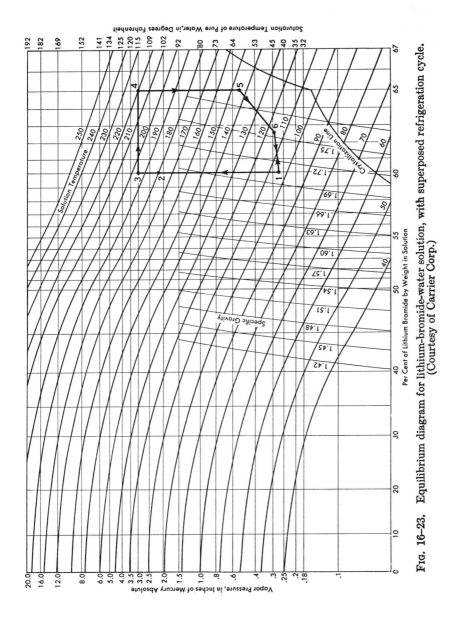

FIG. 16-23. Equilibrium diagram for lithium-bromide-water solution, with superposed refrigeration cycle. (Courtesy of Carrier Corp.)

cessfully developed and has achieved great commercial success. The vapor pressure of an aqueous solution of lithium-bromide high in salt is very low, and if water and solution are placed adjacent to each other in a closed evacuated system, the water will evaporate. For example, a 60

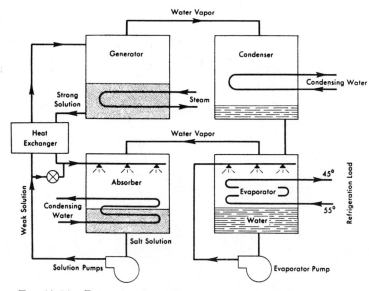

Fig. 16-24. Diagrammatic arrangement of water and lithium-bromide circuits for Carrier absorption system.

per cent lithium-bromide-water solution at 110 F has a vapor pressure of 0.27 in. Hg, which is sufficiently low to cause water at 43 F, or slightly lower, to boil. Figure 16-23 is an equilibrium diagram of lithium-bromide-water solutions, and use will be made of it and of the numbered cycle in describing the absorption system. Figure 16-24 and Fig. 16-25 show the system first in diagrammatic arrangement, and then with the vessels as actually placed in the machine.

Absorber. In the absorber, a circulating pump takes solution at roughly 60 per cent from one end and recirculates it back into the absorber to mix with strong salt solution (point *5*) at 65 per cent, supplied to the absorber from the heat exchanger. The mixed solution at an average temperature of 115 F at 62.5 per cent (point *6* in Fig. 16-23) has a sufficiently low vapor pressure that it can readily absorb water vapor from the evaporator, and the concentration further reduces with the addition of vapor toward point *1*. As the water vapor enters the salt solution to enter the liquid phase, the heat of condensation must be removed as must also the heat of dilution (solution). This heat removal is accomplished by cooling water, which is circulated through the absorber. The main solution pump takes diluted salt solution at 60 per cent and 105 F (point *1*) and delivers it to the *heat exchanger*. In the heat exchanger the weak salt solution is warmed to 180 F (*1* to *2* on Fig. 16–23). From the heat exchanger this weak solution then passes to the *generator*. In the generator it warms to

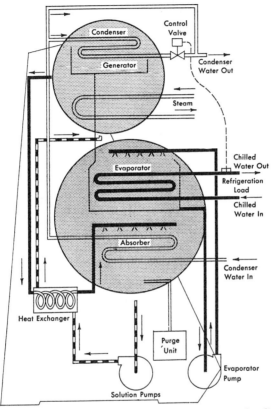

Fig. 16–25. Placement of units and circuit arrangement for Carrier absorption refrigeration unit.

220 F (*2* to *3* to *4*). As the solution receives heat from the steam coils in the generator, excess water is driven off from the solution to enrich it to a salt concentration of 65 per cent (point *4*). The strong solution, which is at 220 F and at the generator pressure of 3 in. Hg abs then enters the *heat exchanger*, where it cools to 145 F (points *4* to *5* in Fig. 16–23) as it warms the counterflowing weak solution. This strong solution then sprays into the *absorber*, dropping in pressure to 0.28 in. Hg abs, where the previously described absorption process takes place.

The water vapor of the refrigerant circuit, when distilled from the solution in the generator, passes to the *condenser*, where circulating water removes heat and changes the water vapor into liquid at 115 F, 3 in. Hg abs. From the condenser, water, the refrigerant of the system, then passes to the *evaporator*, where it drops in pressure to 0.28 in. Hg abs, and a portion of it flashes into low-temperature steam. Water in the evaporator is continuously recirculated by a pump and sprayed over the refrigerant load coils. Because a portion of the water is continuously

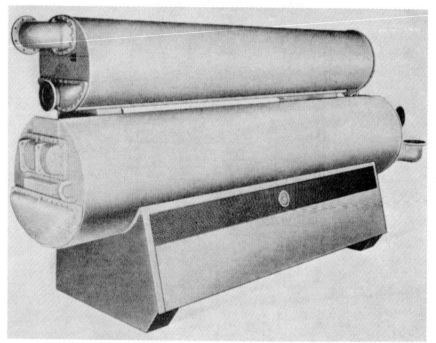

Fig. 16-26. Absorption refrigeration machine using lithium bromide (Courtesy of Carrier Corp.)

evaporated, the evaporator maintains a useful low temperature of approximately 43 F. The water vapor produced passes to the absorber.

The system, although apparently complex, is relatively simple, and in the actual design of the unit the absorber and evaporator are placed in the common lower shell (Figs. 16-25 and 16-26), which operates at the low pressure of the system. The generator and condenser are in the separate upper shell, at the high pressure of the system (3 in. Hg abs). The circulating (condensing) water passes through the absorber, where it rises in temperature in picking up the solution load, following which it moves to the condenser and picks up additional load in condensing the generated vapor. Steam is supplied to the generator essentially at 12 psi gage at a condensing temperature of approximately 245 F. The major energy

Refrigeration Equipment and Arrangement 591

input into the system is from the steam, since the pumps require relatively trivial horsepower.

A Carrier Corporation machine of this type is pictured in Fig. 16–26, where the lower vessel represents the absorber-evaporator unit, and the upper vessel the condenser-generator unit. The pumps are not shown in place. Units of this type are built in sizes ranging from approximately 100-ton capacity to over 700-ton capacity. In performance, a 100-ton machine would need approximately 400 gpm of circulating water and use 19.4 lb of steam per ton hr. Three pumps, all of 2 hp or less, would be required in the 100-ton machine. For the larger size machine the figures can be prorated. A purge unit is required to remove any air which might leak into this highly evacuated system and also to remove any gases which might arise from chemical breakdown. With the advent of relatively cheap gaseous fuel and wherever excess steam is available, units of this type are particularly advantageous for producing refrigeration at air-conditioning levels.

In addition to the Carrier Corporation machine other salt-solution machines have been built; in particular, the small-size machines of 5 to 25 tons that have been designed and marketed by the Servel Corporation.

PROBLEMS

16–1. In testing a cooling unit the refrigerating effect was measured by brine passed through the evaporator. Calcium-chloride brine, of sp gr 1.180 when measured at 60 F, entered the evaporator at 31 F and left at 21 F. During the test the brine rate of flow was 80 lb per min. Find the (a) specific heat of brine (Fig. 16–1), and (b) tons of refrigeration developed. (c) What is the freezing point of this brine? (cf. Fig. 16–1.) *Ans.* (a) 0.731; (b) 2.9; (c) −0.4 F

16–2. The bulb of a thermostatic expansion valve is filled with refrigerant-12 (Freon-12) and controls an evaporator which is to be held at 20 F in the coils. The expansion valve is actuated and causes refrigerant flow when 6 deg of superheat exist in the return line at the point of attachment of the bulb. Under these conditions, find what pressures exist in the bellows chamber for thermostatic gas (Table 15–5) and near the valve outlet to the evaporator and thus, the pressure difference causing the valve to operate. *Ans.* $\Delta p = 4.3$ psi

16–3. A refrigeration system using ammonia (refrigerant-717) condenses at 80 F, and the evaporator refrigerant temperature is 30 F. The liquid at the expansion valve is not subcooled, and dry saturated vapor enters the compressor. Find (a) the enthalpy of the liquid at the expansion-valve inlet; (b) the useful refrigeration in the evaporator, in Btu per pound; (c) the isentropic work of compression, in Btu per pound; (d) the probable shaft horsepower required by the compressor per ton of refrigeration; (e) the probable condenser heat load per ton of refrigeration.
Ans. (a) 132 Btu/lb; (b) 488.5; (c) 56.5; (d) 0.76 shp; (e) 13,930 Btu/ton hr

16–4. A two-cylinder, single-acting compressor, with $2\frac{1}{4}$-in. bore and $1\frac{13}{16}$-in. stroke, runs at 1500 rpm. The condensing temperature is 82 F and the evaporating temperature is 30 F. A heat exchanger which warms to 65 F the 30 F dry saturated vapor from the evaporator is used in subcooling the 82 F condensed liquid to a lower temperature before the liquid enters the evaporator. Refrigerant-12 is used. (a) Compute total piston displacement, in cubic feet per minute. (b) Compute the temperature of the liquid before it enters the expansion valve—after it has passed through the heat exchanger and has been cooled down by the vapor warming from 30 to 65 F. Note that the decrease in enthalpy of the liquid equals the increase in enthalpy of the vapor. (c) Compute the pounds of refrigerant required per ton of refrigeration per minute. (d) For a volumetric efficiency of 88 per cent, compute the pounds of refrigerant handled per minute by the compressor. (e) Find the refrigerating capacity, in tons and in units of 1000 Btuh. (f) Find the isentropic work of compression per pound. (g) Find the theoretical horsepower and specify a motor size for this unit.

Ans. (a) 11.65 cfm; (b) 59 F; (c) 3.4 lb/ton-min; (d) 9.8 lb/min; (e) 3.0 tons, 36,500 Btuh; (f) 7.3 Btu/lb; (g) 1.77 hp and 2.5

16–5. A 100-ton refrigerant-12 (Freon-12) compressor runs with a discharge pressure of 93.3 psig and a suction pressure of 11.79 psig. This liquid is subcooled to 70 F before expansion, and the vapor entering the compressor is superheated 15 deg. Find (a) the heat absorbed in the evaporator per pound of Freon circulated; (b) the weight of Freon flowing per minute per ton of refrigeration; (c) the theoretical piston displacement per minute per ton; (d) the isentropic horsepower; (e) the probable actual horsepower; (f) the probable actual piston displacement per minute per ton.

Ans. (a) 56.0 Btu/lb; (b) 3.54 lb/ton-min; (c) 5.4 cfm/ton; (d) 0.9 hp/ton; (e) 1.26 shp/ton; (f) 6.2 cfm/ton

16–6. In an evaporating condenser assume that 12,000 Btuh, plus 2500 Btuh in the form of work from the compressor, constitute the heat removed in the condenser each hour per ton of refrigeration. (a) If the latent heat of vaporization of water is taken as 1050 Btu per lb, compute, in lb per hr, the water evaporated per ton of refrigeration in the evaporative condenser, and express the quantity of water evaporated also in gpm. (b) In a particular evaporative condenser used on a 5-ton unit, city water supplied has a hardness of 15 grains per gallon. It has been found undesirable to operate evaporative condensers when the hardness exceeds 75 grains per gallon. How much water should be bled (wasted) from this evaporative condenser every hour to keep the hardness at a safe level?

Ans. (a) 13.8 lb/ton hr, 0.0277 gpm/ton; (b) 2.08 gal/hr

16–7. An evaporative condenser serving a 25-ton capacity unit is located on the roof of a building. Compute (a) the probable heat load in Btu per min removed from the condenser, and (b) the gpm evaporated if the latent heat of water is considered to be 1050 Btu per lb. If on a given day the outside air conditions are 80 F dry bulb and 70 F wet bulb and air leaves the condenser essentially saturated at 84 F, compute the air flow in (c) lb per min, and (d) cu ft per min.

Ans. (a) 6250 Btu per min, (b) 0.717 gpm, (c) 488 lb per min, (d) 6750 cfm

16–8. Work problem 16–7 for conditions of a 10-ton evaporative condenser and of outside air at 95 F dry bulb and 75 wet bulb, inlet makeup water at 70 F, and air from the condenser at 100 F dry bulb, at 90 per cent relative humidity.

Ans. (a) 2500 Btu/min, (b) 0.287 gpm, (c) 93.5 lb per min, (d) 1342 cfm

16–9. Consider that the two-stage centrifugal compressor described in example 16–1 is recharged with refrigerant-113 (trichlorotrifluoroethane) for which $k = 1.09$ and the mean specific heat of its vapor can be taken as 0.15. Consider adiabatic compressor efficiency as 70 per cent. A ph diagram for this refrigerant appears in Fig. 15–7. For the same running conditions find the probable maximum pressure ratio and top condensing temperature.

Ans. Total pressure ratio, 8.5; 126 F

Refrigeration Equipment and Arrangement

16–10. A two-stage centrifugal compressor, having two impellers each with a maximum tip diameter of 28 in., is driven by a steam turbine rotating at 4500 rpm. Refrigerant-12 is used, and the suction temperature is -60 F. The compressed refrigerant is liquefied in a condenser cooled by refrigerant from a separate system (cascade arrangement). The refrigerant flow under design conditions at 16 F condensation is 152 lb per min. (a) Compute the probable maximum pressure-rise ratio for this machine, and find the top condensing pressure for -60 F evaporation (consider $\eta_{ic} = 0.72$ and $\mu = 0.9$). (b) Compute the probable horsepower needed from the turbine to run the compressor under design conditions if the power factor compared to isentropic can be considered as 1.38. (The isentropic work can be found by considering the inlet enthalpy and entropy of dry saturated vapor from Table 15–5 and then reading the compressed vapor enthalpy from Fig. 15–4 at the -60 F entropy value.)

Ans. (a) 7.14, 38.3 psia; (b) 65 shaft hp

REFERENCES

1. American Society of Heating, Refrigerating, and Air-Conditioning Engineers, *Fundamentals and Equipment Guide and Data Book* (1965–66), Chapter 38.
2. B. H. Jennings and W. L. Rogers, "Centrifugal Compressors," *Gas Turbine Analysis and Practice*, (New York: McGraw-Hill Book Company, Inc., 1953), pp. 223–41.
3. A. A. Berestneff, "A New Development in Absorption Refrigeration," *Refrigerating Engineering*, Vol. 57 (1949), pp. 553–57, 606–09.
4. E. P. Whitlow and McNeely, "Absorption Air Conditioning," *Refrigerating Engineering*, Vol. 59 (1951), pp. 38–43, 100.
5. S. M. Miner, "Centrifugal Refrigeration," *Refrigerating Engineering*, Vol. 58 (1950), pp. 877–81, 918–20.

17

Cryogenics and Gas Liquefaction

17–1. CRYOGENICS

The word cryogenics, if we consider its sources in the Greek language, means "creation or production by means of cold." The word is presently used in more than one sense but primarily refers to the utilization of low-temperature processes to produce appropriate changes in gases, liquids, or solids to achieve desired objectives. Mediums that are used in the production of low temperatures are described as cryogenic fluids. In one sense also, cryogenics can be considered a synonym for refrigeration but it differs from the latter by virtue of the fact that it occupies a lower-temperature range of utilization which extends from -240 F down to almost -460 F.

It has been shown that ordinary refrigerants are gases which after compression could readily be condensed in the ambient atmospheric-temperature range. Cryogenic fluids, on the other hand, are usually gases which possess critical temperatures much lower than normal atmospheric temperatures and thus cannot be liquefied by conventional compression and heat rejection. Air and its major constituents, nitrogen and oxygen, are the most common cryogenic fluids along with the two more permanent gases, hydrogen and helium.

Various processes are used in the liquefaction of gases. An early and simple process involved first the compression of the gas to high pressure. Following this, after any pre-cooling of the gas that might have been possible, the gas was then allowed to expand either in an expander engine or it was throttled and use made of the Joule-Thompson effect to lower the gas temperature. Because of its importance in cryogenics, Joule-Thompson (Kelvin) cooling is considered in detail in the next section.

17–2. JOULE-THOMPSON EFFECT

When a gas expands freely, without doing work, it carries out a throttling process. As it expands its temperature falls, stays constant, or increases. These widely variable patterns are indicated by the so-called

Cryogenics and Gas Liquefaction

TABLE 17-1
DATA ON CRYOGENIC GASES

Gas	Mol Weight	Boiling Temperature at 14.696 psia		Critical Temperature		Critical Pressure		Representative Inversion Temperature		Triple Point	Latent Heat @ 1 atm	Freezing Point @ 1 atm
		°K	°R	°K	°R	atm	psia	°K	°R	°K	Btu/lb	°R
Helium (He4)	4.003	4.22	7.6	5.3	9.5	2.24	33.2	40.0	72.0	None	8.9	
Helium (He3)		3.19	5.74	3.35	6.03	1.17	17.2			None		
Hydrogen (n)	2.016	20.4	36.7	33.2	59.8	12.98	190.8	177	319	13.9	194	25.1
Hydrogen (para)	2.016	20.3	36.5	32.9	59.4	12.7	187.7	171	308	13.8	193	
Neon	20.183	27.2	48.9	44.4	79.9	26.86	394.7			24.6	37.1	44.2
Nitrogen	28.016	77.3	139.2	126.3	227.3	33.5	492.9	621	1118	63.1	85.8	113.9
Air	28.966	78.8*	141.9	132.4	238.3	37.2	546.6	>600	>1080		88.2	
Argon	39.944	87.3	157.1	150.9	271.6	48.3	710.4	710.4	1279	89.3	70.2	150.4
Oxygen	32.00	90.2	162.4	154.8	278.5	50.1	736.3	893	1607	54.4	96.7	97.8
Krypton	83.70	121.3	218.3	209.4	376.9	54.2	796.5			116	46.5	208.7
Methane	16.04	111.6	200.8	191.1	343.9	45.8	673.0	>500	>900	89.2	219.2	163.3
Ethylene	28.05	169.3	304.7	282.5	508.5	49.8	731.8	>500	>900	103.7	207	187.2
Ethane	30.07	184.6	332.2	305.7	549.8	48.2	708.3	>500	>900	89.9	210	162.0
Carbon dioxide	44.01	194.7ₛ†	350.4ₛ†	304.2	547.5	72.9	1071.1	>500	>900	216.6	246ₛ†	350.4ₛ†
Refrigerant-14	88.01	145.1	261.2	227.6	409.8	36.9	542.3				58.6	161.1

*Bubble point temperature.
†ₛ = Sublimation temperature and energy.

Joule-Thompson coefficient. In ordinary throttling where a gas drops in pressure, kinetic energy effects frequently occur, as the gas accelerates in passing through a small opening. Usually when kinetic energy is thus created, the temperature is lowered. Thus, in throttling where the temperature effect is of interest, it is necessary to eliminate the effect of velocity so that its magnitude before and after expansion is the same and relatively low. A descriptive term used in connection with such expansion is the porous plug, since by such a device the pressure can be reduced and velocity effects essentially eliminated. It has already been shown before that during a throttling process, the enthalpy stays constant, or more specifically the gas moves from one state to another at each of which the enthalpy of the gas has the same value.

If a gas were truly perfect, there would be no change in its temperature as the gas throttles or flows through a porous plug. For real gases, however, two conditions must be considered. The first relates to the fact that real gases do not adhere to the simple relationship (sometimes known as Joule's Law), which states that the internal energy of a gas is independent of its pressure and volume at a constant temperature. This implies that there are essentially no attractive forces between the molecules. The other condition is related to deviations from Boyle's law for gases. This law states that the product of pressure times volume for a gas is constant at a given temperature. Deviations related to the first condition produce cooling during expansion because when the molecules move further apart and gain potential energy, this occurs at the expense of kinetic energy so the temperature decreases. Boyle's law deviations may produce either an increase or decrease in temperature. Particularly at low pressures and temperatures and also at conditions not far from condensation, gases are extremely compressible. Such gases would cool as their pressure is reduced in a throttling expansion. Thus, with two basic and diverse patterns applying to a gas at parts of its range, we must expect variability in the Joule-Thompson coefficient.

In mathematical terminology, the Joule-Thompson coefficient is expressed as

$$\mu = \frac{\partial T}{\partial p}\bigg)_h \tag{17-1}$$

when under conditions of throttling (h = constant)

∂T = change in °K   or   in °R

∂p = change in atmospheres or occasionally in psi

In Fig. 17-1, Joule-Thompson inversion curves for a number of gases are plotted. To the left of the curve for each gas at a given temperature

Cryogenics and Gas Liquefaction

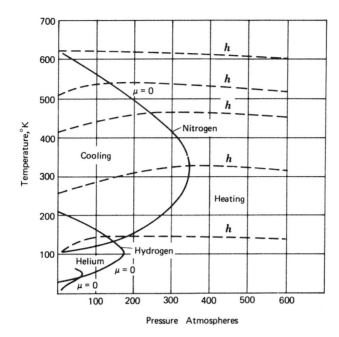

FIG. 17-1. Inversion curves for representative gases including isenthalpic lines for nitrogen.

and pressure, the Joule-Thompson coefficient for that gas is positive. On the curve itself, the coefficient is zero and represents the locus of the inversion points for that gas. To the right of each curve, the Joule-Thompson coefficient is negative and would produce an increase in temperature as the gas throttled (expanded). On the particular curve for nitrogen, a number of constant enthalpy lines are plotted in to show the general shape of isenthalpic lines (h = constant). It will be noted that these rise to the left of the inversion curve, have zero slope at the inversion curve and fall at the right side of the curve. Thus, if we had nitrogen at 200 atmospheres pressure and 300 K and allowed the pressure to drop isenthalpically to 100 atmospheres, the temperature would fall by some 10 degrees. Quantitative computations using the Joule-Thompson coefficient will be shown later in this discussion. It will be noted that air and, of course, its constituents, oxygen and nitrogen, have positive Joule-Thompson coefficients at normal temperatures. The same is true for such easily liquefied gases as carbon dioxide and ammonia. On the other hand, hydrogen and helium have negative Joule-Thompson coefficients under normal conditions and cannot be liquefied by Joule-Thompson expansion unless precooled. In fact, hydrogen would have to be dropped to a temperature of approximately 180 K at 50 atmospheres pressure before it

TABLE 17-2
HELIUM (REFRIGERANT 704), SPECIFIC HEAT OF LIQUID AT SATURATION (C_s), HELIUM VAPOR AT CONSTANT PRESSURE (C_p), AND PROPERTIES OF SATURATED LIQUID AND VAPOR (REFS. 3, 4, AND 5)

°R	C_s for Liquid Cal per gram K (Btu per lb R)	\multicolumn{6}{c}{C_p for Vapor at Varying Pressures (Cal per gram K)}					
		°K	°R	3 atm	5 atm	10 atm	30 atm
3.24	0.672	6	10.8	2.91		1.18	0.77
3.40	0.906						
3.60	1.24	6.5	11.7	2.14	3.40	1.35	
3.78	1.80	7	12.6	1.84	2.83	1.53	0.90
3.87	2.23	8	14.4	1.55	1.96	1.93	1.02
3.89	3.01	9	16.2	1.46	1.67	1.96	1.14
3.96	0.951	10	18.0	1.40	1.53	1.81	1.25
4.68	0.542	12	21.6		1.42	1.59	1.40
5.04	0.559	14	25.2		1.37	1.47	1.47
5.40	0.595	16	28.8		1.34	1.40	1.48
6.30	0.741	18	32.4		1.32	1.36	1.46
7.20	0.953	20	36.0			1.34	1.44
8.10	1.315			1 atm	3 atm	10 atm	100 atm
9.00	2.75	47.4	84.7	1.24	1.27	1.27	1.38
9.09	3.23	102.6	184.7	1.24	1.24	1.24	1.28
		186.0	334.7	1.24	1.24	1.24	1.25
		227.6	409.7	1.25	1.24	1.24	1.25
		283.2	509.7	1.24	1.24	1.24	1.24
		394.3	709.7	1.24	1.24	1.24	1.24
		505.4	909.7	1.24	1.24	1.24	1.24

PROPERTIES OF SATURATED LIQUID AND VAPOR

Temperature		Pressure	Specific Volume		Enthalpy		Entropy	
			\multicolumn{2}{c}{cm^3/g}	\multicolumn{2}{c}{joules/g}	\multicolumn{2}{c}{joules/g K}
K	R	atm	Liquid	Vapor	Liquid	Vapor	Liquid	Vapor
3.0	5.4	0.241	7.085	224.1	5.30	28.96	2.356	10.247
3.2	5.8	0.320	7.185	174.8	5.85	29.48	2.527	9.911
3.4	6.1	0.416	7.303	138.7	6.47	29.93	2.690	9.606
3.6	6.5	0.529	7.439	111.5	7.18	30.34	2.883	9.322
3.8	6.8	0.661	7.597	90.50	7.93	30.59	3.059	9.021
4.0	7.2	0.814	7.779	73.86	8.81	30.72	3.255	8.736
4.2	7.6	0.990	7.989	60.61	9.85	30.79	3.460	8.443
4.4	7.9	1.190	8.231	49.80	10.96	30.65	3.674	8.150
4.6	8.3	1.417	8.547	40.72	12.33	30.32	3.920	7.845
4.8	8.6	1.672	9.033	32.87	13.88	29.49	4.188	7.473
5.0	9.0	1.959	9.940	25.67	16.00	28.02	4.598	7.000

would cool as it expanded in a throttling process. Similarly, helium requires a temperature around 40 K before it can be liquefied under throttling expansion.

Cryogenics and Gas Liquefaction 599

For those interested in carrying out a more-mathematical analysis of the Joule-Thompson process, the following development is presented. Let us first write enthalpy, h, as a function of pressure and temperature and make use of the relationships involved with partial derivatives,

$$d\mathbf{h} = \left.\frac{\partial \mathbf{h}}{\partial P}\right)_T dP + \left.\frac{\partial \mathbf{h}}{\partial T}\right)_P dT \quad (17\text{-}2)$$

If enthalpy is kept constant in a given process $d\mathbf{h} = 0$, and this permits a rearrangement of equation 17-2 to show

$$\left.\frac{\partial T}{\partial P}\right)_\mathbf{h} = -\left.\frac{\partial \mathbf{h}}{\partial P}\right)_T \bigg/ \left.\frac{\partial \mathbf{h}}{\partial T}\right)_P \quad \text{and} \quad \left.\frac{\partial T}{\partial P}\right)_\mathbf{h} = -\frac{1}{C_P}\left(\frac{\partial \mathbf{h}}{\partial P}\right)_T \quad (17\text{-}3)$$

by reference to equation 2-26.

Next combine the first law, equation 2-60, with enthalpy in its differentiated form, equation 2-22,

$$Tds = du + Pdv \quad [2\text{-}60]$$

$$d\mathbf{h} = du + Pdv + vdP$$

then

$$d\mathbf{h} = Tds + vdP \quad (17\text{-}4)$$

and

$$\frac{d\mathbf{h}}{dP} = T\frac{ds}{dP} + v \quad (17\text{-}5)$$

Equation 17-5 is equally valid if T is constant and thus

$$\left.\frac{\partial \mathbf{h}}{\partial P}\right)_T = T\left.\frac{\partial s}{\partial P}\right)_T + v \quad (17\text{-}6)$$

This expression can be modified to eliminate the entropy-pressure term if use is made of the fact that thermodynamic properties are point functions and have exact differentials. For such functions second derivatives of the partial differential coefficients must be equal. For example in equation 17-2, it must follow for the partial coefficients that

$$\frac{\partial^2 \mathbf{h}}{\partial P \partial T} = \frac{\partial^2 \mathbf{h}}{\partial T \partial P}$$

since the order of differentiation is immaterial. Let us now apply this procedure to the coefficients T and v of the differentials in equation 17-4 and differentiating we find that

$$\left.\frac{\partial T}{\partial P}\right)_s = \left.\frac{\partial v}{\partial s}\right)_P \quad (17\text{-}7)$$

This is one of the Maxwell relations, here worked out to show how the method is applied. To improve equation 17-6 let us make use of the

Gibbs function (Z), defined as

$$Z = \mathbf{h} - T\,s$$

$$d\,Z = d\mathbf{h} - T\,ds - s\,dT$$

Substitute equation 17–4 in this, and we find

$$d\,Z = -s\,dT + v\,dP \tag{17-8}$$

Now carry out second order differentiation and there results

$$\left.\frac{\partial s}{\partial P}\right)_T = -\left.\frac{\partial v}{\partial T}\right)_P \tag{17-9}$$

Substitute equation 17–9 in equation 17–5

$$\left.\frac{\partial \mathbf{h}}{\partial P}\right)_T = -T\left.\frac{\partial v}{\partial T}\right)_P + v \tag{17-10}$$

and substitute this in equation 17–3 to show

$$\mu = \left.\frac{\partial T}{\partial P}\right)_\mathbf{h} = \frac{1}{C_p}\left[T\left.\frac{\partial v}{\partial T}\right)_P - v\right] \tag{17-11}$$

This is the basic relationship for μ, the Joule-Thompson coefficient, for all gases. Note the following when throttling at constant enthalpy takes place:

$\mu = 0$ no temperature change occurs

$\mu = +$ temperature falls as pressure falls, cooling effect

$\mu = -$ temperature rises as pressure falls, heating effect

For the perfect gas, $Pv = RT$, note that for equation 17–11

$$\left.\frac{\partial v}{\partial T}\right)_P = \frac{R}{P} \quad \text{and} \quad v = \frac{RT}{P}$$

thus $\mu = 0$, as would be expected since C_p is finite. The values of μ for a real gas must be determined experimentally unless an equation of state for that gas can be expressed in mathematical terminology. With such an equation the value of μ could be computed at various pressures and temperatures with the help of equation 17–11.

Insight into the Joule-Thompson mechanism can be gained from the fact that during throttling, enthalpy is a constant set of equivalent values, such that

$$\mathbf{h}_1 = \mathbf{h}_2 \quad \text{and} \quad u_1 + P_1 v_1 = u_2 + P_2 v_2$$

also

$$u_2 - u_1 = P_1 v_1 - P_2 v_2 \tag{17-12}$$

Note that if we have an expanding gas, *maintained at the same temperature* by heat addition or otherwise, u_2 after expansion is greater than u_1 before

Cryogenics and Gas Liquefaction

expansion because with real gases there is an attractive force between the molecules and with separation taking place, increased potential energy must be stored in the molecules. Thus, during a throttling expansion under adiabatic conditions the above reasoning would show that considering the u effect alone, the temperature of the expanding gas would fall. Now observe the other side of equation 17–12 where the Pv terms appear. For a real gas there is no assurance that the Pv product remains constant and the following possibilities exist:

$$a) \quad P_1v_1 = P_2v_2$$

$$b) \quad P_1v_1 < P_2v_2$$

$$c) \quad P_1v_1 > P_2v_2$$

For case a, temperature falls during expansion since this side cannot offset the overriding effect of $u_1 < u_2$.

For case b, temperature falls during expansion but even more than in case a.

For case c, temperature rises during throttling expansion, except for very small inequalities in the Pv terms.

The integral Joule-Thompson effect appears when equation 17–1 is written in integral form as

$$\Delta T = \int_{P_1}^{P_2} \mu dP \qquad (17\text{–}13)$$

For arithmetic usage with this equation, it is customary to break the pressure drop into a series of steps using an appropriate average value of μ for each pressure range.

Example 17–1. Air at a pressure of 400 psia and at 70 F is throttled to 14.7 psia into an insulated chamber under conditions of trivial kinetic energy. Compute the temperature decrease resulting from the integral Joule-Thompson cooling effect making use of the data of Fig. 17–2. Recompute if the high-pressure air is regeneratively cooled to 200 K before expansion.

Solution: 400 psia = $\dfrac{400}{14.7}$ atm = 27.2 atm and 530 R = $\dfrac{530}{1.8}$ or 294.4 K. The value of μ is not extremely sensitive to pressure in this range and so Fig. 17–2 can be used in a series of steps. Select these by trial, primarily in terms of the temperature range.

Write $\Delta T = \int_{P_1}^{P_2} \mu dP$ in the form

$$\Delta T = \Sigma \, (\mu_1 \Delta P_1 + \mu_2 \Delta P_2 + \mu_3 \Delta P_3 + \cdots)$$

A preliminary trial might show that for a 10° drop an average value is 0.24. Then

$$\Delta T = -(0.24)(27.2 - 1) = -6.3°$$

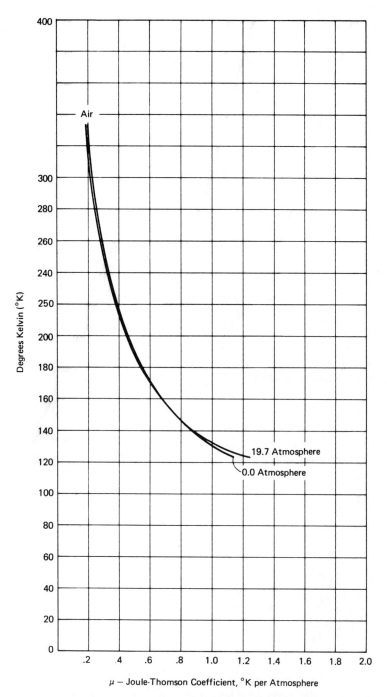

Fig. 17-2. Joule-Thompson coefficients for air. (ref. 1)

Cryogenics and Gas Liquefaction

Thus less than a 10° drop is produced and if we now try 2 steps

$$\Delta T = (0.23)\left(-\frac{27.2 - 1}{2}\right) + (0.24)\left(-\frac{27.2 - 1}{2}\right) = -6.0°$$

Thus T, final, is about $294.4 - 6.0 = 288.4$ K or correspondingly a 10.8° drop on the Rankine scale gives

$$530 - 10.8 = 519.2 \text{ R} \quad \text{or} \quad 59.2 \text{ F}$$

Consider now that the expander is provided with a counterflow regenerative heat exchanger such that after continuing expansion and back flow the equilibrium temperature before the expansion valve reduces to 200 K. To compute the ΔT under these conditions, we observe higher values for μ and using 2 steps

$$\Delta T = (0.46)\left(-\frac{27.2 - 1}{2}\right) + (0.47)\left(-\frac{27.2 - 1}{2}\right) = -12.3°$$

Here the final temperature would be 187.7 K after expansion.

17-3. LIQUEFACTION OF AIR BY THE HAMPSON SYSTEM (JOULE-THOMPSON EXPANSION)

Fig. 17-3 shows in diagrammatic form an early method, often called the simple-Linde or the Hampson system, used for the liquefaction of air or other gases. In it air is cleaned, dried, and then compressed to high pressure with reciprocating compressors. It is further purified and dried and then passes through the Joule-Thompson expander (valve). After this expansion, the now cold air returns through a counterflow heat exchanger and is led back to the compressor. In the heat exchanger the high-pressure air flowing to the expander valve progressively cools to such an extent that after passage through the valve, a fraction of the air becomes liquid while the very cold non-liquefied air moves back through the counterflow exchanger. The throttling process is one of constant enthalpy and thus it is easy to write an equilibrium heat-balance equation in terms of the enthalpies of the air streams. These enthalpies are indicated at appropriate points on the expansion diagram and the apparatus can be sufficiently insulated to assume it to be adiabatic. The very simple equations follow:

$$h_1 - h_2 = (1 - \lambda)(h_4 - h_3) \qquad (17\text{-}14)$$

$$h_2 = \lambda h_f + (1 - \lambda)(h_3) \qquad (17\text{-}15)$$

$$\lambda = \frac{h_4 - h_1}{h_4 - h_f} \qquad (17\text{-}16)$$

here:
h_1 is the enthalpy of the high pressure air entering the heat exchanger;
h_2 is the enthalpy of the high pressure air at the expansion valve;
h_3 is the enthalpy of the non-liquefied low-pressure air;
h_4 is the enthalpy of the low-pressure air leaving the heat exchanger;

Cryogenics and Gas Liquefaction

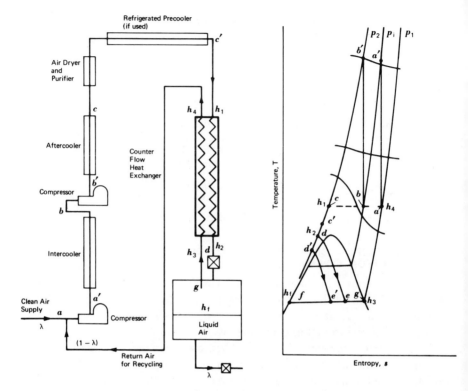

FIG. 17-3. Simple Joule-Thompson expansion for liquefaction of air and Ts diagram showing intercooling for 2-stage compression.

h_f is the enthalpy of liquid air;
λ is the fraction of the inlet air flow which is liquefied.

For liquefaction of air by this method, high pressures are required, reaching 200 or more atmospheres. For compressions of this extent, reciprocating compressors are used and the work is done in 2, 3, or 4 stages with intercoolers between each stage and an aftercooler following the last stage. Two-stage compression is shown in Fig. 17-3 and on the Ts diagram of the cycle. Compression is indicated from a to a', intercooling at an intermediate pressure from a' back to b, with compression in the second stage occurring from b to b'. A similar pattern would be followed if more than two stages were used. Following aftercooling, the highly compressed air enters the well-insulated heat exchanger in which the cold gas, returning from the liquefaction chamber, flows counterflow to the warm air moving to the expansion valve. This counterflow heat exchange chills the air in the heat exchanger to point d on the Ts diagram. In the throttling process at the valve, from d to e the pressure drops to atmospheric or essentially atmospheric pressure, and a fraction of the air liquefies along

Cryogenics and Gas Liquefaction

with a large amount of air which remains in gaseous state. The state point after expansion is indicated at e on the Ts diagram, and the fraction of liquid formed is represented by the length eg with the fraction of gas formed represented by the length fe. A definite improvement in the amount of liquid produced can be brought about if a precooler, shown in Fig. 17-3, is also employed. Such a precooler uses the cooling effect of a separate refrigeration system to bring the temperature of the air to an appreciably lower value for entry into the counterflow heat exchanger. In the Ts diagram this new temperature is indicated as c', and with a lower temperature on inlet to the heat exchanger, a lower temperature on exit from the heat exchanger indicated as d' would also be obtained. It would thus follow that the final state point in the liquid chamber would be at e' and, as can be seen, a much larger yield of liquid air would be produced.

Equation 17-16 was developed as a heat balance for the countercurrent heat exchanger of the system. Another approach to analyzing the performance of the system can be carried out by making an energy and mass balance of the whole system. This balance will assume that λ pound of liquid leaves the system for each pound flowing through the system. It is obviously necessary that λ pound of make-up air gas must also be provided. Energy is added in the form of work at the compressors and heat is rejected at the intercoolers, aftercooler, and precooler, if the latter is used. Some heat gain (Q) also occurs because the temperatures at the cold end are sufficiently low that even with effective insulation, some heat leaks into the system. The energy balance, using the notation of Fig. 17-3, takes the following form:

$$\lambda h_a + (W_c - q_{\text{int}} - q_{\text{aft}} - q_{\text{pre}}) + Q = \lambda h_f \qquad (17\text{-}17)$$

Note that for the term in parentheses, the energy added as work is either transferred out from the system as heat to appropriate cooling mediums or if not transferred out, resides in the gas and is measured by an enthalpy increase. If this term is expressed as a change in enthalpy, it would appear

$$W_c - q_{\text{int}} - q_{\text{aft}} - q_{\text{pre}} = h_1 - \lambda h_a - (1 - \lambda) h_4 = h_1 - h_4 \qquad (17\text{-}18)$$

The final simplification was made because h_4 and h_a have closely the same value with an effective heat exchanger, and thus equation 17-18 can be written:

$$\lambda h_a + h_1 - h_4 = \lambda h_f - Q$$

$$\lambda = \frac{h_4 - h_1 - Q}{h_4 - h_f} \qquad (17\text{-}19)$$

The units of equation 17-17 are the same as those for equation 17-16. Note that with enthalpy h, in Btu per lb, Q would represent the insulation-leakage heat gain in Btu per lb of air entering the hot end of the heat exchanger.

In the operation of an air liquefaction system, many problems must be solved to keep the system in continuous operation. Atmospheric air, for example, contains dirt, extraneous gases, such as water vapor, CO_2, the inert gases, argon, neon, xenon, etc. Fortunately, most of the water vapor can be condensed in the intercoolers and aftercoolers, and the remainder can be taken out with desiccators. The carbon dioxide, CO_2, can also be removed chemically, if desired, but usually the CO_2 is allowed to freeze out as a solid in the heat-exchanger system where the temperature is sufficiently low to produce this effect. Continual freezing out of the CO_2 would ultimately block the heat exchanger passages so this solid CO_2 or "dry ice," in turn, must be removed. Various ingenious arrangements have been developed to bring this about, one of which, the intermittent regenerative heat exchanger, is discussed later.

The Hampson cycle, although described for air, is applicable to the liquefaction of other gases, provided the temperature of the gas is sufficiently low that its Joule-Thompson coefficient is positive on leaving the precooler and at entry to the heat exchanger. It should, however, be mentioned that this essentially basic process for gas liquefaction leaves much to be desired, since the Joule-Thompson expansion is a highly irreversible process and high pressures are required to produce a liquid product. More effective methods will be described in the sections which follow.

Example 17-2. A simple Linde (Hampson) system at the warm end of its countercurrent heat exchanger shows compressed air at 200 atm and 20 C (68 F). The return gas from the liquid-air storage tank on arrival at the warm end of the heat exchanger shows a temperature of 14 C. On the basis of these data, compute the probable weight of liquid air produced from each pound supplied to the Joule-Thompson throttling valve. Make use of the Ts diagram for air. Disregard heat gain to the heat exchanger.

Solution: Refer to Fig. 17-3 and use Fig. 17-5 for the data on air.

Read h_4 = 3.2 cal per gram (k cal per kg) at 14 C and 1 atm

h_1 = -4.0 k cal/kg at 20 C and 200 atm

h_f = -97.0 k cal/kg at 1 atm (by extrapolation)

By equation 17-19

$$= \frac{3.2 - (-4.0)}{-4.0 - (-97.0)} = \frac{7.2}{93.0} = 0.073 \text{ kg of liquid air produced per kg of compressed air supplied or lb per lb compressed.}$$

17-4. IDEAL GAS LIQUEFACTION

The production of liquid from a gaseous medium requires minimum work if the processes involved are reversible both for the heat dissipation

Cryogenics and Gas Liquefaction

to the warm environment and for the work interchange during compression and expansion. The reversible path shown in Fig. 17-4 involves isothermal compression and heat rejection from 0 to 1; isentropic expansion in an expansion engine from 1 to 2 with work delivery, ending with the state point on the liquid line. If one wished to consider the gas undergoing a cyclic process, constant-pressure addition of heat from 2 to 3 as the liquid vaporizes, with further heat added at constant pressure from 3 to 0 as the gas superheats, would bring the medium back to its original state.

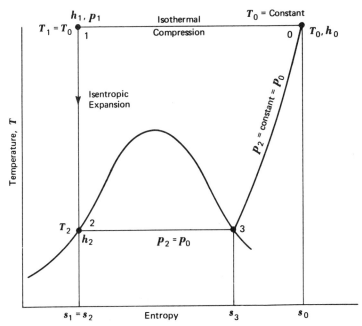

Fig. 17-4. Reversible paths for ideal gas liquefaction on the Ts plane.

The cycle, which brings the medium back to its original state, is not of interest since we wish a continuous product. Thus, for this analysis only the reversible isothermal compression (0–1) and the isentropic expansion (1–2) are required to produce a liquid product at 2.

For steady-flow isothermal compression from 0 to 1, both the heat added and work done by the gas are negative, that is heat is rejected and work is done on the gas, thus

$$-Q_{(0-1)} = \mathbf{h}_1 - \mathbf{h}_0 - \mathbf{W}_{(0-1)}$$

Since the process is isothermal, it follows that

$$-Q_{(0-1)} = T_0 (s_1 - s_0) = -T_0 (s_0 - s_1) \qquad (17\text{-}20)$$

$$-\mathbf{W}_{(0-1)} = -T_0 (s_0 - s_1) - \mathbf{h}_1 + \mathbf{h}_0 \qquad (17\text{-}21)$$

The work delivered by the expansion engine under reversible adiabatic conditions is

$$W_{(1-2)} = h_1 - h_2 \qquad (17\text{-}22)$$

Combine equations 17-21 and 17-22, and we get the total work of the process, really the net work

$$\text{Net work} = -T_0(s_0 - s_1) + h_0 - h_2$$

The net work is negative in sense and should be so indicated by a change in sign. This change is also needed to give positive numerical results.

$$-\text{Net work} = T_0(s_0 - s_1) - (h_0 - h_2) \qquad (17\text{-}23)$$

Notice that $s_1 = s_2 = s_{f2}$ where the subscript f refers to the saturated liquid line, thus:

$$(-)\,\text{Net work} = T_0(s_0 - s_{f2}) - (h_0 - h_{f2}) \qquad (17\text{-}24)$$

$$(-)\,\text{Net work} = T_0\,\Delta s - \Delta h \qquad (17\text{-}25)$$

The minus sign is merely a symbolic indication that net work is needed to produce the required cooling, and the above expressions are a measure of the minimum work required to produce a continuous supply of liquefied gas.

Equation 17-24 measures the minimum work required for gaseous liquefaction, nevertheless no actual system is designed to produce liquefaction by the process illustrated from 0 to 2 in Fig. 17-4. This is true for many reasons but primarily because isothermal compression of a gas cannot be readily achieved in actual compression and secondly, extremely high pressures, reaching unattainable limits from many original conditions, would be required to produce the indicated liquefaction at state point 2. However, the use of expander engines (or turbines) for the gas in other process arrangements is a most important means of lowering temperature even though no engine can actually perform under isentropic conditions.

Example 17-3. Compute the theoretical minimum work required to change dry air at 68 F (20 C) and 14.7 psia (1 atm) to liquid at 141.7 R. Compute by using enthalpy and entropy values by making use of appropriate values from air tables or charts.

Solution: Make use of equation 17-24. In this case $T_0 = 20 + 273.16 = 293.16$ K and from Fig. 17-5 read

$s_{f2} = -0.912$ at 1 atm on the saturation liquid line, $s_0 = 0.019$;
$h_0 = 4.9$ kcal/kg, $h_2 = h_{f2} = -96.8$ kcal/kg.

Net work $= T_0(s_0 - s_{f2}) - (h_0 - h_{f2}) = 293.16\,[0.019 - (-0.912)] -$
$$[4.9 - (-96.8)] = 171.3 \text{ kcal/kg}$$

$$171.3 \times 1.8 = 308.4 \text{ Btu/lb}$$

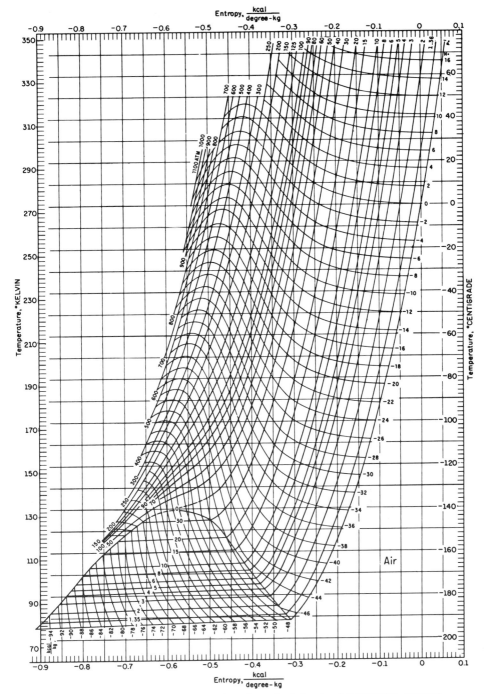

Fig. 17-5. Ts diagram for air (based on data of ref. 1 and ref. 2, NBS File C-3663). Note $h_f = -96.8$ and $h_g = -47.5$ kcal per kg at 1.0 atm.)

Thus 308.4 Btu per lb of liquid air produced is the minimum work required. Expressed in terms of kwhr, since 3413 Btu equal 1 kwhr

$$\frac{308.4}{3413} = 0.0904 \text{ kwhr}$$

= 90.4 watt hr minimum work required per lb of air liquefied.   *Ans.*

17–5. TEMPERATURE-ENTROPY DIAGRAM FOR AIR

Air is a mixture of gases and its representative composition appears on page 85. It is the most common of all gases, yet, because it is a mixture of several gases, its properties have not been as rigorously determined as is the case for its component gases. Since it is a multi-component gas, with every constituent gas having a different boiling point at each pressure, the boiling temperature of air at the liquid line (bubble point) is not the same as the dew-point temperature at the saturated vapor line. Thus the constant-pressure lines in the liquid region on a Ts diagram are not parallel to the abscissa base but slope upward from liquid to vapor line. Because the components boil at different temperatures, it is possible to separate them by fractional distillation as the lower boiling components vaporize at greater concentration in the vapor phase than exists for them in the liquid phase.

Figure 17–5 has been constructed making use of the data of Claitor and Crawford (ref. 1) in the low-pressure region and the data of Michels, Wassenaar and Walkers (ref. 2) in the high-pressure region. The zero value of enthalpy and entropy is arbitrarily chosen at 0 C (273.16 K). The high-pressure lines above the critical-pressure of 37.2 atm can be seen approaching each other in the compressed liquid region. From the peak value for each constant-enthalpy line (where μ the Joule-Thompson coefficient is zero) enthalpy values slope downward. This indicates positive values for μ everywhere on the chart except at extremely high pressures of 300 to 400 atm. Enthalpy is expressed in k cal per kg and can be converted to Btu per lb by the simple multiplying factor 1.8.

17–6. THE LINDE DUAL-PRESSURE CYCLE

An early improvement to the previously-discussed Hampson cycle was brought into use by Linde and was known as the dual-pressure cycle. Using it, the air still had to be highly compressed to about 200 atmospheres but the Joule-Thompson expansion was carried out in two steps. As shown in Fig. 17–6, the first step allows the air to expand to approximately 30 atmospheres and some liquid is created. This liquid then undergoes a second expansion where usable liquid product along with flash gas is produced at essentially atmospheric pressure. Air in gaseous state exists at both expansion points. However, the returning gas, from the first expansion point, has to be compressed only from 30 to 200 atmospheres instead

Cryogenics and Gas Liquefaction

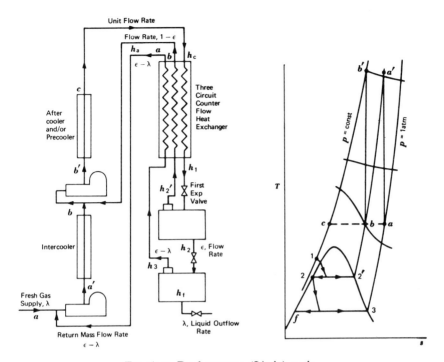

Fig. 17-6. Dual-pressure (Linde) cycle.

of from 1 to 200 atmospheres as is the case with the second expansion step. The result of this is a decrease in the amount of work required for compressing, since the work of compression is directly proportional to the pressure ratio of compression. However, the cooling effect is proportional to the pressure difference, and for this the full 200 atmosphere pressure range is available. This cycle shows a significant lowering of work input over the Hampson process.

Let us write a mass and energy balance in terms of λ, the mass of liquefied gas produced per unit mass flowing from the last compressor. Use conditions at the warm end of the heat exchanger and at exit from the last liquid storage tank.

$$(1)h_c = \lambda h_f + (1 - \epsilon)h_b + (\epsilon - \lambda)h_a$$

$$\lambda(h_a - h_f) = (h_b - h_c) + \epsilon(h_a - h_b) \qquad (17\text{-}26)$$

$$\lambda = \frac{h_b - h_c}{h_a - h_f} + \frac{\epsilon\,(h_a - h_b)}{h_a - h_f} \qquad (17\text{-}27)$$

A comparison of equation 17–27 and 17–19 shows an additional term in the latter equation. However, this does not greatly increase the yield of λ since ϵ is far from unity. The magnitude of ϵ is set in terms of the selected intermediate pressure return flow. The main effectiveness of the cycle lies in the work reduction which develops because only a part of the gas is compressed through the entire pressure range.

17-7. WORK COMPUTATIONS FOR GAS COMPRESSION

For a gas, the minimum work for the actual compression phase occurs when the compression can be carried out isothermally (at constant temperature). For isothermal compression to occur, the heat of compression has to be removed continuously at a rate sufficient to keep the temperature from rising during the compression process. This is not actually possible but the isothermal process does provide a convenient basis for comparison of cyclical arrangements. Real compression is more closely approximated by essentially adiabatic compression in one stage followed by intercooling to the isothermal temperature, with the compression and intercooling pattern repeated in subsequent stages. On the Ts diagram of Fig. 17-3, the idealized isothermal compression follows the line $a\ b\ c$ while the reversible adiabatic pattern for 2-stage compression would be compression a to a', intercooling a' to b, compression b to b' and aftercooling b' to c. Both the isothermal pattern and the adiabatic with intercooling will be used in subsequent analyses.

The first law when applied to a steady-flow compression process with negligible kinetic effects appears

$$W_{(1-2)} = Q_{(1-2)} + h_2 - h_1 \qquad (17\text{-}28)$$

here the signs are chosen as positive for work added (compression) and for heat rejected.

For the reversible isothermal case, referring again to Fig. 17-3, for compression from a to c we can write equation 17-28, noting that $T_0 = T_a = T_b = T_c$, as

$$W_{(a-c)} = T_0 (S_a - S_c) + h_c - h_a \qquad (17\text{-}29)$$

Values resulting from the use of equation 17-29 are significantly lower than the required work of an actual compression. However, for multistage systems with effective intercooling equation 17-29 used with a compression efficiency in the area of 65 per cent gives data comparable with real performance.

For two-stage reversible-adiabatic compression, the work would appear

$$W_{(1-2)} = (h_a' - h_a)_s + (h_b' - h_b)_s \qquad (17\text{-}30)$$

The heat dissipation at constant pressure for intercooling and aftercooling is

$$Q_{\text{total}} = (h_a' - h_b) + (h_b' - h_b) \qquad (17\text{-}31)$$

Cryogenics and Gas Liquefaction

For approximation of true work input, a compression efficiency in the region of 72 per cent is suggested. This makes some allowance for pressure losses in the intercoolers and connecting piping as well as for compressor inadequacies.

An examination of Fig. 17-5, the Ts diagram for air, shows that pressure compression ratios of more than some 2 to 6 are out of chart range for reversible-adiabatic compressions when initial temperature conditions are those of the ambient atmosphere. Thus, compression to 200 atm from 1 atm cannot be read from the chart even for 2-stage operation. In fact, considering air as a perfect gas, use of equation 2-47 would show for a 30 to 1 pressure-compression ratio starting at 68 F, that a final temperature of 935 F would be reached. This is an extremely high temperature for compression and would be objectionable from an operating viewpoint. Consequently it is customary to use 3 or preferably 4-stage compression to reach 200 atm with intercooling provided between stages.

For approximately equal work in each stage, use can be made of the following expressions to determine intermediate pressures with p_1 considered the initial pressure and p_2 the final discharge pressure.

2-stage compression
$$p' = \sqrt{p_1 p_2} \qquad (17\text{-}32)$$

3-stage compression
$$p' = \sqrt[3]{p_1{}^2 p_2} \quad \text{and} \quad p'' = \sqrt[3]{p_1 p_2{}^2} \qquad (17\text{-}33)$$
$$(17\text{-}34)$$

4-stage compression
$$p' = \sqrt[4]{p_1{}^3 p_2} \qquad (17\text{-}35)$$

$$p'' = \sqrt{p_1 p_2} \qquad (17\text{-}36)$$

$$p''' = \sqrt[4]{p_1 p_2{}^3} \quad \text{or} \quad p''' = \sqrt{p'' p_2} \qquad (17\text{-}37)$$
$$(17\text{-}38)$$

Any consistent units of absolute pressure are applicable.

For example, with 4-stage compression from 1 to 200 atm equations 17-35 to 17-38 would show intermediate pressures of 3.76 atm, 14.14 atm, 53.2 atm. If an arbitrarily selected middle pressure of 30 atm was taken, with 2-stage operation up to 30 atm and 2 stages following, p' would be 5.48 atm and p''' would be 77.4 atm.

For accurate evaluations of work required use should be made of Ts, hs or ph diagrams of gas properties. However, since such diagrams are not always available, it is often necessary to use equations of a type similar to 14-17 to 14-19 with k, an appropriate ratio of specific heats for the pressure and temperature range under consideration, since with all real gases k is not a true constant and varies with changes in C_p and C_v.

For non-isentropic compression, which is usually non-adiabatic as well, appropriate values of the polytropic n may be used instead of k to yield good approximations for actual work. However, it is preferable in most cases to compute the isentropic work and modify this by an appropriate compression efficiency to determine the actual work.

Equation 14–19, which is set up as a power equation, can be shown to take the following form per stage for reversible 2-stage compression with intercooling to initial conditions

$$\text{Hp} = \frac{144}{33,000} \frac{k}{k-1} p_1 V_1 \left[\left(\frac{p_2}{p_1}\right)^{(k-1)/2k} - 1 \right] \quad (17\text{–}39)$$

For horsepower, note that V_1 represents the free gas flow in cfm at inlet conditions and p_1 is expressed in psia.

If we express v_1 as cu ft per lb, express pressures in atmospheres, and wish to find the work per lb of inlet gas entering, then the work in ft-lb per pound of air compressed in each stage is

$$W_{(1-2)} = \frac{144\,k}{k-1} (14.7\ p_1)\, v_1 \left[\left(\frac{p_2}{p_1}\right)^{(k-1)/2k} - 1 \right] \quad (17\text{–}40)$$

Note that in both expressions p_2 is the discharge pressure from the final compression stage. The 2 in the denominator of the exponent term of the pressure ratio appears with 2-stage compression. Similarly 3 would be used in this position were 3-stage compression to be employed. For single-stage compression k appears alone in the denominator of the exponent.

Example 17–4. A Linde Dual-Pressure cycle for air liquefaction employs 200 atm top pressure, 30 atm intermediate pressure and receives purified air at 1 atm and 20 C. The return flow $(1 - \epsilon)$ at the intermediate pressure is 0.8 based on unit outflow from the last compressor and thus ϵ is 0.2. Liquid air is delivered at 1 atm pressure. Compute: (a) the liquid produced, λ, for unit air flow from the last compressor, (b) the isothermal compression work required, and (c) the work requirement by isentropic computation if 2 stages are used to the intermediate pressure level and 1 stage operation is involved for the high pressure-compression. Assume intercooling and aftercooling to 20 C. (d) Evaluate the previous results to give the probable actual work requirement for the system described.

Solution: Refer to Fig. 17–6 and make use of the Ts air chart of Fig. 17–5, at

200 atm, 20 C read h_c = $-$ 4.0 $s_c = -0.374$
30 atm, 20 C read h_b = 3.1 $s_b = -0.219$
1 atm, 20 C read h_a = 4.9 $s_a = 0.019$
1 atm, 78.8 K read h_f = -96.8 $s_f = -0.912$

values of h are in k cal per kg or cal per gram, values of s are in k cal per kg K or cal per gram K.

$$T_0 = T_a = T_b = T_c = 20 + 273.16 = 293.16\ K$$

a) By equation 17–27 $\lambda = \dfrac{3.1 - (-4.0)}{4.9 - (-96.8)} + 0.2\,\dfrac{4.9 - 3.1}{4.9 - (-96.8)} = 0.073$ kg of liquid air produced per kg of compressed air leaving final compressor.

Cryogenics and Gas Liquefaction

b) Note that for unit mass flow, 1 kg, from the final compression phase, only ϵ kg have to be compressed to the intermediate pressure. Using these mass values in equation 17-29 there results

$$W_{a-c} = \epsilon \left[T_0 (s_a - s_b) + \mathbf{h}_b - \mathbf{h}_a\right] + 1 \left[T_0 (s_b - s_c) + \mathbf{h}_c - \mathbf{h}_b\right]$$
$$= 0.2 \{293.16 \,[0.019 - (-0.219)] + 3.1 - 4.9\}$$
$$+ 293.16 \,[-0.219 - (-0.374)] - 4.0 - 3.1$$
$$= 0.2 \,(67.8) + 45.4 - 7.1$$
$$= 51.9 \text{ k cal for each kg of air delivered by the last compressor.}$$

$$W_{(a-c)} = \frac{51.9}{0.073} = 711 \text{ k cal per kg of liquid produced}$$

$$= 711 \times 1.8 = 1280 \text{ Btu per lb of liquid produced}$$

or

$$\frac{1280}{3413} = 0.375 \text{ kwhr per lb of liquid air produced} \qquad Ans.$$

c) Fig. 17-5 does not give specific volumes for air but these can be found by equation 2-30 with R for air = 53.34.

$$(144) \,(1) \,(14.7) \,v_1 = (1) \,(53.34) \,(460 + 68)$$

$$v_1 = 13.28 \text{ cu ft per lb}$$

For the work computation to 30 atm from 1 atm in 2 stages with intercooling use equation 17-40

$$W_{(1-2)} = \frac{144 \,(1.4)}{1.4 - 1} \,(14.7) \,(1) \,(13.28) \left[\left(\frac{30}{1}\right)^{(1.4-1)/2(1.4)} - 1\right]$$

$$= 7{,}410 \,(1 \times 13.28) \,[30^{0.1427} - 1] = 62{,}900 \text{ ft lb per stage}$$

For the two stages

$$W = \frac{(62900)2}{778} = 161.5 \text{ Btu per lb of air compressed to 30 atm}$$

At 30 atm and 20 C the air at entry to the final stage of compression at intermediate pressure has an inlet volume of

$$v_i = (13.28) \frac{p_1}{p_2} = (13.28) \left(\frac{1}{30}\right) = 0.443 \text{ cu ft per lb}$$

By equation 17-40 for the intermediate to final pressure in single stage compression

$$W_{(i-f)} = (7410) \,(30 \times 0.443) \left[\left(\frac{200}{30}\right)^{(1.4-1)/1.4} - 1\right] = 70{,}900 \text{ ft lb}$$

$$W_{(i-f)} = \frac{70{,}900}{778} = 91.1 \text{ Btu per lb compressed 30 to 200 atm.}$$

However, note that only ϵ lb is compressed to 30 atm while unit mass is compressed to 200 atm. Thus for each lb of liquid produced the total work would be

$$W_{\text{Total}} = \frac{\epsilon \,(161.5)}{\lambda} + \frac{(1) \,(91.1)}{\lambda} = \frac{32.3}{0.073} + \frac{91.1}{0.073}$$

$$= 1692 \text{ Btu of work required per pound of liquid air produced}$$

$$= 0.495 \text{ kwhr} \qquad Ans.$$

The last computation shows that significantly greater work is indicated, when using isentropic work methods, than by the theoretical isothermal compression. The results would have been closer if 4 stages of compression had been employed. When the isentropic approach is modified by a factor to bring it close to actual work requirements, using the suggested efficiency of 72 per cent, there results

$$W = \frac{1692}{0.72} = 2348 \text{ Btu per lb of liquid air}$$

$$W = \frac{2348}{3413} = 0.688 \text{ kw-hr per lb of liquid air, required as work input.} \quad Ans.$$

17-8. THE CLAUDE SYSTEM

The name of Claude is usually associated with systems which make use of an expander engine (turbine) to drop the temperature of part of the air moving through the system. It is obvious that if work is removed from a gas during its expansion, its temperature drops much more rapidly than is the case with a gas merely throttling in a Joule-Thompson expansion device. The arrangement shown in Fig. 17-7 shows a Claude cycle which produces λ pounds of liquid for unit mass flow in the major circuit of the system. The gas is compressed by multistage compression with inter-

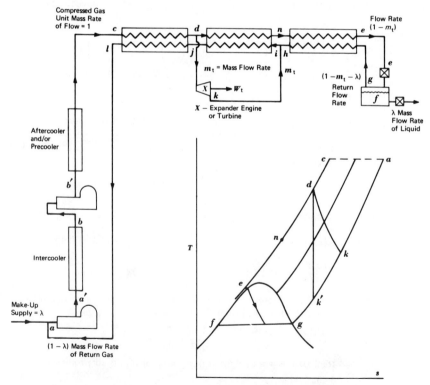

FIG. 17-7. Claude system for gas liquefaction using an expander engine for a fraction of the compressed gas and Ts diagram of process.

Cryogenics and Gas Liquefaction

coolers and aftercoolers to a lower required pressure than for the Hampson system. Satisfactory operation is possible for air when the compression is as low as 40 atm. Varying amounts of air ranging from 30 to 90 per cent of the major flow are taken out after some precooling and expanded from the high pressure of the system to the low pressure of the system where the expanded gas enters the heat exchanger system and returns to the compressor.

Referring to Fig. 17-7 a mass flow and energy balance will be written, assuming unit mass flow rate from the final compressor, λ the mass flow rate of liquid delivered and also the rate flow of make-up gas, with m_t the mass flow rate to the expander turbine (engine). Make the balance between the liquid outflow point and the warm end of the heat exchanger.

$$1\,(h_c) - (1 - \lambda)\,h_l - W_t = \lambda h_f \tag{17-41}$$

For an adiabatic expander and system

$$W_t = m_t\,(h_d - h_k) \tag{17-42}$$

$$h_c - (1 - \lambda)\,h_l - m_t\,(h_d - h_k) = \lambda h_f$$

$$h_c - h_l - m_t\,(h_d - h_k) = \lambda(h_f - h_l)$$

$$\lambda = \frac{h_l - h_c}{h_l - h_f} + m_t\,\frac{h_d - h_k}{h_l - h_f} \tag{17-43}$$

Notice that the first term of equation 17-43 is equivalent to equation 17-19 for the Hampson system if this is also considered adiabatic ($Q = 0$). The second term of equation 17-43 is very sizeable and increases the output of liquid appreciably by virtue of the significant cooling produced by the expander turbine or engine. In large systems the expander output can justifiably be used for reducing some of the work of compression but in small systems the work is absorbed by an artificial load such as a brake or blower fan.

The input temperature to the expander is selected near the warm end of the heat-exchanger. It should be sufficiently high that liquid will not be produced in the expander engine or turbine since the presence of liquid could be damaging to reciprocating or rotating parts. The performance of the Claude system is much better than that of the simpler systems and if the work produced by the expander engine is employed to minimize the compressor work, a very satisfactory performance can be obtained.

In choosing a flow rate to the expander, care must be exercised to ascertain that the value of $\lambda + m_t$ is always less than unity.

Example 17-5. A Claude-type system operates to produce liquid air from ambient temperature conditions of 20 C (68 F) and atmospheric pressure at a head pressure of 50 atm. Reciprocating compressors in three stages are used with intercooling and aftercooling to 20 C. The mass-flow rate to the expander engine m_t is taken as 50

618 Cryogenics and Gas Liquefaction

per cent of the compressor delivery and the temperature at inlet to the expander is −30 C. Consider the return air temperature at the warm end of the exchanger to be 18 C. Compute: (a) the liquid-air production for the given conditions, (b) the idealized isothermal work of compression making use of real air properties and unit weight of gas, (c) the idealized 3-stage isentropic work of compression with perfect intercooling, (d) the isentropic work possible in the expander engine (turbine), (e) the net work required per lb of liquid air produced, and (f) the probable net work required if the overall compressor performance can be considered 72 per cent to account for pressure loss in intercoolers, etc. and considering 70 per cent of the expander work to be realized.

Solution: From Fig. 17–5 read enthalpies in k cal/kg and entropy in k cal/kg K as follows:

h_c = 2.0 at 20 C and 50 atm
h_l = 4.3 at 18 C and 1 atm
h_a = 4.9 at 20 C and 1 atm
h_d = −11.3 at −30 C and 50 atm
$h_{k'}$ = −47.0 at 1 atm after isentropic expansion from conditions at C
h_f = −96.8 liquid air at 1 atm
s_a = 0.019 at 1 atm, 20 C
s_c = − 0.310 at 50 atm, 20 C

Reference will be made to the Fig. 17–7 although this shows 2-stage compression instead of the 3-stage compression specified.

a) By equation 17–43

$$\lambda = \frac{4.3 - 2.0}{4.3 - (-96.8)} + 0.5 \left(\frac{-11.3 - (-47.0)}{4.3 - (-96.8)} \right) = 0.199$$

b) By equation 17–29

$W_{(a-c)}$ = (273.16 + 20) [0.019 − (−0.310)] + 2.0 − 4.9
 = 93.5 k cal/kg of air compressed

$W_{(a-c)}$ = 93.5 × 1.8 = 168.3 Btu/lb of air compressed
 disregarding the benefit of expander engine work. *Ans.*

Probable actual work = $\frac{168}{0.65}$ = 258.9 Btu/lb of air compressed input during compression

c) Make use of equation 17–40 employing 3 in the denominator of the pressure-ratio exponent for 3-stage compression with intercooling. The specific volume of the air at a is taken as at 20 C for both streams considering a slight warm up in the line from l to a and can be found by equation 2–30 as 13.28 cu ft per lb

$$W_{(a-b)} = \frac{144\,(1.4)}{1.4 - 1} (14.7 \times 1)(13.28) \left[\left(\frac{50}{1}\right)^{(1.4-1)/3(1.4)} - 1 \right]$$

 = (7,410) (13.28) [1.451 − 1] = 44,380 ft lb per lb per stage
 = 57.0 Btu per lb of gas compressed per stage

Work for the 3 stages = (3) (57.0) = 171.0 Btu per lb of gas compressed or

171.0 × 0.5556 = 95.0 k cal per kg *Ans.*

The result by this method approximates the result found in *b*.

d) For the expander engine the isentropic work can be found by use of the Ts dia-

Cryogenics and Gas Liquefaction 619

gram Fig. 17–5. Read $h_d = -11.3$ and $h_{k'} = -47.0$ after following down the isentropic line from inlet conditions at -30 C and 50 atm.

$$W_t = 0.5\,[-11.3 - (-47.0)] = 17.85 \text{ k cal/kg} \qquad Ans.$$

e) Net work $= W_{(a-c)} - W_t = 95.0 - 17.85 = 77.15$ k cal/kg of air compressed or $77.15 \times 1.8 = 138.9$ Btu/lb *Ans.*

Based on the weight of liquid air produced these values become

$$\text{Net work} = \frac{77.15}{0.199} = 387.7 \text{ k cal/kg}$$

or

$$\frac{138.9}{0.199} = 697.9 \text{ Btu/lb of liquid air produced}$$

or

$$\frac{697.9}{3413} = 0.204 \text{ kwh per lb of liquid air produced, based on isentropic calculations}$$
Ans.

f) Considering inefficiencies in the cycle

$$\text{Net work} = \frac{W_{a-c}}{0.72} - 0.7\,(W_t) = \frac{95.0}{0.72} - 0.7\,(17.85)$$

$$= 131.9 - 12.5 = 119.4 \text{ k cal/kg of air compressed}$$

$$\text{Net work} = \frac{119.4}{0.199} = 600.0 \text{ k cal/kg of liquid air produced}$$

or

Net work = 1080 Btu per lb or 0.316 kwh per lb of liquid
air produced based on isentropic calculations
and making allowance for inefficiences. *Ans.*

17–9. CASCADE SYSTEM

The cascade system for gas liquefaction is thermodynamically very effective for the liquefaction of gases because it utilizes the liquid-vapor refrigeration system and can interchange heat at essentially constant temperature levels. However, for the large temperature ranges required, it is necessary to employ two or more refrigerants in the cascade and this increases the complexity of the system. Figure 17–8 illustrates an air liquefaction system using three refrigerants and working through a range of 400 degrees. To serve the temperature lift of about 100 degrees in each stage, it becomes necessary to multi-stage the compressors for each refrigerant. Except for the air part of the cycle this is not illustrated in the figure but multi-stage operation is needed. It is also necessary to have effective counter-flow heat exchangers (HE) since these permit compressors to operate with gases at essentially room temperature. Condensation of one refrigerant (gas) adds heat to the evaporator of the next refrigerant. Although there is a temperature difference at every heat exchange point, such a step arrangement is thermodynamically preferable to employing throttling alone. Even in the cascade system, throttling is used but with the throttling involving liquids expanding through small pressure ranges.

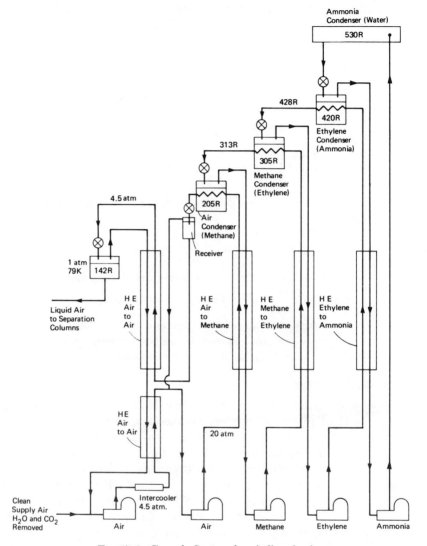

Fig. 17-8. Cascade-System for air liquefaction.

Cooling here is more effective than using Joule-Thompson cooling with a gas. Although cascade systems are not widely used for air liquefaction, the process is used extensively for the liquefaction of other gases, particularly where liquid hydrocarbons are required in petrochemical plants or for distribution as liquid fuels.

In Fig. 17-8 the air is compressed to 20 atmospheres at which pressure it is possible to liquefy the air in a methane evaporator. The liquid air is then expanded to an interstage pressure producing some flash gas, which

Cryogenics and Gas Liquefaction

is taken off by the second-stage air compressor. The remaining cool liquid is then expanded to just above atmospheric pressure for distribution. The compression work is appreciably reduced by carrying out the process in two steps, since the flash gas at intermediate pressure has only to be compressed from some 4.5 atm to the 20 atm head pressure of the air system with an appreciable saving in work. It should be noted that if a lower compressed-air pressure, say 10 atm had been used instead of the 20 atm, the air could not have been liquefied in the methane evaporator but would have merely been cooled to slightly above the evaporator methane temperature. This high-pressure, cool, gaseous air when throttled to the low pressure of the system would produce some liquid air but the arrangement would not be as effective as the one indicated in the diagram. The arrangement of the equipment, the pressure ranges employed, the evaporator and condenser operating points can be adjusted through wide ranges, and the designer's problem is one of optimizing the design to minimize fixed investment while obtaining lowest operating cost. Maintenance and reliability of plant are very important and cannot be overlooked when considering the numerous possibilities and arrangements that can be used. One design alternate involves the possibility of compressing air to a pressure lower than the 20 atm shown and making use of an expander turbine to bring the air to the low pressure of the system.

17–10. GENERAL CONSIDERATIONS ON GAS LIQUEFACTION

The earlier part of this chapter described various methods that could be followed for the production of cryogenic temperatures, and for the liquefaction of gases. A summary of these approaches would include the following:

1. Joule-Thompson cooling resulting from expansion of a compressed gas.
2. Use of an expander engine (turbine) with the energy delivered as work from the system serving to cool the gaseous medium.
3. Cascading of systems with use of multiple refrigerants.
4. Use of reversed Stirling or Brayton Cycle engines (turbines), employing regenerative exchangers. (Method 4 is a special application and will be considered later.)

Most systems fall into one or more of the first three classifications. Thus, the designer of a liquefaction system would arrange to use his choices in such manner as to optimize the yield of liquid product or of cooling for minimum cost in terms of power and equipment. The development of better heat exchange materials, more versatile expansion engines (turbines), and the availability of suitable metals and plastics for low-temperature usage has put many new possibilities at the disposal of designers and the older standardized systems are undergoing drastic changes. However, many of the early investigators in the cryogenic area made contributions of such significant import that their work should be noted here.

P. Kapiza in the 1930's made many significant cryogenic innovations and developed an effective helium liquefier. In particular, he made improvements to reversed-flow regenerators. These are used in pairs in air liquefaction systems to freeze out water vapor, carbon dioxide and hydrocarbons from the make-up air for later elimination from the system. Figure 17-9 shows a typical arrangement. Compressed warm air enters

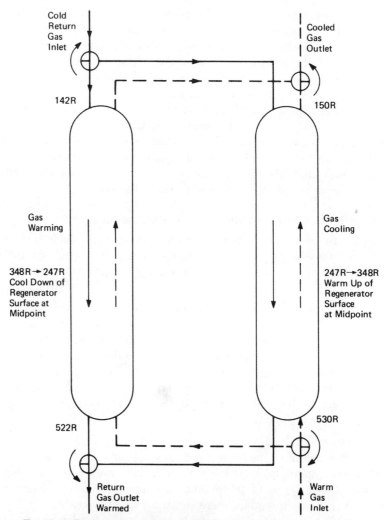

FIG. 17-9. Packed alternating regenerative heat exchangers for cooling and cleaning air supply stream and warming discharge stream. (Representative temperatures shown for packing material at center during the 2-minute air-warming and -cooling cycles.)

Cryogenics and Gas Liquefaction

at the bottom and, as shown, is valved to pass up through the right-hand heat exchanger. As the air stream is cooled, the contaminants freeze on the surfaces to solid form. In this heat exchanger the extended-surface packing remains progressively cooler from the bottom to the top and the air moving upwards and contacting ever cooler surface can be cooled to within 6 to 10 degrees of the packing temperature at the top. At the same time, the left-hand heat exchanger is receiving cold waste gas, which in chilling the exchanger, progressively warms as it moves to the bottom outlet and the gas sublimes or melts the contaminants which then pass out to discharge. The two heat exchangers, having been respectively cooled and warmed, then have the four-valve rotors turned to reverse the gas flows. This is done on a periodic basis, usually repeated at from 45 seconds to 2 minutes. The regenerative process is an effective means of cleaning the supply gas while at the same time the gas streams are cooled and heated. In contrast to chemical gas cleaning, there is some admixture of the gas streams at changeover so that contaminated gas may cause trouble at cryogenic temperatures. Instead of the four valves, shown in the illustration, automatic combination valves operating on a timed cycle are usually employed. The temperatures at each end change very little but at the middle and other places in the exchanger a sizeable temperature change, as indicated in Fig. 17–9, occurs during the alternating cycles.

C. W. P. Hylandt prior to World War I developed an expander engine using a very long piston in the expander cylinder by means of which it was possible to have the warm end work at essentially atmospheric conditions while only the cold end of the piston worked at the very low temperatures involved. More recently Professor S. C. Collins in the late 1940's developed a flexible, small unit for the liquefaction of helium or hydrogen, making use of two or more expander engines. This system is described in a later section.

17–11. HYDROGEN

Hydrogen is more difficult to liquefy than air and its components, nitrogen and oxygen, because its inversion temperature lies in a low range, 171 K to 177 K (308 R to 319 R). This means that gaseous hydrogen has to be cooled to some 300 R before Joule-Thompson cooling becomes effective, and it is difficult to produce temperatures as low as 300 R with conventional refrigerants unless these are cascaded. However, liquid nitrogen, which can be used in an evaporator, is effective for pre-cooling the hydrogen and has been widely employed. More recently, both reciprocating- and turbo-expanders, which are able to work at temperatures in the range of 65 K to 20 K have been developed and can reduce the horsepower required for production of the liquid hydrogen. Hydrogen, with its atomic weight of 1.008, would be expected to have a relatively simple character but such is not the case, since it has two other forms,

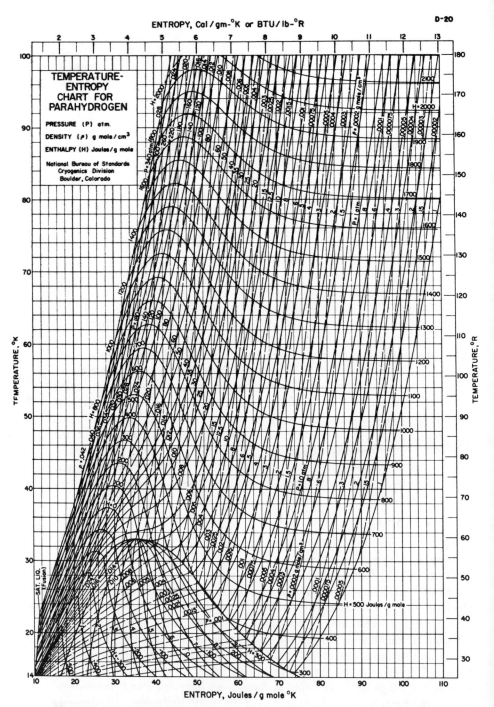

Fig. 17–10. Ts diagram for parahydrogen for low temperature. NBS File D-5881.

Cryogenics and Gas Liquefaction

deuterium with an atomic weight of 2 and the rare tritium with a weight of 3. These, however, are not significant for hydrogen in the cryogenic field, but real complications arise because the diatomic hydrogen exists in two molecular states called orthohydrogen and parahydrogen. These relate to the manner in which the one orbital electron of the atom spins around the spinning proton. Since it is possible for the spins of the two to be in the same or opposite directions, different combinations occur when two atoms join to form the diatomic hydrogen molecule. If the nuclear spins of the atoms are parallel, orthohydrogen results, or if antiparallel, parahydrogen results. When hydrogen gas is liquefied, its composition is approximately 25 per cent parahydrogen and 75 per cent orthohydrogen. Unfortunately, this liquid is not stable and the liquid undergoes a slow conversion, changing from the ortho to the para form and releasing heat (about 305 Btu per lb) during the process. Some gas is lost as a result of the heat release as this takes place. By use of a conversion catalyst, hydrous ferrous oxide, the conversion takes place rapidly at any temperature. The resulting liquid consists of better than 99 per cent parahydrogen with the remainder orthohydrogen. This liquid is stable and can be stored for use in propulsion systems or for other cryogenic operations. A Ts diagram of parahydrogen appears as Fig. 17–10. Parahydrogen requires effective insulation for storage since its boiling temperature at atmospheric pressure is 36.5 R. Its heat of vaporization at that temperature is only 193 Btu per pound and consequently vaporization can take place readily. Hydrogen is a dangerous material because of its ready combustibility and high explosive limits. In fact, it will explode in air through a range of 4 to 75 per cent by volume. Its high heat of combustion makes it extremely valuable as a propellant. Figure 17–11 is an enthalpy-entropy diagram for parahydrogen.

17–12. THE PRECOOL LIQUEFACTION SYSTEM

In the 19th century there were those who believed that hydrogen and helium were truly permanent gases and could not be liquefied. The reason for this rested on the fact that both gases had low inversion temperatures and when attempts were made to liquefy these gases starting from temperatures higher than their inversion temperatures, no liquefaction occurred. However, adequate precooling to a low temperature made liquefaction readily possible. It is, of course, also true that the use of expander engines for either of these media is very effective.

Figure 17–12 is a diagramatic layout of a system that can be used for liquefaction of those gases which have low inversion temperatures and liquefy at very low temperatures. The system is thus suitable for hydrogen or helium. Compression of the make-up and return gas is carried out in as many stages of compression as appear desirable. In the diagram, two

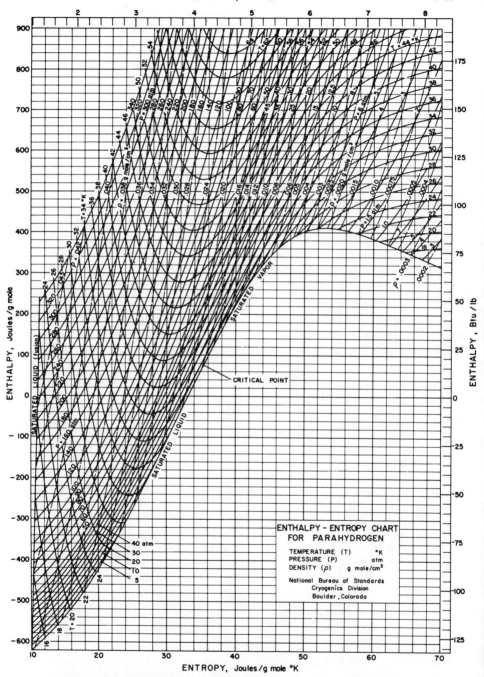

Fig. 17–11. Enthalpy-entropy (hs) diagram for parahydrogen. NBS File D-5880.

Cryogenics and Gas Liquefaction

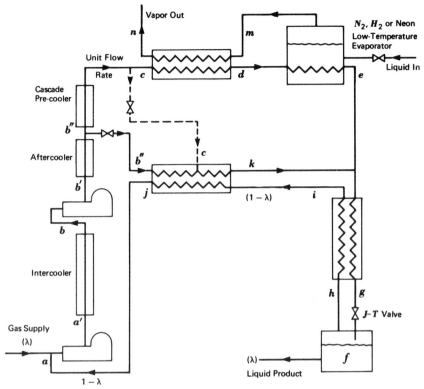

Fig. 17-12. Precool liquefaction cycle for hydrogen (using nitrogen) or for helium (using H_2 or neon).

stages are indicated. Frequently the aftercooler from the final stages is supplemented by a cascade refrigeration arrangement, using a conventional refrigerant such as R-12 or R-22 or even nitrogen to precool the gas. In the case of hydrogen, it first passes through the heat exchanger from c to d and then is led to a flooded evaporator served by a suitable low-temperature refrigerant. Liquid nitrogen is frequently employed in this evaporator since the boiling temperature of nitrogen (139.2 R) is sufficiently low to cool the hydrogen much below its inversion temperature of 177 K (319 R). The hydrogen, now cooled below its inversion temperature, passes through the second heat exchanger and then expands through the Joule-Thompson valve. Liquid hydrogen results at f along with an appropriate fraction of flash gas. The diagram uses the symbol, λ, for the fraction of liquid produced in terms of unit flow leaving the last compressor. Thus $(1 - \lambda)$ represents the fraction of flash gas passing through the heat exchanger from i to j and returning to the compressor inlet for recycling. In starting the system the parallel circuit through the heat exchanger ck is not turned on until the system is effectively underway. It is obvious that in a system of this type, where expander engines are not used, that it is most desirable to have the temperature at point e reduced to a very low value.

For helium, having an inversion temperature, 40 K (72 R), there are few mediums that evaporate at sufficiently low temperatures to cool the helium to its inversion range. The one medium which has been most extensively used is liquid hydrogen. With neon having become available in larger quantities from high-production oxygen plants, this refrigerant, which boils at 27.2 K (48.9 R) at atmospheric pressure, can be employed instead of liquid hydrogen. The amount of hydrogen or neon required is reduced if a liquid-nitrogen precooler is used prior to the heat exchanger before the hydrogen bath.

17-13. HELIUM

Helium is a remarkable material possessing such extremely unique characteristics as to make it an important scientific tool. It does not combine with other elements since it is relatively inert. It possesses a widely variable specific heat, both in the liquid and gaseous phases and is a good heat-transfer medium. A positive Joule-Thompson effect does not exist for helium until it is cooled to a temperature below 40 K (72 R). Its most significant characteristic is that it possesses the lowest boiling point of any material.

It is a rare gas but it is found in recoverable quantities, from 0.3 to 2 per cent by volume, in natural gas fields located in Kansas, and the Oklahoma-Texas panhandle region. It also occurs in trace amounts in the atmosphere.

Two isotopes of helium exist. The one, which carries the number 4 (He^4), is widely available and this isotope boils at 4.2 K (7.6 R). The isotope, He^3, is very rare and difficult to isolate from He^4. Because of this, almost all of the helium used is of type 4. The critical temperature of helium is close to its boiling point at atmospheric temperature, Table 17-1, and it has a very low vapor pressure as well. Helium cannot be solidified by reducing its vapor pressure and thereby lowering its temperature but must have a high pressure applied to it to produce the solid phase. For example, liquid helium at 3.5 K, if put under a pressure of 100 atm, transforms to the solid phase. At 4.2 K, 140 atm are required to cause solidification.

The specific heat of helium also evidences extreme variability, as can be seen from Table 17-2. The specific heat, C_s, of liquid helium along the saturation curve drops with temperature to a low point, then rises to a peak condition at 2.1735 K, after which it falls again. Because of the odd

Cryogenics and Gas Liquefaction

shape of the specific heat curve, this peak is known as the lambda (λ) point because it resembles the top of the Greek letter. The specific heat, C_p, of helium vapor also shows a diverse pattern at low temperatures and varies greatly, both with temperature and pressure.

The lambda point of helium is of interest because of the phenomena that take place at temperatures below the lambda point. The first of these is known as superfluidity, wherein the liquid helium, He^4, shows decreasing values of viscosity. In fact, below the lambda-point temperature, there comes into existence a type of helium called He II which has almost zero viscosity. This appears mixed with the normal helium, called He I, which is similar in characteristics to helium above the lambda point. This odd characteristic of vanishing viscosity as the temperature lowers can be evidenced by rotating a disk or sphere in the fluid; when such rotation is started it continues indefinitely because of the almost absence of viscosity in the fluid. A related effect is that films of helium below the lambda point have the ability to climb up the walls of a containing beaker and travel to the outside, or if a second beaker is placed inside a partially-filled beaker, the smaller beaker fills to the same level of the large beaker by film-travel. The conductivity of liquid He I above the transition point is essentially linear, ranging from 0.011 to 0.016 Btu per hr ft R between 4.5 R and 7.6 R.

In contrast to this, the conductivity of liquid helium II below the lambda point is phenomenally high, reaching 4600 Btu per hr ft R at 2.7 R (1.5 K). This value is many times greater than the k value for copper, 230 Btu per hr ft R at 77 F (537 R), or even than the maximum k value for copper which occurs around 54 R (30 K) and reaches some 810 Btu per hr ft R. With such high values for helium II, it happens that k is not completely independent of the thermal gradient and k also bears a relationship to the size and shape of the vessel in which the helium is placed.

The thermal conductivity of gaseous helium He I is relatively normal and at about 1 atm pressure ranges from 0.0273 to 0.0838 Btu per hr ft R in the interval 90 R to 500 R.

Helium boils like other liquids with ordinary bubble formation, provided the temperature is above the lambda point. Below this, it boils by undisturbed surface evaporation and shows no visible signs of ebullition. This effect is related to unusually high thermal conductivity which apparently develops with the appearance of He II at temperatures below 2.1753 K (3.91 R).

A temperature-entropy diagram for helium appears in two parts as Fig. 17–13 and Fig. 17–14. These diagrams were prepared at the Cryogenic Data Center of the National Bureau of Standards, Reference: Technical Note TN 154 (January, 1962). This was also the source of the enthalpy-entropy diagram, Fig. 17–15.

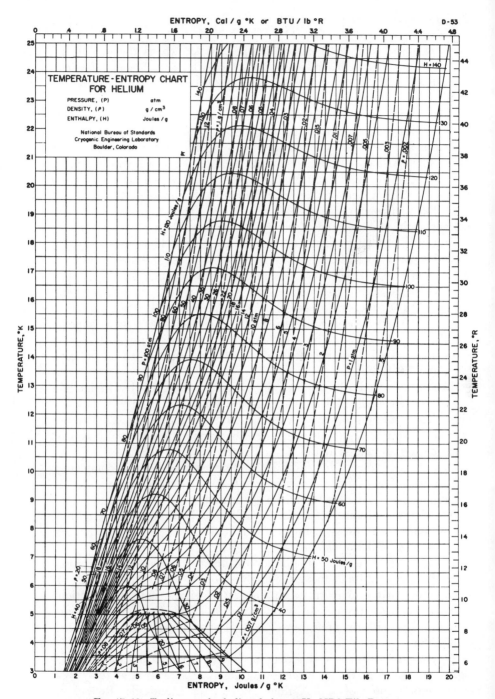

FIG. 17–13. Ts diagram for helium below 25K. NBS File D-5638.

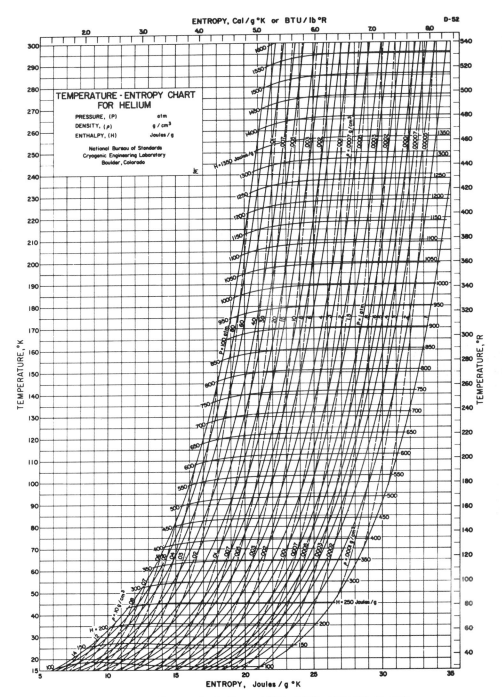

FIG. 17–14. Ts diagram for helium—15K to 300K. NBS File D-5637.

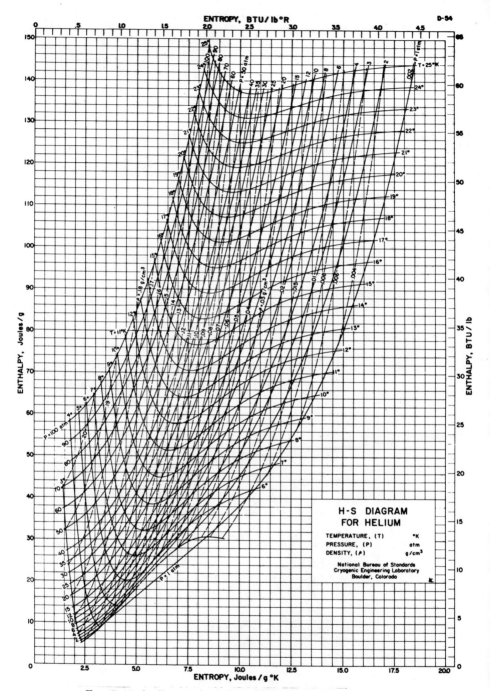

Fig. 17–15. hs diagram for helium. NBS File D-5639.

Cryogenics and Gas Liquefaction

17-14. COLLINS HELIUM LIQUEFIER

Professor J. C. Collins of MIT from about 1946 and into the 1950's pioneered in making small helium and hydrogen plants available for low-temperature usage, on an almost universal scale. Arthur D. Little Company put his designs into production and these ranged in capacity from one liter of liquid helium per hour to 50 or so. All of the designs used two or more expansion engines and most of the later models used nitrogen precooling. The same machines could be used for liquefying other gases by using them in the expander-engine cycle but more easily by using helium in the expander-engine cycle and liquefying the other gases on surfaces cooled by the super-cold helium.

Figure 17-16 is a diagram of a Collins helium liquefier of advanced design which is also applicable for liquefaction of hydrogen by contact cooling. It uses a 3-stage, 90-hp helium compressor and can produce up to 45 liters of liquid helium or up to 50 liters of liquid hydrogen per hour. Helium is the refrigerant gas and hydrogen is liquefied by contact at essentially atmospheric pressure. About 25 liters of liquid nitrogen are required per hour.

The helium circuit will first be described and in this connection the hydrogen circuit 1 to 5 can be disregarded. Valves 1 and 5 are always closed unless hydrogen is being produced. For helium the supply gas enters the 3-stage compressor and is compressed to 16 atm. It is after-cooled and then is divided into 4 streams, the first c-d-e passes to the nitrogen precooler (7) and then to the expansion engine X-1, the second c-h-i passes through heat-exchanger (8) and then to the second expansion engine X-2, the third c-k-l passes through two heat exchangers (2 and 3) into the third expansion engine X-3, the fourth stream passes through the shut-off valve O then moves through four heat exchangers and the Joule-Thompson expansion valve. At this point liquid helium is produced and the flash gases at about atmospheric pressure return through the four heat exchangers for recycling. Typical performance data would show cooling of the helium stream to 80 K in the evaporating nitrogen heat exchanger (7). The expander engine X-1 drops the temperature to the range 40 K to 45 K and this cool gas in 8 chills the counterflowing gas to 45 K to 50 K. In the second expander engine the gas drops to 25 K to 30 K. The third stream cooled in exchangers 2 and 3 to about 18 K drops to 7 K in passing through expansion engine X-3 and then joins the stream of flash gas from the helium liquefier storage tank after Joule-Thompson expansion. The final stream passes through valve O and the heat exchangers 2, 3, 4, and 6 to reach some 5 K to 6 K before the expansion valve. The output work of the expander engines is not used but is expended on an energy-absorbing brake arrangement.

When hydrogen is in production the helium system is operated with

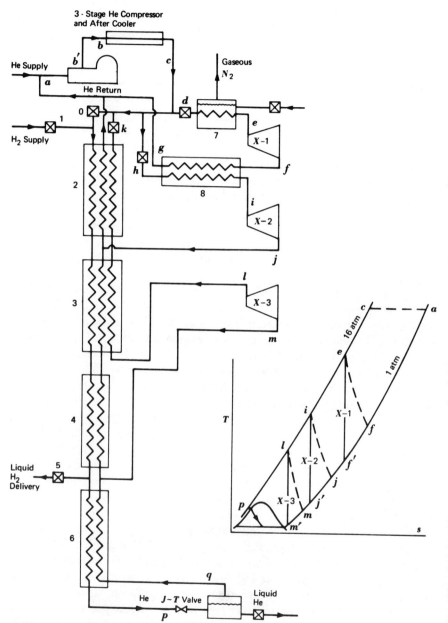

FIG. 17–16. Collins System for the liquefaction of helium using expander engines.

Cryogenics and Gas Liquefaction

the helium being recycled in a closed cycle. The valves O and p are closed and valves 1 and 5 are opened. Gaseous hydrogen at atmospheric pressure entering at 1, passing over the helium-cooled surfaces in the heat-exchangers 2, 3 and 4, passes out as liquid through valve 5 in subcooled form. Most of the condensation occurs in exchanger 4.

A temperature-entropy, Ts, diagram representing the cycle also appears in Fig. 17-16. It is not accurately to scale but does show the events involved. Isothermal compression is indicated from a to c, even though the three-stage compression would be more closely represented by three isentropics with intercooling and aftercooling. The ideal action of the three expander engines is represented by the isentropic lines $e\text{-}f'$, $i\text{-}j'$, and $l\text{-}m'$. However, because of irreversibilities and heat transfer during expansion, the final state points are more accurately represented by f, j, and m respectively and it is with these temperatures that the heat transfer in the exchangers is associated.

17-15. ATMOSPHERIC HELIUM, NEON AND XENON

The high tonnage use of oxygen in industrial processes, particularly for the steel mills, has recently become a widespread practice. Prior to this it was economically unfeasible to consider atmospheric air as a source for helium since in standard air only 0.000524 per cent by volume (1 part per 190,000) of helium exists in the atmosphere and only 0.001818 per cent by volume (1 part in 55,000) of neon is present. Yet with the enormous amounts of air handled, it is now feasible to separate in the liquid-air fractionating column significant amounts of a helium-neon mixture. This mixture also contains nitrogen and hydrogen, which are first removed, following which the helium and neon are separated from each other by selective adsorption. The resultant products can be of very high purity and it is further true that helium from the atmosphere has a helium-3 content, approximately ten times greater than that which is found in the helium associated with underground natural gases. Thus atmospheric air has become the major source for the very small amounts of helium-3 that are available for study and has made neon available in commercial quantities at reduced cost.

Neon. Neon itself is of significant cryogenic value because its boiling point at atmospheric pressure is extremely low, 48.9 R or -410.8 F. This boiling temperature is only slightly higher than that of hydrogen and great use can be made of this low-boiling, non-explosive, inert gas. As more neon becomes available as a by-product from oxygen production, it is felt that its usage in cryogenic service will greatly increase.

Xenon. The same approach has also made this rarest of the atmospheric gases more available than formerly, and it is possible that additional uses will now develop for this gas. Xenon is unique in that it has a high

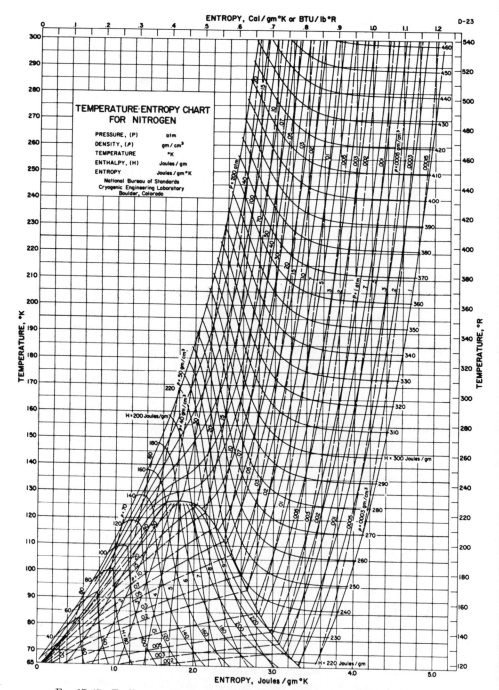

Fig. 17–17. Ts diagram for nitrogen in the range 65K to 300K. NBS File D-5441.

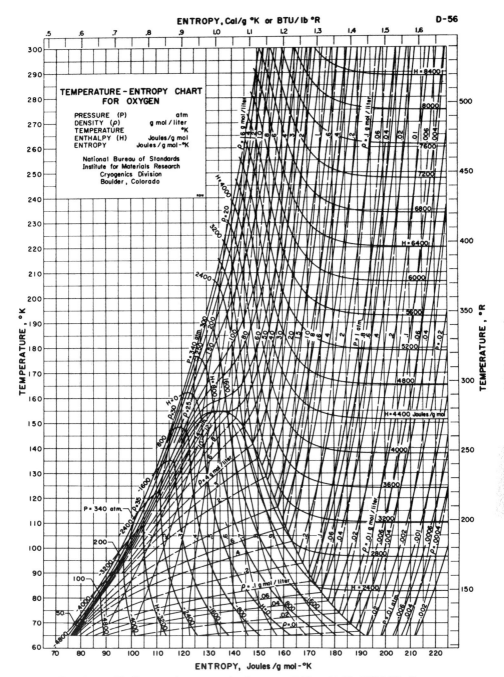

Fig. 17–18. Ts diagram for oxygen in the range 60K to 300K. NBS File D-6098.

molecular weight for a monatomic gas, 131.3. It boils at −162.5 F (259.2 R) and has an extremely high liquid density, 190.8 lb per cu ft at atmospheric temperature and pressure. Up to this time the main use of xenon has been as a filler in certain electronic tubes.

Argon. Argon is a significant constituent of air with a concentration of 0.934 per cent by volume or 1 part in 107. It can readily be separated from air by fractionation and has been used as an industrial gas for some time. Because of a suitable pressure-temperature relationship, boiling at 157.1 R at atmospheric pressure, it is available for cryogenic use but has not been widely employed. It is monatomic, very inert and frequently appears as part of the nitrogen mixture of air unless attention is given to separating this gas from the nitrogen.

PROBLEMS

17-1. Making use of the Ts chart for air find the average values for μ, the Joule-Thompson coefficient, at 30 atm throughout the temperature range 340 K to 160 K. Use values 10 atm, above and below 30 atm, to determine values.

Ans. At 0° C (293.16 K), $\mu = 0.12$ °K per atm, etc.

17-2. Making use of the Ts chart for air (Fig. 17-5) find the pressure and temperature values at which $\mu = 0$ for the approximate temperature range 160 K to 320 K. Note that $\mu = 0$ at the peak temperature-pressure value for each line of constant enthalpy (isenthalp). *Ans.* $\mu = 0$ at 400 atm, $T = 252$ K, etc.

17-3. Using the cycle of Fig. 17-12 draw a T_s diagram (low-temperature end) for the liquefaction of hydrogen (use the chart for parahydrogen although for the gaseous phase normal hydrogen would be in supply). For convenience show isothermal compression at 68 F from 1 to 60 atm replacing intercooling and aftercooling and assume -R-22 precooling to −40 F. The nitrogen evaporator cools the hydrogen to 144 R and in the final heat exchanger the hydrogen is cooled to 83 R. Call the temperatures at e and k essentially the same. Indicate on your diagram temperatures at the various heat exchanger points allowing for a reasonable Δt at each point.

17-4. Making use of the Ts chart for air compute values of the Joule-Thompson coefficient, μ, along the 2.0 k cal/kg isenthalpic line in the range 200 to 150 atm, 50 to 30 atm, 15 to 5 atm. *Ans.* $\mu = 0.08$ °K/atm, etc.

17-5. Helium at 6.5 K and 16 atm passes through an expansion valve on its way to the liquefaction chamber which operates at 1.19 atm. Return flash gas passing out from the heat exchanger is warmed to 9.0 K. Compute the probable kg of liquid produced per kg of helium entering the expansion valve if heat gain can be disregarded. Find the temperature of the compressed medium entering the heat exchanger just prior to the expansion valve. Make use of Fig. 17-13 and Table 17-2.

Ans. $\lambda = 0.26$, $T = 10$ K.

17-6. Compute the minimum work required for ideal gas liquefaction of air from 68 F and 2.0 atm to liquid at 2.0 atm. Express answer in k cal per kg, Btu per lb and watt hours per lb. *Ans.* 149 k cal per kg, 267 and 78.

17-7. Compute the minimum work required for the ideal liquefaction of gaseous

Cryogenics and Gas Liquefaction

helium from 68 F and 1.0 atm to liquid at 1.0 atm. (The entropy of 4.2 K saturated liquid at 1.0 atm is 3.5 joules per gram K and the enthalpy is 9.9 joules per gram.)
$Ans.$ 6622 j per gr, 2846 Btu per lb.

17-8. Compute the minimum work required for idealized liquefaction of gaseous nitrogen from 77 F (25 C) and 1 atm to saturated liquid at 1.0 atm. Express answer in joules per gram and Btu per lb. $Ans.$ 764 j per gram and 328 Btu per lb.

17-9. Solve problem 17-8 except that oxygen is the gas involved.

17-10. Solve problem 17-8 except that parahydrogen is the gas involved.

17-11. In the simple Linde Cycle air enters the heat exchanger without prior refrigeration at 200 atm and 77 F. Liquid air is produced at 1.0 atm and gas leaves the heat exchanger at 68 F and 1 atm. Compute the lb of liquid air produced per lb of compressed air entering the heat exchanger, assuming no heat gains to the system. Make use of the Ts diagram for air. $Ans. \lambda = 0.07$.

17-12. A simple Linde (Hampson) cycle uses a refrigerated precooler which chills the compressed an at 200 atm to -40 F (-40 C) prior to its entry to the heat exchanger in which the gas is cooled and then throttled to 1.0 atm. Gas leaves the heat exchanger at -58 F (-50 C). Liquid and vapor leave the separating and storage chamber in saturated condition. Compute the ratio of liquid to compressed gas produced. Use Fig. 17-5 and refer to Fig. 17-3 for layout. $Ans. \lambda = 0.14$.

17-13. A simple Linde (Hampson) cycle operates between 1.0 atm to 200 atm pressure in an atmospheric environment at 77 F (25 C). Consider minimum temperature differentials at every heat exchange point so that $t_a = t_c$ and also assume that idealized isothermal compression is possible at 77 F. For air as the flow medium, find: (a) the liquid production rate, λ, for unit warm gas flow into the heat exchanger, (b) the isothermal work of compression per lb of air compressed and delivered to the heat exchanger, and (c) the isothermal work required per lb of liquid air produced at 1.0 atm.
$Ans.$ (a) $\lambda = 0.0826$; (b) 108.4 kcal per kg, 195.1 Btu per lb;
(c) 1312 kcal per kg, 2360 Btu per lb

17-14. Solve problem 17-13 for nitrogen.

17-15. Solve problem 17-13 for oxygen.

17-16. A dual-pressure Linde cycle operates from 1.0 atm pressure to an interexpansion pressure of 30 atm and final top pressure of 200 atm. Consider the atmospheric temperature environment at 25 C (77 F) with minimal temperature differentials required so that t_c, t_b, and t_a are the same at all junction points. The liquid flow, ϵ, from the interexpansion chamber to the second throttling valve is 20 per cent of unit warm gas flow entering the heat exchanger. For air as the flow medium, find: (a) the liquid production rate, λ, for unit warm gas flow into the heat exchanger with saturated liquid at 1.0 atm produced, (b) the isothermal work of compression (ideal) to 200 atm for the dual-pressure conditions per lb of air, and (c) the isothermal work of compression per lb of liquid produced. $Ans.$ (a) $\lambda = 0.0694$; (b) 52.5 cal/gr, 94.5 Btu/lb;
(c) 757 cal/gram, 1362 Btu per lb

17-17. Solve problem 17-16 for nitrogen.

17-18. Solve problem 17-16 for oxygen.

17-19. A simple Linde (Hampson) cycle operates between 1.0 atm and 200 atm in an atmospheric environment at 77 F (25 C). Consider perfect intercooling and aftercooling to 77 F as possible. Compute the isentropic work of compression per lb of gas if this is carried out in 4 stages of compression with air as the gas to be liquefied. Consider the air to perform as a perfect gas. Compare your answer with the corresponding answer found in problem 17–13.

17–20. Solve problem 17–19 for nitrogen.

17–21. Rework problem 17–16 on the basis that the compression from 1.0 to 30 atm is carried out under 2-stage isentropic conditions followed by one-stage compression from 30 to 200 atm, with perfect intercooling and aftercooling to 77 F. Use the proper mass flow rates for the 2 pressure regimes.

Ans. 1 to 30 atm is 32.2 Btu, 30 to 200 atm is 92.8 Btu per lb

17–22. Clean dry nitrogen at 77 F (25 C) is compressed from 1.0 atm to 10.5 atm in a 4-stage centrifugal compressor. It is then aftercooled to 77 F and later in a heat exchanger further cooled to −40 F. At this temperature and at 10.4 atm it enters a radial-inflow turbine expander dropping to 1.1 atm pressure. (a) Compute the isentropic work of compression per lb of nitrogen. (b) Estimate the actual work of compression if the internal efficiency of the compressor is 78.0 per cent. (c) Compute the actual temperature leaving the compressor if it can be considered essentially adiabatic. (d) Find the work produced by the expander under isentropic conditions. (e) Find the expander work per lb of nitrogen if a 75 per cent realization of the isentropic work is obtained. (f) Find the temperature of the nitrogen leaving the expander turbine. Use the Ts diagram for nitrogen to solve parts d, e, and f. For part a where nitrogen is being compressed at elevated temperature, use k as 1.395.

Ans. (a) 127 Btu/lb; (b) 162.8 Btu/lb; (c) 731 F; (d) 113 joules/gram, 50.4 Btu/lb; (e) 84.7 j/gr, 36.2 Btu/lb; (f) $h_k = 303.1$, $t_k = 268$ R, −192 F

17–23. Assume that clean dry air, instead of nitrogen, is carried through the processes outlined in problem 17–22 and obtain the results asked for in that problem for air.

17–24. Helium cooled by previous expander engines and heat exchangers enters a final turboexpander at 6 atm and 9.5 °K. It leaves the expander turbine at 1.0 atm and 5.5 °K. The expander turbine exhaust helps cool the pressure stream at 6.0 atm to 6.2 °K at entry to the Joule-Thompson valve where expansion to 1.0 atm takes place and some helium liquid is produced. From the data given, and making use of the Ts chart for helium, find: (a) the fraction of liquid (λ) produced per unit mass of gas entering the J. T. Valve and (b) the efficiency of the turboexpander engine based on isentropic performance. NOTE: The enthalpy of helium at 1.0 atm is $h_f = 9.9$ and $h_g = 30.78$ joules per gram.

Ans. (a) 0.42; (b) 69 per cent

17–25. Making use of a fundamental property of exact differentials, as illustrated in section 17–2, start with the entropy relationship $Tds = du + pdv$, from which $du = Tds - pdv$ and derive the Maxwell relationship

$$\left.\frac{\partial T}{\partial v}\right)_s = -\left.\frac{\partial p}{\partial s}\right)_v$$

REFERENCES

1. L. C. Claitor and D. B. Crawford, *Thermodynamic Properties of Oxygen, Nitrogen and Air at Low Temperatures*, Trans. ASME, Vol. 71 (1949), pp. 885–895.

2. A. Michels, T. Wassenaar, and G. J. Walkers, *Thermodynamical Properties of*

Air between 75C and −170C and Pressures up to 1200 Atmospheres, Applied Scientific Research, Sec. A, 5 (1954), pp. 121–136.

3. V. J. Johnson (ed.), *Properties of Materials at Low Temperature* (Phase I), *A Compendium* (New York: Pergamon Press, 1961).

4. *Design Guide for Pressurized Gas Systems,* Vol. II, NASA Contract NAS7-388, IITRI (Chicago, 1966).

5. National Bureau of Standards, Tech Note No. 154 (January, 1962).

6. J. Hilsenrath, et al., *Tables of Thermodynamic and Transport Properties of Gases* (New York: Pergamon Press, 1960).

7. F. Din, *Thermodynamic Functions of Gases,* Vol. 2 for Air, Butterworth's Scientific Publications, 1956, London, U. K.

8. F. Din and A. H. Crockett, *Low Temperature Techniques* (New York: Interscience Publishers, Inc., 1960).

9. S. C. Collins and R. L. Cannaday, *Expansion Machines for Low-Temperature Processes* (London, England: Oxford University Press, 1958).

10. M. Ruhemann, *The Separation of Gases* (Oxford, England: Clarendon Press, 1949).

11. Russell B. Scott, *Cryogenic Engineering* (New York: D. Van Nostrand Co., 1959).

12. R. L. Barron, *Cryogenic Systems* (New York: McGraw-Hill Book Company, Inc., 1966).

13. J. H. Bell, Jr., *Cryogenic Engineering* (Englewood Cliffs, N. J.: Prentice Hall, Inc., 1963).

18

Gas Separation and Cryogenic Systems

18–1. SEPARATION OF GASES

Discussion has already been directed at the characteristics of gaseous mixtures but little has been said about how to separate specific components from a mixture of gases. Air, for example, contains primarily nitrogen and oxygen but it is almost impossible to separate these components from each other in their gaseous phase except through selective adsorption or by chemical combination of one or more of the components. For example, the oxygen in air can be combined by combustion processes, leaving only nitrogen and combustion products, and if the combustible is hydrogen, the water vapor resulting could readily be condensed from the mixture leaving only nitrogen. Or it is possible to absorb preferentially certain gases from mixtures by use of suitable absorbents. In the case of air, however, where chemically active oxygen is usually the gas desired, the only practical method of separating the oxygen from the air is liquefaction of the air followed by boiling off the different constituents at their respective boiling temperatures to leave the residual ones desired. In the case of air, since nitrogen boils at 139.2 R (-320.5 F), argon at 157.1 R (-302.6 F), and oxygen at 162.4 R (-297.3 F), there is enough differential between the boiling temperatures of these three gases to enable the gases to be separated by fractional distillation. The nitrogen, boiling at the lowest temperature, will of course volatilize most readily and leave the other gases in their liquid phases. However, when this happens nitrogen does not leave alone but carries oxygen and argon with it. The concentration of the nitrogen in the vapor phase exceeds its concentration in the liquid phase but always bears a definite relationship to it.

The relationship of the concentration of a component in the vapor phase to the concentration of the same component in the liquid phase can be found by making use of Raoult's Law. This law states that the partial pressure of any component of a liquid mixture is equal to the product of its vapor pressure in the pure state multiplied by its molal concentration in the liquid mixture

$$p_i = p_{is} x_i \qquad (18\text{--}1)$$

Gas Separation and Cryogenic Systems

where x_i = molal concentration (mol fraction) of any component in a liquid mixture (solution)
p_{is} = saturated vapor pressure of the pure component at the temperature of the liquid
p_i = vapor pressure of the component

Raoult's Law is precisely true only with liquid mixtures (solutions) which have molecules of similar size which mix without association or chemical combination and which yield vapors which follow perfect-gas relationships. It is thus not applicable for many solutions and at higher pressures it may be inaccurate. For liquid-air and nitrogen-oxygen mixtures it is extremely useful and reasonably accurate at the usual pressures of a few atmospheres. Let us consider liquid air primarily as a binary mixture of nitrogen and oxygen. We can write

$$p_{N_2} = p_{N_2 s}\, x_{N_2} \tag{18-2}$$

$$p_{O_2} = p_{O_2 s}\, x_{O_2} = p_{O_2 s}\, (1 - x_{N_2}) \tag{18-3}$$

Here p_{N_2} and p_{O_2} are partial pressures, in units consistent with $p_{N_2 s}$ and $p_{O_2 s}$, the saturation pressures of nitrogen and oxygen, at the liquid mixture temperature and x_{N_2} and x_{O_2} are molal concentrations of the components in the liquid phase.

If we now consider the vapor phase it is obvious that the total pressure (p_t) of the vapor mixture is the sum of the partial pressure of the components, and for air

$$p_t = p_{N_2} + p_{O_2} \tag{18-4}$$

$$p_t = p_{N_2 s}\, x_{N_2} + p_{O_2 s}\, x_{O_2} = p_{N_2 s}\, x_{N_2} + p_{O_2 s}\, (1 - x_{N_2}) \tag{18-5}$$

Let us designate concentrations in the vapor phase by y_{N_2} and y_{O_2} then

$$y_{N_2} = \frac{p_{N_2}}{p_t} = \frac{p_{N_2 s}\, x_{N_2}}{p_t} \tag{18-6}$$

$$y_{O_2} = \frac{p_{O_2}}{p_t} = \frac{p_{O_2 s}\, x_{O_2}}{p_t} = \frac{p_{O_2 s}\, (1 - x_{N_2})}{p_t} \tag{18-7}$$

In a liquid solution the concentration of the more volatile component will be greater in the vapor than in the solution. This is true for solutions following Raoult's Law and is true in general for all solutions. For liquid air and its nitrogen-oxygen vapor this can be seen in Fig. 18-1 which shows experimentally-determined temperature-concentration graphs at several pressures. The liquid line appears as the lower line and the vapor line is the upper line in each graph. For example at 1.0 atm pressure, a 100 per cent nitrogen (zero oxygen) composition boils at 139.2 R, whereas a 21

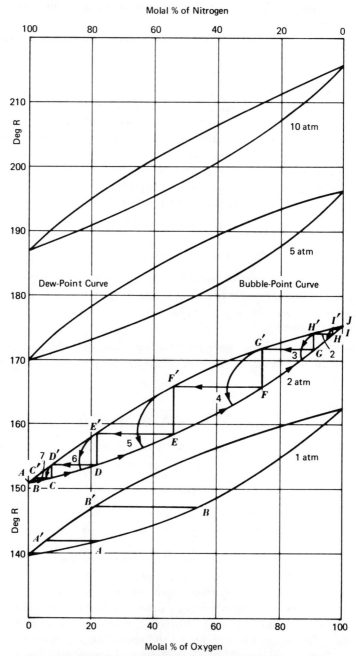

Fig. 18-1. Liquid-vapor phase concentrations for the oxygen-nitrogen system at varying pressures with a rectification-column layout superposed on the 2 atm plot. (ref. 3.)

Gas Separation and Cryogenic Systems   645

per cent oxygen concentration (point A) boils at 141.9 R and at this temperature produces a vapor (point A') having a concentration of about 96 per cent nitrogen and 4 per cent oxygen. Thus the liquid, with greater delivery of nitrogen taking place, becomes enriched in oxygen and its boiling temperature also rises until at a representative concentration of 55 per cent oxygen, the temperature and concentration are as indicated at B and B'.

If the problem is then to separate oxygen from a liquid air mixture, this can be done by employing plates in a fractionating column on which the nitrogen is removed from mixtures progressively richer in oxygen. If a sufficient number of plates are provided, it is theoretically possible to end up with pure liquid oxygen at the bottom of the column.

Example 18–1. Air is primarily a binary mixture of nitrogen and oxygen in the ratio of 79% N_2, molal and 21% O_2, molal. For a total pressure of 14.7 psia on a liquid nitrogen-oxygen mixture determine approximately by Raoult's Law boiling and vapor condensing temperatures over a range of liquid concentrations from 0 to 100 per cent for the two components. The plot of this condition appears in Fig. 18–1 and a point can be found as indicated below.

Solution: The temperature range must run between the boiling points of pure nitrogen at 14.7 psia, 139.2 R, $x_a = 100$ per cent, and pure oxygen 162.4 R, $x_b = 100$ per cent. Then at any selected temperature between the two use equation 18–5 to find x_{O_2}. Choose for example a temperature of 80 K (144 R) at which the saturation pressures of O_2 and N_2 are

$$p_{O_2} = 0.296 \text{ atm} \qquad p_{N_2} = 1.341 \text{ atm}$$

(These can be read from Table 18–1). For 14.7 psia = 1.0 atm by equation 18–5

$$1.0 = 1.341 \, (x_{N_2}) + 0.296 \, (1 - x_{N_2})$$

$$x_{N_2} = 0.674 = 67.4\% \text{ in liquid phase}$$

$$x_{O_2} = 0.326 = 32.6\% \text{ in liquid phase, by difference.}$$

Use equation 18–6 to find the concentration in the vapor phase

$$y_{N_2} = \frac{(1.341)\,(0.674)}{(1.341)\,(0.674) + (0.296)\,(0.326)} = 0.904 = 90.4\% \; N_2$$

$$y_{O_2} = 1. - 0.904 = 0.0896 = 8.96\% \; O_2$$

18–2. RECTIFYING COLUMNS

The operation of a rectifying column depends upon the interchange of energy and of composition between the components on each plate (tray) of the column. As equilibrium is approached between the liquid and vapor phases on a plate, a greater fraction of the more volatile component moves upward as vapor while the less volatile component builds up in the liquid and the latter moves to a lower plate. The temperature composition characteristics of a two component mixture are illustrated for oxygen and

TABLE 18-1
Vapor Pressure of Saturated Liquid-Phase Nitrogen and Oxygen
(Ref. 1)

Temperature		Nitrogen		Oxygen	
K	R	atm	psia	atm	psia
60.0	108.0			0.00716	0.105
62.5	112.5	0.1096	1.61		
65.0	117.0	0.170	2.50	0.0229	0.34
65.57	118.0	0.1889	2.78		
70.0	126.0	0.404	5.94	0.0616	0.905
71.43	128.6	0.465	6.83		
75.0	135.0	0.745	10.95	0.1430	2.10
75.47	135.8	0.793	11.65	0.1550	2.28
77.33	139.2	1.0	14.696		
80.00	144.0	1.341	19.70	0.2964	4.36
85.0	153.0	2.24	32.92	0.5597	8.23
86.96	156.52	2.693	39.58	0.6818	10.02
90.0	162.0	3.53	51.88	0.9803	14.41
90.17	162.4			1.0	14.7
93.02	167.44	4.568	67.14	1.320	19.4
95.0	171.0	5.17	75.98	1.608	23.65
97.56	175.61	6.47	95.09	2.014	29.6
100.0	180.0	7.70	113.2	2.507	36.84
110.0	198.0	10.45	153.6	5.359	78.76
111.1	200.0	15.47	227.4	5.647	83.0
120.0	216.0	25.01	367.6	10.077	148.09
120.48	216.87	25.42	373.6		
126.3	227.3	33.50	492.9	← Critical	
130.00	234.0			17.239	253.4
150.0	270.0			41.620	611.7
154.8	278.5		Critical	→ 50.1	736.3

nitrogen in Fig. 18–1. The lower curve is the boiling (bubble) point curve and represents the temperature at which liquid of a given composition will boil or deliver vapor while the upper curve, known as the dew-point line, represents the temperature at which vapor of a given composition will start to condense. Thus evaporation always causes partial separation with the more-volatile (lower-boiling-temperature) component leaving in greater quantity thereby making the liquid richer in the less volatile component. In rectifying columns heat is added at the base of the column and vapor is generated while from the upper part of the column heat is extracted either directly or by means of a cold reflux liquid which as it evaporates it absorbs heat and produces a cooling effect.

Usually the feed to the column occurs at an intermediate point on the column with the liquid moving over plates toward the bottom, progressively increasing the concentration of the less-volatile component. The vapors move upward both above and below the feed point. In order for equilibrium to exist on a plate, it is necessary to have good admixture of the gas and the liquid involved. This means that the vapor must be brought into intimate contact with the liquid. Two methods in common use are illustrated in Fig. 18–2. In the bubble-cap arrangement, the vapor

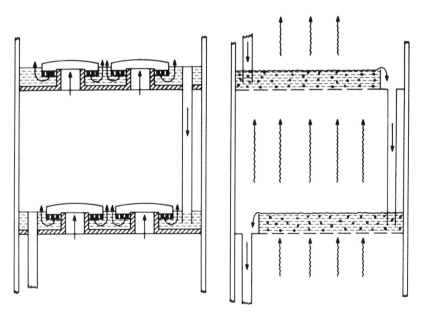

Fig. 18–2. Rectification-column plates (trays) with bubble-cap type at left, perforated plate-type at right.

moving upward, at a temperature higher than the liquid on the upper plate it contacts, comes into intimate contact with the liquid and, as it drops in temperature to reach equilibrium, more of the volatile gas is driven off. In towers with bubble-cap trays, the bubble caps are made of reasonably small size so that a large number of them cover the total area of the plate. The downcomer standpipe transfers the liquid now partially stripped of the less volatile component to a lower plate on which further stripping takes place. In turn the vapor, now enriched with more of the volatile component moves upward to carry out interchange on a higher plate and in turn to become further enriched in the volatile component. In addition to the bubble-cap arrangement, perforated plates are also used as illustrated. These plates are perforated with holes of sufficiently small size that the vapor, under slightly higher differential pressure from beneath, continuously forces its way through the holes and thereby prevents liquid from draining downward. Because the holes are small, in the neighborhood of $\frac{1}{32}$ of an inch or somewhat less, little or no liquid passes downward under the action of gravity. However, in the event that vapor drive from the lower part of the column is not adequate to support the liquid on the plates, the column becomes ineffective as streams of liquid merely drain through the holes to lower plates.

The performance of each plate in a column can readily be analyzed by making a mass and enthalpy balance of the components moving upward and downward on the plate. Perfect equilibrium in composition, however, does not exist on any plate and it is necessary to use more plates than the minimum ideal number. Figure 18–1, in the second diagram at 2 atm, shows the stepwise pattern which is approximated by the components on the plates of a column. Ideally the warmer vapor reaching a given plate moves to reach equilibrium with the liquid on that plate while an increasingly greater fraction of the more-volatile component is driven off to move to the plate next above. There must be a definite temperature difference between the plates, sufficient to maintain equal vapor pressure above each plate each with its liquid of different composition. This must follow because the pressure in the column is essentially constant except for the driving differential of heat addition from below and heat removal above.

To relate these statements to Fig. 18–1 let us consider liquid air feed to the column at a temperature, $D = D'$, entering at the 6th plate from the bottom of the column. Since the liquid feed and its vapor are at the same temperature and in equilibrium the liquid has the concentration shown at D, and the vapor the concentration shown at D'. When the column is in running operation the vapor rising to this plate is at the

Gas Separation and Cryogenic Systems

temperature and concentration indicated by E'. The liquid at D provides the down-flow feed for plate 5 and in turn is warmed to temperature E by the rising vapor from plate 4 at its temperature F'. Similarly the down feed from plate 5 at temperature E is warmed to temperature F by the vapor G' from plate 3. The feed to the first plate is warmed by vapor from the oxygen reservoir and in turn delivers almost pure oxygen I into the storage reservoir (sump). There the liquid is warmed from I to J and most of the small residue of nitrogen is driven out in the oxygen boil off from the liquid oxygen reservoir. However, if greater purity is desired another plate could be added at the bottom of the column between I and J.

Pure nitrogen reflux at A flows onto plate 8 where the up-flowing warm vapor from plate 7 contacts it, driving off essentially pure N_2 gas, while the small remainder of oxygen in the vapor is absorbed in the resulting liquid. This liquid at temperature B flows to plate 7 and is warmed by vapor at D' from plate 6. An extra plate could be provided if greater nitrogen purity is required. It should also be noted that too much reflux or too little reflux can alter the satisfactory operation of the column and change the quality of the product. Notice that the plates located lowest in the column represent the high temperature end of the column and thus appear at the right in Fig. 18-1.

To aid in understanding the step diagram in Fig. 18-1, a set of extra lines, and arrow markings have been provided. These lines and the arrows merely indicate flow patterns and no other interpretation should be made of them. For example, the arrow to the right of D indicates the down flow from plate 6 to plate 5 and the line and arrow from F' indicates that this vapor provides the means of heating the down-flow liquid. This vapor as it bubbles through the plate liquid heats the down flow while condensing out part of the less volatile component in exchange for delivery of a vapor higher in the more volatile component. Thus by following the arrows we would see that the liquid from D combines with the vapor from F' to create the liquid E and deliver a vapor E'. A similar pattern occurs on each plate.

It should also be obvious that this idealized step arrangement does not truly represent column conditions for several reasons. There is always an imperfect interaction of the vapor with the liquid since some bubbles move upward through the liquid path before reaching equilibrium. The liquid down flow at gradually rising temperatures must also deliver vapors at other than the step equilibrium temperature. The balance between raw inlet feed, reflux and main column warming are not always in adjustment. Nevertheless the procedure described does give a ready means of understanding a complex process. Methods of designing rectifying and fractionating columns are well described in the technical literature, references 4, 5, and 6.

18–3. AIR SEPARATING SINGLE AND DOUBLE COLUMNS

Carl Von Linde in 1902 developed a simple method of separating oxygen from air, which employed a fractionating column. As shown in Fig. 18–3 compressed air at from 30 to 200 atm pressure, passes through

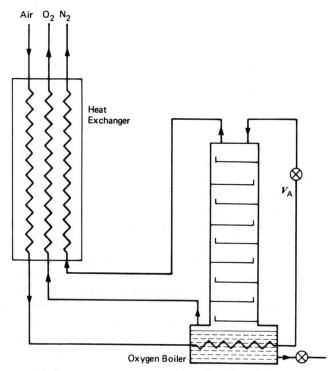

FIG. 18–3. Simple Linde rectification-column for the production of gaseous (or liquid) oxygen and impure nitrogen gas.

the heat exchanger and is further cooled as it moves through a liquid oxygen bath at the base of the column. The further cooling brought about in evaporating some of the oxygen liquefies the high-pressure air. This air then moves through the expansion valve, V_A, and drops to the essentially-atmospheric, low pressure of the system. The liquid resulting from passage through the expansion valve, moves downward over the trays of the column, gradually enriching itself in oxygen while the vapor stream, moving upward, increases in nitrogen content. It is possible to produce

Gas Separation and Cryogenic Systems 651

almost pure oxygen in the liquid bath at the base of the column. However, with no reflux of liquid nitrogen in the upper part of the column, it is not possible to produce a pure nitrogen product. From the bath at the base of the column, a portion of the gaseous oxygen produced is led back through the heat exchanger and contributes to the precooling of the compressed air. If, however, gaseous oxygen is not the product desired, it is possible to deliver liquid oxygen (LOX) from the bath. To deliver liquid oxygen requires more refrigeration than for a gaseous product. This means that the air has to be compressed to a high pressure, around 200 atm, in contrast to the lower values of 30 to 60 atm for the gaseous product.

Liquid air, considered as a binary mixture, of 21 molal per cent O_2 and 79 molal per cent N_2, at atmospheric pressure produces a vapor having a concentration of about 94 per cent nitrogen and 6 per cent oxygen (Fig. 18-1). In the simple Linde column, with oxygen the desired product this 6 per cent of the original oxygen content going away with the waste nitrogen stream represents a sizeable fraction of the oxygen of the original compressed air and thus the column wastes a large fraction, over 20 per cent of the total air compressed. It is possible to improve this performance so that only a trivial amount of oxygen is lost in the waste stream and it is also possible to produce a pure nitrogen product by extending the column upward and providing pure nitrogen as reflux. The so-called Linde double column, which can accomplish this end is illustrated in Fig. 18-4. Since liquid nitrogen is required to carry out the reflux operation a means of condensing nitrogen, using only the liquid air as the basic refrigerant is needed. Reference to Fig. 18-1, or Table 18-1, shows that at 5 atm pressure, nitrogen condenses at 169.8 R and thus if the pressure in the lower column is maintained at 5 or more atmospheres, it would be possible to condense nitrogen if oxygen evaporating at 162.4 R (at about 1.0 atm pressure) serves as a cooling medium to condense the nitrogen. This process is carried out at a middle section of the double column in the heat exchanger, named the nitrogen condenser and oxygen boiler. In it condensed liquid nitrogen is caught in a trap below the condenser to be used as a reflux product, and in addition as output from the unit, if nitrogen is a desired product.

To explain the whole system, let us start with compressed air at between 30 to 200 atmospheres delivered to the sump at the bottom of the column. There in evaporating some oxygen, it is further cooled and essentially liquefied, and it then throttles through the air valve, V_A, to a pressure of about 5 atm. On the lower plates of the column the liquid part of the expanded air drops downward enriching itself in oxygen while the vapor stream moves upward enriching itself in nitrogen until at the top of the

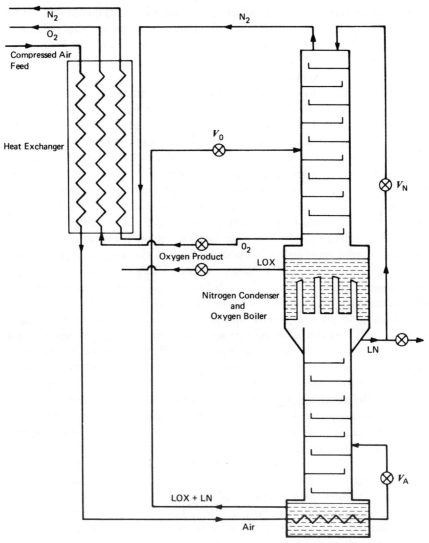

Fig. 18–4. Linde double-column for air separation to almost pure oxygen and nitrogen.

lower column it contacts the cold surfaces of the oxygen boiler. Part of the nitrogen which is liquefied falls back to serve as reflux for the lower column.

Because the reflux rapidly becomes almost pure nitrogen, it is possible to arrive at a continuing operating condition which will produce nitrogen (exclusive of argon) of better than 99 per cent quality. The lower plates

Gas Separation and Cryogenic Systems

of the column are limited in number since they need merely enrich the oxygen content of the liquid product in the sump. This liquid which is controlled at approximately 45 molal per cent oxygen and 55 molal per cent nitrogen, is then led to the mid point of the upper column where after expanding through the valve, V_o, it drops to about 1.0 atm pressure. This resulting product moves down the lower plates of the upper column to enrich in oxygen until an almost pure product is produced. The vapors, in turn, move upward in the column gradually enriching in nitrogen. This is possible because liquid nitrogen, from the nitrogen condenser is also throttled to atmospheric pressure and serves as reflux for the upper column. With pure nitrogen reflux it is easily possible to produce essentially pure gaseous nitrogen. Thus, the double column can produce either liquid or gaseous oxygen at the oxygen boiler section of the apparatus and at the same time it can produce either gaseous nitrogen from the top of the column or liquid nitrogen from the nitrogen condenser part of the column. Moreover by putting a side-arm rectifying column at a suitable location on the upper column, it is possible also to separate argon and produce this as an additional product. More energy is required to produce liquid products than is needed for gaseous products.

18–4. AIR SEPARATION PLANT

Numerous plant designs and arrangements have been made for air separation and the ultimate production of gaseous or liquid oxygen, nitrogen, and argon. The problem faced in these plants is to produce as pure a product as possible with minimum power consumption in relation to a reasonable-low investment cost for the plant facilities. Added to this is the desirability of maintaining continuing operation with a minimum of maintenance. The plant of Liquid Carbonic Corporation in operation at Calumet City, Illinois is a representative modern plant and will be described in detail. Referring to Fig. 18-5, incoming air is seen entering a four-stage centrifugal compressor where it is compressed to about 105 psia. The air is then after-cooled with circulating water, following which it moves through the reversing heat exchangers HE-1 and HE-2. In these the remaining water and the CO_2 are frozen out by means of the counterflow cooling action of low-temperature streams. These exchangers are installed in pairs to work on ten minute cycles, where in part of the cycle the flow is reversed in such manner (see Fig. 17–9) that the frozen CO_2 and H_2O, when sublimed into the waste stream circuit, can be exhausted from the system. Leaving heat exchanger, HE-2, the air, now at almost liquefaction temperature is throttled into the medium pressure column T–1 at a low point in the column.

In this medium-pressure column three streams are produced, the first a crude LOX (liquid oxygen stream) containing 36 to 40 per cent oxygen

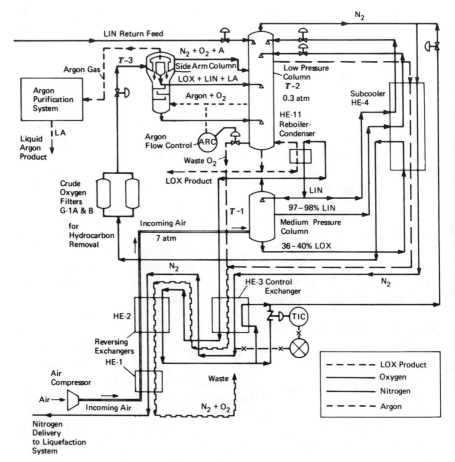

Fig. 18-5. Diagrammatic layout of air-separation plant of Liquid-Carbonic Corporation at Calumet City, Illinois. (ref. 7.)

flows from the bottom of the column. Next a pure nitrogen stream leaves from the top and finally a mixed stream of liquid nitrogen and some oxygen leaves from near the middle of the column. The pure N_2 overhead stream splits two ways. One passes to the heat exchanger, HE-3, called the control heat exchanger. It then passes through the non-reversing passages of heat exchanger HE-2 to maintain the temperature there in the right range for proper freezing and later removal of CO_2, when the exchanger is put on reversed stream. The N_2 returns again through HE-3, following which it is throttled to the low pressure of the system, to pass through subcooler HE-4, again through HE-3, HE-2, and finally HE-1 on its way to the nitrogen-cycle compressors.

The other portion of the pure N_2 overhead stream is looped into and through the oxygen reboiler $-N_2$ condenser, HE-11. In giving up heat to the liquid oxygen, the nitrogen condenses to serve as a liquid reflux for T-1, the medium pressure column, with its remainder passing through subcooler HE-4 finally throttling into the low-pressure column, T-2, to serve as a pure nitrogen reflux feed at the top. It is possible to condense the nitrogen in the oxygen reboiler because the nitrogen, at some 7 atm abs, is at a higher pressure and temperature than the oxygen in the reboiler at approximately 1.3 atm.

The liquid mixture of nitrogen and oxygen which leaves near the middle of the medium-pressure column, passes through subcooler HE-4 and then expands through an expansion valve to enter the low-pressure column, T-2, where it serves as reflux feed to one of the trays in the column. Increasingly pure gaseous nitrogen moves to the top of the column and the gradually enrichening liquid oxygen moves to the bottom of the column.

The crude liquid oxygen from the bottom of the medium-pressure column passes to the subcooler heat exchanger HE-4 where it is partly cooled and then passes to the crude liquid-oxygen filters G-1A and G-1B. These contain silica gel which absorbs the hydrocarbons from the oxygen. The crude liquid oxygen then throttles into the side-arm column, T-3, where it condenses argon vapor to have this serve as a reflux. Gaseous oxygen, nitrogen, and argon vent to the low-pressure column, T-2, as does also a liquid stream to a midpoint of the column T-2. Relatively pure oxygen collecting in the bottom of the side-arm column feeds by gravity to the low-pressure column. In the lower part of the low-pressure column gaseous oxygen and argon are present, while at the very bottom of the column the pure liquid oxygen collects. This liquid oxygen leaves the column in two streams, one forms the LOX product and the other passes through the reboiler condenser, as mentioned before.

Near the bottom of the low-pressure column a side stream of oxygen high in argon is taken off. Part of the stream, which is called the oxygen waste, flows downward to pass through the exchangers while the remainder passes over to the side-arm column, carrying its associated argon. This operation is controlled by the argon-control (AR-C) analyzer which adjusts the rate of flow of this waste oxygen and argon line. An increase in the flow of this line provides more argon for the side-arm column but if the flow is too great, an excess of nitrogen also passes over, which must be avoided. The side-arm column carries out the rectification of the argon product so that gaseous 98 per cent argon is delivered from the top with the other 2 per cent being largely oxygen with some nitrogen. The argon stream then moves to a purification system.

The waste oxygen stream, later joined by the waste nitrogen stream from the low-pressure column, passes through heat exchangers HE-3, HE-2 and HE-1 for ultimate discharge to the atmosphere. It should be mentioned that the low-pressure column which operates at about 4 psig, is also fed liquid nitrogen from the nitrogen system, described later. This nitrogen serves as a refrigerant and the relative amounts of the oxygen, nitrogen, and argon produced, bear a relationship to the amount of liquid nitrogen supplied.

The nitrogen liquefaction system, shown in greater detail in Fig. 18–6 starts with clean N_2 gas entering the four-stage nitrogen compressor C-$4s$ where the gas is compressed to about 155 psia. The gas is then cooled by an aftercooler using water. Part of the compressed gas then passes through heat exchanger HE-7 in which it is cooled and it is then allowed to expand in the expander turbine X-$1A$. The power output of this turbine is absorbed by a booster compressor C-3. The cool expanded gas then moves through heat exchangers HE-9 and HE-7, finally to return for recycling through the four-stage compressor C-$4s$. The other part of the nitrogen stream from C-$4s$ passes to a two-stage compressor C-$2s$ where it is raised in pressure to about 355 psia. The major portion of this compressed gas then moves through heat exchanger HE-8, followed by expansion through the expanders X-$1B$ and X-$1C$. In these it cools to about -310 F. This very cold gas now at low pressure is warmed in heat exchangers HE-10 and HE-8 and also is returned to compressor C-$4s$ for recycling. The other part of the stream from C-$2s$ at 355 psia now passes to the booster compressor C-3, where it is raised in pressure to about 595 psia (40.3 atm). This stream is then cooled by a two-stage R-22 refrigeration system to -40 F which it leaves to pass through the low-temperature heat exchangers HE-9 and HE-10 to leave as a high-pressure, highly-subcooled liquid at approximately -305 F. These relatively low pressures make it possible to employ light-weight, extended surface of brazed aluminum in the heat exchangers. This LIN (liquid-nitrogen) product can then be led to storage after it is dropped in pressure or it can be directly throttled to serve as a refrigerant and reflux feed for the low-pressure column of the air separation system.

The vent condensers on the argon and oxygen storage tanks use this liquid nitrogen supply as a refrigerant to recondense the vapors for restorage.

Flash N_2 gas from all sources is conserved and returned to the system for recycling. To complete the system it should be mentioned that the argon stream is purified by adding gaseous hydrogen to it for the purpose of counteracting the oxygen associated with the argon. For this purpose the argon is compressed to 40 psig and the oxygen then combines with the H_2 in a palladium catalyst reactor. The gas is dried, cooled, liquefied,

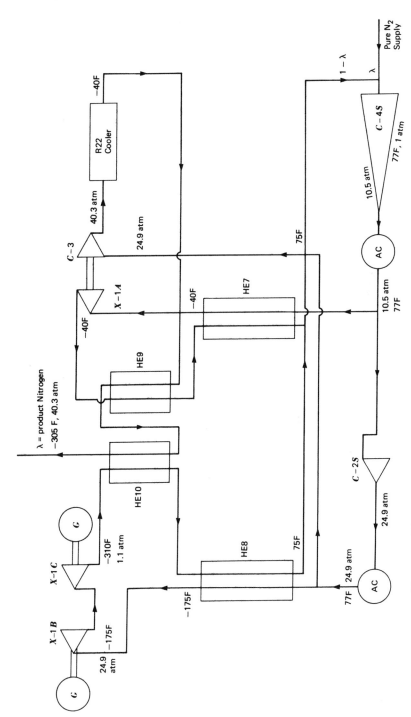

Fig. 18-6. Diagrammatic layout of nitrogen liquefaction system for plant shown in Fig. 18-5.

and then distilled to make a pure product. About 65 per cent of the argon in the original air stream is conserved as final product. This plant is very efficient and on test has been found to require only 31 kwhr per standard million feet of air compressed. It has produced on test 224 tons per day of separated products showing approximate figures of 187 tons of nitrogen, 52 tons of oxygen and 4.8 tons of argon. Moreover the plant is flexible and can be operated to produce more oxygen and less nitrogen. Another test showed for a 24-hour period 113 tons of nitrogen, 100 tons of oxygen and 5.1 tons of argon. In terms of capacity it may be mentioned that the centrifugal air compressor is driven by a 2,000 hp induction motor. The nitrogen cycle compressors are driven by a 7500 hp double-ended synchronous motor. The expander turbines X-$1B$ and X-$1C$ are each loaded with 400 horsepower electric generators.

18-5. STORAGE AND HANDLING OF CRYOGENIC LIQUIDS

Because of the low temperatures involved, the storage of cryogenic liquids is a complex problem. Unless these mediums are kept at very low temperatures, high pressures are involved and a phase change may occur if the temperature exceeds the critical. For liquid-phase air, nitrogen, and oxygen with the temperature not having to go below -320 F, it is possible to hold the liquids in insulated chambers, using fiber, foamed or powder insulation, or vacuum as insulation. However, for very-low temperature cryogenic liquids such as helium and hydrogen, multilayer vacuum insulation with or without powder or with a liquid nitrogen shield is desirable. For such very low-temperature liquids, unless the insulation is effective, the temperatures are sufficiently low to cause ambient air to liquefy or even freeze on the cold surface of the container and this magnifies the heat gain to the cryogenic liquid.

For many years the dewar has been in cryogenic service. This device was named after Sir James Dewar, the pioneer in vacuum insulation who developed it in 1892. The dewar is an evacuated double-wall vessel with the inside surfaces of the double walls silvered to reduce radiation. In diagrammatic form dewars are sketched in Fig. 18-7. It can be seen that the simple dewar, which resembles an elaborate thermos bottle, can be modified to have an inter space provided for liquid nitrogen. The liquid nitrogen is evaporated by the heat which flows in from the external surroundings and in turn holds the inner jacket surface at the temperature of boiling nitrogen so that reduced evaporation loss of contained liquid helium or hydrogen takes place.

In addition to the use of evacuated space as an insulator for cryogenic liquids, use is also made of insulating powders, fibers and foam-type insulation. The powder-type insulations owe their insulating ability not so

Gas Separation and Cryogenic Systems

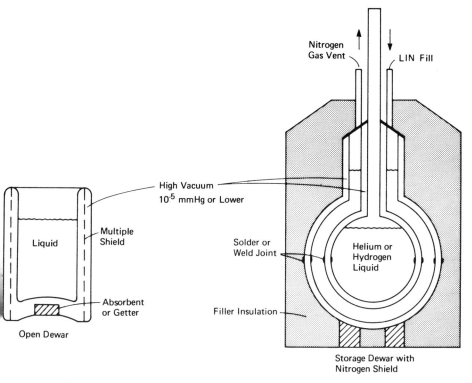

Fig. 18-7. Sectional diagram of dewars for cryogenic liquid storage.

much to low inherent conductivity of a powder but to the fact that with the powder in place, there is a relatively large volume of gas-filled voids compared to the volume of the solid (powder) material so that the apparent conductivity lies in the range of the thermal conductivity of the gas which fills the voids. Convection and radiation losses are minimized in powder insulation of this type. The performance of powder insulation can be further improved if the gas in the interstitial spaces is reduced to high vacuum conditions. For this to be effective, however, the evacuation must be carried down until the pressure is of the order 0.0001 millimeters of mercury or lower. It should also be noted that for low-temperature insulation, the value of k decreases with temperature so that insulation effectiveness is improved as temperatures are lowered.

Foam insulations are made from suitable chemicals which, when mixed in a liquid or semi-liquid state, can be poured into place, after a short delay a reaction starts and a large amount of gas is produced. The resultant product is a solid material consisting of numerous gas pockets of small size. These foams, with their cellular structure, are usually strong structurally and are nearly always impervious to further gas flow through them. A number of different types of foam insulation are in use with the greater number of them being made from the polyurethanes and the polystyrenes.

For large-storage cryogenic tanks, running into thousands of gallons of liquid, use is made of the foam insulators, fiber insulators and also of powder. The solid insulants can be built to several inches thickness and thereby provide the insulating effectiveness required even though these materials are not as effective as the high-vacuum or powder-type insulants. One suspension system for a storage tank is illustrated in Fig. 18–8.

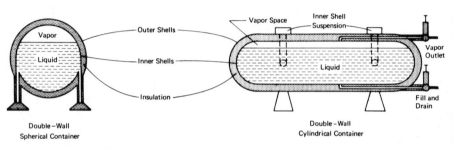

FIG. 18–8. Diagrammatic arrangement of containers for storage of cryogenic liquids in the 5,000 to 1,000,000 gallon range.

An appreciable fraction of the heat gain through the better insulations occurs by means of radiation. Silvering the inner surfaces of dewars represents one of the means of minimizing the radiation heat gain under cryogenic conditions. A more significant method involves the use of low-conductivity insulation material sandwiched between sheets of highly reflective material. With a large number of reflective surfaces and good insulating fillers, containers having extremely low conductivities can be constructed for cryogenic insulations. Such composite insulations, are called super insulations, if vacuums lower than 10^{-4} mm Hg are employed with them.

Table 18–2 presents the properties of a number of insulating materials and the benefits of evacuated space for the insulation can be noted. For vessels, where the insulation space can be effectively sealed but with some air or N_2 trapped in it, when helium is in storage at or near atmospheric pressure, it is obvious that cryogenic pump-down of the insulation space occurs as the air or N_2 freezes on the cold surface. Then, if further air does not enter, a high vacuum in the insulation space develops giving the benefits of vacuum insulation. For a non-vacuum-tight insulation space such leakage would be most undesirable and pre-filling of the space with helium gas would be desirable.

TABLE 18-2
PROPERTIES OF LOW-TEMPERATURE INSULATING MATERIALS

Material	Character	Density lb per cu ft	Pressure mm Hg	Interstitial Gas	Mean Thermal Conductivity (500 R–130 R range)	
					Btu ft/hr ft² °R	Milliwatt cm per cm² K
Diatomaceous earth	Powder, 1–100 micron size	15.6	10^{-4}	.....	6×10^{-4} to 9×10^{-4}	0.0104 to 0.0155
Silica Aerogel	Powder	6.2	10^{-4}		12×10^{-4}	0.0208
	2.5×10^{-2} micron size	6.2	628*	Nitrogen	113×10^{-4}	0.196
		6.2	628	Helium	358×10^{-4}	0.620
		6.2	628	Hydrogen	462×10^{-4}	0.818
Lampblack	Powder	12.0			6.4×10^{-4}	0.011
Perlite	30 mesh	6.6	628		9.8×10^{-4}	0.017
(expanded)	30 to 80 mesh	8.4	628		6.9×10^{-4}	0.012
	80 mesh	8.7	628		5.8×10^{-4}	0.010
Granulated cork	20 mesh		760		144×10^{-4}	0.25
Vermiculite	10–14 mesh		760		260×10^{-4}	0.45
Polystyrene foam (Styrofoam)	Solid	2.4	760		190×10^{-4}	0.33
Glass (Foam Glass)	Solid	9.0	760		200×10^{-4}	0.35
Gases (Independent of convection)						
Helium	Gas	.....	760		6.65×10^{-2}	1.15
Hydrogen	Gas	.....	760		7.51×10^{-2}	1.30
Air and N_2	Gas	.....	760		1.04×10^{-2}	0.18
					Btu per hr ft²	Milliwatts per cm²
High Vacuum						
300 K to 77 K range	0.02 emissivity of facing surfaces	.....	10^{-6}		2.88	0.91
77 K to 20 K range		.....	10^{-6}		0.127	0.004

*Atmospheric pressure at U.S. National Bureau of Standards at Boulder, Colorado.

18-6. REVERSED REGENERATIVE CYCLE FOR AIR LIQUEFACTION OR REFRIGERATION

During the early part of this century, many hot-air engines were in use. These worked on modified Stirling or Ericsson cycles to produce power. Since they were not efficient in their use of fuel and not well adapted for large powers, they disappeared as the use of internal-combustion engines expanded and electric power became available in remote areas. Although the basic cycles of hot-air engines were theoretically reversible and could thus reach high efficiency, the performance of the engines was poor because of the difficulty of transferring heat from the fire (combustion source) through the walls of a cylinder into the low-pressure air, which constituted the working medium of the system. A similar problem existed for the heat sink, where it was difficult to dissipate heat, and the regenerators operating between the hot and cold regions of the engine also left much to be desired. Recently there has been a revival of interest in regenerative cycles arising for several reasons, the most important being: operating with highly compressed gas not only improves heat-transfer performance but also makes it possible to develop more power with a given piston displacement, while at the same time very effective regenerative heat-transfer material can now be found which has good heat capacity combined with low pressure loss for the gas flow.

Since regenerative cycles are typically reversible, it is possible to design a reversed system to which power is supplied for the purpose of producing refrigeration. For this latter usage, great progress has been made, particularly at cryogenic levels. Historically a Scottish engineer, Alexander Kirk, developed a successful refrigeration machine operating on a reversed Stirling cycle in 1874 and a number of his machines were built and put into use. More recently a great step forward was taken when W. L. Köhler and C. L. Jonkers (ref. 8) working at the Philips Laboratories at Eindhoven in the Netherlands, in 1954 produced a sophisticated refrigeration machine to operate at cryogenic temperatures.

The Stirling cycle for power presumes isothermal addition of heat to a gas with increasing volume followed by falling temperature and pressure at constant volume with the gas flowing through the regenerative heat exchanger. Heat is then isothermally rejected at the sink temperature of the system while the volume of the gas returns to its initial value. The gas then moves back through the regenerator, where it picks up heat and warms to the initial temperature of the system. The difference between the heat added at the high temperature of the system and the heat rejected at the low temperatures of the system represents the work produced. The heat interchange in the regenerator is internally adiabatic since heat is given to the regenerator in one part of the cycle and rejected at a later period.

Gas Separation and Cryogenic Systems 663

A diagram showing the reversed Stirling cycle for refrigeration appears as Fig. 18–9. The PV diagram appears in the middle with the sequence diagram shown on the left. The power piston at pressure, P_1, and volume, V_1, rejects heat isothermally and rises in pressure to a value P_2. This is represented by the piston moving from position 1 to position 2. With the left hand piston moving outward while the right hand piston moves inward and the volume remains constant, the gas is transferred through the regenerator, giving up heat and cooling to temperature T_3 and pressure P_3. From 3 the gas then absorbs heat isothermally at the low temperature of the system and expands to the initial volume condition $V_4 = V_1$. The pistons then reverse direction with the gas at 4 moving through the regenerator at constant volume and absorbing heat to return to point 1. On the Ts plane also shown in Fig. 18–9, the heat rejection at high temperature is shown as the area 1-2-7-8 and the heat absorption or useful cooling effect is represented as the area 3-4-6-5. The constant volume lines appear as 2-3 and 4-1. In this cycle the same mass of gas continuously repeats a cyclical process. The regenerator is the critical part of the operation since without an effective regenerator, the system is inoperable. The regenerator must extract heat, represented by the area 2-3-5-7, from the high-temperature gas to cool it to the low temperature of the system and hold the heat in storage temporarily. During reverse flow, it adds heat, represented by the area 4-1-8-6, to the cool gas to return it to the high temperature of the system. The energy absorbed at low temperature (useful refrigeration) plus the work added by the driving motor is dissipated at the sink or high temperature of the system. This is essentially atmospheric temperature or is related to the temperature of available cooling water. In the real case no regenerator is 100 per cent effective but whatever heat it does absorb reduces the temperature of the gas to the low temperature of the cycle. It should also be realized that neither of the external heat transfers can be truly isothermal since there is necessarily a temperature change of most gaseous media when they are compressed or expand. Performance is also lowered because of the large temperature differential which has to exist across the metal surfaces from gas to gas.

The Working Cycle. The actual cycle, in addition to the deviations just mentioned, requires for a refrigerator continuously running at high speed that the constant-volume part of the cycle be modified. One arrangement that can be used is to have the two cylinders working at a crank displacement of about 90°, alternately pulsing gas back and forth through the regenerator. The arrangement that the Philips unit employs, involves the use of a main piston and a displacer piston. These two pistons, each driven off of the common crankshaft, carry out a harmonic variation in their travels which only approximates the constant volume aspects of the Stirling cycle pattern but involves its other charac-

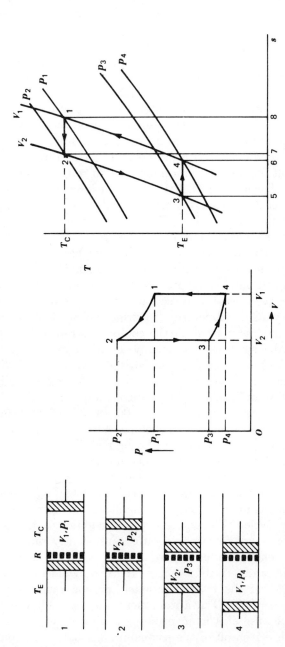

Fig. 18-9. Schematic Stirling cycle for refrigeration using 2 pistons; heat absorbed at low temperature, T_E or $T_3 = T_4$, and heat delivered at atmospheric conditions, T_C or $T_1 = T_2$, with a regenerator R between the temperature regions.

Gas Separation and Cryogenic Systems 665

teristics. The engine itself, which is illustrated in Fig. 18–10, uses as a working medium hydrogen or helium. The reason for selecting these gases is that either of them can operate at very low temperatures and both have good heat-transfer characteristics. Air can readily be liquefied on the refrigeration surface of the unit. In this engine the main piston, numbered 1, moves in the working cylinder 2 and working with the displacer piston

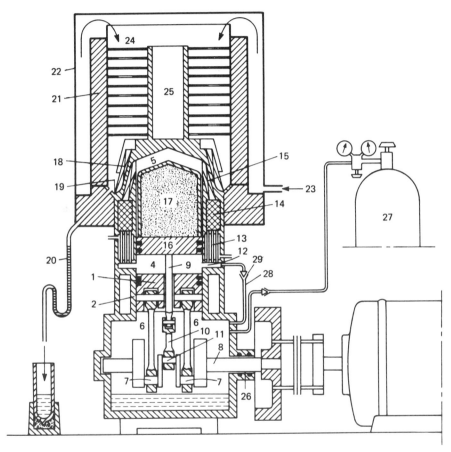

Fig. 18–10. Gas refrigerating machine of N. V. Philips—Eindhoven, as operated for liquefaction of air. (ref. 8.)

varies the volume in spaces 4 and 5. The piston, 1, is driven off the crankshaft at 8 by dual connecting rods, 6. The displacer piston, 16, and its insulated cap, 17, is similarly driven by its piston rod, 9, a short connecting rod and a crank. The cold gas space lies above the displacer piston, 5. The spaces 4 and 5 are in communication with each other through the annular heat sink, 13, operating at normal cooling water temperature, which lies next to the regenerator, 14. The piston rod for the displacer piston passes completely through the main piston. The displacer carries out a harmonic pattern as it rotates out of phase with the main piston at an appropriately chosen angle (slightly more than 90°). The gas flows out of the compression space through the ports at 12, first through the heat exchanger at atmospheric conditions, then through the regenerator and into the cold space. The cap of the displacer, which is filled with an insulating material, is made slightly smaller than the cylinder so as not to touch it and cause heat transfer. The outside of the cold section above 5 performs the useful refrigeration and as arranged for producing liquid air is shown in the illustration. The air to be liquefied enters at 23, passes around the outside cover for the refrigerator surfaces and first comes into contact with the plates, 24. These are thermally tied into the cold block and are so cold that moisture, carbon dioxide and hydrocarbons, present in the air, readily freeze out and solidify on them. The air, now cooled and cleaned of contaminants, moves to the bottom and hits progressively colder surfaces to finally complete its liquefaction and drain out through pipe 20. The crankcase is closed and contains the refrigerant gas at approximately the minimum working pressure, and with a good shaft seal, minimum loss of the working gas occurs. The machine is of heavy construction because, in order to have good capacity in small physical size, a high pressure of the working gas is required. The crankshaft and glands are lubricated but the upper working parts of the machine are completely free of oil as they must be, lest the oil vapor solidify on the cold surfaces at 5. The air to be liquefied is brought in by the condensing action of the cold surface and the pressure unbalance is maintained by means of the looped U-tube, 20. In the particular machine illustrated, the cylinder bore is 70mm, the piston stroke 52mm, crankshaft speed 1440 rpm, maximum pressure 500 psi, minimum pressure about 230 psi.

Machine performance shows that this unit can produce 5.8 kg per hour of liquid air, using 5.8 kw of shaft power or 1.0 kwh are required per kg of air. Liquid air is produced about 13 minutes after start-up. The machine can run for some 20 to 30 hours on moist air before the plates clog with ice, dry ice (CO_2) and other contaminants. Defrost or desublimation requires about one hour. With this machine, as with all cryogenic equipment, the power consumption increases rapidly as temperature lowers. For example, at -50 C, 2.5 kw can produce 2.3 kw of refrigeration but these values change so that at -125 C, 3.8 kw are required to produce 1.7 kw, and at -200 C, 6.2 kw are required to produce 0.4 kw of refrigeration. This is a very good performance for any refrigerator unit and speaks well for regenerative direct-power input cryogenic units. This same manufacturer, N. V. Philips of Eindhoven, makes other units of this general design for lower temperature applications to 20 K, such as

Gas Separation and Cryogenic Systems 667

for recondensing hydrogen or neon, direct hydrogen liquefaction or as a low temperature cryogenic source for cold transfer usage. For such usage, it should be noted that a cryogenic coolant or refrigerant is required in the heat sink at 13 of Fig. 18–10. Other manufacturers are now also producing regenerative cycle cryogenic units.

Example 18–2. A gas refrigeration machine operating on the Stirling cycle, with regenerative heat-exchangers, employs hydrogen as its working medium. The minimum pressure of the system is 13.0 atm abs and the maximum pressure is 34 atm abs. The heat absorber of the machine operates at 135 R and heat is dissipated at 530 R. Assume that the hydrogen acts as a perfect gas, that the regenerator is 100 per cent effective and that the system is adiabatic except at the source and sink points. The designed volume ratio $v_4/v_3 = v_1/v_2 = 1.6$. Compute, per lb of hydrogen circulated: (a) the heat absorption and work done at low temperature, (b) the heat delivered and work input at heat sink conditions, (c) the net work, and (d) the coefficient of performance.

Solution: The heat, added isothermally at the low temperature of the system 135 R, is exactly equal to the work done (for a perfect gas) and can be computed by use of equation 2–45. Using the symbols of Fig. 18–9 and taking the isothermal volume ratio as 1.6

$$W_{3-4} = R\,T\,\log_e \frac{v_4}{v_3} = 2.3\left(\frac{1545.3}{2.016}\right)(135)\log_{10} 1.6$$

$$= 49{,}250 \text{ ft lb per lb of hydrogen}$$

a) $W_{3-4} = Q_{3-4} = 49{,}250 \text{ ft lb} = \dfrac{49{,}250}{778} = 63.3$ Btu absorbed per lb of hydrogen circulated and work delivered. Note that in this part of the system, work is done by the gas.

b) During the isothermal compression phase, work is done on the gas from 1 to 2 and heat is rejected.

$$W_{1-2} = (2.3)(766)(530)\log_{10} 1.6 = 193{,}200 \text{ ft lb}$$
$$= 248.5 \text{ Btu per lb of hydrogen circulated}$$

c) The net work required from an external source $= 248.5 - 63.3 = 185.2$ Btu

d) $\text{C P} = \dfrac{Q_R}{W_{net}} = \dfrac{63.3}{185.2} = 0.34$

On the basis of perfect gas considerations the pressures at 3 and 1 are

$$p_3 = p_4 \times \frac{v_4}{v_3} = 13 \times 1.6 = 20.8 \text{ atm}$$

$$p_1 = p_2 \times \frac{v_2}{v_1} = 34 \times \frac{1}{1.6} = 21.2 \text{ atm}$$

Note that for this reversible cycle the C P must be the equivalent of Carnot-cycle performance. See equation 14–5. This can be shown by writing out the energy terms

$$\text{C P} = \frac{Q_R}{W_{net}} = \frac{RT_R \log v_4/v_3}{RT_c \log v_1/v_2 - RT_R \log v_4/v_3}$$

$$= \frac{T_R}{T_C - T_R} = \frac{135}{530 - 135} = 0.34$$

PROBLEMS

18-1. A liquid mixture of nitrogen and oxygen enters a rectification column at a measured temperature of 90 K. The column is operating under a total pressure of 3.0 atm. Consider that the liquid and its vapor on delivery to the column are in equilibrium. Assume that Raoult's Law is applicable and compute (a) the composition of the liquid phase and (b) the composition of the vapor phase.

Ans. (a) $N_2 = 0.792$; (b) $N_2 = 0.932$

18-2. A liquid mixture of 40 per cent O_2 molal and 60 per cent N_2 molal enters a rectifying column at a temperature of 85 K. Assuming that Raoult's Law holds, find the pressure at which the column is operating and the concentration of the equilibrium vapor for this liquid mixture. *Ans.* $p_t = 1.57$ atm, $x_{N_2} = 0.858$

18-3. In a Linde double-column evaporator, oxygen in the upper boiler at 2 atm is boiled off by condensing nitrogen from the lower column. (a) Find the temperature at which boiling takes place and the heat absorbed by each mol of oxygen evaporated, in joules per gram mol and joules per gram. Nitrogen condenses on the lower surfaces of this boiler at 10 atm. (b) Find the temperature of condensation and the heat of condensation per gram of nitrogen. (c) How many pounds of oxygen are evaporated per lb of nitrogen condensed?   *Ans.* (a) 97.5 K, 6400 j/gram mol, 200 j/gram;
(b) 104 K, 149 j/gram; (c) 0.745

18-4. Sketch a liquid-air rectification column operating at 1.0 atm. Use 6 plates with feed onto the 5th plate above the bottom reservoir. The oxygen reservoir at the bottom is heated by the cool compressed air feed before it enters the *J-T* valve which it leaves as liquid air. Nitrogen reflux enters at the top of the column. Refer to Fig. 18-1 for values and mark on the tower all temperatures of liquid on the plates and rising vapor and letter the tower with the same notation used in Fig. 18-1.

18-5. Reproduce and sketch the 5 atm equilibrium diagram for air using data from Fig. 18-1. On this diagram sketch in a 7 plate rectification column for air constructed in step form (in the fashion employed for 2 atm, Fig. 18-1) designed to produce almost pure liquid O_2 and gaseous N_2. Nitrogen reflux is available and a wet feed, in equilibrium at 36 molal per cent oxygen, is supplied to the 5th plate above the oxygen reservoir at the base of the column. Mark temperatures of rising vapor and liquid for each plate and label points in the column, using the same nomenclature employed in the equilibrium plot. Start the first step on the equilibrium concentration chart at 36 per cent oxygen on the liquid (bubble-point) line.

18-6. Refer to Fig. 18-6 and note that nitrogen at 77 F (25 C) and 10.5 atm is compressed to 24.9 atm in a 2-stage centrifugal compressor having an internal efficiency of 78 per cent. The compressed nitrogen is aftercooled to 77 F and then in passing through a heat exchanger drops in temperature to −175 F. At this low temperature it is supplied to two work-loaded expander turbines in series which expand the nitrogen to 1.1 atm pressure. The internal efficiency of the expander turbines is 80 per cent. The resulting cold nitrogen is warmed in subsequent heat exchangers. (a) Compute the isentropic work of compression in the 2-stage (adiabatic) compressor by means of perfect-gas relations, $k = 1.395$, no intercooling between stages. (b) Find the actual work of compression and the final temperature on entry to the aftercooler. (c) Making use of the Ts chart, compute the isentropic work of expansion for the −175 F nitrogen at 24.9 atm expanding to 1.1 atm in the 2 turboexpanders. (d) Find the probable temperature on exit from the second expander turbine. Pressures are in absolute values, express answers in Btu per lb.

Gas Separation and Cryogenic Systems

18-7. Refer to Fig. 18-6 and note that nitrogen gas at 24.9 atm at 77 F is compressed to 40.3 atm. It is aftercooled by water and then refrigerated by R-22 to −40 F following which it passes through 2 counterflow heat exchangers to reach a temperature of −305 F. (a) Compute the isentropic and actual work of compression if the internal compressor efficiency is 79 per cent. (b) Find the final temperature on exit from the compressor. (c) Find the heat removed in cooling the 40.3 atm-pressure nitrogen from −40 F to −305 F. Express in k cal per kg and in Btu per lb. (d) What is the condition of nitrogen at −305 F and 40.3 atm pressure? Use Fig. 17-17 for data.

18-8. Work problem 18-6 on the basis that the medium is clean dry air instead of nitrogen.

18-9. Work problem 18-7 on the basis that the medium is clean dry air instead of nitrogen.

18-10. A spherical container has a thick layer of insulating material completely surrounding it. The inside layer of the insulation lies at radius r_i from the center of the sphere and the outside surface of the insulation is at radius r_o from the center of the sphere. The surface area of a sphere of any radius is $4\pi r^2$. Employ a procedure similar to that used in section 4-9 and by integration find the average area of the insulation for use in heat-transfer computations, where the conductivity of the insulation is not greatly influenced by the temperature change through the insulation.

$$Ans. \ A_m = \sqrt{A_o A_i} \quad \text{or} \quad A_m = 4\pi r_o r_i$$

18-11. A horizontal double-shell vessel of cylindrical shape is insulated with a polystyrene foam insulation having a k value of 190×10^{-4} Btu ft per hr ft ^{2}R. The outer diameter of the inner shell is 60 inches and the inner diameter of the outer shell is 76 inches. Both shells have almost-flat slightly-rounded ends and the overall length of the cylindrical inner shell is 120 inches and the outer shell 136 inches. Under ambient conditions of 90 F outside, oxygen is stored at one atmosphere pressure. (a) Compute the heat gain per hour through the insulation. (b) Compute the pounds of oxygen vaporized and lost per hour because of heat gain. (c) When the inner tank is exactly half full, how many gallons and how many pounds of LOX are in storage? *Note:* LOX weighs 71.2 lb per cu ft at 90.2 K.

18-12. A nitrogen storage tank is constructed of double spherical shells. The outer diameter of the inner shell is 20 feet and the inner diameter of the outer shell is 22 feet. The space between the two shells is filled with vermiculite in coarse powder form having a k factor of 260×10^{-4} Btu ft per hr ft ^{2}R. Assuming that the outside ambient is at 90 F and the nitrogen is stored at 1 atmosphere pressure, compute the heat gain through the spherical shell. Make use of the answer to problem 18-10 to compute the mean area of the insulation for the spherical shell. (a) Find the heat flow to the nitrogen in Btu per hour, and (b) compute the pounds of nitrogen lost by evaporation per hour because of this heat gain. Disregard heat gain through supports and external piping. (c) When this tank is operated with 10 per cent of its volume in vapor space, how many pounds and how many gallons of liquid nitrogen are in storage? *Note:* LIN weighs 50.4 lb per cu ft at 139.2 R.

18-13. Rework example 18-2 in the text for conditions of holding a source temperature of 160 R and dissipating heat at 530 R.

18-14. Rework example 18-2 if helium gas instead of hydrogen is used but with no other change in the operating conditions.

REFERENCES

1. J. Hilsenrath, et al., *Tables of Thermodynamic and Transport Properties of Gases* (New York: Pergamon Press, 1960).
2. F. Din, *Thermodynamic Functions of Gases*, Vol. 2 for Air, Butterworth's Scientific Publications, 1956, London, U. K.
3. B. F. Dodge and A. K. Dunbar, "Oxygen and Nitrogen System," Journal of American Chemical Society, Vol. 49 (1927), p. 591.
4. M. Ruhemann, *The Separation of Gases*, 2d ed. (New York: Oxford University Press, 1949).
5. J. H. Perry, *Chemical Engineers' Handbook*, 3d ed. (New York: McGraw-Hill Book Company, 1950).
6. B. F. Dodge, *Chemical Engineering Thermodynamics* (New York: McGraw-Hill Book Company, 1944).
7. S. Shaievitz and S. Markbreiter, "High Efficiency Air Separation Plant," *Cryogenic Engineering News*, Vol. 3 (Feb., 1968), pp. 25–28.
8. J. W. L. Kohler and C. O. Jonkers, I. "Fundamentals of the Gas Refrigerating Machine and II. "Construction of a Gas Refrigerating Machine," *Philips Technical Review*, Vol. 16 (Sept., 1954), pp. 69–78 (Oct., 1954), pp. 105–115.

19

Controls and Control Systems

19-1. BASIC PRINCIPLES OF CONTROL

At other places in this text various types of controls for specific purposes have been described. However, up to this point the broader principles underlying the application of automatic controls have not been considered.

From a functional viewpoint, automatic controls serve heating, cooling, and air-conditioning equipment for the purpose of maintaining temperature, humidity, and air motion, and possibly pressure, within close ranges of desired conditions.

The general action of a control system is represented in diagrammatic form in Fig. 19-1.

In Fig. 19-1 the responsive element, or *controller*, which can also be called the *actuating source*, is in a key position. The controller is responsive to a change in temperature, pressure, humidity, or whatever other characteristic is being controlled.

The reaction of the controller is transmitted to the *actuating control* or, as it is sometimes called, the *actuator*. The actuating control starts and stops (or adjusts) the operation of the controlled equipment in response to orders from the controller. Actuating controls are devices such as valves and dampers, and relays or electronic packs, with their attendant circuits.

The *equipment under control* consists of direct heating or cooling equipment, or fans or pumps, which act directly to bring about a desired end result.

The equipment itself must be provided with *limit controls*, which come into play when the equipment reaches too high or too low a temperature (or pressure) or when lack of proper working medium shows that protection is needed. The limit control, in turn, feeds back its instructions to the actuating control and the operation is stopped.

Feedback from the equipment itself is an end result and indirectly causes the controller device to operate. For example, when a room is brought to a desired temperature the controller signals the actuating control to cut off or reduce the output of the heating equipment, because the space requirement has been satisfied.

The example of a simple heating device in the preceding paragraph represents essentially what happens in any completely automatic system. In summary, we find in every such system (a) a sensing device (controller), (b) an actuating device (actuating control, or actuator) controlling the equipment, and (c) a limiting, or protective, device (limit control), with its elements arranged to feed back an effect to other parts of the system.

19-2. HEATING AND COOLING LOAD

In air conditioning, the control of equipment to satisfy loads is critical. The system designer must provide sufficient capacity to meet the maximum demand of a system; and, on the other hand, he must provide control means for adapting oversize units to satisfy load conditions which are usually less

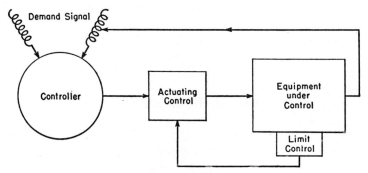

FIG. 19-1. Functional diagram of control-system elements.

than the maximum. In addition to this, a control device may also be required to operate a system equally well whether cooling is required or whether heating is required. This may make it necessary to provide additional controls to superpose on the basic controls in certain systems.

The designer of controls faces a further problem known as system override. In heating, for example, once a system is set into operation so much energy may be supplied to a given space that the space temperature "coasts" to an undesirably high temperature after the control has shut off the equipment. On the other hand, in such a system the space may cool to too low a temperature before the delayed action of the heat supply is felt in the control space. Thus there is a definite need in control systems for some type of anticipative action. This is particularly true in systems where heat-storage effects exist, as the storage effect is not usually synchronized with the heating equipment and the controls are never exactly in phase with controlled action. Examples of typical systems will be given later in this chapter.

Controls and Control Systems

19-3. CONTROLLERS

Before attention is given to the details of control systems, a discussion of controller (sensing) elements is desirable.

Most controllers or controller elements are thermostats, hygrostats, or pressure-responsive devices. A *thermostat* is a control element which is directly responsive to a change in temperature. Its response can be such that it sets in action other circuits which can reposition valves, dampers, or other flow equipment. Thermostats are installed as room thermostats, duct thermostats, and immersion thermostats. Each one of these subdivisions in turn has many variations. For example, a day-night (two-temperature) thermostat controls a cooling or heating source in such a way as to maintain one temperature during a certain period of time and a dif-

Fig. 19-2. Humidistat with cover removed to show biwood humidity-response element. (Courtesy Johnson Service Company.)

ferent temperature during another period. The particular temperature under control at each time period may be set manually or may be under the control of a clock mechanism. Summer-winter thermostats are similar to the day-night type except that their functioning occurs under manual action or under the action of indexing devices. The response may take place when the temperature increases over a certain range or when the temperature decreases over a certain range. *Submaster thermostats* can cause an operating-temperature range to be increased or decreased by a given amount in terms of a temperature change of a different type. For example, the water temperature in a hot-water heating system may be increased by predetermined amounts in relation to the amount by which the outdoor temperature decreases. Simple room and equipment thermostats are illustrated in Figs. 1-1 and 1-2.

Hygrostats, or *humidistats,* are controller devices which respond to changes in relative humidity in a given space or in a supply duct. Hygrostatic elements frequently consist of fine human hair, porous wood, or composition materials, which respond to changes in humidity usually by changing in length, or certain hygroscopic salts, which respond by virtue of a change in electrical conductivity.

The humidity-control device illustrated in Fig. 19-2 makes use of a biwood element for response to humidity change. The biwood element is similar in its action to the bimetal element of a thermostatic unit in that the moisture-responsive strip of wood elongates more or less than the attached base strip, thereby causing the strip to deflect and thus operate equipment.

FIG. 19-3. Electronic humidistat element in position in controller, with cover removed. (Courtesy Minneapolis-Honeywell Regulator Company.)

The sensing element of the humidistat shown in Fig. 19-3 consists of a gold-leaf grid embossed on a plastic base and coated with a special salt which is responsive to humidity changes. The humidistats can effectively control humidities over a range of 12 to 93 per cent. Generally, however, the grids themselves are effective only for small humidity changes and therefore the range of a particular grid is small—from 6 to 10 relative-humidity percentage points (although a range of 25 points is possible). To maintain a relative humidity of (say) 62 per cent in a space, therefore, a grid element covering the range of 53 to 63 per cent might be selected. Close control is possible, with the element actuating the circuit of an electronic relay amplifier for on-off control of humidifying or dehumidifying systems. Figure 19-4 is a diagram of the circuit. The sensing element is shown at the top of the diagram. Current entering the circuit at *1* divides, and then part of the current flows through the element to *3* and part flows

Controls and Control Systems

through the control point adjusting circuit ("C. P. Adj.") to *2*. If the system conditions are as desired, the current flow (or the potential) will be balanced in the two circuits and no demand will be made for the humidification (dehumidification) equipment to respond through the relay amplifier to the signals received at B and R. For humidification the circuit is connected as shown, while for dehumidification the connections *2* and *3* are reversed and connect with R and B respectively.

Pressurestats and other pressure-responsive devices function in relation to changes in pressure. The pressure in question may be a static or total pressure in a space, duct, or pipe. Such controllers may also be made responsive to a differential pressure existing between two other points.

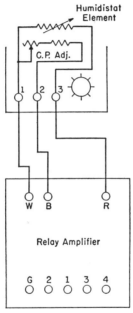

FIG. 19-4. Circuit diagram of electronic humidistat. (Courtesy Minneapolis-Honeywell Regulator Company.)

19-4. ACTUATING CONTROLS (ACTUATORS)

The actuator in a control system is the device which actually makes it possible for the fluid medium to act and bring about a desired end result. A valve (or a damper) is most frequently used to control the flow of a fluid. The valve controls the size or opening of a variable orifice or passage and, by this means, the amount of fluid flowing into or from a device. A control valve (or damper) is usually operated by an electric motor, or an electric solenoid, or a pneumatic bellows or diaphragm. *Normally-open (direct-acting) valves* assume the open position when the imposed (control) operat-

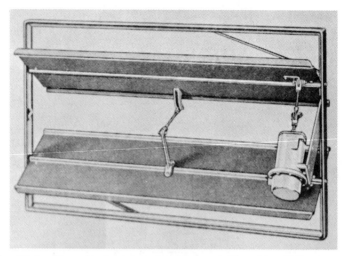

FIG. 19-5. Proportioning louver damper operated by piston-type pneumatic damper operator. (Courtesy Johnson Service Company.)

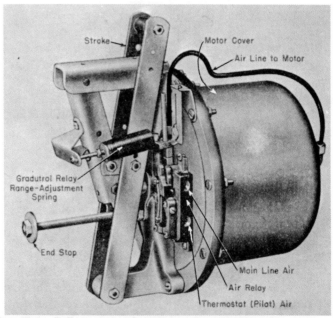

FIG. 19-6. Bellows-operated pneumatic Grad-U-Motor for operation of dampers. (Courtesy Minneapolis-Honeywell Regulator Company.)

Controls and Control Systems   677

ing force is removed. *Normally-closed (reverse-acting) valves* move to the closed position when the imposed (control) operating force is removed. Various valve-positioning elements and numerous design arrangements are employed. For example, three-way valves have three connections, and, depending upon the manner in which the valve passages are arranged, various interconnections for several fluid flows can be planned. Dampers, in contrast to valves, are used to control the flow of gases and air. However, functionally there is no difference between a valve and a damper. Dampers are made in single-blade form or in multiblade, or louver, form. Figure 19-5 is an illustration of a multilouver damper with its damper operator, or damper motor, in position.

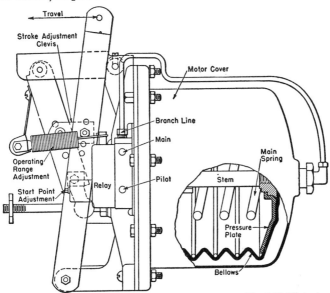

FIG. 19-7. Bellows-operated pneumatic motor (Grad-U-Motor). (Courtesy Minneapolis-Honeywell Regulator Company.)

Figure 19-6 shows the Grad-U-Motor, designed for operation of large air dampers. This particular unit has an over-all length of approximately 13 in. and a maximum diameter of 8 in. Its stroke, or travel, ranges from 2 to 4 in., and a force of 42 lb can be exerted when the unit is operating at 15 psig air pressure. The linkage drive to the damper is connected at the extreme top of the stroke arm, which moves through a slight arc around its pivot, at the bottom, to operate the damper. The extremes of travel of the arm can be controlled by adjustable stops. Referring to Fig. 19-7, it can be seen that in operation of the Grad-U-Motor the air enters the motor chamber through the air tubing at the extreme right of the chamber and acts on the piston end of the contained bellows. The piston end of the bellows is slightly smaller than the inside diameter of the cylinder. Inside the bellows is a strong spring located around the spring stem. This stem (rod) moves outward to position the previously-described driver (stroke) arm. By making use of a bellows the power air pressure acts on but one face of the piston, and leakage of power air around the stem cannot occur. An air pressure of 3 psig starts compressing the spring, and at 13 psig, with no external load on the rod, the spring is compressed to the limit of its travel. Usually an air relay (Fig. 19-8) controls the operation. Thus, under the action of the control thermostat, the relay, in turning at the lever

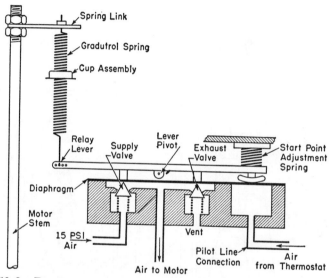

FIG. 19-8. Diagrammatic layout of Gradutrol air relay for pneumatic motor. (Courtesy Minneapolis-Honeywell Regulator Company.)

pivot, deflects the diaphragm to direct supply air into the power cylinder to act on the piston. When the thermostat calls for reverse action, the air is shut off and a vent valve in the air relay opens, which permits the air pressure in the power cylinder to fall as air is vented. The power spring then repositions the piston, with its connected stem and linkage, and consequently the damper.

Figure 19-9 illustrates a pneumatically operated steam valve with a pilot positioner arranged to open or close it by desired amounts.

In Figs. 19-10 and 19-11 is shown a solenoid-operated gas valve. When power is supplied to the solenoid coil, the iron solenoid plunger is pulled up into the hollow center of the solenoid. The valve rod and its seat, which are attached to the solenoid plunger, rise with the plunger, and the valve is thereby opened. A spring partially resists the opening of the valve but

Controls and Control Systems

provides positive closure of the valve when the power is taken off the solenoid.

19-5. TRANSMISSION POWER SYSTEMS

The original action from the controller to the actuator must be brought about by an interconnecting means of power transmission. Both electrical and pneumatic (air) systems are employed for air-conditioning controls. In industrial operations, hydraulic systems—usually operated by oil under

FIG. 19-9. Steam valve with pneumatic operator and pilot positioner. (Courtesy Johnson Service Company.)

pressure—are also used. Electrical systems frequently make use of the *relay*. A relay, when sensitized by the low-current flow from the controller, sets into operation auxiliary higher-output energy sources to operate the actuating controls in the system. Relays may be solely electrical—or they may be electrical-pneumatic, in which case an electrical impulse from the controller operates the small air valves which, in turn, direct air flow to the required circuits. Electronic controllers can serve a function similar to that produced by a relay system. In such control devices, electronic

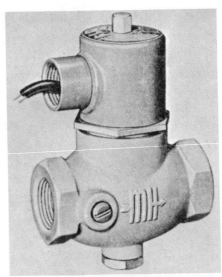

Fig. 19–10. Solenoid-operated gas valve. (Courtesy Minneapolis-Honeywell Regulator Company.)

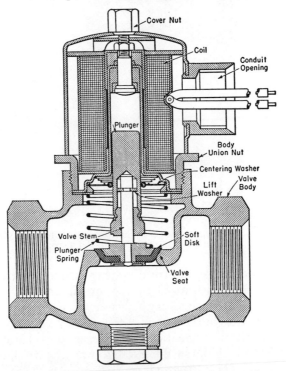

Fig. 19–11. Schematic section through solenoid-operated gas valve. (Courtesy Minneapolis-Honeywell Regulator Company.)

circuits are used to amplify the small currents from the controllers into the full currents used by the actuating equipment.

In operation, the actuator elements, whether electrically or pneumatically operated, are of three common types. (1) The two-position (on-off) controller is the simplest type. As its name indicates, it sets a damper or valve either in the full-open position or in the full-closed position. (2) A proportional, or multiple-position, controller operates to set the control device over a range of settings so as to control the flow more uniformly than is possible with the simple two-position controller. (3) Automatic

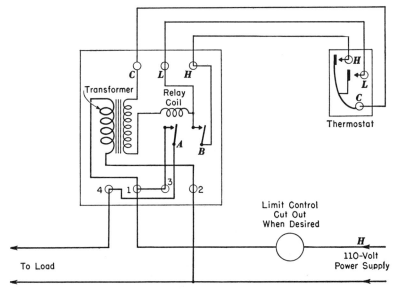

FIG. 19-12. Diagrammatic circuitry of control relay responsive to action of sequence-contact thermostat.

reset controllers change the position of a valve (or damper), just as in the proportional type, but the effect of a change in the valve position is transmitted back to the original controller so that the effect is noted at the basic control point.

19-6. RELAYS AND CONTROL CIRCUITS

Mention has already been made of the purpose of a relay in a control system. Pneumatic or electrical relays, or combinations of both types, can serve a variety of control functions, and the relays vary through a wide range of design arrangements.

Electrical relays can operate on full line voltage throughout, or on full voltage in one part of the relay and reduced voltage elsewhere. To illustrate the workings of one type of relay, Fig. 19-12 has been drawn. This illustrates a simple relay acting under the control of a thermostatic (temperature) element to serve a load which might be represented by such things as a motorized valve, a solenoid valve, or a motorized damper. In this relay, power from the line, usually at 110 v, is supplied to a built-in transformer connected directly across the line through contacts *1* and *2*. The secondary of this transformer provides power at reduced voltage to the thermostat-operating circuit. Power is always available in the secondary circuit and is put to use whenever the contacts close to energize the

Fig. 19–13. Electronic panel controller, with cover removed to show essential control elements. (Courtesy Minneapolis-Honeywell Regulator Company.)

relay coil. When this coil is energized the relay armature closes contacts A and B and, with A closed, the load circuit of the relay causes the load motor or solenoid to function.

The thermostat, as shown, consists of a thermostatic element rigidly anchored at point C. For heating service under the action of falling temperature, the thermostatic element anchored at C moves the contact point to the right so that closure takes place at H. This action, however, does

Controls and Control Systems

not energize the relay coil in series with the secondary of the transformer. As the temperature drops further, the contact at L closes. This immediately energizes the relay coil circuit, and the armature of the relay is pulled to close both the circuit A, for the load, and the holding circuit, B. It will be noticed that now both H and L act to keep the circuit closed. This apparent duplication of effort is provided in order to set up a holding circuit through H so that the relay does not continuously open and shut and thus cause chattering or other undesirable effects under the effect of slight temperature changes. This is accomplished in the following manner: The contact at L is mounted at the terminus of a stiff arm; and when the temperature has sufficiently changed to bring L into contact, the flexible arm of H bends to permit this contact. If a slight rise in temperature now occurs, the contact at L will open, but this action does

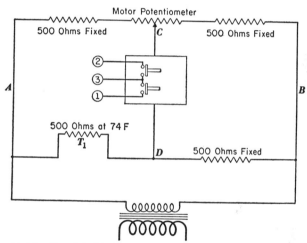

Fig. 19-14. Simplified bridge circuit employed with electronic panel. (Courtesy Minneapolis-Honeywell Regulator Company.)

not disturb the relay circuit and thus load power is still maintained. If, however, the temperature rises appreciably, then the contact H is also broken, the relay-contacts open, and the load power is turned off.

Some relays do not make use of sequence contacts as indicated by H and L as just described, but use a single contactor such as H in a double-position circuit. With this arrangement, when the thermostat calls for heating (or cooling) the thermostat moves to close the relay circuit, and in so doing the relay armature pulls in and closes a holding circuit. This stays closed when the thermostat contact breaks the circuit and until the thermostat contact, in moving, closes an auxiliary circuit which serves a second coil in the relay. This second coil "bucks" the effect of the first coil, to trip the relay armature and open the load circuit.

Increasing use is being made of control circuits which use potentiometer circuitry, and some of these also employ electronic packs. A potentiometer circuit with an electronic pack is illustrated in Fig. 19-13. Figure 19-14 is a simplified diagram of part of the circuit of such a unit, applied to a temperature-sensing element indicated by T_1. Here the transformer voltage from a power system is impressed across the upper and lower legs of a bridge circuit. The potential difference between A and B is always the transformer voltage, which may be taken as 15 v. To assist in this discussion, let us consider the voltage at A to be 0, and the voltage at B to be 15. When the motor potentiometer contact is centered, as at C, half of the resistance of the upper leg appears on either side of it, so that in each case the potential drop from C to A and from C to B amounts to 7.5 v. The temperature-sensing element T_1 has a fixed resistance of 500 ohms at a datum temperature, in this case 74 F; and if the temperature is 74 F, there is no voltage impressed across the amplifier circuit *1-3-2*. If, however, the temperature T_1 should drop, the resistance of its element decreases and the voltage at D is slightly *less* than 7.5 v. Since the contact C is at *exactly* 7.5 v, a voltage difference exists between C and D. This voltage, in acting on the amplifier *1-3-2*, causes the relay to act and to drive the motor or other element constituting the load. This in turn moves the contact point C until the system readjusts itself and the amplifier *1-3-2* is no longer energized. If the temperature of the element T_1 had increased, a similar action, but opposite in sense, would have taken place to drive the motor and the contactor in opposite directions.

With the simple circuit in Fig. 19-14, the resistance of the motor potentiometer sets the temperature change required at T_1 to drive the motor through full range, used for throttling or otherwise metering a heating or cooling medium. When it is desired to change the throttling range with a simple circuit of this type, it is necessary to have a separate motor potentiometer for each range desired. However, by placing a rheostat in parallel with the motor potentiometer, the throttling range can be varied by a simple adjustment. The flexibility of such circuits is unlimited but this cannot be described here in detail. When repair is required, it is possible to replace a complete power pack, thus reducing repair work in the field.

19-7. GENERAL CONSIDERATIONS ON CONTROLS

It is well to consider some of the general purposes of controls. The most significant functioning of control apparatus is directed toward satisfying the following items:
1. Load factor (variation in the thermal load within a space).
2. Varying process requirements.
3. Load variations in terms of climate changes.
4. Specific requirements to meet the comfort of individual occupants.

Controls and Control Systems

The first of these, *load factor*, represents that fraction of the full-load capacity of the system which must be employed. Characteristic load factors which apply to different types of buildings are given in the following tabulation:

$$
\begin{aligned}
&\text{Hotels} \ldots \ldots \ldots \ldots \ldots \ldots \ldots 0.35\\
&\text{Office buildings} \ldots \ldots \ldots \ldots \ldots 0.60\\
&\text{Stores} \ldots \ldots \ldots \ldots \ldots \ldots \ldots 0.50\\
&\text{Residences} \ldots \ldots \ldots \ldots \ldots \ldots 0.75\\
&\text{Restaurants} \ldots \ldots \ldots \ldots \ldots \ldots 0.40
\end{aligned}
$$

The system must be flexible enough to meet variations of the load factor as well as the daily and annual climatic changes which always occur. Among the important objectives which must be met by a control system are the ability to (a) heat and humidify in winter, (b) cool and dehumidify in summer, and (c) perform either of the foregoing functions in spring and fall.

The two basic control instruments are thermostats and humidistats, in one form or another. These, in turn, are used to bring into operation dampers, valves, relays, or motors. Dampers and valves control flow of air, vapor, or liquid, and can operate in three general patterns:
1. Two-position (closed or open).
2. Modulating (over a range of positions).
3. Multiple-position (using a fixed set of operating points).

19–8. AIR CONDITIONING SYSTEMS

A number of air-conditioning-system arrangements were discussed in sections 6–2 to 6–4. The satisfactory operation of each of these basic systems involves the use of control elements and the reader may find it desirable to review them as examples of control practice. However, it should be realized that numerous variations in system arrangement and the control of systems are available, subject only to the ingenuity of the designer. Examples of additional design patterns follow in the next two sections.

19–9. PRECOOL AND REHEAT CONTROL SYSTEM

Figure 19–15 shows a precool and reheat system, in which the mixed air from outside and return are passed through a precooling coil. The partly cooled air then enters a low-temperature dehumidifying coil which condenses moisture from the air as required. The humidistat in the space acts to control the dehumidifier temperature. The water used in the precool coil, having been warmed by the air, is pumped back to the reheat coil, where it can rewarm the air which has been chilled to low temperature in the dehumidification coil. In this way the air is partly warmed and the actual load on the final heating coil is reduced to a minimum amount. This idea of using the same water to cool the air at one point of a cycle and to

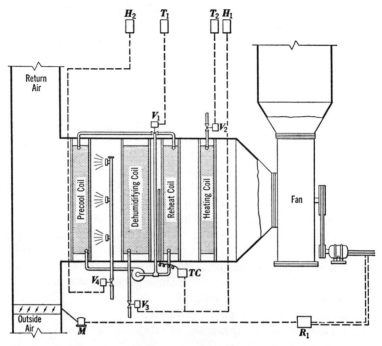

Fig. 19-15. Typical manufacturer's drawing of precool and reheat system.

warm the air at another part of the cycle is known as a run-around system and can, in many cases, result in economy where the dehumidification load is particularly heavy. By close examination of the diagram, one can observe the places at which the thermostats, indicated as T_1 and T_2, and the humidistats, indicated as H_1 and H_2, act on different valves in the system. Notice that a damper motor (M) is placed so as to close the outside-air inlet in the event that the fan power goes off. The humidistat H_2 is available for humidification if the system requires it. A limit temperature control (TC) operates in case the dehumidifier becomes too cold.

19-10. ZONE CONTROL

Figure 19-16 shows an elaborate system in which various zone controls are employed, each under the action of its own thermostat. For winter operation the main air supply, which is returned from the zones and from outside, first passes through a humidifier warming coil, which operates in connection with the humidifier sprays. A heating coil follows the sprays. For summer operation the air is chilled by a cooling coil and then passes through the heating coil, which is under the action of the limit temperature control TC_2. Here, if the air temperature falls to less than 60 F, heat is

Controls and Control Systems

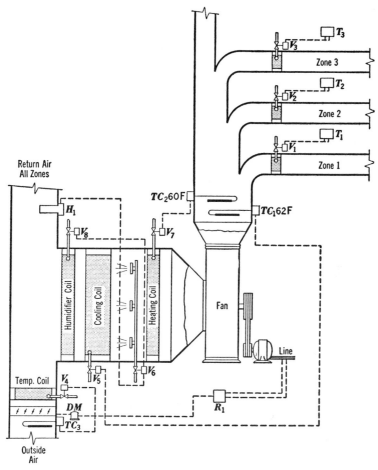

Fig. 19-16. Heating and cooling system using zone control.

supplied. If the temperature rises above 62 F, the limit temperature control TC_1 increases the cooling. The zone thermostats reheat as necessary. The tempering coil is employed under the action of TC_3 whenever the outside air temperature drops below a predetermined point (perhaps 45 F).

PROBLEMS

19-1. The relay shown in Fig. 19-12 is connected to a thermostat which uses sequence contacts. Some thermostats, however, have a three-wire control circuit with a common point C and with the bimetal strip so arranged that under a drop in temperature the bimetal moves, let us say, to the left to close the low-temperature contact. In so doing it energizes the relay coil, which closes the relay contacts of the holding and load circuits. However, when the temperature rises and the thermostat bimetal opens its contact, the relay holding and load circuits stay closed. This condition continues until the thermostat bimetal strip moves far enough to the right to close a high-temperature contact in the thermostat. This action energizes another circuit in the relay which contains a second coil set so as to "buck" (counteract) the action of the first relay coil. When this occurs the relay armature drops free, and in so doing it opens the holding circuit and turns off the load. Draw in diagrammatic form the circuits of such a relay. Show the transformer connected across the line, with one relay armature coil connected to the common and low-temperature contact points of the thermostat. Also show the armature, with its holding and load contacts. Connect the high-temperature (right) contact and the common point of the thermostat with the relay bucking coil and with the transformer. Also show the load connection from the power supply line through the load switch in the relay.

19-2. The pressure face of the bellows in Fig. 19-7 is approximately 6.4 in. in diameter. (a) Compute the respective forces on the stem (rod) when the air pressures in the motor chamber are 10 psig and 15 psig. (b) When the external force at 15 psig air pressure is 42 lb, what compressive force is being exerted on the main spring?
Ans. (a) 321 lb, 482 lb; (b) 440 lb

19-3. In a system using face and bypass dampers, it is found that with the face damper almost closed, the air delivered by the cooling coil is at a temperature of 55 F db and 53.2 F wb, and that for every pound of air passing through the coil 4 lb of air at 82 F db and 70 F wb pass through the bypass dampers. Compute the dry-bulb and wet-bulb temperature of the resultant air being delivered by the fan to the conditioned space. *Ans.* 76.6 F db, 67.2 F wb

19-4. For the face and bypass system described in problem 19-3, rework the problem under the assumption that half of the air passes over the coil and the other half moves through the bypass dampers, and that the conditions of the air leaving the coil and dampers are the same as indicated in problem 19-3. *Ans.* 68.5 F db, 62 F wb

19-5. In a precool and reheat system, air leaves the dehumidifying coil with a temperature of 53 F and a relative humidity of 90 per cent. It then passes through the reheat coil where it is raised in temperature to 63 F. For each pound of water circulating through the precool and reheat system, 4 lb of air pass over the coils. (a) What temperature drop takes place in the water of the reheat coil? (b) If the mixture of outside and return air enters the precool coil at 84 F db and 73 F wb, how much is the air cooled in passing through the precool coil? Refer to Fig. 19-15 for a diagram of a typical system. *Ans.* (a) 9.8 deg; (b) 10 deg (to 74 F)

19-6. In the zone control system of Fig. 19-16, it is found that for the air being supplied the temperature in a zone is 67 F db and 62 F wb. Under these circumstances, the room thermostat calls for heat from the zone reheat coil and the supply temperature is brought to 70 F. If 5000 cfm of air at supply conditions of 67 F db and 62 F wb enter the zone reheater, how much reheat is being employed? *Ans.* 16,300 Btuh

BIBLIOGRAPHY

AMERICAN SOCIETY OF HEATING REFRIGERATING, AND AIR-CONDITIONING ENGINEERS, *ASHRAE Guide and Data Book, Systems and Equipment.* New York, 1967. [Chapters 30 and 44].

HAINES, JOHN E. *Automatic Control of Heating and Air Conditioning.* New York: McGraw-Hill Book Company, Inc., 1953.
LAJOY M. H. *Industrial Automatic Controls.* New York: Prentice-Hall, Inc., 1954

20

Panel (Radiant) Heating and the Heat Pump

20-1. CONDITIONS EXISTING WITH PANEL HEATING

In panel heating (also called radiant heating) the panels of the building structure are heated by pipe coils or warm-air ducts, and these panels in turn transmit heat into the space. Whether heating is done by normal convection or by radiant methods, the objective is to supply enough heat to compensate for the space heat load while maintaining, in the space, conditions satisfactory for occupant comfort.

An occupant at rest dissipates approximately 400 Btu during one hour, because of metabolic body processes, although under conditions of activity the dissipation is necessarily at a higher rate. Of the dissipation loss at rest, that portion associated with moisture evaporation is relatively constant and amounts to some 100 Btuh. Consequently the remainder must be dissipated by convection to the ambient air or by radiation to cooler surfaces. In a room with an average surface temperature of 70 F and with its air temperature at 70 F, the convection loss from a human body to air is approximately 140 Btuh and the radiation loss is 160 Btuh. The net energy transferred by radiation always flows from the warmer surfaces to the cooler surfaces; and for an individual wearing conventional indoor winter clothing the average surface temperature of the body, including clothed and exposed surface, is in the neighborhood of 85 F. Thus when the average temperature of enclosure (room) surfaces is 70 F, a definite radiation loss from the body to the room surface takes place. It is naturally desirable to know how the radiation and convection heat losses vary as the air temperature and average surface temperature change from 70 F. It is difficult to predict the amount of such losses accurately for a change in temperature of more than a few degrees; but in a narrow range (± 8 deg) above and below 70 F, it has been found that each 1-deg rise in the average enclosure temperature decreases the radiant heat loss by very nearly the same amount as a 1-deg decrease in air temperature increases the convection heat loss. This statement can be formulated as follows:

$$\frac{t_a + t_e}{2} = 70 \tag{20-1}$$

Panel (Radiant) Heating and the Heat Pump

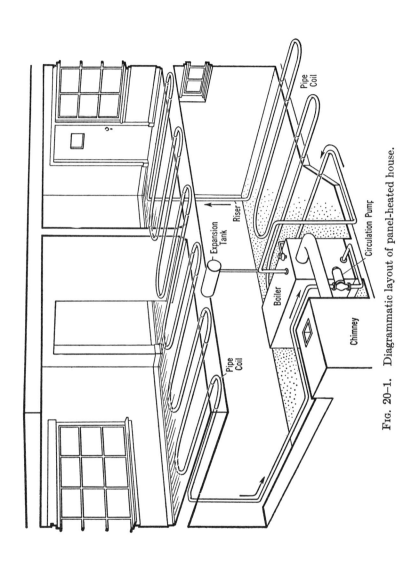

FIG. 20-1. Diagrammatic layout of panel-heated house.

where t_a = air temperature, in degrees Fahrenheit;
t_e = average surface temperature of the enclosure, in degrees Fahrenheit;
70 = operative temperature for office and residence conditions, in degrees Fahrenheit.

Thus, if t_a = 68 F and t_e = 72 F for one case, whereas t_a = 73 F and t_e = 67 F for another case, the conditions in both cases are equivalent for body heat loss or body comfort.

For a store with continuous activity, an operative temperature of 66 should be substituted for the 70 in equation 20–1; for a manufacturing plant with production activity, 64 should be used for the operative temperature.

20-2. PANEL LOCATION

Panels are placed in the floor or in the ceiling but rarely in the side walls. Although panels are thought of as radiant heat devices, a large portion of the energy supplied from the heating source to the panel is actually delivered into the enclosure by convection methods. It is customary to consider the fraction of the heat transferred by convection as approximately 30 per cent from a ceiling panel, 40 per cent from a wall panel, and 48 per cent from a floor panel. Because of the large convection loss from floor panels and because for occupancy requirements the floor should not have a surface warmer than 85 F in walking spaces, interest is being directed to greater use of ceiling panel surface. Panels in this location can be operated to 115 F without discomfort if the average mean radiant temperature is not too high—say 77 to 78 F or less.

Figure 20–1 is a diagrammatic layout of a hot-water panel system, with the floors of the building structure holding embedded pipe coils. This diagram illustrates the general principles of heat circulation and is not intended to be used as a detailed installation guide. In section 11–13 there is a discussion of an arrangement of a forced-warm-air system in which a large portion of the ceiling area is heated from warm-air ducts, and Fig. 20–2 shows a representative layout for a warm-air ceiling-panel system for a one-story house. In this arrangement the warm air is delivered into the attic through a centrally located trunk riser, from which ducts lead over to the seven inlets that supply air to the passage spaces between the true room ceiling and a second continuous closure directly under the joists. The warm air flows in this inner space, following guided passageways as indicated in the illustration, and finally returns to the furnace through the six return ducts shown. The passage space between the ceiling and the gypsum (rock) lath under the joists should be $3\frac{1}{4}$ in. deep, and insulation of not less than 2 in. of rock wool or its equivalent must be provided over the passage. In fact, insulation should always be provided on the back

Panel (Radiant) Heating and the Heat Pump

side of panel surfaces in order that the uncontrolled effect of the panel heat will not pass into spaces above or below that which the panels serve.

20-3. PANEL OUTPUT

To find the heat output of a panel by radiation effects alone, reference should be made to equation 4-34, which shows that the heat transferred by radiation is proportional to the difference between the fourth powers of the emitting and receiving sources. Moreover, in section 4-10 it was stated

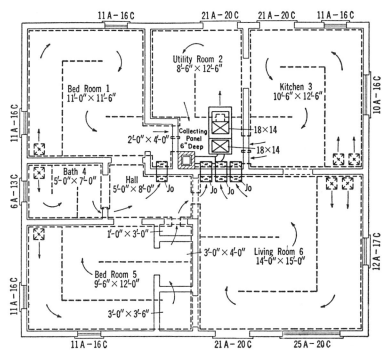

FIG. 20-2. Air-heated-panel plan for one-story house, showing ceiling arrangement. A = full area for window, in square feet; C = running feet of crack for window; Jo = jumpover. Ceiling height is 7 ft 6 in.

that the effective absorptivity factor for parallel planes and large enclosed surfaces appeared as

$$e = \frac{1}{(1/e_s) + (1/e_r) - 1} \quad (20\text{-}2)$$

If, in equation 20-2, representative emissivities of 0.9 are used to represent probable surface absorptivity values in representative buildings, it will be found that the numerical magnitude of equation 20-2 works out to be 0.82. Also, in the case of large plane surfaces such as are met in a representative room enclosure, the geometrical configuration factor F_a

is equal to 1, and when $e = 0.82$ and $F_a = 1$ are inserted in equation 4–34 it is found that it reduces to the form

$$q'_r = 0.142A\left[\left(\frac{T_s}{100}\right)^4 - \left(\frac{T_r}{100}\right)^4\right] \qquad (20\text{–}3)$$

or to the following form when a unit panel area (1 sq ft) is considered:

$$q_r = 0.142\left[\left(\frac{T_s}{100}\right)^4 - \left(\frac{T_r}{100}\right)^4\right] \qquad (20\text{–}4)$$

where q_r = net heat transmitted by radiation per square foot of panel surface, in Btuh, when
T_s = panel surface temperature, in degrees Fahrenheit absolute ($t_s + 460$), and when
T_r = temperature of absorbing wall surfaces, usually considered as the wall *unheated mean radiant temperature* (UMRT), in degrees Fahrenheit absolute.

TABLE 20–1

Heat Delivery by Radiation from Panels
(Btu per Hr Sq Ft)

Temperature of Panel Surface	Unheated Mean Radiant Temperature of Room (UMRT)								
	50 F	55 F	60 F	65 F	70 F	75 F	80 F	85 F	90 F
60 F	8	4	...	...	...	...	...	...	...
70 F	16	12	8	5	...	...	...	...	...
80 F	26	22	17	13	9	5	...	...	...
85 F	30	26	22	18	14	10	5	...	...
90 F	35	31	27	22	18	14	10	5	...
100 F	44	40	36	32	28	24	19	15	10
105 F	49	45	41	37	32	28	24	20	15
110 F	53	50	46	42	37	33	28	24	20
115 F	60	55	52	47	43	40	35	31	16
120 F	65	62	57	53	50	45	40	37	32

This equation is not rigorously precise because of the assumptions involved. However, under most conditions the radiant output of a room panel computed by this formula is correct to within 8 to 12 per cent. Table 20–1 gives radiant heat delivery from a panel, expressed in terms of panel surface temperature and average temperature of surroundings and facing surfaces, the so-called unheated mean radiant temperature (UMRT).

Heat transfer by convection per unit area from a flat surface to air moving by natural methods takes the general form

$$q_c = f(t_s - t_a)^n \qquad (20\text{–}5)$$

Panel (Radiant) Heating and the Heat Pump

where q_c = heat transferred by convection, in Btu per hr sq ft of panel surface;

t_s = surface temperature of panel, in degrees Fahrenheit;

t_a = representative bulk temperature of air adjacent to the panel, in degrees Fahrenheit;

n = an exponent whose value, although related primarily to plate location, is also affected by temperature difference (for heat flow upward from horizontal surfaces, $n = 1.12$; for heat flow downward, as for example from ceiling panels, $n = 1.25$);

f = coefficient of surface conductance, in Btu per hr sq ft deg F temperature difference between surface and air (the value of f varies with panel location; it can be taken as 0.81 for upflow panels and 0.22 for downflow panels).

For floor panels, equation 20–5 becomes

$$q_c = 0.81(t_s - t_a)^{1.12} \qquad (20\text{–}6)$$

and for ceiling panels,

$$q_c = 0.22(t_s - t_a)^{1.25} \qquad (20\text{–}7)$$

TABLE 20-2

HEAT DELIVERY BY CONVECTION FROM PANELS TO AIR
(Btu per Hr Sq Ft)

Temperature Difference between Panel Surface and Air (deg F)	Panel Location		
	Ceiling	Wall	Floor
5	1.8	3.0	5.0
10	4.0	6.8	11.0
15	6.8	11.2	17.5
20	9.2	16.0	23.0
25	12.5	21.2	31.0
30	16.0	26.7	40.0
35	18.9	32.6	52.0
40	22.4	38.4	...
45	25.7	44.6	...
50	29.0	50.4	...
55	33.0	56.7	...
60	36.6	62.4	...

Table 20-2 gives values computed from equations 20-6 and 20-7 for a range of temperature differences.

Example 20-1. For a ceiling panel with a surface temperature of 100 F, find the heat delivery by radiation and by convection into a room having a UMRT of 60 F and an air temperature of 76 F under the ceiling.

Solution: By equation 20-4, the heat addition by radiation from this panel is

$$q_r = 0.142\left[\left(\frac{460+100}{100}\right)^4 - \left(\frac{460+60}{100}\right)^4\right]$$
$$= 35.9 \text{ Btu per hr sq ft}$$

By Table 20-1, the corresponding value is 36.

By equation 20-7, for the heat added by convection,

$$q_c = 0.22(100 - 76)^{1.25} = 11.7 \text{ Btu per hr sq ft}$$

By Table 20-2, the corresponding value is found to be 11.8.

In a room or enclosure the surface temperature is related to the over-all wall resistance R_w, to the temperatures on both sides of the wall, and to the inside film factor R_f. For most cases the film factor of the inside wall is considered to be 1.65 and the corresponding resistance is 0.606, or approximately 0.6. By making use of resistance concepts, the inside surface temperature can easily be found (see sec. 4-4). Designate the outside design temperature as t_o, and the inside air temperature as t_a; then, as indicated previously in equation 4-23,

$$\frac{\Delta t_f}{R_f} = \frac{t_a - t_o}{R_w}$$

and the temperature drop through the air film Δt_f is

$$\Delta t_f = (t_a - t_o)\left(\frac{R_f}{R_w}\right) \quad (20\text{-}8)$$

or

$$\Delta t_{\text{film}} = (t_a - t_o)(R_{\text{film}})(U_{\text{wall}}) \quad (20\text{-}9)$$

And when $t_a = 70$ F, and $R_f = 0.6$,

$$\Delta t_f = (70 - t_o)\frac{0.6}{R_w} = \frac{42 - 0.6 t_o}{R_w} \quad (20\text{-}10)$$

Then the wall surface temperature t_w is

$$t_w = t_a - \Delta t_f \quad (20\text{-}11)$$

and, for $t_a = 70$ F and $R_f = 0.6$,

$$t_w = 70 - (70 - t_o)\frac{0.6}{R_w}$$

$$= 70 + \frac{0.6 t_o - 42}{R_w} \quad (20\text{-}12)$$

Example 20-2. An exposed side of a room is 16 ft by 8 ft and has two 3-ft by 5-ft single-glazed windows. The wall has an over-all coefficient of heat transfer U of 0.25 Btu per hr deg F, considering film effects; and for the glass, $U = 1.13$. When it is 0 F outside and at 70 F design inside air temperature, what is (a) the surface temperature of the wall and of the glass, and (b) the average weighted surface temperature of the exposed wall?

Panel (Radiant) Heating and the Heat Pump

Solution: (a) The resistance of the wall is

$$R_w = \frac{1}{U} = \frac{1}{0.25} = 4.0 \frac{\text{hr ft}^2 \text{ F}}{\text{Btu}}$$

and the resistance of the glass is

$$R_g = \frac{1}{1.13} = 0.885$$

For the wall, by equation 20-12,

$$t_w = 70 + \frac{(0.6)(0) - 42}{4.0} = 59.5 \text{ F} \qquad Ans.$$

and for the glass,

$$t_w = 70 + \frac{(0.6)(0) - 42}{0.885} = 22.6 \text{ F} \qquad Ans.$$

b) Of the 16 ft by 8 ft = 128 sq ft of surface, there are $2(3 \times 5) = 30$ sq ft of glass, and thus the remaining 98 sq ft is wall surface.

We then make use of the areas and their respective temperatures to find

$$t_{\text{avg}} = \frac{(98)(59.5) + (30)(22.6)}{128} = 50.8 \text{ F} \qquad Ans.$$

20-4. PANEL HEATING CALCULATIONS

1) In the layout of a panel heating system it is first necessary to compute the design heat loss from the room or space. In computing the heat load, one uses the same method that is used in computing the heat load for a convection heating system. Allowance for air change should also be made, with at least one air change per hour being considered. Although it is possible to operate radiant systems at air temperatures somewhat below 70 F, from a viewpoint of design 70 F is employed except for unusual applications.

2) After the approximate location and extent of the heating-panel area have been selected from a plan of the space, the unheated mean radiant temperature (UMRT) of the room surface should be found. The UMRT represents an average temperature for the unheated surfaces of the enclosure, with each surface weighted according to its area and temperature. Whenever the surfaces of inside walls, floors, and ceilings are not exposed to unheated space on their far side, the surface temperature is usually considered to be 70 F, although this figure should be modified for the case of adjacent spaces definitely operated at lower design temperatures. The method of finding the surface temperature of an outside (exposed) wall was discussed in section 20-3.

3) Two approaches can be made for finding the panel surface. One is to select the maximum or desired surface temperature which might be used for the panel in question. This temperature is, of course, related to whether the panel is in the floor, the ceiling, or a side wall. Then, in terms

of this surface temperature and the UMRT, Table 20–1 is used to find the radiant heat output; and, employing the respective surface and air temperatures, Table 20–2 is used to find the convection heat output. The sum of these two items represents the total heat output per square foot of panel surface. This unit figure can now be used in connection with the total room heat loss to find the required panel area. If this area is larger than the space available in the ceiling or floor, it is then necessary either to raise the operating temperature of the panel surface or, if this is not possible, to supply supplementary heat from another source or from an additional panel surface in another location.

The other approach is to make use of tabular data based on confirmed field and test data and listed here in Table 20–3. The manufacturer's literature also contains a variety of data of this type for use in design.

TABLE 20–3
APPROXIMATE HEAT DELIVERY OF PANELS SERVING SPACES DESIGNED
FOR NORMAL COMFORT OCCUPANCY
(Btu per Hr per Panel)

PANEL AREA* (SQ FT)	LOCATION AND SURFACE TEMPERATURE			
	Ceiling at 100 F	Ceiling at 110 F	Ceiling at 120 F	Floor at 85 F
	Tube Spacing and Approximate Average Water Temperature			
	$\frac{3}{8}$-In. Tube 4 In. C to C at 121 F 6 In. C to C at 133 F	$\frac{3}{8}$-In. Tube 4 In. C to C at 137 F 6 In. C to C at 153 F	$\frac{3}{8}$-In. Tube 4 In. C to C at 155 F	$\frac{1}{2}$- and $\frac{3}{4}$-In. Tube 9 In. C to C at 120 F 12 In. C to C at 130 F
100	5000	6500	8000	4000
200	10000	13000	16000	8000
500	25000	32500	40000	20000

* For other panel areas, prorate the base values listed.

4) The coil layout and spacing should be designed to produce the desired surface temperature and heat output. The necessary flow through the system is then computed.

Figure 20–3 shows a few of the arrangements which are employed in using copper-tubing or steel-pipe coils for panel heating. For floor and structural slabs, $\frac{3}{4}$-, 1-, or $1\frac{1}{4}$-in. sizes, of steel or wrought-iron pipe, and $\frac{3}{4}$-in. or 1-in. sizes of copper water tubing, are most common. For ceilings, $\frac{3}{8}$-in. or $\frac{1}{2}$-in. copper tubing, or $\frac{1}{2}$-in. steel (or wrought-iron) pipe sizes are most common. The permissible top water temperatures used are determined, to a large degree, by the necessity of limiting the surface temperature below

Panel (Radiant) Heating and the Heat Pump

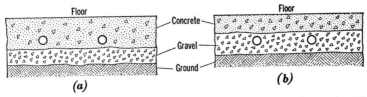

(a) Coil embedded in concrete poured on bed of packed gravel.

(b) Coil laid in packed gravel, with concrete floor placed above.

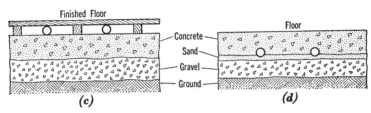

(c) Coil resting on top of concrete and covered by wood flooring supported on sleepers.

(d) Coil placed in sand resting on bed of packed gravel, with concrete flooring placed above.

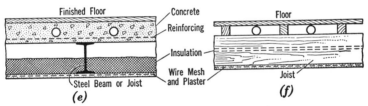

(e) Coil embedded in concrete floor panel. Floor supported by steel beam, with insulation on far side of panel.

(f) Coil laid on joists beneath wood flooring supported on sleepers. Insulation on far side of coil.

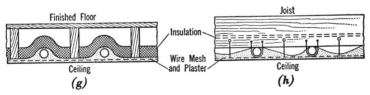

(g) Ceiling heating, with coil laid directly on metal lath and plaster. Insulation in place above coil.

(h) Coil supported by joists and metal lathing wired to coil. Insulation applied above coil and plaster.

Fig. 20-3. Typical pipe-coil arrangements of radiant-heating coils installed in floor or ceiling.

values which could cause discomfort to occupants and, to a lesser degree, by the necessity of preventing damage to the plaster, concrete, or other surface employed. During the initial drying-out (stabilizing) period, care must be exercised to keep temperatures below 90 F for at least forty-eight hours after the plaster or concrete has set.

Surface temperatures for *floor panels* should hardly ever exceed 85 F, with 80 F being preferable for many installations. Water temperatures in floor panels reach 120 F for conventional design, and go to 130 F for an acceptable maximum. For *ceiling panels* the height of the room, and the room arrangement, dictate the surface temperatures acceptable. Up to 120 F is considered possible; but for residential application, temperatures of 110 F or below are usually preferable. Top water temperatures of from 155 F to 165 F are used with ceiling panels. *Wall panels* are governed by

TABLE 20-4

Tube Spacing and Panel Output for Copper-Water-Tube Panels at High Ratings

Panel Position	Nominal Tube Size (in.)	Tube Center-to-Center Spacing	Panel Output (Btu per hr sq ft)	
			Nominal Design	Maximum Output
Ceiling...........	$\frac{3}{8}$	$4\frac{1}{2}$	65	75
	$\frac{3}{8}$	6	50	60
	$\frac{3}{8}$	9	35	40
Wall.............	$\frac{3}{8}$	$4\frac{1}{2}$	65	75
	$\frac{3}{8}$	6	50	60
	$\frac{3}{8}$	9	35	40
Floor.............	$\frac{1}{2}$ or $\frac{3}{4}$	9	40	50
	$\frac{3}{4}$ or 1*	12	40	50

* Because of the low surface temperature for floor panels (80 F to 85 F) which can be tolerated, the coil output is not limited by tube size.

considerations similar to those which hold for ceiling panels if the wall panels are built high in the side wall; if low, they are in a category similar to floor panels in regard to temperature effects on occupants. Top water temperatures employed range from 150 F to 160 F. Table 20-4 gives approximate output values for panels using copper pipe when operated with water temperatures in the range indicated in this paragraph.

Example 20-3. Compute the unheated mean readiant temperature (UMRT) for the combined living-dining room of the residence in Figs. 11-7 and 11-8. The overall U for the walls is 0.23, and the windows are single glass. The outside design temperature is -10 F, and the assumed air temperature is to be taken as 70 F. Take all inside walls as being at 70 F, but find the temperature of each of the other surfaces before

Panel (Radiant) Heating and the Heat Pump

starting the computation. Exclude the ceiling in this calculation, as it is desired to add a panel surface there. The basement is heated.

Solution: The dining room has a gross outside wall area of

$$12.5 \times 9 + 16.5 \times 9 = 261 \text{ sq ft}$$

and a glass area of

$$(2.5 \times 4)(4) = 40 \text{ sq ft}$$

There are 221 sq ft of net outside wall area in the dining room.

The living room has a gross outside wall area of

$$(19.5 + 16.5)(9) = 324 \text{ sq ft}$$

and a glass and door area of

$$\frac{37 \times 38.375}{144}(3) + (2.67)(7) = 48.3 \text{ sq ft}$$

The net outside wall area of the living room is 275.7 sq ft (note that the room has two exposed sides, since the garage is largely at a lower level than the room). Also,

Floor area = $16.5(12.5 + 19.5) = 528$ sq ft

Inside wall area = $9(12.5 + 19.5) = 288$ sq ft

The inside surface temperature of the outside wall can be found by equations 20–9 and 20–11. Recall that the inside film factor is $f_i = 1.65$ and $R_{film} = 1/1.65 = 0.606$. Then,

$$\Delta t_{film} = [70 - (-10)](0.606)(0.23) = 11.5 \text{ deg}$$

and the wall surface temperature is therefore

$$70 - 11.5 = 58.5 \text{ F}$$

For the glass surface temperature,

$$\Delta t_{film} = [70 - (-10)](0.606)(1.13) = 54.8 \text{ deg}$$

and the glass temperature is therefore

$$70 - 54.8 = 15.2 \text{ F}$$

A summary of the results is shown in the accompanying tabulation. Thus the unheated mean radiant temperature (UMRT), exclusive of the ceiling and its panels, is 62.4 F.

Surface Location	Net Area (sq ft)	Surface Temperature (deg F)	Area Times Surface Temperature
Outside area of dining room..........	221	58.5	12,930
Outside area of living room..........	275.7	58.5	16,130
Glass area of dining room............	40	15.2	608
Glass area of living room............	48.3	15.2	734
Floor area.........................	528	70	36,960
Inside wall area....................	288	70	20,160
Total........................	1401.0	[62.4]	87,522

Example 20-4. For the combined living-dining room of the residence illustrated in Figs. 11-13 and 11-14 and considered in example 20-3, prepare a design using (a) ceiling-panel surface and (b) coils imbedded in the floor slab.

Solution: The heat loss from this space, computed by conventional methods and allowing for one air change per hour, was found to be 23,920 Btuh with 70 F inside air temperature and −10 F outside air temperature.

a) Assume that the ceiling panel surface temperature will be held at 100 F for design conditions. Since the UMRT from example 20-3 is 62.4 F, the radiant heat output, by Table 20-1, is 34 Btuh per sq ft. Table 20-2, for 100 F − 70 F, or 30 deg difference, shows the convection output to be 16.0 Thus there is a total output of 34 + 16 = 50 Btuh per square foot of heated ceiling panel. The panel area needed can now be found as

$$(A_p)(50) = 23,920$$

and $\qquad A_p = 478$ sq ft $\qquad$ *Ans.*

This panel area requires the use of a large portion of the ceiling area (528 sq ft available) but the requirement is not excessive. The mean radiant temperature can now be computed with the help of the tabulated data which is already available in example 20-3 and to which the ceiling data can now be added:

$$
\begin{aligned}
1401 \times 62.4 &= 87{,}522 \\
478 \times 100 &= 47{,}800 \\
50 \times 70 &= 3{,}500 \\
\hline
1929 \times [\,72.0] &= 138{,}822
\end{aligned}
$$

Thus the mean radiant temperature of the heated room under design heat loading is 72 F. This is a satisfactory temperature and it might be possible and even desirable to reduce the air temperature to slightly below 70 F.

b) For the floor panel, if 80 F is selected as the surface temperature for a UMRT of 62.4 F, Table 20-1 shows a radiant output of 15.5 Btuh per sq ft and, for a 10-deg floor-to-air temperature difference, Table 20-2 shows a convection heat delivery of 11.0. The total is thus 26.5 Btuh per sq ft, which is too small to carry the design heat loss. If 85 F is tried, the corresponding figures are 20.0 and 17.5, or 37.5 Btuh per square foot of floor area. For this,

$$(A_f)(37.5) = 23{,}920$$

and thus $\qquad A_f = 637$ sq ft

Thus more panel surface is required than the available floor area of 528 sq ft. A floor panel, then, will not suffice unless supplementary heat from another source is provided—or unless the floor temperature is raised above 85 F, which is not a desirable solution.

As the ceiling panel design is satisfactory, reference to Table 20-4 shows that the design conditions can be met using $\frac{3}{8}$-in. copper water tubing spaced on 6-in. centers. A similar result could also be reached using $\frac{1}{2}$-in. nominal-size steel pipe on 8-in. centers, with possibly a slightly higher water temperature employed.

20-5. WATER REQUIREMENTS FOR PANELS

In previous sections, methods were developed for finding the required panel area to offset the heat loss of a room and to maintain desired mean radiant and air temperatures in the space. It is obviously necessary to circulate adequate amounts of water at proper mean temperatures, in

Panel (Radiant) Heating and the Heat Pump

order to hold the panel surfaces at desired temperatures. The amount of water and its allowed temperature drop must be sufficient not only to offset the heat losses of a space but also to offset the heat loss through the far side of the panel—as, for example, from a floor panel into the ground, or from a ceiling panel into a room above or into an unheated attic. The latter type of loss is called the reversed heat loss. To provide for reversed heat loss, up to 10 per cent additional input energy to the panel should be allowed in the case of ceiling panels, and with floor panels up to 25 per cent should be allowed.

For a given room design, when the heat loss is increased by an adequate factor to compensate for reversed loss, water calculations can then be made. Call the heat loss from the panel, with its associated reversed loss, Q_p. The water supplied to the panel in cooling must exactly balance this loss. Thus we can write

$$Q_p = 60 W C_p(t_{in} - t_{out}) = 60 W(1)(t_{in} - t_{out})$$
$$= (60)(\text{gpm})(8.1)(t_{in} - t_{out})$$
$$= 490(\text{gpm})(t_{in} - t_{out}) \qquad (20\text{--}13)$$

where Q_p = total required input from water, including reversed heat loss, in Btu per hour;
W = water flow, in pounds per minute;
gpm = flow equivalent to W, in gallons per minute;
t_{in} = temperature of water in, in degrees Fahrenheit;
t_{out} = temperature of water out, in degrees Fahrenheit.

TABLE 20–5

APPROXIMATE LENGTH OF TUBING PER SQUARE FOOT OF PANEL SURFACE FOR VARIOUS CENTER-TO-CENTER SPACING OF TUBES

Spacing (in.)	Length (ft)
4	3.0
$4\frac{1}{2}$	2.7
6	2.0
8	1.4
9	1.3
12	1.0

With the required pump flow to the coil determined, it is next necessary to find whether one or whether more than one circuit will be required for the panel in question. Two factors enter into this consideration: (1) the frictional loss of the water in flowing through its circuit and (2) whether it is possible to maintain the desired temperature difference and mean water temperature in the flow circuit. Table 20–5 furnishes the approximate length of tubing required to serve a given panel area. For example, if a panel requiring 500 sq ft of surface had tubes spaced on 6-in. centers, the factor 2.0 from Table 20–5 would show that 2 × 500 = 1000 lin ft of tubing are required. Unquestionably, such a length is too long to put into one circuit—particularly if the tubing is small, such as $\frac{3}{8}$ or $\frac{1}{2}$ in. In general, circuit lengths should not exceed 200 ft from header to header unless tubing $\frac{3}{4}$ in. or larger is employed. For ceiling panels, it is generally considered that with $\frac{3}{8}$-in. tubing, circuits should not greatly exceed 120 to 140 ft with standard pumps, or 160 to 200 ft with high-head pumps.

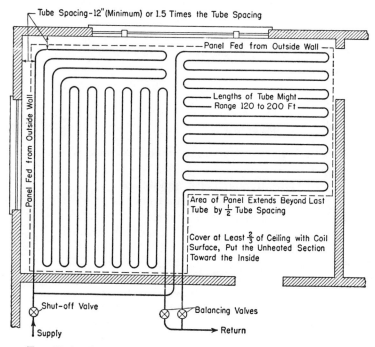

Fig. 20–4. Typical arrangement for layout of two radiant ceiling panels with balancing valves.

For floor panels with larger-size tubing, the corresponding figures might be 200 ft for standard pumps and up to 350 ft for high-head circulating pumps.

In laying out the friction design for the hot-water circuit, the methods developed in Chapter 8 can be employed. The friction-loss graphs of Fig. 8–14 for copper tubing, and of Fig. 8–13 for steel pipe, are applicable. Data on circulating pumps are given in Figs. 8–11 and 8–12.

Figure 20–4 shows an arrangement in which two coil circuits (zones) are employed to serve a ceiling panel in a room of large size. With the arrangement indicated in Fig. 20–4, the flow through the different panels

Panel (Radiant) Heating and the Heat Pump

can be adjusted (balanced) so that one portion of the room is at a lower temperature than the other. This is accomplished by means of the balancing valves. In addition to the simple arrangement shown in Fig. 20-4, much more elaborate zoning arrangements can be set up for large buildings. These can also lead to elaborate control arrangements, with one or more pumps and with mixing valves used to mix portions of the return water with the supply water to give the exact temperatures required in a given panel. Such an arrangement, using one pump and deflecting tee fittings, is shown in Fig. 20-5.

Coils are most usually sinuous in character, looping back and forth. However, sometimes headers on each side of a panel are used with tubing running across the space. It is always advisable to insulate the reverse side of a coil, not only to control the reversed heat loss but also to avoid the possibility of undesired and uncontrolled heat leakage into adjacent spaces. In laying out the coils, care must always be taken to see that adequate vent points are provided so that air caught in the system can be vented and not interfere with the free flow of water in the various circuits.

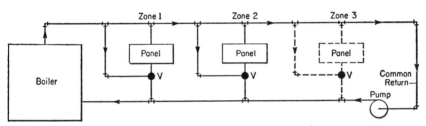

FIG. 20-5. Zoning arrangement for multiple panels, using one pump and adjustable or automatic mixing valves to control temperature of panel supply water.

20-6. PANEL HEATING CONTROL

Radiant heating can be controlled by ordinary thermostatic controls, but under certain conditions the space may be overheated or underheated. This condition is particularly prevalent, with floor coils embedded in heavy masonry slabs, during periods of rapid fluctuation in outside temperature. It should be realized that, as the outside temperature falls, a radiant panel can provide only for the additional heat load required by the space because of an increase in surface temperature. The end result is to bring the room temperature up to the desired point but at the expense of having a panel temperature which is so high as to cause overheating and discomfort. With a thermally heavy structure, even after the thermostat has cut off the heat supply, the energy stored in the structure will maintain the space temperature for a delay period and prevent the heat supply from coming on sufficiently soon, producing a period of low-temperature discomfort. Consequently, with a simple thermostat and a heavy masonry-panel structure, "hunting" will result. On the other hand, with a light panel system (small thermal storage), a simple indoor thermostat with some manual adjustment available can be perfectly satisfactory.

Because of thermal lag it is necessary, in a fully automatic control system, that provision for change in the temperature of the panel be anticipated before the space conditions change, if the space conditions are to be held essentially constant. As the space internal load does not greatly vary, a control responsive to change in outdoor temperature might be used to actuate a change in panel temperature. One commercial control system employs an outdoor bulb which in turn sets the control-point limits of a heating-medium bulb. Thus the panel water temperature is set at predetermined values and the heating medium is controlled in response to outside temperature rather than the temperature of the space itself. Actually, a room thermostat may be superposed on a system like that just described, to take care of zone problems or to make desired modifications.

20-7. SNOW-MELTING

Active interest has recently arisen in the use of panels for melting snow from roadways, pavements, runways, and the like. Most snow-melting systems of this type consist of ferrous or nonferrous coils embedded in the concrete at the time it is poured. Instead of water, an antifreeze mixture, frequently ethylene glycol, is employed in the snow-melting circuit. The antifreeze mixture in turn is warmed through a steam or hot-water heat exchanger. The problem with these devices is to maintain the surface at a sufficiently high temperature, at least above 32 F, so that snow falling on the surface will melt and later evaporate. Space does not permit a complete discussion of the design of snow-melting panels, but this matter has been extensively discussed in the engineering literature.

20-8. SUMMARY

In this chapter we have developed methods for use in designing radiant (panel) heating systems. However, it should be mentioned that throughout the industry there is somewhat of a divergency of opinion as to the best over-all design methods for all conditions. This is true not only in design but also in the selection and installation of controls. Many successful installations using the design methods indicated in this chapter are in use.

Both copper tubing and steel (or wrought-iron) piping are used for panels. Corrosion is not a serious problem, since the water in the system is continuously recirculated, and when, during the process of recirculation, it has once been deoxygenated, the basis for corrosion disappears. The coefficient of linear expansion of steel or iron (0.0000068 ft per ft per deg F) is close to that of concrete or plaster. The expansion coefficient for copper is somewhat higher, but for the temperature ranges employed the cracking

Panel (Radiant) Heating and the Heat Pump

of plaster because of unequal expansion is rare for any of the metals unless a very sudden heating (temperature rise) is imposed on the panel. Surface coverings over panel surfaces do not greatly limit output unless the coverings are of high insulating capacity. Rugs over panels do reduce the output somewhat. The radiation emissivity of woods, rugs, plaster, painted surfaces, etc.—in fact, of almost everything except polished metal—lies in a range of 0.85 to 0.90, and the type of surface is not too significant from a radiation viewpoint.

Initial costs of panel heating systems are usually somewhat higher than costs of most other good heating systems, but in many cases the difference is not great. A study made of typical residences showed panel-system costs ranging from 6.8 to 11.0 per cent of the cost of the residence. In contrast, conventional systems run from about 5 to 10 per cent.

20-9. THE HEAT PUMP

Panels for effective operation can be designed for use with moderate-temperature water, and for this purpose the heat pump can be very effective. The heat pump can also supply warm air at moderate temperatures with good efficiency.

An analysis of the discussions on refrigeration would show that by supplying mechanical work to a refrigeration system, energy absorbed at a low temperature level can be made available for use at higher temperature. A device which uses energy produced in this way is called a *heat pump*. Interest in commercial development of the heat pump has increased in recent years, as it is becoming apparent that the operating cost of the heat pump is competitive with other fuels wherever electric-power costs are low or moderate. Except in regions of mild winter temperature the outside winter air is not in general an economical low-temperature source for the heat-pump sytsem. In addition to the operating costs for power and maintenance, and possibly for water, there exist the fixed charges on the equipment of the heat-pump system. These latter costs are generally higher than for a conventional heating system and it is difficult to justify them unless there is some need of using the same equipment for cooling in summer.

Figure 20–6 shows in schematic arrangement how a heat-pump system can be substituted for the furnace of a residence or building which employs forced-warm-air circulation for heating. The liquid refrigerant from the condenser and receiver enters the expansion valve and expands into the cold evaporator. Here vaporization of the refrigerant is carried out as it absorbs heat from the cold outside air passing over the coils. The vaporized refrigerant then enters the compressor, where mechanical energy is added to raise the pressure and temperature of the refrigerant. The hot refrigerant then passes to the condenser, where it gives up superheat and

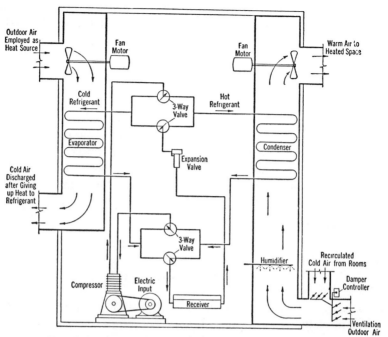

Fig. 20-6. Diagram of heat pump used to heat residence.

enthalpy of vaporization in useful heating. For example, if 200 Btu are added in the evaporator and 100 Btu are supplied by the compressor, then 300 Btu are delivered to the condenser and the *heating coefficient of performance* (HCOP) is $\frac{300}{100}$, or 3. This figure is representative of what might actually be experienced with a temperature of 30 F outside. At 55 F outside temperature the corresponding value for the HCOP would be 4. Because of the temperature difference required for heat transfer, the refrigerant is at least 10 deg colder than the air passing over it.

Notice that by changing the four three-way valves in Fig. 20–6 the same system can be used for cooling. The evaporator in this case becomes the condenser, and the former condenser serves as the evaporator for cooling the residence air.

Equation 14-3 showed that the coefficient of performance (CP) of a refrigeration system was expressed as the ratio of the refrigeration produced (Q_R) to the work (W) required to produce it. Equation 14-3 can also be written as

$$\text{CP} = \frac{Q_R}{W} = \frac{Q_C - W}{W} \qquad (20\text{--}14)$$

where Q_C represents the heat transfer at higher temperature during condensation. The useful output that could be derived from a heat pump

Panel (Radiant) Heating and the Heat Pump

(Q_C) is greater than the refrigeration effect Q_R to the extent of the work addition W; that is, $Q_C = Q_R + W$. In equational form, the heating coefficient of performance is expressed as

$$\text{HCOP} = \frac{Q_C}{W} \quad (20\text{-}15)$$

Example 20-5. A certain residence with a design heat loss of 60,000 Btuh employs a heat pump. The pump has a refrigeration coefficient of performance of 3 when it is 50 F outside and when 70 F is maintained inside. Under these conditions the condenser and evaporator coil temperatures are, respectively, 80 F and 40 F. Disregarding power used by auxiliaries, compute (a) the kilowatts required from the power lines when the over-all motor-compressor efficiency is taken at 0.82, and (b) the heating coefficient of performance.

Solution: (a) Subsituting in equation 20-14, the shaft work is found as follows:

$$3 = \frac{60{,}000 - W}{W}$$

$$W = 15{,}000 \text{ Btuh}$$

Therefore the number of kilowatts supplied is

$$\frac{15{,}000}{(0.82)(3413)} = 5.35 \text{ kw} \qquad Ans.$$

where 3413 represents the heat equivalent of a kilowatthour. (If power can be purchased at 2.5 cents per kilowatthour, the 5.35 kw represents an operating cost of 13.4 cents per hour.)

b) The heating coefficient of performance, by equation 20-15, is thus

$$\text{HCOP} = \frac{Q_C}{W} = \frac{60{,}000}{15{,}000} = 4 \qquad Ans.$$

Table 20-6 gives specific data on two small heat-pump units of one manufacturer.

20-10. HEAT SOURCES FOR HEAT PUMP

Air. The most obvious energy source for the heat pump is the air. Where the air temperature does not go to low extremes this source is quite adequate. Certainly in climates where winter temperatures are in general about 32 F, air can be considered an effective and useful source of low-temperature energy. In regions where the temperature does drop to values appreciably below 32 F, not only is the temperature lift of the system greatly increased, but the problem of frost formation on the heat-transfer coils may become serious.

In general, however, this has not been the case, because as the temperature lowers, the amount of moisture carried by the air decreases.

The main disadvantage of air-to-coil transfer lies in the fact that the maximum heat-loss loading on the building occurs simultaneously with the lowest temperature available for supplying energy to the coils. This lowers the HCOP; and not only is more power needed per unit of heating output, but this condition occurs when the demand is greatest. The widespread use of heat pumps within a given utility system might cause a very heavy loading on the system with every extreme drop in atmospheric temperature. If such a peak coincided with other peak loads of the system, generating capacity might be overextended or even inadequate.

TABLE 20-6

HEAT-PUMP SPECIFICATIONS

Item	3 Hp	5 Hp
Winter heating output, Btuh*	32,500	54,200
Summer cooling rating, Btuh†	30,000	50,000
Winter: Heat delivered to	Building air	Building air
Summer: Heat dissipated to	Outdoor air	Outdoor air
Winter heat source	Outdoor air	Outdoor air
Winter supplementary heat	Electric strip heaters	Electric strip heaters
Indoor fan, cfm	1200	2000
Outdoor air fan, cfm	1200	2000
Compressor type	Hermetic	Hermetic
Size	50" × 29" and 75" high	71" × 29" and 75" high
Design for house having	5-6 rooms	6-8 rooms

* Based on 70 F inside, 35 F outside.
† Based on 80 F db and 67 F wb inside; 95 F outside.

Well Water. The temperature of water in wells does not deviate greatly from an annual minimum value; consequently, when such water is available, it forms an excellent source of energy for heat-pump coils. However, the cost of sinking a well and pumping the water must be considered, and unfortunately the water-level table in the United States has been dropping in almost all sections of the country, because of greater usage and because of less-effective rain absorption into the soil (resulting from intensive land use and bare fields).

Some double wells have been drilled as a means of water conservation. Water is pumped from one of the wells, and the cooled water, after passing through the system, is forced back into the other well. The second well must be placed at a sufficiently great distance from the first well to permit the water, in percolating back to the source point, to be warmed by contact with the subsurface of the earth. The additional complications of this system, and the cost of the second well, are deterrents to its extensive use.

Underground Water. In the case of the double-well system, underground water is employed by bringing it to the surface of the earth and then sending it back. It is sometimes possible to accomplish the same result by placing a coil below ground in such manner that a moving underground current passes over the coil. It is desirable, of course, that the water be continuously moving; otherwise the coil will freeze a layer of ice around it as heat is abstracted, and the useful process of heat transfer will be retarded as the insulation effect of the ice becomes greater and greater. The problem of burying such a coil and finding a source of underground water is, of course, difficult.

Earth Coils. Instead of depending on continuous movement of water below ground, it is possible to bury a network of pipe coils so arranged that they come into intimate contact with the earth itself. Heat can then flow from the earth into the colder coil. In turn, the temperature of the earth adjacent to the coil becomes reduced. It is necessary that the coil remove heat at a slow enough rate to prevent an excessive localized lowering of ground temperature. Earth has such widely different conductivities, ranging from 0.2 to 2.0 Btu per hr sq ft deg F per foot, that designs of this nature are somewhat uncertain. If evaporator operation is carried out continuously at low temperatures, the moisture in the earth around the buried pipe freezes. This is not a serious disadvantage, since frozen ground with good moisture content has a higher thermal conductivity than unfrozen ground, namely from 1 to 2.5 Btu per hr sq ft deg F per foot. In the winter season the temperature of an underground area supplying coils would be greatly reduced, and therefore it might be desirable in summer to reverse the system and put heat back into the ground, with the former evaporator acting as the condenser of the system. The ultimate source of ground heat of this type is largely the sun, heat being obtained by conduction from the sun-warmed surface of the earth during summer periods. A relatively small amount of heat passes from the core of the earth outward. If sufficient investment is made in burying an adequate surface of pipe coils, the earth can be an effective source of heat for pumps.

With ¾-in. copper tubing buried at a depth of 6 ft, heat-absorption rates have ranged from 50 to 25 Btuh per lineal foot over a winter season when the coil temperature was about 21 F and the minimum temperature of the uncooled ground not adjacent to the coil was 41 F.

Miscellaneous Sources. Natural lakes or even artificial water-storage reservoirs can be considered as possible heat-pump sources. Certainly such storage could be used to carry peak demands in the case of a normal air-to-air operating system. Even the possibility of using city water for peak demands may be feasible in some cases. The possibility of freezing water and utilizing this latent heat of fusion might also be considered, but the inflexibility of rigid ice, and the insulating effect of ice, make this scheme appear not particularly feasible.

20-11. ROOM-COOLER HEAT PUMPS

Increasing interest is arising in connection with room-cooler units which can have their operation reversed so as to heat the room as a heat pump when outside conditions make heating necessary. Such dual-purpose devices can be of particular use where climates are inherently mild and also during fall and spring in more-rigorous areas. However, although these are sometimes of the window type, it is still true that most of the heat pumps in use are of the central type, with the distribution system sending either cooled or warm air to the rooms of a building.

PROBLEMS

20-1. Making use of basic equations, compute the heat output per hour of a 10-ft by 10-ft ceiling panel having a surface temperature of 105 F in a room with a UMRT of 61 F and an air temperature of 68 F. Compare your answer with tables 20-1 and 20-2.
Ans. 6010 Btuh

20-2. If the panel surface of problem 20-1 is extended along the top side-wall instead of on the ceiling, what is the heat output in Btu per hour? Assume that the panel-surface and room-temperature conditions are the same as in problem 20-1. Use Table 20-1 and Table 20-2.
Ans. 7500 Btuh

20-3. Rework problem 20-1 by any method, assuming the following conditions: a ceiling-panel surface temperature of 95 F, a UMRT of 60 F, and an air temperature of 68 F.
Ans. 4540 Btuh

20-4. Rework problem 20-2 for a panel surface temperature of 95 F, a UMRT of 60 F, and an air temperature of 68 F.
Ans. 5490 Btuh

20-5. Compute the unheated mean radiant temperature for the combined living-dining room of the residence shown in Figs. 11-7 and 11-8, with no change except that the outside design temperature is + 5 F. Refer to example 20-3 for the area and for other data on the room.

20-6. Consider the room described in problem 5-1 to use panel heat in its ceiling, and find the UMRT for the conditions indicated. In a radiant-heated room the temperature gradient from floor to ceiling is small and no correction for vertical temperature gradient need be made. Except for the exposed walls, all other surfaces are adjacent to heated spaces.
Ans. 61.7 F

20-7. A recreation lounge in a resort hotel is 60 ft long, 30 ft deep, and 10 ft high. Along one of the long sides is a glass area amounting to 300 sq ft. The space is heated by panel coils in the ceiling, which are designed to operate at a surface temperature of 100 F. Surface temperatures are as follows: the glass, 50 F; the exposed wall, 60 F; the end walls, 64 F; and the back wall, the floor, and the unheated ceiling, 70 F. The design heat loss, including infiltration, amounts to 19,000 Btuh. (a) Compute the unheated mean radiant temperature, considering that half

Panel (Radiant) Heating and the Heat Pump

the ceiling is not provided with panel surface. (b) What area of panel surface is needed to heat the space if a 73 F air temperature is maintained?

Ans. (a) 67.2 F; (b) 430 sq ft

20–8. (a) For 430 sq ft of panel area in problem 20–7, at an average temperature of 100 F, refer to tables 20–3 and 20–5 and select appropriate tubing. (b) Find the gpm required to serve the panel with a 20-deg drop in water temperature. (c) Find an appropriate number of circuits through the panel, and a probable pumping head.

Ans. (a) 1290 ft of $\frac{3}{8}$-in. tubing on 4-in. centers; (b) (c) 1.94 gpm in 7 circuits each 185 ft long or 0.28 gpm per circuit, 3.4 ft head loss

20–9. The room of problem 20–6, under the design conditions indicated and under certain infiltration conditions, has a total loss of 6150 Btuh. (a) Find the panel area which would be required at surface temperatures of 115 F and 110 F and (b) estimate the lowest surface temperature which would suffice with the 80 sq ft of ceiling area available. If this temperature is above 110 F, supplementary heat should be used.

Ans. (a) Available area inadequate (81.4 sq ft); (b) 115 F

20–10. Refer to example 20–3, and find the room UMRT for a location where -5 F is the outside design temperature.

20–11. Carry through the ceiling panel design on the basis of the data in problem 20–10, with a recomputation showing the heat load to be 22,400 Btuh. Find (a) the panel area, (b) the length of the $\frac{3}{8}$-in. copper tubing on 6-in. centers which is to be used, and (c) a desirable number of circuits to use.

20–12. (a) For example 20–3, compute the gpm of water required in the coil if it enters at 145 F and cools to 121 F. (b) Assume that this flow is handled in 20 parallel circuits each 48 ft in length using $\frac{1}{2}$-in. copper tubing, and find the friction drop. Refer to Chapter 8 for friction-loss charts.

20–13. A two-cylinder, single-acting compressor, with $2\frac{1}{4}$-in. bore and $1\frac{1}{8}$-in. stroke, runs at 1500 rpm. The condensing temperature is 82 F and the evaporating temperature is 30 F. A heat exchanger is used which warms to 65 F the 30 F dry saturated vapor from the evaporator, in subcooling the 82 F condensed liquid to a lower temperature before the liquid enters the evaporator. Freon-22 is the refrigerant. (a) Compute total piston displacement in cubic feet per minute. (b) Compute the temperature of the liquid before it enters the expansion valve—after it has passed through the heat exchanger and has been cooled down by the vapor warming from 30 F to 65 F. Note that the decrease in enthalpy of the liquid equals the increase in enthalpy of the vapor. (c) Compute the pounds of refrigerant required per ton of refrigeration per minute. (d) For a volumetric efficiency of 88 per cent, compute the pounds of refrigerant handled per minute by the compressor. (e) Find the refrigerating capacity, in tons and in units of 1000 Btuh. (f) Find the isentropic work of compression per pound. (g) Find the theoretical horsepower and specify a motor size for this unit.

Ans. (a) 11.65 cfm; (b) 65.7 F; (c) 2.5 lb/ton-min; (d) 11.93 lb/min; (e) 4.77 tons; (f) 9.4 Btu/lb; (g) 3.7 shp and 5-hp motor

20–14. A heat-pump system must supply a design heat load of 50,000 Btuh. Freon-12 is used with a minimum evaporator temperature of 30 F, and the condenser must work at not less than 106 F (143.2 psia). (a) Find the isentropic work of compression and the probable actual work input per pound of refrigerant (use a factor of 1.4). (b) Find the pounds of refrigerant which must be circulated per hour for the heating load. (c) Find the shaft horsepower required by the compressor. (d) What is the kilowatt input (assume a motor efficiency of 90 per cent) and the hourly operating cost (assume a cost of 1.2 cents per kilowatthour)? (e) Find the over-all HCOP in terms of input electric power.

Ans. (a) 9.5 and 13.3 Btu/lb; (b) 802 lb/hr; (c) 4.2 hp; (d) 3.5 kw, 4.2 cents; (e) 4.18

20–15. Work problem 20–14 for a top pressure condition of 110 F (151.1 psia) and an evaporator at 38 F.

20–16. A heat-pump system in a large office building in Portland, Oregon, uses well water, being supplied with 150 gpm at 64.5 F from one well and with 450 gpm at 62.5 F from a second well. For heating, the water is chilled to 50.4 F and disposed of in a third well of greater depth. The heat-pump equipment consists of four centrifugal refrigerating units having a total rated capacity of 540 tons under summer cooling conditions. The water for warming the air in winter is heated to 101 F in the condenser and cools to 86 F. Various energy-saving and heat-saving features are employed in this system, but they will not be considered here. (a) Compute in Btu per hour the heat that is absorbed from the water when the full capacity of both wells is absorbed. (b) The evaporators operate at 46 F and condensation occurs at 105 F. The refrigerant used is Freon-11. Estimate the shaft horsepower that is required if the compressor efficiency is 77 per cent, by computation from isentropic power. (c) Assume the motor efficiency is 93 per cent and find the power, in kilowatts, drawn from the power lines. (d) For these data find the over-all HCOP.
Ans. (a) 3,780,000 Btuh (68.5 Btu/lb refrigerant, 919 lb/min); (b) 251 shp, 193 isentropic hp; (c) 201 kw; (d) 6.4

20–17. Assume that in problem 20–16 half of the condensers, and their compressors, warm the water from 86 F to 93.5 F, with condensation at 98 F, and that the others warm the water from 93.5 F to 101 F, with condensation at 105 F. On this basis (a) compute pertinent items as called for in problem 20–16 and (b) find two values of over-all HCOP.
Ans. (b) 6.4 at 105 F, 7.2 at 98 F

BIBLIOGRAPHY

AMERICAN SOCIETY OF HEATING, REFRIGERATING, AND AIR-CONDITIONING ENGINEERS, *Guide and Data Book, Systems and Equipment 1967*, Chapter 50, "Heat Pump Systems," Chapter 58, "Panel and Radiant Heating."

BURKE, R. H. "Ramp Snow Melting for Underground Garage," *Heating, Piping and Air Conditioning*, Vol. 27, No. 3 (March, 1955), pp. 105–8.

CHAPMAN, W. P. "Design Conditions for Snow Melting," *Heating and Ventilating*, Vol. 49, No. 11 (November, 1952), p. 88.

——, "Design of Snow Melting Systems," *Heating and Ventilating*, Vol. 49, No. 4 (April, 1952), p. 95.

COOGAN, C. H., JR. "Heat Transfer Rates—Coils in Earth," *Mechanical Engineering*, Vol. 71, No. 6 (June, 1949), pp. 495–8.

SARTAIN, E. L., and W. S. HARRIS. "Heat Flow Characteristics of Hot Water Floor Panels," *Heating, Piping and Air Conditioning*, Vol. 26, No. 1 (January, 1954), pp. 183–92.

SCHUTRUM, L. F., G. V. PARMELEE, and C. M. HUMPHREYS. "Heat Exchanges in a Floor Panel Heated Room," *Trans. ASHVE*, Vol. 59 (1953), pp. 495–510.

SHOEMAKER, R. W. *Radiant Heating*. New York: McGraw-Hill Book Company, Inc., 1948.

21

The Cleaning of Air

21-1. AIR IMPURITIES

Among the impurities which appear in air that is used for ventilating are carbon (from incomplete combustion); bacteria; plant pollen; dust from manufacturing processes; dust from the ground, such as sand, animal excrement, and rubber from tires; and (from wearing apparel) lint, leather particles, and the like. The intake of air-conditioning systems must frequently be placed near horizontal surfaces from which accumulated dust may be swept by the wind, and most outside-air intakes to such systems operate at rather high velocities. The size and number of dust particles which enter such systems therefore vary within a wide range. The greater part of atmospheric dust has particles smaller than 5 microns (1 micron equals 0.001 millimeter, or about 0.00004 in.), but many particles entering the filters of ventilating systems are larger than 800 microns, or $\frac{1}{32}$ in. Dust causing lung damage ranges from about 6 microns to 0.6 micron. Smoke particles average about 0.3 micron. Figure 21-1 shows a comparison of various sizes of dust particles.

Dust is removed from air by the following general processes:

1. Contact with water (washing).
2. Straining the air through orifices smaller than the dust particles (dry filtering).
3. Impact and reaction against viscous-coated barriers having interstices larger than the dust particles (wet filtering).
4. Electrostatic and charged-media units.
5. Centrifugal separation.

The probable maximum amount of dust to be removed by an air filter in a ventilating system is as follows, per 1000 cu ft of air handled by the fan:

Residence and country, 0.2 to 0.4 grains
Congested areas in cities, 0.4 to 0.8 grains
Industrial zones, 0.8 to 1.5 grains

21-2. CONTACT WITH WATER (WASHING)

Washing is one of the oldest methods employed to remove dust from air. It was used for air cleaning many years before the present, more-extensive employment of water for heat-transfer purposes.

In the washer the air meets the water sprays, and many of the dust particles in the air are surrounded by water and fall with the water to the open tank under the spray chamber. Here much of the dust settles to the floor of the tank as sludge, or mud. The water, after being screened, can be recirculated. In actual practice some of the dust, especially if it contains

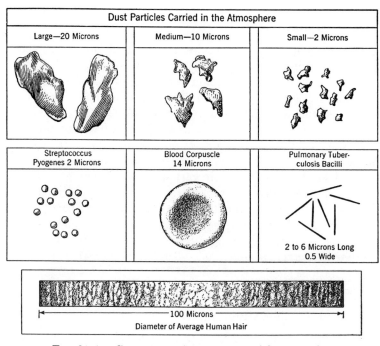

FIG. 21-1. Comparison of various sizes of dust particles.

a trace of oil, passes through the spray chamber. On the leaving side of the spray chamber there are vertical baffles, usually set at an angle to the direction of air flow. The air passes easily around these baffles, but the droplets of water and particles of dust which pass through the sprays impinge against the wetted baffles and are carried by a washing stream of water into the tank.

If the water-circulating pump of an air washer fails, or if any of the spray nozzles should clog, much dust may pass through. This weakness, which may destroy confidence in the device, is a very real one in practice. When sprays are used for direct contact with air in heat-transfer work, as in

The Cleaning of Air

humidifying, a few clogged nozzles may not matter, but when used in the chilling of air, as in dehumidifying and when removing dust, failure of one nozzle among many may mean failure of the whole process.

The velocity of the air through washers is ordinarily 500 fpm, based on gross area.

Figure 21-2 illustrates an air washer. Air enters at the left and passes through two banks of sprays in series. In this case the first bank, b, faces downstream to the air, and the second, c, upstream, but other arrangements may be used. The eliminator plates, with their flooding equipment d and the collecting pan, are also shown.

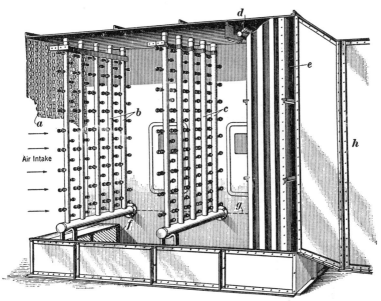

Fig. 21-2. Air washer with side removed.

21-3. DRY FILTERING

There are many successful dust-removing devices of the dry-filtering type. An early type used a number of fabric bags held by draw strings around circular flanged openings in a partition across the airway. The air passed through, while the dust gathered in the bags or was caught in the meshes of the fabric.

The modern successor to this primitive sieve is usually accordion-pleated so as to increase the filter area without too great enlargement of the main airway. The filter medium may be specially-woven cloth or may be a cellulose composition. Such filters may have exceedingly high dust-arrestance, which increases as the interstices clog with dust; but they also have a rapidly increasing resistance to air flow as the dust accumulates, and therefore must be reconditioned frequently. The dust may be removed by application of a vacuum cleaner on the windward side, as is done with some of the cloth media, or the dust-loaded elements may be removed, to be replaced with new media, as is usually the case with the cellulose types.

The cleansing and replacing of dry air filters is always a disagreeable task, involving considerable labor, and requires skilful technique to prevent passage of dust to the discharge side of the filter frames if they are removed. Any fan will deliver much more air, other conditions being constant, when sieve-type air filters are clean than when they are nearly clogged. Since the lowered resistance following filter reconditioning causes a higher velocity of air through the ducts, dust accumulations in the ducts may be whirled into the rooms, and drafts may be noticed by the occupants.

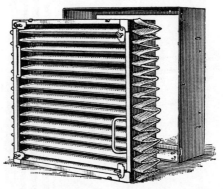

FIG. 21-3. A typical cellulose-type dry air filter.

For this reason the dry-filter area should be in increments of certainly not more than $33\frac{1}{3}$ per cent, and only one-third or less of the total area should be reconditioned at any one time. Under no circumstances should the fan be operated unless the filter barriers are in place, to prevent dust from entering the ducts. The design air velocity through the filter medium is usually from 10 fpm to 50 fpm, depending on the fabric. Dry filters should be processed to minimize the tendency of such substances to absorb moisture, which absorption causes clogging of the tiny air-passage interstices. A permanently mounted differential-pressure gage to show resistance should always be provided to indicate the necessity for filter replacement.

Practically all air filters of this type are made with V-shaped supports (see Fig. 21-3) over which the fabric is clamped so as to give a maximum straining area within a minimum-size frame. The usual resistance, when the filter is clean, is about 0.1 in. of water, and about 300 cfm per square foot of frame area is the maximum allowable air capacity. It is possible to increase the thickness of the medium, but this increases not only arrestance but also resistance.

The Cleaning of Air

21-4. VISCOUS-COATED AIR CLEANERS

Viscous-coated air cleaners operate on an entirely different principle from that of sieves. The air is divided into innumerable fine streams which, at rather high velocity, pass around the barriers, while the dust, because of momentum, is diverted less easily from a straight course and therefore accumulates on the windward sides of the barriers, or loses velocity and piles up behind them. The barriers are coated with nonevaporating adhesive, which holds the dust particles, impregnates them, and thus causes the accumulated dust to catch and hold successive particles regardless of their size. The interstices through the filter are always larger than the particles of dust, and the medium (unlike the necessarily thin close-knit fabric of the dry filters) may be very coarse, and is usually several inches in thickness. The efficiency of dust-arrestance is improved if the density of the packed fibers which comprise the barriers increases toward the discharge side. Thus the larger dust particles are caught as the air passes

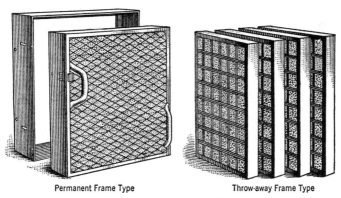

Permanent Frame Type   Throw-away Frame Type

FIG. 21-4. Cell-type viscous-coated air filter.

around the relatively coarse barriers at the entrance side, and the finer dust particles are arrested toward the discharge side in the smaller passages. The dust-holding capacity of an air filter operating on the impingement and reaction principle far surpasses that of any other type, and the arrestance increases with accumulation of dust up to the point when the accumulations of dust commence to break down and re-enter the air stream.

The barriers of such an air filter have been made of many different materials, but commercially they are usually of a nonorganic substance, such as glass wool, iron wire, woven sheet copper, or expanded sheet metal.

The glass-wool, cell-type air filters of the replaceable type usually have cardboard retaining frames and grilles, so that they may easily be burned or otherwise destroyed when they become dust-laden. Others, using treated organic fillers, are furnished in self-supporting pads which may easily be fitted into the metallic frames, which remain in place across the air passages. The old pads are thrown away.

Another type of cell filter is designed for an indefinite number of cleanings and replacements. It generally has a permanent frame and grilles for retaining the filtering elements, which fit dust-tight into a second frame which remains in the airway (frames are shown in Figs. 21–4 and 21–5). The filter cell can be removed from the permanent outer frame for washing in hot water and soda, during which time a spare frame can be used. The cleaned filter cell, after drying, may be dipped in the viscous fluid and allowed to drain for at least twelve hours. The number of spare cells

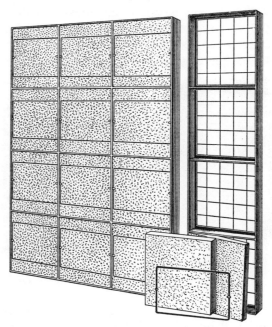

FIG. 21–5. Renewable-pad air filter.

which should be provided depends on the amount of dust in the air, but it is wise to provide one spare cell for not more than six active cells. The usual design air volume allowed per 20-in. by 20-in. cell for ventilating systems is 800 cfm, and the average resistance (when about one-half its ultimate dust accumulation is present) approximates 0.28 in. of water.

All of the cell-type air filters require manual replacement; and since replacement is often delayed too long, some dust is released into the air system. There is also a noticeable tendency toward leakage of dust through the cracks between the cells. For these reasons, air filters which are mechanically renewed are more desirable. In some of these a continuous belt of filtering elements is moved intermittently or continuously across the airway, in a vertical direction, the bottom of the belt dipping into a bath of viscous fluid. The belt is usually moved a predetermined distance periodically, thus placing the dust-laden parts in the fluid (oil) and exposing a cleaned and freshly coated section to the air. While the belt is in the oil the particles of dust slowly separate from the filter and settle as sludge in the bottom of the oil tank, from which the sludge can be dipped or drained at infrequent times, say annually or semiannually.

The only foreign matter which fails to settle out after being accumulated by the filters is lint from manufacturing processes. Lint is common in the air in certain parts of cities. The best solution to the lint problem seems to be a special parallel-bar type of grille to windward of the filter. The lint is then collected from the bars by a revolving brush or accumulator in felt-like masses.

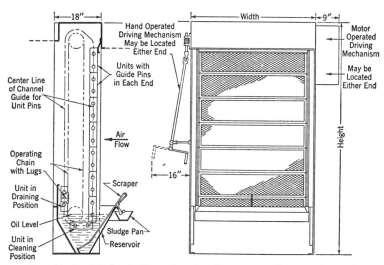

FIG. 21-6. Mechanically-replaced and automatically-reconditioned cell-type air filter.

In Fig. 21-6 the barrier across the airway comprises seven cells. As the lowest cell is moved out of the oil bath to the draining position, the cell above it rises to the top, passes over the upper sheaves, and assumes its working position. The same side of each cell is always to windward, so that the desirable, progressively closer-meshed filter packing is on the discharge side. In Fig. 21-6 the cell-moving mechanism is indicated as a crank, in the right-hand drawing, though either the electric or compressed-air (hydraulic) method of cell replacement is to be preferred. Frequently the cells are moved under control of a Telechron switch. A wide range of sizes in both height and width is available in this type of air filter, though the usual width does not exceed 48 in. For greater widths a number of parallel units are employed. A very important advantage of this type of filter is that the resistance to air flow is practically constant, as compared with the wide variations when cell-type filters are used.

Figure 21-7 illustrates one manufacturer's overlapping, continuous-belt, mechanically-reconditioned air filter. Unlike the type in Fig. 21-6, the air passage is not always in the same direction across the medium, so that

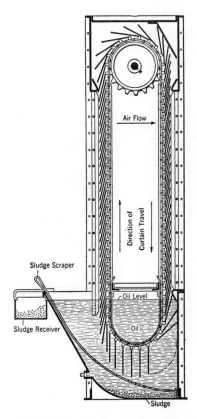

Fig. 21-7. Continuous belt filter.

progressively finer interstices are not practicable. The downcoming dust-laden barriers, on the right side, are separated and cleaned as they very slowly pass through the oil. The hoe, or scraper, shown at the lower left, is used to drag the sludge to the receiver, which may easily be carried out and dumped. The movement of the filter belt may be controlled by a time clock.

The Cleaning of Air

21-5. RELATION OF FILTERS TO OTHER PARTS OF SYSTEM

Figure 21-8 shows the usual, preferred arrangement of an air-conditioning system which can clean, heat (or cool), and humidify (or dehumidify) the air. The winter preheater, when it is located ahead of the filter, must be controlled in order not to warm the air to such an extent as to cause evaporation of the viscous oil of the filter, and in order not to develop over-humidification from the air washer. The air must be warm enough, however, to prevent freezing of the spray water. The air filter acts to trap coarse dust particles which would otherwise cause rapid wear or clogging of the spray nozzles. In winter or summer the reheater is used to warm the

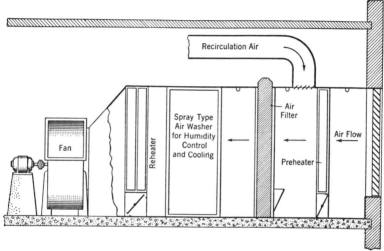

Fig. 21-8. Sectional diagram showing component parts of air-conditioning system.

humidified and cleaned air to a comfortable temperature for delivery to the rooms of the building. There is no serious objection to placing the air filter ahead of the preheater except that in such an arrangement the recirculation duct should also enter ahead of the filter so that no uncleaned air may pass.

21-6. ELECTROSTATIC (ELECTRONIC) AIR CLEANING

The electrostatic (electronic) method of precipitating dust and other particles from air had its inception with industrial precipitators developed early in this century by Cottrell for the control of dusts from industrial chimneys and metallurgical plants. However, modern air-cleaning units hardly resemble the earlier industrial models. The present units use electron-tube power packs which, with alternating-current voltage as low as 115, develop the 13,000- and 6000-dc voltages required by the ionizer and collector circuits. The power required by the units is small—about 40 watts per thousand cubic feet of air cleaned per minute.

An understanding of electronic air cleaners can be gained by a study of Fig. 21-9, which shows the principle of operation of a representative type of unit. Air first passes through an ionizer section, where an electrical charge is imparted to the dust particles in the air. As the air and the

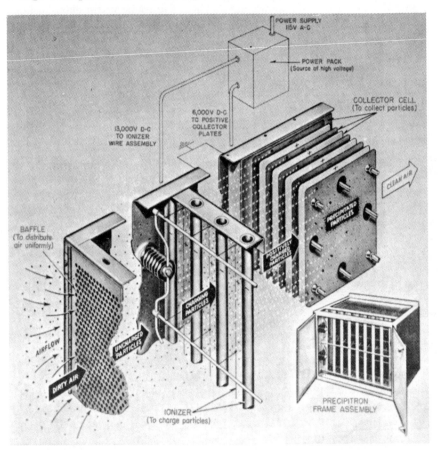

FIG. 21-9. Principle of operation of Precipitron air cleaner. (Courtesy Sturtevant Division, Westinghouse Electric Corporation.)

charged particles flow into the collector cells, the charged particles are drawn from the air to the surface of the collector plates by electrical attraction. The dirt thus separated sticks on the plates with the help of an adhesive coating, while the clean air passes through. To remove the dirt accumulation, the collector cells are periodically cleaned by washing them down with hot water sprayed onto the collector plates by a nozzle washer.

The Cleaning of Air

The dirt is scrubbed out and flushed into a drain. Following the washing and draining period the collector cells are sprayed with adhesive, and the unit is then ready for operation.

Figure 21-10 is a cutaway section of a vertical unit. In this unit, outside and recirculated air enters the bottom of the unit, passes up through the ionizing section, and then passes through the collector plates, and the clean air is delivered from the upper part of the unit. The washing period

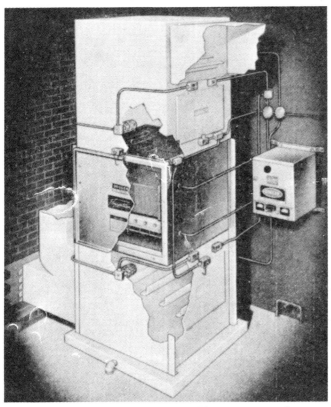

FIG. 21-10. Cutaway view of vertical Precipitron air cleaner.
(Courtesy Sturtevant Division, Westinghouse Electric Corporation.)

for the plates is short, so that the unit is not out of service for any appreciable time, and the messy labor involved in handling conventional filters is completely eliminated.

Electrostatic precipitation of dust from air is especially effective for exceedingly small particles, even for particles of smoke or mist. The space occupied by electrostatic cleaners does not greatly exceed that required by the usual automatic air filter, and is usually less than that needed for a spray-type air washer. It is possible, although objectionable, to deliver air and dust through the device without cleaning, when the ionizing elements are not energized, or to bypass dust-laden air around some of the elements which may be out of order.

Charged-media electronic air cleaners are also in use. These units consist of a nonconductive filtering material placed on supports in the same manner as pleated dry filters. An ionizing circuit is not used, but the supporting grillework, consisting of alternating grounded and charged members, holds the filtering medium in place. The resulting electrostatic

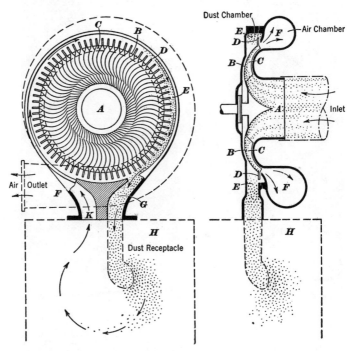

FIG. 21-11. Roto-Clone dust separator.
(Courtesy American Air Filter Company.)

field causes the airborne particles to move into the filter mat. However, like all mat filters, the resistance of the unit rises rapidly as the dust load increases. It is necessary to replace the filter medium completely when the unit is serviced.

21-7. CENTRIFUGAL DUST-SEPARATION

Large particles of dust may be removed from air by proper use of centrifugal force. One device employing this principle is the Roto-Clone dust separator, a high-speed centrifugal air pump having a large number of blades. The dust-laden air enters at A in the cross section shown at the

The Cleaning of Air

right of Fig. 21–11. The heavier dust particles are intercepted by the impeller disk B, and the lighter dust particles are caught by impact on the advancing (forward) surfaces of the scoop-like curved blades C. Both the light and heavy dust particles move toward the periphery of the impeller because of centrifugal force, and converge at its outer edge D. Opposite this edge in the stationary housing, there is a projecting lip which catches

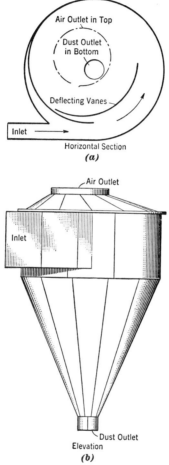

FIG. 21–12. Cyclone dust-separator.

and separates the dust, passing it into the chamber E, while most of the air goes on to the scroll F and is eventually discharged at the air outlet indicated by the dotted lines of the left-hand drawing. There is a secondary air circuit in the annular chamber E substantially parallel to that in chamber F, and from this chamber the dust passes through the port G into the hopper H. This chamber is airtight, and serves as a dust-settling receptacle so that the air from it returns through port K to the annular space E for re-use. Thus the air in E is in constant recirculation.

The *cyclone separator* (Fig. 21–12) comprises a funnel-shaped housing which receives high-velocity air bearing a comparatively heavy burden of solids. The discharge duct from the exhaust fan enters the cylindrical body of the cyclone tangentially near the top, and the air whirls as a vortex around the outer periphery and lowers into the sloping sides of the separator. Still rotating, the air then passes upward through the central section and out at the top. The solids suspended in the air impinge against the sides and spiral down to the outlet, or tailpiece, while the air is escaping near the center of the top. The cyclone diameter should be at least $3\frac{1}{2}$ times the diameter of the fan discharge duct. Sometimes several cyclones may be installed in series, with booster fans picking up the discharge air from successive treatments. Cyclone dust separators are employed especially with exhaust systems used with grinders, polishers, saws, planers, and the like.

It should be realized that the air first travels around in an outer vortex of gradually decreasing radius and then leaves from an inner vortex with much smaller radius but greater speed of rotation. There is a pressure loss involved in the vortices and it is therefore necessary to bring the air to the initial velocity it had at entry to the cyclone. The vortex loss plus the velocity head, equivalent to inlet velocity, may be expressed as

$$P = \frac{3}{2}\frac{\rho v_1^2}{g} + \frac{\rho v_1^2}{2g} \tag{21-1}$$

where P = pressure loss in pounds per square foot;

ρ = density, in pounds per cubic foot;

v_1 = velocity of entering dust-laden gas (air), in feet per second;

g = the standard gravitational constant 32.2.

To change pounds per square foot to inches of water, recall equation 12–18, which gives 5.19 as the factor; thus,

$$\text{Inches of water} = \frac{\text{lb per sq ft}}{5.19}$$

PROBLEMS

21–1. A certain viscous-type filter in a duct system offers pressure drops as follows: 0.07 in. of water at 600 cfm; 0.13 in. of water at 800 cfm; 0.18 in. of water at 1000 cfm. The duct system is designed for 875 cfm, and 0.15 in. of water is allowed for the filter. After two weeks of service, with dirt accumulation, it is found that only 800 cfm are passing through the filter, with 0.15-in. pressure drop. Draw the original characteristic curve for the clean filter, plotting pressure drop against air flow in cubic

The Cleaning of Air

feet per minute, and on the same graph sheet draw in a curve to estimate what pressure drop would take place if 875 cfm were flowing through the dirty filter.

21-2. The pressure drop (resistance) of a filter is roughly proportional to the square of the velocity. A certain filter passes 2000 cfm at a velocity of 150 fpm, and the pressure drop under these conditions is 0.3 in. of water. How much increase in filter area is required if it is desired to cut the filter pressure drop to 0.15 in. of water and deliver the same quantity of air? *Ans.* $18.9 - 13.3 = 5.6$ sq ft

21-3. A certain 20-in. by 20-in. viscous-type filter 2 in. thick has a resistance of 0.085 in. of water when air is flowing through the filter at 300 fpm. As the filter becomes loaded and approaches the end of its life, the resistance increases to 0.25 in. of water for the basic 300-fpm velocity of a clean filter. Estimate the probable decrease in velocity (capacity) of a system with a dirty filter as compared with a clean one if no change in the fan pressure of the system can be made as the filter loads up with dirt.
Ans. 102 fpm (if the relationship to pressure drop is linear; however, this is not the exact relationship and a somewhat higher velocity would hold)

21-4. An electrostatic filter of the ionizing-wire type has a low resistance to air flow—approximating 0.15 in. for the complete unit. Resistance increases by some 20 per cent before cleaning. Velocities through units of this type range from 300 to 370 fpm. Cleaning is usually by washing. (a) What other advantage do units of this type possess which make them somewhat distinctive in performance? (b) For 4000 cfm, find the probable air-flow cross section and the change in capacity between the clean and dirty phases of operation. As a rough approximation, a linear relationship between pressure drop and flow can be considered to exist.
Ans. (a) Fine particles can be removed, such as tobacco smoke; (b) 11.9 sq ft at 335 fpm, 3330 cfm

21-5. An air washer consisting of one bank of spray nozzles facing downstream in the direction of air flow has a saturating effectiveness of 60 to 70 per cent, measured in terms of bringing the dry-bulb temperature of the entering air to the limiting (thermodynamic) wet-bulb temperature. For this it is considered that recirculated water is employed, with trivial heating or cooling of the water. With the bank facing upstream, 65 to 75 per cent saturation effectiveness can be expected. With two banks downstream in succession, or two banks upstream in succession, the corresponding figures are, respectively, 85 to 90 per cent and 90 to 95 per cent. Cleaning effectiveness is not influenced by the effectiveness of saturation but by wetting of the dirt particles and separation of them on the eliminator plates or mats. The washing down of these plates moves the dirt into a collection sump at the bottom of the tank, for settlement and removal. Air velocities of 250 to 650 fpm are used in the spray chamber. Consider an air washer in an industrial area taking air containing 0.8 grains per 1000 cu ft. The washer section is 6 ft by 6 ft and the air velocity is 500 fpm. Assume that some 60 per cent of the solids in the air are removed in the sludge and find the weight of solids discharged to the sewer per 8-hr day of operation. *Ans.* 9.6 lb

21-6. A cyclone separator 4.8 ft in diameter at the air-inlet point is supplied with air at 65 fps. What is the original speed of the vortex in the cyclone in revolutions per minute? *Ans.* 259 rpm

21-7. For the inlet velocity of problem 21-6, and with standard air, calculate the total pressure loss for the cyclone. *Ans.* 3.8 in. water

21-8. A cyclone separator 6 ft in diameter at the air-inlet point is supplied at 70 fps. (a) What is the original vortex speed in the cyclone in revolutions per minute? (b) Calculate the vortex pressure loss and the velocity head (that is, the total pressure loss) for the cyclone. *Ans.* (a) 223 rpm; (b) 4.4 in. of water

21-9. A heavily dust-laden atmosphere moved into a certain area and practically stopped there during a six-hour period. The dusty air was approximately 1000 ft deep and carried 1.6 grains per 1000 cu ft. If approximately 25 per cent of the dust carried dropped out during this period, how many tons of dust fell on a square mile of land surface? *Ans.* 0.795 tons

21-10. To remove obnoxious vapors or fumes from tanks it is customary to build hoods over the tanks. The edges of the hood project slightly past the sides of the tank. In the perimeter around the edge of the tank and up to the hood, a sufficiently great air velocity must be established to draw the fumes into the air stream and carry them up the hood and duct. In a specific case a hood was placed 30 in. above the edge of a fuming vat 4 ft wide by 10 ft long. It is assumed that an air velocity of 50 fpm will satisfactorily carry the fumes over the edge of the tank and into the exhaust pipe. (a) Find the quantity of air which must be moved in over the edge of the tank and under the hood, and (b) assuming that 2000 fpm is the designed velocity, find the exhaust duct area above the hood. *Ans.* (a) 3500 cfm; (b) 1.75 sq ft

22

Industrial Air Conditioning and Food Preservation

22-1. MOISTURE CONTENT

Nearly all substances have the characteristic property of absorbing or releasing moisture under different atmospheric conditions. Substances which are particularly variable in the moisture content which they can possess at different times are known as *hygroscopic materials*. Most of the organic materials, such as wood, leather, foodstuffs, paper, tobacco, cloth, and hair, are quite hygroscopic.

The familiar case of the swelling of wooden doors, furniture drawers, and the like in homes in summer time is known to all. In this case, during the summer the relative humidity is relatively high (around 50 to 60 per cent) and the cells of the wood absorb moisture and the wood increases in both volume and weight. During the winter, the inside of the house is under low relative humidity, often less than 30 or even 20 per cent, and the wood gives up moisture to the dry air, thereby shrinking both in volume and in weight.

A distinction should be made between hygroscopic moisture in a body and moisture content. For example, a piece of cloth plunged in water, removed, and wrung out is obviously wet and in a given case the moisture content may be more than 50 per cent of the total weight. If this cloth is hung up it will eventually dry with no apparent surface moisture. Yet there is still hygroscopic moisture in the cloth, for if it is put in a very dry warm atmosphere it will continue to lose weight until an equilibrium point is reached at which time no further loss in weight will occur. The final constant weight reached by a hygroscopic substance after being dried out is known as its *bone-dry weight*, or *dry weight*, under the conditions maintained in the drying chamber. The drying limits cannot be the same for different substances, because many organic substances are damaged if heated or dehydrated beyond a certain point. *Regain* is the hygroscopic moisture content of a substance expressed as a percentage of the bone-dry weight of the material. *Moisture content* is usually expressed as a percentage of the gross weight of the body, and may refer to both hygroscopic and purely surface moisture. Table 22-1 gives the regain values for various hygroscopic materials.

TABLE 22-1

MOISTURE IN HYGROSCOPIC MATERIALS

(Moisture is expressed as *regain*, per cent of the dry weight of the material. Values are given at various relative humidities and at 75 F temperature.)

Classification	Material	Description	Per Cent Relative Humidity								
			10	20	30	40	50	60	70	80	90
Food-stuffs	Bread	White	0.5	1.7	3.1	4.5	6.2	8.5	11.1	14.5	19.0
	Flour		2.6	4.1	5.3	6.5	8.0	9.9	12.4	15.4	19.1
Miscellaneous Organic Materials	Leather	Oak-tanned	5.0	8.5	11.2	13.6	16.0	18.3	20.6	24.0	29.2
	Paper	Newsprint, wood pulp	2.1	3.2	4.0	4.7	5.3	6.1	7.2	8.7	10.6
	Paper	White bond, rag	2.4	3.7	4.7	5.5	6.5	7.5	8.8	10.8	13.2
	Rubber	Solid tire	0.11	0.21	0.32	0.44	0.54	0.66	0.76	0.88	0.99
	Tobacco	Cigarette	5.4	8.6	11.0	13.3	16.0	19.5	25.0	33.5	50.0
	Wood	Timber (average)	3.0	4.4	5.9	7.6	9.3	11.3	14.0	17.5	22.0
Miscellaneous Inorganic Materials	Activated charcoal	Steam activated	7.1	14.3	22.8	26.2	28.3	29.2	30.0	31.1	32.7
	Domestic coke		0.20	0.40	0.61	0.81	1.03	1.24	1.46	1.67	1.89
	Silica gel		5.7	9.8	12.7	15.2	17.2	18.8	20.2	21.5	22.6
Textile Materials	Cotton	Absorbent	4.8	9.0	12.5	15.7	18.5	20.8	22.8	24.3	25.8
	Cotton	Cloth	2.6	3.7	4.4	5.2	5.9	6.8	8.1	10.0	14.3
	Cotton	Raw fibres	2.5	3.7	4.6	5.5	6.6	7.9	9.5	11.5	14.1
	Hemp	Manila and sisal-rope	2.7	4.7	6.0	7.2	8.5	9.9	11.6	13.6	15.7
	Linen	Table cloth	1.9	2.9	3.6	4.3	5.1	6.1	7.0	8.4	10.2
	Rayon	Fibre	0.8	1.1	1.4	1.9	2.4	3.0	3.6	4.3	5.3
	Silk	Skein	3.2	5.5	6.9	8.0	8.9	10.2	11.9	14.3	18.8
	Wool	Skein	4.7	7.0	8.9	10.8	12.8	14.9	17.2	19.9	23.4

Industrial Air Conditioning and Food Preservation

A study of this table indicates the extreme moisture fluctuations which take place with the materials under varying humidity conditions. The secondary effects of moisture are very significant; for example textiles under proper humidity conditions are soft and pliable, but when too dry, they become brittle and weakened. On the other hand, sustained high humidities are unsatisfactory, at certain temperatures, leading to fungus growth. In the printing industries the change in size of paper under changing humidity conditions may be a serious problem, particularly with certain types of multicolor printing. From a viewpoint of purchasing and selling certain products, standards of regain must be set and adhered to when the products are sold by weight.

Air conditioning in industrial processes is thus concerned with control of temperature and humidity, and with air circulation, to maintain the products in desired condition. In this way frequent adjustments on machinery handling the materials can be avoided and the materials themselves can be kept in the best condition for fabrication or manipulation.

Example 22-1. Raw cotton has been stored in a warehouse at 60 F and at a relative humidity of 50 per cent. (a) What is the probable regain of this cotton and the moisture in 100 lb? The cotton goes through a mill and passes through the weaving room kept at 70 F and 70 per cent relative humidity. (b) For each 100 lb of cotton from the warehouse, how many pounds should appear in the woven cloth, neglecting lintage and thread losses?

Solution: (a) The values in Table 22-1 can be used in absence of specific information on a given cotton, while the fact that the warehouse was at 60 F instead of 75 F does not greatly affect the regain. Thus the regain is taken as 6.6 per cent.
Ans.

A 100 lb hamper of cotton equals the bone-dry weight (bdw) plus the weight of the moisture in the sample or

100 lb = bone dry weight + 0.066 bone dry weight = 1.066 bdw

$$\text{bdw} = \frac{100}{1.066} = 93.8 \text{ lb of dry cotton material}$$

Moisture in 100 lb = 100 − 93.8 = 6.2 lb; also

$$93.8 \times 0.066 = 6.2 \text{ lb} \qquad \textit{Ans.}$$

b) Finished cotton has been fabricated and made into cloth, so its texture has changed somewhat; using the column (Table 22-1) for cloth at 70 per cent relative humidity shows a regain of 8.1 per cent. From part (a) bone-dry weight of 100 lb in warehouse was 93.8 lb. With a regain of 8.1 per cent this weight becomes

$$93.8(1 + 0.081) = 101.4 \text{ lb} \qquad \textit{Ans.}$$

22-2. CONDITIONS FOR CERTAIN PROCESSES

For various industrial processes the maintenance of certain conditions has been found, by experience, to yield good results. Some conditions for process work are found in Table 22-2. For storage of foodstuffs and materials, desirable conditions may be found in Table 22-3.

TABLE 22-2
Temperature and Humidity Conditions for Various Industrial Processes

Industry	Operation	Desirable Temperature, F	Desirable Relative Humidity
Baking	Cake icing	70	50
	Dough fermentation	80	78
	Mixing	75	55–70
	Proof boxes	80–90	80–90
Brewing	Fermentation in vats	44 to 50	50
Ceramic	Molding room	80	60
Confectionery	Chocolate covering	60–65	50–55
	Packing	65	50
	Storage	60–68	50–60
Electrical	Insulation winding	104	5
	Manufacture cotton covered wire	60–80	60–70
Food	Preparation of cereals	60–70	38
	Preparation of macaroni	70–80	38
	Ripening of meats	40	80
Fur	Storage	28–40	25–40
Paint	Air drying lacquers	70–90	25–50
	Air drying paints	60–90	25–50
Paper	Storage	60–80	35–45
Printing	Binding	70	45
	Press room	75	60–75
Soap	Drying	110	70
Textile	Cotton roving	75–80	60–66
	Cotton spinning	60–80	60–70
	Cotton weaving	68–75	70–80
	Rayon spinning	70	85
	Silk spinning	75–80	65–70
	Silk weaving	75–80	60–70
	Wool carding	75–80	65–70
	Wool weaving	75–80	50–56
Tobacco	Cigar making	70–76	55–65
	Softening	90	85

Industrial Air Conditioning and Food Preservation

TABLE 22-3
STORAGE CHARACTERISTICS OF PERISHABLE PRODUCTS

	Products	Range of Storage F		Optimum Rel Hum. Per Cent	Freezing Point F	Composition % Water	Specific Heat Btu per Lb Deg F		Latent Heat of Fusion	Maximum Storage Period
		Short Time Storage	Warehouse Storage				Above Freezing	Below Freezing		
FRUITS	Apples	35–40	30–32	85	28.5	85	0.90	0.49	122	8 months
	Bananas	55–56	55–56	80	26–30	75	0.90	0.61	112	10 days
	Grapes	35–40	30–32	80	28	77	0.90	0.50	130	1 to 6 mo
	Lemons	55–60	50–55	80	28	89	0.94	0.47	124	90 days
	Oranges	40–45	32–34	80	28	86	0.90	0.48	128	2 months
	Peaches	35–40	31–33	80	29.5	88	0.92	0.49	122	30 days
	Pears	35–40	30–32	85	28.5	84	0.91	0.48	131	1 to 7 months
	Strawberries	35–40	31–33	80	30	90.5	0.92	0.48	131	10 days
VEGETABLES	Asparagus	40–45	32–34	90	30	94	0.91	0.49	136	30 days
	Beans (string)	40–45	32–34	85	30	68.5	0.80	0.46	98.5	30 days
	Beets	40–45	32–34	85	27	88.5	0.86	0.48	128	7 to 90 days
	Cabbage	35–40	32–34	90	31	91.5	0.93	0.47	132	4 months
	Carrots	35–40	32–34	90	29.5	88	0.86	0.45	126	2 to 4 months
	Celery	35–40	31–33	90	30	94.5	0.91	0.46	136	2 to 4 months
	Corn (green)	35–40	31–33	85	29	75.5	0.86	0.38	108	10 days
	Corn (dried)	50–60	35–40	60		10.5	0.29	0.24	15	12 months
	Lettuce	35–40	32–34	95	31	94.5	0.90	0.46	136	20 days
	Potatoes	36–50	38–42	85	29	78.5	0.86	0.47	113	6 months
	Tomatoes	50–55	50–55	80	30.5	94.5	0.92	0.46	132	10 days
	Vegetables mixed	40–45	35–40	85	30	90	0.90	0.45	130	

TABLE 22-3 (Continued)
Storage Characteristics of Perishable Products

	Products	Range of Storage F		Optimum Rel Hum. Per Cent	Freezing Point F	Composition % Water	Specific Heat Btu per Lb Deg F		Latent Heat of Fusion	Maximum Storage Period
		Short Time Storage	Warehouse Storage				Above Freezing	Below Freezing		
MEAT AND FISH	Bacon............	40–45	28–30	80		20	0.50	0.30	29	15 days
	Beef (fresh)......	35–40	30–32	84	27	68	0.75	0.40	98	3 weeks
	Fish (frozen).....	10–0	–20–0	80		70	0.76	0.41	101	6–8 months
	Fish (iced).......	34–38	30–32	85	28	70	0.76	0.41	101	15 days
	Hams and Loins..	34–38	28–30	80		60	0.68	0.38	86.5	3 weeks
	Lamb............	34–38	28–30	85	27	58	0.67	0.38	83.5	2 weeks
	Pork (fresh)......	34–38	30–32	80	29	60	0.68	0.30	86.5	15 days
	Pork (smoked)....	40–45	28–30	80	28	68	0.68	0.32		15 days
	Poultry (fresh)....	28–30	28–30	84		57	0.60	0.37	106	15 days
	Poultry (frozen)...	10–0	–20–0	85	27	74	0.79	0.37	106	10 months
	Sausage (fresh)...	35–40	21–27	80	27	74	0.79	0.56	93	15 days
	Sausage (smoked).	40–45	32–40	75	26	65	0.89	0.56	86	6 months
	Veal.............	34–38	28–30	84	25	60	0.86	0.39	91	15 days
					29	63	0.71			
MISCELLANEOUS	Beer..............	35–40	34–38	85	28	92	1.0	0.36	79	6 months
	Butter............	45–40	32–34	80	75–70	15	0.64	0.36	79	10 days
	Cheese, American.	40–45	38–42	80	17	55	0.64	0.56	40	15 months
	Cheese, Swiss.....	40–45	60–75	55	15	55	0.64	0.40	100	60 days
	Chocolate coating.	65–70	30–31	85	95–85	0.5	0.30	0.41	100	6 months
	Eggs (crated).....	40–45	0–5	60	27	73	0.76			12 months
	Eggs (frozen).....	15–20	35	85	27					18 months
	Flowers (cut).....	40	<–10	85	32					1 week
	Ice cream.........	0–10	32–34	80	27–0	60	0.78	0.45	96	6–12 months
	Lard..............	45–50	35–40	70			0.52			6 months
	Milk..............	35–40	30–32	75	31	87.5	0.93	0.49	124	5 days
	Nuts (dried)......	35–40	42–44	90	25	3–10	.21–.29	.19–.24	4.3–14	8–12 months
	Tobacco and Cigars.	42–44								

Industrial Air Conditioning and Food Preservation

A study of these two tables gives some idea of the optimum conditions for many industrial processes and for storage. The wide diversity of humidity and temperature ranges for different processes is sometimes obvious but not necessarily so. Approved practices should always be followed when operating data are available. The reasons for various process humidities and temperature may be (a) biochemical, as in dough fermentation, or fruit and meat ripening or food storage; (b) control of chemical reaction as in drying of paints, where a hard skin must not oxidize before the inside paint has had some chance to dry out; (c) control of regain for purposes of strength for manipulation, for size and weight limitation, or for appearance; (d) control during purely drying processes to prevent such rapid drying as to cause checking or cracking. An example of the latter case is the drying of green wood, where, if the atmosphere surrounding the wood in a kiln or in the open is too dry, moisture will leave the wood surface faster than moisture can travel from the inside of the wood, and cracking may occur as the outside shrinks.

22-3. HEAT AND COOLING LOADS

The calculation of heating and cooling loads during industrial processes follows the same general procedure that has already been outlined. However, several additional items may also enter into the calculations.

Product load. Heat may be required by the product itself to keep the process in action, to bring the product up to desired temperature, or to evaporate and superheat mositure from the product.

Ventilation conditions may be unusual and extreme, sometimes to keep fumes or odors to a minimum consistent with safety or comfort, to distribute air at proper velocity through the product, and to provide adequate air for carrying humidity loads. In storage of apples, oranges, pears, etc., and to a lesser extent with vegetables, there is a respiration action during which the product generates carbon dioxide and uses oxygen. With such products, from four to six air changes per 24 hours are required. Some heat is generated by respiration in any green or living vegetal product and forms an additional load on the cooling system.

Door losses, from people and products entering and leaving, may be high in both heating and cooling processes and should always be considered.

Washwater brought into a conditioned space may cause either a heating or a cooling load. This is often negligible except in some industrial processes.

Basic losses or gains from people working, lights, motors, combustion processes, insulation, and sun loads are calculated as previously outlined.

22-4. COLD STORAGE

Cold storage conditions for food products differ for every product, but certain basic facts are true for all fruits, vegetables, and meats. In the case of fruits and vegetables particularly, the problems faced in storage and handling are (1) *desiccation*, or loss of moisture, (2) *physical breakdown* of tissue, (3) *chemical changes* such as oxidation of certain matter in the product, hydrolysis of fats and esters, and changes in the food proteins, (4) *respiration changes* in which the living cells take up oxygen and give up carbon dioxide (this action takes place along with the enzyme actions), and (5) *bacterial, yeast,* and *mold growth* in or on the product. Such growth is very rapid at room temperature, decreasing as the temperature is lowered and almost ceasing as the temperature drops to about 10 or 15 F.

Several kinds of storage conditions should be recognized, namely (1) long-term warehouse storage in which the product is frozen and maintained in that condition, (2) long-term warehouse storage with the product not frozen, and (3) short-term or retail storage in which the product is usually not frozen, although it may be, as in the case of the so-called quick-frozen foods. Quick-frozen foods are frozen very rapidly to reduce the size of ice crystals formed, to reduce the separation of water from the cells, and to limit the growth of bacterial and mold products during the freezing period. Quick freezing by cold-air blast, brine spray, or contact evaporators usually takes place at temperatures of -15 to -45 F. Ordinary cold-storage freezing takes place at temperatures of from about $+20$ to below 0 F.

Example 22-2. Dressed poultry is brought to a warehouse at 40 F for freezing and temporary storage at 5 F until shipped away. The supply of poultry averages 4000 lb per 24 hr and comes in open-type wooden boxes weighing 10 lb each and holding 50 lb of poultry. The poultry remains in the plant from 2 to 3 weeks, but daily shipment closely equals average daily supply. Approximately 10 tons of refrigeration are used in the rooms where the final chilling and general storage takes place. What additional tonnage is required for the initial chilling and freezing room, assuming a uniform daily supply of the product and that the poultry is moved out after it is chilled to 10 F? Insulation loss to this room, held at -10 F, is 4000 Btu per hr for the 20 by 30 by 10 ft room.

Solution: Supply is 4000 lb of poultry, or $4000/50 = 80$ wooden boxes at 10 lb each. Chilling poultry to freezing point 27 F, from Table 22-3,

$$(W)(C)(\Delta t) = (4000)(0.79)(40-27) = \qquad 41{,}080 \text{ Btu}$$

Freezing poultry, from Table 22-3, $(4000)(106) = \qquad 424{,}000 \text{ Btu}$
Chilling frozen poultry, from Table 22-3,

$$(W)(C)(\Delta t) = (4000)(0.37)(27-10) = \qquad 25{,}160 \text{ Btu}$$

Chilling pine wooden boxes, from Table 4-1,

$$(W)(C)(\Delta t) = (80 \times 10)(0.67)(40-10) = \qquad 16{,}080 \text{ Btu}$$

Industrial Air Conditioning and Food Preservation

Estimate 2 men working in this room, 8 hr per day. Estimate 1000 Btu per man per hr, as this is far out of the comfort zone. Human latent heat is not separated, since the moisture evaporated probably freezes on the evaporator coils and directly adds to the load $2 \times 8 \times 1000 =$ 16,000 Btu

Assume door loss and ventilation loss equal to one half change of outside air per hour for this room. With air outside at 80 F and 50 per cent rel hum, $h = 31.3$ Btu per lb outside air (Plate I) and sp vol 13.83 cu ft. With air inside at -10 F and 80 per cent rel hum, $h = -2.0$ Btu per lb cold room air (Plate II). Weight of air from outside ½ change per hr

$$\frac{20 \times 30 \times 10}{2 \times 13.83} = 217 \text{ lb}$$

Heat removal per 24 hours for ventilation air

$(24)(217)[31.3 - (-2.0)] =$ 173,426 Btu
Assume ten 40-w lights for 8 hr per day $= 400 \times 8 \times 3.413 =$ 10,922 Btu
Insulation loss $4000 \times 24 =$ 96,000 Btu

Total   802,668 Btu per 24 hr period

Refrigeration required for initial chilling and freezer room is

$$\frac{802,668}{24 \times 12,000} = 2.79 \text{ tons}$$

This unit should be increased by some 10 to 15 per cent to take care of defrosting and contingencies, giving

$$2.79 \times 1.10 = 3.1 \text{ tons} \qquad \textit{Ans.}$$

During cold storage of food products a certain amount of dessication always takes place because moisture from the products evaporates into the air moving past the product. This drying out does not increase directly the heat load in the space because the water vapor comes from moisture in the product and the energy for this evaporation can only come from a sensible heat cooling of the air or of the product itself. However, dessication must be kept to a minimum and this can be done best by maintaining relative humidity and temperature at conditions found most suitable by experience. Dessication can take place with both frozen and nonfrozen products although much less with the latter. In the case of nonfrozen meats, if the relative humidity is kept too high, molds start to form. Yet at low humidities the shrinkage or weight loss in meat is excessive and its appearance also becomes displeasing so that it is necessary to strike an optimum storage condition.

In selecting a conditioning unit, which should be done from manufacturer's test data (except for pipe coils), it is necessary to choose a unit which gives a moderate air velocity and does not require a refrigerant temperature very much below room temperature, otherwise too much moisture will freeze on the coils and excessively dry out the air. The same statement applies to pipe-coil temperatures in a cold-storage room, and enough surface should be supplied to make unnecessary the use of large temperature differences between room and coil for the requisite quantity of heat transfer.

The division of the heat removed at the coils as sensible heat or as latent heat of condensation (or of condensation and freezing combined) runs over wide limits. Above freezing, it varies from about 60 to 90 per cent, or more, sensible heat. For rough estimates, in lack of specific information, a value of about 72 per cent sensible can be used for nonfrozen meats, 75 per cent for usual vegetables, and about 80 per cent for fruits. For inherently dry or damp products, these values should be varied by judgment—lower for short storage and higher for long storage.

Cooling of rooms in large cold-storage central plants is usually done by brine cooled at a central point and pumped through pipe lines to the individual floors and rooms. Figure 22–1 shows two typical systems suitable for a multistory storage warehouse. The sketch at the left shows a two-pipe system, which is cheaper to install but takes more throttling in the closer circuits and is not so satisfactory as the three-pipe system shown on the right. In the latter, the length of piping for every circuit is about the same and it is easy to balance loads and vent the system.

Brine-pipe coils, in a given room, are usually placed parallel to and near the center of the ceiling, seldom more than two to four pipes deep. Thermal air circulation down from the coils is set up and the cooled air passes around the product. Table 9–4 can be used to find the required coil surface for given conditions.

22–5. UNIT CONDITIONERS

For cold-storage rooms, unit conditioners are in wide use. In typical unit conditioners, warm air from the room is drawn in at the bottom, passes over the coils and, as chilled air, is blown out in the room through the directing louvres. Such units can use brine or direct refrigerant in their coils. Where the vapor load is very high, the air to be cooled may be passed through a direct brine spray instead of over a coil surface.

Defrosting is necessary for installations in which the temperature of the coils is below 32 F and below the air dew point, because frost will then form on the coils and must be removed periodically. When space temperatures are above 32 F, this can sometimes be done by turning off the refrigerant and letting the room air melt off the frost. Some rise in room temperature occurs during the process. With direct refrigerants, it can often be arranged to turn hot gas from the compressor into the evaporator for defrosting. Brine units have, in a few instances, been

Industrial Air Conditioning and Food Preservation 741

equipped to spray some of the brine itself over the coils during a defrosting period. With brine-spray systems not using coil surface, the brine is progressively diluted by moisture from the air and must be strengthened periodically by adding more salt or by evaporating the surplus water in a heater.

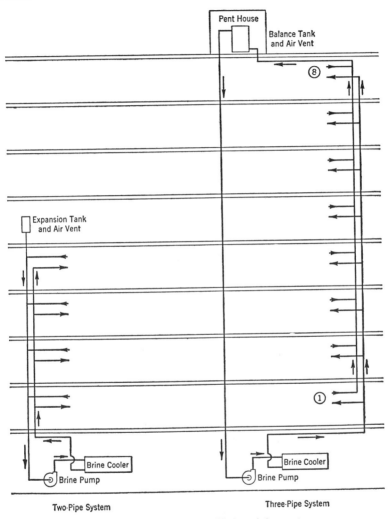

FIG. 22-1. Diagrams of typical brine piping systems.

Small ceiling-mounted unit conditioners consisting of an evaporator or cooling coil and backed by an electric fan which blows the air horizontally or downward into the space are quite common. They are employed for both summer cooling and winter heating.

742   Industrial Air Conditioning and Food Preservation

Figure 22-2 shows a picture of a complete unit conditioner with the front cover removed, and Fig. 22-3 is a diagrammatic view of the unit showing details of construction. This type of unit conditioner is primarily designed both for comfort conditioning and for industrial operation in the range above freezing. Modified forms of such units are also available for food-storage installations in the subfreezing range. The fan, driven by a belt from the motor, draws air from the room through louvers (not

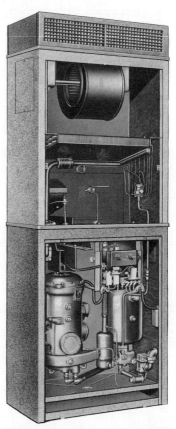

Fig. 22-2. Complete unit conditioner with front cover removed.
(Courtesy Airtemp Division, Chrysler Corporation.)

shown) located just below the cooling coil. The cooled air is discharged from the top of the unit. The design pictured uses a radial compressor directly connected to the motor, with both built integrally in a hermetically sealed housing. The gas is condensed in a water-cooled condenser, shown in upright position at the right of each diagram. A thermal-expansion valve controls the flow of refrigerant. The unit under discussion uses a

Industrial Air Conditioning and Food Preservation

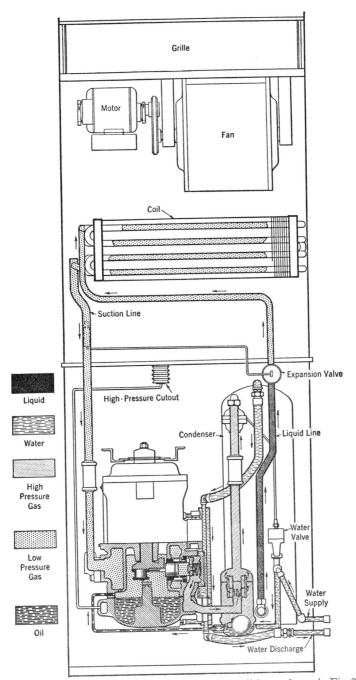

Fig. 22–3. Diagrammatic view of complete unit conditioner shown in Fig. 22–2. (Courtesy Airtemp Division, Chrysler Corporation.)

3-hp motor to drive at 1750 rpm a compressor whose three cylinders have a $1\frac{7}{8}$-in. bore and a $1\frac{1}{2}$-in. stroke. A $\frac{1}{3}$-hp fan motor supplies 1200 cfm, and 375 fpm face velocity is used over the cooling coil. In a packaged unit of this type, water and electric connections alone are required.

Air conditioning is especially necessary where precision machine work is required, such as that found in the manufacture of aircraft engines. In particular, the gaging and inspection rooms, where standards are kept, need controlled temperature and humidity. Several large aircraft plants employ mechanically circulated water as the heat transporting medium. The water is warmed in winter and refrigerated in summer, but the same pipes, pumps, and convectors are used the year round. The heat removal may run into thousands of tons, and for these large loads, centrifugal refrigerating compressors are generally employed. The ultimate heat disposal into the air is through mechanical-draft cooling towers.

Another type of plant is illustrated by a large lens-grinding factory, where dust-free air and definite year around relative humidity control are required. It has four central air-supply stations with electrostatic dust eliminators at each of the four apparatus rooms. Each department or zone demands different conditions from those in other departments and has individual steam reheating convectors which facilitate both temperature and humidity adjustment. Certain critical departments even have secondary local electrostatic precipitators as an aid in preventing cross-contamination by dust. The heat removed from the air is absorbed by water, while that added to the air is delivered by steam. Steam reheaters, during hours when production is stopped, prevent undue cooling of the partly manufactured glass.

Where heavy machining and high-speed grinding or finishing is carried out, as in automotive production, the air of a shop may become laden with oil vapor. The conventional solution has been to increase the amount of ventilation and to install hoods, ducts, and powerful fans.

22-6. FROZEN-FOOD DISTRIBUTION

During the last decade a great increase has taken place in the extent to which foods are prepared, stored, and distributed in a frozen state. This has increased the demand for refrigeration equipment; first, in frozen food production; second, for intermediate short-term storage units in warehouses; and finally, for freezers and locker stores in institutions and homes. For initial freezing of the food products a variety of methods have been developed, which include blast freezers, immersion freezers, and double-contact plate freezers. The latter type of freezer is used on pre-packaged products, such as packaged peas, fish filets, berries, pies, and the like, with the freezer consisting of a horizontal series of refrigerated plates. The packaged food is put between the plates, which are

Industrial Air Conditioning and Food Preservation

brought into contact with the packages, but prevented from crushing the packages by a limit stop. With the packages in contact between the plates, heat is abstracted from the product at top and bottom until it is frozen. The plates are then separated by an automatic mechanism, and the frozen packages are shoved out for removal to storage or to distribution channels. In machines of this type, direct-expansion coils are located in the plates, and ammonia, refrigerant 22, or 12, is customarily used.

Even when foods are frozen, some deterioration takes place, and storage life is not unlimited in extent. In general it can be stated that the lower the temperature at which frozen foods are kept, the better and longer their quality is maintained. In most cases, for extended storage it is desirable to maintain foods at 0 F or less. However, small freezer equipment usually operates in the range of 0 to 20 F, and quality is not always maintained at the highest level.

22-7. ICE MANUFACTURE

Although ice as a basic means of preserving food at the domestic level is little used because of the advent of mechanical refrigerators and food freezers, commercial ice is still important and is used extensively in transportation for prechilling and for food materials which cannot be brought to extremely low temperatures. For example, lettuce for long-distance shipment is frequently packed in containers around which and between which crushed ice is blown. The melting of the ice maintains the product in fresh condition for longer periods than would otherwise be possible. The same type of process can also be used to advantage with many other kinds of fresh garden foods. For use in serving foods and beverages, ice still renders an important function.

Much artificial ice is made by the can system. In this system tapered cans of rectangular cross-section (usually 10 in. by 21 in. at the bottom, 11 in. by 22 in. at the top, and 49 in. high), which hold about 300 pounds of water, are immersed in a moving bath of brine until freezing is completed. Although water for ice making may be clean, it nevertheless contains dissolved gases and solids; and, if frozen quietly, the ice would not be transparent. A common system for manufacturing clear ice is that of air agitation. In this system, compressed air is bubbled down the central axis of the can and, as the ice freezes from the sides of the can inward, the air motion sweeps away dissolved gases and prevents them and solids from freezing into the cake. The small central core of liquid, which remains before freezing is finally completed, is sucked out with its entrained solids, and the cavity is refilled with clean water. Brine velocities of 16 to 30 fpm are used and temperatures are held between 12 and 18 F. Ice made at lower temperatures may be brittle. The time of freezing depends on the brine temperature and velocity but, in general, about 50 hr may be considered representative for a 300-lb can.

In recent years, more and more ice has been produced in unit machines, which make the ice in little pieces called by such descriptive names as "flakeice," "tube ice," and "packice." In one of these unit machines, water is sprayed onto a refrigerated rotating drum, on which it freezes. The ice is scraped off in the form of chips or slivers, in which form it is suitable for ultimate use and is delivered as the final product.

22–8. MOISTURE-ABSORBING DEVICES

The removal of water vapor from air by cooling the air below its dewpoint and condensing out the moisture has been previously discussed. However, it is also possible to dry or remove water vapor from air by the use of appropriate absorbers or adsorbers. An *adsorber* is a solid substance which has the ability to trap and hold certain vapors in its cell structure. *Silica gel,* a chemical which looks very much like sand, has numerous ultramicroscopic pores and can adsorb moisture to the extent of 6 to 40 per cent of its bone-dry weight. The air to be dried is made to come in contact with the gel and the latter adsorbs a large portion of the water vapor. Heat is generated during this process in an amount equal to the latent heat of condensation plus about 200 Btu for each pound of vapor adsorbed. This latter is called the heat of wetting. The air passing through the gel receives, as sensible heat, about 0.174 Btu per lb of dry air per grain of moisture adsorbed. From these data it is possible to calculate the temperature rise of the air leaving a silica-gel dehumidifier.

The silica gel eventually adsorbs so much moisture that it must be reactivated. This operation is performed by exposing the gel to hot air (250 to 350 F), during which process, the gel dries out to a very low residual vapor content (7 to 8 per cent). The gel is then cooled before going into service again. A silica-gel apparatus is always arranged to permit regeneration of the moisture-laden bed of gel, while a previously prepared bed is in service.

The air delivered from such a unit, if it is to be used for drying purposes alone, can be sent directly to the point of utilization. However, for air-conditioning work, the air may first be cooled by passing it over water-cooled convectors, or it may be exposed to water sprays for evaporative cooling. Silica-gel systems are employed extensively where gas heat is available for regenerating the gel. Silica-gel dehumidification is especially successful in manufacturing processes where exceedingly low relative humidities are necessary.

Other adsorbents of water vapor are activated alumina, activated carbon, lamisilite, and certain border-line materials, such as calcium

Industrial Air Conditioning and Food Preservation 747

chloride and lithium chloride, which may operate in the solid state and also in the liquid, or solution, state.

The last two substances are more truly absorbents, since they are usually employed in liquid form as strong brine. The water-vapor pressure over such brines is usually much lower than the partial pressure of the water vapor in the air, and vapor from the air thus condenses into the brine, thereby diluting the brine. The diluted brine is then heated in a separate chamber to expel the surplus moisture and, after cooling, is ready for re-use. The liquid absorbent, which can be pumped through its various phases of treatment, has advantages over the more cumbersome granular media. The latent heat of condensation plus the heat of solution with a brine must be dissipated just as with silica gel, and as a result, the temperature of the air rises during dehumidification.

PROBLEMS

22-1. Tobacco in a warehouse, held at 75 F and 40 per cent rel hum, is placed in a room at 70 per cent rel hum and 75 F preparatory to being worked on. For each 100 lb of tobacco moved from the warehouse, what is its bone-dry weight? What is the actual weight of this amount of tobacco after staying in the working room? *Ans.* 88.5 lb dry; 110.5 lb

22-2. A 10-lb sample from a batch of material under test is found to have a bone-dry weight of 8.5 lb. This material is processed and is then found to have a regain of 20 per cent. How much weight of product appears for each pound of original material? *Ans.* 1.02 lb per lb

22-3. Material enters a dryer containing 60 per cent water and 40 per cent solids and leaves with 5 per cent water and 95 per cent solids. Find the weight of water removed, based on each (a) pound of original product, (b) pound of final product, (c) pound of bone-dry material. *Ans.* (a) 0.58; (b) 1.38; (c) 1.45

22-4. Find the heat that must be removed to cool 2000 lb of apples from 80 to 32 F. Neglect respiration load and containers. *Ans.* 86,400 Btu

22-5. Find the heat removal required to cool, freeze, and chill a ton of fish from 70 to 10 F. *Ans.* 280,600 Btu

22-6. Air in a cold-storage room enters the bottom of a unit conditioner at 36 F and 85 per cent rel hum and leaves at 30 F and 40 per cent rel hum. What heat load is given up per pound of dry air and how much of this is latent? *Ans.* 4.3 Btu; 2.9 latent per lb dry air

22-7. A 20 by 20 by 10 ft cold-storage room at 10 F and 70 per cent rel hum has six changes of air per 24 hr, measured at outside conditions of 90 F and 50 per cent rel hum. Find (a) the cooling load required for ventilation and (b) the weight of moisture brought in by the outside air. (c) Where does this moisture probably go? *Ans.* (a) 59,000 Btu/24-hr total, plus 3750 Btu in freezing water on coils; (b) 25 lb water; (c) freezes on coils

22-8. A freezer room 20 by 20 by 10 ft high and insulated on all sides by the equivalent of 6 in. of corkboard with asphaltic binder (include film effects) maintains its load at 0 F when 95 F outside temperature holds at the ceiling and adjacent to one side of the room. Three air changes are required per hour, measured at outside conditions, with the relative humidity both inside and outside the room at 60 per cent. Although a varying product load exists in this room, the new product added is essentially the equivalent of 1000 lb of beef per 24 hr brought in at 60 F and frozen. The lighting is 200 w and 2 men work in the room for 8 hr a day. The lights are not used when the men are not in the room. Compute (a) the heat gain through the two warm surfaces; (b) the equivalent product load per day per 1000 lb of meat cooled, frozen, and chilled; (c) the ventilation load; and (d) the total load. Find the compressor tonnage required for this room (e) on a basis of continuous operation for 24 hr, and (f) if the compressor carries the load for 16 hr of actual compressor running time in each 24 hr with the unit idle intermittently during 8 hr of each day.

Ans. (a) 70,200 Btu/day; (b) 133,580 Btu/day; (c) 925,000 Btu/day + 67,800 Btu/day from freezing water on coils and cooling ice; (d) 26,260 Btu/day men and lights and 1,222,840 Btu/day total; (e) 4.25 tons; (f) 6.4 tons

22–9. A small cold-storage plant uses refrigerant-12 (Freon-12) in the refrigerant coils of one of its freezer rooms where a temperature of -10 F is maintained. The direct-expansion coils are held 10 deg lower than room temperature and condensation takes place at 96 F in the condenser. The liquid subcools to 90 F on the way to the expansion valve. Essentially dry saturated vapor flows to the compressor. The load on the plant is 15 tons of refrigeration. (a) Sketch a ph diagram showing necessary values required for solution of this problem. Compute (b) the refrigerant circulation in the system in pounds per minute, (c) the theoretical piston displacement in cfm, (d) the cylinder size for a six-cylinder, 1150-rpm compressor with a volumetric efficiency of 80 per cent, if the bore-stroke ratio is unity, (e) compute the compressor shaft horsepower and the probable motor size required.

Ans. (b) 64.6 lb/min; (c) 158 cfm; (d) 3.98×3.98 say 4×4; (e) 36.0 hp; next standard, 40 hp

22–10. Rework problem 22–9 on the basis of refrigerant-22 (genetron-22 or Freon-22) being employed. (b) 46.3 lb/min; (c) 96.7 cfm; (d) $3\frac{3}{8} \times 3\frac{3}{8}$; (e) 36,40

REFERENCE

American Society of Heating, Refrigerating and Air-Conditioning Engineers. *ASHRAE Guide and Data Book Applications,* 1966–67, Chapters 38 and 42.

Appendix

TABLE A-1
Conversion Factors for Pressure Units

(Multiply units of left column by appropriate factor in table to obtain result in units designated at top of each other column.)

	atm	lb/in.²	kg/cm²	in. Hg at 32 F	mm Hg at 0C
atmospheres	1.0	14.6960	1.03323	29.9212	760
lb/sq in.	0.068046	1.0	0.070307	2.0360	51.7147
kg/cm² ≡ metric atm	0.96784	14.2234	1.0	28.9590	735.5592
in. Hg at 32 F	0.33421	0.49116	0.34532	1.0	25.40005
mm Hg at 0C	1.31579×10^{-3}	0.0193369	1.35951×10^{-3}	0.03937	1.0

TABLE A-2
Conversion Factors for Units of Density

(Multiply units of left column by appropriate factor in table to obtain result in units designated at top of each other column.)

	lb/cu ft	lb/cu in.	grams/cu cm	grams/ml	lb/gal
lb/cu ft	1.0	5.7870×10^{-4}	0.016018	0.016019	0.13368
lb/cu in.	1728	1.0	27.6797	27.6805	231
grams/cu cm	62.4283	0.036127	1.0	1.00003	8.34545
grams/ml	62.4266	0.036126	0.99997	1.0	8.34522
lb/gal	7.48052	4.3290×10^{-3}	0.119826	0.119829	1.0

TABLE A-3

Conversion Factors for Energy Units

(Multiply units of left column by appropriate factor to obtain result in the units designated at top of each other column.)

	Btu	k cal	1000* I.T. cal	Int Joules	ft-lb	Int kw hr
Btu	1.0	252.161×10^{-3}	251.996×10^{-3}	1054.89	778.16	2.930×10^{-4}
cal	3.9657×10^{-3}	1.0×10^{-3}	0.99935×10^{-3}	4.1833	3.086	1.1620×10^{-6}
k cal	3.9657	1.0	0.99935	4183.3	3086	1.1620×10^{-3}
1000 I.T. cal*	3.9683	1.00065	1.0	4186.05	3088	1.1628×10^{-3}
Int kw hr	3412.76	860.56	860.0	3.6×10^6	2.6556×10^6	1.0
hp-hr	2544.4	641.617	641.197	2.6840×10^6	1.980×10^6	0.74558
ft-lb	1.2851×10^{-3}	0.3241×10^{-3}	0.3238×10^{-3}	1.3556	1.0	3.7656×10^{-7}
Int Joule ≡ 1 watt-sec	0.9480×10^{-3}	0.2390×10^{-3}	0.2389×10^{-3}	1.0	0.7378	2.779×10^{-7}

*1000 International Steam-Table Calories = 1/860 Int kw hr

Appendix

TABLE A-4

Conversion Factors for Thermal Conductivity

(Multiply units of left column by appropriate factor to obtain result in the units designated at top of each other column.)

	$\dfrac{\text{Btu ft}}{\text{hr ft}^2\text{ F}}$	$\dfrac{\text{Btu in.}}{\text{hr ft}^2\text{ F}}$	$\dfrac{\text{Btu ft}}{\text{sec ft}^2\text{ F}}$	$\dfrac{\text{Btu in.}}{\text{sec ft}^2\text{ F}}$	$\dfrac{\text{kcal cm}}{\text{sec cm}^2\text{ K}}$	$\dfrac{\text{watts cm}}{\text{cm}^2\text{ K}}$
$\dfrac{\text{Btu ft}}{\text{hr ft}^2\text{ F}}$	1.000	12	2.778×10^{-4}	3.333×10^{-3}	4.136×10^{-6}	1.730×10^{-2}
$\dfrac{\text{Btu in.}}{\text{hr ft}^2\text{ F}}$	8.333×10^{-2}	1.000	2.3148×10^{-5}	2.778×10^{-4}	3.447×10^{-7}	1.442×10^{-3}
$\dfrac{\text{Btu ft}}{\text{sec ft}^2\text{ F}}$	3600	43200	1.000	12	14.892×10^{-3}	62.296
$\dfrac{\text{Btu in.}}{\text{sec ft}^2\text{ F}}$	300	3600	8.333×10^{-2}	1.000	1.241×10^{-3}	5.191
$\dfrac{\text{Btu in.}}{\text{sec in.}^2\text{ F}}$	4.32×10^4	5.184×10^5	12	144	178.5×10^{-3}	7.474×10^2
$\dfrac{\text{kcal cm}}{\text{sec cm}^2\text{ K}}$	241.9×10^3	2903×10^3	67.17	0.8058×10^3	1.000	4.183×10^3
$\dfrac{\text{watts cm}}{\text{cm}^2\text{ K}}$	57.788	693.46	1.605×10^{-2}	0.1926	0.239×10^{-3}	1.000
$\dfrac{\text{cal cm}}{\text{sec cm}^2\text{ K}}$	241.9	2903	6.717×10^{-4}	0.8058	0.001	4.183

TABLE A-5

Conversion Factors for Specific Heat Capacity (Specific Heat)

(Multiply units of left column by appropriate factor to obtain result in the units designated at top of each other column.)

	$\dfrac{\text{Btu}}{\text{lb F}}$	$\dfrac{\text{Cal}}{\text{g K}}$ or $\dfrac{\text{k cal}}{\text{kg K}}$	$\dfrac{\text{Joules}}{\text{g K}}$	$\dfrac{\text{Watt-sec}}{\text{g K}}$
$\dfrac{\text{Btu}}{\text{lb F}}$	1.0	1.000654	4.1867	4.1867
$\dfrac{\text{Cal}}{\text{g K}}$	0.999346	1.0	4.1840	4.1840
$\dfrac{\text{Joules}}{\text{g K}}$	0.23885	0.23900	1.0	1.0
$\dfrac{\text{Watt-sec}}{\text{g K}}$	0.23885	0.23900	1.0	1.0

TABLE A-6

Conversion of Energy per Unit Mass and of Enthalpy

$$\dfrac{\text{Btu}}{\text{lb}} \times 0.55556 = \dfrac{\text{cal}}{\text{gram}} = \dfrac{\text{k cal}}{\text{kg}}$$

$$\dfrac{\text{Btu}}{\text{lb}} \times 2.3263 = \dfrac{\text{Joules}}{\text{gram}} = \dfrac{\text{watt-sec}}{\text{gram}}$$

$$\dfrac{\text{cal}}{\text{gram}} \times 1.8 = \dfrac{\text{k cal}}{\text{kg}} \times 1.8 = \dfrac{\text{Btu}}{\text{lb}}$$

$$\dfrac{\text{Joules}}{\text{gram}} \times 0.42987 = \dfrac{\text{watt-sec}}{\text{gram}} \times 0.42987 = \dfrac{\text{Btu}}{\text{lb}}$$

TABLE A-7

EQUIVALENCE OF MISCELLANEOUS UNITS

Lengths

1 ft = 0.3048 m	= 12 in.	= 0.3333 yd
1 m = 3.28084 ft	= 39.37008 in.	= 10^6 μ (microns)
1 mi = 5,280 ft	= 1,760 yd	= 1,609.34 m
1 mi = 0.86898 (nautical) mi	= 1.60934 km	= 320 rd

Areas

1 sq ft = 0.09290 sq m	= 144 sq in.	= 0.11111 sq yd
1 sq m = 1549.99 sq in.	= 10.7639 sq ft	= 1.19599 sq yd
1 Acre = 43,560 sq ft	= 4,840 sq yd	= 0.40469 ha (hectare)
1 Acre = 4046.87 sq m	= 0.001563 sq mi	
1 sq mi = 640 A	= 3,097,600 sq yd	= 2,589,999 sq m
1 sq mi = 2.59000 sq km	= 259.0 ha	
1 sq km = 0.38610 sq mi	= 247.104 Acre	= 10^6 sq m = 100 ha

Masses and Weights

1 lb	= 0.45359 kg	= 16 oz	= 14.5833 oz (troy) = 0.0005 ton
1 lb	= 7000 grains	= 0.000464 long ton	
1 kg	= 2.2046 lb av	= 2.2692 lb tr	= 35.274 oz av
1 kg	= 15,432.4 grains	= 0.00110 ton	= 0.001 m ton
1 ton	= 2,000 lb	= 907.185 kg	= 32,000 oz = 0.90722 m ton

Volume and Capacity

1 cu ft	= 1728 cu in.	= 0.03704 cu yd	= 0.028317 cu m
1 cu ft	= 29.9221 qt (liq)	= 7.4806 gal (liq)	
1 cu ft	= 6.229 Imp, gal (Br)	= 0.80356 bu	
1 cu yd	= 46,656 cu in.	= 27 cu ft	= 0.76456 cu m
1 cu yd	= 807.896 qt (liq)	= 201.974 gal (liq)	
			= 21.6962 bu
1 gal (liq)	= 231 cu in.	= 0.13368 cu ft	= 4 qt
1 gal (liq)	= 0.83268 Imp. gal	= 0.00378543 cu m	
1 cu m	= 61,023 cu in.	= 35.314 cu ft	= 1056.7 qt (liq)
	= 264.18 gal (liq)	= 28.38 bu	= 1.308 cu yd

Index

Adiabatic process, 52
Adiabatic saturation, 102
Air
  atmospheric composition, 85
  building, changes in, 388
  bypassed, 214–16
  cascade system, 619–21
  changes of, assumed, 197, 388
  chart, Ts, 609
  cleaning of, 715–28
  combustion, requirements for, 354–56, 360
  compressibility, 7
  –cycle refrigeration, 500–05
  delivery of, 445–54
  dew point of, 104–05, 124
  distribution of, 445–54
  dry, 364–65
  dry-bulb temperature of, 104
  dust separation with, 725–26
  eliminator, 256–58
  enthalpy of, 99–100
  excess, 356–59
  filtering of, 717–23
  flow of, 475–82
  forced warm, 396–408
  grille location for, 450–54
  heating of, 116–17
  high-velocity distribution of, 454–58
  impurities in, 715–16
  induction and mixing of, 450
  inside design temperatures for, 189, 377
  leakage of, 198
  liquefaction, 603–06, 610–12, 619–22
  mixing, equations for, 119
  moist, properties of, 88–94
  molecular weight, 85
  properties, 79, 85, 88–94
  recirculated, 498
  relative humidity of, 87, 99, 124
  requirements for quality and quantity of, 386–91
  saturation ratio for, 97
  standard, 418
  supply, conditions for, 391
  throw of, 449
  Ts diagram, 609–10, 646
  ventilation, 238
  venting of, 256–58, 295–98

Air—*Continued*
  washing, with water, 716–17
  wet-bulb temperature of, 100, 102, 104, 108–11
Air capacity, maximum, 219–20
Air circulation, velocity of, 449
Air cleaner
  charged-media, 726
  Precipitron, 724–25
Air conditioning, definition of, 2
  systems, 213–19
Air conditions
  inside, 189, 377
  outside, 191–216
Air diffusers, 453–54
Air elimination, boiler, 256–58, 295–98
Air filters, 717–22
Air flow
  measurement of, 475–82
  minimum, 387
Air-fuel ratio, 354–56, 360
Air outlet noise, 451
Air relay, 678
Air temperature
  daily variations, 223
  dry-bulb and wet-bulb, 102, 104, 108–11
  inside design, 189, 377
  outside summer, 191–216
  outside winter, 189–96, 208
Alumel, 11
Aluminum sheet, 445
  gages of, 446, 448
Ammonia, 419–24
Anemometer, 478–81
Anemostat, 453–54
API, degrees, 352
Apjohn equation, 110
Apparatus dewpoint, 245–47
Argon, 595, 655, 683
Aspect ratio, 424, 453
Atmospheric air, 17, 85
Azeotrope refrigerant, 541

756   Index

Babcock formula, 279
Bare-iron pipes, heat loss from, 180
Barometer, 17
Baseboard radiation, 330–33, 400–01
Basement, heat loss from, 169, 173–75
Bimetal strip, 5–7
Bismuth telluride, 27
Bituminous coal, 349–51
Body, human; see Human body
Boiler, 38, 346–49
   blowing down of, 347
   energy equation for, 37–38
   fire-tube, 347–49
   furnace
      heat balance with, 361
      loss in, 357–62
   gas-fired, 348, 365–66
   horizontal-return-tubular, 349
   horsepower of, 42
   hot-water, accessories and controls for, 346–69
   H. R. T., 349
   low-pressure, 347
   safety devices for, 369–71
   water level in, 347, 370
Bourdon-tube pressure gage, 18
British thermal unit, 34–35
Buildings
   air changes in, 197–98
   fuel consumption in, 204–09
   warm-up of, 199
Burners, controls for, 363–70
Bypass damper, 215
Bypass factor, 245–47
   coil, 214–16, 238
Bypass system, 214–16

Calorie, 42
Carbon black, 166
Carbon dioxide, content, 389–90
   properties, 484, 489–91, 531
Carrier equation, 111
Cascade system, 619–21
Central systems in air conditioning, 338
Chimneys, 367–69
   draft of, 367–68
   effect in buildings, 196–98
   height of stack for, 368
Chromel in thermocouples, 11
Circuits
   relay, 678, 681
   thermocouple, 9–13, 366–67
Coal
   analyses of, 350–51
   Dulong formula for, 351
   heating value of, 351

Coefficient of performance (COP), 24, 26, 486–88
Coil efficiency, 238
Coke, 351
Collins, J. C., helium liquifier, 633–35
Column
   gas separation, 645–53
   Linde, 650–53
Combustible losses in refuse, 361
Combustion
   calculation of, 353–57
   definition of, 349
   incomplete, loss from, 360
Combustion chamber, 364
Comfort air conditions, 374, 377, 379
Comfort chart, 376–79
Compression
   isentropic work of, 496–97, 499–500
   and shaft work, 497, 509
Compressor
   calculation of work, 495–97, 499, 502, 509
   centrifugal, 569–77
   characteristics of, 567, 577
   charge efficiency of, 498–502, 505–09
   clearance, 505–08
   control, 506, 566–68
   high-pressure cutout for, 570
   mechanical efficiency of, 509
   radial, 564–65
   reciprocating, 559–68
   steam jet, 582–84
   stopping and starting of, 567
   volumetric efficiency of, 497–98, 505–09
Condenser, heat removed in, 498–99
Condensers, 577–83
Conductance
   of air space, 140
   thermal, 151
Conduction
   definition of, 134
   through walls, 138
Conductivity, thermal
   of building and insulating materials, 142–49
   of miscellaneous substances, 136–38
Conduit air-distribution systems, 454–58
Constantan in thermocouples, 11
Contaminants, air, 391–92
Control
   actuating, 671, 675–79
   general considerations, 684–85
   limit, 369–70
   precool and reheat, 686
   principles of, 671–72
   reciprocating compressor, 567–68
   refrigerant valve, 558–59

Index

Control—*Continued*
  relay, 678, 681–83
  zone, 687
Control systems
  damper, 215
  all-year, 216–19
Controller, 671, 673–74
  electronic panel, 682
Convection
  definition of, 134
  forced, 135
  free, 135
  heat delivery by, 695
  heat transferred by, 692, 694
  natural, 135, 329
Convectors
  capacities of, 330
  definition of, 326
  details of, 329–30
  output of, 330
  stack height of, 328–29
Cooling, convector for, 334–35, 339
Cooling load, 213, 223–49
  air arrangements for, 213–28
  air quantities in, 245–49
  calculations for, 240–44
  heat sources for, 223
  internal, 244–45
  sensible-heat ratio for, 127, 245, 249
Cooling unit, location of, 339
Critical pressure, 68
Cryogenics, 594–667
  air liquefaction plant, 653–58
  charts of gases, 609, 624, 626, 630–32, 636–37
  compression of gases, 495–97, 499–500, 612–16
  definition, 4, 594
  gas data, 595
  gas liquefaction, 603–35
  insulants for, 658–61
  Joule Thompson
    coefficient, air, 602
    effects, 594–605
  liquid storage, 658–61
  regenerative cycle for, 662–67
  separation of gases by, 642–58
Cyclone dust-separator, 726–27

Damper, 675, 685
  bypass, 685
  control system, 675
  pneumatic, 677–78
Data sheet for heating, 200–01
Degree days, 204
  by months, 208
  table of 191–96
Dehumidification, 117–18, 124, 246
Dehumidifier, 68

Dehumidify, 3
Dehumidifying system, 685–87
Density, 62
Design conditions, outside, 191–96
  for summer, 377
Design temperatures in heating, 189, 191–96
Dew point, 104–05
  apparatus temperature for, 595
Dewar, 658–59
Diathermous media, 135
Dichlorodifluoromethane, 525–27, 533–35
Diffusing passage, 466
Diffusion, 430
Dry-bulb temperature, 104
Dry return, 260
Dual-duct systems, 217–21
Duct
  circular, frictional loss in, 423
  construction of, 445
  equipment in, 449
  fittings, pressure loss in, 424, 426–27
  fricton-factor chart, 415
  heat gain to, 442–44
  heat loss to, 442–44
  joints, 447
  losses in, 417–27
  noncircular, 417
  pressure loss from air flowing in, 430
  rectangular, frictional loss in, 423
  regain equations for, 430, 435–37
  return, 442
  return-air, 399–400, 402
  roughness, correction factors for, 423
  seams used in, 445
  temperature drop in, 443–44
  temperature rise in, 443–44
  trunk, 399
  turbulent flow in, 410
  velocities, recommended, 428
  warm-air design of, 398–402
Duct design
  procedure for, 427–37
Duct sizes
  aluminum, 446–48
  steel, 448
Duct system
  characteristics of, 471
  equipment, losses for, 428
Dulong formula, 351
Dust
  removal of, from air, 715
  separation of, centrifugal, 725–27
  size of particles, 715–16
Dust separator
  cyclone, 726–27
  Roto-Clone, 725

EDR
  definition of, 255
  hot water heating, 302
Effective piston displacement (PD), 497–98, 507
Efficiency
  of coal-fired boiler, 206
  of gas-fired furnace, 206
  motor, 237, 319
  of oil-fired boiler, 206, 361
  volumetric, 505–06
Elbow equivalents, 303–04
Elbows
  rectangular, loss in, 426
  round, loss in, 426
  vaned, pressure loss in, 427
Electromotive forces, 11, 23
Electronic packs, 684
Eliminator, air, 256–58, 295–98
Energy, 32–34, 39
  internal, 34, 51
  kinetic, 33
  potential, 36
  steady flow of, 35–39
Energy equation, 35–39
Enthalpy, 37, 57–58, 752
  of air, 99–100
  of wet steam, 66
  pressure correction, 107
Enthalpy-humidity-difference ratio, 125–26, 245
Entropy, 43–45
  gases, 58–59
Equivalent bypass, 246–47
Equivalent length, 278–79, 303–04
Estimates, hearing, 200–03
Evaporation, heat loss by, 375–76, 384
  from skin, 379–80
Evaporative cooling, 123
Evaporators, 553–57
  air-warmed, 557
  pressure in, 486
  refrigerant, 344
Expansion
  coefficient of, 5, 19–20
  of pipe, chart, 272–73
  thermal, 5–7, 19–20
Expansion tank, 294–97
Expansion valve, 486, 493–95, 558–59
Exposure factor, 203

Factory building, heating of, 337–38
Fans, 461–83
  axial-flow, 470–72
  backward-tip, 465
  centrifugal, 461–62, 466–70
  characteristics of, 469

Fans—*Continued*
  efficiency of, 464
  forward-tip, 464
  horsepower characteristic of, 468
  noise in, 469, 475
  operating velocities for, 467
  pressure-capacity table for, 474
  pressure rise in, 468
  propeller, 463
  radial-tip, 465
  selection of, 473
  speed of, 472
  static work of, 39
  tip speeds for, 467
  total work of, 39
  tubeaxial, 463
  types of, 461–62
  vaneaxial, 461
  velocity diagrams for, 466
  and system characteristics, 471–72
Fan laws, 472–75
Feedback, 671
Feeder, boiler-water, 369–70
Ferrel equation, 110
Film coefficient, resistance, 140–41, 344
Film factor, 140–41, 179–80
  forced-convection, 345–46
Filter
  continuous belt, 722
  mechanically-replaced, 721
  medium, air velocity through, 716–17
Filtering, dry, 717–18
Fireplace, heat loss from, 197, 199, 202
Fittings
  compression-type, 274
  flared tube, 274
  resistance of, 278
  solder-type, 274
  tee, 299–300
Flow
  air-duct, 417–27
  fluid, 409–60
  steady, 136
  streamline, 409
  turbulent, 410
  viscous, 409–16
Flow nozzles, 482
Flow work, 36, 418
Flue gas, 353–59
  analyzer, Orsat-type, 357
  carbon dioxide in, 359
  loss due to, 360–61
Fluid film, conditions of, 139
Food storage, 735–40
Foot-pound, 32–35
Fourier's equation, 135
Freon-11, 531, 537
  pressure-enthalpy chart for, 536

Index

Freon-12, 525–27, 533–35
  pressure-enthalpy diagram for, 534
Freon-22, 518, 528–30, 535, Plate V
Friction factor, 414–16
Friction loss
  air, chart for, 420
  in copper tubing, 310
  in steel pipe, 309
Frozen foods, 744–46
Fuel, requirements of, 205–07
Fuels, solid, 349–51
Furnace
  coal-fired, 206
  gas-burning, 365–67
  gas-fired, efficiency of, 206
  location of, 401
  oil-fired, 206, 363–65
  warm-air, 7, 396–97
Fusion, heat of, 33

Gage
  Bourdon-tube, 17–18
  pressure, 17
Gas, 352–53
  automatic shutoff valve for, 366
  burning, boilers for, 365–67
  compressibility, 77–78
  draft hood with boilers, 365, 369
  exhaust, analyses of, 357–60
  firing, control for, 366
  as fuel, 352–53
  liquefaction, 603–35
  mixtures, 70–72
  perfect, 48–49, 50, 56
  separation, 72–76, 642–58
  specific heat of, 46–48
  universal constant, 49
  velocity heads for, 415
  viscosities of, 412
  work, perfect gas, 53–55
Genetron-22, 518, 528–29, 535, Plate V
Gibbs-Dalton law, 70
Glass, window
  convection and radiation from, 224–30
  shading effect upon, 229
  trap effect of, 226
Grain, defined, 39
Grille
  location of, 450–54
  throw through, 450

Hampson system, 603–06
Hartford loop, 261, 272
Head, 418
  total, 419
  velocity, 418, 466
Heat, 32–34, 37
  balance, boiler, 361
  of fusion, 33

Heat—*Continued*
  humid, 117
  latent, 66, 254
  seasonal requirements, 207–08
  specific, 46–48, 136–37
  vaporization, 34
Heat dissipation from individuals, 380–85
Heat exchangers, 342–46
Heat ratio, sensible, 127, 245
Heat gain
  from body evaporation, 382, 384
  from occupants, 381–82
  temperature differentials for, 231–34
Heat load from equipment, 237–38
Heat loss
  from attics, 173
  from bare-iron pipes, 180
  from basements, 169, 174–75
  from evaporation, 382, 384
  through glass surfaces, 188, 190
  from ground floors, 173–74
  from human body, 380–85, 690
Heat losses
  convection, distribution of, 690
  radiation, distribution of, 690
  reversed, 703
Heat-moisture ratio, 125–26, 245–46
Heat pump, 707–10
  diagram of, 708
  heat sources for, 709–10
Heat transfer
  coefficients for heat exchangers, 343–44
  coefficients, values of, 136–38, 142–50, 152–65, 344
  by conduction, 134
  by convection, 134–35, 340
  counterflow, 341
  devices, 326–39
  equations, 135, 138–39, 151
  insulating materials, characteristics of, 166–67
  liquid elements for, 340
  through metal surfaces, 340–46
  modes of, 134–35
  over-all coefficient of, 151
  parallel-flow, 342
  pipe coverings for, 177–83
  through pipes, 177
  by radiation, 135, 183–84
  symbols used in, 151
  thermal conductivity, units, 136
Heaters
  floor-mounted, 335
  operation of, 334–37
  unit, 333–38
Heating, definition of, 1, 2
Heating load, 188–203
Heating systems, 2

Heating value, fuels, 349–53
Helium
  charts, 630–32
  compressibility
  liquefaction, 633–35
  properties, 77, 598, 628–35
High-velocity distribution, 454–58
Hot-water heating, 289–322
  advantages of, 295
  air in, 296–98
  baseboard (perimeter), 305–06
  boiler arrangements, 295
  copper tubing for, capacities of, 312
  design practice, 307
  diversion tees for, 299–300
  flow required for, 302
  friction charts, 309–10
  high-temperature systems, 322
  in large buildings, 321
  pressure-differential graph, 292
  radiator heat emission, 302
  rate of circulation, 293
  steel pipes for, capacities of, 311
  steps in design of, 305, 307–09
Hot-water heating systems, 2, 289
  automatic air valve for use on, 298
  circulating arrangements, 293, 315–16
  one-pipe, 299
  temperature drop in, 303
  thermal system, circulation in, 291–292
Human body
  heat control within, 1, 375–76
  heat loss from, 380–85
  metabolic energy in, 375
  metabolic processes of, 690
  moisture evaporated from, 381–86
  physiological responses of, 376
Humidification of air, 121
Humidifying, 3, 121–23
Humidistat, 674–75, 686
Humidity, 87
  percentage of, 97
  ratio, 87, 96
  relative, 86, 99
  specific, 96
  system, controlled, 636
Hydraulic radius, 410, 417
Hydrogen, 623–27
  charts for, 624, 626
Hydronic heating, 289–322
Hygroscopic material, 731–32
Hygrostat, 674

Ice, 484–745
  manufacture, 745–46
Ideal gas liquefaction, 606–08
Industrial heating, 336–39

Infiltration, 196–98, 234, 236
  through window and door cracks, 197–99, 236
Instrumentation, 4–13, 16–18
Insulating materials, density and other characteristics of, 136–38, 143–50
  low temperature, 658–61
Insulation loss, 188
Invar, 7
Isothermal process, 55, 606–08

Joint
  duct, 447, 449
  solder, 274
  sweated, 274
Joule's law, 596
Joule-Thompson
  coefficient, 596–603
  effect, 603–06, 610–11
  inversion, 595, 597

k, units of, 136
Kata thermometer, 480
Köhler, W. L., cryogenic unit, 662–67

Leader, 398
Lights, heat gained from, 200
Limit control, 7
Linde system
  column, 650–53
  dual, 610–12
  simple, 603–06
Liquids
  subcooled, 583
  viscosities of, 583
Log mtd, 341–43

Main
  connection from, to riser, 272
  dry-return, 260
  return, 260
  steam, pitch of, 281
  supply and return, 279–83
  wet-return, 260
Manometer, draft-gage type, 16
Materials, properties of, 4, 19–20, 136–38, 143–50
Maxwell relation, 599
Mean temperature difference, logarithmic, 341–42
Mercury, 8, 14
Metabolic process, 1, 375
Metric units, 39, 41
Midgley, Thomas, 513
Moist air
  problems of, 117–23
  tables, 88–94
Moisture absorbing devices, 746–47

Index

Moisture content, 731-32
Moisture, transfer for condensing, 345
Mole, 49, 70
Monochlorodifluoromethane, 518, 528, 529, 535
Motors
  bellows-operated, 676
  heat gained from, 200, 237
MRT
  definition of, 380, 694

Neon, 595, 628, 635
Nitrogen, 595, 597, 636, 642-49, 656-57
Noise, air outlet, 451
Nozzles, flow, 482
Nusselt number, 435

Odors, concentration of, 388
Oil fuels, 351-52
Oil burners, 363-65
Orifice plates, radiator, 283-84
Orifices, 480-82
  fixed, 283-84, 320-21
Orsat analyzer, 357
Outlets
  ceiling, 451
  location of, 450
  pan-type, 451
Oxygen, 78, 642-56
  chart, 637

Panel
  arrangement for, 704
  ceiling, 700
  convection from, 695
  emissivity of, 706
  floor, surface temperatures for, 700
  heat delivery of, 698
  location of, 692
  multiple zoning arrangements for, 705
  output, 693-97
  radiant, 705
  radiation delivery from, 694
  surface temperature of, 692, 694-96, 700
  warm-air, 404, 692-96
  water requirements for, 692, 702-03
  water supplied to, 703
Panel heating, 690-706
  calculation form, 697-99
  control of, 705
  heat delivery from, 694-95, 698
  heat losses, distribution of, 690
  mean radiant temperature from, 694
  water requirements, 702-04
  zone arrangement, 705
Peltier, Jean, 18, 20-23
Perfect gas, 48

Performance, coefficient of, 488
Perimeter, 169
Perimeter heating
  hot water, 330-33
  warm-air, 400-03
Pipe
  Babcock formula for, 279
  bare-iron, 275
  coils for, 330
  copper, dimensions of, 274-76
  equivalent length of, 278-79, 303-04
  expansion of, 272-73
  galvanized, 274
  red-brass, dimensions of, 277
  resistance of, 304, 310
  roughness of, correction factors for, 423
  steam, sizes, 279-84
  steel, characteristics of, 273-74
  steel data, table, 275
Pipe fittings, 278, 303-04
Piston displacement, effective, 497-98, 507
Pitot tube, 475-76
  ten-point method, 477
Plenum chamber, 434
Pneumatic damper, 676
Poiseuille equation, 409
Polytropic
  values of n, 61
  work, 60
Potentiometer circuitry, 684
Power, 42
Prandtl number, 345
Pressure, 13, 86-87, 95
  absolute, 14
  barometric, 14, 17
  conversion units, 749
  enthalpy charts, 491-93
  fluid column, 14-15
  gage, 14
  impact, 477
  loss (balanced), 434
  manometer, 14-15
  static, 463
  total, 418, 477
  total dynamic, 418
  vacuum, 14
  velocity, 424
Pressure drop
  constant, 433
  fluid, 489
Pressurestat, 675
Propeller fan, 461
Psychrometers
  aspiration, 109
  sling, 108

Psychrometric charts, 113–17, Plates I, II, III, IV
  effect of pressure on, 107, 115
  pressure corrections for, 127–29
Psychrometric equations
  adiabatic saturation, 102–04
  Apjohn, 110
  Carrier, 111
  comparison of, 111
  Ferrel, 110
Pump
  boiler-feed, 265–66
  capacity characteristics of, 308
  condensate-return, 265–68
  forced-circulation, 308, 318–19
  vacuum, 265–68

R, for gases, 49, 50
Radiant heating, 690–706
Radiation
  baseboard, output capacity, 331–33
  definition of, 135, 182–84
  equivalent direct, 225
  heat transfer by, 302, 327–28
  square feet of, 225
Radiators
  air-vent valve for, 257–58
  definition of, 255, 326
  free-standing, 328
  heat transfer from, 327
  heat-transmitting capacity of, 306, 327
  hot-water, 302
  latent heat in, 254
  location of, 327, 329
  packless valve in, 262, 271
  rating of, 326–27
  representative connections of, 328
  test temperature of, 327
  types of, 328
Radius, hydraulic, 345, 410, 417
Radius taps, 481
Rankine, 45
Ratio
  enthalpy-humidity difference, 125–26, 245
  sensible heat, 127, 245–49
  specific heats, 47, 52
Receiver, alternating, 264–65
Recirculation, 213–16
Refrigerants, 512–44
  ammonia, 519–24
  calculations, 491–99
  carbon dioxide, 484, 489–91, 531
  desirable qualities of, 512–14
  dichlorodifluoromethane, 525–27, 533–35
  fluorine group of, 513–19
  Freon, 516–19, 525–29
  hydrocarbon, 541
  methyl chloride, 532, 537–38

Recirculation—*Continued*
  monochlorodifluoromethane, 518, 528–30, Plate V
  sulfur dioxide, 531, 540
  temperature-pressure chart for, 516
  testing for leaks of, 543–44
  thermodynamic characteristics, 514
  toxicity of, 533, 542
  trichloromonofluoromethane, 518, 531, 536–37
  water-steam, 582–85
Refrigeration
  absorption, 585–91
  air-cycle, 500–05
  brine, 553–54
  calculations, 491–99
  coefficient of performance in, 491
  cycle, 493–99
  diagrams
    pressure-enthalpy, 491–93
    temperature-entropy, 488–91
  dry ice (CO_2), for, 484
  horsepower expended for, 496–97, 509
  ice for, 484–85
  mechanical vapor, 485–86
  ranges, 4
  ton of, 42
Regain
  design pressure drops in, 435
  equations, 430–37
  static, 436
    method of design for, 431, 437–44
  static pressure, 430
Regenerative heat exchanger, 622–25
Register box, 398
  types of, 399
Relay, 678, 681–84
Relay, air, 678
Relay control, 681
Resistance
  electrical, 25
  pipe fittings, 303–04
  thermal, 139–40, 151
Return-air, 400
Return trap, 264–65
Reversibility, 43–45
Reynolds number, 410, 413–14
Risers, 262, 272
Roto-Clone dust separator, 725

Saturation, 85
  adiabatic, 102–04
  degree of, 97
Saturation ratio, 97, 99
Saybolt viscosimeter, 413
Season, length of heating, 204
Second law, 43
Seebeck, Thomas, 18, 20–22

Index

Sensible-heat ratio, 127, 245–49
Sigma function, 103
Silica gel, 746
Skin, evaporation from, 375, 377–78
Slug, 40
Snow-melting, 709
Soil, thermal conductivity of, 167
Solar effects heat from, 224–35
Solar heat gain
　through glass, 224–30
　through walls and roof, 230–34, 235
Specific heat, 46, 51–52
　tables of, 47, 93–95, 598, 735
Stacks
　definition of, 367
　forced-warm-air, 398
Standard air, 418
Static pressure, 463
Static regain, 429
Steam
　air-vent valve for, 257–58
　basic circulation of, 255
　consumption of, 206–07
　flow requirements of, 280, 282, 285
　pipe expansion, 272–73
　process, 267
　properties of, 63–65
　saturated, 87
　seasonal requirements for, 204, 207
　superheated, 65
　temperature-entropy, 67–69
　trap, purpose of, 262–63
　wet, 66–67
Steam flow, 255–56, 278–81
Steam heating, 254–85
　contraction in, 255
　expansion in, 272–73
　high-pressure systems for, 284–85
　orifices for, 283
　pitch of main in, 260, 262, 281
　radiators for, 326–28
　square feet of radiation in, 255
　vacuum system for, 265–68
　vapor system in, 262–65
Steam mains, 260
　piping arrangement for, 268–71
Steam-pipe arrangements, 259, 261, 263, 266–68
Steam systems, 258–70
　circulation in, 255–56
　definition of, 259
　Hartford loop in, 272
　high-pressure, 284
　one-pipe, air-vent, 259
　return-pipe capacities for 30 psig, 285
　steam capacities for 30 psig, 284
　two-pipe, air-vent, 260–62
　types of, 259
　vacuum, 265–70, 262–64

Steel sheet, 445
　gages, 448
Stem correction, 8–9
Stokers, 362–63
Storage, cold, 738–40
Summer
　daily temperature variations in, 223
　door infiltration in, 236
　heat transmission of flat roofs in, 233
Surface roughness, recommended values
　of, 414
Surface temperature, 285
Systems
　air-vent, pipe sizes for, 280
　all-year, 216–19
　central fan, 338–39
　circulating, 299–301
　damper-control, 215–16
　direct-return, 298
　dual-duct, 217–19
　forced-circulation
　　hot air, 396–400
　　water, 293–301
　gravity-circulation, 291–93
　high-temperature, 322
　hot-water, 289–322
　low-pressure
　　return-pipe capacities for, 282, 285
　　steam-pipe capacities for, 280, 284
　natural-circulation, 291–93
　override in, 672
　precool, 685–86
　reheat, 685–86
　return-trap, 264–65
　reversed-return, 298
　thermal-circulation, design for, 319–21
　transmission-power, 679–81
　two-pipe, layout of, 260–62
　vapor, 262–65
　warm-air heating, 396–407

Tank, expansion, 294–97
　sizes, 296
Temperature, 7–9, 45
　absolute, 8, 45
　air, limit setting, 404
　apparatus dew-point, 245
　average computed, 208
　breathing-line, 188–89
　centigrade, 8
　design, inside, 189, 377
　dew-point, 104–05
　drop in hot-water heating, 292–93, 303
　dry-bulb, 104
　effective, 377–79
　Fahrenheit, 7–9
　inside air, 188–89
　inside design, 189, 377
　lowering of, at night, 208
　mean radiant (MRT), 380

Temperature—*Continued*
    operative, 692
    outside, 191–96
    Rankine, 8
    scales, 7–9
    sol-air, 227
    stem correction, 8–9
    summer comfort, inside design
        conditions for, 377
    surface, 285
    thermocouple, 7, 9–10
    unheated mean radiant (UMRT), 694
    vapor-pressure, 9
    wall-surface, 169
    wet-bulb, 102–03, 108–11
    winter design, 176, 189, 191–96
Ten-point method, 477
Therm, definition of, 352
Thermal circuit, 167
Thermal conductance, 151
    of air spaces, 141
Thermal conductivity, 136–38, 142–50
Thermal expansion, 19, 20, 272–73
Thermal resistance, 96, 121
Thermocouple circuits, 13
Thermocouples, 7, 9–10
    copper-constantan, 11, 12
    control devices for, 10–13
    EMF chart for, 11
Thermodynamics, laws of, 33, 43
Thermoelectric
    effects, 18, 20
    equations, 21–27
    performance, 28–29
    refrigeration, 22–23
Thermometer, 9
    mercury-in-glass, 8
Thermostat, 5, 6
Ton of refrigeration, 42
Transmission
    through brick- and stone veneer
        masonry walls, 155
    coefficients of, 152–65
    through concrete-construction floors
        and ceilings, 158
    through flat roofs, 159–60, 235
    through frame-construction ceilings
        and floors, 157
    through frame partitions, 156
    through frame walls, 152–53, 231
    through masonry walls, 154, 231
    through pitched roofs, 161–62
    through power systems, 679–81
    through solid wood doors, 163
    through windows, skylights, and
        glassblock walls, 164, 224–30

Traps
    definition of, 262–63
    inverted-bucket, 269–70
    thermostatic, 269, 328
Threshold limit, 392
Trichloromonofluoromethane, 576
Tubeaxial fan, 463
Tube shapes, 330
Tubing, copper, 244, 276–77
    sizes of, 276–77

Unit conditioner, 742–43
Unit systems, 81
Units, conversion of, 749–50
    English and metric, 42
    energy, 750
    pressure, 749
    specific heat, 752
    thermal conductivity, 751

Vacuum, 14, 16
Vacuum system, 265–69
Valves
    air, 257–58, 298
    air-vent, 257–58, 298
    automatic air, 258–98
    control types, 685
    expansion, 466–67, 558–59
    packless radiator, 271
    pneumatic steam, 679
    pressure-relief, 295, 370–71
    reducing, 284
    resistance of, 304
    solenoid-operated, 680
    vacuum-type, 257
    vent, 258, 298
Vapor density, 56
Vapor systems
    definition of, 262–64
    operation of, 264
    pipe sizes for, 279–83
    return trap in, 264–65
    sketch, 2
    steam heating in, 262–65
Velocity
    air, ASHRAE comfort chart for, 379
    assumed, 429–34
    high-system, 222, 454–58
    wind, 151
Velocity head, 418
    pressure loss in, 424
Velometer, 480
Ventilation
    definition of, 1, 238
    standards, 386–89
Viscosity
    absolute, 410
    coefficient of, 410

Index

Viscosity—*Continued*
  kinematic, 413
  liquid, graph of, 411
Volume, specific, 48, 66
  steam, 66
Volumetric efficiency, 505–08, 540

Walls, heat gain through, 579–80
Warm-air heating, 396–408
  forced-circulation, 396–400
    basement plan of, 402–03
  panels with, 404
  perimeter, 400–04
  systems, types of, 396
Washer, air, 716

Water
  circulation of, 290–93
  high-temperature, for heating, 322
  make-up of, 295
  pressure-temperature relations in, 63–64, 322
Water leg, 262
Wet-bulb temperature, 100, 102, 104, 108–11
Wet return, 260, 262
Wind, 190–98
Window, crack, 190, 196
  infiltration of air, 198
Winter design temperature, 189, 191–96
Work, 32–34, 36–39